WATER

A COMPREHENSIVE TREATISE

Volume 4

Aqueous Solutions of

Amphiphiles and Macromolecules

WATER
A COMPREHENSIVE TREATISE

Edited by Felix Franks

WATER
A COMPREHENSIVE TREATISE

Edited by Felix Franks
Unilever Research Laboratory
Sharnbrook, Bedford, England

Volume 4
Aqueous Solutions of
Amphiphiles and
Macromolecules

PLENUM PRESS • NEW YORK AND LONDON

Library of Congress Cataloging in Publication Data

Franks, Felix.
Aqueous solutions of amphiphiles and macromolecules.

(His Water: a comprehensive treatise, v. 4)
Bibliography: p.
1. Macromolecules. 2. Solution (Chemistry) I. Title.
QD169.W3F7 vol. 4 [QD381] 553'.7'08's [547'.7] 74-17244
ISBN 0-306-37184-7

© 1975 Plenum Press, New York
A Division of Plenum Publishing Corporation
227 West 17th Street, New York, N.Y. 10011

United Kingdom edition published by Plenum Press, London
A Division of Plenum Publishing Company, Ltd.
4a Lower John Street, London W1R 3PD, England

Preface

This volume of "Water—A Comprehensive Treatise" is devoted to aqueous solutions of macromolecules and aqueous systems of small molecules which exhibit reversible aggregation in solution. This latter type of behavior is closely linked with the hydrophobic interaction and a chapter devoted to this subject is therefore included in the volume.

As the volumes constituting this treatise progress from pure water through solutions of simple molecules to the present volume, the level of our understanding of observed phenomena decreases and the experimental and theoretical techniques available become less adequate to provide the detailed information required. Thus, for example, we have so far been unable to include the generally acknowledged solvent effects in conformational calculations on biopolymers. On the credit side, however, some of the information gained from the study of simple aqueous solutions and water by itself is now beginning to find application in the elucidation of the properties of the systems discussed in this volume.

Perhaps an editorial explanation is required for lack of uniformity of approach by the various contributing authors, but this is primarily due to the historical development of the subjects covered. Thus the study of lipids, nucleotides, proteins, and polysaccharides originated in the biological disciplines where water has always been accepted as the natural substrate, even though its role in promoting the various *in vivo* processes has only recently been recognized. On the other hand the subject of aqueous solutions of synthetic macromolecules has grown out of classical polymer physics and chemistry where water as a solvent is generally avoided, and for good reasons. It is nevertheless my view that water-soluble and water-sensitive polymers have considerable technological potential and it is my hope that the material collected in this volume may help in the exploitation of these interesting substances.

All the subjects covered by this book are still in a state of rapid development so that much of what is written here may look out of date two years from now. It is nevertheless my hope that the approaches taken by the authors will provide new slants on well established phenomena.

Once again my thanks are due first and foremost to Mrs. Joyce Johnson for her invaluable and cheerful help with preparation of manuscripts, indexing, checking references, and communicating with authors and publishers. I should also like to thank my colleagues and many friends in the "International Water Fraternity" for advice and guidance.

October, 1974

F. FRANKS

Contents

Chapter 1

The Hydrophobic Interaction

F. Franks

Chapter 2

Surfactants

G. C. Kresheck

Chapter 3

Dyestuffs

D. G. Duff and C. H. Giles

Chapter 4

Lipids

H. Hauser

Chapter 5

Nucleic Acids, Peptides, and Proteins

D. Eagland

Chapter 6

Polysaccharides

A. Suggett

Chapter 7

Synthetic Polymers

P. Molyneux

Contents of Volume 1: The Physics and Physical Chemistry of Water

Contents of Volume 2: Water in Crystalline Hydrates; Aqueous Solutions of Simple Nonelectrolytes

Contents of Volume 3: Aqueous Solutions of Simple Electrolytes

Contents of Volume 5: Water in Disperse Systems

The Hydrophobic Interaction

F. Franks

Biosciences Division
Unilever Research Laboratory Colworth Welwyn
Colworth House, Sharnbrook, Bedford, England

1. INTRODUCTION

The peculiar features exhibited by aqueous solutions of some very simple molecules have been studied for some 30 years and various attempts have been made to account for the nature of such solutions in terms of conceptual models which take particular account of the observed solution behavior. The solutes in question include the rare gases, alkanes, and alkyl and aromatic compounds with only one, or at most two, polar functional groups. However, since we are dealing with a special type of interaction between water and apolar residues, a whole range of polymers must be included in the discussion, and indeed it is the influence of these particular interactions on polymer conformation and aggregation behavior which provides the most interesting manifestation of the phenomenon under discussion. The role of apolar group interactions (hydrophobic bonding) in maintaining the tertiary (native) structures of proteins was reviewed by Kauzmann,[759] who suggested that comparatives studies of aqueous and nonaqueous solutions of simple analogs of apolar amino acid side chains, ideally alkanes, might provide a valuable insight into the peculiarities of water as a solvent. His suggestion was taken up by many physical chemists and biochemists, and thanks to their efforts, the properties of simple and, more usually, not so simple molecules in aqueous solution are well documented. Although hydrophobic hydration and solute association are now accepted phenomena, it must be emphasized that, as yet, we have no knowledge of the origin of such behavior. Thus, whereas electrostatic, dipole, dispersion, core repulsion, and even hydrogen bonding types of interactions have been described in terms of more or less credible pair potential functions, relating the potential to the distance and mutual orientation of the interacting molecules, this has not yet been achieved for the hydrophobic interaction.

2. PHENOMENOLOGY AND NOMENCLATURE

The fact that hydrocarbons are very sparingly soluble has been known for a long time ("oil and water don't mix"), but the cause of such low solubilities is related to entropic effects and this distinguishes such aqueous solutions from the more normal liquid mixtures in which mutual solubility is determined by enthalpic factors. Thus the transfer of, say, a rare gas atom or a simple hydrocarbon molecule from the ideal gas state into water, as shown in Fig. 1a, is accompanied by a negative entropy of hydration $\Delta S_h°$, where $\Delta S_h° = \bar{S}_2° - S_2^{\ominus}(g) < 0$. This observation led to the suggestion that inert solutes promote the structuring of the neighboring water molecules, i.e., the "iceberg" effect,[477] so that the hydrophobic bond, or the association of apolar molecules, according to Fig. 1b, could then be regarded as a partial reversal of the thermodynamically unfavorable process of solution.

This reasoning also been extended to much more complex molecules, the only two requirements being substantial apolar residues and an aqueous environment. Thus the reversible aggregation of molecules or ions with long apolar chains or aromatic groups, as in the case of surfactants, phospholipids, glycerides, and dyestuffs, is believed to depend mainly on the phenomenon of hydrophobic association.* In this way complex aggregates such as micelles, bimolecular layers, and lamellar structures can exist, depending on the nature of the aggregating species. These structures and the processes giving rise to them are treated in detail in the following chapters of this volume.

On a more complex level, interactions between apolar residues attached to a common polymeric backbone can help to promote conformational changes of which the folding of globular proteins is probably the best-

---

* It should, of course, be emphasized that entropy-driven association is in no way unique to hydrophobic interactions; processes dominated by electrostatic interactions, e.g., ion pairing, also take place primarily as a result of favorable entropic factors. This can be shown by simple calculations based on the Born model.

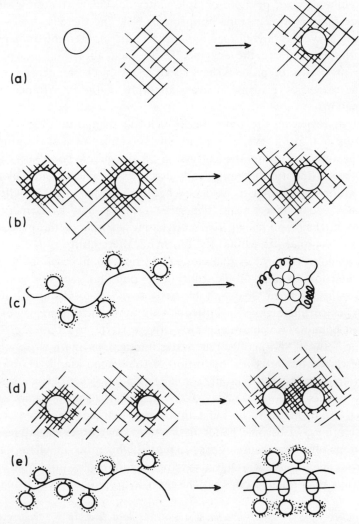

Fig. 1. Diagrammatic representation of (a) hydrophobic hydration and (b)–(e) hydrophobic interactions; (b) Kauzmann–Nemethy–Scheraga contact interaction, (c) globular protein folding, (d) proposed long-range interaction (see text), and (e) possible stabilization of helix by interaction as in (d).

known example.[1463] This is shown diagrammatically in Fig. 1c in its simplest form, i.e., an ordering process which results in the removal of apolar residues from the external aqueous environment to the interior. Hydrophobic interactions are also believed to contribute to the maintenance of ordered macromolecular structures, whether they be simple helices, such as polyglutamic acid, or more complex intermolecular structures, such as the blood plasma or membrane lipoproteins,[1399] and there is also evidence that some enzyme-catalyzed reactions involve as one step the apolar binding of the substrate to the enzyme.[271] In fact, apolar or hydrophobic binding properties of proteins have been the subject of intensive studies since the review by Kauzmann drew attention to the importance of this phenomenon.

Before reviewing the properties which are diagnostic of the type of interaction outlined above, a word about the subject of nomenclature. As far as the author is aware, the expression "hydrophobic bond" was coined by Kauzmann and it rather graphically describes the phenomenon by which apolar groups are driven to associate because of their ordering effect on the local aqueous environment. The same term was used by Nemethy and Scheraga in their well known thermodynamic analysis of the process[1092, 1093] and it has since passed into the biochemical literature.

By analogy, the process that causes hydrophobic interactions to occur, i.e., the entropically unfavorable solution of apolar molecules or residues in water, has been termed "hydrophobic hydration"[486] to distinguish it from other types of interactions in aqueous solutions, such as ion-dipole and hydrogen bonding, which are enthalpy driven. Hertz, in an attempt to differentiate between "normal" solute–water interactions and hydrophobic effects, described the latter as "hydration of the second kind" (Hydratation zweiter Art) but this rather nongraphic description never became popular. Imaginative descriptions of newly discovered physical phenomena, or new concepts, are of necessity simplifications, and as these phenomena receive closer scrutiny and become better understood, the earlier "headline"-type descriptions sometimes cause offense to purists who attach much importance to rigorous definitions.* This has also happened to the "hydrophobic bond." Thus it has been pointed out that the term hydrophobic bond is inappro-

---

* A classic example is Frank and Evans' use of the word "iceberg"[477] to describe the possible origin of the negative hydration entropies of rare gases and hydrocarbons in aqueous solution. The authors not only put this word in quotes but explained carefully the usefulness and limitations of the "iceberg" concept. Later generations, when citing Frank and Evans, first of all omitted the quotes and eventually ascribed to the originators of this concept interpretations which cannot be found in the original paper. Some authors have even gone to the length of searching for icebergs almost in the literal sense, i.e., long-lived, quasicrystalline assemblies of water molecules in the neighborhood of apolar groups, and after satisfying themselves that such permanent crystalline hydration shells do not appear to exist, have concluded that Frank and Evans were incorrect in their interpretation of the thermodynamic properties of aqueous solutions of apolar compounds.

priate[661] since (a) alkyl groups do not aggregate because of their phobia for water, and (b) it is not a bond in the accepted sense. To meet the former objection, one could consider replacing the word "hydrophobic" by "apolar," although it will then no longer emphasize that the phenomenon is unique to water and highly aqueous solvent media.* On balance, therefore, and accepting the limitations, "hydrophobic" seems to describe the situation well enough. The other objection would appear to be trivial, since the word "bond" has been used in the physical sciences to describe all manner of interactions other than chemical (covalent) bonds. For instance, the generic term hydrogen bond is still somewhat ill defined and includes interactions with a variety of characteristic lengths, orientations, and energies. Nemethy in his more recent publications[1089] and Ben-Naim[117,118] in his statistical thermodynamic analysis of hydrophobic effects have adopted the term "hydrophobic interaction" to replace "hydrophobic bond."

Although those "skilled in the art" will know fairly precisely what is meant by the term "hydrophobic bond," we shall follow the present trend and refer to the phenomenon as hydrophobic interaction. Quite possibly, by the time this volume is published, this term will itself have come under attack.

3. PHYSICAL PROPERTIES OF INFINITELY DILUTE AQUEOUS SOLUTIONS OF "EFFECTIVELY HYDROPHOBIC" MOLECULES

3.1. Thermodynamic Characterization

Simple apolar molecules, notably the alkanes, are so sparingly soluble that their aqueous solution properties are not easily investigated experimentally. Most of the available data are based on solubility determinations over a range of temperatures and pressures. Notable exceptions are a few direct (i.e., calorimetric) estimations of the enthalpies of solution[820,1239] and a direct (i.e., volumetric) estimate of partial molar volumes,[978] although the latter data have not yet been confirmed. Because of the inherent difficulties in the solubility determinations, the agreement between published data from different sources is only moderate, and hence there are

---

* A more general "solvophobic" effect has been identified in solutions of other associated solvents, e.g., methanol and N-methyl-formamide, and a superficial comparison might suggest that the phenomenon is not peculiar to water. However, a closer scrutiny reveals that the thermodynamic features discussed in Sections 3 and 4 have no counterpart in the nonaqueous systems.

large uncertainties in the van't Hoff enthalpies computed from such solubilities, and this is of course even more true for the heat capacities, obtained from the van't Hoff enthalpies by a further differentiation.

The experimental problems associated with the study of aqueous solutions of alkanes can, however, be circumvented, since it has been realized that there is a whole range of alkyl derivatives whose dilute aqueous solutions exhibit the same features as those of the unsubstituted hydrocarbons. These compounds, as a first approximation, consist of such alkyl derivatives that possess only *one* simple polar site capable of participating in specific hydrogen bonding with water, i.e., alcohols, ethers, amines, ketones, esters, but *not* necessarily amides or carboxylic acids.[485] Such compounds in dilute aqueous solution therefore behave as "soluble hydrocarbons" and it has been shown that the functional group provides only a minor contribution to the measured thermodynamic properties of a given homologous series.[485] In the case of the partial limiting heat capacities of solution $\bar{C}_{p_2}^\circ$, the contributions from a whole range of polar groups, such as $-OH$, NH_2, $-COOH$, $-CONH_2$, are identical,[799] so that by performing calorimetric measurements with solutes of increasing chain length or chain complexity, one is truly determining the effect of CH_3, CH_2, or CH groups in normal, branched, or cyclic configurations. A more detailed discussion of the aqueous solution thermodynamics of such "effectively hydrophobic" compounds is presented in Chapter 5 of Volume 2 of this treatise, and for present purposes it suffices to summarize the observed trends.

In all cases free energies of solution are positive, i.e., $\Delta G_s^\circ > 0$, resulting from large, negative entropies, whatever the sign of ΔH_s°. On ascending a homologous series, ΔG_s° increases for each additional CH_2 group, as do also $-\Delta H_s^\circ$ and $-\Delta S_s^\circ$. In the case of $\bar{C}_{p_2}^\circ$ the increment per CH_2 appears to be constant.[799,839] A comparison of normal and branched chain isomers shows that as a general rule ΔG_s° increases with the number of internal degrees of freedom i.e., due to entropic factors. However, on occasions, enthalpic effects may also be of importance. This is particularly well illustrated by comparing the aliphatic alcohols with the cyclic ethers. Thus ΔG_s° for n-BuOH and tetrahydrofuran is 2352 and 1715 cal mol^{-1}, respectively, at 25°. This difference is due not so much to the lower ΔS_s° of tetrahydrofuran as to its much more negative ΔH_s° (see Chapter 5, Volume 2 of this treatise). Various investigations are on record of the role played by differences in the configurational entropies of cyclic and normal compounds in determining the solution behavior, but the matter remains to be clarified. That the alkyl chain configuration does affect ΔG_s° is demonstrated by the solubility of the higher alkanes ($>C_{10}$) which, on the basis of a more or less constant $\Delta G_s{}^\circ$ increment, should be much less soluble than in fact they are.[484] It appears that chain folding (intramolecular hydrophobic interaction?) makes these long-chain hydrocarbons appear to the solvent as their lower-chain homologs.*

A comprehensive study of $\bar{C}_{p_2}^\circ$ of the lower alcohols by Arnett[50] also bears out the effects of alkyl chain conformations. Unfortunately, the calorimeter used was not of the same high precision as that used by Wadsö,[799] and the actual experimental results do not show too good an agreement with those of Alexander and Hill,[14] but an analysis of the data shows that the errors are of a systematic type. The trends reported by Arnett et al. provide interesting information about alcohol–water interactions: Although the $\Delta C_{p_s}^\circ$ increments per CH_2 residue decrease on ascending the homologous series, corresponding increments in $\bar{C}_{p_2}^\circ$ remain constant at 25 cal(deg mol)$^{-1}$. Ring closure and possible chain branching lead to a decrease in $\Delta C_{p_s}^\circ$, but here standard-state problems obscure the significance of the experimental results.

On a more general basis, the recent calorimetric studies of Cabani and his colleagues on a variety of homologous series have thrown further light on the $\Delta C_p{}^\circ$ effects produced by differences in alkyl chain configuration.[220–224] They find that a clear difference exists between the $\bar{C}_{p_2}^\circ$ contributions of the $>CH_2$ group in open-chain [22 ± 2 cal(mol deg)$^{-1}$] and cyclic [18 ± 2 cal(mol deg)$^{-1}$] derivatives. However, they point out that this difference should not be ascribed to a loss of flexibility in the cyclic derivatives but should be discussed in terms of the "hydrophobic surface" exposed to the solvent.

All the recent calorimetric determinations of $\bar{C}_{p_2}^\circ$ per CH_2 group show good internal agreement[14,50,224,489,799,839] but are of course obtained from experiments on polar alkyl derivatives. These results are at variance with estimates from alkane solubility measurements. A recent statistical treatment[1553] suggests "most probable" values for $\Delta C_p{}^\circ$ per CH_2 group which increase with the alkane chain length and appear to be significantly lower than the calorimetric estimates. Wadsö discusses this discrepancy[839] and reaches the conclusion that the precision of the solubilities is not such that the van't Hoff differentiations can be applied with a high degree of confidence (see also Ref. 1239).

The volumetric properties of "effectively hydrophobic" compounds in aqueous solution also show some interesting features which distinguish these mixtures from their more normal nonaqueous counterparts. We con-

---

* See also Lal on the statistical mechanics of alkane chain conformations.[847]

sider, in particular, the limiting partial molar volumes, expansibilities, and compressibilities, as given by

$$\bar{V}_i^\circ = \partial\mu_i^\circ/\partial P \tag{1}$$

$$\bar{\alpha}_i^\circ = \partial^2\mu_i^\circ/\partial P\,\partial T = \partial\bar{V}_i^\circ/\partial T \tag{2}$$

$$\bar{\beta}_{i_T}^\circ = (\partial^2\mu_i^\circ/\partial P^2)_T, \qquad \bar{\beta}_{i_S}^\circ = (\partial^2\mu_i^\circ/\partial P^2)_S \tag{3}$$

If the pure organic substances are adopted as standard state, then a comparison of the limiting excess properties of members of homologous series again shows fairly regular increments for additional CH_2 groups, but also the effects of chain branching and cyclization are quite pronounced.

As is also the case for thermal properties, in the absence of extrapolation procedures based on credible solution theories, very precise experimental measurements on very dilute solutions are required, and even then the exact magnitudes of limiting values are in some doubt.[492,499]

The following features seem to bear directly on the phenomenon of hydrophobic hydration:

1. $\bar{V}_2^{\circ E}$ is always negative, i.e., the solute molecule appears to occupy less volume in an aqueous solution than it does in its own pure liquid state at the same temperature.* As a first approximation, $\Delta\bar{V}_2^\circ = -15\ \mathrm{cm^3\ mol^{-1}}$ of CH_2 groups.[12,486,487]

2. In spite of some very precise density data (e.g., one part in 5×10^7) at very low concentrations ($x_2 = 10^{-4}$) there is still considerable doubt as to the limiting value of $\partial\bar{V}_2/\partial x_2$. By comparison with better behaved solutes, such as polyhydroxy compounds, for which $(\partial\bar{V}_2/\partial x_2)_{x_2\to0} = 0$,[489] it seems that in solutions of effectively hydrophobic compounds, solute–solute interactions persist to unexpectedly low concentrations.[492] Similar trends have been observed with solutions of tetraalkylammonium halides,

---

* Although pure solutes at the same temperature are favorite standards in comparisons of thermodynamic properties, the validity of such an approach is by no means certain. Thus, at 25°, say, methanol is not far from its boiling point, whereas tert-butanol has only just melted. Some workers utilize the concept of transfer from an "apolar environment" to obtain information about an aqueous environment, but this only raises the question of what constitutes the standard "apolar environment." Morcom and Smith have performed ΔV^E measurements of 1,4-dioxane in hexane, cyclohexane, and benzene and have shown significant differences in $\bar{V}_2^\circ$ and $\Delta V^E(x_2)$ of these three systems.[1041] It therefore appears that a *standard* apolar environment does not exist. It has also been suggested that, rather than referring all measurements on pure compounds to one temperature, the use of reduced temperatures might be more realistic, but there are not nearly enough available experimental data to validate such a suggestion.

where, depending on the size of the alkyl groups, deviations from limiting Debye–Hückel behavior can occur at lower concentrations than would be predicted on the basis of electrostatic ion pairing.[491] Futhermore, the deviations are opposite in sign to those observed for salts, e.g., $MgSO_4$, where ion pairing has been shown to occur at low concentrations.

3. Like the solvent water, very dilute solutions of effectively hydrophobic solutes exhibit negative coefficients of thermal expansion and this is reflected in $\bar{\alpha}_2^{\,\circ}$, as defined by eqn. (2). The effect is even more pronounced when one examines $\bar{\alpha}_2^{\circ E}$. Figure 2 shows $\bar{V}_2^{\circ E}(T)$ for the four isomeric butanols.[492] The surprising feature is that tert-butanol, at limiting concentrations, is able to maintain the negative excess expansibility of its aqueous solution up to relatively high temperatures, more so than other solutes which have been studied, although the phenomenon appears to be common to all monofunctional alcohols, ethers, amines, and ketones.[487,489]

4. Closely linked to the negative excess thermal expansibility is the well-known maximum density phenomenon observed for pure H_2O at 3.98°. This may be enhanced by the addition of low concentrations of effectively hydrophobic solutes. This can be regarded as a stabilization of the order (structure) which exists in water against the effect of temperature. An analysis of the maximum density effect predicts that, on the basis of ideal mixing, *all* solutes will lower θ, the temperature of maximum density, and the power of various solutes to reduce θ can be expressed in terms of a molal lowering (Despretz constant) analogous to the molal cryoscopic constant. Since $\Delta\theta$ is a function of the solute concentration, the effect discussed here depends not only on the effect of isolated solute molecules on the intermolecular order of water, but on the effects of pairs, triplets,

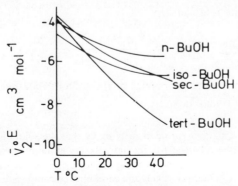

Fig. 2. Temperature dependence of $\bar{V}_2^{\circ E}$ for the isomeric butanols in aqueous solution. From Ref. 492.

etc. of solutes on this order. The fact that $\Delta\theta(x_2)$ is not linear suggests a cooperative mechanism. With effectively hydrophobic solutes the observed $\Delta\theta$ ($= \theta - 3.98$) is always less negative than that predicted by the Despretz constant, and in some cases can actually assume positive values[493,1545,1546]; the relationship between $\Delta\theta$ and the volumetric properties of the solution is given by

$$\Delta\theta = \theta_{\mathrm{id}} - \theta^{\mathrm{E}} = \frac{x_2}{1 - x_2} \frac{a_2}{2a_1} \frac{V_2^{\circ}}{V_1^{*}} - \frac{\partial(\Delta V^{\mathrm{M}})/\partial T}{2(1 - x_2)a_1 V_1^{*}} \tag{4}$$

where a_1 and a_2 are the coefficients of expansion of the two pure components and V_1^{*} is the molar volume of water at 3.98°. Thus, only when $\theta^{\mathrm{E}} > \theta_{\mathrm{id}}$ will an elevation of θ be observed.* One interesting consequence of the negative expansibilities and the maximum density effects is that for each solute there is a characteristic concentration, which can be calculated, at which the coefficient of thermal expansion of the solution is zero, i.e. the partial molal volume is independent of temperature.[49] Some of the characteristic volumetric anomalies of "effectively hydrophobic" compounds are summarized in Table I.

5. The limiting partial molal compressibilities [eqn. (3)] of the solutions under discussion in this section also show some interesting features, suggesting the existence of the unique solute–water interactions which we have termed "hydrophobic hydration." Although the isothermal compressibility is able to provide more direct information about this phenomenon, its measurement is laborious, and therefore, for reasons of expediency, the isentropic (adiabatic) compressibility is usually determined from measurements of the velocity of sound through the solution and application of the Laplace equation. If the necessary heat capacity and expansibility data are available (which is seldom the case, but see Ref. 13), then $\bar{\beta}_{2_T}^{\circ}$ can be computed from the relationship

$$\bar{\beta}_{2_T}^{\circ} = \bar{\beta}_{2_S}^{\circ} - \frac{T(V_1 a_1)^2}{C_{p_1}} \left(\frac{\bar{C}_{p_2}^{\circ}}{C_{p_1}} - \frac{2\bar{a}_2^{\circ}}{a_1 V_1} \right)$$

where a_1, V_1, and C_{p_1} are properties of the solvent (coefficient of expansion, molar volume, and heat capacity).

---

* The converse, $\Delta\theta < 0$, does *not* necessarily imply that the solute is a "water structure breaker," as has been claimed for the tetraalkylammonium halides.[319] All it does mean is that $\theta_{\mathrm{id}} > \theta^{\mathrm{E}}$, but θ^{E} may still be negative.[1304] In the case of tetraalkylammonium salts this is in fact very likely the case, but the computation of θ^{E} requires the assumption of a common standard state *and* knowledge of the coefficient of expansion of the substances in the standard state [see eqn. (4)].

TABLE I. Characteristic Volumetric Properties of Some "Essentially Hydrophobic" Solutes in Dilute Aqueous Solution[a]

	MeOH	EtOH	PrOH	i-PrOH	BuOH	t-BuOH	Tetrahydrofuran	Tetrahydropyran
$-\bar{V}_2^{\circ E}$ at 5°, cm³ mol⁻¹	—	2.5	3.5	3.5	4.1	4.1	4.1	5.3
$-\bar{V}_2^{\circ E}$ at 25°, cm³ mol⁻¹	—	3.7	4.5	5.5	5.0	7.0	4.7	6.5
$10^2 \bar{\alpha}_2^{\circ E}$ at 15°, cm³ mol⁻¹ deg⁻¹	—	−5	−6.7	−10	−5	−13	−3	−6
$-\bar{V}_{min}^E$ at 5°,	3	6	8	10	Insoluble	>12?	—	Insoluble
at x_2 cm³ mol⁻¹	0.12	0.07	0.05	0.06		0.02?		
Concentration x_2 at which $\bar{\alpha}_2^E = 0$	—	0.035	0.02	0.02	—	?[b]	—	—
θ_{max}^E, deg,	~6	4.6	2.5	~5	1	6	>1	—
at x_2	<0.10	0.035	0.015	~0.04	0.01	0.02	>0.001	—
$-\bar{\beta}_2^\circ$ at 5°, 10⁴ cm³ mol⁻¹ bar⁻¹	—	—	—	—	—	20	12	20

[a] From Refs. 12, 49, 486–488, 492, 493, 1080, 1545, 1546.
[b] Experimental density data appear to be too unreliable for evaluation of $\bar{V}_2^E$.

At low temperatures ($<30°$), $\bar{\beta}^{\circ}_{2_T} = \bar{\beta}^{\circ}_{2_S}$, within the limits of experimental error.[13] For effectively hydrophobic solutes $\bar{\beta}_2^{\circ}$ may be negative at low temperatures (e.g., for propanol $\bar{\beta}_2^{\circ} = -3 \times 10^{-4}$ cm^3 bar^{-1} mol^{-1} at $0°$) but becomes positive at higher temperatures. For a homologous series, increasing the size of the alkyl residue leads to a lowering of $\bar{\beta}_2^{\circ}$, but it is not quite clear whether there is a constant decrement in $\bar{\beta}_2^{\circ}$ per CH$_2$ group.[277,1525] Thus for the symmetric tetraalkylammonium halides $\bar{\beta}_2^{\circ}$ decrements (per CH$_2$ group) for the transitions Et$_4$N$^+$–Me$_4$N$^+$, Pr$_4$N$^+$–Et$_4$N$^+$, and Bu$_4$N$^+$–Pr$_4$N$^+$ are 0.6, 0.9, and 2.5×10^{-4} cm^3 bar^{-1} mol^{-1}, respectively,[277] and a similar behavior is observed for the alkylamine halides R$_2$NH$^+$X$^-$.[1525] Conway and Verrall have discussed the contributions to $\bar{\beta}_2^{\circ}$ that might be assigned to the intrinsic ionic compressibility and to structural (hydrophobic hydration) factors. That the latter effect is significant can be deduced from the marked temperature dependence of $\bar{\beta}_2^{\circ}$.

The limiting thermodynamic excess properties that distinguish aqueous solutions of "effectively hydrophobic" solutes from their polar counterparts and from nonaqueous solutions are summarized in Table II.

The total thermodynamic picture which emerges suggests that apolar groups are able to promote in their vicinity an aqueous environment which differs from that in ordinary bulk water in such a way that the volumetric

TABLE II. The Thermodynamic Characterization of Hydrophobic Hydration and a Comparison with "Normal" Solution Behavior

	Essentially hydrophobic solutes	Polar solutes (and nonaqueous solutions)
ΔG_s°	>0	$\lesssim 0$
ΔH_s°	$\gtrsim 0$	$\gtrsim 0$
ΔS_s°	<0	$\lesssim 0$
	$T\,\lvert \Delta S^{\mathrm{E}} \rvert > \lvert \Delta H^{\mathrm{E}} \rvert$	$T\,\lvert \Delta S^{\mathrm{E}} \rvert < \lvert \Delta H^{\mathrm{E}} \rvert$
$\Delta C_{p_s}^{\circ}$	>0	≤ 0
$\bar{V}_2^{\circ\mathrm{E}}$	<0	$\gtrsim 0$
$\bar{\alpha}_2^{\circ\mathrm{E}}$	<0	>0
θ^{E}	>0	<0
$\bar{\beta}_2^{\circ}$ (at 25°)	$\gtrsim 0$	<0 (for ionic solutes $\ll 0$)
$\partial\bar{\beta}_2^{\circ}/\partial T$	>0	>0 (?)

anomalies characteristic of water (e.g., the temperature of maximum density) are enhanced. The structural and/or dynamic modifications induced by apolar residues are thermally very labile, as evidenced by the large, positive $\Delta C_{p_s}^{\circ}$, but appear to extend over long distances, i.e., evidence of solute–solute interactions at extremely low concentrations.

To obtain further information about the nature of these changes, other experimental techniques need to be employed which measure space- and orientation-dependent correlations between water molecules and between water and solute molecules. Unfortunately such techniques, e.g., based on diffraction and scattering of radiation, are not yet sufficiently sensitive for experiments to be carried out at limiting concentrations where solute–solute effects are eliminated. Such information as has been obtained has therefore been based on experiments at higher (in some cases very much higher) concentrations and the placing of faith into extrapolations to infinite dilution conditions. Thus Narten and Lindenbaum, from an X-ray diffraction study of a concentrated solution of Bu_4NF, concluded that the radial distribution function was rather similar to that for pure water, but the O–O distance was shown to be 2.80 Å (cf. 2.85 Å for water), and the coordination number 3.8 (cf. 4.4 for water). These results contrasted with those obtained for solutions of NH_4F, in which the O–O distance was identical to that for pure water. However, the authors stated that their results were incompatible a clathrate hydrate-like geometry but could be accounted for on the basis of the ice I geometry,[1084] in spite of the fact that the salt is known to form a clathrate hydrate of ideal composition $Bu_4NF \cdot 32 \cdot 8H_2O$. It is, however, not at all certain whether small differences in calculated radial distribution functions (rdf) for different models and their comparison with the rdf obtained by processing the experimental scattering intensity functions (of a concentrated solution) allow one to discriminate between the validities of similar models. (See also Ref. 1570.)

3.2. A Formal Description of the Hydrophobic Hydration Phenomenon

Much has been written about "water structure" and the effects of various types of solutes on this structure. The terms "structure making," "structure breaking," positive and negative hydration, electrostriction, hydrophilic hydration, and hydrophobic hydration are found in the scientific literature, but few attempts have been made to specify these effects more closely, e.g., in terms of molecular distribution functions. In 1964 Hertz attempted to define more exactly the qualitative descriptions applied to aqueous solutions.[650] He described in a very graphic manner the possible

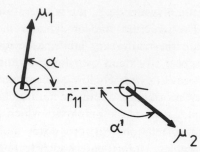

Fig. 3. Distances and orientations (in two dimensions only) which define the molecular pair distribution function for two water molecules.

influences of temperature, ions, and apolar particles on the orientation-dependent molecular pair distribution functions f_{ij} (i.e., water–water and water–solute) and tried to correlate the changes in f_{ij} so produced with the measured physical properties of such aqueous solutions.* At that time, of course, he did not have at his disposal the wealth of experimental data which has become available since the impetus given to studies of dilute aqueous solutions of nonelectrolytes by the three famous publications by Nemethy and Scheraga[1091–1093] in which a quantitative description of the hydrophobic interaction was attempted.

Before one speaks of the "structural" changes in the solvent, brought about by the introduction of a solute molecule, it is as well to be clear about the specification of the initial or unperturbed "structural" state and it is for this reason that we discuss recent developments in the molecular description of liquid water in some detail.

The position and orientation of a water molecule may be specified in terms of the coordinates of the mass center (components of a vector $\mathbf{r}$), and the three Euler angles α, β, and γ. The position and orientation of one such molecule relative to another may be similarly described in terms of the vector $\mathbf{r}_{11}$ joining the two mass centers, and differences in the Euler angles, α_{11}, β_{11} and γ_{11}. This situation is illustrated in Fig. 3 in which, for simplicity, both molecules have been confined to a common plane. In this hypothetical case, water molecules require only one angular coordinate (α) to specify absolute orientation.

* Unfortunately, this quite fundamental study of Hertz, which is really a significant development in the interpretation of solution properties, has not become one of the "popular," i.e., much cited, publications, probably for language reasons. It therefore seems opportune to summarize the main arguments and conclusions, most of which are as relevant today as they were in 1964.

The pair distribution function f_{11}° $(\mathbf{r}_{11}, \alpha_{11}, \beta_{11}, \gamma_{11})$ for pure water, is a measure of the probability that two molecules will simultaneously take up particular spatial positions and orientations. This distribution may be formally defined in terms of the configuration integral appropriate to the canonical ensemble, but if W_{11}° $(\mathbf{r}_{11}, \alpha_{11}, \beta_{11}, \gamma_{11})$ is the potential of average force, a succinct expression may be written for f_{11}°. This is,

$$f_{11}^{\circ}(\mathbf{r}_{11}, \alpha_{11}, \beta_{11}, \gamma_{11}) = (\varrho_1^2/8\pi^2) \exp[-W_{11}^{\circ}(\mathbf{r}_{11}, \alpha_{11}, \beta_{11}, \gamma_{11})/kT] \quad (5)$$

where ϱ_1 is the number density of water molecules. The potential of average force is related to the vacuum potential for a pair of water molecules, $U_{11}^{\circ}(\mathbf{r}_{11}, \alpha_{11}, \beta_{11}, \gamma_{11})$ and is the potential energy of two water molecules in specific positions, averaged over all configurations taken up by the remaining molecules in the liquid sample.

The radial distribution function $g_{11}^{\circ}(\mathbf{r}_{11})$ may be obtained from f_{11}° by integration over all angular coordinates. It is accessible experimentally, and has been calculated by several authors, always on the assumption of specialized structural models, and usually with the aim of using the agreement between the experimental and calculated distribution functions as support for the credibility of the chosen model. Recent Monte Carlo[85,1565] and molecular dynamics[1209] computer simulations of liquid water appear to be the first instances where the radial distribution function has been calculated from first principles without the assumption of some structural model, degree of hydrogen bonding, distinct molecular species, etc. On the other hand, these simulation experiments have utilized a drastically simplified model for the H_2O molecule, i.e., a hard sphere (in fact, a neon atom) with four embedded point charges. This in turn has led to the use of a simple vacuum pair potential $U_{11}^{\circ}(\mathbf{r}_{11}, \alpha_{11}, \beta_{11}, \gamma_{11})$ based on a Lennard-Jones potential for neon and a "pseudo-hydrogen bond" potential.[1427]

In fact a realistic specification of U_{11}° has been, and still is, the main stumbling block to the application of integral equation methods to the evaluation of f_{11}°. Apart from complexities introduced by the asymmetry and anisotropy of the water molecule, there are good indications that the interactions between water molecules are not pairwise additive.[332,600]

While W_{11}° and hence f_{11}° cannot be evaluated with any degree of certainty, it is difficult to specify exactly what will be the effect of a solute molecule on the "water structure." The effect of introducing solute molecules must be expressed in terms of three pair distribution functions, f_{11}, f_{12}, and f_{22}, and in general, if N_1 and N_2 are the numbers of solvent and solute molecules, respectively, and ϱ_1 and ϱ_2 are corresponding number densities, these distributions are given by the equations,

$$f_{11}(\mathbf{r}_{11}, \alpha_{11}, \beta_{11}, \gamma_{11})_{N_1,N_2} = (\varrho_1{}^2/8\pi^2) \exp[-W_{11}(\mathbf{r}_{11}, \alpha_{11}, \beta_{11}, \gamma_{11})/kT] \qquad (6)$$

for solvent alone, and

$$f_{12}(\mathbf{r}_{12}, \alpha_{12}, \beta_{12}, \gamma_{12})_{N_1,N_2} = (\varrho_1\varrho_2/(8\pi^2)^2) \exp[-W_{12}(\mathbf{r}_{12}, \alpha_{12}, \beta_{12}, \gamma_{12})/kT] \quad (7a)$$

$$f_{22}(\mathbf{r}_{22}, \alpha_{22}, \beta_{22}, \gamma_{22})_{N_1,N_2} = (\varrho_2/8\pi^2)^2 \exp[-W_{22}(\mathbf{r}_{22}, \alpha_{22}, \beta_{22}, \gamma_{22})/kT] \qquad (7b)$$

for solvent–solute and solute–solute pairs.

Again W_{11}, W_{12} and W_{22} are potentials of average force, and if required, the angular dependences may be removed by integration. Although f_{22} must be intimately connected with the phenomenon of the hydrophobic interaction, we shall not discuss it at this stage but limit ourselves to the very dilute solution ($N_1 = N, N_2 = 1$).

In order to determine the structure-making potential of a solute, we need to definine f_{soln} as

$$f_{\text{soln}} = f_{11} + f_{12} \qquad (8)$$

where f_{soln} refers to a configuration space whose dimensions are made up of the dimensions of the configuration spaces that describe water–water and water–solute distributions in the solution. Thus f_{soln} is not a probability but the sum of two probabilities. Equation (8) can serve as an index of the effect of concentration on the molecular distribution functions. Thus, as the solute is added, the shape of f_{11}° can be expected to undergo a change (to f_{11}). Simultaneously a peak will probably develop in f_{12} and also in f_{soln}. If the f_{soln} peak narrows as a function of concentration, this indicates structure promotion.

Several possible types of behavior can be distinguished (see Table III). If hydrophobic hydration does not exist, aqueous solutions of, say, alkanes will show type I behavior. Hydrophobic hydration would be expected to give rise to type IV or V behavior. In the case of type IV the effect is confined to changes in f_{11} and could be expected to disappear with increasing concentration as a peak in f_{22} develops. Type V seems to be more consistent with the experimental evidence, but even if specific water–solute correlations do not exist, the foregoing treatment still suggests that hydrophobic hydration will always be associated with a narrowing of the water–

TABLE III. Effect of Solute Concentration on the Molecular Pair Distribution Functions

Type	f_{11}	f_{12}	f_{soln}	Type of solution
I	Broader than f_{11}^{0}	Isotropic	Broad-structure breaking	Weakly hydrated, e.g., I^-
II	Broad	Sharp maxima	Narrow	Strongly hydrated ions, Li^+, Mg^{2+}, F^-
III	No change, $= f_{11}^{0}$	Well-developed maxima	Slight structural promotion	H_2O–dipole interaction, e.g., hydration of polar groups in organic molecules
IV	Narrow	Isotropic	Narrow	
V	Narrow	Well-developed maxima	Narrow-structure promotion	Hydrophobic hydration?

water pair distribution function (per mole of dissolved solute), referred to pure water.

The above discussion combined with the data in Table III will show that terms such as "structure making" and "structure breaking" are meaningless if they are not carefully defined, since f_{soln} can be narrow, for a combination of reasons. Thys type II behavior leads to a very narrow f_{soln}, while at the same time the water–water correlations that exist in the pure liquid are largely destroyed. Whatever structure is made, therefore, has little in common with the mainly tetrahedral order in pure water. Types IV and V, on the other hand, are indicative of *water* structure promotion, although here again the actual values of r_{11}, α_{11}, β_{11}, ... at which peaks are observed in f_{11}° may differ somewhat from the corresponding values in solution (f_{11}).* We are therefore led to suggest that descriptions such as "structure maker," etc. might well be limited to situations where a narrow f_{soln} is due mainly to the fact that f_{11} is narrower than f_{11}°.

According to accepted views, hydrophobic interactions involve a partial (or even complete, as in the case of micelles) reversal of the hydration process (see Fig. 1b).[759,1093] In line with the foregoing arguments that this would involve the emergence of a new f_{22} peak and a broadening of f_{11} and f_{12} it is to be expected therefore that f_{soln} would undergo a broadening. There exists, however, another possibility, namely that hydrophobic interactions take place with f_{11} remaining either unchanged or actually exhibiting a further narrowing. Although this is incompatible with the concept of the hydrophobic interaction as a partial reversal of the solution process, there is experimental evidence for such an interpretation, and this will be discussed in Section 4.5.2, and is shown diagrammatically in Fig. 1(d).

3.3. The Microscopic Nature of Hydrophobic Hydration

Following on from the discussion in the previous section, it is now appropriate to enquire into the molecular nature of the processes which might give rise to a sharpening of f_{11} and/or f_{12} in solutions of effectively hydrophobic solutes. Intuitively, type V in Table III can be taken to describe the phenomenon of hydrophobic hydration and would certainly be consistent with the entropy and heat capacity changes observed for the process

$$X(\text{ideal gas}) \rightarrow X(\text{aqueous, infinite dilution})$$

---

* This is easily seen from a comparison of r_{11}, α_{11}, β_{11}, ... in the ice I lattice and in the various clathrate hydrate structures.

Ideally, one would like to know the difference between ω_{11}° and ω_{11} in eqns. (5) and (6) and the distance and orientation dependences of ω_{12} and ω_{22} in eqn. (7). Unfortunately, there are as yet no experimental techniques which will directly provide such detailed structural information for complex poly-atomic systems. However, f_{12} and any changes in f_{11} due to the presenc of a solute molecule can be monitored by the diffusional motions of the two species in solution. Over the past decade Hertz and Zeidler, together with their colleagues, have made detailed studies of the phenomenon of hydrophobic hydration by the application of nmr relaxation techniques.[539, 540] As is well known, one of the practical weaknesses of these methods is their low sensitivity, which necessitates the use of relatively concentrated solutions $(x_2 \geq 0.05)$ for which thermodynamic measurements indicate pronounced solute–solute interactions (i.e., f_{22} cannot be neglected). How-ever, in the absence of other, more informative spectroscopic data, the conclusions derived from such experiments must serve as the best presently available insight into the microscopic nature of hydrophobic hydration.

Basically three time-dependent correlation functions are used, express-ing the velocity, position, and orientation of a molecule (or a pair of molecules) as a function of time. The accessible experimental quantities are the time integrals of these correlation functions, e.g., the self-diffusion coefficient D_i of molecular species i is given by the integral of the velocity correlation function:

$$D_i = \tfrac{1}{3} \int_0^\infty \overline{\tilde{v}_i(0)\tilde{v}_i(t)}\, dt$$

where $v_i(0)$ and $v_i(t)$ are the velocities at zero time and after time t. The other two correlation functions employed are measures of the molecular rotational and translational motions. Intramolecular and intermolecular correlations can now be described, as shown in Fig. 4.* The simplifying assumption is made that solvation effects are limited to the primary (near-est neighbor) hydration sphere. Whether in the case of hydrophobic effects such an assumption is realistic must still remain open to doubt. In Fig. 4, τ stands for τ_c, the rotational correlation time, which is related to the reorientation time τ_r of the vector under consideration by

$$\tau_r/3 \gtrsim \tau_c < \tau_r$$

* The notation adopted in Fig. 4 and in the text is as follows: D and τ are the measured diffusional properties (translational and rotational). The preceding subscripts w, h, and s refer to the species bulk water, water in the hydration sphere, and solute, respectively. The subscript following τ refers to the vector whose rotation is being considered; this can be intramolecular or intermolecular, as shown in Fig. 4.

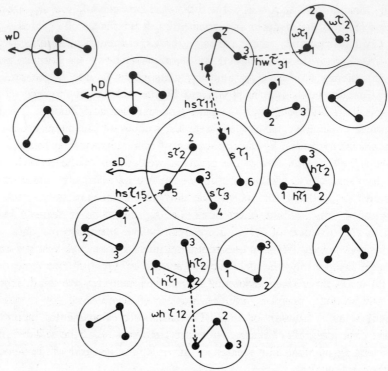

Fig. 4. Model of an infinitely dilute aqueous solution showing atomic positions and the various translational and rotational correlation terms used in the text. The nearest-neighbor water molecules of the solute make up the hydrate shell and are included in n_h. After Goldammer and Hertz.[539]

Goldammer and Hertz, by considering the nature of the hydration process, have derived certain rules which limit the permissible range of values which D and τ_c may adopt and their experimental results provide tests for these rules. Thus, it is evident that for long-lived (compared to the observed reorientation time) hydration states to exist, then for $x_2 \rightarrow 0$,

$$_hD = {_sD} \tag{9}$$

and if the solute molecule is rigidly incorporated into the hydration lattice, then

$$_h\tau_1 = {_h\tau_2} = \cdots = {_s\tau_1} = {_s\tau_2} = \cdots = {_{sh}\tau_{11}} = {_{sh}\tau_{12}} = \cdots \tag{10}$$

Of course if long-lived hydration states exist, but the solute is free to rotate within, say, a clathratelike cavity, then eqn. (9) will still hold as a necessary condition, but there will be no simple relationship among the various rotational correlation times.

For stoichiometric, pairwise hydration complexes one would expect eqns. (9) and (10) to hold for $x_2 \sim 0.5$, but even if these equations are obeyed, self-diffusion coefficients and correlation times must have "reasonable" values before the existence of identifiable hydration shells can be postulated. As useful guidelines, the Debye equation for rotational orientation

$$\tau_i = (4\pi a^3/3kT)f(\eta) \tag{11}$$

and the Stokes–Einstein equation for self-diffusion

$$D = kT/6\pi a f(\eta) \tag{12}$$

can be employed. Here a is the radius of the diffusing unit (molecule or molecular aggregate) and $f(\eta)$ is a function of the viscosity.*

Another criterion of "structure promotion" in the solvent relates f_{ij} (as defined on p. 16) to τ_i and D_i. It was suggested (see Table III) that structure promotion changes the shapes and/or peak positions of some of the pair distribution functions in such a manner that there is a net narrowing in f_{soln}. It can be shown that this in turn would lead to a decrease in $_h D$ (and possibly $_s D$) and to a lengthening of $_h \tau_i$ (and possibly of $_s \tau_i$).

Finally, one of the correlation function integrals P (see footnote on p. 60) that can be derived from nmr measurements provides an estimate for f_{ii} and f_{jj} in terms of the closest distance of approach of two protons on different molecules of the same species (a_{ii}), and the mean square displacement $\langle r^2 \rangle$ associated with the primary diffusive step length and D. For a uniform distribution of solute molecules throughout the solution, $_s DP_{22}$ does not deviate much from a constant value. If, as x_2 increases, the product $_s DP_{22}$ decreases, then this can be related to a nonuniform solute distribution, i.e., microheterogeneity.

The above criteria for particular solute–water (and water–water) correlations have been examined by means of very detailed nmr experiments, involving aqueous solutions of acetone, methanol, ethanol, tert-butanol, tetrahydrofuran (THF), pyridine, and acetonitrile. In terms of the ther-

---

* The approximation of equating $f(\eta)$ with the bulk viscosity of the medium is usually made, but since we are here suggesting that the possibility that $_w \tau_i \neq _h \tau_i$, etc., a viscosity term would presumably reflect more accurately the situation in the hydration sphere. The use of such "microviscosity" term in eqns. (11) and (12) has been suggested several times,[513] but there are obvious difficulties in assigning numerical values to $f(\eta)$.

TABLE IV. Diffusional Properties of Water and Effectively Hydrophobic Solutes at 25°, As Determined by Nuclear Magnetic Relaxation Methods[a]

	n_h'' [b]	$\left(\dfrac{{}_hD}{{}_sD}\right)_{c=c^*}$	$\left(\dfrac{{}_h\tau}{{}_w\tau}\right)_{x\to0}$	$\dfrac{({}_s\tau_{\text{alkyl}})_{x_2\to0}}{({}_s\tau_{\text{alkyl}})_{x_2\to1}}$	$\dfrac{({}_s\tau_{\text{polar}})_{x_2\to0}}{({}_s\tau_{\text{polar}})_{x_2\to1}}$
Acetone	23	1.3	1,3	1.3	0.8
Acetonitrile	19	1[c]	1.1	1.05	1.6
Methanol	17	—	1.4	1.2	1.2 ?
Ethanol	20	1.5	1.5	0.68	0.68?
tert-Butanol	25	2.4	1.7	0.53	0.53?
Tetrahvdrofuran	24	1.8	1.4	2.4	—

[a] See text for meaning of symbols. From Refs. 539 and 540.
[b] These values have been obtained from consideration of molar volumes.
[c] This does not indicate the existence of a long-lived hydration sphere, but arises from the fact that D of the two liquids is the same.

modynamic criteria already discussed, all these molecules, with the possible exception of acetonitrile, can be considered as "effectively hydrophobic." There is some uncertainty about the quantification of the conclusions reached, because the interpretation of the experimental nmr results partly depends on the assignment of hydration numbers, n_h, defining the primary hydration sphere. In the absence of independent information, such estimates

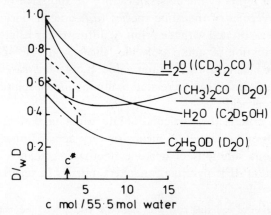

Fig. 5. Self-diffusion coefficients of solvent and solute in dilute, binary aqueous mixtures normalized to ${}_wD$; c^* corresponds to $n_h/55.5$ (see text). To allow for the difference between H_2O and D_2O, the measured ${}_sD$ values should be increased to fall on the broken lines as shown. From Ref. 539.

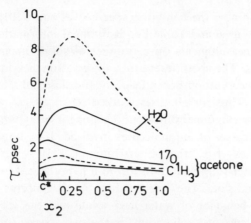

Fig. 6. Rotational correlation times of water and acetone (^1H and ^{17}O) in binary mixtures at 5° (– – –) and 25° (—). Note the sensitivity of $_h\tau$ to temperature and acetone concentration, which is almost completely missing in $_s\tau_{CH_3}$. Data from Ref. 539.

of n_h have been based on molar volume considerations. In this way n_h values shown in Table IV have been obtained. It can be argued (see Section 3.4) that a better way of assigning hydration numbers might be based on the clathrate hydrate geometry, especially for those compounds that are known to form such hydrates. In this manner acetone and THF would have $n_h \gtrsim 28$, this being the number of vertices of the hexakaidecahedra typical of the class II hydrate structure.[337]

Some of the D and τ data are summarized in Figs. 5 and 6. In each case the data refer to the substance which is underlined, measured in the presence of the other component, shown in parentheses. Since $_sD$ was measured in D_2O, allowance was made for the greater mobility of H_2O and the $_sD$ values were adjusted as shown by the dashed lines in Fig. 5. The vertical lines indicate solute concentrations $c = c^* = 55.5/n_h$. Thus for the concentration range $0 < c < c^*$ it might be meaningful to distinguish between unperturbed water and the hydration shell. For $c > c^*$ the experimental observations yield $_wD = {}_hD$.

The results in Fig. 5 show clearly that $_hD \neq {}_sD$, so that by the criterion of eqn. (9), rigid, long-lived hydration structures do not exist in the neighborhood of c^*, although the possibility of such structures at very low concentrations cannot be ruled out by consideration of D alone, particularly in the case of tert-butanol.*

* The fact that for acetonitrile $(_hD/_sD)_{c=c^*} = 1$ (see Table IV) is fortuitous and due to the fact that the self-diffusion coefficients of the two pure liquids are equal.

Considering now the condition specified in eqn. (10), we turn to the information in Fig. 6 and Table IV. Rotational information has been derived from τ_c measurements on a number of solute nuclei and a detailed picture emerges. The most interesting observation lies in the differences between the concentration dependence of translational and rotational diffusive motions. Thus for all cases studied $_sD < {}_wD, {}_hD$, whereas the opposite is true for rotational diffusion. This might be expected from a comparison of the masses of the molecules involved. It is, however, of interest that in some cases—e.g., ethanol, tert-butanol, THF— the solute molecule can perform faster rotational diffusion in a hypothetical hydrate cage than in the pure liquid state. The other striking result concerns the marked differences in the behavior of water and solute and the very different temperature coefficients of τ_c observed for the two species (see below). From Fig. 6, which is typical for all solutes studied, it is apparent that eqn. (10) does not apply, so that correlated rotation of solute and those water molecules composing the hydration cage does not take place. Indeed, both $_s\tau_{CH_3}$ and $_s\tau_{polar}$ are shorter than $_w\tau$, and very much shorter than $_h\tau$. The data in Fig. 6 indicate that the addition of acetone (or any of the other solutes) to water results in a considerably perturbed solvent, $_h\tau > {}_w\tau$, whereas the solute molecule is not greatly affected by the solvent medium, since $_s\tau_{CH_3}$ hardly changes over the whole composition range.

The probability of pairwise solute–water interactions (via the polar group) is indicated for pyridine, methanol, and tert-butanol, since $_s\tau$ describing the motion of the vector pointing in the direction of the polar group–water bond is longer than $_s\tau$ for any other vector of the solute–water pair. For acetone, acetonitrile, and THF such a pairwise hydration effect is not very pronounced.

From the temperature dependence of D and τ, activation energies $\Delta E^{\ddagger}$ corresponding to the various types of motion can be calculated and the results for acetone and acetonitrile are shown in Fig. 7. The differences between translational and rotational diffusion are again apparent, in that $\Delta E_D{}^{\ddagger}$ for water and solute are similar in magnitude, whereas the rotational motion of the solute, as described by $\Delta E_{s\tau}^{\ddagger}$ is facilitated. In other words, there is a very weak coupling between the solute and water molecules. The opposite is true in the case of deuteropyridine, where in equimolar mixtures $_h\tau = {}_s\tau_{C-D}$ and $\Delta E_{h\tau}^{\ddagger} = \Delta E_{s\tau_{C-D}}^{\ddagger}$, showing strong correlation between the motions of the two molecular species.

Having established the existence of solute–water correlations, it would be interesting to learn something about the mutual orientation of solute and water molecules and the configuration of the hydration shell about an

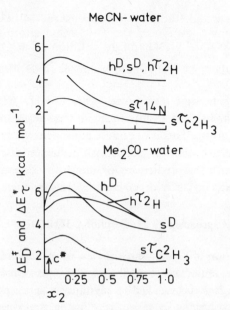

Fig. 7. Activation energies of self-diffusion $\Delta E_D^{\ddagger}$ and molecular reorientation $\Delta E_\tau^{\ddagger}$ for different nuclear species in aqueous mixtures of MeCN and Me₂CO. Note the complex behavior in the latter system, which is characteristic of an "effectively hydrophobic" solute. From Ref. 539.

apolar group in general. Such information, which will also enable us to define f_{11} and f_{12} more precisely, can be obtained from dipole-induced (intermolecular) nuclear magnetic relaxation rates $(T_1)_{\mathrm{inter}}^{-1}$. It can be shown that

$$(T_1)_{\mathrm{inter}}^{-1} \sim a^{-n}a^2/\bar{D} \tag{13}$$

where a is the distance of closest approach between two interacting magnetic dipoles and $\bar{D}$ is the mean self-diffusion coefficient of the molecule containing the magnetic dipole whose relaxation rate is being observed; n assumes the value of six or three, depending on whether or not tight binding exists between the sites containing the two dipoles. Thus a single T_1 measurement, say on a ¹H nucleus in a solute molecule that is interacting with a ¹H nucleus in H₂O, yields an estimate of the proton–proton distance. Now the solvent molecule also contains an ¹⁷O nucleus and the measurement of T_1 (solute) as affected by ¹⁷O will provide another internuclear distance and hence the orientation of the solvent molecule with respect to the proton site of the solute. In order to determine the orientation of water molecules in the primary hydration sphere with respect to the alkyl group in methanol, Hertz and Rädle[653] have measured T_1 in the systems

$$CD_2H \cdot OD + HDO \text{ (dilute solution of } H_2O \text{ in } D_2O)$$
$$CD_2H \cdot OD + D_2{}^{17}O \text{ (dilute solution of } D_2{}^{17}O \text{ in } D_2{}^{16}O)$$

Similar experiments were performed with $CD_2H \cdot CD_2 \cdot COOD$ in the same solvent systems. The results indicate that water protons point away from the methyl group. This is of course consistent with the clathrate model of aqueous solutions and the rotational behavior discussed above. It is also consistent with the predictions of molecular dynamics computer simulation experiments on the hydrophobic hydration effect (see Section 3.4.2).

3.4. Theoretical Approaches to Hydrophobic Hydration

The publication of Kauzmann's review on protein stability[759] in which he discussed the important contribution of hydrophobic interactions to nonbonded tertiary structures led to significant developments in studies of simple aqueous solutions in terms of "water structure" and changes in "water structure" induced by simple nonpolar solutes. Apart from new experimental approaches to the problem, several attempts were made to develop statistical thermodynamic descriptions of dilute aqueous solutions. Formally, the starting point has usually been pure water, and the common feature of the various treatments is the assumption of a locally perturbed water resulting from the introduction of an idealized inert, hard-sphere

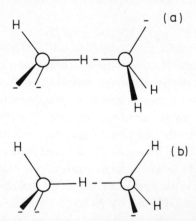

Fig. 8 (a) Favored orientation of H_2O molecules in liquid water (other orientations are possible); (b) required orientation for clathrate cage formation; protons and lone-electron pairs are in eclipsed positions.

molecule or a real molecule. The adequate description of the aqueous solution can, of course, be no better than the specification of the "ground state," i.e. pure water, and it is therefore not too surprising that the experimental features of infinitely dilute aqueous solutions of apolar molecules can be accounted for satisfactorily almost irrespective of the model chosen to represent pure water. This has led to the criticism that the various theoretical attempts to describe the properties of simple solutions are meaningless while the problems surrounding pure water are still awaiting a clearer insight. It is our aim to show that considerable advances have nevertheless been made in developing an understanding of the nature of the hydrophobic interaction, even though its origin is still shrouded in mystery but must be closely linked with the peculiar intermolecular nature of water.

The position as regards pure water has been reviewed in detail in Volume 1 of this treatise, and in Volume 2 the merits of various models for aqueous solutions of nonelectrolytes have been compared. We shall therefore confine ourselves to those features of the theoretical approaches that help to improve our understanding of the hydrophobic interaction. As regards the appreciation of pure water, the application of molecular dynamics simulation methods[1209,1428] constitutes an important new development since the publication of Volumes 1 and 2 of this treatise. The findings throw some new light on previously developed concepts and will certainly provide a new point of departure for studies of solute perturbations of water structure. For this reason we summarize the findings briefly, although these must be taken to be of a preliminary nature.

3.4.1. *Statistical Thermodynamic Treatments of Hydrophobic Hydration*

Theoretical investigations dealing explicitly with the hydrophobic effect date from the early 1960's, although previously Frank and Evans' statistical thermodynamic analysis of aqueous solutions[477] really treated the same problem, and the peculiar nature of aqueous solutions of the rare gases and simple alkyl derivatives had already been the subject of earlier studies.[218,219,392,393]

The theoretical studies of Nemethy and Scheraga, including pure water, simple aqueous solutions, and solutions of biopolymers[1091–1093] have become well known and were the first attempts to quantify the observed solution phenomena in terms of statistical thermodynamics. The calculations of the hydrophobic hydration effect are based on the setting up of a partition function to describe the perturbation of liquid water caused by the introduction of a simple (or not so simple) solute molecule. The method depends on the grouping together of water molecules according to the number of hydrogen bonds (zero, one, two, three, or four) in which they participate and the assignment of certain energy levels to these various states. Thus the properties of the four-bonded species were made to resemble those of the ice lattice and the unbonded species was characterized by the vibrational modes of water in the vapor state.

While several of the assumptions and mathematical procedures used by Nemethy and Scheraga in the calculation of the partition function for water have subsequently been criticized,[387,1090] their method of comparing pure water with dilute solutions has much to commend itself.* Thus it was argued that in the neighborhood of an apolar solute molecule the distribution of water molecules among the various hydrogen-bonded species would suffer a change. The main assumption was that the presence of a solute molecule would stabilize the four-bonded water network because of additional favorable van der Waals contacts. On the other hand, a nonpolar molecule next to an unbonded water molecule was considered unfavorable with respect to pure water, since water–water dipole interactions are stronger than dipole–induced dipole interactions between water and solute. These effects were described quantitatively in terms of shifts of the energy levels of the various hydrogen-bonded water species. These energy shifts were treated as adjustable parameters (i.e., their magnitudes differ for different hydrocarbons) and enabled $\Delta G_s{}^\circ$ to be calculated with reasonable success. Surprisingly, the agreement between theory and experiment was found to be better for butane and benzene than for methane. The theory has since been extended to dilute aqueous solutions of simple alcohols[845] with the further assumption that the alcohol hydroxyl groups can be treated in a manner analogously to water, i.e., they can participate in hydrogen-bonded clusters and exist in three, two, one, or unbonded states. Once again $\Delta G_s{}^\circ$ for the process

$$ROH(\text{apolar solution}) \rightarrow ROH(\text{aq})$$

could be reproduced well for higher homologs but not for methanol. The Nemethy and Scheraga type treatments are not so successful in accounting for the temperature and pressure derivatives of $\Delta G_s{}^\circ$, which for much more stringent criteria for any model. This may be partly due to the shortcomings of some of the assumptions and to the mathematical procedures used in the calculations of the partition functions.

---

* A revised treatment has recently been published[587] but this is confined to pure water.

Another approach to the problem of hydrophobic hydration was adopted by Frank and Quist at about the same time.[479] This took as the point of departure the well-known clathrate hydrate of methane, $CH_4 \cdot 7.66H_2O$.* The assumption was then made that apolar molecules occupy interstitial sites in a regular water network. The model is discussed in detail in Chapter 1 of Volume 2 of this treatise. It can account reasonably well for the thermodynamic properties of aqueous solutions of methane and there is no reason why it should not be just as successful when applied to other types of solutes. The major limitation is its inability to account for the behavior of solutions at elevated temperatures at which there are no longer sufficient interstitial sites to accommodate the solute molecules. This model, although, like many others, based on the concept of water as a mixture of distinguishable molecular species, only makes use of that fraction of the water molecules that is specified to exist as a quasiclathrate hydrate, providing cavities for occupancy by solute molecules. A natural extension of this concept[478] allows solute molecules also to interact with those water molecules that are not in the quasiclathrate environment to form a quasilattice (regular) solution. Thus a distribution of the solute is envisaged between the two solvent environments (for a more detailed description see Chapter 1, Volume 2).

The model has been applied successfully to aqueous solutions of simple alkanes and, in particular, to explain the remarkable effect of urea on the thermodynamic properties of such solutions.[1580] It has provided a novel insight into the origin of the observed negative entropies and the properties of the hypothetical quasiclathrate and quasilattice solutions have been specified in some detail. One of the interesting findings is that the negative $\bar{S}_2°$ values arise mainly from the removal of water molecules from their function as solvent, i.e., by providing interstitial sites and that the water "structural equilibrium shift" referred to above only provides a small contribution to $\bar{S}_2°$. It is not immediately apparent how or whether this result can be reconciled with the previously discussed trends in f_{11} which clearly indicate a "slowing down" of the diffusional processes of water molecules within the sphere of influence of the solute, and such increases in the microviscosity are usually equated to a water structuring process. Certainly the model developed by Frank and Franks contains some drastic simplifications, e.g., it is assumed that the "low-density" environment has a

---

* Evidence supporting the clathratelike geometry of water molecules in the neighborhood of apolar solutes has been discussed by Glew,[534] who compares the thermodynamic properties of aqueous solutions with those of the solid clathrate hydrates.

clathratelike geometry, even in the absence of interstitial solute molecules. Furthermore, the model does not take into account the possibility that water molecules may occupy interstitial sites. Thus, although the number of adjustable parameters has been kept to a minimum, the approximations inherent in the model may lead to numerical values for the various contributions to $\bar{S}_2{}^\circ$ which are at odds with the inferred values from other types of measurements.

The type of approach based on multiple equilibria involving interstitial and regular solutions has also been employed by Mikhailov,[1012,1013] who introduced several refinements, thus enabling him to treat solutions of larger solute molecules. Whereas the equations derived by Frank and Quist and Frank and Franks only apply in the limiting case of infinite dilution, Mikhailov's model successfully describes the rather complex behavior of water–ethanol mixtures over the whole composition range (see Chapter 1, Volume 2); the implications of the calculated thermodynamic properties will be discussed in Section 4.5.

A rather different approach has been adopted by Pierotti,[1170] who examined aqueous solutions of inert solutes on the basis of the scaled particle theory, in an effort to establish whether water was in fact "different" from other solvents.

The scaled particle theory provides for two processes which determine the solubility; the formation of a cavity in the solvent capable of accommodating a solute molecule, and an interaction term. Thus, the Henry's law solubility $\varkappa$ is given by

$$\ln \varkappa = (\mu_c/RT) + (\mu_i/RT) + \ln(RT/V) \tag{14}$$

where the subscripts c and i denote cavity formation and interaction, respectively, and V is the molar volume of the solvent. The first term on the right-hand side can be evaluated in terms of the number density, a solvent molecule "hard-sphere diamater" σ_1, and a composite term σ_{12}, given by $\sigma_{12} = \frac{1}{2}(\sigma_1 + \sigma_2)$, where σ_2 is the diameter of the cavity, i.e., a function of the size of the solute molecule. The second term on the right-hand side of eqn. (14) has been estimated in terms of a solvent–solute potential function which combines a Lennard-Jones function with an inductive energy contribution. The treatment is based on the assumption that the solvent molecules are uniformly distributed around the solute. By adopting certain numerical values (for water $\sigma = 2.75$ Å* and $\varepsilon/k = 85.3°K$) for the Lennard-Jones parameters, a reasonable fit to the experimental Henry's law

---

* This is lower than most estimates.

TABLE V. μ_c, $\bar{H}_c$, and $\bar{S}$, Associated with the Introduction of Argon into Water and Benzene at 25° According to the Scaled Particle Theory[1170]

	Water	Benzene
μ_c, cal mol^{-1}	4430	3610
$\bar{H}_c$, cal mol^{-1}	690	3520
$\bar{S}_c$, e.u.	-12.5	-0.30

coefficients is obtained, even for aqueous solutions. By suitable differentiations of eqn. (14) the corresponding partial entropies, enthalpies, and volumes can be obtained. With regard to the last quantity, $\bar{V}_c$ can be equated to the observed limiting partial molar volume. Table V shows the comparative data for cavity formation required to accommodate an argon atom in water and benzene at 25°. It can be seen that the main difference between aqueous and benzene solutions is that in the former μ_c is determined by the large, negative entropy. Contrary to an earlier view[392,393] that $\mu_c = 0$ for water at 4°, Pierotti points out that it is $\Delta \bar{H}_c$ which is equal to zero, leaving a large, negative $\Delta \bar{S}_c$. The general conclusion is that liquid water behaves like an assembly of hard spheres of 2.75 Å diameter, confined to a volume determined by the observed density of water and that there is no need to postulate special structural features for this liquid. Nelson and de Ligny have tested the scaled particle theory against experimental thermodynamic data for the higher alkanes[1088] and find good agreement only for $\Delta S_s°$. There is a divergence between experimental and theoretical $\Delta G_s°$ values which reaches 100% for hexadecane. They also state that the methylene group additivity in $\Delta G_s°$ is confined to the lower alkanes ($\leq C_8$) and speculate that chain coiling may be responsible for the lower contributions (per CH_2) in the higher homologs (see also Ref. 847).

More recent studies[1584] have indicated that μ_c and $\bar{S}_c$ are very sensitive to the value chosen for σ_1. Thus an increase of 10% in the value for σ_1 decreases $\bar{S}_c$ by 40 e.u. in the case of water, but increases $\bar{S}_c$ by 80 e.u. in the case of nitrobenzene. Although it is again shown that the large μ_c obtained for water can be matched by other solvents (e.g., iodobenzene, nitrobenzene), a unique feature of aqueous solutions is the dependence of $\bar{S}_c$ on σ_2, i.e., $\bar{S}_c$ becomes more negative with increasing σ_2.

The validity of the scaled particle theory has also been tested by Lucas,[923] who finds that in aqueous solutions $\Delta S_s°$ expressed as a simple function of σ_1 corresponds to $\bar{S}_c$ at 4°, except for aromatic solutes. At higher temperatures ΔS_s° also depends on the nature of the solute functional group, e.g., H, halogen, OH, NH_2.

Although it has been pointed out that the scaled particle theory applies to aqueous solutions without the need for any extra adjustable parameters, nevertheless the treatment requires as one of the parameters the experimental molar volume of the solvent, and it can be argued that the rather remarkable tetrahedral packing of the molecules embodies the peculiar structural features of water, which are therefore implicit in eqn. (14) and all other equations obtained therefrom. In addition, as has been mentioned above, the actual numerical values assigned to the Lennard-Jones parameters also determine the quality of the fit to the experimental data. It also seems arguable that such a simple potential function, coupled with the assumption that the solvent is represented as an assembly of close-packed spheres, can have much physical significance in the case of aqueous systems, however good the fit to experimental ΔG_s° values.

Finally, aqueous solutions of hydrocarbons have also been examined in terms of the significant structure theory of liquids.[639] The water molecules adjacent to the solute are considered to be subject to an asymmetric electric field which acts on the water dipoles such as to restrict their motions. A particular function is derived on this basis and once again the calculated ΔG_s° and ΔS_s° show good agreement with the experimental values. It is, however, questionable whether the significant structure theory provides a credible model for water, since in order to match the physical properties of this liquid, three significant structures are required.[419]

3.4.2. Computer Simulation Experiments

There are basically two computer simulation techniques which can provide information about the structural properties of an assembly of molecules. The Monte Carlo method yields canonical ensembles but is restricted to examinations of equilibrium properties of the given molecular assembly. The technique has been applied to water, for which a radial distribution function was calculated.[85] The small number of water molecules (64) included in the sample and the nature of the pair potential function chosen[1276] resulted in only moderate agreement with the X-ray radial distribution function—in particular, the calculated coordination number 6.4 and the O–O distance of 3.5 Å were rather larger than the corresponding values obtained from X-ray diffraction measurements.

A recently reported extension of this work[1565] has highlighted the critical importance of the nature of the potential function used in the simulation. Thus it has been found that the Rowlinson potential function[1278] emphasizes the spherical contribution, while the Ben-Naim–Stillinger potential,[1427] based on the Bjerrum four-point-charge model (see below), overcompensates for the orientational properties of the water molecule.

The technique of molecular dynamics offers the added possibility of probing the dynamic properties of the molecular assembly and its application to a study of liquid water[1209,1428] is probably the most important development since the publication of the studies by Nemethy and Scheraga in 1962.[1091–1093] However, the recently published molecular dynamics studies probably constitute only the beginnings of a renewed effort to obtain a more quantitative description of the intermolecular nature of liquid water.

The central problem is the adequate and realistic specification of the Hamiltonian. Although the sample used by Rahman and Stillinger is large, 216 molecules, these are treated as rigid asymmetric rotors, which precludes any information regarding the nature of intermolecular hydrogen bonding or effects on the intramolecular vibrational modes. Furthermore, although water–water interactions probably contain a significant nonadditive contribution, this has not been allowed for in the calculations, which are based on a pair potential function[1427] according to which the water molecules are treated as neon atoms which have embedded in them four point charges tetrahedrally disposed. (For details see p. 424 of Chapter 11 of Volume 1.) The effective pair potential then consists of a Lennard-Jones type potential function for neon and an electrostatic contribution to simulate the effects of hydrogen bonding.

In spite of the above limitations, the molecular dynamics calculations constitute a tour de force and allow the detailed investigation of a large variety of parameters, some of which are not yet amenable to direct experimental study. Thus the calculations enable three radial correlation functions to be evaluated, corresponding to the three possible nuclear pairs O–O, O–H, and H–H.

An examination of these three pair correlation functions indicates water to be "a random, defective and highly strained network of hydrogen bonds that fits space rather uniformly." There is ample evidence of considerable deviation from the ideal hydrogen bonding lengths and directions, but there appear to be no pronounced density fluctuations, a finding which suggests to the authors that the description of water as a "mixture" of high- and low-density species is probably not realistic. A detailed examination of the molecular distributions shows no completely "free" water molecules, suggesting that the various "interstitial" models which have been proposed (see Volume 1) cannot be reconciled with the results of the computer simulation. At the same time the results do indicate the presence of "free OH" groups which persist much longer than the water molecular vibration periods, and since the concentration of such non-hydrogen-bonded OH groups must be expected to increase (at the expense of the fully hydrogen-bonded molecules) with rising temperature, the description of water as a mixture in terms of different OH environments seems to make good sense.

One other finding of some relevance to the hydrophobic effect is that in the case of pairs of water molecules linked by undistorted hydrogen bonds, all angles of rotation about the bond axis appear to occur with equal probabilities. This is of interest in connection with the formation of cavities, such as exist in clathrates, capable of accommodating solute molecules. The water molecules constituting such cavities *must* be oriented with all the hydrogen bonds directed away from the center of the cavity, i.e. non-bonded protons on adjacent oxygen atoms would necessarily be in the eclipsed positions. The attendant lowering in the configurational entropy of such an arrangement may well be reflected in the observed negative ΔS_s° values of solutions of apolar solutes (see below), and result in a shift and a narrowing of f_{11}.

While the equilibrium properties of water, as given by the molecular dynamics simulation, are in fair agreement with the experimentally derived values, the agreement between experimental and calculated diffusive properties of water is quite remarkable, showing the promise of this technique.

There are of course several features of water which, although of a second- or third-order nature, are symptomatic of its unique role in geological, biological, and technological processes. These features, e.g., the density maximum and the complex pressure and temperature dependence of viscosity, will need to be accounted for by available simulation techniques if these are to become accepted as credible methods for studying aqueous systems. Recently a modified potential function has been developed, still based, however, on the above principle, by means of which an even more faithful simulation of the equilibrium behavior of water can be achieved, including the density maximum, albeit not at 4°.[1429]

The Monte Carlo-type simulation studies have recently been extended to the hydrophobic effect. A combination of statistical mechanical and Monte Carlo calculations has been performed on two systems: a hard phere in aqueous solution and a methane molecule in aqueous solution.[320] For the former case $\Delta G_s^\circ(\sigma_2)$ agreed qualitatively with the values derived by Pierotti, using the scaled particle theory.[1170] However, ΔH_s° is found to be negative, becoming more so as σ_2 increases. This contrasts with the predictions of the scaled particle theory. The results for methane do not agree too well with the experimental results. A solubility minimum is predicted at $\sim 70°$ (actually observed at $80°$), but the shape of the solubility–temperature curve does not reproduce the experimental data.

The molecular dynamics simulation of water[1209] has recently been extended to the system one neon atom $+$ 215 water molecules.* The radial distribution function (measured from the position of the neon atom) indicates about 12 nearest-neighbor water molecules in the first peak. This must be compared with 4–5 in pure water. In addition the Ne–O nearest-neighbor distance is longer by about 10% than is the O–O distance in water. Of particular relevance to speculations that hydrophobic hydration can be likened to clathratelike dispositions of water molecules is the finding that the nearest-neighbor water molecules tend to be reoriented such that no OH vectors of neighboring water molecules lie in the same plane and gives rise to the observation that first-layer water molecules surrounding the neon atom show a pronounced tendency to form hydrogen-bonded pentagons such as commonly occur in ice clathrate structures. This result lends credence to the speculations in Table III, where it is suggested that f_{11} is narrower than f_{11}° because of a restriction in the number of permitted angles and/or O–O distances in the neighborhood of the apolar particle.

The molecular dynamics simulation also indicates that the rate of exchange of kinetic energy between neon and its neighboring water molecules is much slower than that between water molecules. Such a finding is in accord with the nuclear magnetic relaxation results discussed in Section 3.3. It would be of interest to ascertain whether the computer simulation also finds the rotational modes of water molecules in the neighborhood of the neon atom to be affected in the manner suggested by the relaxation measurements.[539,540] At the time of writing it certainly seems that the computer simulation results support a clathratelike water environment of apolar molecules, as was first postulated by Glew.[531]

How far the analogy with the crystalline clathrate can be taken is still open to conjecture, but there is a mounting weight of experimental evidence that water molecules in the vicinity of apolar groups are "different" from water in the bulk and if this difference can be specified in more quantitative terms, then it is implicit that a general description of aqueous solutions

---

* F. H. Stillinger, personal communication. As mentioned above, the "water" molecules used in the computer experiments were neon atoms with embedded tetrahedrally oriented charges. In order to examine the effect of a hard-sphere apolar solute, the charges were removed from one of these water molecules. The only change required in the pair interaction potential was the removal of the electrostatic contribution for neon–water pairs whose interaction could be described adequately by the Lennard-Jones function.

must be based on the concept of water as a mixture of distinguishable species, characterized, for instance, by f_{11}° and f_{11}.

3.5. Summary and Conclusions

Although the twin problems of "inert" solute–water interactions and any resulting local changes in the water–water correlations cannot be regarded as solved, nevertheless the last ten years have seen good progress, mainly on the experimental aspects, which are now beginning to be utilized in the development of a general theoretical framework. In summary of the preceding sections, the following observations are characteristic of the phenomenon which we term "hydrophobic hydration."

1. The transfer of an essentially hydrophobic solute from the gas state into aqueous solution is accompanied by a positive free energy change which originates mainly from the unfavorable entropy associated with such a transfer. The magnitudes of the thermodynamic transfer functions depend almost entirely on the size and configuration of the apolar residue of the solute molecule, the polar head group only contributing in a very minor way. Second- and higher-order temperature and pressure derivatives of the free energy also exhibit abnormal behavior (in terms of what is expected for liquid mixtures).

2. The diffusion modes, both translational and rotational, of water are markedly affected by essentially hydrophobic solutes, but in different ways. Since both the self-diffusion coefficient and the rotational correlation time are related to the molecular pair distribution function, it can be inferred that the solute causes a narrowing of the water–water pair distribution function; i.e., in terms of Table III, f_{11} is narrower than f_{11}°. The situation regarding f_{12} is not quite so clear.

3. Various statistical thermodynamic models have been proposed to account for the unique thermodynamic properties of infinitely dilute solutions of simple solutes. Their common basis is the postulate that water can be treated as a mixture, i.e., that water molecules exist in two or more distinguishable environments. The manner in which this distinguishability is defined differs for the various models, e.g., one can discriminate between –OH groups, that are hydrogen bonded, giving rise to intermolecular vibrational modes, or "free," or one can consider separately those water molecules that have a solute molecule as a nearest neighbor, provided that some measurable property of such water of hydration is sufficiently different from the corresponding property of bulk water for such a distinction to be made. The basic postulate common to all mixture models is that the

proportions of the various subgroups of water molecules depend on pressure, temperature, and the nature of the solute. Although a few attempts are on record to explain the properties of aqueous solutions of essentially hydrophobic solutes in terms of other models, these have not been too successful in the case of infinitely dilute solutions and even less so in accounting for the unique concentration dependences of thermodynamic and diffusional properties of such systems.

4. THE MANIFESTATION OF HYDROPHOBIC ASSOCIATION EFFECTS IN AQUEOUS SOLUTIONS OF SIMPLE ALKYL DERIVATIVES

4.1. Thermodynamics of Moderately Dilute Solutions

The thermodynamics of aqueous solutions of essentially hydrophobic solutes has been treated in detail in Chapter 5 of Volume 2 of this treatise and therefore we only require a summary of those thermodynamic properties that reflect the unique interactions in dilute aqueous solutions of monofunctional solutes.

For the purposes of this discussion such solutions can be conveniently divided into three or possibly four composition ranges, each characterized by a different type of behavior. At the extreme water-rich end is the "infinitely dilute solution," the properties of which depend on solvent–solute interactions (hydration) only. The actual concentration where this applies may be extremely low. Thus, it is claimed that solute–solute interactions can be identified in aqueous solutions of gas mixtures, e.g., $H_2 + N_2$,[576] $CH_4 + O_2$, $N_2 + O_2$, and $N_2 + CH_4$.[960] For the type of molecule of interest to this discussion the composition which can be described as "infinitely dilute" decreases as the number of carbon atoms in the alkyl residue increases. At the other end of the composition scale one is concerned with a dilute solution of water in the organic component; actually this concentration range usually extends from $x_2 \simeq 0.3$ to $x_2 = 1$. This type of solution shows none of the peculiarities associated with the unique properties of water as a solvent. We are then left with a relatively narrow range of composition, $0 < x_2 \gtrsim 0.3$, within which phenomena such as maximum density effects, hydrophobic association, etc. can be observed. In terms of the molecular pair distribution functions in Table III it seems that water loses its peculiar solvent properties as f_{11} is gradually broadened by an increasing solute dilution effect.

For operational reasons the dilute solution range can be further sub-divided into two regions, denoted by x_2' and x_2'', in which different concentration and temperature dependences are observed for a variety of physical properties. For ethanol, $x_2' = 0.08$ and $x_2'' = 0.25$, whereas for tert-butanol, $x_2' = 0.04$ and $x_2'' = 0.12$. In solutions for which $x_2 > x_2''$ water behaves as an "ordinary" polar solvent, i.e., aqueous solutions can be described reasonably well by the various solution theories which have been successful for other polar solvents. For the composition range $x_2 < x_2''$, however, none of the solution theories can successfully account for the observed structural, thermodynamic, and transport properties.

Since the publication in 1966 of a review of the structural properties of alcohol/water mixtures[486] some of the experimental data and hypotheses put forward have received closer scrutiny and, as is often the case, it has been found that the situation is more complex than it appeared to the reviewers at that time. In Volume 2 of this treatise comment was made that reliable thermodynamic information about dilute aqueous solutions of non-electrolytes is still surprisingly sparse. For instance, over the past ten years hardly any new activity coefficient data have been reported, in spite of the remarkable findings by Knight[797] (quoted in Ref. 486). A little more progress has been made in calorimetric studies of excess enthalpies and heat capacities of mixing, and Fig. 9 shows $\Delta H^E(x_2)$ for the system ethanol–water at 25°.[162] The rather complex shape of this curve is now well corroborated, but is at variance with data published prior to 1966. Presumably the two inflections in the higher concentration range will eventually also be established for other alcohol–water systems, and in view of the known

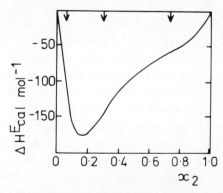

Fig. 9. $\Delta H^E(x_2)$ curve for water–ethanol mixtures at 25°. At the inflection points indicated $\bar{H}_2^E(x_2)$ undergoes extrema, showing the complex nature of the system. Data from Refs. 162 and 848.

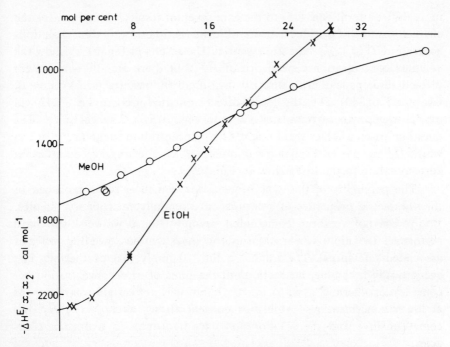

Fig. 10. $\Delta H^{\mathrm{E}}/x_1 x_2$ vs. x_2 curves for aqueous MeOH and EtOH at 25°. Data from Refs. 162 and 848. The significance of the inflection point (well developed in the EtOH curve) to the interpretation of the hydrophobic interaction is discussed in the text.

trends in $\Delta H^{\mathrm{E}}(x_2, T)$ their existence might have been expected. The interpretation of $\Delta H^{\mathrm{E}}(x_2)$ in terms of possible molecular interactions is complicated since it requires an equation of the type

$$-\Delta H^{\mathrm{E}} = x_1 x_2 \sum_{n=0}^{5} B(n)(1 - 2x_2)^n \qquad (15)$$

to fit the data. Figure 9 serves to illustrate the point that different types of interactions predominate in the various composition ranges.

From the point of view of hydrophobic interactions, the initial part of the curve ($x_2 \rightarrow 0$) is of greatest interest, and it is here that there is still considerable uncertainty. This is illustrated in Fig. 10 by the concentration dependence of $\Delta H^{\mathrm{E}}/x_1 x_2$ for two alcohol–water mixtures. It now appears more than probable that previous extrapolations from higher concentrations are invalid and that $(\partial^2/dx_2^2)(\Delta H^{\mathrm{E}}/x_1 x_2)$ changes sign, possibly several times at values of x_2 which depend on the nature of the solute. This makes it even more difficult to fit eqn. (15) to the experimental results.* However, for the characterization of possible hydrophobic interactions in very dilute solutions accurate $\Delta H^{\mathrm{E}}(x_2)_{x_2 < x_2'}$ data are essential. The results in Fig. 10[848] show the complex concentration dependence of $\bar{H}_2{}^{\mathrm{E}}$, but there are still some unexplained discrepancies with directly measured $\bar{H}_2^{\circ\mathrm{E}}$ values (see Volume 2, Chapter 5, p. 340). Thus the "best" direct estimate for ethanol is -2420 cal mol^{-1}, whereas the extrapolated value in Fig. 10 is -2300 cal mol^{-1}. The question arises whether there might be a concentration range $(x_2 < x_2')$ in which $\bar{H}_2{}^{\mathrm{E}}(x_2)$ passes through a minimum, i.e., characterized by an inflexion corresponding to the first arrow in Fig. 9.†

The possibility of this type of behavior, which is well established in the volumetric properties of solutions of essentially hydrophobic solutes, [486,758] has not yet been commented upon by those workers who have performed the more recent calorimetric investigations (see Fig. 10 and associated references). The results in Fig. 10 imply that exothermic, i.e., energetically favorable, interactions in the limit of zero concentration become less exothermic, as $x_2' < x_2 < x_2''$, but it is not certain what happens at lower concentrations. At higher concentrations, as $x_2 > x_2''$, $\bar{H}_2{}^{\mathrm{E}}$ becomes positive and the solution loses its hydrophobic hydration character.

This rather sudden transition from (presumably) hydrophobic hydration to "normal" interactions is well illustrated by $C_p(x_2, T)$.[797] As is well known, C_p of liquid water is remarkably insensitive to temperature and exhibits a shallow minimum at $35°$. It is found that dilute solutions of alcohols behave in a similar way, i.e., C_p is dominated by the aqueous component, but quite dramatic transitions are observed at $x_2 > x_2''$ to the behavior expected for a normal liquid, i.e., C_p increases with temperature.

It might be expected that $\bar{C}_{p_2}(x_2)$ would provide a sensitive criterion for the effects discussed above, but reliable data are only now beginning to appear in the literature. Leduc and Desnoyers have reported apparent heat capacities, ϕ_{C_p} and volumes ϕ_v for a series of related electrolytes, in particular $\mathrm{Bu_4N \cdot COOC_7H_{15}}$ (TBOc), $\mathrm{Bu_4N \cdot COOC_3H_7}$ (TBBu), $\mathrm{Bu_4NBr}$, and $\mathrm{C_3H_7 \cdot COONa}$.[866] These results supplement earlier measurements of activity coefficients and apparent molal heat contents ϕ_L on similar sys-

---

* For the systems ethylene oxide and tetrahydrofuran in $\mathrm{H_2O}$ and $\mathrm{D_2O}$ a satisfactory fit to the low-concentration data can only be obtained with eqn. (15) if $n = 10$.[535]
† At even higher concentrations, according to the results represented in Fig. 9, $\bar{H}_2{}^{\mathrm{E}}$ exhibits a maximum.

tems.[901] The latter studies had revealed a very pronounced concentration dependence of ΔH^E and ΔS^E for the series Bu_4NX, where X represents carboxylate anions with increasingly longer alkyl chains. The results also showed that for TBBu, $\Delta H^E(x_2) = T\,\Delta S^E(x_2)$, but that for the larger anions valerate, heptanoate, pelargonate, and caprate the difference between entropy and enthalpy became increasingly larger.

No clear-cut pattern has so far emerged from the ϕ_{C_p} results, represented in Fig. 11. Sodium octanoate, TBBu, and TBOc clearly show the steep decrease in ϕ_{C_p} which can be associated with micellization, but at even lower concentrations ϕ_{C_p} appears to *increase* (in the case of TBOc this increase is particularly marked. Similar "premicellization effects" have been observed in thermodynamic properties of surfactants[141,490,1049] and are so far unexplained, although the possibility of hydrophobic association, as distinct from micellization, has been discussed.

Unfortunately, no mention is made by Leduc and Desnoyers of the

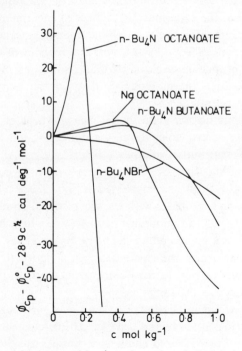

Fig. 11. Apparent molal heat capacities ϕ_{C_p} of a series of alkyl derivatives, corrected for interionic effects. Note the increase in ϕ_{C_p} at low concentrations, which is not in agreement with the predicted $\phi_{C_p}(c)$ behavior for hydrophobic interactions. Reproduced from Ref. 866.

possible effects of hydrolysis on ϕ_{C_p}, especially at the very low concentrations under discussion. Hydrolysis makes the interpretation of the behavior of associating sytems difficult,[374] and we may therefore have to wait for a more complete analysis of these rather startling data.

Not very much is known about $\Delta S^{E}(x_2)$ and hence about $\bar{S}_2^{E}(x_2)$ of nonelectrolytes for the low concentration range of interest. The investigations by Knight[797] were based on freezing point determinations and therefore cannot yield Raoult's law activity coefficients and hence $\Delta G^{E}(x_2)$, and no accurate vapor pressure results are available. However, in the limit of $x_2 \to 0$ all solutions of essentially hydrophobic solutes exhibit the type of behavior known as Barclay–Butler behavior[84] or compensation,[927] i.e., ΔS_h° is a linear function of ΔH_h° (the subscript h denotes hydration, i.e., we are considering the transfer of solute from the ideal gas phase to an aqueous medium). It therefore seems safe to predict that $\bar{S}_2^{E}(x_2)$ follows a similar trend to that tentatively established for $\bar{H}_2^{E}(x_2)$.

The above account emphasizes the considerable uncertainty which still exists about the true concentration dependence of ΔH^{E} and ΔS^{E}. No doubt this is partly due to the experimental difficulties encountered in high-precision calorimetric and vapor pressure measurements, especially in connection with the achievement of equilibrium, and the analysis of the vapor phase. However, for a proper description of dilute solutions in terms of solute *pair* interactions, such thermodynamic data in the concentration range $0 < x_2 < x_2{}'$ are essential.[810,1634]

These problems do not arise in the measurement of volume, and reliable information is available about the concentration and temperature dependence of partial molar volumes.[487,491,492,534,1080] Although the exact nature of the limiting slope of $\bar{V}_2^{E}(x_2)$ is still open to conjecture,[492] the general features of $\bar{V}_2^{E}(x_2)$, as represented in Fig. 12, are (1) negative slopes, the steepness of which increases with the size of the alkyl residue for a given homologous series, (2) the minimum at $x_2 = x_2{}'$, which becomes progressively more developed on ascending the homologous series, (3) the inflection at $x_2 \simeq x_2{}''$, beyond which the curve takes on a "normal" appearance (cf. H_2O–H_2O_2 included in Fig. 12 for comparison), and (4) the complex temperature dependence in the range $x_2 < x_2{}''$.

It was pointed out in Section 3.1 that $\bar{V}_2^{E}(x_2)$ isotherms cross over in a narrow concentration range and that $\bar{V}_2^{E}$ at low concentrations and temperatures is quite insensitive to changes in temperature. This observation is of course related to the maximum density phenomenon and to the capability of essentially apolar solutes of raising the temperature of maximum density [see eqn. (4)].

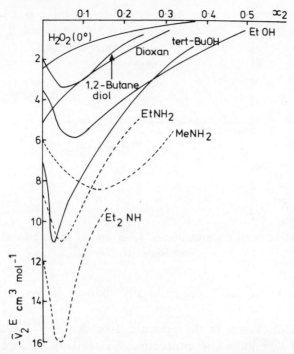

Fig. 12. $\bar{V}_2^{\mathrm{E}}(x_2)$ curves for representative alcohols, amines, 1,2-butane diol, dioxan, and hydrogen peroxide at 25°. The minima that are characteristic of essentially hydrophobic solutes are not observed for dioxan or H_2O_2. The positions and depths of the minima depend on temperature and size and configuration of the hydrophobic group (see Table I). Data from Refs. 486, 758, 1080.

A somewhat different approach to the study of the volumetric characterization of the hydrophobic interaction has been adopted by Wen and his associates, who measured ΔV^{E} accompanying the mixing of two electrolyte species with a common anion. Thus for the species R_4NBr and KBr, ΔV^{E} is written in terms of pair effects:

$$\Delta V^{\mathrm{E}}/[I^2x(1-x)] = 2\bar{v}(R_4N^+\text{–}K^+) - \bar{v}(R_4N^+\text{–}R_4N^+) - \bar{v}(K^+\text{–}K^+) \cdots$$

where I is the ionic strength.[1574,1575] It was found that for $R = Pr$ and Bu, $\Delta V^{\mathrm{E}} = 0.6$ and $0.7\ \mathrm{cm^3\ mol^{-1}}$, respectively, at $I = x = 0.5$. By far the largest contribution to ΔV^{E} was provided by the second term in the above equation, i.e., 9 and $11\ \mathrm{cm^3\ mol^{-1}}$, thus implicating akylammonium ion pairing as the major factor in the nonideal behavior of the mixture. This conclusion has since been confirmed by means of a more complete

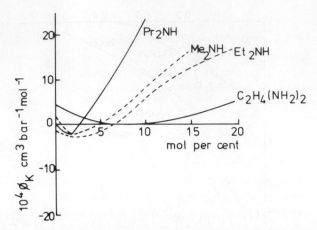

Fig. 13. Apparent molal compressibilities $\phi_\varkappa$ of amines in aqueous solution at 25°. Data from Ref. 758.

characterization of the hydrophobically induced cation–cation interaction.[1573]

A brief discussion of the pressure dependence of $\bar{V}_2$ completes the summary of thermodynamic properties of essentially hydrophobic solutes in dilute aqueous mixtures. Although from a theoretical point of view the isothermal compressibility $\varkappa_T$ is required for the interpretation of thermodynamic quantities in terms of molecular interactions, the isentropic (adiabatic) compressibility $\varkappa_S$ is more easily accessible experimentally—via measurements of sound velocity. Most published work is therefore based on ultrasound velocity measurements and expressed in terms of $\varkappa_S$. The conversion to $\varkappa_T$ requires a knowledge of heat capacities and expansivity.

The corresponding apparent molal compressibility $\phi_\varkappa$ is obtained from $\varkappa_S(x_2)$ as

$$\phi_\varkappa = (V - n_1 V_1^\circ \varkappa^\circ)/n_2$$

where V is the molar volume of the solution and the superscript zero refers to the properties of the pure solvent.

Representative $\phi_\varkappa(x_2)$ curves are shown in Fig. 13 for aqueous solutions of amines[758] (see also Refs. 277, 278, 1525). It is seen that $\phi_\varkappa$ becomes more negative as $x_2 \to x_2'$. As in the case of $\bar{V}_2^{\mathrm{E}}$, the temperature dependence of $\phi_\varkappa$ is complex and changes sign in the region of x_2'.

The qualitative implications of the observed thermodynamic properties can be summed up as follows:

1. The *limiting* properties (thermodynamic, kinetic, etc.) as discussed in Section 3 are consistent with the picture that the main effect of the essentially hydrophobic molecule is reflected in a modification of the water–water interactions, described as hydrophobic hydration.

2. The hydrophobic interaction is generally treated as a partial reversal of the process[759,1093]

$$\text{solute(apolar solvent)} \rightarrow \text{solute(aqueous)}$$

If this approach is taken literally, then the effect of solute concentration should be gradually to remove the anomalous physical properties observed in the limit of "infinite dilution."

3. This does in fact happen when $x_2 > x_2''$, but in the range $x_2 \leq x_2'$, several measured properties change in the *opposite* direction, i.e., the anomalies, presumably originating from the hydrophobic hydration phenomenon, become more pronounced and the effect is of a cooperative nature, i.e., it depends on the concentration of the solution.

4. The concentration dependence of some thermodynamic properties of simple, dilute aqueous solutions does therefore not appear to be compatible with the classical model for solute association by hydrophobic interactions.

4.2. Evaluation of Molecular Interactions from Thermodynamic Data

Whereas the phenomenon of hydrophobic hydration is reflected in the *limiting* thermodynamic solution properties, the *concentration dependence* of the various thermodynamic quantities provides information about solute association effects. A general way of formalizing such interactions is by means of power series expansions of thermodynamic quantities in terms of solute concentration. Thus the second virial coefficient is a measure of molecular pair interactions, etc. Since the aqueous solution properties of essentially hydrophobic solutes are concentration dependent even at very low concentrations, the evaluation of reliable second, let alone third, virial coefficients requires experimental data of the highest quality and such data are still almost nonexistent. Kozak[809] has made a thorough analysis of the available experimental results and the following paragraphs summarize his methodology and findings (see also Ref. 810).

Basically the solvent activity coefficient γ_1 is expressed as a power series in terms of the solute mole fraction concentration:

$$\ln \gamma_1 = Bx_2^2 + Cx_2^3 + Dx_2^4 + \cdots \tag{16}$$

where the coefficients $B, C, D, \ldots$ can, via the existing solution theories, be related to solvent–solute interactions, solute size, solute–solute interactions, etc. Thus the signs and magnitudes of these coefficients are of importance in an understanding of solution properties. There are several experimental methods available for the determination of activity coefficients and the choice of a given technique will largely depend on the nature of the solute and its solubility. For volatile solutes freezing point measurements provide a direct method for the determination of B at $0°$, B(0), although for the evaluation of $C(0)$ corresponding enthalpy of dilution (or mixing) data are also required. To obtain $B(T)$ and $C(T)$ at any other temperature T, one further requires $\bar{C}_{p_2}$, the partial molal heat capacity as a function of concentration and temperature, and it is here that reliable data for nonelectrolytes are almost nonexistent. For nonvolatile solutes $B(T)$ and $C(T)$ can be obtained directly from isopiestic measurements. There are several ways of calculating $B(0)$ and $C(0)$ from the experimental freezing point depression and calorimetric data and Table VI shows a selection of data evaluated by Kozak from Knight's[797] experimental results. It is seen that even with up to date techniques the uncertainties in $B(0)$ and particularly $C(0)$ are considerable. This is also the case for $\partial B/\partial T$ and $\partial C/\partial T$ in Table VII. The coefficients are, however, reliable enough to be compared with the predictions of various existing lattice solution theories, and Kozak et al. have carried out such a comparison.[810] The dependence of the free energy of mixing $\varDelta G^{M}$ on the solute size can be expressed in terms of the well-known Flory–Huggins theory,* according to which N lattice sites are occupied by N_1 solvent molecules and N_2 solute molecules each of which occupies r sites, so that $N = N_1 + rN_2$. The total number of distinguishable configurations for the molecules on the lattice $g(N_1, N_2)$ is calculated[454,455] in terms of r and various other parameters characterizing the lattice. The entropy of mixing is then given by

$$\varDelta S^{M} = k \ln g(N_1, N_2) \tag{17}$$

Now

$$\mu_1 = \mu_1{}^{\circ} + kT \ln \gamma_1 x_1 = \mu_1{}^{\circ} + (\partial \varDelta G^{M}/\partial N_1)_{N_2} \tag{18a}$$

and

$$\ln \gamma_1 x_1 = -(\partial/\partial N_1)[\ln g(N_1, N_2)]_{N_2} \tag{18b}$$

---

* This theory was developed to account for the behavior of polymer solutions $(r \gg 1)$, and cannot be expected to apply equally well to mixtures of molecules of approximately the same size.

TABLE VI. Virial Coefficients $B(0)$ and $C(0)$ in Eqn. (16)[a]

	$B'(0)$	$B(0)$	$B''(0)$	$C'(0)$	$C(0)$	$C''(0)$
Methanol	-0.10	-1.00	-1.50	-6.2	-5.8	$+11.5$
Ethanol	$+1.20$	$+1.00$	$+0.70$	-32	-24	-17
1-Propanol	-2.60	$+2.30$	$+1.90$	-53	-43	-33
2-Propanol	$+1.80$	$+1.40$	$+1.10$	-65	-57	-48
2-Methyl-1-propanol	$+6.90$	$+5.20$	$+3.70$	-260	-150	-80
2-Methyl-2-propanol	$+0.98$	$+0.30$	-0.28	-110	-90	-70
1-Butanol	$+4.70$	$+4.30$	$+3.90$	-120	-100	-90
2-Butanol	$+3.70$	$+1.60$	$+0.80$	-200	-80	-60
1,4-Dioxane	-1.40	-1.50	-1.90	$+28$	$+31$	-37

[a] From Ref. 809. Primes indicate the extreme values bearing in mind the scatter of experimental data. Columns 3 and 6 give the "most probable" values.

Differentiation of eqn. (18b) yields

$$\ln \gamma_1 = -\tfrac{1}{2}(r-1)^2 x_2^2 + \tfrac{2}{3}(r-1)^3 x_2^3 \cdots \tag{19}$$

which can be compared directly with eqn. (16) so that B and C are expressed in terms of r. According to eqn. (19), B is always negative and C is always positive, and both coefficients increase in magnitude with solute

TABLE VII. Temperature Dependence of Virial Coefficients B and C in Eqn. (16) at 25°[a]

	$\left(\dfrac{\partial B}{\partial T}\right)'$	$\left(\dfrac{\partial B}{\partial T}\right)$	$\left(\dfrac{\partial B}{\partial T}\right)''$	$\left(\dfrac{\partial C}{\partial T}\right)'$	$\left(\dfrac{\partial C}{\partial T}\right)$	$\left(\dfrac{\partial C}{\partial T}\right)''$
Methanol	0.019	0.018	0.017	0.18	0.54	0.96
Ethanol	0.018	0.017	0.015	1.50	2.20	3.10
1-Propanol	0.036	0.035	0.032	1.50	1.90	2.40
2-Propanol	0.024	0.023	0.021	1.20	1.70	2.10
2-Methyl-1-propanol	0.081	0.074	0.071	-0.80	$+1.70$	$+3.20$
1-Butanol	0.091	0.086	0.082	1.40	3.40	6.50

[a] From Ref. 809. Primes indicate the extreme values bearing in mind the scatter of experimental data. Columns 3 and 6 give the "most probable" values.

size. Reference to Table VI shows that the signs of the experimentally derived values do not agree with eqn. (19). Huggins introduced further refinements into the lattice treatment of solutions.[693] These lead to the following modifications of eqn. (19):

$$\ln \gamma_1 = \left[\left(\frac{1}{z} - \frac{1}{2}\right)(r - 1)^2\right]x_2^2 + \left[\left(\frac{2}{3} - \frac{1}{z} + \frac{4}{3z^2}\right)(r - 1)^3\right]x_2^3 \quad (20)$$

where z is the coordination number of the lattice. Equation (20) reduces to eqn. (19) as $z \to \infty$. The Huggins treatment also predicts that B becomes more negative and C more positive with increasing solute size, in contradiction to the results presented in Table VI. It can therefore be concluded that, at least with the athermal mixing approximation, the lattice theory cannot account for the observed trends in B and C with solute size in aqueous solutions of alcohols.

Stokes and Robinson have suggested that deviations from ideal behavior can be accounted for by hydration (specific solvent–solute interactions), and as a first approximation this hydration model can account reasonably well for the activity coefficients of aqueous solutions of polyhydroxy compounds.[488,1262] A hydration number n_h is introduced and eqn. (16) now becomes

$$\ln \gamma_1 = -n_h x_2^2 - (n_h^2 - n_h)x_2^3 \cdots \quad (21)$$

predicting that both B and C should always be negative. This is clearly not the case for homologous series of alkyl derivatives.

A combination of the hydration and the cell theories has been developed by Kozak, who considers two cases: (1) hydration with water sharing, where two solute molecules are placed on lattice sites such that they are separated by only one solvent molecule, and (2) hydration with no water sharing, where sites occupied by solute molecules are separated by at least two solvent occupied sites. The third case, i.e., where two different solute molecules (or parts thereof) occupy adjacent sites, is considered to represent "no hydration."

For the "no sharing" situation the lattice model yields

$$\ln \gamma_1 = [-\tfrac{1}{2}n_h(n_h + 2)]x_2^2 + [-\tfrac{1}{3}n_h(n_h^2 + 3n_h + 3)]x_2^3 \cdots \quad (22)$$

showing that hydration tends to make both B and C negative [cf. eqns. (19) and (20)]. For "water sharing" the corresponding relationship is

$$\ln \gamma_1 = -n_h x_2^2 - \tfrac{1}{2}n_h(n_h + 1)x_2^3 \quad (23)$$

It thus appears that specific solute–water interactions *always* lead to negative deviations from ideal behavior, although it must again be emphasized that the above results were obtained assuming that $\Delta H^M = 0$.

The calculation of B and C for the more realistic situation when $\Delta H^M \neq 0$ has been carried out by defining an interchange energy w which is equivalent to the energy change produced by substituting a solvent molecule on a pure solvent lattice by a solute molecule taken from a pure solute lattice. By assuming a random distribution of solute and solvent molecules on the lattice (known as the zeroth approximation), it can be shown that

$$\ln \gamma_1 = (w/kT)x_2^2$$

so that B can now adopt positive or negative values, depending on the relative strength of solute–solute and hydration interactions. However, since we are now considering specific interactions, the zeroth approximation of a random distribution is hardly realistic. Guggenheim therefore introduced the "quasichemical approximation" by weighting the distributions according to the sign and magnitude of w, so that, for instance, when w is large and negative, the probability of solute–water pairs is greater than that calculated on purely statistical considerations.[575] The quasichemical approximation yields the following equation:

$$\ln \gamma_1 = (w/kT)x_2^2 - (4/z)(w/kT)^2 x_2^3 \cdots \tag{24}$$

from which it is seen that $B \gtrless 0$ and $C < 0$. The observed trends in Table VI compared with the prediction of eqn. (24) therefore suggest that in solutions of alcohols, solute–solute interactions predominate and that these become more pronounced as the solute size (as reflected by r) increases.

Another, perhaps more suitable approach to the interpretation of thermodynamic properties is provided by the McMillan–Mayer theory,[953] which expresses the osmotic pressure π of a solution as a power series in the number density ϱ:

$$\pi = kT(\varrho + B_2^* \varrho^2 + B_3^* \varrho^3 \cdots) \tag{25}$$

The virial coefficients B_2^* and B_3^* are related to certain integrals of the radial distribution function via $u^{(2)}$, $u^{(3)}$, etc., the potentials of average force for molecular pairs and triplets, etc., respectively. The osmotic pressure virial coefficients are also related to the constants B and C in eqn. (16) via the limiting partial molar volumes $\bar{V}_i^\circ$ and their concentration dependence $\bar{V}_i(x_2)$. Thus B_2^* can be written in terms of molecular pair interactions

$$B_2^* = -(2V)^{-1} \int \{\exp[-u(2)/kT] - 1\}\, d\{2\} \tag{26}$$

where V is the total volume of the system and $d\{2\}$ means that the integration is carried out over all molecular pair distances and orientations. Similarly, B_3^* is expressed in terms of pair and triplet interactions.

Unfortunately, neither B_2^* nor B_3^* can be calculated, because at the present time we have no reliable ways of deriving $u^{(2)}$ or $u^{(3)}$. The position is thus similar to that discussed earlier for solute–water pair potentials. Kozak et al. proceed by decomposing B_2^* into attractive and repulsive components.[810] Thus the repulsive interaction is obtained from the integration of eqn. (26) within the limits $0 \leq r \leq \sigma$, where σ is the distance of closest approach and is calculable approximately from the molecular geometry (it is assumed that $u^{(2)}$ is positive and much greater than kT). The attractive contribution to B_2^* is obtained by the integration of eqn. (24) for $\sigma \leq r \leq \infty$, so that the exponential term is positive and hence the attractive contribution to B_2^* is negative. By making various assumptions about the molecular geometry, the minimum repulsive contribution to B_2^*, denoted by $R_{2\mathrm{min}}$, is estimated. The minimum attractive contribution A_2 is then simply given by

$$A_2 = R_2 - N_{A_2}^* = +[\bar{V}_2^\circ + V_1^\circ(\tfrac{1}{2} + B_2)] \tag{27}$$

where N_A is the Avogadro number and V_1° the molar volume of water. B_3^* can be similarly derived, although the results are even more uncertain, bearing in mind the complex orientation dependence of $u^{(3)}$ when the molecules do not approximate to hard spheres. Finally, one can derive a measure of the relative importance of pair versus triplet interactions by writing

$$\Gamma = 12(N_A B_2^*)^2 - 3N_A^2 B_3^*$$

where $\Gamma > 0$ signifies that triplet clustering is more pronounced than would be predicted from the summation of three pairwise interactions.

Table VIII provides a summary of estimates for A_2, A_3, and Γ based on the values for B_2 and B_3 and their temperature derivatives, shown in Tables VI and VII.

While bearing in mind that the exact magnitudes of the virial coefficients and their temperature dependences are subject to considerable uncertainties, nevertheless some trends are identifiable in the results cited in Tables VI–VIII. For a given homologous series $A_{2\mathrm{min}}$ becomes more nega-

TABLE VIII. Computed Pair and Triplet Interactions of Solutes in Aqueous Solutions[809]

	Temperature, °C	$-A_{2min}$, cm^3 mol^{-1}	$-1000A_{3min}$, (cm^3 mol^{-1})^2	1000Γ, (cm^3 mol^{-1})^2
Methanol	0	110 ± 12	15 ± 2	$+28$
Ethanol	0	192 ± 4	22 ± 2	-14
1-Propanol	0	261	38 ± 2	-30
2-Propanol	0	250 ± 5	34 ± 2	-37
2-Methyl-1-propanol	0	360 ± 30	—	—
2-Methyl-2-propanol	0	265 ± 12	—	—
1-Butanol	0	347 ± 7	—	—
2-Butanol	0	$296 + 30$	—	—
Glycine	25	235 ± 10	-8.3 ± 8	-38
α-Alanine	25	170 ± 20	32 ± 11	-47
Glycylglycine	25	425 ± 15	—	—
Alanylalanine	25	355 ± 15	—	—
Glucose	0	270 ± 35	—	—
Sucrose	0	558 ± 6	360 ± 10	-710

tive with the size of the alkyl group, and derivatives of the same alkyl chain length but different functional groups are subject to similar attractive potentials. Different functional groups therefore do not appear to affect molecular pair interactions to different extents, a conclusion which is in harmony with the $\bar{C}_{p_2}^\circ$ results[799] already discussed in Section 3.1. Chain branching seems to contribute to the magnitude of A_{2min}, but it does so to a lesser extent than does the number of CH$_2$ groups. It is further found that A_{2min} becomes more negative with rise in temperature, which is in accordance with the view that increase in temperature favors hydrophobic interactions.

Triplet interactions, measured by A_{3min}, follow similar trends, but for all compounds studied, $\Gamma < 0$ indicates that pair interactions predominate. The exception to this general finding is methanol.

Kozak et al. make the point that higher-order terms in eqn. (16) are large and positive (except for methanol) and they suggest that within the framework of their application of the McMillan–Mayer theory this might indicate the existence of large clusters or "micelles." It is, however, not easy to reconcile the highly cooperative nature of micelle formation with the finding that association between alcohol molecules is predominantly a molecular *pair* interaction.

Finally, there is one aspect of this otherwise attractive theory which casts doubt on its value, and that concerns the magnitude of A_{2min} and A_{3min} for molecules that are polyfunctional and do not exhibit hydrophobic interactions, e.g., urea and polyhydroxy compounds. In every case documented, A_{2min} is negative and of a similar magnitude as for essentially hydrophobic molecules of the same size. The authors suggest that pair interactions here take place via hydrogen bonding (A_{2min} is related to the number of donor sites); since for sucrose and glycerol $dB/dT > 0$, it is suggested that pairwise (hydrogen bonding) interactions increase with increasing temperature. Such conclusions cannot, however, be reconciled with what is known about hydrogen bonds in general and the detailed physical chemistry of aqueous solutions of sugars in particular.[488,1262,1458] It is possible, therefore, that the particular way in which B_2^*, B_3^*, etc. are decomposed into A_{2min} and R_{2min} [eqn. (27)] and the various assumptions made in the numerical evaluation of R_{2min} place undue emphasis on the size (rather than the chemical nature) of the solute molecule. This in turn will be reflected in A_{2min}, suggesting that pair interactions depend mainly on molecular size, i.e., this particularly theory, as applied, is incapable of distinguishing between hydrophobic and hydrogen bonding interactions.*

Suffice it to say that of the solution theories that have been applied to the study of aqueous solutions, the McMillan–Mayer treatment as used by Kozak et al.[810] shows the most promise and with generally more reliable experimental information, coupled with the use of more realistic pair potential functions, might well provide valuable information about the microstates of liquid mixtures. One limitation is its total neglect of the solvent component of the mixture. The properties of the solvent are only introduced into the treatment via the solute–solute parameters characterizing the pair potential.

The usefulness of the method could perhaps be extended by applying it to ternary aqueous systems of two essentially hydrophobic solute species. This would yield solute–solute "hydrophobic interaction coefficients." The value of this type of experimentation and analysis has been well demon-

---

* It is also possible, however, that some of the rather surprising results originate from insufficiently precise activity coefficient and calorimetric data. This is well illustrated by Kozak.[809]

strated by Wood and his colleagues for aqueous solutions of electrolytes (Chapter 2, Volume 3). Up to the present time no very high-precision activity or enthalpy data on such ternary systems have been reported (see, however, ref. 1634), but the interesting features of such mixtures have been highlighted by studies of alkane and rare gas solubility in "modified aqueous media," i.e., aqueous mixtures of alcohols,[114–116,119,120] tetraalkylammonium halides,[1574] and urea.[1582]

4.3. Direct Investigation of Time-Averaged Structures in Solution

Direct structural studies on solutions of essentially hydrophobic solutes would be expected to yield the most unambiguous evidence of hydrophobic interactions if such aggregates do in fact exist as time-averaged structures. Aggregated structures such as micelles or lipid bilayers have been extensively studied by diffraction and scattering techniques and some of these experiments are described in Chapters 2 and 4. However, such aggregates are geometrically well defined and consist of at least 50 individual ions or molecules.* Exchange problems do not interfere in their study by X-ray or light scattering.

The interactions between pairs or small groups of, say, alcohol or ether molecules or tetraalkylammonium ions in moderately dilute aqueous solution pose more subtle problems since long-time correlations between the molecules or ions constituting the aggregates probably do not exist, certainly not to the same extent as in micellar type systems. Thus the application of scattering techniques might not be expected to yield direct structural information. It is also questionable how much, if any, internal order can exist in such aggregates composed of a small number of essentially spherical molecules or ions. All known micellar or lamellar structures consist of molecules or ions with polar head groups and long, extended apolar chains or rigid aromatic residues. There have been suggestions[902] that molecules such as symmetric tetraalkylammonium halides or tert-butanol form micelles, but this must be regarded as most unlikely.

To the author's knowledge, only one X-ray diffraction study of simple aqueous solutions of essentially hydrophobic molecules has ever been pub-

---

* One very important difference between such structures and any hydrophobic aggregates of small molecules which might exist in aqueous solution is that in the former the alkyl groups are not in an aqueous environment. Indeed, it is usually stated that the alkyl groups in micelles, etc., behave as liquid hydrocarbon. The aqueous component is in contact only with the polar or ionic head group. In a dilute aqueous solution of, say, ethanol the position is very different.

lished[232]: The results clearly indicate the existence of time-averaged "structure" in aqueous solutions of ethanol, with x_2' and x_2'' very marked. On the other hand, for solutions of 1,4-dioxan the results indicate the lack of such "structure" (as compared to pure water): Once again x_2'' is clearly defined as that concentration ($x_2'' = 0.2$) at which the diffraction band is observed in the position which is characteristic for pure 1,4-dioxan; therefore beyond this concentration the water contributes no more to the X-ray intensity than would "dense water vapor" at the same temperature.

It has been suggested that association in simple solutions could be studied by X-ray methods in a manner similar to that employed for pure liquids and that the radial distribution function for the solute could be obtained by subtraction of the known (pure) water value from the experimental radial distribution of the solution. Such procedure implicitly assumes that the presence of the solute molecules has no effect on $g_{H_2O}(r)$, and this is clearly unrealistic. We have tried in Section 3 to summarize how $g_{H_2O}(r)$ might be affected by the phenomenon of hydrophobic hydration, and by now it seems clear by reference to Table III that $f_{11}^\circ \neq f_{11}$. Hence the subtraction of a "pure water" radial distribution function will not provide a useful indication of the distribution of solute molecules in the solution.

Hydrophobic dimers or higher aggregates, if they exist, will give rise to a microheterogeneous solution, i.e., there will be domains rich in, and others poor in, solute and water molecules, respectively. Such concentration and density fluctuations, if they are of a sufficient magnitude, should lend themselves to study by light and/or small-angle X-ray scattering methods.

Parfitt and his colleagues have investigated the light scattering by aqueous solutions of methanol, ethanol,[1137] and acetone,[1136] while Vuks et al. have studied solutions of propanol and dioxan.[1542,1543] In principle, scattering due to microheterogeneities can arise from density, concentration, and molecular orientation effects, all of which can be identified in the properties of the scattered light. Thus the observed scattering, in terms of the Raleigh ratio $R(\theta, \lambda)$ at a given angle θ and wavelength λ, is given by

$$R(\theta, \lambda) = R_d + R_c + R_o \tag{28}$$

The third term on the right-hand side of eqn. (28) represents the anisotropic contribution to $R(\theta, \lambda)$ and is a measure of any orientation ordering of the molecules. It can be obtained from $R(\theta, \lambda)$ and the depolarization ratio Δ. It can also be calculated from the anisotropy of the polarization tensor of the various molecular species. The terms R_d and R_c measure the density

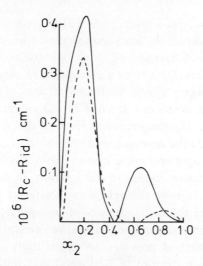

Fig. 14. Nonideal contribution to Raleigh scattering ratio due to concentration fluctuations in H_2O–EtOH (– –) and H_2O–Me_2CO (—) mixtures at 25°; $\lambda = 4358$ Å. From the data of Refs. 1136 and 1137.

and concentration fluctuations and can also be calculated. The latter is related to the chemical potentials, and hence to the activity coefficients, of the solution components. It can therefore be referred to an "ideal" value R_{id} which is characteristic of a solution that obeys Raoult's law. An analysis of $R(\theta, \lambda)$ indicates that $R_d \gg R_o > R_c$. Thus, for instance, for a 10 mol % solution of ethanol ($\theta = 90°$, $\lambda = 435.8$ nm), $R(\theta, \lambda) = 3.5 \times 10^{-6}$ cm^{-1}, of which R_0 amounts to 22% and R_c to 17%. The quantity $R_c - R_{id}$ is shown in Fig. 14 for two aqueous mixtures. Whereas the observed light scattering $R(\theta, \lambda)$ is due mainly to fluctuations in the density, the results suggest that composition fluctuation (i.e., solute clustering) largely accounts for the observed thermodynamic nonideality of the mixtures.* Unfortunately the measurements were only performed at a single temperature. If the concentration scattering is indeed caused by hydrophobic aggregation, one would predict R_c to increase with temperature.

Not much can be inferred from the experimental data about possible orientational microheterogeneities. The measured depolarization values,

---

* The fact that both curves in Fig. 14 exhibit a concentration for which $R_c - R_{1d} = 0$ does not mean that these mixtures are ideal. The opposite is in fact true—they show maximum deviations from ideal behavior. Both R_c and R_{1d} show a dependence on the refractive index change with concentration and at the concentrations in question the refractive index undergoes a maximum, so that $R_c = R_{1d} = 0$.

although subject to a large scatter, nevertheless seem to show a systematic concentration dependence. R_0 has been calculated on the assumption of a random distribution of nearest-neighbor orientations and compared with the experimental values. Deviations are found in all three aqueous systems, but hardly in the binary system methanol/ethanol. The authors are uncertain whether such deviations are due to (1) experimental error in the determination of Δ, (2) the inapplicability of the theory to hydrogen-bonded liquids, or (3) the nonrandom distribution of mutual orientations.

The method of light scattering cannot provide information about the size of any heterogeneous regions in the solution. In principle, such information is obtainable from small-angle X-ray scattering, but here the theory is not well developed for fluid mixtures of nonspherical molecules where the intermolecular forces depend on the relative orientations of the particles.

However, the method provides an opportunity of testing the thermodynamic fluctuation theories of Kirkwood and Buff[799] and McMillan and Mayer.[953] For a two-component system the scattered intensity $I(h)$ $[h = (4\pi/\lambda)\sin(\theta/2)$, where λ is the wavelength of the X rays and θ the angle between the incident and scattered beams] contains three radial distribution functions which are generally not known. The expression for the zero-angle intensity $I(0)$, however, only contains the *integrals* of these radial distribution functions

$$G_{ij} = \int_v [g_{ij}(r) - 1]\, dV \qquad (29)$$

which in turn are related to the thermodynamic properties of the solution. Thus for pure water the zero-angle scattering intensity is given as

$$\frac{I_1(0)}{I_e(0)V} = \varrho_1^\circ n_1^2 \varkappa_1^\circ kT \qquad (30)$$

where ϱ_1° and $\varkappa_1^\circ$ are the number density and isothermal compressibility of water, n_1 is the number of electrons in a molecule, $I_e(0)$ is the zero-angle intensity scattered by a single electron, and V is the volume.[1368]

For a binary solution $I(0)$ has three terms containing the $G_{ij}(r)$ expressions in eqn. (29). By substituting the "pure water" result in eqn. (30), G_{11} and G_{12} can be eliminated, and it can be shown that for the binary solution

$$\frac{I(0)}{I_e(0)V} = c_2\left(n_2 - \frac{n_1 V_2}{\bar{V}_1}\right)^2 (1 + c_2 G_{22}) + \frac{c_1 \varkappa I_1(0)}{c_1^\circ \bar{V}_1 \varkappa_1^\circ}\left[1 + \left(\frac{2l_2}{l_1} - \frac{\bar{V}_2}{\bar{V}_1}\right)\frac{w_2}{w_1}\right]$$

where $l_i = n_i/$molecular weight, c_i is the molar concentration, and w_i is the weight fraction.

Zimm has shown that for a condensed system (where terms involving $\varkappa$ can be neglected) G_{22} can be conveniently expressed by

$$\frac{G_{22}}{\bar{V}_2} = -\phi_1 \left(\frac{\partial(a_2/\phi_2)}{\partial a_2} \right)^{-1}$$

where ϕ_i and a_i are the volume fraction and activity of the ith compo-nent.[1622] The quantity $\phi_2 G_{22}/\bar{V}_2$ (Zimm cluster integral) is a measure of the excess of solute molecules in a given volume element over and above the mean value expected for a random mixture. Similarly $\phi_2 G_{12}/\bar{V}_2$ repre-sents the excess of solute molecules which are in the neighborhood of a given solute molecule. Also

$$(\phi_2/\bar{V}_2)G_{12} = - (\phi_2/\phi_1)(1 + c_2 G_{22}) \tag{31}$$

Thus if like molecules tend to aggregate, then $(\phi_2/\bar{V}_2)G_{12} < 0$.

A comparison of experimental and calculated zero-angle scattering intensities from benzene, ethanol, and water shows that $I(0)_{\exp}$ is $\sim 11\%$ in excess of the calculated value.[1368] This discrepancy may arise from the neglect of the orientation-dependent parts of the radial distribution func-tions. For solutions of sucrose and univalent electrolytes $I(0)$ agrees with values calculated from the Kirkwood–Buff theory.

Only one experimental study appears to have been reported on an aqueous solution of an essentially hydrophobic solute, namely tert-butanol.[77] Here $I(h)$ shows a complex angle dependence in the concen-tration range $0 < x_2 < 0.4$. It increases with concentration and passes through a maximum at $0.1 < x_2 < 0.2$ (corresponding to x_2'' for this sys-tem). The value of $I(0)$ also increases sharply with rising temperature. This behavior is characteristic of a system in which lower critical demixing takes place and this is of course consistent with the solution thermodynamics of water–alcohol mixtures.[486] The combination of the scattering and thermo-dynamic data allows the computation of the left-hand term in eqn. (31) and this is indeed found to be negative for all tert-butanol concentrations. Conversely the cluster integrals for solvent and solute show sharp peaks such that $\phi_1 G_{11}/\bar{V}_1 \simeq 30\phi_2 G_{22}/\bar{V}_2$.*

Unfortunately, the authors conclude that not much can be deduced from the $I(h)$ data regarding the probable sizes of clusters, although it is

---

* Maxima in the same composition range have also been observed in the ultrasound absorption of these mixtures (Chapter 9, Volume 2), but the origin of the acoustic relaxation process is not yet very clear. One important difference between the acoustic absorption maxima and the X-ray scattering intensity is that the former effect decreases with rising temperature so that the mixture becomes more "normal," whereas the scat-tering results clearly indicate the existence of a lower critical solution temperature.

suggested that in a mixture of $x_2 \simeq 0.11$ the cluster diameter might well be ≥ 20 Å.

Since the above results were reported the compressibilities of tert-butanol–water mixtures have been measured,[911] so that all the information required for the calculation of $I(0)$ is now available. The scattering experiments have also been repeated in greater detail[60] [see Fig. 15 for $I(0)$ as a function of x_2 and temperature] and the general trends agree with those discussed in this section. The pronounced shape scattering (i.e., dependence on scattering angle) of solutions in the composition range $x_2 \cong 11$ mol % ($= x_2''$) is surprising. This is of course absent for the pure components and also for solutions lying outside the critical concentration range.

The inset in Fig. 15 shows the temperature dependence of $\tilde{R}_g$, the radius of gyration, as obtained from Guinier plots. Here again the indications are that the size of the microheterogeneous regions increases with rising temperature.

At the time of writing the new data have not yet been fully analyzed, but with the new information available it is hoped that the three G_{ij} terms in the scattering equation can be separately evaluated and examined without the necessity of making assumptions of doubtful validity.

In summary we can say that, although the above approach to the study of hydrophobic aggregation has not yet been worked out in detail, neverthe-

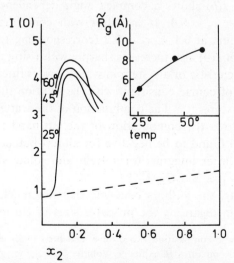

Fig. 15. X-ray zero-angle scattering intensities $I(0)$ of aqueous solutions of tert-BuOH at 25, 45, and 60°. The broken line indicates the "ideal" $I(0)$. The inset shows the temperature dependence of $\tilde{R}_g$, the radius of gyration calculated from $I(0)$ for a 10 mol % mixture. From Ref. 60.

less it has been shown that clustering by essentially hydrophobic molecules in aqueous solution is a real effect and the scattering methods should be able to probe more deeply into the conditions which govern hydrophobic clustering phenomena. Its particular advantage is that the scattering intensities are related via the Kirkwood–Buff and McMillan–Mayer theories to the solution thermodynamic properties.

4.4. Time-Dependent Properties

In principle, the line of reasoning presented in Sections 3.2 and 3.3 for the infinitely dilute solution can be extended to describe moderately dilute solutions in which solute interactions become important, but not to the extent of unduly altering the nature of the solvent interactions. Referring once again to Table III, an additional molecular pair distribution function f_{22} must now be considered to take account of solute–solute pair interactions, so that f_{soln} is now given by a combination of $f_{11} + f'_{12} + f_{22}$. It is seen that the solvation term f_{12} in Table III has now been changed to f'_{12}, because it is likely that the nature of the molecular hydration sphere will undergo a change when two hydrated solute molecules interact in solution. For the commonly accepted model of the hydrophobic interaction (see Fig. 1b) f'_{12} would be broader than f_{12}, since the formation of the solute pair destroys hydration contacts. For the same reason f_{11} would exhibit a broadening, i.e., a trend toward the value for pure water, f°_{11}.

As was seen earlier, direct information about f_{ij} can in principle be derived from scattering methods. There are, however, experimental and other problems which make the use of spectroscopic methods more attractive, although any information about f_{ij} will be of a more circumstantial nature. This is well illustrated by the various reports of investigations into the water proton nmr chemical shifts in solutions of essentially hydrophobic solutes. Since the proton chemical shift δ in water is an indicator of the degree and nature of the hydrogen bonding (see Chapter 6, Volume 1), any shift in δ resulting from the presence of solutes should in principle provide information about changes in the state of hydrogen bonding. The theory of chemical shifts predicts that upfield shifts in δ correspond to a decrease in the degree of hydrogen bonding and vice versa. Prior to 1968 the indications were that all solutes gave upfield shifts* and this was equated to a "water structure breaking" effect which could not easily be reconciled with the current interpretations of nuclear magnetic relaxation rates (see

---

* For hydroxylic solutes, the observed δ will correspond to an averaged hydroxyl proton resonance absorption.

Section 3.3). Glew *et al.*, on the basis of some very careful measurements on solutions of effectively hydrophobic solutes, discovered downfield shifts in δ at low concentrations and temperatures.[532,533]

These results were later confirmed,[26] and have since been subjected to thorough scrutiny and analysis.[769,790,1571] An "excess" chemical shift δ^E can be defined which measures the deviation of δ from the "theoretical" value obtained by the additivity principle. From the available data it appears that maxima in the $-\delta^E$ for different solutes occur at concentrations corresponding to their particular x_2' values, suggesting the implication of hydrophobic hydration and interactions. In addition, however, acid–base proton transfer between water and solute provides another possible mechanism for changes in the degree of hydrogen bonding. These factors, together with the rather rudimentary state of the theory of chemical shifts, make it difficult to derive hard information from the experimental data.

Considerable progress in the definition of the molecular nature of the hydrophobic interaction between simple molecules has been and is still being made by Hertz, Zeidler, and their colleagues at Karlsruhe by the methods of intermolecular nuclear relaxation techniques, already referred to in Section 3.3. Essentially the method involves the application of eqn. (13), the aim being to determine a, the distance between two protons situated on different molecules. The comparison of the distance with the expected value calculated on the basis of a uniform molecular distribution then indicates the presence or absence of solute aggregation in solution.*

Reference has already been made in Section 3.3 to a correlation function which measures time-dependent correlations between diffusive modes of two molecules and can be related to $(T_1)_{\text{inter}}^{-1}$. Thus $(T_1)_{\text{inter}}^{-1}$ is a linear combination of a number of spectral intensities, i.e.,

$$(T_1)_{\text{inter}}^{-1} = K \sum J(\omega_i) \tag{32}$$

and $J(\omega)$ is the Fourier transform of the correlation function P under discussion.†

---

* For a description of the theoretical background to the relationships between f_{ij} and the nuclear magnetic (dipole) relaxation rate, the reader is referred to Refs 539, 540, 542, 653, and 654.

† Actually

$$P = \int_0^\infty \overline{\{Y_{2m}(0)Y_{2m}(t)/[r(0)]^3[r(t)]^3\}}, \qquad m = 0, \pm 1, \pm 2, \ldots$$

where Y_{2m} are the normalized spherical harmonics of the second order at times zero and t, respectively, and $r(0)$ and $r(t)$ are the relative positions and orientations of the two nuclei under discussion at times zero and t.

It can be shown that for the organic component $(T_1)_{\overline{\text{inter}}\,s}^{-1} D/N_I$ is proportional to $P_{22\,s}D$, where N_I is the number of nuclear spins per cm³ in the solution. Similarly for the aqueous component $(T_1)_{\overline{\text{inter}}\,h}^{-1} D/N_I$ $\propto P_{11\,h}D$. Goldammer and Zeidler first observed that in aqueous mixtures of tetrahydrofuran and tert-butanol $(T_1)_{\overline{\text{inter}}\,s}^{-1} D/N_I$ fell sharply in the concentration range $0 < x_2 < 0.2$, whereas for a uniform distribution of molecules this quantity should be independent of concentration. The observed results, together with the analysis of rotational correlation times $_h\tau$ and $_s\tau$ discussed in Section 3.3, led to the suggestion that "for some mixtures the organic component, as well as the water associate preferentially among themselves."[539]

Having established the possibility of microheterogeneities over quite narrow concentration ranges in dilute aqueous solutions, the phenomenon has been subjected to a much more detailed and systematic scrutiny.[542,654] From measurements of intermolecular proton relaxation rates in D_2O solutions of CH_3CD_2COOD and CD_3CH_2COOD the results shown in Fig. 16 were obtained. Once again it is apparent that positive correlations exist between alkyl groups in dilute solution, and the effect between methylene groups appears to be more marked than between the terminal methyl groups. By allowing for the nonspherical nature of the acid molecules (the correlation functions were derived for spherical particles) and making various simplifying assumptions, the intermolecular proton–proton distance is calculated as 2 Å and the interproton correlation exists for $\sim$10 psec $(= {}_s\tau_{CH_3}, {}_s\tau_{CH_2})$.

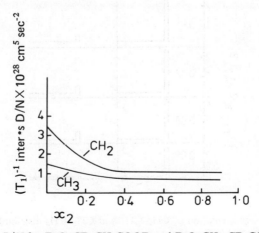

Fig. 16. $(T_1)_{\overline{\text{inter}}\text{-}s}^{-1}D/N$ for D_2O–CD_3CH_2COOD and D_2O–$CH_3 \cdot CD_2COOD$ mixtures, describing intermolecular proton relaxation; from Ref. 542.

The pair distribution functions f_{22} for CH_3-CH_3 and CH_2-CH_2 pairs in propionic acid can now be evaluated approximately by first calculating the probability that an acid molecule (actually a proton) is somewhere in a sphere of 2 Å radius from the reference proton and then converting to a probability density (radial distribution function) $G(r)$ of molecular pairs. Figure 17 shows the summarized results for methylene pairs and methyl pairs in solutions of propionic acid of increasing concentration. Even with the various simplifying assumptions which had to be made in the calculation of $G(r)$, the results clearly indicate (1) that pair (or multiple) interactions between nonpolar residues are real and most pronounced in relatively dilute solution, and (2) that CH_2-CH_2 correlations are more pronounced than CH_3-CH_3 correlations for the same molecular species. This latter finding is surprising and has yet to be explained.

It has been mentioned in several places in this chapter that aqueous solutions of the symmetric tetraalkylammonium halides provide good models for studies of the hydrophobic interaction, and their properties have recently

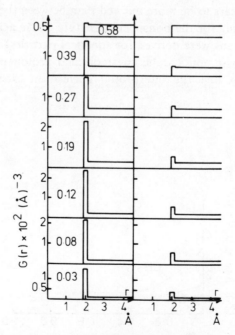

Fig. 17. Schematic representation of CH_2-CH_2 and CH_3-CH_3 distribution functions for propionic acid in D_2O solution. Numbers indicate mole fractions of acid. For method of calculation see text and Ref. 542.

been reviewed.[1570]* Intermolecular nuclear relaxation measurements have been performed on such solutions in an attempt to analyze the concentration dependence of $(T_1)^{-1}_{\text{inter}}$ in terms of cation pair correlations.[651] Ideally such investigation, in a manner analogous to that descried above for the carboxylic acids, would involve the use of derivatives of the type $(CH_3 \cdot CD_2)_4N^+$, $(CD_3CH_2)_4N^+$, etc. Although these were in fact used for self-diffusion measurements, the relaxation studies were performed on alkyl derivatives with either ^1H or ^2H substitution throughout. From measurements of $(T_1)^{-1}_{\text{inter}}$ and D, the closest distance of approach of cation pairs is calculated as 3.7 and 1.6 Å for Me_4N^+ and Et_4N^+, respectively. This trend is unexpected, unless one assumes that cation pairing does not take place in the case of the smaller ion.

At the time of this investigation the cation pair distribution functions for alkylammonium halides had not yet been calculated, but since then Friedman and his colleagues have developed methods for calculating $g_{++}(r)$ from activity coefficient data[1214] (see also Chapter 1, Volume 3) so that a reevaluation of the nuclear magnetic relaxation data is now possible. Certainly the shape of the plot of $(T_1)^{-1}_{\text{inter}}D$ versus concentration for Et_4NCl is unexpected and suggests a nonuniform cation distribution (see also Section 4.5.2).

A related method for studying interactions between essentially hydrophobic particles in aqueous solution makes use of the effect of a diamagnetic hydrophobic molecule on the electron paramagnetic resonance (epr) spectrum of a hydrophobic paramagnetic probe on another (diamagnetic) hydrophobic molecule.[734,736] Spin probes which have been used in such

TEMPO   OTEMPO   DTBN

---

* However, two slight objections against their use should be mentioned: (1) Counterion effects may influence the experimental measurement in such a way that they cannot be unambiguously isolated from the experimental results. Of course, counterion effects can also be turned into an advantage, as witness ^{79}Br and ^{81}Br nmr relaxation studies in solutions of tetraalkylammonium bromides.[904,1635] (2) Although on a macroscopic scale the symmetric R_4N^+ ions may be successfully treated as spherical particles,[375] this is certainly not the case where the experimental point of reference is in a particular location within the ion, e.g., in a study of intermolecular nuclear magnetic relaxation effects with reference to a given proton within an alkyl chain, the motion of the proton–proton vector cannot be described on the basis of spherical symmetry.

experiments include 2,2,6,6-tetramethylpiperidine nitroxide (TEMPO), 2,2,6,6-tetramethylpiperidone nitroxide (OTEMPO), and di-tert-butyl nitroxide (DTBN).

A comparison was made between the observed and simulated spectra; in particular the linewidths W_H were analyzed in terms of the following equation:

$$T_2^{-1} = \pi(3W_H)^{1/2} = a_0 + a_1 m_N + a_2 m_N^2 \tag{33}$$

where $m_N = 0, \pm 1$. Essentially two correlation times can be extracted from W_H [(777)]:

$$W_H(m = 0) = K + L\tau_\theta + [M\tau_\theta/(1 + \omega^2\tau_\theta^2)] + N\tau_J \tag{34}$$

Here τ_θ is the reorientational correlation time of the probe and τ_J is the correlation time for the angular velocity of the probe; L, M, and N are constants which are readily obtainable from the line shape analysis; and K is a constant of more uncertain origin. This makes it impossible to calculate τ_J directly and rigorously from the experimental data and eqn. (34). Jolicoeur and Friedman therefore describe three independent methods for estimating τ_J from the observed linewidths. One method depends on the proposition that for Brownian rotational motion both τ_θ and τ_J can be readily expressed as

$$\tau_\theta = 4\pi r^3 \eta/3kT \qquad \text{[see eqn. (11)]}$$

$$\tau_J = I/8\pi r^3 \eta$$

where r is the hydrodynamic radius of the probe, I is its moment of inertia, and η is the viscosity of the medium.* Thus

$$\tau_J = (I/6kT)(1/\tau_\theta)$$

so that for $\omega\tau \ll 1$, W_H is given by

$$W_H = K + (kT/\eta)\alpha + (\eta/kT)\beta \tag{35}$$

where α and β are positive and related to the constants in eqn. (34). $W_H(\eta, T)$ undergoes a minimum at

$$\eta/kT = (\alpha/\beta)^{1/2}$$

_____

* See p. 21 for a discussion of the uncertainties regarding the viscosity term in the equations for Brownian diffusion.

Jolicoeur and Friedman have shown[735] that the proportionality between τ_θ and η is very close, i.e., Brownian motion is a good approximation for the reorientation of the probe. Thus any deviation of W_H from the value predicted by eqn. (35) has been attributed to non-Brownian behavior of τ_J.

The analysis of W_H as function of η and T shows several interesting and unexpected features which are peculiar to aqueous solutions, i.e., they are not features of polar solvents in general. Thus W_H shows a minimum at low values ot T/η. After eliminating other possible origins for this effect, the authors conclude that the observations can be accounted for by the postulate that τ_J increases as the temperature decreases from 25°. This is consistent with the view that hydrophobic hydration is akin to clathration. It must therefore appear that the rotational diffusion of the hydrophobic TEMPO probe is highly non-Brownian.

Application of the three methods for calculating τ_J leads to the general result that $\tau_J \sim 0.01\tau_\theta$. By adopting the Brownian rotation approximation for τ_θ, eqn. (34) can be rewritten as

$$W_H{}^\circ = W_H(\text{exp}) - W_H(\tau_\theta) = K + N\tau_J$$

and the effects on $W_H{}^\circ$ produced by increasing concentrations of essentially hydrophobic molecules can then be studied. Figure 18 shows two such sets

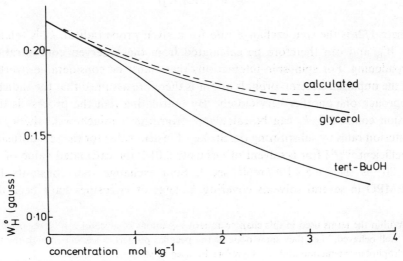

Fig. 18. $W_H{}^\circ$ of TEMPO spin probe in aqueous solutions of glycerol and tert-BuOH at 25°. Probe concentration = 100 μM. The broken line shows $W_H{}^\circ$ with calculated values of τ_θ and τ_J, as described in text. From Ref. 735.

of results for aqueous solutions of glycerol and tert-butanol.* It is found that pronounced deviations are observed in $W_H{}^{\circ}$ and for isoviscous solutions the essentially hydrophobic solute produces a greater deviation from the behavior to be expected for Brownian rotation and a calculated τ_θ. The observed trends for DTBN are similar but qualitatively different from OTEMPO, the spin-rotation interactions of which do not appear to be sensitive to the proximity of a hydrophobic molecule.

Employing their intepretation of W_H and $W_H{}^{\circ}$ in terms of τ_θ and τ_J, Jolicoeur and Friedman have developed a two-state model according to which the probe may exist in two solution environments, either in an isolated state or interacting with another hydrophobic molecule. With the aid of this model they calculate equilibrium constants for the hydrophobic interaction and show that the degree of hydrophobic association depends on the size of the alkyl group.

Another way of using the epr technique to study the hydrophobic interaction has been employed by Barratt et al.[91] and is based on line broadening by electron spin exchange. This is a process in which unpaired electrons on two colliding probe radicals exchange their spin states. It can only be detected when the electron spin states involved are different. The second-order rate constant k_2 for electron spin exchange is given in terms of the radical concentration [R] by

$$k_2 = 1/(2t[\text{R}])$$

where $1/2t$ is the spin exchange rate for a given probe radical. k_2 is related to W_H and can therefore be calculated from the observed concentration broadening. For spin–spin interactions there must be considerable overlap of the unpaired electron orbitals and it is therefore assumed that the radicals approach one another very closely. By postulating that the process is diffusion controlled, k_2 can be calculated from the Smoluchowski theory of diffusion rates by substituting the Stokes–Einstein value for the self-diffusion coefficient.[1019]† For a solvent of viscosity 0.01 P the calculated value of k_2 at 20° is then 6.5×10^9 $l\,\text{mol}^{-1}\,\text{sec}^{-1}$. Spin exchange rate constants for TEMPO in several solvents covering a range of viscosities have been de-

---

* Within the terms used in this chapter aqueous solutions of glycerol can be regarded as "well behaved," i.e., they show none of the peculiar properties associated with the hydrophobic interaction and can therefore be used as reference solutions.

† It would be preferable to determine D by an independent technique, e.g., spin echo nmr, but the low aqueous solubility of hydrophobic spin probes makes this a difficult task.

termined and reduced to a common viscosity of 0.01 P. Experimental $10^{-9}k_2$ values are 4.0 (pyridine), 6.3 (ethanol), 4.4 (n-decane), 2.9 (acetone), and 2.1 (water). The low spin exchange rate ($\sim$30% of the calculated value) in water can be accounted for in a number of ways, all of them relying on the existence of a hydrophobic (or other?) interaction between probe radicals. If pairs or cluster aggregates have a considerable half-life, then the calculated Smoluchowski collision rate would be in excess of the spin exchange rate. On the other hand, if the rms distance between two probe radicals taking part in hydrophobic pairing is greater than the sum of the "hard-sphere" or hydrodynamic radii, then the efficiency of the interaction in terms of spin exchange would be reduced.

Finally, mention should be made of the application of infrared spectroscopy to the study of hydrophobic interactions. This has been confined to an analysis of the effects of temperature and concentration on the 0.97-μm band of water in solutions of tetraalkylammonium bromides.[210] This band is mainly determined by O–H stretching frequencies and is therefore an indicator, albeit a qualitative one, of the degree and/or strength of hydrogen bonding.

An analysis of the band shifts indicates that similar effects are produced by lowering the temperature of water and raising the concentration of R_4NBr, with shifts becoming more pronounced as the size of R is increased. There is, however, a limiting temperature, $\sim$60° for Bu_4NBr, beyond which no further shifts from the pure water value can be observed.

It is interesting to note that in the case of Et_4NBr the hydrophobic and charge contributions cancel out, giving this substance the appearance of an "ideal" electrolyte. A similar conclusion has been reached by Kay (see Chapter 4, Volume 3) on the basis of a variety of physicochemical measurements.

4.5. Theoretical Studies of the Hydrophobic Interaction between Simple Molecules

The main interest in the hydrophobic effect centers around its involvement in biopolymer conformational states and in the formation of complex molecular aggregates, e.g., lipoproteins. The theoretician attempting to define the energetics of the interaction is likely to have only a nodding acquaintance with the details of biochemical structures and processes, and he will feel intuitively that the available theoretical techniques cannot yet be applied to such systems. He will therefore try to simulate the effect to be studied by substituting manageable small-molecule analogs for such naturally occurring molecules as phospholipids, glycerides, proteins, cholesterol, nucleotide bases, etc. It is, however, up to the user of models to convince himself and those at whom he directs his theories that the particular model used does exhibit the effects which are to be investigated.

Nemethy has sounded a timely note of warning in this respect.[1090] He rightly points out some of the problems which arise when the hydrophobic interaction is compared to the transfer of small molecules from a nonpolar solvent to water. Here differences in the nature of the solute may be of great importance, e.g., an ethyl or isopropyl side chain on a polypeptide backbone may not be strictly comparable with ethane and propane, or indeed with ethanol or isopropanol.

Then again, if the "real" process involves the transfer of a molecule or a group from an aqueous medium into an effectively nonpolar medium (e.g., micelle formation, phospholipid bilayer formation, globular protein folding), the water → hydrocarbon transfer may constitute a permissible model. Where, on the other hand, the hydrophobic interaction takes place in water *without* removal into a nonpolar environment (e.g., clusters of small molecules, α-helix formation by polypeptides, binding of small molecules by biopolymers, sol–gel transition of biopolymers), then the above model experiment does not reflect the real situation.*

4.5.1. Statistical Mechanical Treatments Based on Hydrocarbons

As was the case for hydrophobic hydration (Section 3.4.1), the studies of Nemethy and Scheraga also constitute the first attempt to quantify the phenomenon of hydrophobic bonding along the lines suggested by Kauzmann. From a historical point of view, therefore, they should be discussed first. On the other hand, the objective of Nemethy and Scheraga was the elucidation of protein stability and they therefore employed amino acids rather than alkanes as model compounds. Their results and conclusions are reviewed in Section 4.5.2.

More recently Ben-Naim has devoted a considerable effort to the problem of a quantitative description of the hydrophobic interaction. In a series of publications[117,118] he treats the problems of pairwise and multiple interactions of alkanes† within the framework of classical statistical mechanics. For a system composed of N water molecules and two simple solute

---

* Attempts to "force" such processes to fit the results of the model transfer experiments can lead to paradoxical situations. Some of these are examined in Chapter 5.
† See also Chapter 11 of Volume 2.

molecules the Helmholtz free energy can be written as

$$A_{N+2}(\mathbf{r}_2, \mathbf{r}_{2'}) = A^\circ + U_{12}(\mathbf{r}_2, \mathbf{r}_{2'}) + A^{H\phi}(\mathbf{r}_2, \mathbf{r}_{2'}) \qquad (36)$$

Here $\mathbf{r}_2$ and $\mathbf{r}_{2'}$ denote the distances and orientations of the two hydrophobic molecules with respect to some reference point; A° is a term describing the properties of the solvent; $U_{12}(\mathbf{r}_2, \mathbf{r}_{2'})$ is a solute pair potential isolating the contributions that are solvent independent; and the last term is the hydrophobic part of the pair interaction. Omitting the orientation-dependent contributions to the pair interactions, $A^{H\phi}$ measures the work done in allowing two solute molecules to approach each other *in water* from an infinite distance to a distance $r_{22'}$. The term $U_{12}(r_{22'})$ describes the same process carried out in vacuum. A special case of eqn. (36) can be defined where $r_{22'} = \sigma'$, an effective solute molecular diameter. For the case of two methane molecules, σ' is approximated by the length of the C–C bond in ethane, so that

$$\delta A^{H\phi}(\sigma') = A^{H\phi}(r_{22'} = \sigma') - A^{H\phi}(r = \infty) = \Delta\mu^{\ominus}_{C_2H_6} - 2\Delta\mu^{\ominus}_{CH_4}$$

where the $\Delta\mu^{\ominus}$ terms are the standard free energies of solution. The quantity $\delta A^{H\phi}(\sigma')$ is shown for a series of solvents in Fig. 19, which illustrates the unique nature of water as solvent. The difference in $\delta A^{H\phi}$ between water

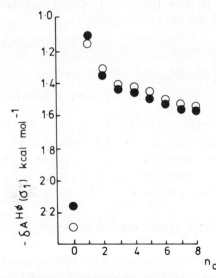

Fig. 19. $\delta A^{H\phi}(\sigma_1)$ for two simple hydrophobic particles in a number of solvents at 25° (●) and 35° (○). From Ref. 117.

and the alcohols, $\sim$600 cal mol^{-1}, is of the same order as the thermal energy of the molecules at the same temperature (25°), so that, in spite of the thermal energy, the excess attraction in water enables the two methane molecules to approach one another.

The temperature dependence of $\delta A^{\mathrm{H}\phi}$ in different solvents, as shown in Fig. 19, is also worthy of note. It is pronounced for water and much less so (but of opposite sign) for other, less polar solvents.

Two other methods for the evaluation of $\delta A^{\mathrm{H}\phi}(r_{22'})$ are considered by Ben-Naim: In one case $r_{22'}$ is put equal to zero. This corresponds to a process where the two solute molecules are brought from infinity to zero distance and the distribution of water molecules around the new *single* particle is chosen such that its field of force is equivalent to twice the field of force produced by the original single particle. As a tentative example, the process $2CH_4 \rightarrow Xe$ is put forward. Finally the case is considered where $r_{22'}$ is equal to the "contact distance" σ. This is taken to be the value of $r_{22'}$ at which $U_{12}(r_{22'})$ [see eqn. (36)] becomes repulsive. Estimates of $\delta A^{\mathrm{H}\phi}(\sigma)$ can be obtained by employing the Percus–Yevick approximation for hard-sphere fluids, or by numerically solving the Percus–Yevick equation for real liquids.

Comparison of the above three methods leads to the result that calculated values for $\delta A^{\mathrm{H}\phi}(r_{22'})$ decrease in the order $r_{22'} = 0, \sigma', \sigma$. Of the three approaches, the first can lend itself to comparison with experiment; the other two are of intrinsic theoretical interest, but the necessary experimental data for their practical exploitation are not yet available.

The treatment has been extended to the interaction between M hydrophobic particles, and in a manner analogous to the formation of the pair interaction described by eqn. (36) one can write

$$A_{N+M}(\mathbf{r}^M) = A_N{}^\circ + U_M(\mathbf{r}^M) + A_M^{\mathrm{H}\phi}(\mathbf{r}^M) \qquad (37)$$

Here $\mathbf{r}^M$ is a given configuration of the M particles ($\mathbf{r}^M = \mathbf{r}_2 \mathbf{r}_{22'} \ldots \mathbf{r}_M$) and the other symbols have the same significance as in eqn. (36). The following questions are then posed: (1) Is there any configuration $\mathbf{r}^M$ for which $A_{N+M}(\mathbf{r}^M)$ has a minimum and, if so, what is its minimum value? (2) Does $A_{N+M}(\mathbf{r}^M)$ have any particular low minimum in water, compared to other solvents?

Once again a useful approximation is to put $\mathbf{r}^M = \sigma$, where σ is a configuration for which the M particles might be situated at short (e.g., hard sphere) distances from one another. For instance, for the process $2CH_4 \rightarrow C_2H_6$ discussed above, $\mathbf{r}^2 = \sigma(\text{ethane})$ signifies two methane molecules at a configuration consistent with the equilibrium distance in ethane.

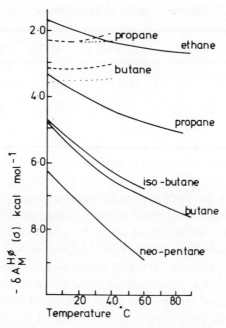

Fig. 20. $\delta A_M^{H\phi}(\sigma)$ as computed from eqn. (39) for alkanes in water (—), methanol (– –) and ethanol (...). From Ref. 118.

This process is accompanied by a change in Helmholz free energy as follows:

$$A_{N+M}(\mathbf{r}^M = \boldsymbol{\sigma}) - A_{N+M}(\mathbf{r} = \infty)$$
$$= U_M(\mathbf{r}^M = \boldsymbol{\sigma}) + A_M^{H\phi}(\mathbf{r}^M = \boldsymbol{\sigma}) - A_M^{H\phi}(r^M = \infty) \tag{38}$$

where the first term on the right-hand side is independent of the nature of the solvent. The process visualized in eqn. (38) can be simulated by allowing a number of methane molecules to approach one another to a configuration $\boldsymbol{\sigma}$ to form higher members of the homologous series. Thus a triplet CH_4 interaction gives rise to propane, etc.* Then

$$\delta A_M^{H\phi}(\boldsymbol{\sigma}) = \Delta\mu_M^{\ominus} - M \, \Delta\mu_{CH_4}^{\ominus} \tag{39}$$

Figure 20 shows the computed $\delta A_M^{H\phi}(\boldsymbol{\sigma})$ values according to eqn. (39) for $2 \leq M \leq 5$. Included in the figure are some comparative data for the "solvophobic" interaction in methanol and ethanol. The most striking

---

* A differentiation is made between situations where σ is or is not affected by internal degrees of freedom, e.g., n-butane and iso-butane, respectively.

TABLE IX. Thermodynamics of Hydrophobic Pairing at 20° [a]

	$2CH_4$	$2C_2H_6$	$4CH_4$	$2C_6H_6$
$-\delta G^{H\phi}$, cal mol^{-1}	2112	1500	5724	1136
$\delta H^{H\phi}$, cal mol^{-1}	1300	3000	5600	3200
$\delta S^{H\phi}$, cal (mol deg)$^{-1}$	11	15	37	14

[a] From Ref. 118. See text for method of calculation.

difference between the alcohols on the one hand and water on the other lies in the magnitude of the effect with increasing M. With some imagination one could say that the trend might be similar for the three solvents, although the initial steep decrease in $\delta A_M^{H\phi}(\sigma)$ is not paralleled by the alcoholic solvent. Interesting features of the aqueous solution data are the constant increase (in a negative sense) of $\delta A_M^{H\phi}(\sigma)$ with increasing M for the n-alkanes and the increasingly more pronounced temperature dependence. The branched chain derivatives (no internal degrees of freedom) do not appear to conform to the same pattern, and this is also borne out by results from experimental thermodynamic studies on a number of homologous series of alkyl derivatives (Chapter 5, Volume 2).

A systematic study of alkane solubilities in H_2O and D_2O has been performed to test the above theoretical predictions.[121] For $\delta A^{H\phi}(\sigma')$ the experimentally more accessible quantity $\delta G^{H\phi}(\sigma')$ has been substituted, and the hydrophobic interaction has been studied with the aid of the following hypothetical process:

$$2CH_4 \rightarrow C_2H_6$$
$$2C_2H_6 \rightarrow C_4H_{10}$$
$$4CH_4 \rightarrow C_4H_{10}$$
2-benzene → biphenyl

Table IX summarizes the thermodynamic quantities associated with hydrophobic pairing at 20°. Similar measurements in D_2O were performed and the results indicate that the hydrophobic interaction in H_2O is stronger,*

* The authors report that this agrees with a conclusion reached by Kresheck *et al.* based on $H_2O \rightarrow D_2O$ transfer studies with alkanes, alcohols, and amino acids.[819] However, the opposite appears to be true in that Kresheck *et al.* predict that the hydrophobic interaction should be stronger in D_2O than in H_2O.

although the reverse is the case for benzene. Both $\Delta H^{H\phi}$ and $\Delta S^{H\phi}$ are positive in all cases but it is not clear why the hydrophobic interaction between two C_2H_6 molecules should be less favorable than that between two CH_4 molecules and particularly why the interaction between four CH_4 molecules appears to be so much more favorable.

In a recent reexamination of the aqueous solution behavior of benzene Green and Frank have remarked upon the implications of the data in Table IX.[565] By writing $\Delta G^{H\phi}$ as

$$\Delta G^{H\phi} = -RT \ln [K_D^{(aq)} - K_D^{(gas)}]$$

where K_D is the dimerization equilibrium constant, they obtain an estimate of $K_D^{(aq)}$ from a knowledge of $\Delta G^{H\phi}$ (see Table IX) and the gas second virial coefficient. This leads to the result that benzene is 24% dimerized in aqueous solution, a finding which is quite incompatible with the observation that the solutions obey Henry's law reasonably well up to saturation. Thus the $\Delta G^{H\phi}$ values (at least for benzene) in Table IX cannot be reconciled with a "real" benzene dimer.*

It is too early to gauge the success of the Ben-Naim formalism as embodied in eqns. (36) and (37). It is certainly arguable that it has not yet achieved the degree of refinement that would make the numerical estimates realistic, e.g., one can conceive that cyclobutane would be a better model than n-butane for the hydrophobic interaction between two ethane molecules. Also, a comparison, where this is possible, with other numerical estimates of $\Delta G^{H\phi}$ shows up discrepancies. Thus, for the maximum hydrophobic interaction between pairs of alanine molecules, namely

$$H_2N \diagdown NH_2$$
$$CH\!-\!CH_3 \quad CH_3\!-\!CH$$
$$HOOC \diagup COOH$$

which should be similar to Ben-Naim's pairs of CH_4 molecules at a distance of σ_1 (see above), Nemethy and Scheraga[1093] have calculated a value of -700 cal mol^{-1} at 25° compared to -2168 cal mol^{-1} reported by Ben-Naim et al. for methane.[121] The inconsistency between the calculated $\delta G^{H\phi}$ value for benzene and its ideal (Henry's law) behavior has already

---

* It can also be argued that diphenyl is not a particularly good model for the study of hydrophobic interactions between two overlapping aromatic ring systems.

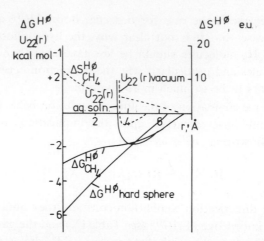

Fig. 21. $\Delta G^{H\phi}(r)$ and $\Delta S^{H\phi}(r)$ for hard spheres and methane molecules. Included also is the pair potential $u_{22}(r)$ for methane in the gas phase (vacuum) and the corresponding calculated value in aqueous solution. From Ref. 320.

been touched upon.* Nevertheless, as a first attempt to apply the rigorous methods of statistical thermodynamics to a well-documented important physical phenomenon, the studies of Ben-Naim deserve the close attention not only of physicists and physical chemists, but also of biophysicists and biochemists who profess an interest in problems relating to molecular conformations of biopolymers.

A slightly different approach to the evaluation of $\Delta G^{H\phi}$ has more recently been taken by Dashefsky et al.,[320] who based their calculations on a combination of statistical mechanics and Monte Carlo simulations. Atom–atom potential functions of the Kitaygorodsky type[776] were used, together with a semiempirical hydrogen bond potential. The $\Delta G_s{}^\circ$ values had previously been calculated (see Section 3.4.2). Perhaps one disadvantage of the procedure was that the sample was limited to 64 water molecules (i.e., almost 2 M aqueous methane). On the other hand, the advantage of the method compared to the approach taken by Ben-Naim was that the distance dependence of $\Delta G^{H\phi}$ could be evaluated. Figure 21 shows $\Delta G^{H\phi}(r)$ for hard spheres (cavities) and methane molecules. It is shown that two cavities of diameter 4 Å interact in a similar manner as two CH_4 molecules. It is also of interest to note that at $r = 0$, $\Delta G^{H\phi}$ does not differ much from

---

* It is interesting to note, however, that for the hydrophobic pairing of alanine with phenylalanine, Nemethy and Scheraga quote $\Delta G^{H\phi} = -1400$ cal mol^{-1},[1093] which is in very good agreement with the Ben-Naim estimate of -1200 cal mol^{-1}.

$-\Delta G_s{}^\circ$ for methane. The entropic origin of the hydrophobic interaction becomes apparent from $\Delta S^{\mathrm{H}\phi}(r)$, included in Fig. 21. On the other hand, $\Delta H^{\mathrm{H}\phi}$ (not shown) is positive, but small and almost independent of r.

Also included in Fig. 21 is the effective pair potential of methane $u_{22}(r)$ in the gas phase and in aqueous solution. It is seen that the potential of average force is much larger in aqueous solution (compare 0.3 kcal mol^{-1} in vacuum with 1.8 kcal mol^{-1} in solution; the equilibrium distance also appears to be somewhat longer for the aqueous solution).

The very important point is made by Dashefsky et al. that in conformational energy calculations on proteins, etc. it is the solution potentials which should be employed, rather than the vacuum potentials.

4.5.2. Statistical Thermodynamic Treatments Based on Other Small Molecules

Although as credible models for hydrophobic particles the alkanes or rare gases can hardly be improved upon, the experimental tests of any theoretical predictions are all but impossible. Furthermore, in most "real" systems the hydrophobic groups are attached to highly polar residues, be it in monomers or polymers. Thus each hydrophobic "molecule" contains at least one dipole, but practical experience has shown that molecules which possess only one such dipole, e.g., —O—, —OH, —NH, >C=O, etc., exhibit properties which are, to a first approximation, characteristic of truly apolar molecules. It has also been shown conclusively that any specific hydration effects between such a dipole and water do not affect materially the hydrophobic interactions of the remainder of the alkyl chain, but are localized and experienced only by the carbon atom to which the dipole is bonded.[489] It is doubtful whether this can also be true for ionic head groups. Thus, it has been reported that for n-alkyl sulfates the terminal sulfate group affects the hydrophobic interactions of the chain as far as the fifth carbon atom.[261] On the other hand, the symmetric tetraalkylammonium halides, apart from the first member of the series Me$_4$N$^+$, behave more like "soluble hydrocarbons" than symmetric electrolytes, and their physical properties exhibit the same chain length dependence as is observed for the homologous series of alcohols and ethers.

The first systematic investigation into the phenomenon of the hydrophobic interaction was that of Nemethy and Scheraga.[1093] The treatment was developed from their statistical thermodynamic model for water and aqueous solutions of hydrocarbons and therefore benefits from the merits and suffers from the shortcomings of this model. The main features of the hydrophobic interaction between, say, the apolar side chains of two amino acid molecules in solution is the substitution of solute–water contacts by solute–solute contacts, i.e., "... the approach of two or more amino acid side chains until they touch (within their van der Waals radii) and thereby decrease the number of water molecules in contact with them. This can be considered as a partial reversal of the solution process..." (see Fig. 1b). From the quantities which characterize pure water and the aqueous solution of the corresponding hydrocarbon, Nemethy and Scheraga were therefore able to calculate $\Delta G^{H\phi}$ with a minimum number of further assumptions. If Δn_h solute–water contacts are removed (n_h is the number of nearest neighbors of the apolar group at infinite dilution) and the water so "liberated" is assumed to revert to its bulk state, then for the process depicted in Fig. 1b

$$\Delta G^{H\phi} = \Delta n_h [G(\text{pure water}) - G(\text{water in hydration shell})]$$
$$- \Delta G(\text{side chains})$$

ΔG(side chains) is itself composed of several terms, specifying the number Z_R of $CH_2 \cdots CH_2$, $CH_3 \cdots CH_2$, etc., pair contacts, their energies, and effects of restrictions on internal bond rotations. This last term only provides a small contribution to $\Delta G^{H\phi}$ in the case of free amino acids, although it is likely to become important where the interacting apolar groups are joined to the same polymer backbone. For the most favorable type of interaction (two groups are completely free to approach each other), Δn_h varies from four (for pairs involving ala), to ten (for ileu–ileu, ileu–met), and twelve (for phe–phe).* Under the same conditions Z_R covers a range from two (ala–any other amino acid) to five (ileu–ileu, ileu–met, etc.) and eight (phe–ileu).

From the temperature and pressure dependences of $\Delta G^{H\phi}$ the other thermodynamic quantities characterizing the hydrophobic interaction can be calculated and Table X shows a selection of such data at 25° for a number of amino acid side-chain pairs. The model predicts that $\Delta C_p^{H\phi} < 0$, $\partial(\Delta C_p^{H\phi})/\partial T < 0$, and that at higher temperatures, as "water structure" breaks down, $\Delta H^{H\phi} < 0$, so that the term hydrophobic interaction loses its meaning because the solubility relationships will then be governed by the component interaction energies rather than by entropic factors.

The above discussion has dealt with the "ideal" hydrophobic interaction, i.e., both apolar moieties are free to approach each other. In polymeric systems this situation is of course rare. Nemethy and Scheraga there-

---

* ala = alanine, ileu = isoleucine, met = methionine, phe = phenylalanine.

TABLE X. Thermodynamics of the Hydrophobic Interaction at 25° between Pairs of Isolated Amino Acid Side Chains[1093]

Composition of pair	$\Delta G^{H\phi}$, cal mol^{-1}	$\Delta H^{H\phi}$, cal mol^{-1}	$\Delta S^{H\phi}$, cal mol^{-1} deg^{-1}	$\Delta V^{H\phi}$, cm^3 mol^{-1}
ala–ala	− 700	700	4.7 ⎫	
ala–ileu	− 600	700	4.4 ⎬	0.74
ileu–ileu	−1500	1800	11.1 ⎭	
phe–phe	−1400	800	7.5	0.65

fore distinguish between several variants of the hydrophobic interaction. For instance, for the case where the two side chains cannot approach each other to make direct van der Waals contact (e.g., an α-helix) the interacting apolar moieties may interact through a layer of water molecules* (see Fig. 1e). By making the necessary adjustments in the various parameters, as dictated by the nature of the model, $\Delta G^{H\phi}$, etc., can be calculated for the water-separated hydrophobic interaction. Nemethy and Scheraga conclude that "the (contact) hydrophobic interaction is strongly favored over the structure involving a single water layer. Therefore it will form whenever this is sterically possible." It is also stated that, because van der Waals forces fall off rapidly with distance, long-range hydrophobic interactions (i.e., through several water–molecular layers) do not contribute significantly to $\Delta G^{H\phi}$.

Several other cases of relevance to more complex situations have been discussed by Nemethy and Scheraga, e.g., the transfer of a nonpolar side chain from an aqueous to a nonpolar environment.† For this particular case, $\frac{1}{2}n_h = Z_R$ and the model predicts the following $\Delta G_t^{\ominus}$ (cal mol^{-1})

_____

* This actually corresponds to the situation in the simple clathrate hydrate crystals. These structures are stable only if the guest molecules are separated by a layer of water. Double occupancy of cages (hydrophobic interaction?) is unknown. The distance between the centers of neighboring clathrate cages in 10–12 Å. Nemethy and Scheraga discarded the possibility that nonpolar side chains on the periphery of a protein might be stabilized by clathratelike interactions and make the point that the favorable accommodation of water molecules around a propane molecule cannot be achieved around a molecule of valine.

† Such transfer studies have become popular and have helped considerably in advancing our understanding of the differences between various solvent systems.[122,821,1104] A full discussion can be found in Vols. 2 and 3 of this treatise (Chapters 1 and 5 of Volume 2; Chapters 1, 2, and 4 of Volume 3).

values for nonpolar side chains: ala -1300, ileu -1900, phe -1800. These estimates might be used in calculations of the energetics of micelle formation but not in predictions of conformational transitions involving biopolymers, since $\Delta G_t^\ominus$ does not include contributions due to the transfer of the polar part of the molecule (or residue) into the nonpolar environment.

A comparison of the Kauzmann–Nemethy–Scheraga model for the hydrophobic interaction with some of the more recent experimental evidence discussed in this chapter raises doubts whether the process depicted in Fig. 1b ("partial reversal of the solution process") can account for all, or indeed most, of the experimental observations. We are concerned here with the concentration dependence in dilute solutions of thermodynamic and transport properties. Having established in Section 3 a description of the solution process in the absence of solute–solute interaction (i.e., hydrophobic hydration), then the process shown in Fig. 1b should be characterized by reductions in the magnitudes of the various measurable physical quantities. In other words, the trends should be toward "normal" behavior. As has been discussed in Section 4.1, the recent detailed thermodynamic results, both from thermal and volumetric measurements, are not compatible with such a model for the hydrophobic interaction. All the indications support the existence of solute–solute interactions in *very* dilute aqueous solutions, but these interactions seem to occur with a further ordering of the water. This is particularly well illustrated by the $\bar{V}_2^E(x_2)$ isotherms in Fig. 12. The Nemethy and Scheraga theory predicts $\Delta V^{H\phi} > 0$ (see also Table X), whereas the experimental evidence indicates that $\Delta \bar{V}_2^E < 0$ in the concentration range $x_2 < x_2''$. Similar effects are indicated in the concentration dependence of the enthalpies, compressibilities, and viscosities, the nmr rotational correlation times for water (Fig. 6), the proton chemical shifts of water, etc.

In the light of the available evidence, it can therefore be concluded that in terms of the molecular pair distribution functions (Table III), increasing concentrations of essentially hydrophobic molecules in aqueous solution give rise to a peak in f_{22} but, at the same time, a further narrowing takes place in f_{11}, thus showing the simple hydrophobic interaction not to be a partial reversal of the solution process, but rather a process which involves a further ordering of water over and above that which exists in the hydrophobic hydration shell of an isolated solute molecule; this is shown diagrammatically in Fig. 1d.

The experimentally determined complex concentration dependence of $\bar{V}_i^E$ can be fitted by the aqueous solution model of Mikhailov,[1012,1013] already referred to in Section 3.4.1, and discussed fully in Chapter 1,

Volume 2. Furthermore, the model also predicts the nature of the $\bar{H}_2^E(x_2)$ isotherms, which have yet to be determined in the concentration range $0 < x_2 < x_2'$. The Mikhailov formalism implicitly confirms the hypothesis of a qualitative change in the nature of the solute–solute interactions in the range $x_2' < x_2 < x_2''$ and is consistent with the suggestion advanced above, i.e., that the hydrophobic interaction is not necessarily a contact (van der Waals) but a longer-range interaction. Such hydrophobic stabilization of solute molecules at some distance greater than the collision diameter could also account for the unexpectedly low spin exchange rates (see Section 4.4) of hydrophobic spin probes in aqueous solution.

In this context the calculations by Friedman and his colleagues on ionic interactions in electrolyte solutions are of special interest.[1214] The model treats the ionic pair potential $u_{ij}(r)$ in terms of four contributions as follows:

$$u_{ij}(r) = (e_i e_j/\varepsilon r) + u_{ij}^{\text{core}}(r) + u_{ij}^{\text{diel}}(r) + u_{ij}^{\text{overlap}}(r) \qquad (40)$$

The first three terms on the right-hand side refer to the electrostatic contribution, the core repulsion, and a cavity effect in a dielectric medium the nature of which is well known. The final term arises from the overlap of the hydration spheres of the two ions and is again based on the assumption that, upon interaction at a distance r, some of the solvent will revert to its "bulk" state, but the assumption is not made that r is the sum of the ionic crystallographic radii. For purposes of computation the following equation is used:

$$u_{ij}^{\text{overlap}}(r) = A_{ij} V_{\text{H}_2\text{O}}^{-1} V_{\text{mu}}(r_i^* + w, r_j^* + w, r) \qquad (41)$$

where A_{ij} is the molar free energy associated with the partial relaxation of the hydration sphere to "bulk" water, r_i^* is the ionic radius, w is the thickness in molecular layers of the hydration sphere, $V_{\text{H}_2\text{O}}$ is the molar volume of water (in $\text{Å}^3 \text{ mol}^{-1}$), and V_{mu} is a mutual volume function determined by the geometry of the ion pair. Since A_{ij} is the only adjustable parameter in eqn. (41), it can be obtained by comparison with experimental data (osmotic or activity coefficients). Clearly the actual numerical value will depend on the type of potential function used for $u_{ij}^{\text{core}}(r)$ in eqn. (40), and also on the assignment of numerical values to r^*, w, etc. The calculated values for A_{++} (i.e., cation pairs) are of the order of -130 cal mol^{-1} (water) for Me_4N^+ and become less negative as the alkyl chain length is increased. For "normal" ion pairs A_{+-} decreases in magnitude for the series F′, Cl′, Br′, I′.

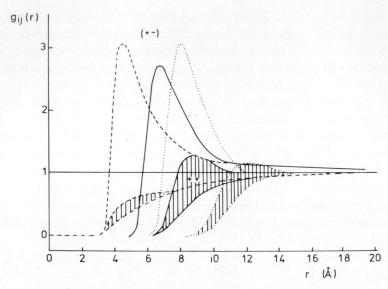

Fig. 22. $g_{+-}(r)$ and $g_{++}(r)$ for 0.2 M KCl (--), Me$_4$NCl (—), and Bu$_4$NCl (...), as calculated from eqn. (40). See Ref. 1214. Shaded areas show the contribution from the hydration overlap term, as defined by eqn. (41). Redrawn from Ref. 1570.

From eqn. (40) the various pair correlations $g_{ij}(r)$ (radial distribution functions) can be calculated. In physical terms $c_i g_{ij}(r)$ is an estimate of the concentration of species i at a distance r from a given j ion, where c_i is the stoichiometric concentration of species i. Figure 22 shows $g_{+-}(r)$ and $g_{++}(r)$ for 0.2 M aqueous KCl, Me$_4$NCl, and Bu$_4$NCl.[1572] The shaded areas show the hydration contribution as calculated from eqn. (41). Positive correlations are not obtained for $g_{--}(r)$, since this is in accordance with the expected behavior for ions with like charges. The peaks in $g_{++}(r)$ for the tetraalkylammonium ions arise from the fact that $u_{++}(r)$ is negative over a certain range of r and this in turn appears to be due to the magnitude of the last term in eqn. (40). The excess concentration of cations at a distance b from any given cation is calculated from

$$n_+^+(b) = c_+ \int_0^b g_{++}(r) 4\pi^2 r dr$$

For Et$_4$N$^+$ in Et$_4$NCl with b equivalent to the peak in $g_{++}(r)$, $n_+^+ = 0.7$. In terms of cation pair interactions calculated only on the basis of electrostatic effects this number is significant and probably originates directly from the hydrophobic interaction.

A particularly interesting feature of $g_{++}(r)$ concerns the value of b, i.e., the distance at which the pair correlations exhibit their maxima. This occurs at 9 and 12 Å for Me_4N^+ and Bu_4N^+, respectively. Although there is some doubt about the absolute magnitudes of the ionic radii, even for the highest available estimates[1263] (namely 3.47 and 4.94 Å) it is found that $b > 2r_+^*$. Other estimates for r_+ are: Me_4N^+ 2.56 or 2.85 Å; Bu_4N^+ 3.75 or 4.37 Å (see Ref. 1214). The results of these calculations are therefore compatible with the view that hydrophobic pairing in an *aqueous* environment might not involve van der Waals contact between solute particles, but is a longer-range interaction involving one or several layers of water molecules (see Fig. 1d). It has already been shown (p. 35) that a molecular dynamics computer simulation for a *single* inert particle in water suggests a clathratelike environment for the solute with little exchange of kinetic energy between the solute molecule and its nearest water neighbors. We therefore look forward with impatience to the extension of such calculations to two inert particles and their hydration shells. At this stage it is perhaps permissible to speculate that the equilibrium configuration of such a system might still approximate to a clathrate type of environment, and that it is only when the number of water–water pairs required to maintain such a configuration under the given conditions is no longer available that contact hydrophobic interactions replace the solvent-separated ones. This appears to occur at a solute concentration which we have characterized by x_2'', beyond which f_{11} can be expected to become progressively broader.

Finally, the model for the hydrophobic interaction developed here can also account for the interesting small-angle X-ray scattering data discussed in Section 4.3 and the phenomenon of lower critical demixing which is characteristic of aqueous solutions of essentially hydrophobic solutes. Thus, the pronounced clustering which is inferred from the excess scattering intensities should not be seen as the formation of microdroplets of "oil" or micelles, but as complex aggregates in which the hydrophobic moieties are still hydrated and therefore prevented from coalescing, until the critical solution temperature is reached, where the thermal energy suffices to disrupt the water framework which is stabilized by the presence of the hydrophobic groups.

5. THE HYDROPHOBIC INTERACTION IN COMPLEX AQUEOUS SYSTEMS

The foregoing discussions may well leave the reader with the impression—and correctly so—that much remains to be done before a rigorous description of the hydrophobic interaction can be formulated, even if only in terms of the various molecular *pair* distribution functions (and it still remains to be shown that sums of pair interactions provide a permissible approximation of the *real* situation, i.e., multiple interactions).

What, then, has been achieved during the past few years in furthering our understanding of the role of the hydrophobic interaction in such complex processes as biopolymer folding and unfolding equilibria, the reversible aggregation of amphiphilic species, small-molecule binding to polymers in vitro and in vivo, the specific aggregation of polymers to form supramolecular structures, etc.? These various aspects are either discussed in detail or at least referred to in subsequent chapters of this volume. It is very clear that, in most cases, the simple Kauzmann–Nemethy–Scheraga treatment is employed directly to calculate the magnitude of $\Delta G^{H\phi}$, and the thermodynamic transfer data (apolar solvent $\rightarrow$ water) for low-molecular-weight model compounds form the basis for such estimates. In most cases such experimental data are limited to free energies $\Delta G_t^{\ominus}$, although it has been shown quite convincingly that, at least for processes in aqueous systems, the Gibbs free energy function hides more than it reveals (see Chapter 1, Volume 2). This phenomenon, known as thermodynamic compensation,* manifests itself by trends in the enthalpies being almost exactly matched by similar trends in the entropies, leaving the free energy almost unaffected. Thus free energy measurements are not a very potent tool to highlight differences between processes taking place in slightly different solvent media or involving a range of related substrates. Nevertheless most speculation is based solely on ΔG measurements at single temperatures.

There is certainly no lack of experimental information implicating the hydrophobic interaction in a variety of biochemical and biological processes of varying complexity. It is impossible to provide a complete catalog and even less to discuss them in any detail. It may be useful, however, to provide a small representative sample, if only to demonstrate the present gulf between those who think in terms of pair potentials for methane molecules and those who are attempting to gain an insight into drug function, membrane processes, enzyme–substrate interactions, etc.

Thus over a number of years Hansch and his colleagues have shown that the partition coefficient between water and octanol of a series of related pharmaceutical compounds can be correlated with their biochemical and physiological properties.[602,603] This is claimed to be due to differences in the "hydrophobicity" of the related drugs.

---

* The phenomenon is being referred to with increasing frequency as "Lumry's law," from the well known review of compensation phenomena in biochemical systems.[927]

A similar linear correlation between the logarithm of the partition coefficient and the efficiencies of various compounds as inhibitors or substrates for chymotrypsin has also been established.[128,604]

The effects of alkyl chain branching on hydrophobic binding to proteins and possible relationships to disease conditions has been investigated by means of the binding of palmitic acid and 3,7,11,15-tetramethyl palmitic (phytanic) acid to serum albumin.[55] Another investigation of chain branching makes use of the anticholinesterase activity of trialkyl phosphates or phosphorothiolates.[164] It appears that added CH_2 groups only act by contributing to the total hydrophobic surface. Where alkyl groups are oriented away from the enzyme, they provide no contribution to the activity, from which it is concluded that folding of the enzyme does not take place during phosphorylation.

The implication of hydrophobic bonding in local and general anaesthesia has been suggested for several years. A recent study based on the effect of paramagnetic shift reagents on the $-N^+Me_3$ protons in liposomes (phospholipid bilayer systems; see Chapter 4) shows that the induced proton shifts are reversed in the presence of the local anaesthetic tetracaine, but not by Me_4NCl, procaine, or lidocaine, none of which interacts hydrophobically with lecithin. It is suggested that the anaesthetic effect may be related to a displacement of divalent metal ions from the phosphate group and that this requires the (hydrophobic) incorporation of the anaesthetic into the phospholipid bilayer.[439]

$Me(CH_2)_3NH$⟨⬡⟩$—COO \cdot CH_2 \cdot CH_2 \cdot N^+HMe_2 \, Cl'$ tetracaine hydrochloride

⟨⬡⟩$—COO \cdot CH_2 \cdot CH_2 \cdot N^+HEt_2 \, Cl'$ procaine hydrochloride

Me
⟨⬡⟩$—NHCO \cdot CH_2 \cdot N^+HEt \, Cl'$ lidocaine hydrochloride
Me

On a more technological note, it has been suggested that the insolubilization of soya milk during heating and drying cannot be due solely to disulfide formation but that "cross-linking" by hydrophobic groups contributes significantly.[502] Thus, if the reactive $-SH$ groups are blocked, soya milk can still be insolubilized by heat, but the effect is reversible.

The existence of well-defined conformational states in various synthetic copolymers, stabilized by hydrophobic interactions, is receiving increasing attention. Thus for a series of alternating copolymers of maleic anhydride and alkyl vinyl ethers of varying chain lengths, it has been found that the polymer containing ethyl vinyl ether behaves upon hydrolysis as a normal weak poly acid, but the butyl and hexyl copolymers give rise to hypercoiled states at low pH which undergo conformational transitions to extended states with increasing degree of neutralization.[366] From the difference in the free energy of transition between the butyl and hexyl copolymers, the contribution of a CH_2 group to the hydrophobic stabilization of the hypercoiled state is estimated as 400 cal mol^{-1}. This may be compared with Nemethy and Scheraga's estimate for $\Delta G^{H\phi} = 900$ cal mol^{-1}.

The importance of hydrophobic binding in some catalytic processes, and probably in enzyme catalysis, has been convincingly demonstrated by Klotz and his colleagues.[786,789,790] The catalysts consist of polyethylene-imine in which a proportion of the terminal amino groups carry dodecyl chain substituents which act as binding sites. The polymer also has a number of methyleneimidazole side chains which function as nucleophilic catalytic sites. These modified water-soluble polymers catalyze the hydrolysis of p-nitrophenyl esters of acids with reasonably long alkyl chains, e.g., caproic acid, and are more than 100 times as effective as imidazole. More recently similar synthetic enzymes (synzymes) have been found to enhance the rate of hydrolysis of aromatic sulfates by two orders of magnitude over that observed for aryl sulfatase enzymes and by 12 orders of magnitude over that achieved by free imidazole.[766]

The importance of the primary binding step is stressed by Klotz, and can be shown experimentally by the rate dependence on the size (and stereochemistry?) of the hydrophobic residues both on the synzyme and the substrate.

The catalytic action of polyelectrolytes carrying hydrophobic residues is becoming more generally recognized. Thus, Okubo and Ise have reported a marked effect of various synthetic copolymers on the alkaline fading of triphenylmethane dyes,[1110] and related the observed results to a balance between electrostatic and hydrophobic interactions.

The above examples, drawn from different fields of study, illustrate both the importance of hydrophobic interactions and also the present rather empirical state of our knowledge. This also becomes clear from a study of Tanford's recent monograph,[1470] in which he discussed the properties (mainly thermodynamic) of chemical, biochemical, and biological systems (e.g., micelles, lipid bilayers, cell membranes) the integrity of which depends largely on hydrophobic interactions.

5.1. Thermodynamic Studies in Perturbed Aqueous Media

5.1.1. *Urea and Guanidinium Chloride As Perturbants of Hydrophobic Interactions*

A common method of studying the hydrophobic interaction in biopolymer systems makes use of the modification of the aqueous solvent. This can be achieved by ionic or nonelectrolyte additives, of which latter class urea is the favorite. The use of such modified solvents, also known as denaturing solvents, allows thermodynamic comparisons to be made between the interactions of given segments of the biopolymer (or sometimes the synthetic polymer, see Chapter 7) in the native and the perturbed or denatured states. The importance of this technique will become very clear in subsequent chapters of this volume. Urea (and guanidinium chloride, GuHCl) act as disaggregants in systems where the hydrophobic interaction is believed to be the main driving force for association, e.g., micelles. Similarly, in biopolymer systems that exhibit specific tertiary and/or quaternary structures, urea and GuHCl exert a general disordering effect, which is believed to originate from a weakening of the hydrophobic interaction.

It is impossible here to review the literature dealing with urea–water mixtures and the effects of urea on the ordered or aggregated aqueous systems which form the subject matter of this volume. The reader is referred to Volume 2 of this treatise, where the various physicochemical aspects of urea–water mixtures are fully discussed. Suffice it to say that the question remains: How does urea, which forms near-ideal solutions with water, act in such a chaotropic manner toward water itself and hydrophobically associated or ordered systems?

Before returning to this question, it is opportune to digress briefly to show the unique effects of urea and GuHCl among water-soluble additives. Tanford has discussed the transfer thermodynamics of protein backbone (i.e., peptide) groups and apolar amino acid side chains from water to various solvent media, including aqueous electrolyte, alcohol, glycol, urea, and GuHCl solutions.[1466] Figures 23a and 23b summarize the $\Delta G_t^{\ominus}$ data for —CO—NH$_2$—CH$_2$ (or —CH$_2$—CO—NH—) and —CH$_2$—CHMe$_2$ (leucyl) groups, respectively. As might be expected from a consideration of purely electrostatic, dipole, and dispersion interactions, electrolytes "salt in" peptide groups and "salt out" apolar residues. Again in a predictable manner, organic additives act in the opposite sense. Urea and GuHCl "salt in" both nonpolar *and* polar groups. Tanford comments as follows: "(GuHCl) behaves like other electrolytes towards peptide groups but clear-

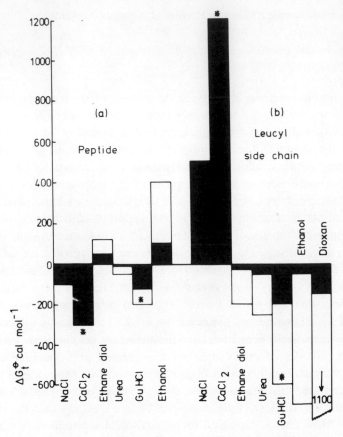

Fig. 23. Standard free energies of transfer of (a) peptide groups and (b) leucyl groups from water to the media shown at concentrations 2 M (dark areas) and 8 M (white areas). Asterisks refer to extrapolated values. From data reviewed in Ref. 1466.

ly acts quite unlike inorganic salts towards hydrophobic groups. The reasons for this difference are not understood at the present time."

However, data such as illustrated in Fig. 23 are used in calculations of protein unfolding mechanisms, etc., but, to quote Tanford again, "It has not been possible to account for the stability of the (protein) native state over all denatured states under native conditions. Were it necessary to make a prediction in the absence of experimental knowledge, one would probably conclude that the native state should not exist." He then continues to show that the calculation of the free energy change associated with the unfolding of ribonuclease is a futile undertaking. However the calculated group contributions are adjusted (within reason), the final result still predicts that the random coil should be the stable conformation under conditions where this is patently not the case.*

We now return to the question of the chaotropism of urea and GuHCl. Its possible origin has been fully discussed in Volume 2, and in the present context all that need be said is that these reagents, since they "salt in" apolar groups (and that includes hydrocarbons),[1580] probably reduce the property which makes water such a bad solvent for apolar species (hydrophobic hydration), i.e., they turn water into a more "normal" liquid. This, however, is exactly the type of process which will perturb a folded polymer, the stability of which depends in some measure on hydrophobic interactions. It is suggested therefore that urea and GuHCl can act quite nonspecifically through their action on the degree of intermolecular order in water, which is itself a fundamental requirement for the maintenance of certain naturally occurring conformations. Apart from this effect, urea and GuHCl may also act in a more specific manner, e.g., by dipole interactions with peptide groups, but such effects may well be less important [cf., for instance, the magnitude of $\Delta G_t^\ominus$(GuHCl) in Figs. 23a and 23b] than the role of urea and GuHCl as chaotropic agents.

Finally, in support of the above views, it is well to emphasize that the observed effects of urea and GuHCl are by no means confined to the unfolding of globular proteins (see particularly Chapter 5). Similar effects are observed with polysaccharides, nucleotides, and even with some water-soluble synthetic polymers. Furthermore, urea either inhibits completely the formation of many aqueous gels, or at least reduces their mechanical strength considerably. As discussed in Chapter 2, urea also raises the CMC of surfactants; in fact it interferes with all those phenomena that are believed to rely on the peculiar intermolecular order in liquid water and therefore on hydrophobic interactions. Another citation from Tanford's protein denaturation review[1466] is pertinent: "The point is being made... that the answer to this question... (why hydrophobic groups have a lower free energy in aqueous GuHCl solution than in water alone)... cannot come from studies of the effect of GuHCl on the properties of water. It requires an investigation of the molecular organization in a three compo-

---

* In all fairness, it should be pointed out that the alternative methods for calculating protein conformational states, i.e., potential energy minimization techniques, have not been any more successful in accounting for the existence of "native" structures. One of the main shortcomings of such techniques is of course the present inability to incorporate solvation interactions in the computations, although it has been convincingly demonstrated by crystallographic and other techniques that water plays a dominant part in the maintenance of native conformations, even in the crystalline state.

nent system: hydrocarbon, GuHCl, and water." We would contend strongly that before such advice is taken, the properties of the binary GuHCl–water and the hydrocarbon–water systems should be fully considered (and for hydrocarbons, read also biopolymers, surfactants, etc.) since otherwise the necessary base of reference is missing and important effects may be overlooked.

5.1.2. Essentially Hydrophobic Molecules As Perturbants of Hydrophobic Interactions

We should now like to develop further the points raised in the last paragraph of the preceding section. It has already been noted that urea and GuHCl form near-ideal aqueous solutions and this is also the case with many polyhydroxy compounds, although minor differences in the behavior of, say, hexose sugars may be associated with quite fundamental and specific biochemical and biological effects[511] (see Volume 2 for the properties of such systems).

The physicochemical behavior of aqueous solutions of essentially hydrophobic solutes is much more complex; qualitatively different types of behavior are encountered in the different concentration ranges $x_2 = 0 \rightarrow x_2'$, $x_2' \rightarrow x_2''$, $x_2'' \rightarrow 1$. Unless this fact is fully appreciated, the potential of this type of molecule in studies involving hydrophobic interactions in complex situations cannot be fully exploited, and indeed can lead to paradoxical situations (see Chapter 5). An example will illustrate this point: It is commonly observed that the presence of alcohols raises the critical micelle concentration (CMC) of surfactants compared to their CMC in water at the same temperature. This effect is fully discussed by Kresheck in Chapter 2. It is immaterial to the present argument whether the observed results are due to an increase in the solubility of the monomer species or to the partial destabilization of the micellar state. What is relevant is that *at low alcohol concentrations* the observed effects are reversed, i.e., the CMC is lowered[373] and that the alcohol concentration at which the CMC effect is reversed corresponds to x_2' for that particular alcohol. Similar alcohol effects have been reported for such widely different phenomena as the flocculation of hydrophobic sols by electrolyte[1530] and the thermal stability of ribonuclease.[173]

Operationally, the effects of additives on two-state or multistate equilibria are considered in terms of "binding" of the additive to one or several "sites" (e.g., Ref. 1466). Thus a two-state equilibrium

$$N \rightleftharpoons D$$

where N might refer to a native and D to a denatured conformational state of a biopolymer, is characterized by an equilibrium constant K.* An additive X is added such that its activity in the aqueous solution is a_X. It is proposed that X (and water) can *bind* to both the N and D states and that the binding of X is *preferential* when the mole ratio $X:H_2O$ on the surface of the polymer is larger than the mole ratio in the bulk of the solvent. It can then be shown that

$$\partial(\ln K)/\partial(\ln a_X) = \Delta \bar{\nu}_{X,pref} \qquad (42)$$

where $\Delta \bar{\nu}_X = \bar{\nu}_{X,D,pref} - \bar{\nu}_{X,N,pref}$.

Preferential binding can be positive or negative, with $\bar{\nu}_{X,pref} < 0$, signifying that water is preferentially bound. Similarly $\Delta \bar{\nu}_{X,pref}$ can adopt positive or negative values. Indeed, for one and the same aqueous solvent mixtures, as x_2 is increased from zero to unity, $\Delta \bar{\nu}_{X,pref}$ may undergo complex changes such that the N and D forms, respectively, may be stabilized in aqueous solvent mixtures of different compositions. It might, for instance, be postulated that the N form has few binding sites for X but those few are "high-energy" sites, i.e., the binding process is accompanied by a large decrease in free energy. On the other hand, the D state may have many X binding sites of a low-energy kind. In this situation low concentrations of X would stabilize the N state but high concentrations of X would favor the conversion to the D state.

The generality of this type of treatment coupled with the very loose specification of the term "binding" severely limits its usefulness in attempts to elucidate possible mechanisms leading to stability or instability of particular states under a given set of experimental conditions. Let us consider a hypothetical case[†] where it is found from experiment that a given additive X at low concentration, $x_2 < x_2''$, promotes state N, but that higher concentrations of X favor state D. The arguments put forward in the previous paragraph may now be advanced to account for the observed results. However, if it were now known that the properties of the binary mixture H_2O-X

---

* The treatment can equally well be applied to other processes, such as monomer $\rightleftharpoons$ micelle.

† This may not be nearly as hypothetical as it may appear. The indications are, and careful experimentation will confirm this, that this is quite a common occurrence. The trouble with much of the literature devoted to this subject is that experimenters who are unaware of the properties of the binary aqueous mixtures themselves add the organic cosolvent in steps of 20% and thus completely miss the composition range $0 < x_2 < x_2''$ discussed at length in Section 4 in which all the interesting phenomena associated with the hydrophobic interaction are observed.

underwent some drastic change at the same concentration x_2'' and further-more that below this same concentration the aqueous mixture depressed the CMC of surfactants, promoted the aggregation of dye stuffs, and stabilized hydrophobic sols against flocculation, but that all these effects were reversed above this critical concentration x_2'', then eqn. (42) would lose some of its credibility as an informative criterion for the effect of X on the N $\rightleftharpoons$ D equilibrium. As suggested earlier in this chapter, $\Delta G_{N \rightarrow D}(x_2)$ is likely to be a monotonic function, so that the experimental data can probably be adequately fitted with the available adjustable parameters $\bar{\nu}_{i,N}$ and $\bar{\nu}_{i,D}$ (where there are i solvent species available for binding). No phys-ical significance can, however, be attached to the $\bar{\nu}$ data and one must doubt whether the relevant binding sites in the N and D states could be identified, say, by spectroscopic methods. This is of course not to imply that binding sites do not exist, or that macromolecule–small-molecule com-plexes have not been identified or do not play an important role in chemical and biological processes. The careful work of Tanford,[1466] Klotz,[786] and many others has also shown that hydrophobic interactions are often directly involved in such binding processes. The point which is made, however, is that a fit of eqn. (42) alone tells us nothing about binding sites, mechanisms, etc. and may well arise from the curious behavior of the binary or ternary solvent itself.*

More stringent tests of any model are the pressure and temperature derivatives of the free energy. Unfortunately, there are very few data for complex systems the stability of which is affected by hydrophobic inter-actions with small molecular species in aqueous solution. The reader is referred, however, to the work of Brandts and Hunt,[173] whose calorimetric studies of the effects of alcohol on the unfolding of ribonuclease beautifully illustrate the arguments advanced in this section.

In closing this discussion, we therefore refer again to the quotation on p. 87. As suggested by Tanford, the answers to many of the problems touched upon are not likely to come from studies of binary aqueous sol-vents, but we would add the rider that the *right* answers are not likely to come from any studies that ignore the properties of such solvent mixtures.

---

* In this context it is opportune to mention that the adsorption of moisture onto a pro-tein can very often be described quite adequately by the BET isotherm (another free energy function). This can hardly mean that the exterior of a protein has the property of a typical BET solid, nor that the assumptions inherent in the BET treatment apply to the protein–water systems. It does mean that the free energy is a very non-discriminating property. This becomes apparent when heats and entropies of adsorption are examined. These do not follow the predicted BET behavior.

5.2. Spectroscopic Identification of the Hydrophobic Interaction in Complex Systems

In Sections 3.3 and 4.4 we touched upon the application of spectroscopic techniques to identify and elucidate the phenomena of hydrophobic interactions. Of the available techniques, nmr is particularly well suited to such studies because the chemical species directly involved in the interactions, i.e., water molecules and methylene chains, can be observed directly. The incorporation of "labels," such as free radical spin labels or fluorescence lables into an alkyl chain, although popular with the practitioners of esr and fluorescence spectroscopy, always raises doubts about the effects of such bulky foreign groups on the motions of the remainder of the molecule.*

These problems do not arise with nmr spectroscopy and "substitutions" only involve exchange of ^1H by ^2H, ^{16}O by ^{17}O, ^{12}C by ^{13}C, etc., none of which is likely to affect the interactions under study. We have emphasized the sophisticated studies by Hertz and his colleagues at Karlsruhe of the relaxational properties of small, essentially hydrophobic species in aqueous solution and the wealth of information which has been and is still being derived from these experiments.

The use of different spectroscopic techniques in connection with studies of hydrophobic interactions in aggregated or macromolecular systems is touched upon in several chapters of this volume. As concerns a better insight into the detailed nature of the hydrophobic interaction in such complex systems, spectroscopic techniques have hardly begun to contribute significantly. Rather, the methods are used to detect and monitor interactions which are then treated as Kauzmann–Nemethy–Scheraga hydrophobic bonds, during complex changes such as micellization, small-molecule binding to proteins and nucleotides, unfolding reactions, etc. In terms of nmr, for instance, the widths and intensities of signals associated with alkyl chain protons are followed during processes which are believed to involve changes in the hydrophobic interactions between a polymer and a smaller molecule or between hydrophobic groups attached to the same polymeric backbone. An example of such a process is presented in Fig. 24, which illustrates the pmr spectrum of lysolecithin (a) in aqueous solution, and (b) in the presence of the protein $\varkappa$-casein with which it is believed to

---

* These doubts do not apply to the exchange of diamagnetic ions by paramagnetic ions for esr purposes, provided that it can be shown that the substituted paramagnetic ion acts in a similar manner and does not promote any specific effects in the complex system under study.

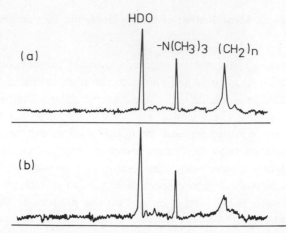

Fig. 24. Pmr spectra lf (a) 1% lysolecithin in D_2O and (b) as in (a) in the presence of 1.5% $\varkappa$-casein. Hydrophobic interactions are indicated by the differential broadening observed in the methylene proton signal. From Ref. 92.

interact. The spectrum contains two prominent signals arising from the phospholipid alkyl chain protons and the protons associated with the choline end group.[92] On addition of the protein, the choline proton signal is unaffected, while the $-(CH_2)_n$ signal is differentially broadened, the effect being proportional to the protein concentration. Such results, which are typical of many protein–lipid or protein–surfactant systems, indicate therefore that the interaction between the two species involves the alkyl chains. Much more detailed studies are, however, needed if we want to derive information about distances, orientations, rates, energetics, and intramolecular dynamics of the interactions involved, and in time such investigations will no doubt be undertaken.

Much the same is true for "label" experiments. There are many reports of the use of fluorescing probes to identify hydrophobic binding sites in biological systems. Thus 8-anilinonaphthalene-1-sulfonate (ANS) fluoresces strongly when "bound" to hydrophobic groups, whereas the dye in aqueous solution is virtually nonfluorescent. This effect is utilized in the identification of such hydrophobic regions in a variety of complex systems, e.g., immunoglobulins[1138] or cell surfaces.[371] Esr spin probes can be used in similar ways, either in studies of the interactions of the isolated probe with its environment, or by covalently attaching the probe to a molecule whose interactions are to be studied. Several esr parameters provide indicators about the environment (i.e., aqueous or apolar) of the probe. In particular the hyperfine coupling is used as such an indicator, whereas the reorientation correlation time τ_0 describes the mobility of the probe. However, here again, apart from identifying a hydrophobic effect, the methods have not yet provided additional information about the nature or mechanism of the hydrophobic interaction.

5.3. Summary and Prospects

In this chapter we have tried to trace the somewhat tortuous development of the currently held views on the hydrophobic interaction. It originates from the realization that aqueous solutions of nonpolar molecules have some rather remarkable properties which are inextricably bound up with the physical/chemical behavior of water itself. The fact that a more general solvophobic effect has since been studied in other associated solvents in no way detracts from the unique nature of the hydrophobic effect[371,821] (see Chapter 1, Volume 3).

The so-called "structure promotion" effect in dilute aqueous solutions of hydrophobic groups is now firmly established, although its dependence on concentration is not yet well understood. Once the importance of hydrophobic association to protein stability and other phenomena was realized, the natural way of specifying such an interaction was by considering it as "a partial reversal of the solution process." We have attempted here to review the available evidence in the light of this specification and find many unresolved problems. At high concentrations in substantially nonaqueous environments, such as exist in micelles, lipid bilayers, or the interior of globular proteins, this model no doubt accounts well enough for the observed behavior, but in predominantly aqueous environments, the suggestion that hydrophobic groups are stabilized not by contact interactions, but by longer-range interactions (water acting as "cement") must be taken seriously and receive closer scrutiny. Such a model of a longer-range hydrophobic interaction might well be able to account for the α-helical structures of certain homopolypeptides in aqueous solution (see Fig. 1e), the existence of which is still something of a puzzle.* It might also help in the future to explain the function(s) of the extensive hydrophobic regions which are found on the *surfaces* of many proteins and the rather more subtle hydrophobic interactions which occur in the complex conformational states of polysaccharides.[1086]

---

* The famous quotation, "Poly-L-glutamic acid forms a very stable α-helix, whereas poly-L-aspartic acid forms almost no helix, and no one has the vaguest notion why,"[1123] is as relevant today as it was in 1968. Most of the attempts to rationalize this puzzling situation have been at best only partially successful.[1126]

As regards the hydrophobic interaction in complex situations, one can be confident that detailed experiments (and this includes computer experiments) will eventually provide the necessary quantitative information which will help to place this interaction in its right context so that it becomes amenable to calculation. Hopefully, this in turn will materially assist our understanding of the "native states" of biopolymers, their solution structures, dynamics, and functions.

ACKNOWLEDGMENTS

It is a pleasure to thank the following for making available manuscripts and other material prior to publication: Prof. V. G. Dashevsky, Dr. F. H. Stillinger, Dr. H. L. Friedman, Dr. W. Y. Wen, Dr. H. S. Frank, Dr. H. G. Hertz, Dr. M. Lal, Dr. S. Cabani, Dr. M. Lucas, Dr. R. Lumry, Dr. I. Wadsö, Dr. J. E. Desnoyers, and Dr. M. V. Kaulgud.

In particular, I owe a debt of gratitude to those of my colleagues who, by frequent discussion, have helped to shape my ideas along the lines discussed in this chapter: David Reid, Allan Clark, Elliot Finer, Martin Barratt, David Atkinson, Alan Suggett, and Michael Phillips.

CHAPTER 2

Surfactants

Gordon C. Kresheck

Department of Chemistry
Northern Illinois University
DeKalb, Illinois

1. INTRODUCTION

The inclusion of a volume dealing with the properties of solutions in a comprehensive treatise on water is understandable in view of the known effects of water "structure" on intermolecular interactions in aqueous solutions. It is worthy to mention at this point that Frank[476] recently suggested that the study of aqueous solutions in general (and one might include nonaqueous solutions) will likely increase our presently incomplete understanding of water itself. Surface-active molecules, which characteristically contain a polar and a relatively large nonpolar portion, are known as surfactants, and their solutions in particular are attractive for study since they include a wide variety of solute molecules which may be conveniently studied by a wide variety of physical techniques. The ability of the polar portion of the surfactant molecule to increase the solubility of the nonpolar side chain makes it possible to determine experimentally the effects of normally sparingly soluble groups on the properties of water itself, subject to a determination of the contribution of the polar portion (head group effect). An estimation of the importance of the head group effect can be made by comparison of the solution properties of nonpolar solutes and surfactants. Finally, a study of several fairly well-defined types of molecular association with surfactants may allow a distinction to be made between the contributions of the solute and the solvent to the observed bulk properties of the solutions.

2. CRITICAL MICELLE CONCENTRATION

The most characteristic and thoroughly studied property of surfactant solutions is the cooperative self-association of the solute within a fairly narrow concentration range in dilute solution to form high-molecular-weight aggregates known as micelles. This topic has been thoroughly considered in several recent reviews.[282,434,746,1049,1067,1321,1387] The present discussion will attempt to provide a physical description of the process and, where possible, discuss the contribution of the solvent. Water will be implied as the solvent throughout this chapter unless stated otherwise.

Considerable attention has been given to developing methods for determining the solute concentration at which micelle formation first occurs, known as the *critical micelle concentration* (CMC). The CMC for a particular surfactant solution represents a physical quantity of no less significance than a melting point, boiling point, or refractive index, for example, of a pure substance. As such, the CMC can be correlated with many other properties of the solution which are related to the surface activity of the solute. The literature from 1926 through 1966 was critically reviewed and a collection of CMC values for over 700 compounds compiled in Ref. 1060. As such, it serves as the most complete single source of critical micelle concentrations and represents a valuable contribution to the literature of surfactants. A brief description of the usefulness of the CMC value, methods for the determination of the CMC, and the effect of additives was also included.

2.1. Methods of Determination

A cursory review of the literature reveals that a large variety of methods has been used for the determination of the CMC. In general, the depletion of ions from the bulk solution or the binding of ions to the charged micelle (counterion binding) can be measured for ionic surfactants, or changes in the environment of the nonpolar side chains for all surfactants can be detected either directly or indirectly through the effect on the solvent. Finally, appropriate indicators can be used to function as chemical amplifiers or transducers to facilitate the determination. Various dyes have been widely used for this purpose, although many other types of indicators can be imagined. Due to the generality of the above-mentioned changes, it then comes as more of a surprise when a particular physical method cannot be used to measure the CMC than when another new method is reported. A few of the more interesting recent methods include optical rotatory dispersion,[1063] gel filtration,[1451] side chain fluorine magnetic resonance,[1070] counterion magnetic resonance,[730,903] fluorimetry,[1237] various types of electrochemical determinations,[961,1388,1585] and interference refractometry.[254]

Dependending upon one's purpose and prejudice, CMC determinations by various methods may be considered to be either in poor or in good agreement. In a typical CMC determination, some physical property of the solution is plotted as some function of solute concentration. An apparent discontinuity will result at the CMC, although curvature in the vicinity of the CMC is nearly always found if one searches diligently enough. The

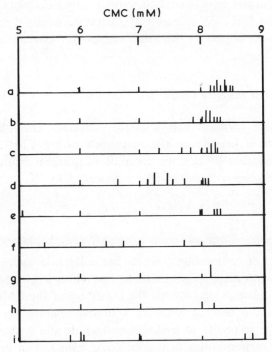

CMC (mM)

Fig. 1. Variation of the CMC of sodium dodecyl sulfate at temperatures between 20° and 30° as monitored by various experimental methods. Each horizontal line represents the results obtained by a given experimental method, with the height of each bar corresponding to the number of independent determinations. The methods employed are identified as follows: (a) specific conductivity, (b) equivalent conductivity, (c) other conductance, (d) surface tension versus the logarithm of concentration, (e) other relationships between surface tension and concentration, (f) absorbance, (g) solubilization of orange OT dye, (h) light scattering, and (i) other methods such as refractive index, emf, vapor pressure, sound velocity, and viscosity. Taken from Ref. 1060 by permission of the U.S. Department of Commerce, NBS.

latter reflects the fact that the CMC is not a precise concentration per se, but an average of a narrow range of surfactant concentrations in which micelles form. Certainly the same order of magnitude is obtained by most methods, but better agreement between the results obtained might be expected considering the precision of the individual methods. For example, Mukerjee and Mysels[1060] give a comparison of several CMC determinations for sodium dodecyl sulfate, which are reported to be at least precise to about 3% and accurate to 5% (Fig. 1). The observed values may be seen to range from about 5 to 9 mM, with a clustering about 8.0–8.5 mM. The variation in the CMC depicted in Fig. 1 is considered by Mukerjee and Mysels to result primarily from the fact that the micelles are polydisperse, that slight impurities exist, and that various empirical plots were used to establish a given CMC. Whenever CMC data are required for some comparative purpose, the use of the same method for internal consistency is clearly desired.

2.2. Importance of the Side Chain

The size and chemical nature of the surfactant nonpolar side chain have an important bearing on the magnitude of the CMC. It has been known for a long time[779] that the variation of the CMC with chain length for a given homologous series can be represented by

$$\log \text{CMC} = a_0 - a_1 n \tag{1}$$

where a_0 and a_1 are constants for a particular experimental condition and n is the number of carbon atoms in the nonpolar side chain. A collection of data for sodium alkyl sulfates that fit eqn. (1) is shown in Fig. 2. The negative linear relationship between the logarithm of the CMC and length of the side chain is seen to resemble a similar relationship between the logarithm of the solubility of n-aliphatic alcohols and hydrocarbons and the number of carbon atoms in the molecule, also shown in Fig. 2. It is likely that the two processes are similar though not identical, because of the common role of the solvent, water. Typical values of a_0 range from 1.0 to 3.5 depending upon the nature of the head group. However, as can be seen from Table I, values of a_1 cluster about 0.3 and 0.5 for ionic and nonionic surfactants, respectively. In view of the existence of a well-established empirical relationship such as eqn. (1), it becomes an immediate challenge to give physical meaning to the coefficients a_0 and a_1. In view of the irregular manner in which the values of a_0 change with different head

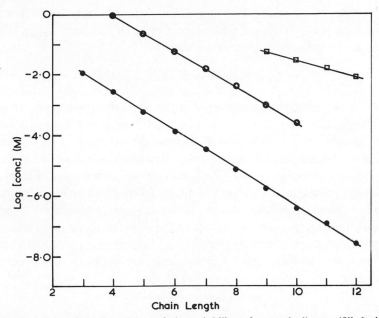

Fig. 2. Variation of the logarithm of the solubility of normal alkanes (filled circles). normal alcohols (open circles), and the CMC °F sodium alkyl sulfates (open squares) with chain length. Data were taken from Refs. 770, 853, and 937 respectively.

TABLE I. Values of a_0 and a_1 Defined According to Eqn. (1) for Various Surfactants[a]

Surfactant	Temperature, °C	a_0	a_1
Sodium carboxylates	20	2.41	0.341
Potassium carboxylates	25	1.92	0.290
Potassium carboxylates	45	2.03	0.292
Alkane sulfonates	40	1.59	0.294
Alkane sulfonates	50	1.63	0.294
Alkyl sulfates	45	1.42	0.295
Alkylammonium chlorides	25	1.25	0.265
Alkylammonium chlorides	45	1.79	0.296
Alkyltrimethylammonium bromides	60	1.77	0.292
Alkyltrioxyethylene glycol monoether	25	2.32	0.554
Alkylglucoside	25	2.65	0.531
Alkylhexaoxyethylene glycol monoether	25	1.81	0.488
Alkyldimethylphosphine oxide	30	2.31	0.466
Alkyldimethylamine oxide	27	3.3	0.5

[a] Values for the ionic surfactants were taken from Ref. 898 and those for the nonionic surfactants were calculated from the data in Ref. 1060.

groups, it is not surprising that at present they are rarely used. However, the situation is quite different with respect to a_1. It has been shown[1383,1384] that a_1 can be defined for ionic surfactants as

$$(a_1)_{\text{ionic}} = [1/(1 + K_g)](W/2.303kT) \tag{2}$$

where k is the Boltzmann constant, T is the absolute temperature, W is the free energy change for transferring one methylene group from the micelle to an aqueous environment, and K_g is related to the degree of counterion binding to the micelle. The term W was described as a cohesive energy in early theories of micelle formation, in keeping with the notion of Debye[329] that it reflected the favorable van der Waals interactions between the hydrocarbon chains in the micelle interior. Using modern terminology,[1188] which acknowledges the importance of the solvent in nonpolar interactions in aqueous solutions, it would be described as the contribution of hydrophobic bonding to micelle formation.* For nonionic surfactants, K_g disappears and a_1 becomes

$$(a_1)_{\text{nonionic}} = W/2.303kT \tag{3}$$

At this point, it is possible to use eqn. (3) and data from Table I for nonionic surfactants, and obtain values of W ranging from $1.08kT$ to $1.28kT$. It is interesting to note for comparison that a value of $3.2kT$ is associated with the solution of alcohols in water at $25°$.

Using the data for nonionic surfactants to estimate W and an average of 0.29 for a_1 from data for ionic surfactants, K_g as given by eqn. (2) is found to range from 0.6 to 0.9. The expected value of K_g may be compared with experiment by making use of another equation,[1384] which describes the effect of added salt on the CMC of ionic surfactants. This relationship is given as

$$\log \text{CMC} = \text{const} - K_g \log C_i \tag{4}$$

where C_i is the total concentration of counterions in solution. Values of K_g are reported to range from 0.4 to 0.6.[1384] Therefore, consistency between eqn. (2) and eqn. (4) is obtained with W approximately equal to $1.1kT$ and K_g equal to 0.6.

Lin and Somasundaran[899] have chosen to use an equation proposed by Shinoda[1383] from which eqn. (2) was derived, namely

$$\ln \text{CMC} = -(Wn/kT) + E_{\text{el}} + c_0 \tag{5}$$

---

* The free energy of hydrophobic bond *formation* would be $-W$.

where E_{el} represents the electrostatic contribution to ionic micelle formation and c_0 is another constant, and assume that E_{el} does not vary with
counterion concentration. They represent eqn. (5) as linear in n, except
the slope is now formally the same as given by eqn. (3) for nonionic surfactants, since it does not contain the term $1/(1 + K_g)$ found in eqn. (2).
They clearly obtain a different apparent value for W in this case. They
then introduce a term which indirectly accounts for the dependence of E_{el}
on counterion concentration with specific account taken of the effect of
chain length. Their final result is

$$\ln \text{CMC} = (n/kT)(W_{el} - W) + c_0 \qquad (6)$$

where $nW_{el}/kT = E_{el}$. The term W_{el} was not evaluated and a theoretical
discussion of it was not given. Since the slope of eqn. (6) still contains
two unknowns, Lin and Somasundaran resort to eqn. (2) and eqn. (4) to
evaluate W. In either case, the same interpretation is given to W by Shinoda
and by Lin and Somasundaran, with a numerical assignment of about
$1.08kT$ for ionic surfactants.

Thus, subject to the development of more refined theories of micelle
formation, the effect of changing the chain length for a homologous series
on micelle formation can be understood. One apparent weakness of existing theories of micelle formation is their failure to predict changes in thermodynamic parameters other than the changes in free energy with chain
length, although these could be derived in principle from the temperature
and pressure dependences of the CMC if the temperature and pressure
dependences of the various constants are known.

Data which illustrate the influence of the number of ionic groups, branching, unsaturation, and substitution of various groups on the CMC have
been collected by Shinoda.[1384] Of these, the effects of unsaturation appear
to be the least thoroughly studied, although of some potential interest,
since the degree of stiffness in the chain could potentially be altered in a
systematic way in order to determine the effect of rotational restriction. It
has been shown recently that in the limiting case, the CMC of sodium
cyclododecyl sulfate is about ten times higher than for the corresponding
linear surfactant.[660] Shinoda et al.[1386] have demonstrated that eqn. (1)
is valid for perfluoroctanoate surfactants except that W for a —(CF_2)—
group is about $1.6kT$ as compared to $1.08kT$ for a —(CH_2)— group. This
indicates that the CMC of a fluorinated surfactant will be the same as for
a hydrocarbon surfactant with a chain length about 1.5 times larger than
for the fluorinated compound. Data that illustrate the dramatically lower

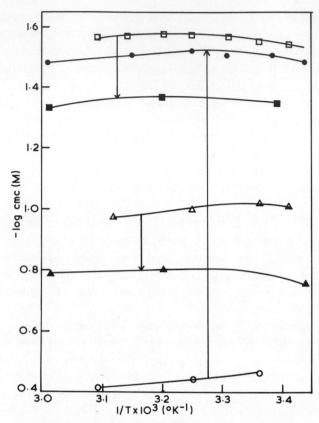

Fig. 3. Plot of $-\log$ CMC versus $1/T$ ($^{\circ}\mathrm{K}^{-1}$) for fluorinated and protio sodium carboxylates. The open symbols represent the protio octanoate (circles), decanoate (triangles), and dodecanoate (squares); the filled circles represent perfluoroctanoate, and the filled triangles and squares represent decanoate and dodecanoate, respectively, with only the ultimate carbon atom fluorinated.

CMC due to complete fluorine substitution for sodium octanoate[1074] are shown in Fig. 3, along with related data for sodium decanoates and dodecanoates.[1060] Perfluoroctanoate may be seen to have nearly the same CMC and dependence of the CMC on temperature as protiododecanoate. On the contrary, Muller and co-workers[1070,1073] have observed that a terminal (CF_3)— group about doubles the CMC for ionic and nonionic surfactants, as shown in Fig. 3 for sodium decanoate and dodecanoate. These apparently conflicting effects of fluorination can be understood if it is imagined that fluorination of only the ultimate carbon atom resembles branching of the side chain with the concomitant increase in the CMC due to steric factors. However, with complete fluorination, perhaps a better situation exists for packing the chains in the micelle interior. This interpretation would be consistent with Muller's observed differences in solvation of the interior of the two types of micelles.[1073] An alternate interpretation would be to allow for coiling of the side chain so that the fluorinated group was in fact in contact with the solvent rather than imbedded in a hydrocarbon interior. The increased CMC would then parallel the influence of covalent ring formation previously mentioned for the dodecylsulfate results. The large value of W for the $-(CF_2)-$ group suggests the interesting possibility that hydrophobic interactions between fluorocarbons might be stronger than hydrophobic interactions between hydrocarbons in aqueous solutions, as suggested by Rochester et al.[1264–1266] This difference may be related to the probable alignment of the water molecules in the first solvation layer around the solute molecules, being $(-CH_2-) \cdot OH_2$ for hydrocarbons and $(-CF_2-) \cdot H_2O$ for fluorocarbons. Interesting differences between aqueous solutions of hydrocarbons and fluorocarbons may be expected on the basis of these results, with the fluorocarbon solutions perhaps exhibiting physical properties which reflect greater immobilization of the solvent than commonly noted with aqueous hydrocarbon solutions, discussed in Chapter 1 of this volume.

2.3. Temperature

The effects of temperature changes on the CMC are so small that the best obtainable precision for CMC measurements is required to determine differences for quantitative purposes, yet large enough to provide the basis for a calorimetric method to determine the CMC.[111] The effect of changing the number of carbon atoms in the side chain has a much greater effect on the CMC than changing the temperature, as can be noted from the data presented in Fig. 3. Numerous workers have used the temperature dependence of the free energy of micelle formation to find the enthalpy and entropy of micellization. The latter two thermodynamic quantities serve as useful links between experiment and theory of micellization, and will be discussed separately later. For the moment, only the temperature dependence of the CMC itself will be considered.

The variation of the CMC with temperature may be compared with the variation of the solubility of nonpolar substances in water with temperature. The solubility behavior from 10 to 70° of several gases in water,[100, 662] shows that the solubility of every gas decreases initially as the temperature is increased from 10° and seems to approach a minimum which is in fact realized in a few instances. This tendency of gases to exhibit a minimum solubility at some temperature is paralleled in the variation of the CMC with temperature. The generalization for surfactants can be made more consistent with experiment by noting that the temperature at which the minimum occurs, T_{min}, may be outside the experimentally accessible range so that in some instances only an ascending or descending limb is observed. It has been the author's experience that a maximum in the temperature dependence of either the solubility of nonpolar molecules in water or the CMC of a surfactant is not to be expected, and, if found, represents some unusual state of solute association. Values of T_{min} for various surfactants[4-6,301,453,1060,1065,1452] are presented in Table II. Although the number of observations is limited, two general effects can be noted. First, the influence of the size of the counterion is just opposite for the cationic and the anionic surfactants. The lowest T_{min} is observed with the largest and less highly solvated counterion for the pyridinium halides and for the smallest and more highly solvated counterion for the alkyl sulfates. This difference between cationic and anionic surfactants may be general and could be related to other known fundamental differences in the physicochemical properties of the two types of surfactants.[896]

TABLE II. Temperature at Which the Temperature Dependence of the CMC Exhibits a Minimum, T_{min}, for Various Surfactants

| Head group | Chain length | | | | | | | Ref. |
	C_6	C_8	C_9	C_{10}	C_{11}	C_{12}	C_{14}	
N-Betaine	—	—	—	41	35	$\lesssim 10$	—	1452
Lithium sulfate	—	—	—	—	—	$\lesssim 10$	—	1060
Sodium sulfate	—	29	—	29	—	26	21	453, 1452
Octylphenoxyethoxyethanols[a]	47	51	49	50	—	—	—	301, 1452
Pyridinium iodide	—	—	—	—	—	0	—	1065
Pyridinium bromide	—	—	—	—	—	15	—	4
Pyridinium chloride	—	—	—	—	—	25	—	1060
Alkylammonium bromide	—	—	—	25	—	20	—	6
α-Picolinium bromide	—	—	—	20	—	20	≤ 5	5

[a] A chain length for this entry refers to the number of oxyethylene groups in the polar head group.

The second feature to note from the data in Table II is the effect of chain length for a given homologous series. A maximum value for T_{min} seems to exist near C_{10} for the compounds studied, although it is certainly not clearly established. From empirical considerations, some more regular variation of T_{min} with chain length would be expected for the following reason. Since the logarithm of the CMC relates in a regular manner to the number of carbon atoms in the side chain according to eqn. (1) and the CMC is proportional to the standard free energy of micellization, it is also proportional to the standard enthalpy and entropy changes $\Delta H_m^{\ominus}$ and $\Delta S_m^{\ominus}$, so that

$$\Delta H_m^{\ominus} - T \Delta S_m^{\ominus} = a_0' + b_0' n \tag{7}$$

where a_0' and b_0' are constants related to a_0 and b_0 defined by eqn. (1). Since a linear relationship between $\Delta H_m^{\ominus}$ and $\Delta S_m^{\ominus}$ is observed for many transfer processes when dealing with homologous series in aqueous solutions,[218,819,927] such a relationship could reasonably be expected for micelle formation in view of the linear variation of the CMC with n. Therefore, if

$$\Delta S_m^{\ominus} = \text{const} + \text{const } \Delta H_m^{\ominus} \tag{8}$$

substitution of eqn. (8) for $\Delta S_m^{\ominus}$ into (7) and setting $\Delta H_m^{\ominus} = 0$ yields $T = T_{min}$, and the result is

$$T_{min} = \text{const} + \text{const} \times n \tag{9}$$

which predicts a linear relationship between T_{min} and n. The data in Table II clearly cannot be described by eqn. (9), and possible contributions to either the enthalpy or entropy of micelle formation from a source in addition to solvation effects or some very unusual solvation effects are suggested.

2.4. Salts

Consideration of the influence of different types of salts on the CMC will deal with ionic and nonionic surfactants separately, since different empirical relationships have been found to hold for the two types of surfactant. The nonionic surfactants will be discussed first since electrostatic factors do not complicate the interpretation. The effectiveness of salts in altering the CMC of nonionic surfactants approximately follows the lyotropic series,[963,1229] which for anions and cations,[1229] respectively, are

$$\tfrac{1}{2}SO_4^{2-} > F^- > Cl^- > ClO_4^- > Br^- > NO_3^- > I^- > SCN^- \tag{10}$$

and

$$Na^+ > K^+ > Li^+ > \tfrac{1}{2}Ca^{2+} \tag{11}$$

In most cases the observed relationship between the CMC and salt concentrations C_i, at least up to 1.0 M salt, can be described as

$$\log \text{CMC} = \text{const} + k'C_i \tag{12}$$

where the constant k' may be identified as a specific salt effect constant. It has been shown[1048] that the salting in and salting out of nonionic molecules by electrolytes can serve as a basis for understanding the effects of salts on the CMC of nonionic surfactants. For this purpose it is convenient to recall the Setschenow equation,[1367] which has been found to relate the solubilities of nonpolar solutes in the absence and presence of salt, S_i^* and S_i, respectively, as

$$\log(S_i^*/S_i) = KC_i \tag{13}$$

where K is an empirical constant. The constant K can be identified[100] with a solubility parameter k_s given by McDevit and Long[943] as

$$K = k_s = \bar{V}_i^{\circ}(\bar{V}_s^{\circ} - V_s)/2.3\beta^*RT \tag{14}$$

where $\bar{V}_i^{\circ}$ and $\bar{V}_s^{\circ}$ are the limiting partial molal volumes of the nonpolar solute and electrolyte, respectively, V_s is a derived quantity which represents the molar volume of the pure liquid salt and is discussed by McDevit and Long and by Harned and Owen,[608] and β^* is the compressibility of pure water.* A thorough thermodynamic discussion of this equation has been presented.[573] The effect of temperature at constant pressure P on k_s has been given as[915]

$$\frac{dK}{dT} = \frac{dk_s}{dT} \simeq \frac{-\bar{V}_i^{\circ}}{2.3RT} \left(\frac{d\bar{V}_s^{\circ}}{dT}\right)_P \tag{15}$$

A recent review by Millero[1021] provides a useful collection of partial molal volumes of electrolytes at various temperatures which permits the determination of $(d\bar{V}_s^{\circ}/dT)_P$ for a variety of salts ranging from those that typically salt in to those that salt out nonelectrolytes in aqueous solutions for use with eqn. (15).

It is also possible to view the effects of temperature on k_s by employing the enthalpy of solution of the nonpolar solute in water and electrolyte solutions. If eqn. (13) is rewritten as

$$\ln S_i^* - \ln S_i = 2.3KC_s \tag{16}$$

---

* The application of the McDevit–Long equation to problems relative to protein stability are discussed in Chapter 5.

the temperature derivative of eqn. (16) is

$$\frac{d \ln S_i^*}{dT} - \frac{d \ln S_i}{dT} = 2.3 C_s \frac{dK}{dT} \qquad (17)$$

Since the temperature derivative of the solubility yields the enthalpy of solution $\varDelta H_i$, eqn. (17) becomes

$$(\varDelta H_i^* - \varDelta H_i)/RT^2 = 2.3 C_s \, dK/dT \qquad (18)$$

Upon rearrangement and integration, eqn. (18) gives

$$(T_2 - T_1)/T_1 T_2 = 2.3 R C_s (K_2 - K_1)/\varDelta H_t \qquad (19)$$

where $\varDelta H_t$ is the usual enthalpy of transfer of a solute from an aqueous to a nonaqueous or salt solution, given in the present case as

$$\varDelta H_t = \varDelta H_i - \varDelta H_i^* \qquad (20)$$

The derivation of eqn. (19) also follows from the usual definition of transfer parameters and assumes that the enthalpies of solution are constant over the small temperature interval $T_2 - T_1$, but provision can be made for the situation where the enthalpy changes are temperature dependent by use of the change in heat capacity $\varDelta C_{p_t}$ by means of the equation

$$(\varDelta H_t)_{T_2} = (\varDelta H_t)_{T_1} + \int_{T_1}^{T_2} \varDelta C_{p_t} \, dT \qquad (21)$$

Finally, an equation which may be used to estimate the temperature dependence of $\varDelta H_t$ is obtained from a combination of eqns. (15) and (18) to give

$$(\varDelta H_t)_{T_2} - (\varDelta H_t)_{T_1} \simeq \frac{\bar{V}_i^\circ C_s [(\bar{V}_s^\circ)_{T_2} - (\bar{V}_s^\circ)_{T_1}]}{\ln(T_2/T_1)} \qquad (22)$$

It may be assumed for the discussion of eqns. (14)–(22) that provision could be made for the size of the nonelectrolyte solute and electrolyte by inclusion of a correction factor proposed by Long and McDevit.[915]

Since the effect of temperature on the salting in and salting out of aliphatic hydrocarbons is known,[1044,1572] it is of interest to determine the extent that aliphatic hydrocarbons can substitute for benzene, using eqn. (14) at temperatures other than 25°. For this purpose eqn. (14) can be equated for various pairs of hydrocarbon–benzene solutes to give

$$k_s^{\text{hydrocarbon}} = \frac{\bar{V}_{\text{hydrocarbon}}^\circ}{\bar{V}_{\text{benzene}}^\circ} k_s^{\text{benzene}} \qquad (23)$$

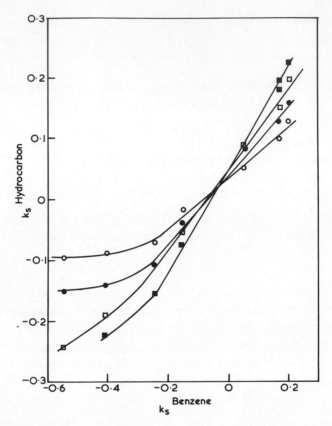

Fig. 4. Plot of $k_s^{\text{hydrocarbon}}$ versus k_s^{benzene} defined by eqn. (23) for methane (open circles), ethane (filled circles), propane (open squares), and butane (filled squares) in the presence of various electrolytes at 25°. The data were taken from Refs. 987, 1044, 1572 as described in the text.

A plot of eqn. (23) for methane, ethane, propane, and butane at 25° is given in Fig. 4 for electrolytes that typically salt in[1572] and salt out[1044] nonelectrolytes. The curves do not pass through the origin and seriously depart from linearity for $k_s^{\text{benzene}} < -0.2$. However, the slopes of the linear portions, 0.42, 0.57, 0.73, and 0.83, compare well with the predicted slopes of 0.45, 0.62, 0.81, and 1.0 for methane, ethane, propane, and butane, respectively, using the molar volumes reported by Masterson[987] and Traube's rule to estimate $\bar{V}_i^\circ$ for butane. It would be tempting to propose that deviations from linearity for the tetrabutylammonium and tetrapropylammonium bromides reflect some type of solute–solute or solute–solvent interaction that does not exist with the other salts. The tentative conclusion is that it does not appear encouraging to attempt to extrapolate data on the basis of eqn. (23) to various temperatures for the larger alkylammonium salts, although it may be useful in some cases with the other salts.

An alternate approach to the use of eqn. (23) to predict the temperature dependence of k_s would be to employ either eqn. (19) or (22). Therefore, the same hydrocarbon data used previously[987,1044,1572] and data for argon[120,412] were plotted according to eqn. (22) and these results are shown in Fig. 5. Although the data are decidedly nonlinear, the correct trend exists even with the alkylammonium bromides. Deviations from the predicted slope of unity could be ascribed to (a) rather large errors in ΔH_t for the hydrocarbons, (b) the inexactness of eqn. (15), and (c) the simplicity of the model. It may be concluded that the ability of eqn. (15), from which

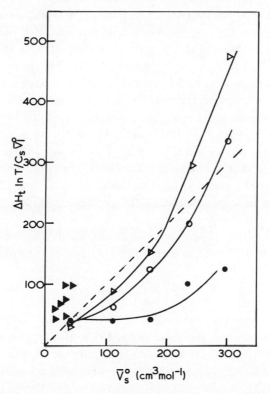

Fig. 5. Plot of $\Delta H_t \ln(T)/C_s \bar{V}_i^\circ$ versus $\bar{V}_s^\circ$ according to eqn. (22), using the average data contained in Refs. 987, 1044, 1572 for methane, ethane, propane, and butane at 5° (filled circles), 25° (open circles), and 35° (open triangles), and argon (120, 412) at 25° (filled triangles). The broken line with a slope of 1.0 passes through the origin, and is included for reference.

eqn. (22) was derived, to account for the effect of temperature upon K is approximate within a factor of about two. Clearly, the use of eqn. (19) is preferred.

That eqn. (14) may also describe salt effects on the CMC of nonionic surfactants was originally proposed by Mukerjee,[1048] who demonstrated that the specific salt effect constant k' given by eqn. (12) could be identified with the solubility parameter k_s in eqn. (14). The specific salt effect constant k' was decomposed into additive effects for the hydrophobic and hydrophilic portions of the surfactant molecules as

$$k' = k_a + k_b - k_{bm} \qquad (24)$$

where k_a, k_b, and k_{bm} reflect contributions to micelle formation in the presence of salts for the nonpolar and polar portion of the monomer and the polar portion in the micelle, respectively. It was assumed by Mukerjee that the interior of the micelle was not in contact with the solvent, and since the polar portion of the micelle is in contact with the solvent, the terms k_b and k_{bm} are expected to nearly cancel. Situations where k_b and k_{bm} fail to cancel may be understood in terms of a large effective concentration of head groups on the micelle surface as opposed to the low concentrations of monomer. Therefore, it would appear to be reasonable to rewrite eqns. (12), (14), (15), (20), and (24), substituting k_a for k', K, and k_s to relate salt effects on the CMC with temperature changes. One important difference to note would be that the enthalpy change would correspond to the enthalpy of micelle formation in the presence and absence of salt rather than the heat of solution of the nonelectrolyte as in the previous discussion.

An extensive test of Mukerjee's analysis has been performed by Ray and Némethy.[1229] They combined eqns. (14) and (23) to obtain the equation

$$k_a = (\bar{V}^{\circ}_{\text{surfactant}} / \bar{V}^{\circ}_{\text{benzene}}) k_s \qquad (25)$$

Reasonable agreement between experiment and eqn. (25) at 25° was demonstrated for the nonionic p-tert-octylphenoxy(polyethoxy) ethanols, OPE_{30} and OPE_{9-10} (Triton X-305 and Triton S-100), as shown in Fig. 6. The observed slope of 3.5, indicated by the dashed lines, approximated the expected slope of 2.8, which passes through the origin. The observed abscissa intercepts are found to be positive for the inorganic salts and negative for the tetraalkylammonium halides, in agreement with differences observed in Figs. 4 and 5 between the two types of salts. Deviations between observed and expected behavior were ascribed to (a) failure of k_b to cancel

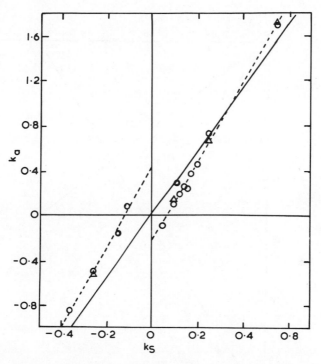

Fig. 6. Plot of k_a versus k_s according to eqn. (25) as given in Ref. 1229 (by permission of the American Chemical Society) for OPE_{30} (circles) and OPE_{9-10} (triangles).

k_{bm}; (b) specific interactions between the salts and the polar head group; (c) head group–head group interactions; and (d) failure of the oversimplified theory of Long and McDevit.

The previous discussion of the effects of temperature on the solubility of nonelectrolytes may be extended to micelle formation. It the data of Schick[1060,1319] for the condensate of n-dodecanol and ethylene oxide (30) in the presence of NaCl are plotted as $(\bar{V}_s^{\circ})_T$ versus $k_a T$ according to eqn. (15), a linear relationship would be expected with a *negative* slope equal to $2.3R/\bar{V}_i^{\circ}$. This plot is shown in Fig. 7, and the slope is seen to be *positive*. Also shown for comparison is the same plot for butane in the presence of NaCl,[1044] where the slope is of the expected sign. These results reflect the fact that k_a becomes more positive and K becomes less positive with increasing temperature. The origin of this difference between micelle formation and solubility is not known, but it could reflect a salt effect on the large ionic head group of the surfactant. This explanation has some support in that methoxy–polyoxyethylene dodecyl ether with a shorter head group

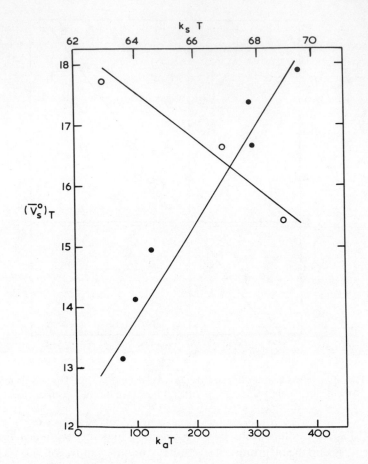

Fig. 7. Plot of $(\bar{V}_s^\circ)_T$ versus k_aT or k_sT according to eqn. (15) for a nonionic surfactant (filled circles) and butane (open circles) in the presence of NaCl with the data from Refs. 1044, 1060, 1319.

consisting of about 12 ethylene glycol units[836,1060] shows a negative trend for k_a with increasing temperature between 30° and 50° with 0.2 m NaCl as the electrolyte.* Clearly, more data are needed to test these relationships.

Finally, the heat of micelle formation ΔH_m can be used to obtain ΔH_t from the equation

$$(\Delta H_m)^{\text{H}_2\text{O}} - (\Delta H_m)^{\text{salt}} = \Delta H_t \qquad (26)$$

* It could also indicate that the solubility of hydrocarbons is not a good model for micelle formation and vice versa.

if the micelle is assumed to have the same partial molal heat content in the two solvents. The value of ΔH_t obtained from eqn. (26) would yield the dependence of k_a on temperature according to eqn. (19).

The CMC of ionic surfactants in the presence of added univalent salts, CMC′, was shown in 1947 by Corrin and Harkins[289] to be described by the log–log relationship

$$\log \text{CMC}' = a \log C_i + b \tag{27}$$

where C_i represents the concentration of counterions and a and b are constants. The counterion concentration can be expressed as the sum of the concentration of surfactant CMC plus the concentration of added salts. The values of a and b were first estimated to be about -0.5 and -3.0 by use of the Debye–Hückel theory.[670] The coefficient a was later identified as $-K_g$ by Shinoda as given in eqn. (4), by again focusing upon the primary electrostatic effects in the solutions. The effect of added NaCl on the CMC of CF_3—$(CH_2)_8$—CO_2Na has also been shown to be described by eqn. (27) with values of -0.58 and -1.24 for a and b, respectively.[1070] An alternate empirical equation was suggested by Yan[1602] and has the form

$$\text{CMC}' = a'(\text{CMC}' + C_i)^{1/2} + b' \tag{28}$$

A modified form of eqn. (28) was able to describe the effects of polyvalent salts on the CMC. It has recently been shown that the nature of the similion has a measurable effect on the CMC[1070] at higher electrolyte concentrations. Also, the importance of the nature of the counterion on the CMC of n-dodecyl sulfates was demonstrated by Schick,[1320] although this is best described as a second-order effect. Values of a from eqn. (27) of -0.68, -0.66, and -0.57 were found for Li^+, Na^+, and Me_4N^+ chlorides, respectively. These differences were attributed to differences in binding affinities of the counterions to the micelle as a result of differences in hydration of the ions. It may also be possible to interpret the effects of counterion size on the basis of water structural properties such as water-structure-enforced ion pairing[349] or structural salting in or out,[474] but in the author's opinion such attempts seem premature. However, data of this type are very attractive areas for future study in view of the quantitative relationships which exist between the CMC and ion type. It is these second-order effects, after consideration of primary electrostatic factors, that will likely have solvation implications related to water structure.

2.5. Nonelectrolyte Additives

The addition of low-molecular-weight organic additives to aqueous solutions of surfactants is known to alter the CMC.[1060] The most commonly used additives have been urea and its derivatives, alcohols, dioxane, and sugars. Each of these types of compounds would be expected to alter the bulk dielectric constant of the medium and influence the state of solvation of both the polar and nonpolar portion of the surfactant molecule. It is assumed for the present discussion that the additives are not transferred into the micelle interior, for these situations will be considered with respect to mixed micelle formation. Such a distinction has been made previously by Emerson and Holtzer[409] in classifying additives as either penetrating or nonpenetrating. The apparent relative unimportance of the change in dielectric constant per se due to the presence of the additives is demonstrated by the fact that both urea, which increases the dielectric constant,[1597] and dioxane, which decreases the dielectric constant,[608] consistently increase the CMC.[192,409,1060,1071,1323] Also, examples exist where low concentrations of added alcohols reduce the CMC, but high concentrations tend to increase the CMC for nonionic[111] and ionic surfactants[373,409] alike. Therefore, to understand the effect of various organic additives on the CMC, it becomes necessary to consider how a range of concentrations of additives affect the relative stability of the monomer and micelle. For example, an increased CMC would accompany either a decreased chemical potential of the monomer or an increased chemical potential of the micelle, or both.

To determine the relative effect of the additives on the polar head group and the nonpolar side chain, it may be instructive to examine the data in Table III, which demonstrate the effect of $6 M$ urea on the CMC of various nonionic and ionic surfactants. To aid in this evaluation, a relative energy diagram is constructed in Fig. 8, which is similar to that presented by Klotz and Farnham[787] and used by Kresheck and Klotz[818] to estimate the contribution of solvation effects to the self-association of N-methylacetamide in aqueous and nonaqueous solutions. A similar approach has been taken by Lin and Somasundaran[899] to assess the importance of solvation in aqueous micelle formation. The energy coordinate represented in Fig. 8 could in principle be replaced by a variety of thermodynamic parameters. The model compound could also be chosen so as to represent the polar portion of the surfactant molecule. Additivity of the contributions of two portions of the surfactant molecule is assumed for this type of analysis, which should only be considered to be approximate, due to the possibility of head group effects, although it represents the best

TABLE III. Effect of 6 M Urea on the CMC for Various Surfactants at 25°

Surfactant	Type	(CMC)urea/(CMC)water	Ref.
Dimethyldecylamine oxide	Nonionic	2.50	111
Dodecanol-(ethylene oxide)$_7$ condensate	Nonionic	2.50	1320
Dodecanol-(ethylene oxide)$_{30}$ condensate	Nonionic	3.13	1320
Hexadecanol-(ethylene oxide)$_{30}$ condensate	Nonionic	1.82	1320
Branched nonylphenol-(ethylene oxide)$_{10}$ condensate	Nonionic, branched	3.2	1323
Branched nonylphenol-(ethylene oxide)$_{20}$ condensate	Nonionic, branched	3.4	1323
Branched nonylphenol-(ethylene oxide)$_{30}$ condensate	Nonionic, branched	4.0	1323
Dodecyltrimethylammonium bromide	Cationic	3.1	192, 409
Sodium decylsulfate	Anionic	1.61	1320
Sodium dodecylsulfate	Anionic	1.67	1320
Lithium dodecylsulfate	Anionic	1.25	1320
Tetramethylammonium dodecylsulfate	Anionic	2.08	1320
Sodium dodecyl (ether)$_3$ alcohol sulfate	Anionic	2.50	1320
Sodium dodecyl (ether)$_{17.5}$ alcohol sulfate	Anionic	3.67	1320

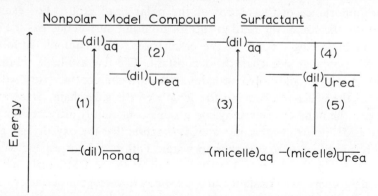

Fig. 8. Energy diagram representing the transfer of model compounds from water to urea solution, and micelle formation in water and urea solutions.

that can be done at present. The various steps represented in the figure can be identified as follows: Step 1 is the transfer of the model compound from a nonaqueous reference state to water to produce a standard solution with the properties of infinite dilution; step 2 corresponds to the transfer of the model compound from a standard dilute aqueous solution to a standard dilute urea solution; step 3 represents the reversal of micelle formation to a standard state of dilute solution in water; step 4 is a transfer step for the surfactant which is the same as step 2 for the model compound; and step 5 refers to disruption of micelles in urea solution. All of the steps are reversible and can be treated accordingly. The micelle may not be at the same energy level in the aqueous and urea solutions.

The above approach has the advantage that each process can be measured. For example, if standard free energy changes are considered, step 2 can be evaluated from the solubility of the model compound in water and urea solution,[1104] and likewise for step 4 with the surfactant molecule. Steps 3 and 5 can be found by means of the CMC in water and urea solution, respectively.[111] It is not necessary to obtain a value for step 1 for the present purposes, however; this can also be done from solubility measurements.[1580] Inherent problems in this analysis exist with respect to accounting for the free energy of dilution of the appropriate saturated solutions.[567] It is generally assumed that the activity coefficients are the same in aqueous and urea solutions for a given concentration of a particular chemical species. This assumption is most serious in the determination of a value for step 4 in view of the complexity of concentrated surfactant solutions; an alternate procedure, however, may be to use step 2 to obtain an estimate for step 4.

Perhaps the most striking thing to be noted from the data in Table III is the importance of the nature of the polar head group. For example, increasing the size of the head group for nonionic surfactants has the same effect as increasing the size of the counterion for the anionic surfactants with the nonpolar side chain held constant. The few available results for the alkyl sulfates and polyoxyethylene alkanols are contradictory with respect to the effect of changing the length of the side chain. Since an increase in the number of oxyethylene groups consistently increases the ratio of the CMC in urea to that in water, it suggests that the oxyethylene group is more soluble in 6 M urea than in water. This would correspond to steps 2 and 4 in Fig. 8. The fact that the increment in the CMC ratio per oxyethylene group is not constant for the various homologous surfactant series included in Table III may be cited as evidence of varying micelle structure, which is represented as step 5 in Fig. 8.

In spite of the scattered reoccurring suggestions of possible clathrate formation in aqueous surfactant solutions containing urea, the bulk of the recent evidence tends to support a general structure-breaking role.[447,1323] One recent example was the demonstration[1002] that the effect of urea on micellar catalysis follows the Setschenow equation identified earlier as eqn. (13). This type of observation makes it tempting to propose that an expected relationship between CMC and chain length for a nonionic surfactant in the presence of urea can be obtained by combining eqns. (1) and (3) to give

$$\log \text{CMC} = a_0{}^u - (W_{[\text{urea}]}/2.3kT)n \tag{29}$$

where $W_{[\text{urea}]}$ corresponds to the hydrophobic bonding contribution to micelle formation, which includes the effect of urea. The superscript in the coefficient $a_0{}^u$ is intended to provide for the effect of urea on the polar head group. It is difficult to discuss the relative importance of this term in the presence of urea or in its absence, although it would probably be sensitive to the changes in solvation of the polar head group that accompany micellization. Data to test eqn. (29) are not available; however, $W_{[\text{urea}]}$ would likely be about 10% more than for micellization in the absence of urea using model compound data[1104,1580] to estimate step 4 and assuming that the difference between steps 3 and 5 is independent of chain length. In addition, an equation of the same form as eqn. (29) may also be expected to represent the effect of urea on the CMC of ionic surfactants to a first approximation, although the effect of urea upon the $a_0{}^u$ term of eqn. (29) is counterion dependent, as evidenced by the data included in Table III.

Finally, a few remarks may be made with respect to the temperature dependence of the effect of urea on the CMC. The data of Schick[1320] reveal similar trends for ionic and nonionic surfactants, in that the CMC ratio with 6 M urea decreases for three different nonionic surfactants and two anionic surfactants. These observations are consistent with the calorimetric results of Benjamin,[111] which demonstrate that the heat of micelle formation of dimethyl decylamine oxide is less positive in the presence than in the absence of urea. Benjamin interpreted his results in terms of the effect of urea on the solvation of the monomer, which corresponds to the transfer process designated as step 4 in Fig. 8. Related studies with model compounds[816] support this view.

Shirahama et al.[1390,1391] have determined the effect of urea on the solubility and CMC of potassium dodecyl and hexadecyl sulfate. In general, the ratio of the CMC in the urea solutions to the CMC in water followed the ratio of the solubility of the surfactant in the same two solvents. However, marked differences between the two ratios occurred at low concentrations of urea. These results suggest that solubility measurements represented by step 2 of Fig. 8 do not provide a consistent estimate for step 4, and/or the micelle structure varies in some irregular way with urea concentration. It is interesting to note with respect to this latter point that the temperature dependence of the CMC of sodium dodecyl sulfate in 3 M urea has been shown[1169] to exhibit anomalous behavior around 10°.

Finally, the addition of urea to surfactant solutions containing a nonionic fluorine-labeled surfactant is reported to increase the micelle size,[1073] although it decreases the micelle size for ionic sodium trifluorododecyl sulfate.[1071]

A comparison of the effect of urea and various amides on the CMC of several surfactants has been reported.[409,967,1071] Whereas the various reports consistently show that urea and various related amides increase the CMC differently for anionic and cationic surfactants, only Malik and Verma[967] report that the effects of acetamide and formamide level off after addition of 1.0 M additive. This point would apparently warrant clarification. Differences between the effects of urea and its substituted derivatives on the CMC are to be expected in view of the known difference in solution properties of the compounds.[90,788,815,944,1283]

In concluding a discussion of the influence of urea on the CMC, it may be said with little reluctance that its primary action is to increase the CMC by increasing the stability of the monomer species. Exceptions clearly exist and have been mentioned in the preceding discussion. Again, it is the second-order effects that are perhaps the most interesting, though poorly understood at the present time.

The addition of sucrose to nonionic surfactant solutions was found to promote a small lowering of the CMC.[562] The observed effect was quantitatively much less than what might be expected from the fact that the free energy of transfer of argon from water to 0.5 m sucrose is about 100 cal mol^{-1}.[115] This suggests an enhanced solubilization of the polar polyether head group in the presence of the sugar. Adderson and Butler[3] have studied the temperature dependence of the CMC of several alkylammonium bromides in the presence and absence of glucose and sucrose. It was found that the addition of sucrose increased the CMC at all temperatures, and glucose may promote either CMC increases or decreases, depending upon the temperature. They interpreted the observed effects on the basis of a modification of water structure and a reduction of the dielectric constant of the medium due to the added sugars.

Understanding the mechanism of CMC alterations requires knowledge of whether a particular additive has its greatest effect on the monomer or the micelle. Emerson and Holtzer[409] have proposed three experimental criteria to distinguish between penetrating and nonpenetrating additives, making use of the concentration and temperature dependence of the effect of the additives on the CMC. An additional experimental factor which is characteristic of solutions containing penetrating (or mixed micelle forming) additives is a reduction in the sharpness of the CMC. This latter fact is suggestive of the importance of step 5 in Fig. 8 to micelle stability. Therefore, in view of the added complexity of the importance of altered micelle structure in the presence of penetrating additives, mainly qualitative conclusions can be drawn. However, solubility data can always be used for either model compounds[115] or surfactants[1390,1391] to establish the contribution of step 4 of Fig. 8 to the altered CMC. Indeed this approach has been the basis of the studies of Tanford and his associates in order to estimate the hydrophobic contribution to protein stability.[1466] A notable exception which led to quantitative data for penetrating solutes was the study by Metzer and Lin[1006] of the effect of ethane and propane on the CMC of dodecylamine hydrochloride. The hydrocarbon gases lowered the CMC in a manner which was similar to the effect of increasing chain length on the CMC described by eqn. (11). This study provides a reference point for comparison of the effect of complete solute penetration upon the CMC,[1588] which is a CMC reduction. The action of methanol, which is probably a nonpenetrating additive,[409,899] provides a reference point at the other end of the spectrum, and is responsible for CMC increases. The action of other additives is likely to lie between the effects demonstrated by methanol and the hydrocarbons, and may show both effects at different concentration levels. To this must be added the complexity of the structural properties of alcohol–water mixtures as reviewed by Franks and Ives.[486]

2.6. D_2O

One final type of additive that one can consider is the substitution of D_2O for H_2O as the solvent. The suggestion was made[819] that hydrophobic bonds may be expected to be stronger at 25° in D_2O than in H_2O, based on the relative solubility of amino acids and the CMC of dodecylpyridinium iodide in the two solvents. Subsequent studies have confirmed the observation that the CMC is lower in D_2O than H_2O for other surfactants.[409,1059] However, difficulties in the above interpretation exist in that hydrocarbon gases[819] and argon[113] are more soluble in D_2O than in H_2O at low temperatures, although this behavior becomes less pronounced with increasing temperature. Thus, the differences between the surfactants and amino acids on the one hand and the hydrocarbon gases and argon on the other may be attributable to a "head group effect." The surfactant and amino acid solubility data have been used by Berns and co-workers[1354] to interpret the effect of D_2O on protein structure. The effect on the polar portion of the surfactant has not been established. Perhaps the determination of heat capacities and comparison with model compound data[1159] will clarify this question.

3. AGGREGATION NUMBER

In addition to the CMC, the size of the micelle or number of monomeric units in the aggregate (known as the aggregation number m) is most important in describing the physical state of the system. A quantitative expression for m can be obtained from the simplest case with a closed association model

$$m\mathrm{S} \rightleftharpoons \mathrm{S}_m; \qquad K = [\mathrm{S}_m]/[\mathrm{S}]^m \tag{30}$$

where S corresponds to the monomeric surfactant and S_m is the micelle. Typical values of m range from about 20 for very small micelles to several hundred for large ones.[1079] Since individual surfactant monomers may be expected to have molecular weights of several hundred, the determination of m becomes a problem of colloid physical chemistry, and thus dictates the methodology employed. In general, the problem is to determine how m varies with experimental conditions once it can be found with precision. Another important factor is to determine the degree of polydispersity in the system, which can be estimated from a comparison of the number-average and weight-average aggregation numbers.[1054] The latter can be easily obtained by dividing the number-average and weight-average molecular weights, respectively, by the molecular weight of the surfactant monomer. Examples of the most commonly used methods to determine the molecular weight of micelles include light scattering,[24,103,408,645,723] sedimentation equilibrium,[24,762,781] membrane osmometry,[63,142] magnetic resonance,[1070,1073] gel filtration,[268,370,1451] dye solubilization,[723,1342] potentiometric titrations,[1598] and hydrodynamics.[103,290] The references cited are by no means intended to be exhaustive, but rather illustrative. Perhaps the most unifying approach to relating the aggregation number to experimental parameters was that by Barry and Russell.[98] They combined eqn. (1) with a similar empirical relationship that exists between micelle molecular weight

M and chain length n,

$$\log M = \text{const} + \text{const} \times n \tag{31}$$

to obtain the result

$$\log \text{CMC} = \text{const} \times \log M + \text{const} \tag{32}$$

They also combined eqn. (4) with eqn. (32) to show that

$$\log M = \text{const} \times \log C_i + \text{const} \tag{33}$$

Thus linear log–log relationships are expected between the molecular weight M and CMC and between M and the counterion concentration according to eqns. (32) and (33), respectively.

The temperature derivative of eqn. (32) could be used to find the temperature dependence of the aggregation number. The latter was also derived from a theory of hydrophobic bonding for similar, spherical, non-ionic micelles[1188] which suggests a cube root dependence of the most probable micelle size with the reciprocal of temperature. A maximum value of the most probable micelle size in the vicinity of $50°$ was predicted. Additional evidence in support of an increase in m with an increase in temperature for nonionic surfactants has been found[62,1073] in addition to evidence of a maximum value of m with temperature.[843] Consideration of the second-order processes which relate to the aggregation number reveal some interesting effects. An unusual threshold temperature has been noted for a few nonionic surfactants which corresponds to a temperature at which an abrupt change in m occurs.[63,103] This has been interpreted in terms of a change in geometry or shape of the micelle.[103]

Strict adherence to eqn. (33) is to be expected for a given electrolyte, since the CMC, which is related to m by eqn. (32), is subject to the lyotropic series as discussed previously. Indeed, the salts that are most effective in lowering the CMC of nonionic surfactants[1229,1322] give rise to the largest values of m, with the exception of the tetramethylammonium chloride, which is anomalous in this respect. In a study of the effects of divalent metal ions on the molecular weight of sodium dodecyl sulfate[1108] it was found that Pb^{2+} increased m more than did Zn^{2+}, Cu^{2+}, or Mg^{2+}. The possible biological importance of this observation could be related to the ability of Pb^{2+} to replace Na^{+} in membranes.

The importance of the structure of the polar head group was demonstrated for various homologous series of cationic surfactants.[22,23] The aggregation number was larger when Br$^-$ was the counterion rather than Cl$^-$. Also, the largest micelles occurred with the smallest head group, e.g., NH$_4^+$ versus Me$_4$N$^+$. These results were interpreted on the basis that the more tightly bound counterion would promote the largest aggregation number. Determinations of the approximate degree of counterion binding to the micelle support this view. However, the finding that a decreased CMC due to a structural change of the head group is not necessarily accompanied by an increase in m, as eqn. (32) would predict, leads to the conclusion that no simple relationship between head group structure and aggregation number is presently known to exist.

Organic additives which increase the CMC also decrease m[645,1071,1073] in qualitative accordance with eqn. (32). In this case, the CMC is less abrupt, suggesting the existence of polydisperse aggregates. The question of monodisperse versus polydisperse micelles represents an inconsistent situation in the literature. Some nonionic surfactants apparently form monodisperse micelles,[288,1451] while others give rise to polydisperse ones.[62,63,288,370] Micelles of sodium dodecyl sulfate are monodisperse at 0.4 M NaCl and polydisperse at 0.5 M NaCl, and micelles of dodecyltrimethylammonium bromide are monodisperse in 0.1 M NaBr but polydisperse in 2.0 M NaBr.[24] A general consideration of the problem of the size distribution of micelles in general, with a special emphasis on light scattering data, which helps to explain some of these observations was recently presented by Mukerjee.[1054] He found that many small ionic or nonionic spherical micelles appear to be fairly homogeneous. However, large micelles are expected to have a wide size distribution, with the ratio of the weight- to number-average molecular weights equal to two. Also, a linear relationship between the weight-average molecular weight and square root of the surfactant concentration was predicted. The importance of taking nonideal effects into consideration was also emphasized.

4. SHAPE AND HYDRATION

Since the questions of the shape and hydration of micelles are closely related, they will be discussed together. As in the determination of the aggregation numbers, the techniques applicable to colloid physical chemistry are involved. In general, either hydrodynamic or light scattering data are required to obtain information regarding the shape of the micelles, and the hydration can then be estimated. Alternatively, information regarding hydration of the micelles can be found from various spectroscopic and thermodynamic studies. A rather complete presentation of the idealized shapes that may exist in aqueous nonionic surfactant solutions ranging from dilute solution to the neat mesomorphic phase was given by Corkill and Goodman.[282] Also, Muller[1067] has discussed the use of spectroscopic data to obtain information regarding the hydration of micelles.

An examination of recent data based both on light scattering and parallel studies by another method were conducted for a given set of compounds in the same laboratory and quite good agreement between the correspondence of the two sources of data was found.[62,103,288,1252,1322,1598] A variety of evidence in support of the idealized spherical shape in the vicinity of the CMC has been reported for cationic,[1252,1598] anionic,[251,252,290,1626] and nonionic[103,290,391,1322,1564] micelles; although there appear to be exceptions for some nonionic surfactants[62,103, 1385] and perhaps minor distortions.[1469] In general, an increase in either surfactant or salt concentration promotes asymmetry.[103,206,251,252,391,1252,1322,1385]

Becher and Arai[103] have used light scattering and hydrodynamic techniques to show how the log-boom or Philippoff model is consistent with the changes in shape and hydration that accompany the change in molecular weight for nonionic surfactants. A change from small, heavily hydrated, oblate ellipsoid micelles to very large, slightly hydrated, prolate ellipsoid micelles was shown to occur via a spherical form of intermediate hydration. This interpretation of the gradual change in shape of the micelle depending on the aggregation number receives support from an examination of thermodynamic data for the octaoxyethylene dodecylether–water system.[1385]

The variation of shape and hydration with size of nonionic micelles just described may be contrasted with that of ionic micelles studies by Ekwall and his co-workers.[391] They have found that small ionic micelles tend to be spherical and very highly hydrated in the vicinity of the CMC. With an increase in concentration, asymmetric micelles form. However, unexplained differences between the manner in which the specific viscosity changes with concentration above the CMC for anionic and cationic micelles were noted. These differences in viscosity appear to be related to differences in counterion binding. The importance of taking the virial coefficients into consideration for more concentrated solutions was again noted,[646] although this alone does not describe or account for all of the changes in light scattering behavior under associating conditions.[62] Ionic micelles have been classified as being either (a) spherical at all concentrations, (b) rod-shaped at all concentrations, or (c) spherical at low concentrations and rod-shaped at high concentrations, from the results of X-ray scattering studies.[1241]

The extent and nature of hydration of micelles and, more particularly, the hydrocarbon interior represent aspects of micellization in which progress has recently been made.[1067] There does not seem to be any question that the polar surface of the micelle is more heavily hydrated than the interior,[287,290,1047,1177,1583] but the extent of penetration of the aqueous region into the nonpolar core is uncertain. Convincing evidence in support of the existence of a hydrocarbon core has come from solubilization studies.[404,1078] Recent investigations of this type have dealt with the relative distribution of solubilized molecules in either the hydrocarbon core or the more hydrophilic mantle.[1052,1053] The solubilization of p-xylene in sodium dodecyl sulfate micelles was studied[469] by a nmr method by which paramagnetic counterions caused line broadening of the p-xylene signal. The results were consistent with a model in which p-xylene was allowed to range freely throughout the hydrocarbon core. The dynamic fluid, though perhaps viscous nature of the hydrocarbon core, has also been suggested on the basis of fluorescence[627] and fluorescence polarization,[1568] differential ultraviolet spectroscopic,[1237] spin labeling,[569,1105] nmr,[549,1070,1182] and optical rotatory dispersion[1063] studies.

Another approach to obtain information regarding the hydration of the micelle interior has been to compare some physical property of the surfactant molecule above the CMC or the solubilized molecule with the same property in a nonaqueous solvent. The reference point for the aqueous environment is usually the infinitely dilute aqueous solution. In this manner, Muller and Birkhahn[1070] have defined a parameter Z obtained from nmr studies which ranges from zero for a hydrophilic environment to 1.0 for a hydrophobic environment. The use of this parameter was discussed by Muller[1067] with respect to micelles formed from surfactants containing a terminal trifluoromethyl group in the alkyl side chain. Values of Z close to 0.5 were found for cationic, anionic, and nonionic surfactants of various chain lengths. Interpretation of these values becomes somewhat difficult since the extent of solvation of the hydrocarbon core is not precisely known. Results from a variety of studies suggest that the first four to six hydrocarbons retain their hydration in the micellar state.[111,285,1182,1422] Additional support for this view comes from nmr studies of sodium perfluorooctanoate[1074] which yielded values for Z of 0.37, 0.66, 0.63, 0.83, and 0.84 for substitution on the second, fourth, sixth, seventh, and height positions of the chain, respectively. Considerably lower values of Z were found in solubilization studies involving the distribution of benzene between the bulk solvent and micelle interior.[549] This was considered to reflect the dynamic nature of micelles, although it is also consistent with a partial hydration of the hydrocarbon core. Ericksson and Gillberg[414] concluded that benzene along with nitrobenzene is solubilized at the surface of cationic micelles, whereas cyclohexane and cumene are located in the hydrophobic interior. From the study of ^{19}F chemical shifts and shielding parameters for a series of solubilized para-substituted fluorobenzene,[52] it was concluded that the probes are more exposed to the solvent with anionic and cationic than with zwitterionic micelles.

5. COUNTERION BINDING

A moment's reflection on the structure of ionic micelles quickly leads to a recognition of the necessity for some type of neutralization of electrostatic repulsions between the charged surfactant head groups at the micelle–solvent surface. This is accomplished by the "binding" of some fraction of the total number of counterions in the solution. To represent this interaction, the following equilibrion can be proposed:

$$mM^- + pX^+ \rightleftharpoons [M_mX_p]^{(m-p)-} \tag{34}$$

where M^-, X^+, and $[M_mX_p]$ correspond to the concentration of the monomer, counterion, and micelle, respectively, and the aggregation numbers of the monomer and counterions are m and p. The quantity $m - p$ corresponds to the residual charge on the micelle, and leads to negatively charged micelles for an anionic surfactant according to eqn. (34) since $m > p$, and positively charged micelles for cationic surfactants, for the same reason. The ratio $(m - p)/m$ is the degree of dissociation of the counterions, and the degree of association, or counterion binding ,α is given as

$$\alpha = 1 - (m - p)/m \tag{35}$$

It is of interest to determine the quantity α from experiment, and Anacker[21] has cited several examples of different methods which have been used previously, and described the use of light scattering, ultracentrifugation, colligative properties, emf measurements, electrophoretic mobility, and conductivity in some detail. In addition he has included a description of the use of eqn. (4) as indicated previously in this chapter. More recently, it has been possible to use nmr, employing paramagnetic ions[903,1250] and ^{85}Rb counterions,[903] and esr[1614] to study counterion binding, and the use of magnetic resonance in general for this type of study appears to be quite promising. The binding of counterions to the micelle may be viewed in chemical terms as a mass action type of ion pairing or in physical terms as a distribution of counterions in a Gouy–Chapman type of double layer. A calculation of the free energy of micelle formation by either approach leads to values which are of the correct order of magnitude upon comparison with experimental data for sodium dodecyl sulfate, although the mass action model was favored by Mukerjee[1049] in spite of various uncertainties in both interpretations.

A summary of values for α for various surfactants under different experimental conditions is given in Table IV. It can be seen from the first five entries that α decreases slightly with increasing temperature. A similar trend has been reported for other surfactants.[5,6,98] Although a change in aggregation number with temperature may occur, it would not likely produce a large change in α since Evans[417] has shown that α is nearly independent of aggregation number over a fourfould range for various alkyl sulfates. If counterion binding is compared with ion pair formation[323] or ion binding to proteins,[917] it may be considered to be an exothermic process for at least some counterions, and on this basis the degree of association would be expected to decrease with an increase in temperature. Indeed, Ingram and Jones[708] have calculated the enthalpy of dissociation of counterions from dodecyltrimethylammonium bromide, dodecylpyridinium bromide, and sodium dodecyl sulfate micelles to be 2.02, 1.90, and 2.51 kcal mol^{-1}, respectively, and Eatough and Rehfeld[379] have interpreted the heat of dilution of sodium dodecylsulfate above the CMC in terms of counterion binding. They obtained a binding enthalpy of about -11 kcal mol^{-1} and a binding constant of about 300 for ion pairing at the micelle surface. It would be of interest to study the influence of supporting electrolyte on the heat of dilution in order to test their treatment and extend it to other ion types such as those collected by Davies.[323]

An increase in length of the nonpolar side chain is accompanied by an increase in α, as seen from a comparison of entries 1, 6, and 7; 8 and 9; 11, 13, and 15; 17, 18, and 19. The acidified solutions of carboxylate salts (entries 31, 32, and 33) do not appear to exhibit the same behavior; however, their alkaline solutions (entries 26, 27, and 29) do. The differences between the amount of counterion binding of the carboxylates is likely due to the existence of mixed micelles containing the undissociated acids at lower pH values, as shown by potentiometric studies, with the result that competition between protons and counterions at the micelle surface exists.[428,1273,1482,1483] An increase in counterion binding with an increase in chain length can be understood in terms of an increased micelle size for the surfactants with a longer chain, and a higher charge density at the

TABLE IV. Summary of Values of the Degree of Counterion Binding α for Various Surfactants

Surfactant	Temperature, °C	Additive	Method	α	Ref.
1 Decyl α-picolinium bromide	15	—	Conductance	0.69	5
2 Decyl α-picolinium bromide	25	—	Conductance	0.70	5
3 Decyl α-picolinium bromide	35	—	Conductance	0.68	5
4 Decyl α-picolinium bromide	45	—	Conductance	0.67	5
5 Decyl α-picolinium bromide	55	—	Conductance	0.67	5
6 Dodecyl α-picolinium bromide	25	—	Conductance	0.78	5
7 Tetradecyl α-picolinium bromide	25	—	Conductance	0.80	5
8 Decyltrimethyl-ammonium bromide	25	—	Conductance	0.72	6
9 Dodecyltrimethyl-ammonium bromide	25	—	Conductance	0.82	6
10 Dodecylbenzyldimethyl-ammonium bromide	25	—	Conductance	0.78	6
11 Decyltrimethyl-ammonium bromide	25	—	Light scattering	0.78	98
12 Decyltrimethyl-ammonium bromide	25	—	Equation (4)	0.34	98
13 Dodecyltrimethyl-ammonium bromide	25	—	Light scattering	0.80	98
14 Dodecyltrimethyl-ammonium bromide	25	—	Equation (4)	0.62	98
15 Tetradecyltrimethyl-ammonium bromide	25	—	Light scattering	0.86	98
16 Tetradecyltrimethyl-ammonium bromide	25	—	Equation (4)	0.70	98
17 Sodium decylsulfate	25	—	emf	0.70	1371
18 Sodium dodecylsulfate	25	—	emf	0.78	1371
19 Sodium tetradecylsulfate	25	—	emf	0.87	1371
20 Sodium dodecylsulfate	25	—	emf	0.84	157

TABLE IV. (*Continued*)

	Surfactant	Temperature, °C	Additive	Method	α	Ref.
21	Sodium dodecylsulfate	25	—	Light scattering	0.82	1160
22	Dodecylaminoethanol hydrochloride	25	—	Conductance	0.73	1251
23	Dodecylaminoethanol hydrobromide	25	—	Conductance	0.80	1251
24	Dodecylaminoethanol hydroiodide	25	—	Conductance	0.84	1251
25	Dodecylaminoethanol nitrate	25	—	Conductance	0.89	1251
26	Potassium decanoate	23	pH 12.5	emf	0.21	427
27	Potassium dedecanoate	24	pH 12.5	emf	0.31	427
28	Potassium tetradecanoate	25	pH 12.5	emf	0.76	427
29	Sodium decanoate	26	pH 12.5	emf	0.43	427
30	Sodium dodecylsulfate	24	pH 12.5	emf	0.86	427
31	Potassium dodecanoate	25	—	emf	0.77	966
32	Potassium tetradecanoate	25	—	emf	0.73	966
33	Potassium hexadecanoate	25	—	emf	0.74	966
34	Lithium dodecylsulfate	25	—	emf	0.83	708
35	Sodium dodecylsulfate	25	—	emf	0.77	708
36	Potassium dodecanoate	25	—	emf	0.73	708
37	Sodium dodecanoate	25	—	emf	0.70	708
38	Dodecyltrimethyl-ammonium bromide	25	—	emf	0.75	708
39	Dodecylpyridinium chloride	25	—	emf	0.75	708
40	Dodecylpyridinium bromide	25	—	emf	0.77	708
41	Dodecylpyridinium iodide	25	—	emf	0.77	708
42	Dodecylpyridinium chloride	25	0.1 M KCl	emf	0.73	708
43	Dodecylpyridinium bromide	25	0.01 M KBr	emf	0.81	708
44	Dodecylpyridiniumiodide	25	0.01 M KI	emf	0.91	708
45	Hexadecyltrimethyl-ammonium bromide	25	—	emf	0.77	991

TABLE IV. (*Continued*)

	Surfactant	Temperature, °C	Additive	Method	α	Ref.
46	Hexadecyltrimethyl-ammonium bromide	25	0.1 M NaBr	emf	0.67	991
47	Hexadecyltrimethyl-ammonium chloride	25	0.1 M CCl$_4$	emf	0.77	991
48	Hexadecyltrimethyl-ammonium chloride	25	—	emf	0.76	991
49	Hexadecyltrimethyl-ammonium chloride	25	0.1 M NaCl	emf	0.82	991
50	Hexadecyltrimethyl-ammonium chloride	25	0.1 M CCl$_4$	emf	0.77	991
51	Decylammonium chloride	30	0.1 m NaCl	Light scattering	0.84	23
52	Decylmethylammonium chloride	30	0.1 m NaCl	Light scattering	0.84	23
53	Decylethylammonium chloride	30	0.1 m NaCl	Light scattering	0.83	23
54	Decylpropylammonium chloride	30	0.1 m NaCl	Light scattering	0.83	23
55	Decylammonium bromide	25	0.5 m NaBr	Light scattering	0.93	23
56	Decylmethylammonium bromide	25	0.5 m NaBr	Light scattering	0.93	23
57	Decyldimethylammonium bromide	25	0.5 m NaBr	Light scattering	0.92	23
58	Decyltrimethylammonium bromide	25	0.5 m NaBr	Light scattering	0.87	23
59	Decyldiethylammonium bromide	25	0.5 m NaBr	Light scattering	0.87	23
60	Decyltriethylammonium bromide	25	0.5 m NaBr	Light scattering	0.77	23
61	Decylquinuclidinium bromide	25	0.5 m NaBr	Light scattering	0.88	23
62	Decyldiazabicyclo-[2.2.2.]octanium bromide	25	0.5 m NaBr	Light scattering	0.88	23
63	Sodium trifluorodecanoate	35	0.6 M NaCl	nmr	0.58	1070

micelle surface would exist for the larger micelles since the volume of a sphere increases faster than the area with an increase in radius. An increase in counterion binding would accompany an increase in electrostatic potential at the micelle surface. Mukerjee has described the use of surface potentials in calculations of the electrostatic contribution to the free energy of micelle formation.[1050,1067]

The nature of the counterion has an important bearing upon α for cationic surfactants, as can be seen by comparing entries 22, 23, 24, and 25; 36 and 37; and 39, 40, and 41. These results consistently reveal that the ions that have the lowest energy of hydration and the largest amount of nonsolvational coordinated water[155] are associated with the largest values of α. The nature of the counterion for anionic surfactants does not seem to influence α as much as for cationic surfactants, as seen by comparing entries 34 and 35, and 36 and 37. The relative differences between the amounts of nonsolvational coordinated water for these ions are not as large as for the anions. A trend similar to that observed with anions is observed with the tetraalkylammonium counterions.[1062]

A kinetic description of the Stern layer of dodecyl sulfate and trimethylammonium micelles was obtained from a nmr investigation with ions.[1250] The rate of exchange of water molecules between the hydration layer of the ion and bulk water was not found to change when Mn^{2+} was incorporated into the micelle Stern layer. However, the rate of Mn^{2+} rotation was found to decrease by about one-third of its value in bulk solution. Finally, it was shown that ion pair formation between Gd^{3+} and SO_4^{2-} is stronger than the attraction of SO_4^{2-} to a cationic micelle surface. The contribution of charge transfer to the interaction between the pyridinium and halide ions[1064] probably also contributes to α for entries 39, 40, and 41 and is responsible for the differences between this series and entries 22, 23, and 24.

The effect of added salt is to increase α, as can be seen for entries 39 and 42; 40 and 43; 41 and 44; and 48 and 49. An exception seems to exist for entries 45 and 46. An increase in counterion binding with an increase in ionic strength can be explained on the basis of mass action, whereas Mathews et al.[991] rationalized the difference for entries 45 and 46 on the basis of a salt-induced increase in micelle size with a concomitant reduction of charge density. The solubilization of organic solutes by hexadecyltrimethylammonium bromide micelles (entries 47 and 50) did not alter α within the limits of experimental error. This result is rather surprising in view of the known influence of chain length on counterion binding previously discussed, and warrants further investigation. Finally, the effects of the structure of the polar head group are described by entries 9 and 10, 51–54, and 55–62. For all three of these groups, the larger degree of ion binding occurs with the smaller head group. Other possible interpretations of the influence of subtle changes in the structure of the polar head group on counterion binding have been discussed by Anacker et al.[23] but unambiguous interpretations are not possible at this time.

Quantitative differences exist between the values of α reported for a given compound using the same method (entries 17 and 20), and also between results for the same compound when studied by different methods (entries 8, 11, 12, and 58; 13, 14, and 38; and 15 and 16). Discrepancies of the latter type have been noted previously by Anacker[21] and Mukerjee et al.[1062] The reason for the observed differences is probably similar to that mentioned for variation in the reported CMC values, namely one is not dealing with an easily defined physical process and different methods are sensitive to different aspects of the process. Again the general statement can be made that was used when discussing CMC determinations, that the values of α obtained by different methods, except for the use of eqn. (4), may be considered to be in either good or poor agreement depending upon the use one makes of the data. Shinoda[1384] has indicated that experimental values of K_g, which is conceptually similar to α, range from 0.4 to 0.6, but nearly all the values of α given in Table IV tend to range from 0.7 to 0.9. Since K_g evaluated from eqns. (2) and (3) ranges from 0.6 to 0.9, in agreement with observed values included in Table IV, it may be that the use of eqn. (4) to determine α is not reliable because of the oversimplified model. The origin of the failure of eqn. (4) is not known, but Mukerjee et al.[1062] have shown this equation becomes a more complex function if the activity coefficients of the ionic species and polydispersity are taken into consideration.

An unconventional way of viewing ion binding to micelles was recently proposed by Mijnlieff,[1010] who suggested that the micelle and its counterlayer out to a Debye distance $1/\varkappa$ be viewed as an electroneutral micellar unit, by regarding all of the counterions as being bound to the micelle. A thermodynamic derivation from this model allowed the calculation of the negative adsorption of added electrolyte, equal to the number of univalent–univalent electrolyte molecules excluded from the electroneutral micellar units. Calculations for sodium dodecyl sulfate micelles in the presence of NaCl were in satisfactory agreement with previous light scattering experiments. The concept of negative adsorption together with double layer theory has been used previously by Mukerjee and Banerjee[1055] to account for the variation of the pK_a of a solubilized indicator dye with micelle concentration.

6. PRE-MICELLAR ASSOCIATION

Results from a variety of recent thermodynamic,[9,104,141,207,208,315,374,452, 490,810,815,1344,1422] transport,[217,262,1049] and kinetic studies[851] of aqueous solutions of substances containing a nonpolar group have appeared which suggest self-association by means of hydrophobic interactions. Therefore, it can be said with confidence that pre-micellar association as a phenomenum is to be fully expected, and it is the details of the process that are presently incompletely understood (but see Chapter 1). Extensive thermodynamic analysis of aqueous solutions of a variety of nonelectrolytes indicates that pairwise interactions are more probable than triplets, etc., except for concentrated alcohol solutions, which appear to form aggregates.[810] Indeed, pairwise interactions are in accord with results with other systems,[315,490,1049,1344] although alkyl carboxylates show evidence of both dimer formation and higher aggregates as well.[315,374]

Studies of the effect of temperature on dimer formation appear to lead to contradictory results. Kozak et al.[810] found that hydrophobic interactions became stronger with an increase in temperature, whereas Butler et al.[217] observed that dimerization was more pronounced at low temperatures. A similar difference between the effect of temperature on hydrophobic bond formation with homologous series of alcohols and carboxylic acids was observed.[1329] Perhaps detailed studies will show that the influence of temperature on dimer formation is similar to its influence on the CMC, in that dimerization may be expected to generally approach a temperature of minimum stability, and the nature of the polar head group would dictate the temperature at which the minimum is observed. The effects of temperature on hydrophobic interactions has been discussed in detail by Franks in Chapter 1 of this volume. Recently, Birch and Hall[141] have observed deviations from Debye–Hückel theory by means of heat of dilution measurements on several submicellar sodium alkylsulfate and alkyltetramethylammonium bromide samples. They concluded that the data may be equally well interpreted on the basis of either dimerization due to hydration effects or changes in the distance of closest approach of the hydrophobic ions.

The structure of the head group also has an important bearing on the stability of the dimer. Dimerization has been reported for short- and long-chain fatty acids,[315,1049,1344] whereas only long-chain alkylsulfonates and alkylammonium surfactants exhibit deviations from the Onsager limiting law which can be attributed to dimerization.[262] However, frictional factors can also be responsible for deviations from the Onsager limiting law, as evidenced by the fact that both positive[1061] and negative[217] deviations have accompanied dimer formation. It is very possible at a given temperature for a given surfactant that the conductance of a monomer and dimer could be equal within experimental error, due to a cancellation of effects. Variable temperature studies may be one way in which to separate the effects. Apparent increases in the conductivity of micelles in concentrated salt solutions have recently been attributed to an obstruction effect of the colloidal particles on the conductance of the electrolyte.[1076]

Finally, another detail of the dimerization process which has received some comment in the past[282,1067] is the question of degree of side chain extension or curling of either the free monomer or the dimer in aqueous solution. Both recent reviewers of this topic favor the view that hydrocarbon side chains shorter than about C_{16}–C_{18} are likely in an extended configuration; this opinion is based upon an analysis of partition coefficients,[282] nmr data,[1067] and enthalpy of transfer data.[1067]

7. POST-MICELLAR ASSOCIATION

The continuous changes in the physical state of micelles with changes of temperature, surfactant concentration, and solvent composition have been discussed previously. In addition, there are several fairly well-defined experimental conditions which give rise to abrupt changes in the micellar properties of isotropic solutions. One such example is the *Krafft point*,[404] which corresponds to the temperature, or small range in temperature, at which the solubility of a surfactant is greatly increased. The onset of micelle formation with increasing temperature occurs at the Krafft point and is regarded as the basis for the enhanced solubility. This transition from a less-ordered to more-ordered form with increasing temperature resembles the well-known coil $\rightarrow$ helix transition of poly-γ-benzylglutamate in nonaqueous solvents.[361] The unusual behavior of the polypeptide is attributed to enhanced solvation of the low-temperature form, and in this way resembles the behavior of the surfactant. Sometimes another abrupt change in the physical properties of nonionic surfactants above the CMC occurs at what is known as the *cloud point*. Examination of the phase diagram of a two-component (surfactant–water) system reveals a region where two immiscible isotropic liquid phases exist, and they are described as conjugate solutions.[646] One solution will be richer in surfactant than the other, and the one with the higher concentration may be referred to as the concentrated phase, and the more dilute solution as the dilute phase. When a maximum or minimum occurs in the phase boundary, the temperature at which phase separation occurs is called an upper or lower consolute temperature, respectively.[646] The excess thermodynamic functions over the entire range of composition for a surfactant–water system have been obtained from vapor pressure and calorimetric experiments.[1366] It was shown that phase separation must be accompanied by a favorable entropy change, with the solvent playing an important role. Increasing the hydrophilic chain length of a series of octylphenyl polyoxyethylene ethers progressively raises the cloud point,[541] and the dilute phase is found to be enhanced with surfactant by the addition of NaSCN and slightly depleted by the addition of NaCl.[357] Shinoda has shown that a pseudophase separation model is consistent with the thermodynamic properties of the process.[1385] Extreme nonideality exists in the vicinity of the consolute boundaries and must be accounted for when interpreting thermodynamic data in this region.[646]

Finally, abrupt changes in the physical properties of surfactant solutions have been noted above the CMC, between the Krafft and cloud points with either increased concentration[370,389,817,1026,1451,1594] or temperature.[103,288] In some cases these changes have been termed a second[389,1026] and third[389] CMC. Both ionic and nonionic surfactant solutions exhibit these effects. The transitions undoubtedly reflect changes in size, shape, polydispersity, and degree of counterion binding to the micelles and perhaps also changes in hydration.[103,391,1385] It may be that systems will be found which exhibit various combinations of these effects. However, this aspect of the physical chemistry of micelles represents an interesting though incompletely developed area.

8. MIXED MICELLES

The formation of micelles from more than one chemical species gives rise to what is known as *mixed micelles*. In the simplest case, binary or ternary mixtures of surfactants of similar, but not identical chain lengths may be studied, and the thermodynamics of this type of micelle formation has been described.[102,1384] Alternately, micelles may be formed from compounds which are either heterodisperse or polydisperse, and Gibbons[512] has identified important differences between the two types of compounds. Polydisperse substances may be expected to exhibit average properties which would not differ very much from those of a homogeneous compound, with the properties corresponding to the mean, whereas this behavior would not be expected for a heterodisperse sample, which contains an uneven distribution of components. Némethy and Ray[1230] have recently taken advantage of this important property of polydisperse compounds in a thermodynamic study of micelle formation by nonionic surfactants in ethylene glycol–water mixtures. Another class of mixed micelles results when low-molecular-weight molecules are *solubilized* by micelles formed from surfactants containing a relatively larger nonpolar side chain. The solubilized substance, also called a penetrating additive,[409] may be located in the hydrocarbon core[1052] or the hydrophilic mantle[1053] as discussed previously. Several recent studies have been concerned with this aspect of micelle formation.[404]

By assuming that the CMC of a mixed micelle is given by the sum of the contributions of the components, Shinoda[1384] was able to show that the CMC of the mixture C_{mix} is given by the equation

$$C_{\text{mix}}^{1+K_g} = \sum C_{m_i}^{1+K_g} \frac{X_i' \exp(n_i W/kT)}{\sum X_i' \exp(n_i W/kT)} \tag{36}$$

where n_i is the number of carbon atoms in the side chain of the ith component and C_{m_i} and X_i' correspond to the CMC of the n_i component and mole fraction of nonmicellar material, respectively. The CMC of the mixture will fall within the highest and lowest individual CMC values of the pure components, as shown for binary and ternary mixtures.[1384] The validity of eqn. (36) for commercial samples of long-chain alkyltrimethylammonium bromides containing up to seven (though only three major) components was reported.[97] This work was extended to obtain an expression for the micellar weight-average molecular weight of a surfactant mixture,[98] and good agreement between experimental and theoretical molecular weights was found for the commercial samples previously described.[97] Alternatively, Attwood *et al.*[64] applied the Hall and Pethica[592] treatment of mixed micellar solutions of nonionic surfactants to calculate the number–average micellar molecular weights of various compositions of a binary nonionic surfactant mixture. The molecular weights that were calculated by assuming perfect mixing of the micellar components closely approximated the experimental values, whereas large deviations between the experimental and calculated molecular weights were found by assuming the surfactants did not interact to form mixed micelles. Hall[591] has recently developed a general thermodynamic theory of solutions containing interacting multicomponent charged aggregates. He indicated that examples of the application of this theory to multicomponent systems will be forthcoming.

The surface properties of ionic micelles have been shown to be altered by mixed micelle formation. For example, Tokiwa and Ohki[1483] have found that the addition of an anionic surfactant increases the apparent dissociation constant of micelles of a cationic–nonionic surfactant, and the addition of a cationic surfactant produces the opposite effect. It has also been shown that the degree of counterion binding by mixed micelles formed from anionic and nonionic surfactants decreases as the proportion of the nonionic component increases and as the nonionic polyoxyethylene head group is lengthened.[1482] These observations can be understood in terms of an altered charge density at the micelle surface as the result of mixed micelle formation, and possible interaction between the anionic and nonionic head groups. The latter has been demonstrated by nmr studies[824,1484] in which the aromatic portion of the anionic surfactant shifts the proton resonance signal of the polyoxyethylene group upfield. The extent of the upfield shift depends on the mole ratio of the mixture and the length of the polyoxyethylene chain.

Studies involving organic additives that are solubilized by the micelle have primarily dealt with their effect on the CMC and aggregation number and their location in the micelle interior. It has been known for a long time that micelles enhance the solubility of hydrocarbons in water,[404,1588] but only recently has it been shown that hydrocarbons alter the CMC.[1006] The CMC of dodecylamine hydrochloride solutions saturated with ethane and propane at 25° was found to be lower by an amount which was equivalent to lengthening the surfactant side chain by $(0.35–0.46)n$ for ethane and $(0.72–1.80)n$ for propane, where n is as defined in eqn. (1). It would be of great interest to extend these studies to other surfactants and temperatures to determine the thermodynamic parameters that accompany this process. The solubilization of benzene, ethylbenzene, octane, and cyclohexane in sodium carboxylate micelles from 20–60° decreased the CMC, and the greatest effect was observed at the lower temperatures.[805,806] The addition of low concentrations of trialkyl phosphines[1604] and short-chain alcohols[1392] to sodium dodecyl sulfate solutions also produced CMC reductions. In both cases the depression was more marked as the size of the hydrophobic group in the additive increased. The CMC reduction was attributed to a reduction of the micelle surface charge density and hydrophobic interaction between the additive and micelle hydrocarbon core.[1392] Finally, two studies of the effect of additives with nonionic surfactants may be contrasted. The addition of short-chain alcohols to dimethylamine oxide solutions lowered the CMC,[111] whereas the addition of ethylene glycol to alkyl polyoxyethylene solutions increased the CMC.[1230] However, in the study with the alcohols it was found that the effect on the CMC increased in the order n-butanol > n-propanol > ethanol > methanol. It seems that mixed micelle formation favors a CMC reduction, as with the hydrocarbons, and less penetrating additives like methanol and ethylene glycol increase the CMC by lowering the free energy of the monomer by enhancing its solubility. However, at higher concentrations of the additive the CMC may be raised as previously described. A similar though not identical view has also been given.[409] It has been noted[268,991] that penetrating additives, or those that are solubilized by the micelle, increase the aggregation number in addition to lowering the CMC as previously discussed. These results are in qualitative agreement with eqn. (32), but are too incomplete at this time for further comment.

Recent attemps to locate the site of penetrating substances that have been incorporated in the micelle interior[404,1078] have utilized nmr,[52,411, 469,549,722,1067,1105] esr,[1105,1191] UV absorption,[1105,1237,1568] fluorescence spectroscopy,[359,1568] and solubilization[1052] techniques. The results of these investigations are summarized in Table V. The Z values are given for several entries in the table, and the interpretation of this quantity is the same as described previously when discussing hydration of the micelle interior. All of the solutes represented contain an aromatic moiety, but they differ with respect to the amount of hydrophilic character they possess. If benzene is used as a reference point, it can be seen that solutes that are more hydrophobic than benzene, such as pyrene, perylene, 2-methylanthracene, cyclohexane, isopropylbenzene, and NFR-612, in general have properties when solubilized by the micelle that resemble those properties observed when dissolved in hydrocarbon solvents, and for this reason are identified as being located in the hydrocarbon core. However, polar derivatives such as DNPHTEMPO, TEMPO, nitrobenzene, N,N-dimethylaniline, and phenol appear to be located on the surface with the polar portion of the molecule aligned toward the solvent and the nonpolar portion directed toward the hydrocarbon interior. It is interesting that the apparent location of benzene (or the extent of hydration) depends upon the nature of surfactant, as shown by the range of values for Z, and also the relative concentration. For example, it has been reported that benzene[414] becomes a penetrating solute when it is present at an about 1:1 mole ratio with the surfactant.[414]

The concentration dependence of Z values can be difficult to interpret, and Muller has suggested that additional contributions to the solvation effect may arise.[1067] These were attributed to (a) swelling of the micelles with concomitant changes in internal organization, (b) interaction between the solubilized molecules, (c) unequal binding affinities for the various sites,

TABLE V. Summary of the Results from Some Recent Studies Dealing with the Location of Penetrating Solutes in the Micelle Interior

	Solute	Micelle	Method	Location	Ref.
1	Pyrene	Dodecyltrimethylammonium bromide	UV abs. and fluorescence	Hydrocarbon interior	359
2	Pyrene	Hexadecyltrimethylammonium bromide	UV abs. and fluorescence	Hydrocarbon interior	359
3	DNPHTEMPO[a]	Sodium dodecyl sulfate	UV abs, esr, and nmr	Near micelle surface	1105
4	TEMPO[b]	Sodium dodecyl sulfate	UV abs, esr, and nmr	Near micelle surface	1105
5	Benzoic acid and derivatives	Alkylpolyoxyethylenes	Solubilization	Mostly mantle	1052
6	2-Methylanthracene	Dodecyltrimethylammonium bromide[c]	Fluorescence	Hydrocarbon interior	1568
7	Perylene	Dodecyltrimethylammonium bromide[c]	Fluorescence	Hydrocarbon interior	1568
8	Xylene	Sodium dodecyl sulfate	nmr	Uniformly distributed	469
9	Benzene	Dodecyl polyoxyethylene (23) ether	nmr	$Z = 0.06-0.16$	549
10	Benzene	Hexadecyltrimethylammonium bromide	nmr	$Z = 0.18$	549
11	Benzene	Sodium dodecyl sulfate	nmr	$Z = 0.20-0.22$	549
12	Benzene	Sodium dodecanoate	nmr	$Z = 0.52$	549
13	Benzene	Sodium decanoate	nmr	$Z = 0.29-0.33$	549
14	p-FBDA[d]	Sodium dodecyl sulfate	nmr	Surface	52

15	p-FBDA[a]	Dodecyltrimethylammonium bromide	nmr	Surface	52
16	p-FBDA[a]	Dodecyldimethylammoniapropane sulfonate	nmr	Surface[c]	52
17	M-O,p-FB[e]	Sodium dodecyl sulfate	nmr	Surface	52
18	NFR-612[f]	Hexadecyltrimethylammonium bromide	esr	Hydrocarbon interior	1191
19	NFR-612[f]	Sodium dodecyl sulfate	esr	Hydrocarbon interior	1191
20	Benzotrifluoride	Sodium decanoate	nmr	$Z = 0.5$	1070
21	Benzotrifluoride	Sodium dodecanoate	nmr	$Z = 0.5\text{–}0.6$	1070
22	Phenol	Sodium dodecyl sulfate	nmr	Surface	722
23	Cyclohexane	Hexadecyltrimethylammonium bromide	nmr	Hydrocarbon interior	414
24	Isopropylbenzene	Hexadecyltrimethylammonium bromide	nmr	Hydrocarbon interior	414
25	Benzene	Hexadecyltrimethylammonium bromide	nmr	Surface	414
26	Nitrobenzene	Hexadecyltrimethylammonium bromide	nmr	Surface	414
27	N,N-dimethylaniline	Hexadecyltrimethylammonium bromide	nmr	Surface	414

[a] DNPHTEMPO = 2,4-dinitrophenylhydrazone of 2,2,6,6-tetramethyl-4-piperidine-1-oxyl.

[b] TEMPO = 2,2,6,6-tetramethyl-piperidine-1-oxyl.

[c] Also investigated were tetra- and hexadodecyltrimethylammonium bromide and octadecyldimethylbenzylammonium bromide.

[d] p-FBDA = p-fluorobenzaldehyde diethyl acetal.

[e] M-O, p-FB = methyl ortho-p-fluorobenzoate.

[f] NFR-612 = 17β-hydroxy-4',4'-dimethylspiro[5α-androstane-3,2'-oxazolidin]-3'-yloxyl.

and (d) penetration of water into the micelle interior. It is interesting in this regard that the solubilization of an oil-soluble dye in a mixed anionic–nonionic micelle depends upon the relative composition of the two surfactants.[1481] The amount of dye solubilized increased upon the addition of the nonionic component until a mole ratio of about 4:1 of nonionic to anionic surfactant was reached. Finally, the line broadening caused by paramagnetic counterions was used to study the solubilization of p-xylene in sodium dodecyl sulfate micelles, and it was found that an average of six molecules per micelle were uniformly distributed throughout the micelle interior.[469]

Although all the results obtained to date with penetrating solutes indicate that the micelle interior is generally fluid, there are several indications that it is less fluid than a comparable hydrocarbon solvent of the same chain length as the surfactant at the same temperature.[359,1051,1105,1191,1568] Thus the sodium dodecyl sulfate micelle interior has been described as being more restricted than water or dodecane, by magnetic resonance studies,[1105] or having long-range order to give it some solidlike character, by studies of the variation of the CMC with chain length.[1051] The microviscosity of the micelle interior was determined for substituted methylammonium bromide micelles by polarization of fluorescence and found to be characteristic of a liquid, although less fluid than hydrocarbon solvents of similar chain length,[1568] and quantum yield studies showed that the micelle interior became less rigid as the temperature was increased.[359] The fluidity of the hexadecyltrimethylammonium bromide increased markedly upon incorporation of cholesterol or hexadecanol.[1568] Only one investigation suggests that the hydrocarbon core is completely solidlike,[1191] and in this instance the results of an esr investigation found that both the sodium dodecyl sulfate and hexadecyltrimethylammonium bromide micelle interior resembled hexadecane at $-22°$ (mp 18°). In summary, it can be said that interesting spectroscopic data are just beginning to be obtained and at present do not provide a consistent view of the nature of the micelle interior. Part of the problem may reside in obtaining suitable references for comparison with studies involving micelles, which leads to alternate interpretations of a given set of data by different workers, as in the case of UV differential spectroscopy studies.[438,1238]

9. NONAQUEOUS MICELLES

Perhaps one of the more interesting though less thoroughly studied aspects of micellization is the formation of micelles in nonaqueous solvents. The existence of micelles in a nonaqueous solvent may seem a bit surprising in view of the emphasis normally placed on hydrophobic interactions when discussing micelle formation in aqueous solutions. However, evidence was obtained as early as 1902 for the formation of micelles in nonaqueous solutions.[751] Recent results have been reviewed,[102,404,774,1384] and it is now possible to make a few generalizations. The nature of the nonaqueous solvent is most important for the formation of nonaqueous micelles and their structure. Nonaqueous solvents can be classed[1227,1228] as (a) those that give rise to solvophobic interactions, (b) those in which "inverted" micelles are formed, and (c) those in which micelles do not exist.

Solvents of the first type are those that have two or more potential hydrogen-bonding centers and are capable of forming a highly hydrogen-bonded structured solvent similar in many respects to water.[1228] Micelles formed in solvents of this type such as glycol, 2-aminoethanol, formic acid, 2-mercaptoethanol, 1,2-propanediol, etc., are assumed to have the same structure as those formed in water, by analogy with the limited solubilities of hydrocarbons in this type of solvent. However, micelle formation is not as favored in a nonaqueous solvent as in water for a given surfactant. For example, the CMC at 27.5° of dodecylpyridinium bromide, tetradecyltrimethylammonium bromide, and hexadecylpyridinium chloride changes from 1.22×10^{-2}, 3.84×10^{-3}, and 9.2×10^{-4} M in water to 0.55, 0.25, and 0.23 M in ethylene glycol, respectively. Conversely, micelles of an "inverted" type are formed in typical hydrocarbon solvents such as benzene, cyclohexane, heptane, etc., in which the structure of the micelle is reversed —the polar head groups are at the center and the nonpolar side chains are directed toward the solvent. Finally, polar solvents with a single hydrogen-bonding center such as methanol, ethanol, and dimethylformamide do not support micelle formation. Since the demonstration of micelle formation by solvophobic interactions is very recent, the micelles that are formed by this process have not been characterized. Therefore, the following discussion will mainly deal with micelles of the inverted type that form in hydrocarbon solvents.

In many respects, the problems dealing with inverted micelles in hydrocarbon solvents are similar to those encountered when characterizing aqueous micelles. Thus, a discontinuity in some physical property of the solution can be used to identify the CMC, and techniques such as light scattering, ultracentrifugation, and viscosity are used to determine the size and shape of the micelle. Specific examples of techniques which have been used recently to determine the CMC include dye solubilization,[186,1576] water solubilization,[1286] nmr,[402,435,436,437,1603] solubility,[999] and surface tension.[186,998,1351] Additional methods which have been used in the past have been collected by Shinoda.[1384] Of the various methods which have been employed, the solubilization methods appear to be the most objectionable in the light of the observed fourfold CMC reduction of dodecylammonium propionate and dodecylammonium benzoate in benzene due to the presence of a solubilizate (2,3,4,6-tetramethyl-α-D-glucose).[437] The importance of eliminating impurities when dealing with inverted micellar systems was emphasized by Peri, who noted differences in the size and shape of Aerosol OT (di-2-ethylhexyl sodium sulfosuccinate) micelles due to the presence of very minor amounts of an impurity.[1151] The CMC of several alkylammonium carboxylates in a variety of solvents has been studied[402, 435,436] and examples were found where the CMC either increased, decreased, or remained the same with increasing number of carbon atoms in the alkyl chain. In general, it was observed that the CMC values increased with increasing solvent polarity or reciprocal of the dielectric constant.

Only a few thorough quantitative studies of the effect of temperature on the CMC in nonaqueous solvents have been made.[186,944a,1286] Marked differences appear to exist between the changes in thermodynamic parameters that accompany micelle formation of the inverted and solvophobic types. Whereas the former are accompanied by moderately small, negative enthalpy and entropy changes,[186,1286] comparable enthalpy and very large, positive entropy changes are associated with the formation of solvophobic micelles. Clearly, more data would be useful and the magnitude of the enthalpy change indicated above suggests the usefulness of calorimetric techniques. The existence of a Krafft point has been demonstrated for copper soaps in various organic solvents[999] and for solvophobic micelles.[1228] The organic solvents employed were benzene, chlorobenzene, xylene, 1-propanol, 1-butanol, and 1-pentanol. The CMC was much larger with the three alcohol solutions than with the other solvents.[998,999]

The aggregation numbers of inverted micelles tend to be much lower than those of their aqueous counterparts, with typical values seldom exceeding about 25.[1384] Inverted micelles are also less polydisperse. Recent studies are consistent with those general statements.[402,435,436] The aggregation number of sorbitan monostearate micelles in o-xylene was recently found to increase rapidly with increasing temperature,[186] as described previously for several aqueous solutions of nonionic surfactants. However, the molecular weight of Aerosol OT in n-nonane[1151] only changed slightly between 25° and 75°. Therefore, this aspect of nonaqueous micelle formation would appear to be an attractive one for further study.

The nature of the solvent has an important bearing on the aggregation number, as it does on the CMC.[402,435,436] Solvents that do not support micelle formation also reduce the aggregation number. Thus molecular weight determinations by ultracentrifugation of barium dinonylnaphthalene sulfonate yielded aggregation numbers of ten in pure toluene, but this was reduced to four by the addition of less than 1% methanol.[501] The aggregation number of sodiodiethyl butylsodiomalonate[88] was about 50 in benzene, but only monomers were found in 1,2-dimethoxyethane by sedimentation equilibrium studies. The nature of the polar head group and size of the nonpolar side chain also have an important bearing upon the aggregation number.[800] The aggregation number of surfactants having nonpolar groups of comparable size was anionics $\gg$ cationics $>$ nonionics in various solvents. The size of the micelle formed by both cationic and anionic surfactants generally decreased with an increase in the size on the alkyl group. Peri[1151] attempted to correlate micelle size with properties of the solvent for Aerosol OT, and found that the molecular weight decreased markedly in solvents of lower molar volume but was less affected by solvents of a high molar volume, in contrast to the behavior of alkali metal dinonylnaphthalene sulfonates, which exhibit a linear correspondence between aggregation number and the solubility parameter of the solvent.[662] The analysis used by Little and Singleterry,[908] in analogy with the use of salting-out parameters to treat the effects of salts on the CMC with aqueous solutions, seems to be the most promising to date, although Chung and Heilweil[257] have recently presented a statistical treatment which reproduces some of the thermodynamic aspects of micelle formation in nonassociated solvents characteristic of inverted micelles. The existence of pre-micellar association also appears to be a factor which must be considered when dealing with nonaqueous association of surfactants.[884]

Finally, the existence of two isotropic solutions has been observed for three-component systems containing an alkali soap, water, and a liquid fatty acid or alcohol. Several physical methods and their properties have been studied in detail by Ekwall and Mandell and their associates, and a summary of this work, including citations to early work, was recently presented.[388] One of the two isotropic solutions was water rich, designated L_1, and contained micelles of the ordinary type found in aqueous solutions. The other isotropic solution, designated L_2, was water poor and contained inverted micelles. The characteristics of the inverted micelles were found to be different depending upon the nature of the intermicellar liquid. In fatty acid solutions, the micelles are composed of a hydrogen-bonded molecular compound of soap and fatty acid. When the water content exceeds the amount that corresponds to the hydration of the alkali soap, a gross change in structure occurs. In alcohol solutions, no molecular compounds are formed and the relative amounts of alcohol/alkali soap in the micelle depend upon the water content of the system. At high water contents phase separation occurs. A cationic surfactant, decyltrimethylammonium bromide, was substituted for the alkali soap, and the hexanol-rich solutions of the L_2 region were again characterized by a variety of physical measurements.[390] Evidence of reversed mixed micelles with an aqueous interior and intermicellar liquid containing hexanol was found. The bromide ions were found to be located in the aqueous core in a restricted environment, by ^{81}Br nuclear magnetic resonance.[900] The original papers should be consulted for a detailed description of the concentrations employed, since the nature and composition of the micelles changes quite drastically with the composition of the solution. A similar conclusion was reached for the ternary system of n-octylamine–p-xylene–water from nmr and light scattering studies.[844] In these studies reversed micelles were only found when the molar ratio of water to octylamine was less than 2/1. At higher ratios, between 2/1 and 13/1, apparently lamellar micelles with a hydrocarbon core existed. The ability of inverted micelles to solubilize water has been shown by thermodynamic and spectroscopic studies to depend upon the nature of the nonaqueous solvent and the surfactant.[481,523,775,800,844,1605] In all cases, however, the importance of an ion–dipole interaction between the polar core and the water molecules was emphasized.

10. THERMODYNAMICS OF MICELLE FORMATION

Several very thorough discussions of the way to calculate changes in thermodynamic parameters that accompany micelle formation from experimental data have been described.[21,282,592,1049,1384] For most purposes, it is probably sufficient to approximate the standard free energy of micelle formation $\Delta G_m^{\ominus}$ as[21,592,1468]

$$\Delta G_m^{\ominus} = RT \ln \text{CMC} - (RT/m) \ln X_m \tag{37}$$

where X_m is the mole fraction of micelles and the other symbols have been defined previously. The temperature and pressure derivatives of eqn. (37) give the standard enthalpy change $\Delta H_m^{\ominus}$ and volume change $\Delta V_m^{\ominus}$ per monomer, where

$$\Delta H_m^{\ominus} = RT^2 \left(\frac{d \ln \text{CMC}}{dT} \right)_P + \frac{RT^2}{m} \left(\frac{d \ln X_m}{dT} \right)_P \tag{38}$$

and

$$\Delta V_m^{\ominus} = RT \left(\frac{d \ln \text{CMC}}{dP} \right)_T - \frac{RT}{m} \left(\frac{d \ln X_m}{dP} \right)_T \tag{39}$$

It should be noted that the free energy changes obtained using eqn. (38) are unitary values when the CMC is expressed in mole fraction units. Since the aggregation number is often not known, many workers approximate eqn. (37) by omitting the second term, and estimate $\Delta G_m^{\ominus}$ by the relation

$$\Delta G_m^{\ominus} = RT \ln \text{CMC} \tag{40}$$

The second terms are also dropped from the right-hand side of eqns. (37) and (38). Only a relatively small error in the calculated thermodynamic quantities is introduced by this approximation.[1468] The enthalpy of micellization can also be determined by calorimetry, and it is of interest to compare enthalpy changes determined by the two approaches. Finally, the unitary entropy of micelle formation $\Delta S_m^{\ominus}$ is most often obtained from the relationship

$$\Delta G_m^{\ominus} = \Delta H_m^{\ominus} - T \Delta S_m^{\ominus} \tag{41}$$

and the change in heat capacity $\Delta C_{P_m}^{\ominus}$ can also be determined from calorimetry or the temperature dependence of $\Delta H_m^{\ominus}$.

A summary of the thermodynamic results obtained recently for several surfactants in various solvents is presented in Table VI. Nearly all of the values of $\Delta H_m^{\ominus}$ and $T \Delta S_m^{\ominus}$ are plotted in Fig. 9, and it can be seen that the data are roughly consistent with an arbitrary line of slope 1.0, corresponding to a compensation temperature of about 300°K.[927] Considering the approximations in the theory of micellization and the uncertainties in the experimental quantities, the correlation is reasonable. Selected points will be extracted from the figure for more detailed comparisons in the following discussion. The short line with a slope of about 1.5 to the left of the compensation line for micelle formation represents the beginning of a compensation line for hydrogen bonding between phenol and several amines in CCl_4.[1178] A comparison of these two compensation lines at a given enthalpy value reveals the unusual positive entropy change that is characteristic of micelle formation in aqueous solution. Whereas negative entropy changes for hydrogen bond formation may result from the loss of rotational degrees of freedom,[1178] the positive entropy changes that accompany micelle formation could be partly ascribed to a greater freedom of rotation of hydrocarbon chains in a nonpolar than in a polar environment. Structural changes in the solvent that give rise to hydrophobic interactions would also be expected to make positive contributions to the entropy of micellization. However, it is clear from the data included in Fig. 9 that the entropy of micelle formation is also positive for the formation of micelles in ben-

TABLE VI. Summary of Observed Changes in Standard Thermodynamic Quantities That Accompany Micelle Formation

Surfactant	Solvent	Temp., °C	$\Delta H_m^{\ominus}$, kcal mol^{-1}	$T\Delta S_m^{\ominus}$, kcal mol^{-1}	$\Delta C_{Pm}^{\ominus}$, cal mol^{-1} deg^{-1}	$\Delta V_m^{\ominus}$, cm^3 mol^{-1}	Ref.
1 Hexylsulfinyl ethanol	Water	25	2.4	5.6	—	3.2	286
2 Octylsulfinyl ethanol	Water	25	1.2	5.8	—	6.3	286
3 Octylsulfinyl ethanol	2 M NaBr	25	0.7	5.7	—	—	286
4 Octylsulfinyl ethanol	6 M Urea	25	0.8	4.7	—	—	286
5 Hexylsulfinyl propanol	Water	25	2.7	5.9	—	3.0	286
6 Octylsulfinyl propanol	Water	25	1.7	6.4	—	6.1	286
7 Octylsulfinyl propanol	2 M NaBr	25	1.0	6.2	—	—	286
8 Octylsulfinyl propanol	6 M Urea	25	1.1	5.0	—	—	286
9 Hexylsulfinyl butanol	Water	25	3.4	6.7	—	2.7	286
10 Octylsulfinyl butanol	Water	25	2.0	6.7	—	5.5	286
11 Octylsulfinyl butanol	2 M NaBr	25	1.2	6.3	—	—	286
12 Octylsulfinyl butanol	6 M Urea	25	1.3	5.4	—	—	286
13 Dimethyloctylamine oxide	Water	30	4.0	7.5	—	4.5	110, 112
14 Dimethylnonylamine oxide	Water	30	4.4	8.5	—	1.2	110, 112
15 Dimethyldecylamine oxide	Water	30	2.7	7.5	—	6.3	110, 112
16 Dimethyldecylamine oxide	4 M Urea	26.5	1.2	5.7	—	—	111
17 Dimethyldecylamine oxide	4 M GuHCl[a]	26.5	1.0	5.3	—	—	111
18 Dimethyldecylamine oxide	$0.05X_2$ Methanol	26.5	1.4	6.2	—	—	111
19 Dimethyldecylamine oxide	$0.05X_2$ Ethanol	26.5	−0.1	4.9	—	—	111

	Name	Solvent	Temp					Ref.
20	Dimethyldecylamine oxide	$0.05X_2$ Propanol	26.5	−1.6	4.0	—	—	111
21	Dimethyldecylamine oxide	$0.02X_2$ Butanol	26.5	3.8	9.8	—	—	111
22	Dimethylundecylamine oxide	Water	30	—	—	—	9.4	112
23	Dimethyldodecylamine oxide	Water	30	2.6	8.7	—	13	110, 112
24	Dimethyloctylphosphine oxide	Water	30	—	—	—	4.4	112
25	Dimethyldodecylphosphine oxide	Water	30	1.5	8.4	—	—	645
26	Dimethyldodecylphosphine oxide	10% Ethanol	30	−6.5	0.3	—	—	645
27	Dimethyldecylammonio propane sulfonate	Water	30	—	—	—	3.9	112
28	Dimethyldodecylammonio propane sulfonate	Water	30	—	—	—	4.7	112
29	Sodium octyl sulfate	Water	25	0.8	4.4	—	—	379, 1060
	Sodium octyl sulfate	Water	25	1.5	5.1	—	—	110
30	Sodium decyl sulfate	Water	25	0.5	4.9	—	—	379
31	Sodium dodecyl sulfate	Water	25	0.09	5.3	−137	—	1177
	Sodium dodecyl sulfate	Water	25	0.52	5.7	—	—	379
	Sodium dodecyl sulfate	Water	25	−0.3	4.9	—	12.1	112
32	Sodium dodecyl sulfate	0.023 m NaCl	25	−0.15	5.6	−95	—	1177
33	Sodium dodecyl sulfate	3 m Urea	25	−1.44	3.7	−78	—	741
34	Sodium tetradecyl sulfate	Water	26	—	—	—	15.9	112
35	N-Decylbetaine	Water	25	0.92	5.5	—	—	1452
36	N-Undecylbetaine	Water	25	0.95	6.3	—	—	1452
37	N-Dodecylbetaine	Water	25	−1.40	4.6	—	—	1452
38	Decyl α-picolinium bromide	Water	25	−0.2	4.0	−60	—	5
39	Dodecyl α-picolinium bromide	Water	25	−0.4	4.7	−72	—	5

TABLE VI. (*Continued*)

Surfactant	Solvent	Temp., °C	$\Delta H_m^{\ominus}$, kcal mol^{-1}	$T\Delta S_m^{\ominus}$, kcal mol^{-1}	$\Delta C_{Pm}^{\ominus}$, cal mol^{-1} deg^{-1}	$\Delta V_m^{\ominus}$, cm^3 mol^{-1}	Ref.
40 Tetradecyl α-picolinium bromide	Water	25	−0.75	5.1	−87	—	5
41 Decyltrimethylammonion bromide	Water	25	0.1	4.4	−53	—	6
41a Decyltrimethylammonion bromide	1 m Sucrose	25	−0.4	3.5	—	—	3
41b Decyltrimethylammonion bromide	2 m Glucose	25	−0.8	3.3	—	—	3
42 Dodecyltrimethylammonium bromide	Water	25	−0.3	4.6	−67	—	6
42a Dodecyltrimethylammonium bromide	1 m sucrose	25	−1.0	3.8	—	—	3
42b Dodecyltrimethylammonium bromide	2 m glucose	25	−1.4	3.5	—	—	3
42c Dodecyltrimethylammonium bromide	4 m glucose	25	−2.3	2.8	—	—	3
42d Tetradecyldimethylammonium bromide	1 m sucrose	25	−1.4	4.2	—	—	3
42e Tetradecyldimethylammonium bromide	2 m glucose	25	−1.6	4.2	—	—	3
43 Dodecylbenzyldimethylammonium bromide	Water	25	−0.7	4.7	−77	—	6
43a Dodecylbenzyldimethylammonium bromide	1 m sucrose	25	−1.4	4.0	—	—	3

43b Dodecylbenzyldimethylammonium bromide	2 m glucose	25	−1.8	3.8	—	—	3
44 Dimethyldodecylamine oxide hydrochloride	Water	25	−0.3	5.2	—	—	110
45 Dodecylpyridinium bromide	Water	25	−1.0	4.1	−75	—	4
46 Dodecyltrimethylammonium bromide	Water	25	−0.33	4.5	−101	—	416
47 Dodecyltrimethylammonium bromide	0.175 m NaBr	25	−0.38	4.7	−111	—	416
48 Dodecyltrimethylammonium bromide	0.05 m NaBr	25	−0.44	4.9	−106	—	416
49 Dodecyltrimethylammonium bromide	0.1 m NaBr	25	−0.49	5.2	−100	—	416
50 Dodecylpyridinium iodide	Water	25	−2.98	2.5	−98	—	740
51 Dodecylpyridinium iodide	0.01 M KI	25	−2.86	3.0	−91	—	740
52 Dodecylpyridinium iodide	2 M urea	25	−3.1	2.1	−85	—	740
53 OPE_{9-10}[b]	Water	35	0.9	8.5	−105	—	1230
54 OPE_{9-10}[b]	50% EG[c]	35	−4.5	1.4	−160	—	1230
55 OPE_{12-13}[b]	Water	35	0	7.6	−205	—	1230
56 OPE_{12-13}[b]	50% EG[c]	35	−4.0	1.9	−150	—	1230
57 OPE_{30}[b]	Water	35	2.1	9.0	−215	—	1230
58 NPE_{30}[d]	Water	35	2.1	9.9	−185	—	1230
59 NPE_{30}[d]	47.2% EG[c]	35	−2.3	3.8	15	—	1230
60 TFHEG[e]	Water	25	4.9	9.6	−94	—	1073
61 Octylhexaoxyethylene monoether	Water	25	4.8	9.9	−87	—	284

TABLE VI. (*Continued*)

Surfactant	Solvent	Temp., °C	$\Delta H_m^{\ominus}$, kcal mol^{-1}	$T\Delta S_m^{\ominus}$, kcal mol^{-1}	$\Delta C_{Pm}^{\ominus}$, cal mol^{-1} deg^{-1}	$\Delta V_m^{\ominus}$, cm^3 mol^{-1}	Ref.
62 TFHEG[e]	2 M urea	25	4.2	8.6	−54	—	1073
63 TFHEG[e]	4 M urea	25	3.8	8.0	−37	—	1073
64 Octylmethyl sulfoxide	Water	25	1.7	6.3	—	—	283
65 TFOMS[f]	4 M urea	28	1.5	5.3	−17	—	1073
66 Dodecyltetraoxyethylene glycol	Formamide	25	−0.55	3.6	—	—	944
67 Dodecylhexaoxyethylene glycol monoether	Formamide	25	−0.9	3.1	—	—	944
68 Dodecyloctaoxyethylene glycol monoether	Formamide	25	−0.7	3.1	—	—	944
69 Dodecylammonium hexanoate	Benzene	26	−1.9	1.9	—	—	774
70 Dodecylammonium octanoate	Benzene	26	−1.3	2.4	—	—	774
71 Dodecylammonium octanoate	Cyclohexane	40	−7.7	−2.9	—	—	774
72 Dedecylammonium decanoate	Benzene	40	1.2	4.8	—	—	774
73 Octadecylammonium propionate	Benzene	33	−3.0	1.5	—	—	774

[a] Gu-HCl = Guanidinium chloride.
[b] p-tert-octyl phenoxy(polyethoxy)ethanol, where n represents the mean ethoxy chain length.
[c] EG = Ethylene glycol.
[d] NPE$_n$ = p-tert-monyl phenoxy(polyethoxy)ethanol, where n represents the mean ethoxy chain length.
[e] TFHEG = 8,8,8-Trifluoro octyl hexaoxyethylene glycol monoether.
[f] TFOMS = 8,8,8-Trifluoro octyl methyl sulfoxide.

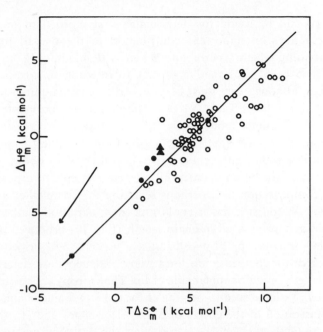

Fig. 9. Compensation plot of data presented in Table VI. The data for aqueous solutions with and without additives are represented by open circles, whereas the filled circles and triangles correspond to benzene or cyclohexane and formamide solutions, respectively.

zene and formamide but the absolute values tend to be 30–50% less than for micellization in water without additives. A few negative entropies of micelle formation were observed with cyclohexane and 60% ethylene glycol as solvents. The negative values could reflect the contribution of interactions between polar groups with the concomitant loss of rotational entropy analogous to hydrogen bond formation as the dielectric constant of the solvent decreases. A more definitive interpretation will await the development of a thorough thermodynamic understanding of the basis for compensation effects in general.[1477] The effects of the addition of ethylene glycol to aqueous surfactant solutions will be discussed in more detail in conjunction with a consideration of temperature-dependent compensation changes.

Since all of the data listed in Table VI were obtained in the vicinity of 25°, it may be instructive to compare the effects of changing the surfactant structure and the solvent composition on the thermodynamics of micellization. The compensation plot in Fig. 9 can serve as a phenomenological basis for the identification of unusual effects, since simultaneous changes in the process with different compensation properties can be distinguished from changes which follow a single compensation line. Such comparisons can be useful for determining that more than one type of interaction makes an important contribution to the overall process, or perhaps deciding which type of interaction is dominating. Compensation plots can also be of assistance in helping to interpolate, extrapolate, and smooth data, although care must be exercised to eliminate unjustified bias.

All of the data in Table VI that refer to the addition of urea and guanidinium chloride or monohydric alcohols to water, except for entry 21, represent negative shifts along the compensation line due to the presence of the additive, within an average experimental error of about ± 0.5 kcal mol^{-1}. The data for butanol, entry 21, offer an example of a positive shift along the compensation line, perhaps reflecting the complex effects of this additive.[373] Although a few irregularities exist, variation of the structure of the head group for a given chain length (Fig. 10) produces changes in the enthalpy and entropy of micellization which can be ascribed to movement up or down compensation lines whose intercepts are determined by the chain length of the nonpolar side chain. The entropy change for a given value of $\Delta H_m^{\ominus}$ can be seen to be greater for the larger side chains.

An illustration of the effects of added electrolytes and changes in the chain length with a given head group on the enthalpy and entropy of micellization is presented in Fig. 11 in view of the common effect these two factors have on altering the CMC. The compensation line identified as a is an arbitrary line with a slope equal to 1.0, which fairly well represents the effect of increasing the chain length by one methylene group on the

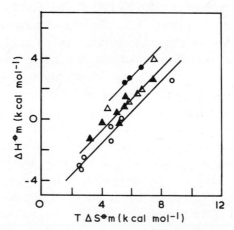

Fig. 10. Compensation plots for various surfactants with different head groups. The filled circles represent the hexanoates, the triangles represent the octanoates and decanoates, and the dodecanoates are identified by the open circles.

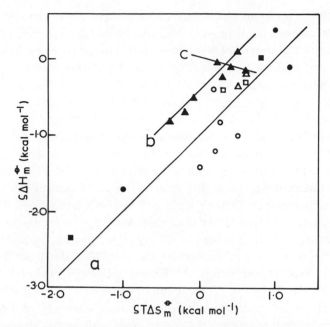

Fig. 11. Compensation plots for different chain length surfactants and in the presence of supporting electrolytes. The three lines, a, b, and c are identified in the text. Data for line a are taken from Table VI and correspond to entries 38–40 (open triangles), 35–37 (filled squares), 29–31 (open squares), 13–15, 23 (filled circles), and 1, 2, 5, 6, 9, 10, and 41–45 (open circles). The salt effects are identified with filled triangles and data for line b are taken from entries 3, 7, 11, 32, and 51, and line c is described by entries 46–49.

enthalpy and entropy of micelle formation, $\delta \Delta H_m^{\ominus}$ and $\delta \Delta S_m^{\ominus}$, respectively. The compensation line in Fig. 11 identified as b is parallel to line a and includes values of $\delta \Delta H_m^{\ominus}$ and $\delta \Delta S_m^{\ominus}$ that result from the addition of salts. The three points with the more negative values of $\delta \Delta H_m^{\ominus}$ and $\delta \Delta S_m^{\ominus}$ correspond to the addition of 2.0 M NaBr to different octylsulfinols,[286] and the other two points represent the results of adding low concentrations of NaCl and KI to sodium dodecyl sulfate[1177] and dodecylpyridinium iodide[740] solutions, respectively. The three points that lie on the line in Fig. 11 identified as c correspond to the addition of low concentrations (0.1 m and less) of NaBr to dodecyltrimethylammonium bromide solutions.[416] Since the data for low NaBr concentrations fall on a compensation line which is quite different from that observed under salting-out conditions, it may be postulated that a different mechanism is dominating to some unknown extent. Increased counterion binding would be expected to be an important factor. As previously mentioned, Eatough and Rehfeld[379] have used the apparent molal heat content curve above the CMC to calculate the changes in thermodynamic quantities that accompany counterion binding. The degree of binding was calculated to be nearly constant above a surfactant concentration of about 0.1 M. Therefore, the addition of salt above this concentration to surfactant solutions would likely exert its effect on the nonpolar side chain through a salting-in or salting-out mechanism of the type previously discussed, and a compensation line with a slope equal to about 1.0, in keeping with the observations for the nonionic surfactants and observed salting-out effects of various salts on argon,[120] would be expected. It would clearly be of interest to have thermodynamic data for intermediate and higher concentrations of several electrolytes at several temperatures with both ionic and nonionic surfactants. The tentative interpretation given to the data that define line c should be regarded with caution, since the salting-out of argon by various concentrations of KI also has a different compensation line than the salting out by various salts at a fixed concentration.[120] Finally, although attention has been focused on the deviations of the compensation data from a slope of 1.0 with a systematic change in salt concentration, it should be emphasized that the individual points are not out of place with respect to line b within the limits of the experimental error for the other points. This latter observation is suggestive of a common general mechanism for micelle formation in the presence of salts, in addition to the specific interaction that follows line c. An identical argument could be presented with respect to the effect of increasing the hydrophobic side-chain size for a particular head group by noting that most of the values presented in Fig. 11 are located in the lower right-hand quadrant of the figure (corresponding to negative values of $\delta \Delta H_m{}^{\ominus}$ and positive values of $\delta \Delta S_m{}^{\ominus}$) for both ionic and nonionic surfactants. However, several exceptions exist, and the importance of specific interactions is suggested in addition to a possible general common mechanism.

The effects of temperature on $\Delta H_m{}^{\ominus}$ and $T \Delta S_m{}^{\ominus}$ represent a rather dramatic example of compensation, and data are plotted in Fig. 12 for several cationic surfactants. Although the compensation lines have different intercepts, they are all consistent with a unit slope. Similar compensation lines for three different nonionic surfactants in water and 60% ethylene glycol[1230] reflect the temperature dependence of the CMC, as shown in Fig. 13. However, the compensation line for water falls about 5 kcal mol^{-1} to the right of the line for ethylene glycol. Several points are also included to illustrate the effect of gradual changes of the solvent composition at a given temperature. The change in compensation from one

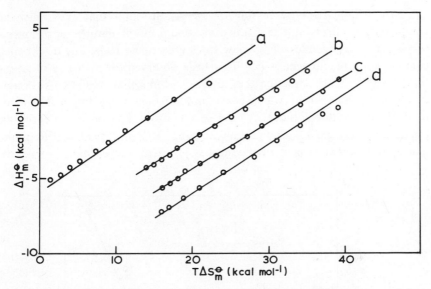

Fig. 12. Variation of $\Delta H_m{}^\ominus$ with $T\,\Delta S_m{}^\ominus$ for several surfactants. Lines *a*, *b*, *c*, and *d* correspond to dodecylpyridinium bromide, decyl α-picolinium bromide, dodecyl-α-picolinium bromide, and tetradecyl-α-picolinium bromide, respectively.

line to the other is nonlinear, with the largest change occurring between 40 and 50% ethylene glycol. Up to that point an increase in temperature and an increase in concentration of ethylene glycol produce similar compensation effects. The departure from the compensation line with water as the solvent can be seen to be from right to left with increasing concentrations of ethylene glycol (Fig. 13) and a decrease in surfactant chain length (Fig. 12), whereas the opposite trend was observed with increasing concentrations of NaBr (Fig. 11).

The effects of sucrose and glucose on the thermodynamics of aqueous alkylammonium bromide micelle formation were investigated by Adderson and Butler.[3] Compensation effects are illustrated in Fig. 14 and it can be seen that negative shifts along a compensation line with a slope of about 1.0 are observed, as previously noted, upon the addition of urea, guanidinium chloride, and most monohydric alcohols. Thus substances such as the sugars, which are said to promote water stabilization around alkyl chains,[760] and urea, which functions as a water structure breaker,[1323] produce the same compensation effects. This suggests either a possible reexamination of the designation given to sugars or an indication of the insensitivity of enthalpy and entropy changes to water structure changes that accompany micelle formation.

In concluding this discussion of the relationships that exist between compensation effects and experimental conditions, it may be appropriate to ask what has been learned. There does not appear to be any doubt that compensation phenomena which have been observed for a variety of situations[927,1178,1477] also describe the behavior of micellar systems. Although the arbitrary manner by which compensation lines were drawn is subject to criticism, it is hoped that they have served a useful purpose in identifying trends that may exist with the data at hand. It is not clear at this point that a statistical treatment of the data is justified. The existence of nearly

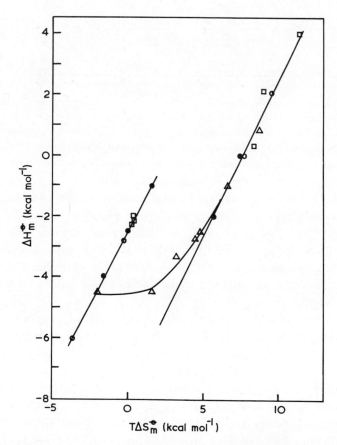

Fig. 13. Variation of $\Delta H_m^{\ominus}$ with $T\,\Delta S_m^{\ominus}$ in water (right-hand line) and 60% ethylene glycol (left-hand line) for three nonionic surfactants taken from Ref. 1230. Data are for OPE_{9-10} (open circles), OPE_{12-13} (filled circles), and NPE_{30} (squares) from Table VI. Data are also presented for OPE_{9-10} (triangles) for the variation of $\Delta H_m^{\ominus}$ with $T\,\Delta S_m^{\ominus}$ at intermediate ethylene glycol concentrations between zero and 60% at 35°.

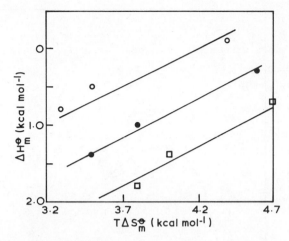

Fig. 14. Effect of the addition of sucrose and glucose on $\Delta H_m{}^{\ominus}$ and $T\,\Delta S_m{}^{\ominus}$ for decyl-trimethylammonium bromide (open circle), dodecyltrimethylammonium bromide (filled circles), and dodecylbenzyldimethylammonium bromide (squares) at $25°$ taken from Ref. 3.

parallel compensation lines for aqueous and 60% ethylene glycol solutions indicates the importance of contributions of enthalpic origin to micelliza-tion that cannot be attributed solely to normal water structural effects. The departure from movement up and down compensation lines was observed in the case of added salts, increased chain length, and high concentrations of ethylene glycol. These departures suggest the possibility of the existence of some unidentified important contribution to micelle formation. The identification of compensation lines that accompany temperature changes may provide a phenomenological basis for detecting common mechanistic changes which may not be clear from a comparison of thermodynamic quantities for different surfactants at a single temperature. Finally, it is hoped that this discussion will stimulate interest in the development of a rigorous understanding of the basis for the observed changes.

Since a large number of the enthalpy changes presented in Table VI were obtained by calorimetry, it is relevant to discuss the methods by which different workers have evaluated the enthalpy of micellization. A hypo-thetical relative apparent molal heat content ϕ_L curve obtained from either integral heats of solution or dilution is shown in Fig. 15 for this purpose. A relative partial molal heat content curve is also shown, calculated from the same data using the relationship

$$\bar{L}_2 = \bar{H}_2 - \bar{H}_2{}^{\circ} \tag{42}$$

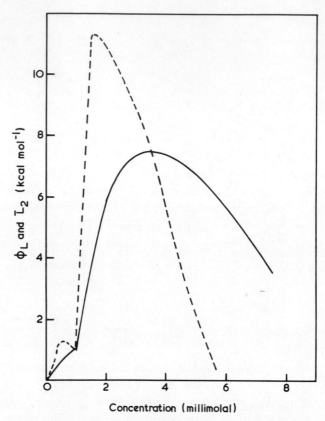

Fig. 15. Variation of the relative apparent molal heat content ϕ_L (solid line) and partial molal heat content $\bar{L}_2$ (broken line) with molality for a hypothetical surfactant.

where $\bar{L}_2$ and $\bar{H}_2$ are the relative partial molal enthalpy and a partial molal enthalpy of the solute at a given concentration, and $\bar{H}_2{}^\circ$ is the partial molal enthalpy of the solute in a standard state, corresponding to the infinitely dilute solution. The partial molal enthalpy of the solute is related to the observed heat of solution ΔH_s by[110,538]

$$\bar{H}_2 = \bar{H}_2{}^s + \Delta H_s + m[d(\Delta H_s)/dm] \qquad (43)$$

where $\bar{H}_2{}^s$ refers to the partial molal enthalpy of the solid surfactant. Alternatively, the heat of dilution from a concentration C' to C, $\Delta H_{(C' \to C)}$ and the apparent molal heat content at C', ϕ_L, can be used to obtain $\bar{H}_2$, where[538]

$$\bar{H}_2 = \bar{H}^\circ - \phi_L{}' + \Delta H_{(C' \to C)} + m[d(\Delta H_{(C' \to C)}/dC] \qquad (44)$$

The use of partial molal quantities when studying solution properties is usually required,[783] although the distinction between apparent and partial enthalpies has not often been made in the past. A recent exception has been the work of Benjamin.[110]

Returning to Fig. 15, the change in molal heat content between 0 and 1.0 mm represents nonideality below the CMC, which can likely be mainly attributed to pre-micellar association, although ion pairing and electrostatic interactions may play minor roles.[141,379] Since the CMC represents a finite range of concentrations during which the formation of micelles is first detected, the arbitrary concentration of about 1.5 mm may be chosen since it represents the point of intersection of the two lines defining the change in $\bar{L}_2$ just below and above the CMC. The change in enthalpy above 1.5 mm can be attributed to counterion binding,[379] post-micellar association, and other thermodynamic nonideal characteristics of micelles.[1177] The standard enthalpy of micellization $\Delta H_m^{\ominus}$ can be conveniently defined as

$$\Delta H_m^{\ominus} = \bar{H}_2(\text{just above the CMC}) \qquad (46)$$
$$- \bar{H}_2^{\circ}(\text{infinite dilution})$$

or

$$\Delta H_m^{\ominus} = \bar{H}_2(\text{just above the CMC}) \qquad (46)$$
$$- \bar{H}_2(\text{just below the CMC})$$

Very different values of $\Delta H_m^{\ominus}$ would result if ϕ_L was used, and the difference between the two definitions was recognized by Goddard and Benson.[538] Conventions based both upon eqns. (45) and (46) are found. Results from titration calorimetry directly give $\Delta H_m^{\ominus}$ defined according to eqn. (45), to a first approximation.[111,645] Unfortunately, most of the data included in Table VI were not obtained by a common method, and it is not known for certain if they even approximate the results which would be obtained by the method of partial molal enthalpies just described. This would therefore somewhat limit one's enthusiasm in using the data at face value for theoretical purposes. A thorough analysis of the data is not always possible, due to the inability to develop apparent molal heat content curves from the published results.

Only a few attempts have been made to compare the enthalpy of micellization obtained by calorimetry with the results from the temperature dependence of the CMC. Pilcher et al.[1052] have compiled a very complete listing of enthalpies of micellization for sodium dodecyl sulfate in water. No systematic difference between the values obtained by the two methods is apparent at any given temperature. However, the calorimetric results appear to show a more pronounced temperature dependence ($\Delta C_P^{\ominus}$), although the experimental errors tend to reduce the differences. The calorimetric results obtained at 25° for dodecyltrimethylammonium bromide[416] and the variation of the CMC with temperature[6,416] are in good agreement. However, poor correspondence was observed with dodecylpyridinium iodide in water, 0.01 m KI, and 2 m urea.[740] In all of these recent comparisons the theoretical treatment of micellization by Ingram and Jones[708] was used and the calculation of $\Delta H_m^{\ominus}$ involved multiplication of the left-hand side of eqn. (38) by the factor $(1 + \alpha)$ to give

$$\Delta H_m^{\ominus} = (1 + \alpha)R \, d(\ln \text{CMC})/d(1/T) \tag{47}$$

Although a comparison of calorimetric results with values obtained from the use of either eqns. (38) or (47) would appear to enable one to determine which equation is preferred, no systematic differences exist between the results obtained from use of the two equations.* Hall and Pethica[592] have noted that eqn. (38) constitutes a good estimate for the enthalpy of micellization when the aggregation number is large and only changes slightly with temperature. However, deviations from calorimetric results are to be expected when the aggregation number is small and when the CMC and aggregation number have a large dependence on temperature.[588,589] The value of $\Delta C_P^{\ominus}$ obtained from calorimetric data[416] is also larger than the value from the temperature dependence of the CMC[6] for dodecyltrimethyl ammonium bromide, but less than a factor of 1.8 which might be expected from the use of eqn. (38) rather than eqn. (47). This point was recognized several years ago by White and Benson.[1582] Caution should be exercised, however, when using second derivatives of the variation of the CMC with temperature to obtain heat capacity changes without a complete error analysis of the results. There is need for additional data with other systems before the situation can be clarified. It may be worthwhile again at this point to mention the possibility of obtaining thermodynamic quantities of polydisperse mixtures which are as valuable as those for pure surfactants, which are often difficult to prepare.[98,1230]

Finally, it may be noted that all the $\Delta V_m^{\ominus}$ and $\Delta C_P^{\ominus}$ values, defined in an analogous manner to $\Delta H_m^{\ominus}$, yield fairly consistent results in that all values of $\Delta V_m^{\ominus}$ are positive, as are nearly all $\Delta C_{P_m}^{\ominus}$ values. A small, positive value of $\Delta C_{P_m}^{\ominus}$ was obtained with 60% ethylene glycol as the solvent. A

---

* Reasons for possible discrepancies between calorimetric and van't Hoff enthalpies are discussed in more detail in Chapter 5 (p. 337) of Volume 2.

positive change in $\Delta C_{P_m}^{\ominus}$ was also observed by Adderson and Taylor[6] for aqueous decyltrimethylammonium bromide above 55°. Positive changes in $\bar{V}_2^{\circ(263)}$ and $\bar{C}_P^{\circ(816)}$ were also observed for the transfer of model compounds from water to 6 M urea solutions, and pronounced head group effects were observed in the volume studies which were not apparent from the calorimetric results. All of these results would appear to be consistent with the interpretations of White and Benson[1582] and Benjamin,[112] who emphasized the role of changes in water structure which would decrease with increasing temperatures and action of cosolvents. However, Benjamin[112] recognized the possible oversimplification of this interpretation due to the possible importance of (a) head group effects, (b) compression effects on the micelle interior, and (c) uncertainties regarding the efficiency of packing in the micelle interior. Since the partial molal volumes and heat capacities of the surfactants in the micellar state are comparable with values for liquid hydrocarbons,[112] a liquidlike hydrocarbon interior was suggested. This is contrary to the suggestion recently made with respect to the nature of the protein interior on the basis of an interpretation of volume measurements based on the scaled particle theory of liquids.[778]* The protein interior has often been likened to the micelle core in the past. An interesting minimum in a plot of $\Delta C_P^{\ominus}$ versus temperature has been observed with aqueous solutions of dodecylpyridinium bromide[4] and decyltrimethylammonium bromide[4] around 55°, which has been interpreted[4] as a possible change in micellar structure. Further study of this effect may provide a better understanding of the nature of the micelle interior.

Another way to view the thermodynamics of micellization is to compare the process with the transfer of model compounds from water to nonaqueous solutions. This approach was taken by Butler[218] several years ago to determine the changes in thermodynamic parameters that accompany the hydration of organic molecules. Némethy and Scheraga[1093] used a similar approach, following the suggestion of Kauzmann,[759] to estimate the contribution of hydrophobic pairwise interactions to protein stability; the interpretation in a preceding section of this chapter of the effects of additives on the CMC also used similar arguments. Finally, Franks has discussed the current state of knowledge regarding apolar hydration and hydrophobic interactions in the preceding chapter of this volume and the significance of transfer data for small molecules is included. Therefore, the present discussion is limited to studies that specifically involve surfactants.

An extensive comparison of the free energy of micelle formation with

---

* For details see Chapter 1, Section 3.4.1 and Chapter 5.

the free energy of transfer from an aqueous to a nonaqueous medium was presented by Mukerjee,[1049] and a schematic representation of various related transfer steps has been given.[899] Mukerjee made the important observation that whereas the change in free energy per methylene group is about $-1.08kT$ for micelle formation, the corresponding change for transfer from an aqueous to a nonaqueous solvent is $-1.39kT$. A definitive explanation for the difference between the processes has not been given, although several probable contributions have been identified. The hydrophilic nature of the polar head group has been shown to have a marked effect on the free energy of transfer/methylene group, with reported values ranging from $-1.06kT$ to $-1.53kT$ for the solubility of sodium alkyl sulfonates and primary amines in water.[6] This head group effect would also be expected to provide an important contribution to the free energy of micellization, and it has also been discussed by Mukerjee.[1049]

The standard enthalpies and entropies of solution of either aliphatic alcohols[14,67,110] or hydrocarbons[218] in water and a nonpolar medium may also be used to estimate the hydrophobic contribution to micelle formation. Emerson and Holtzer[407] proposed the separation of the overall free energy into an electrical and nonpolar contribution, although objections to specific aspects of their model have been raised.[1050] The entropy of micellization in aqueous solutions without additives can be seen from Fig. 9 to be positive without exception, as observed for the transfer of alcohols from an aqueous to a nonaqueous state.[110] Also, the bulk of the data for aqueous solutions without additives included in Fig. 11 represent positive increments in $\Delta S_m^{\ominus}$ and negative increments in $\Delta H_m^{\ominus}$, again in agreement with transfer data involving alcohols with chain lengths in excess of five carbon atoms.[110] Additional comparisons at this time do not seem warrented except to note the similarity in sign between the standard heat capacity of transfer of alcohols from water to a nonpolar solvent[14] and $\Delta C_{P_m}^{\ominus}$. The importance of carefully selecting a reference state for transfer processes may be stressed by noting that the dependence of the enthalpy of hydration on chain length is linear when the transfer is from the gas state,[218] but is nonlinear when the transfer is from the liquid state[110] for normal aliphatic alcohols. The complications resulting from the well-known[486] association of alcohols in the pure liquid can be eliminated by adopting gas as the reference state. The recent mass-spectrometer studies of ion hydration by Kebarle[53,372,761] represent significant progress in the characterization of species in the gaseous state.

In concluding this section on the thermodynamics of micelle formation, special mention should be made of the recent theoretical achievements reported in a series of papers by Hall.[588-591] In the first paper of the series[588] the thermodynamics of ideal multicomponent micelles was discussed as an alternative to the application of the thermodynamics of small systems to nonionic micelles previously reported. Several useful expressions were obtained which relate the temperature dependence of the CMC to the enthalpy of micellization as determined by calorimetry, give the variation of the aggregation number with temperature, and provide an operational definition of the CMC and a description of the Krafft point. A limiting law for multicomponent micelles similar to Henry's law was derived in a second paper.[589] Expressions which may be used for the determination of various number-average molecular weights and the CMC from turbidity and osmotic pressure data were developed. An equation was also presented which provides a relationship between solubilized water in soluble solutes and the CMC. An exact phenomenological interpretation of the micelle point in multicomponent systems was presented more recently.[590] Expressions were given which account for the mutual dependences on temperature, pressure, and concentration at the CMC. His thermodynamic treatment of interacting multicomponent charged aggregates was cited previously.[591] Finally, Watterson and Elias have investigated the closed association model of micelle formation and obtained functions which define the number- and weight-average molecular weights for various aggregation numbers.[1563] Since a theta system* is only defined at a particular temperature, it was argued that the temperature dependence of the CMC should not be used to determine the enthalpy of micellization.

11. KINETICS OF MICELLE FORMATION

One of the most promising recent developments has been the formulation of possible mechanisms from kinetic studies of micelle formation. Experimental results have been obtained from a wide variety of methods, and several alternate mechanisms have been proposed. This discussion will attempt to provide an account of the current status of the problem.

The existence of a single relaxation process which is found slightly above, but not below, the CMC has been observed by temperature jump,[123, 124,817,851] pressure jump,[1011,1460] and stopped flow[306,724,850,1612] studies. The observed relaxation has been assigned to the complete dissociation of the micelle into the monomeric form on the basis of observed differences in physical properties of the monomer and micelle such as molecular

---

* For a detailed discussion of the significance of theta conditions see Chapter 7.

weight,[123,124,851] absorption coefficient,[817,850] eletrical conductivity,[306,724] and molar volume.[1011,1460] In addition, a relaxation process has been observed by means of ultrasonic absorption measurements under similar experimental conditions,[554,1223,1610,1611,1615] but the relaxation time obtained from ultrasonic measurements is much shorter than obtained by the other methods.[1068] In fact, the relaxation frequency obtained from ultrasonic experiments coincides with the frequency for the exchange of monomers between micelles and the bulk solution determined by nmr[1105] and esr[58,468] spectroscopy studies.

Several mechanisms have been proposed which attempt to relate observed relaxation times to rate constants of chemical meaning,[555,817,1068,1303,1611,1616] and rate constants which differ by as much as 10^3–10^5 have been reported. All workers interpret the observed results in terms of a transfer of the monomer from a micellar to bulk solution environment, except Yasunaga et al.,[1610,1611] who proposed a change in counterion binding; however, this mechanism has not received support. The main difference between the other proposed mechanisms involves the question of the existence of a rate-determining step in micelle formation or dissociation. Muller[1068] identified a questionable mathematical approximation which was used by Kresheck et al.[817] in the derivation of a rate equation for a mechanism which invokes the importance of a single rate-limiting step. Although his criticism may be valid for *some* systems immediately after the perturbation, the equations are quite correct for all systems as they approach chemical equilibrium. Minor modification of the mechanism proposed by Kresheck et al.[817] to include about 5% dimer, trimer, etc. held together by hydrophobic bonds which increase in strength with increasing temperature[1093] would greatly extend the concentration range for which the equations would be mathematically valid. Although other reasonable modifications are also possible, such as provision for the polydispersity of the micellar state, as proposed by Graber and Zana,[555] the mathematical complexity is usually prohibitive and probably unjustified in view of the fact that a single experimental variable, a first-order relaxation time, is being observed.

On the other hand, two alternate tenable approaches have been suggested which view micelle formation as a process which does not involve a single rate-limiting step.[817,1303] Muller[1068] has suggested that dimer formation may be fast and all subsequent steps leading to micelle formation are slow, whereas Sams et al.[1303] propose a two-state model consisting of either monomer or the associated forms. The model of Sams et al.[1303] has the advantage that it yields an expression which shows the reciprocal of the relaxation time to be a linear function of surfactant concentration, in agreement with several experimental results,[123,124,817,851,1460] whereas the expression given by Muller does not show the relaxation time to depend upon surfactant concentration. However, the mechanism given by Muller tends to reduce the discrepancy between the ultrasonic, nmr, and esr results (stationary) on the one hand and temperature jump, pressure jump, and stopped flow (transient) results on the other.

It is possible to provide yet another explanation for the lack of agreement cited above, and this is to propose that different processes are being measured by the stationary and transient methods. A first-order process has been observed by stopped-flow,[1612] temperature jump,[124] and pressure jump[1460] studies on sodium dodecyl sulfate solutions above the CMC, which has a relaxation time of the order of 1–10 msec. The dissociation of the nonionic surfactant, poly(oxyethylene) glycol, is also characterized by a first-order process on a millisecond time scale, although the agreement between stopped-flow and temperature-jump results is not as good as for sodium dodecyl sulfate. This may reflect either basic differences between the two mechanisms, such as the relative importance of intermediate size aggregates [eqn. (44a)] with the relatively small ($m \simeq 11$) nonionic micelles. Since the time range of the usefulness of ultrasonic studies is given[595] as 10^{-10}–10^{-5} sec, the process observed by the transient methods would be outside the time span of the ultrasonic technique. Nevertheless, an ultrasonic relaxation process is observed with sodium dodecyl sulfate[724] and other surfactants[554,1610,1615] which is clearly faster than the one observed by the transient methods. To account for these results, it is postulated that the mechanism for the exchange of a monomer between the micellar state and nonmicellar state observed by stationary methods may be different than for complete dissociation, as in the case of studies involving a stepwise perturbation. These processes could be imagined to proceed as

$$(m - 1)A_1 \rightleftharpoons A_{m-1} \qquad (48)$$

$$A_{m-1} + A_1 \underset{k_{m,m-1}}{\overset{k_{m-1,m}}{\rightleftharpoons}} A_m \qquad (48b)$$

$$A_m + A_1 \underset{k_{m+1,m}}{\overset{k_{m,m+1}}{\rightleftharpoons}} A_{m+1} \qquad (49)$$

where A_i represents the state of aggregation of the various species, $k_{m,m-1}$ is the rate-limiting step in micelle formation and dissociation, and the remaining rate constants given in eqns. (48b) and (49) and those leading to the formation of the A_{m-1} species are all fast.[817] The process depicted by eqn. (49) would be small and could lead to a rapid exchange as observed by stationary methods, whereas the slow step resulting in complete micelle dissociation represented in eqn. (48b) would be responsible for the slow relaxation time measured by the transient methods. Although it may be presumptuous at the time of writing to put forward a mechanism such as indicated by eqns. (48a, b) and (49), it can nevertheless be viewed as a suggestion of an unproven possible attempt to account for the existing kinetic data. The distinction between the kinetic properties of the A_m and A_{m+1} species should not be taken literally since some heterogeneity of micellar sizes must certainly exist. Instead, the intention is to emphasize the notion of a preferred micelle size with thermodynamic and kinetic properties which are different than those for significantly larger or smaller size species. Other kineticly equivalent mechanisms are indeed possible, such as the importance of steps other than dissociation of the first monomer.

Finally, kinetic results have been obtained in three investigations[554, 817,1610] which indicate the existence of altered physical properties of the micelle at concentrations somewhat above the CMC. A possible mechanism for such a process has been proposed recently by Muller et al.[1072] on the basis of nmr studies. All of these studies appear to be related to the second CMC discussed previously, and provide additional methods by means of which this can be characterized.

12. CONCLUSION

Several characteristics of micelle formation generally parallel aspects of the solubility of nonpolar substances in water, such as qualitative dependence on chain length, influence of temperature, and action of certain additives. However, there are enough differences between the two processes which have been observed to indicate that they are by no means identical. Examples of the latter include quantitative dependences on chain length illustrated in Fig. 1, the irregular variation of T_{min} with surfactant head groups found in Table II, deviations from eqn. (9), deviations from eqn. (25) shown in Fig. 6, opposite temperature dependence of salt effects given in Fig. 7, opposite effects of low concentrations of urea,[1390,1391] quantitative differences due to the addition of sucrose,[3,115,562] apparent head group effect upon substitution of D_2O for H_2O,[113,819,1059] temperature dependence of pre-micellar association,[217,810] and irregular compensation effects associated with micelle formation. The basis for these differences between solubility on nonpolar substances and micelle formation may be complex, but they represent experimental parameters that can be studied to provide a more complete understanding of noncovalent interactions in liquids.

Several empirical examples of compensation phenomena which accompany micelle formation have been observed. In general, negative shifts along a given compensation line accompany changes associated with less solvent structure, whereas positive shifts reflect the opposite. When comparing parallel compensation lines the more negative ordinate intercept seems to be associated with the greatest solvent effect. The basis for these effects is not presently understood.

The recent uses of magnetic resonance in conjunction with other spectroscopic techniques for the study of micellar systems should lead to a rather detailed picture of micellar structure. The mechanism of micelle formation may be determined with other specialized techniques for the study of fast reactions. These should be exciting areas to follow.

Finally, although the general physicochemical features of micelle formation may be considered to be well established, many interesting details await further elucidation. This will increase the value of micelles for use as model systems such as for the study of the conformational stability of biological macromolecules,[593] electron transport in biological systems,[1579] photosynthesis and vision,[1549] biological function of heme proteins,[1396] and biological membranes. In addition, developments of practical applications such as detergency,[76] design of flotation agents,[897] solubilization of drugs,[1343,1524] and micellar catalysis[433,434] may also be aided.

CHAPTER 3

Dyestuffs

D. G. Duff

Department of Chemistry
Paisley College of Technology
Paisley, Scotland

and

C. H. Giles

University of Strathclyde
Department of Pure and Applied Chemistry
Cathedral Street, Glasgow, Scotland

1. INTRODUCTION

It has long been known that many dyes aggregate in aqueous solution. The hypothesis was first suggested by Stegner[1420] and the earliest contributions in the field, both concerned with the spectral characteristics of aqueous solutions of dyes, were made by Formánek and Grandmougin[464] and Sheppard.[1374] The latter author concluded that the spectral changes observed were due to reversible molecular aggregation. Subsequent publications have dealt in detail with the state in aqueous solution of dyes both for textile usage and for photographic sensitizing processes. An understanding of the association of dyes in water is of some importance in studies of dyeing systems since practically all textile dyes are still applied from aqueous solutions, while the difference between good and bad sensitizing dyes may be explained in part by the inability of the latter to associate either in aqueous solution or when adsorbed on silver halides. Certain cationic dyes have recently been used to produce continuously tunable lasers and the dimerization of the dyes must be quenched by addition of surfactant. The self-association of dye ions may occur in water at concentrations as low as 1 mg liter^{-1} and is dependent on several factors. Thus reduction in temperature or increase in dye and inorganic electrolyte concentration causes increased association. The effects of dye concentration and temperature are reversible.

The spectral characteristics of dyes in aqueous solution vary widely, and where aggregation induces marked changes in absorption spectra such measurements have been used to study the phenomenon, and the results have indeed been interpreted semiquantitatively. Otherwise, techniques commonly used in polymer studies have been applied to measure the degree of aggregation. While the existence in aqueous solution of dimeric units of dyes is fairly common, aggregates containing up to 10^5 molecules have been described. It appears that the presence of water is very important in inducing the aggregation of dyes, since there is only limited evidence of the phenomenon in organic solvents of low dielectric constant, except at very low temperatures.[1625]

2. THE NATURE OF THE DYESTUFFS

Dyes are organic molecules having molecular weights in the approximate range 300–1000; the precise features that confer their special properties may be summarized as (a) ability selectively to absorb visible radiation (400–700 nm), and (b) limited aqueous solubility.

(a) Color in a dye molecule is associated with the presence in the molecule of an unsaturated system known as the chromophore, and certain electron-donor or electron-acceptor groups called auxochromes. Typical chromophores in dyestuff chemistry include aromatic and heterocyclic polymethine chains and the azo group, while the common auxochromes include the carbonyl, hydroxyl, amino, and substituted amino groups. The combination of chromophore and auxochrome in a dye molecule gives rise to intense color (dyes have molar absorption coefficients in the range 10^4–10^5)

(b) The ratio of water–miscible (hydrophilic) to water-immiscible (hydrophobic) portions of a dye molecule is one of its important characteristics. This ratio is small and, according to Sheppard and Geddes,[1377] the bulk of the dye molecule is readily solvated in organic solvents up to relatively high concentrations, insulated from other dye molecules. In water, however, only the isolated hydrophilic groups will be solvated, and this fact, coupled with the planarity of many dye molecules, favors intermolecular association through the hydrophobic parts of the molecule. It has also been pointed out[712] that the structural features of azo dyes likely to promote aggregation are also those that appear to favor substantivity to cellulose, namely linearity and coplanarity of the molecule, the presence of hydrogen-bond-forming, i.e., polar groups, and a minimum number of water solubilizing groups, preferably arranged along one side of the molecule and not centrally placed. The aggregating tendency of dyes is therefore favored by the existence in the molecule of a high proportion of hydrophobic character, especially if it is not prevented by steric factors from assuming a planar orientation.

Most dyes have ionic character in solution, a notable exception being the "disperse" class used in the dyeing of synthetic polymer fibers. In this review, anionic and cationic dyes will be dealt with separately, since studies of dye aggregation tend to deal exclusively with one or other of these types and the spectral effects, in particular, shown by them are markedly different. In addition to the dissimilarity in the sign of the ionic group, the two types differ in many cases with respect to the location of the ionic charge. Thus in many cationic dyes the positive charge is distributed by resonance throughout the chromophoric system, while in the anionic class the negative charge is localized, most often on an ionized sulfonate group. In some cationic dyes, also, the positive charge is localized on a group attached to the aromatic nucleus by a short alkyl chain. In the reversible aggregation of dye ions in aqueous solution the mutual Coulombic repulsion between similarly charged ions must be overcome, and the forces that oppose this repulsion to cause aggregation will be discussed later. Since a grouping of dyes in accordance with their ionic character cuts across more conventional classifications, the types of dyes included in the two ionic classes are briefly considered.

2.1. Cationic Dyes

In the present context a cationic dye is defined as one in which the positive charge is delocalized throughout the chromophore. This excludes those mentioned above that have localized charges. The dyes of interest here include the early "basic" textile dyes and the large group of photographic cyanine dyes. Little information has appeared on the aggregating properties of the recently developed modified basic dyes used in the dyeing of acrylic fibers, but unpublished work from our laboratories suggests that in this respect they fit into the same pattern as other dyes.

The synthetic dyestuff industry began with the discovery by Perkin in 1856 of the cationic or basic dye Mauveine, which belongs to the azine class. This was followed by basic dyes of the oxazine, thiazine, acridine, xanthene, and triaryl methane classes, and many dyes from these classes still form the basis of studies on dye aggregation. Thus the thiazine dye Methylene Blue [(I), $R = CH_3$] has been the subject of numerous papers over the last fifty years, though, like the rest of the original basic dyes, it is no longer used in the dyeing of textiles, due to inadequate fastness properties, particularly light fastness.

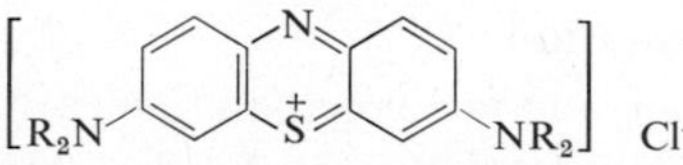

(I)

The cyanine (polymethine) photographic sensitizing dyes comprise a large group of compounds in which generally two nitrogen heterocyclic rings are joined by a conjugated chain of carbon atoms to form the chromophoric system, though a number of related structures are included in the class. Dähne[308] has suggested that all cyanine dyes can be described by the general formula

$$X—(CH)_n—X'$$

where n is an odd integer and $(n + 3)$ π electrons are distributed through the structure between the terminal atoms X and X'. A typical example, much used in aggregation studies, is 1,1'-diethylcarbocyanine chloride (pinacyanol) [(II), $n = 1$]:

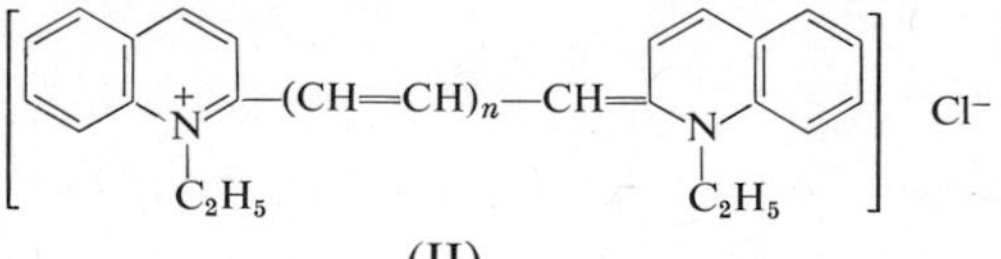

(II)

Since the ability of a cyanine dye to undergo reversible association in water is related to its usefulness as a photographic sensitizer, it is not surprising that numerous publications on the subject have appeared. The properties of these dyes became of technical importance with the discovery in the early 1900's that some members of the class can sensitize photographic emulsions to yellow, orange and red light, a property now essential to the production of almost all modern photographic films.

As will be discussed in Section 3, the absorption characteristics of the basic and cyanine dye classes in aqueous solution show marked concentration effects, and these have been related to the reversible formation of dimers and higher polymers. Because of the magnitude of these spectral effects, very few other properties have been considered in aggregation studies with cationic dyes.

2.2. Anionic Dyes

Chemically, these are sodium salts, mainly of sulfonic acids, occasionally of carboxylic acids, and they include a number of textile dye classes of which the main ones of interest here are the direct dyes for cellulosic fibers and the acid dyes for wool. Azo compounds form the largest proportion of both these classes.

The colloidal properties of azo anionic dyes in aqueous solution have long been recognized, and the direct cotton dye Congo Red [(III), R = H] was used by Donnan[356] in his early investigations of colloidal electrolytes.

(III)

While a large number of direct and acid dyes aggregate at room temperature in aqueous solution, most are applied to their respective fibers at temperatures around 100°C and under these conditions little if any aggregation is likely. The possible effects of the colloidal nature of direct cotton and acid wool dyes in the dyeing process have been investigated by Morton[1045] and Goodall,[544] respectively.

In general, the effect of concentration on the spectral properties of the anionic azo dyes is much less marked than with the cationic type, and, as a result, other techniques have been used to measure the size of aggregates in their aqueous solutions. These are mainly the standard techniques used in polymer studies and the results will be discussed in a later section.

Despite some speculation as to the nature of the process, it is certain that many dyes associate in aqueous solution, and it has been suggested[1207] that this tendency to polymerize in water is a universal property of organic dyes. This is in contrast to certain short-chain, low-molecular-weight organic compounds such as acetic acid, where the dimers, which exist in the vapor state and in nonpolar solvents, dissociate in water.

3. SPECTRAL CHARACTERISTICS OF DYES IN SOLUTION

The electronic spectroscopy of dyes has been comprehensively reviewed by Mason.[986] It has been shown[1313,1315] that for electronic transitions of unit probability, the molar absorption coefficient ε is of the order of 10^5. Since some dyes give ε values around this figure, it is obvious that the absorption of a photon by the dye molecule results in a high-intensity transition.

In the spectrum of an extended conjugated system such as a dye molecule the main absorption band in the visible region is due to a $\pi-\pi^*$ excitation, though the lowest-energy (longest-wavelength) electronic transition is theoretically an $n-\pi^*$ one. This is seldom observed in dyes, where it is submerged by the strong $\pi-\pi^*$ transition.

The visible absorption spectra of cationic dyes with a delocalized charge differ somewhat from those of the anionic type, where the charge is localized and the spectra are in general relatively featureless.

In organic solvents of lower dielectric constant than water, cationic dyes of the cyanine type show a principal long-wavelength band coupled with at least one shoulder, on the short-wave side of the main band. These, and other low-wavelength shoulders when present, represent a progression of vibrational modes in the electronically excited state. Similar auxiliary bands also appear in the spectra of other cationic dyes, e.g., the thiazine class.

Anionic dyes generally exhibit a single, broad absorption band in the visible region. The spectra of both cationic and anionic dyes in polar organic solvents are considered to be due to the monomeric species. The effects on the spectra of reversible aggregation which takes place when these dyes are dissolved in water are now considered.

3.1. Aqueous Solutions

For most dyes in very dilute aqueous solution the visible absorption spectrum is very similar to that of the dye in an organic solvent, i.e., the dye probably exists mainly in the monomeric state. With increase in concentration the aqueous spectrum undergoes reversible changes which are attributed to aggregation. The nature of such changes may be quite marked, as in the case of some cationic cyanine dyes, or hardly noticeable, as with many anionic dyes.

3.1.1. Cationic Dyes

The anomalous spectral properties of cationic dyes in aqueous solution have long been known. Thus, studies of the photographic cyanine dyes[406,

[1313,1315,1376,1578] and the azine,[1617,1618] thiazine,[131,617,873,887,1056,1207,1378] xanthene,[1271] and Rhodamine[879,880,881,1271,1365,1617,1618] dyes have revealed features in the absorption spectra additional to those mentioned in the introduction to this section.

With increasing concentration, the main long-wavelength band (M band), attributed to the monomeric dye ion, loses in intensity, and a new band appears, on the short-wave side of the monomeric band. This is termed the D band, and, as will be discussed below, is attributed to the dimeric dye species. With further increase in concentration a third band or shoulder may appear (H band) on the low-wavelength side of the D band and is associated with larger aggregates. As will be emphasized later, these new bands are not merely intensifications of the shoulders already visible in the M bands resulting from vibrationally coupled transitions associated with the monomer. At very high concentrations ($\sim 10^{-2}$ M) the spectra of certain cyanine dyes in water show a sharp, intense maximum (J band)[725] on the long-wavelength side of the M band, due to polymeric aggregates containing some 10^5 molecules. This is accompanied by a remarkable increase in the solution viscosity. The presence of M, D, H, and J bands in aqueous solutions of varying concentration is shown in Fig. 1 for the cyanine dye 1,1'-diethyl-2,2'-cyanine chloride at 25°C.

The spectral changes associated with the appearance of these bands are reversible with variation in either temperature or concentration and while the reversible changes involving M, D, and H bands are observed with many classes of cationic dye, the presence of J bands appears to be unique to the photographic sensitizing cyanine class.

In a study involving over 100 dyes, anionic as well as cationic, Holmes[675] observed in the aqueous spectra of the latter class the appearance of new bands at lower wavelengths with increasing concentration. This he attributed to the existence of a tautomeric equilibrium between two constitutional forms. Some years earlier, Sheppard[1374] had suggested that some form of association of molecules in water is the cause of the anomaly. That this association resulted in the initial formation of dimeric species was demonstrated by Scheibe[1313,1315] for the cyanine dye pinacyanol [(II), $n = 1$] and by Rabinowitch and Epstein[1207] for the thiazine dyes Thionine [(I), R = H] and Methylene Blue [(I), R = CH$_3$]. Both applied the mass law to the equilibrium

$$2D^+ \rightleftharpoons (D_2)^{2+}$$

and claimed quantitative agreement with theory for the dimerization hypothesis.

Both these sets of results were criticized[1378] on the basis of certain assumptions made in the evaluation of the molar absorption coefficient of the monomer species. Sheppard and Geddes concluded, however, that the assumption of the dimerization process is not in itself wrong, but that the law of mass action is inadequate only because certain activity factors had not been taken into account. The same authors also disagreed with the interpretation of the nature of D and M bands, and concluded that they are due to a series of vibrationally coupled transitions associated with the monomer species, but intensified in the dimer. In support of this, the presence of a subsidiary shoulder in the spectra in organic solvents was cited (e.g., see Fig. 2) since no evidence of association in these media can be obtained under normal conditions. Subsequent workers, however, have claimed evidence of the existence of dimers in organic solvents. (See Section 6.)

In a study of the spectral properties of Methylene Blue [(I), $R = CH_3$] in aqueous solution, Lemin and Vickerstaff[873] also concluded that the simple monomer $\rightleftharpoons$ dimer equilibrium is an oversimplification.

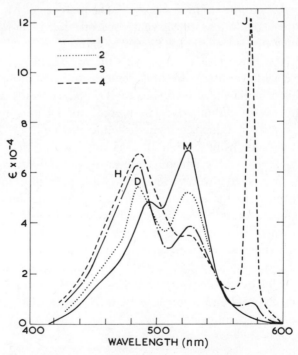

Fig. 1. Absorption spectrum of aqueous solutions of 1,1′-diethyl-2,2′-cyanine chloride at 25°: (1) 1.3×10^{-5} M; (2) 1.3×10^{-3} M; (3) 7.1×10^{-3} M; (4) 1.4×10^{-2} M. (Reproduced with permission from West and Carroll.[1577])

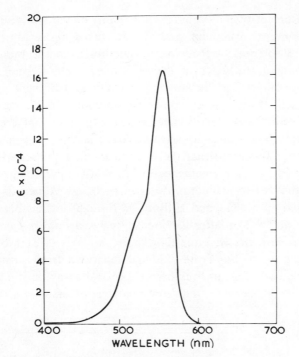

Fig. 2. Absorption spectrum of 3,3'-diethylthiacarbocyanine *p*-toluene sulfonate [dye (IV), $n = 1$] in methanol. (Reproduced with permission from West and Pearce.[1578])

Some years later in a significant study of the dimeric state of cyanine dyes West and Pearce[1578] investigated a series of dyes of general formula

(IV)

and established the presence of an isosbestic point at ~518 nm (Fig. 3). This indicates a two-species equilibrium shown by these authors to be of the form

$$2D^+ \rightleftharpoons (D_2)^{2+}$$

They confirmed the presence of the dimer by a conformity with the law of mass action; thus they assumed that the monomer has a similar ε_{max} value in methanol and in water; then by a series of approximations they calculated the spectrum of pure dimer. They noted an approximately constant displacement of $\sim$1200 cm^{-1} ($\sim$300 nm) of the vibrational shoulder from the monomer maximum, and also an increasing separation of 1200–2200 cm^{-1} (30–50 nm) of the dimer maximum from the monomer maximum throughout the series of dyes. This led them to suggest that the vibrational shoulder and the dimer maximum are different spectroscopic entities. Much of the earlier confusion surrounding the nature of low-wavelength bands and shoulders arises from the fact that the vibrational transition and dimer bands appear at about the same wavelength.

It is generally agreed that the H band in the aqueous spectra of cationic dyes is due to the emergence of other low-polymer aggregates. The possibility that a dye–counterion interaction could explain its appearance was suggested by McKay and Hillson,[951] though this has been refuted by Emerson et al.[406] The latter authors suggest that higher aggregates have a stacked-up structure with the long molecular axis of each dye molecule perpendicular to the aggregate axis and separated from each other by a distance of about 3.3 Å, as found from a crystallographic study by Wheatley.[1581] Using a particularly stable cyanine dye, Emerson et al. also claimed

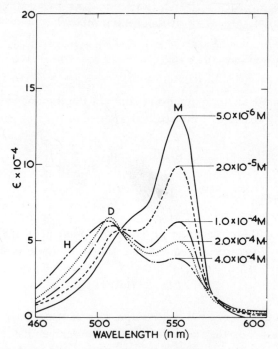

Fig. 3. Absorption spectrum of 3,3'-diethylthiacarbocyanine p-toluene sulfonate [dye (IV), $n = 1$] in water. (Reproduced with permission from West and Pearce.[1578])

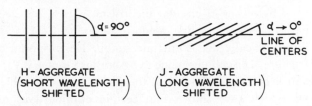

Fig. 4. The orientation of transition dipole moments (or molecular long axes) for linear H and J aggregates. (Reproduced with permission from Emerson *et al.*[406])

to show the first electron micrographs of an H aggregate, and they gave an interpretation of the blue and the red spectral shifts, resulting from the formation of H aggregates and J aggregates, respectively, by use of the molecular exciton model.[956] Thus, as illustrated in Fig. 4 for a stacklike aggregate, the transition intensity accumulates in the long-wavelength transition for a tilt angle α (angle between the long axis of the individual molecules and the line of centers in the aggregate) of less than 54° 44′, i.e., a J-aggregate structure. For tilt angles greater than this value (approaching 90°) the intensity builds up in the short-wave transition and the H band appears.

In the cyanine class, dyes showing J bands at high concentrations in water are in the minority; they generally have a fairly high aqueous solubility. The ability to form a J aggregate is also dependent on molecular architecture, and "compact" types will be more liable to do this than "looser" types of molecule.* The aggregate is considered as an array of molecules having their molecular planes parallel to each other and arranged to form a threadlike aggregate. The nature of the J band in aqueous solutions of cyanine dyes is fully discussed by West and Carroll[1577] and Sheppard.[1376] The band itself is believed to be due to an electronic interaction of π electrons in a direction normal to the stack of planar molecules, such that the whole filamentous aggregate behaves as a conjugated molecule to electronic excitation along the axis of the aggregate. In conformity with this hypothesis it is found that the fluorescence of J aggregates is quenched by exceedingly low concentrations of suitable substances, the dye aggregate here behaving as a single optical unit. The differences in fluorescence efficiency for two geometrically different forms of aggregate have been discussed[986] semiquantitatively in terms of "sandwich" and "end-on" aggregates (see below), and will be referred to in Section 5.1.

---

* The terms "compact" and "loose" describe the steric properties of cyanine dye structures.[1376] Both refer to planar dye molecules, the "loose" type having free space between the nonbonded atoms and "compact" describing these dyes when this free space is taken up. Introduction of further groups then distorts the molecule from planarity.

Mason,[985] by means of circular dichroism spectra, observed induced optical activity of the planar dye pseudo-isocyanine [(II), $n = 0$] in the presence of the optically active tartrate anion. He proposed a helical J aggregate structure. The cationic azine dye Acridine Orange [(V), R = CH$_3$] showed similar properties when bound to optically active soluble polymers. Berg and Haxby[129] have shown that certain cyanine dyes reveal large optical activities when caused to form J aggregates by the addition of simple inorganic salts (NaCl and MgSO$_4$), and they assumed that the dye aggregates form a long-range order system, i.e., a liquid crystal.

(V)

In fact, in his original work on J-aggregating cyanine dyes, Jelley[725] had concluded that the gels of filamentous aggregates were similar to a nematic phase of liquid crystals rather than a stacked-up arrangement of dye ions. It was subsequently suggested[1379] that the mesophase on the threshold of the crystalline state consisted of an extended stack of aggregated planar dye ions. The possible incorporation of water molecules into both dimers and J aggregates is mentioned in Section 6.

We now summarize some of the qualitative and semiquantitative explanations given for the spectral changes associated with the reversible aggregation of cationic dyes in aqueous solution.[406,986]

(a) Bands due to the dimeric species. The qualitative picture of the dimeric system differs according to whether the two equivalent dye molecules form a "sandwich" or an "end-on" dimer. Thus while the monomer has a single excited-state wave function, two possible excited energy states exist for the dimer, one in which the transition moments of the two molecules are parallel, and one where they are antiparallel. In the "sandwich" dimer the higher energy state is that in which the transition moments are parallel and therefore absorption intensity is greatest, while the reverse situation is true for the "end-on" dimer. This is illustrated in Fig. 5. The result in the case of a "sandwich" dimer (by far the commoner type) is a blue shift of the absorption band as typified by Acridine Orange [(V), R = CH$_3$]. The "end-on" system is predicted to show a red shift of the absorption band and it has been suggested that the aggregate of pseudo-isocyanine [(II), $n = 0$] approximates to an "end-on" arrangement. These effects are discussed more fully by Mason.[986]

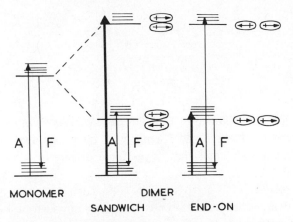

MONOMER   DIMER

          SANDWICH   END-ON

Fig. 5. Energy levels on dimer formation. (Reproduced with permission from Coates.[263])
A = absorption; F = fluorescence.

(b) Bands due to higher aggregates. Aggregation beyond the dimer stage results in the splitting of the first excited singlet into n possible excited energy states of the n-mer. Emerson et al.,[406] using the exciton model, have been able to identify the resolved low-wavelength bands of a particular cyanine dye with the states of aggregation.

3.1.2. Anionic Dyes

Despite the fact that many anionic azo dyes are known to associate strongly in water, the resulting spectral effects are much less marked than those of the cationic dyes discussed above. Since the effects are small, the precision of spectral measurements in aggregation studies with anionic dyes has been questioned[873] and in general other physical methods have been employed to obtain quantitative information. Changes in absorption spectra with mixtures of anionic dyes are much greater than those caused by the self-association of simple dyes and deviations from the theoretical additive spectrum of such mixtures in water have been used to establish the existence of 1:1 aggregates.[339,712,873]

Anionic dyes exhibit small, if any, blue shifts with increase in concentration, and intensity changes are also small. The qualitative effects of dye concentration,[1374,1625] electrolyte concentration,[195,983] and organic solvents[983,1377] have been demonstrated. On the basis of spectral evidence, coupled with potentiometric data, Coates and Rigg[264] showed that the anionic dye C.I. Mordant Violet 5 (VI) in the form DH_2^- dimerizes in the concentration range studied. The form DH^{2-}, resulting from the second ionization stage, does not, however, associate.

(VI)

In later studies on aqueous solutions of reactive dyes used in the dyeing of cellulosic fibers, Padhye and Karnik,[1133] from examination of spectra, suggested that a number of these dyes form dimers in a parallel stacked configuration.

In the absence of the gross effects associated with aqueous solutions of cationic dyes, little quantitative evidence of aggregation is likely from examination of spectra of anionic dye solutions. Recently, however, two semiquantitative methods have been published to evaluate the degree of aggregation of anionic dyes in aqueous solution from spectrophotometric data. Hida et al.[657] describe a "maximum slope" method in which a simple monomer–n-mer equilibrium is first considered. From information so obtained, mathematical adjustments are made to minimize the mean deviation to give computed parameters of an equilibrium system including a monomer and several polymers. This method has also been extended[656] to cationic dyes. The approach of Pugh et al.[1203] is based on a polymeric system heterogeneous with respect to molecular weight, and involves expressions used in polymer studies for evaluating size distributions, these then being related to deviations of the dye solution spectra from Beer's law. The theory has been applied semiquantitatively to a number of anionic dyes in aqueous solution, the anionic type being the more likely type to fit the model on which the theory is based.

3.2. Apparent Deviations from Beer's law

Deviations from linearity are common in the normal "Beer's law plot" (absorbance vs. concentration) for aqueous solutions of all classes of water-soluble dyes.[464,1377] In a theoretical discussion of the deviation of aqueous solutions of dyes and organic ions in general, Kortüm and Seiler[803] argued on the basis of classical dispersion theory that it is not the molar absorption coefficient, but a related quantity involving the refractive index of the solution which is essentially the concentration-independent constant. Other studies upon the validity of the law have been published.[519,606,891,1436] The effects of association on the Beer's law behavior of aqueous dye solutions have been described by Duff and Giles[368] for a number of anionic dyes.

Three types of plot are obtainable:

(a) A linear plot. This is obtained with many dyes at very low concentrations (up to $\sim 10^{-4}\ M$). Apparently linear plots are also observed for dyes that have been shown by other methods to aggregate strongly. It has been demonstrated experimentally[516] that deviations from Beer's law occur only when the *particle size distribution* of the species changes over the concentration range under study. No matter how large the particles, apparently, provided their size distribution remains constant over the concentration range studied, the plot is linear; and hence an apparently linear plot does not indicate absence of aggregation.[225]

(b) Negative deviations. This type of behavior is shown by many dyes in aqueous solution over certain concentration ranges. With certain anionic dyes, the extent of the deviation is time dependent; the aging effect on dye association has been described.[368,875]

(c) Curves consisting of at least two distinct portions: (i) a curved, low-concentration portion, and (ii) a linear or slightly curved, high-concentration portion. This is illustrated in Fig. 6 for the anionic dye C.I. Acid

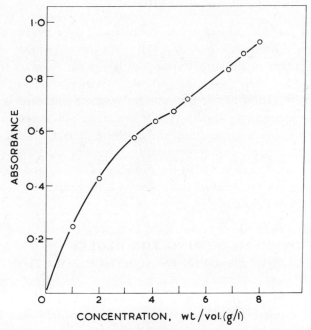

Fig. 6. Beer's law plot for aqueous solutions of C. I. Acid Yellow 29 [dye (VII)]. λ_{max} 385 nm, path length 0.1 mm.

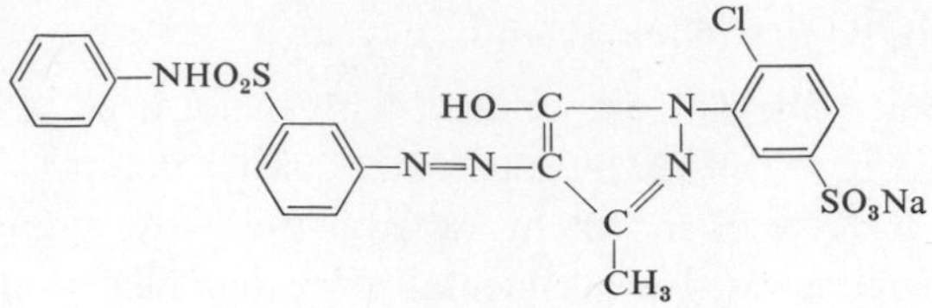

(VII)

Yellow 29 (VII), and the effect at high concentration is characteristic of a small number of anionic dyes showing marked surface activity. For a typical surfactant, hexadecyltrimethyl ammonium bromide, the break in the curve has been shown[367] to coincide with the CMC, independently derived. Little is known of the structure of dye micelles, but it has been suggested[15] that a highly aggregating acid dye, C.I. Acid Orange 63 (VIII), may form colloidal micelles similar in structure to that proposed for surfactants.

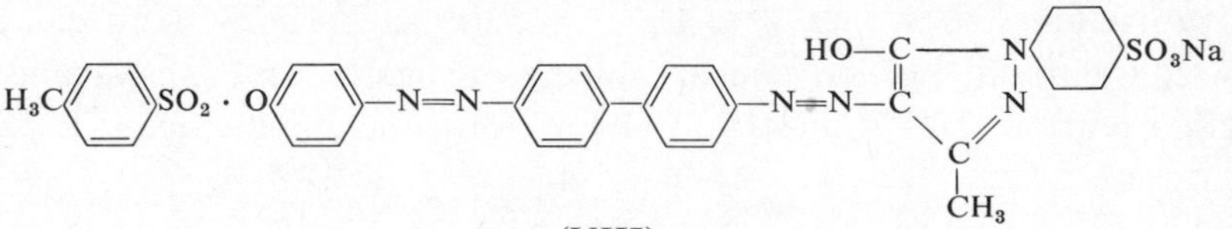

(VIII)

Thus it can be concluded that solutions of dyes in organic solvents generally obey Beer's law (see Fig. 11B), but there are many examples in all classes of dye of deviations from the law when water is the solvent. Such effects have been shown, by various physical methods, to result from reversible aggregation of the dye ions in water. Some care is necessary, however, in interpreting results: While nonlinear absorbance vs. concentration plots indicate aggregation, the converse, namely that a linear plot precludes aggregation, is not necessarily true, and other physical methods may be required to establish the presence of aggregated species. Some of the methods used to measure the degree of aggregation of dyes in aqueous solution are now discussed.

4. METHODS OF MEASURING THE DEGREE OF AGGREGATION OF DYES IN AQUEOUS SOLUTION

In those cases where a simple monomer–dimer equilibrium exists and the system is suitable for spectrophotometric measurements, results are reasonably simple to interpret. Some doubt must exist, however, about the validity of the results from measurements where higher dye polymers, containing up to thousands of molecules, may be present. Thus there are systematic differences between techniques that measure a number-average molecular weight, with high weighting to small particles, and those that give a weight-average molecular weight with consequent weighting to the large particles. In addition, some methods (e.g., diffusion) require the presence of electrolyte, whereas conductometric measurements must be carried out without the interference of other electrolytes.

The following sections briefly describe methods used to measure the degree of aggregation, and some of the factors limiting a quantitative assessment of the results are also considered. Further information can be found in the reviews by Vickerstaff[1526] and Coates.[263]

4.1. Methods Employing Electromagnetic Radiation

As stated above, the spectral characteristics of aqueous solutions of cationic dyes have been widely studied. These are further considered here along with results of the less commonly used techniques, such as nmr and light scattering.

4.1.1. Absorption Spectra and Other Spectral Methods

Spectral methods have indicated the presence of dimers in aqueous solutions of Methylene Blue [(I), R = CH$_3$][131,174,873,1207] Thionine [(I), R = H][617,1207] cyanine dyes,[1314,1578] the Rhodamine dyes of the Xanthene class,[879,880,1365] and some triphenylmethane dyes.[812,1433] The following general comments summarize the results of these studies.

(a) Optical studies are used, mostly on the assumption of a simple monomer–dimer equilibrium

$$2D^+ \rightleftharpoons (D_2)^{2+}$$

Accordingly, over the concentration range studied, the concentrations of the two equilibrium species will adjust themselves according to the law of mass action. Thus $(D^+)^2/(D_2^{2+}) = K$, the dimer dissociation constant.

(b) Mathematical expressions involving the above relationship are employed to validate the assumption of a simple monomer–dimer equilibrium, using measurements from a series of absorption curves. These expressions contain parameters (e.g., ε_{dimer} and the degree of association) not directly measurable and these must be found indirectly by fitting data to some model or by taking arbitrary values of K. In some cases the spectral curves pass through one or more isosbestic points, but in others they do not. The presence of the isosbestic point indicates that two colored species are in equilibrium, but the absence of such a point may in certain cases be due to activity factors.[1377]

(c) The dimer constant is fitted at a single wavelength[880] or over the entire absorption band[1433]; alternatively, it can be fitted by evaluating the monomeric and dimeric concentrations from an iterative procedure involving successive approximations about the overlap of the dimeric band on the monomeric band[1578]; by this means the resolved monomer and dimer spectra have been revealed. Where comparisons are available[880,1433] it is not surprising that the dimeric spectra for the same dye differ and are quite dependent on the method of approximation. The results from such spectral studies do suggest, however, that at least over a limited concentration range, dyes of the types mentioned at the beginning of this section initially form dimers in aqueous solution. In a number of cases[174,1433] optical studies have been accompanied by confirmatory evidence in the form of physical measurements of another sort. At higher concentrations, the evidence suggests that larger aggregates are formed, though quantitative information on higher dye polymers is less common.[406]

It can be concluded, therefore, as a result of the spectral studies that while many dyes form dimers in aqueous solution, this process alone is an inadequate description of their self-associating properties, and the validity of investigations by spectral absorption methods in any system of extended "multimers" has been questioned by Mukerjee and Ghosh.[1057]

Nuclear Magnetic Resonance Spectra. The association of Acridine Orange [(V), $R = CH_3$] has been studied[150] by following the proton chemical shifts as a function of dye concentration in D_2O/H_2O (90:10). The magnitude of the deshielding is interpreted as due to dissociation of aggregates larger than dimers on dilution.

4.1.2. Light Scattering

The measurement of the turbidity τ of dilute solutions is the standard method for obtaining the weight-average molecular weight $\bar{M}_w$ of polymeric species. In this method Hc/τ is plotted against c, where H is given by the Debye equation

$$H = (32\pi^3/3\lambda^4 N)n^2(dn/dc)^2$$

with λ the wavelength of incident light; n the refractive index of the solution; and c the concentration of solute. $\bar{M}_w$ is obtained by extrapolating to zero concentration. This technique has been applied in a number of aggregation studies of dyestuffs.[15,471,1403] Alexander and Stacey[15] examined four azo anionic dyes in aqueous solution using light of $\lambda > 700$ nm to minimize absorption by the dyes. Scattering appeared to be very slight in the absence of added electrolyte, but these authors claimed molecular weights of up to 10^6 for the dye Benzopurpurine 4B [(III), R = CH_3] in solutions of sodium chloride, a figure close to that obtained by Frank[471] for the same dye in 0.04 M KCl, from light scattering measurements. Frank also suggested that large aggregates are only formed in solutions containing added electrolyte and that the aggregates are built up in the manner of a stack of cards, basing his suggestion on results also of viscosity and conductance measurements. In addition to the problem caused by absorption by dyes of the radiation of the light source (often isolated Hg lines at 436 or 546 nm) referred to above, the refractive index shows anomalous behavior near wavelengths of high absorption. The validity of light scattering measurements is critically dependent on the accuracy of the dn/dc term in the Debye equation above.

4.2. Methods Based on Diffusion Properties

The diffusing properties of both cationic and anionic dye aggregates in water have been measured in estimations of the size of these aggregates. The measurements involve the determination of the diffusion coefficient D of the dye either directly[66,297,515,655,874,1254,1510,1512] or by the indirect polarographic method.[263,666,950,962,965] From these values, the particle size is found by use of the Stokes–Einstein equation[515,1254] or from an empirically determined relationship between D and molecular weight.[666]

4.2.1. Direct Measurement of the Diffusion Coefficient

From measurements on Benzopurpurine 4B [(III), R = CH_3] and its meta isomer, Robinson[1254] concluded that the presence of inert electrolyte was necessary to obtain an indication of particle sizes from diffusion coefficients, and all studies using this technique involve the aggregation of dyes in aqueous electrolyte solutions. Valko[1510,1512] examined a large number of anionic dyes and concluded that acid dyes of low molecular weight,

(IX)

like Orange II (IX), are almost monodisperse in aqueous solution, but the higher molecular weight direct cotton dyes such as Benzopurpurine 4B [(III), R = CH₃] and Congo Red [(III), R = H] form ionic micelles. He found that the self-association is temperature reversible and at 90° he found, over normal ranges of concentration, no aggregates containing more than three molecules.

Anthraquinonoid acid wool dyes have also been studied by diffusion methods[297,515] and aggregation numbers from 3 to 18 obtained.

The diffusion coefficient D is evaluated from an equation of the type[938]

$$D = \frac{2}{\lambda + 1} \frac{\beta}{t} \log \frac{C_0}{C_0 - (\lambda + 1)C_1}$$

where λ is the volume of the external solution divided by the volume of internal solution for the diffusion cell; C_0 and C_1 are concentrations of dye solutions in the inner and outer compartments at time t; and β is a cell constant. The Stokes–Einstein equation then relates D to the particle radius r

$$D = RT/6\pi\eta rN$$

where η is the viscosity of the medium. In theory, therefore, evaluation of D will allow the size of dye aggregates to be found in terms of their radius. Two implicit assumptions in the Stokes–Einstein relation make its application in the above form questionable. These are that the aggregate particles are both spherical and uncharged.

In an aqueous solution of a dye, the gegenions will diffuse, under a concentration gradient, ahead of the aggregates, eventually setting up a potential gradient. This gradient accelerates diffusion of aggregates and, as a result, the observed diffusion rate is much increased. In practice, the diffusion is controlled by carrying out the process in an excess of inert electrolyte which minimized the accelerating potential of the dye aggregates.

Early diffusion measurements[66,655] ignored the effects of electrolyte impurities in causing diffusion potentials. Further difficulties arise in the calculation of the aggregate weight from the determined value of r, in that assumptions must be made both in the interpretation of r and in the value taken for the aggregate density. Despite these difficulties, the method has been used quite widely, and as a consequence of the requirement of added electrolyte, quantitative results tend to be higher than those from other methods, such as conductance, where it is essential to avoid interference from other ionic species.

4.2.2. *Polarographic Measurements*

The diffusion coefficients of dyes in aqueous solutions of electrolyte can also be measured by polarography. The excess electrolyte is again used to eliminate electrical potential effects on diffusion. Hillson and McKay[666] first used the technique to measure the aggregation of two dyes, Methylene Blue [(I), $R = CH_3$] and Congo Red [(III), $R = H$], both of which had been extensively studied by other methods. The diffusion current is related, by the Ilkovic equation, to the number of electrons taking part in the reduction, to the diffusion coefficient, to the concentration of the dye, and to the polarograph variables. In this case the instrument was calibrated by the use of Cd^{2+} ions. The reduction of organic dye ions involves an even number of electrons (usually two), and with this knowledge, the diffusion coefficient of the dye in water can be calculated from the limiting polarographic current. Hillson and McKay calculated aggregation numbers by reference to an empirically derived relationship of the form* $D = AM^{-B}$ between diffusion coefficient D and molecular weight M for some 50 high- and low-molecular-weight species. Their results were in good agreement with those obtained by other authors, who had used different techniques. Thus they found that Methylene Blue forms dimers only, while the aggregates of Congo Red vary in size from about three molecules at the lowest to around 10^6 at the highest concentrations.

In a subsequent publication, McKay[950] used the technique to study the cyanine dye, 1,1'-diethyl-2,2' cyanine bromide [the bromide of (II), $n = 0$] and an asymmetric isomer. He attempted to relate the aggregating powers of the two dyes thus measured to changes in their absorption spectra. He found aggregates of up to 100 molecules of the symmetric dye, but the isomer appeared to aggregate only weakly (up to ten molecules) and just below its solubility limit. The latter dye, however, showed larger metachromatic changes. Malik and collaborators[962,965] subsequently investigated the effect of additives of varying dielectric constant on the aggregation of dyes in water using the polarographic technique and found, e.g., that urea, formamide, and methanol all bring about a decrease in the tendency for dye ions to aggregate.

Polarographic measurements suggest that there is a critical aggregate concentration below which the dye is monodisperse in water, though it is unlikely that this simple picture is an accurate one.

---

* This relationship has been criticized by Stork.[1433]

4.3. Methods Based on the Electrical Properties of Dyes in Solution

Conductance measurements have been employed to estimate the degree of aggregation in water of both cationic[350,1025,1255,1376] and anionic[308,471, 674,796,1025,1147,1255-1257] dyes, though it has been pointed out[1511] that the resulting data do not in general provide much information on aggregation. Indeed early studies[796,1147] on the dyes Congo Red [(III), R = H] and Benzopurpurine 4B [(III), R = CH$_3$] indicated little or no tendency toward aggregation. Subsequent work[1257] on the latter dye and its meta isomer, however, suggested that Benzopurpurine 4B, the unhindered dye, formed stable micelles containing a number of associated gegenions and that these showed no tendency to break down on dilution as far as 0.0005 M solutions. A conductance maximum was also observed at very low dye concentrations. Robinson and Garrett[1255] investigated the aggregating properties of several anionic direct cotton dyes and concluded that conductance methods give aggregation numbers lower than those given by other techniques. They also studied the electrochemical properties of the cationic dye Methylene Blue [(I), R = CH$_3$] and obtained an aggregation number of two, in agreement with the results of the numerous subsequent spectral studies. As an alternative to the spectral method for the examination of the aggregation of cationic dyes in aqueous solution, Sheppard[1376] investigated the electrical conductance and the transference numbers as functions of concentration for a series of cyanine dyes, including pinacyanol chloride [(II), $n = 1$]. In all cases the degree of aggregation was low, an average aggregation number of two being the largest obtained, and Sheppard assumed that at this level inclusion of gegenions is unlikely.

Measurements of the above-mentioned electrical properties of dyes in aqueous solution must be carried out in the absence of other electrolyte, and hence results of degrees of aggregation from these techniques are often lower than those from other methods. An aggregate will have[1526] an increased mobility compared with a single dye ion, since this mobility varies directly as the charge and inversely as the particle radius, which is itself proportional to the cube root of the mass.

The aggregation number is determined by measuring the equivalent conductance Λ_0 of the dye solution at infinite dilution. This value is inserted in an expression obtained by equating the electrostatic driving force on the micelle to the frictional resistance. Thus

$$\text{aggregation number} = 0.852(M/\varrho)^{1/2}(\eta/z)^{3/2}(\Lambda_0 - \mu)$$

where M is the molecular weight of the dye ion, ϱ is the density of the particle, η is the viscosity of the medium, z is the number of ionizable groups per dye molecule, and μ is the ionic conductance of the gegenions.

This expression assumes that no genenions are present in the aggregate, though Holmes and Standing[674] give a modified equation containing a factor representing the fraction of total gegenions not included in the aggregate. The value of the particle density is an estimated one, and since the above expression contains an implicit assumption as to sphericity of particles, it is not surprising that the validity of conductance measurements in dye aggregation studies has been questioned.

4.4. Other Methods

A number of early studies involved measurements of the osmotic pressures of aqueous dye solutions[308,356,1258] and as a result of his work on Congo Red [(III), R = H], Donnan[356] developed his famous membrane equilibrium theory. These measurements were normally made in the presence of excess electrolyte, and despite difficulties associated with adsorption of dye on the membrane, at least qualitative agreement was obtained with results from other techniques for direct dyes of the Congo Red type.

Other methods which have been used to indicate the degree of association of dyes in aqueous solution include ultracentrifugation,[1206] low-angle X-ray scattering,[813] calorimetry,[338] surface tension,[566] viscosity,[471] sodium ion electrode response,[1014] capillary analysis[630] and an "iso-extraction" partition technique.[1057] All of these have been used only to a limited extent.

4.5. Summary of Results

Many cationic and anionic dyes in aqueous solution show deviations from Beer's law over an extended concentration range and it may therefore be concluded that they form aggregates in water. Due to their low aqueous solubility, the nonionic disperse dyes have been little studied in this respect, though it has been suggested[947] that association probably also occurs with solutions of some dyes of this class.

The results of studies on solutions of cationic dyes point to the initial formation of dimers, followed, as concentration increases, by the appearance of larger aggregates; subsequently, in the case of certain cyanine dyes, giant stacked-up units appear. Most information has been derived from spectral absorption measurements, assuming adherence to the law of mass action, and although this approach has been criticized, the prediction of dimers, which is a direct consequence of the mass law, is in general quantitative agreement with results from other methods. Some controversy has developed over the question of whether aggregation proceeds beyond the dimer stage with these dyes. In most cases higher polymers are almost certainly formed, but one cannot be authoritative about the results of techniques that afford average values of aggregation numbers. The theoretical interpretation of the dimerization process, based on the formation of head-to-tail, or, more commonly, back-to-back aggregates agrees at least semi-quantitatively with spectral data for cationic dyes in aqueous solution.

Despite general agreement on some features of the aggregation process, the quantitative results of experiments on solutions of anionic dyes only show a limited measure of agreement. However, as pointed out by Coates,[263] an exact comparison of results is not very meaningful due to differences in experimental conditions, particularly with respect to electrolyte concentration. Where environmental conditions are comparable, aggregation numbers resulting from different techniques may either show fair agreement or vary quite considerably. Thus for a $6.9 \times 10^{-5}\ M$ solution of

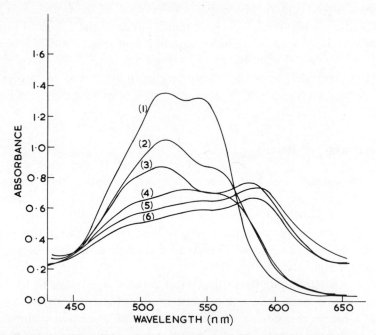

Fig. 7. Spectral characteristics of C. I. Acid Red 138 [dye (X)] in the presence of electrolyte: dye concentration $10^{-4}\ M$; path length 5 mm; NaCl concentration: (1) none, (2) 0.02 M, (3) 0.2 M, (4) 0.5 M, (5) 1.0 M, (6) 2.0 M.

Benzopurpurine 4B [(III), $R = CH_3$] in 0.01 M NaCl, light scattering results[15] give an aggregation number of 40, while diffusion experiments[1510] indicate a value of 5–6. Aging effects could also influence results. The spectral changes associated with increase in concentration of anionic dyes in water are generally too small to be amenable to quantitative measurements, and other physical methods have been employed in determining the degree of aggregation of these dyes. Some anionic dyes, such as C.I. Acid Red 138 (X), do, however, show marked spectral effects (Fig. 7).

(X)

The physicochemical methods used to measure aggregation express the results as some average value, and it would seem likely that many anionic dyes in aqueous solution form a system of aggregates heterogeneous with respect to molecular weight. The self-aggregation of dyes of all types is a reversible process, being promoted by an increase in dye concentration and a decrease in temperature. In both cationic[1577] and anionic[1256] dye classes planar dyes aggregate more readily than their respective nonplanar isomers, as a result of dispersion interaction and steric effects which may be due either to the presence of auxiliary groups on the dye or nonplanarity of the conjugated chromophore system itself.

5. DYE AGGREGATES

The unequivocal proof of the presence of aggregates in aqueous solutions of many dyes has prompted a number of investigators to attempt to gain a more detailed insight into the manner in which the similarly charged ions overcome the mutual Coulombic repulsion forces in the formation of aggregates. Some progress has been made in describing the geometric arrangement of the ions in an aggregate and the nature of the interactions that hold them together. Most success has been achieved with the cationic dyes, where the various steps in the formation of aggregates are defined more precisely and are better understood. Thus the results of optical and X-ray measurements, coupled with a number of thermodynamic studies, have produced information on interionic packing distances and binding forces, though the evidence from different sources is often conflicting. These two aspects of the aggregation process are now briefly dealt with.

5.1. The Nature of the Aggregates

It is clear that two similarly charged dye ions will associate to form a dimer only if some strong attractive forces exist to produce an entity which is both electrostatically and thermally stable. The type of dye ion likely in principle to form the most energetically favored dimer unit by self-association is that of a planar monosulfonated anionic dye. The dimer then consists of ions arranged back to back, allowing maximum overlap of the aromatic systems, together with maximum charge separation. An example is the acid wool dye Orange II (IX), which has been shown to exist as dimers in aqueous solution.[1014] Such an arrangement is described by Coates,[263] who suggests that with increase in concentration larger polymers will be formed from these dimeric species. This buildup of large polymeric units in multiples of two should theoretically ensure maximum charge separation and the mutually repulsive forces may be further reduced by the inclusion of gegenions. One would expect a cation with a multiple charge to be more effective than a singly charged cation in the above situation. Bubser and Eichmanns[196] have demonstrated the importance of the "true" or outwardly effective concentration of the gegenion, this being dependent on its ionic mobility, which decreases with increasing size of the ion. Using the range of alkali chlorides, it was shown that the lithium salt was least effective in promoting aggregation since its cation (in the hydrated state) is the largest of the group and therefore shows the lowest mobility. The above ideas of dimerization and higher polymerization processes are illustrated in Fig. 8, where the planar monomeric (A) and dimeric (B) units are shown, and each line in the polymer (C) represents a dimeric unit.

The important structural features explicit in such a model for dimers and higher polymers formed between dye ions carrying single and localized charges, namely planarity and maximum charge separation, have also been pointed out by Zollinger[1627] and Valko[1511]; and Coates[263] has attempted a semiquantitative description in kinetic and thermodynamic terms. The same author also suggests that the eventual instability of a back-to-back array of ions as outlined above might lead to the formation of an end-on type of aggregate held together by electrostatic forces between layers of aggregates and associated gegenions, as in Fig. 9. There is, however, experimental evidence to suggest that dye (IX) does not form aggregates much bigger than dimers and a calorimetric determination of the heat of dimerization of this dye,[338] while in agreement with the superimposition of the two ions to form a dimer with maximum charge separation and maximum overlap of aromatic surfaces, also suggests that the addition of

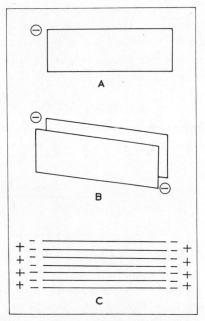

Fig. 8. Suggested aggregation of planar aromatic units (anionic). (A) Monomer, (B) dimer, (C) polymer composed of dimeric units (horizontal lines). (Reproduced with permission from Coates.[263])

a third dye ion with comparable overlap and charge separation is not possible and in any case is energetically too unfavorable to occur to any extent. Higher aggregates of the free acid of this dye may well be formed, however, at higher concentrations due to more efficient shielding of the anionic charge by the smaller hydrogen gegenion.

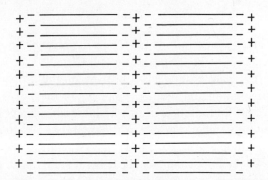

Fig. 9. Suggested formation of higher anionic aggregates by electrostatic union incorporating gegenions. (Reproduced with permission from Coates.[263])

The ratio of hydrophilic to hydrophobic portions (mentioned in Section 2) in a dye of the above type is clearly a determining factor in its ability to form aggregates, and thus polysulfonated anionic dyes might be expected to show less evidence of aggregation in solution than the monosulfonated type. However, the aromatic hydrophobic parts of the former tend also to be extensive and many examples of aggregating polysulfonated dyes are known [e.g., (III), R = H and R = CH₃]. In general these sulfonate groups are situated on terminal positions in the dye molecule, since the same groups centrally placed, as in the direct cotton anionic dye, Chrysophenine G (XI), do not satisfy the condition of maximum charge

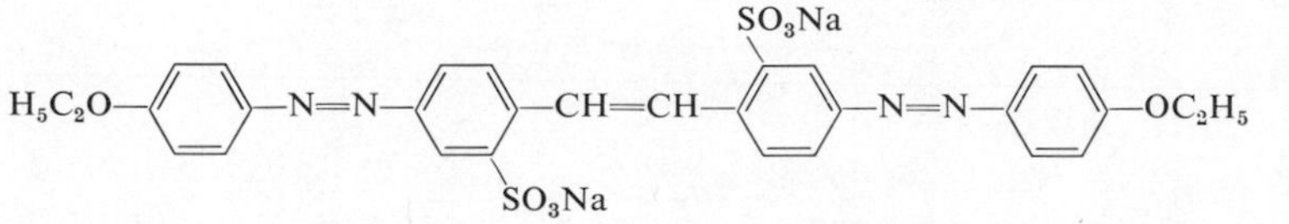

(XI)

separation and do not allow the close approach of hydrophobic parts of neighboring dye ions. Thus Holmes and Standing[674] could find no appreciable evidence of self-aggregation in aqueous solutions of this dye, even at high concentrations or in the presence of electrolyte, though Derbyshire and Peters[339] conclude that a 1:1 mixed aggregate is formed between it and the tetrasulfonated dye C.I. Direct Blue 1 (XII) where all four sulfonate groups are terminally situated:

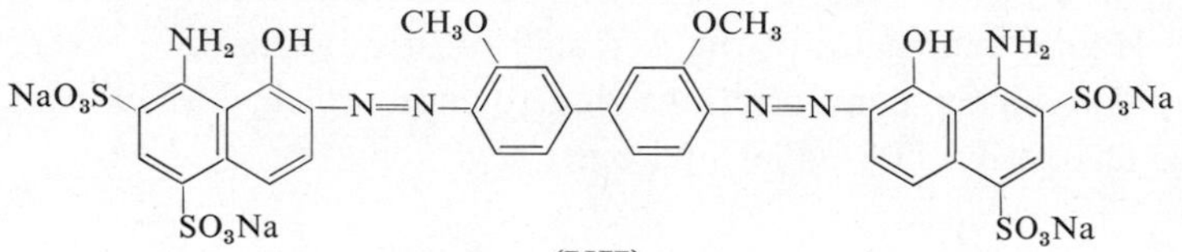

(XII)

This terminal arrangement allows (XII) to aggregate in water.[196,873,1510]

From an extensive study, Speakman and Clegg[1413] concluded that the position of the sulfonate groups has a bearing on the properties of anionic dyes in aqueous solution and that the tendency of these dyes to form colloidal particles in acid solution decreases with increasing degree of sulfonation.

The formation of aggregates of a different type is suggested[15,367,566] for a number of anionic dyes, e.g., (VI), the structures of which somewhat resemble those of the long-chain soaps. It is possible that in aqueous solution these dyes, having the hydrophilic group(s) at one end of the otherwise relatively hydrophobic structure, may aggregate to form structures of similar type to those formed by surfactants in concentrations above their CMC.[1049] These soaplike dyes cause decreases in surface tension of water approaching those of typical surfactants[566] and much greater than those caused by dyes of more usual types.[520] These highly surface-active dyes are all acid wool dyes, and mainly of the milling class, i.e., with high fastness to wet treatments.

In order, then, that in the general case the mutual repulsion of similarly charged dye ions should be as small as possible, the structure of the dimeric or polymeric ion must be such that the charges associated with the individual dye ions have maximum separation. We have seen how, in principle, this condition may be satisfied when the charge on each dye ion is localized as in the anionic sulfonates. In many cationic dyes, however, the charges are delocalized, making their relative separation in a dimer more difficult, and demanding, for example, an alternation of phase of the resonance in each planar ion as suggested by Sheppard and Geddes.[1378] Förster[465] suggested that with Methylene Blue [(I), $R = CH_3$] the dimer structure having minimal repulsion is that where the charge is on the substituted amino groups of the two ions, and that these may lie along opposite edges of the sandwich. There would be four possible resonance structures of this sort. Nevertheless, as a result of a number of theoretical and semi-quantitative treatments the aggregating properties of these dyes are more precisely understood. Optical studies have conclusively established that dimerization is the first stage in the aggregation process in water. As indicated in Section 3.1.1, the spectrum of the dimer shows the appearance of a new band commonly on the blue wavelength side of the monomer band, and more rarely on the red side, the location being determined by the dimer structure, i.e., whether it has the "sandwich" or the "end-on" configuration. This has been qualitatively interpreted on the basis of the molecular exciton model.

Further evidence of the nature of the dimer is often obtained from differences in the emission spectra. It is a consequence of the postulate of the exciton model and the n-fold splitting of the excited singlet on the aggregation of n dye ions that if these aggregates are of the sandwich type, the quantum yield of phosphorescence may be enhanced at the expense of fluorescence due to the accumulation of triplet species.[955] As a result, the fluorescence of such a dimer is weak and shifted toward the red. Both these effects are shown by Acridine Orange [(V), $R = CH_3$]. In a dimer with a head-to-tail configuration, the model predicts more efficient fluorescence as a result of fewer quenching collisions,[986] and this type of behavior, coupled with a red shift of the absorption band, is shown by pseudoisocyanine [(II), $n = 0$].

With increasing concentration, most cationic dyes aggregate beyond the dimer stage (to form H aggregates); eventually, in the case of certain photographic cyanines, they form filamentous J aggregates containing thousands of individual ions. The spectral characteristics of these species have already been discussed in Section 3.1.1. The structure of H aggregates has been discussed by Emerson et al.,[406] and a number of studies reveal details of the structure of J aggregates. These have involved monolayer,[757] X-ray,[679,680,1581] crystallographic,[1581] and spectral reflectance[440,1379] measurements. As a result, the structure of the J aggregate has been revealed as a stack of mutually parallel dye ions at a separation of between 3.3 and 4.5 Å. The individual ions are probably acutely tilted from the normal to the axis of the aggregate. The staggering of successive dye ions has been suggested[1316] in one case. This would give a structure in which the carbon atom of the methine group in the bridge of one ion is almost perpendicularly opposite the carbon atom in the 4-position of the quinoline ring of its neighbor. The optical coupling of such a stacked-up aggregate of ions in producing the so-called J absorption band has already been mentioned (Section 3.1.1). In agreement with this type of assemblage are the fluorescence quenching[755] and the formation from two dyes of mixed aggregates in water which give rise to a single J band. These and other features of J aggregates of cyanine dyes are fully dealt with by West and Carroll.[1577]

It is thus fairly well established that the aggregation of dyes in water results in the formation of stacked-up polymers, the main structural requirements of which are planarity of the individual dye ions and a maximum charge separation between neighboring, similarly charged ions in order to allow the intermolecular forces of attraction to become operative. The following section deals with the nature of these forces.

5.2. Nature of the Bonding in Aggregates

The forces influencing relatively large organic ions such as dyes in solution are briefly considered by Coates,[263] and the more important forces of attraction likely to have a bearing on the formation of dye aggregates are now discussed.

5.2.1. Hydrogen Bonds

The requirement for this type of bonding is that a hydrogen atom, with its high polarizing power, is covalently bound to one of high electron affinity, whereby it is positively polarized. This occurs mainly when hydrogen is bound to N, O, or F atoms and less commonly to Cl, S, and C, and since most dyes satisfy this condition, particularly with respect to the presence of N and O atoms, intermolecular hydrogen bonding between adjacent dye ions to form dimers or higher polymers is, in principle, possible. Some evidence has been put forward in support of this possibility for some Xanthene basic dyes[880,1365] and sodium fluorescein.[1271] Rose (in Ref. 1526) suggested a hydrogen-bonded structure to explain the large aggregates of Congo Red [(III), R = H] but Coates rejected this both on steric and energetic grounds,[263] in favor of the overlapping arrangement for anionic dyes mentioned in Section 5.1. Most of the indirect evidence for hydrogen bonding has been obtained from thermodynamic data derived from spectral studies; in particular values of the (van't Hoff) enthalpies of dimerization ($\Delta H° = -21$ to -46 kJ mol^{-1}). The more negative these values, the more is the dimerization process favored. However, the measured value of this thermodynamic function has a number of contributions and it may be accidental that it happens to be close to that associated with the bond energy of a hydrogen bond. Thus it has been shown[1365] that the value of $\Delta H°$ is decreased when the dye is dissolved in a solvent less capable of forming hydrogen bonds than is water, suggesting that, for some dyes at least, hydrogen bonding is important in the formation of dimers. It is probable, however, that its contribution is indirect, as suggested by Mukerjee and Ghosh[1058] for Methylene Blue [(I), R = CH$_3$]. Levshin and Slavnova[882] studied a number of cationic dyes and concluded that those with no free hydrogen (i.e., those incapable of hydrogen bonding) give absorption spectra in water which are independent of concentration, though certain of the results are not in accord with those of later authors. The balance of evidence seems to be against the hydrogen bond being a major factor in the bonding of dye aggregates in aqueous solution, and indeed it has been pointed out in another context by Meggy[997] that in many dye molecules many suitable groups are already involved in intramolecular hydrogen bonding. Thus in Orange II (IX) the hydroxyl group and a nitrogen of the azo group form such a chelate system. These are particularly strong when a six-membered ring is formed. The enthalpy of dimerization of this dye has been determined calorimetrically to be -43.81 kJ mol^{-1}.[338] Derbyshire concluded that this value is too high to be accounted for by intermolecular H bonding, and that such bonding resulting from the breaking of an intramolecular H bond would not afford the necessary energy. Similar conclusions were reached by Coates and Rigg[264] from a spectral study of a chemically similar dye, and by Giles and D'Silva from adsorption experiments.[515] Rabinowitch and Epstein,[1207] in a spectral study of Methylene Blue [(I), R = CH$_3$] and the related Thionine [(I), R = H], distinguished between the behavior of dye aggregates and that of simple short-chain organic acids that also dimerize readily. The association of the latter in organic solvents is usually attributed to hydrogen bonding and, in contrast with the dimerization of dyes, is reversed in water. In addition, the increase in aggregation of dyes with concentration is not reflected in the behavior of the hydrogen-bonding organic acids. It was concluded, therefore, that the nature of bonding in dye aggregates must be interpreted in some other way.

5.2.2. Van der Waals Forces

These can conveniently be divided into two main groups, known respectively as polar and dispersion forces. The attraction experienced by two molecules as a result of the polar forces will result from dipole–dipole interaction (orientation effect) and dipole–induced dipole interaction (induction effect). Theoretical treatment of the situation for the randomized approach of all molecules leads to two results relevant to stacked dye aggregate systems. These are first that polar van der Waals forces (more accurately, that part due to the orientation effect) have a marked temperature coefficient and second that they fall off with the sixth power of the molecular separation. The dispersion or London forces account for the fact that nonpolar molecules can attract one another since the attraction can arise from the interaction of fluctuating dipoles. The energy of such a system is lowered by the formation of weak bonds, this energy being proportional to α^2/r^6, where α is the polarizability of the molecules and r their separation. The value of α increases with increase in molecular volume, so that for large molecules the dispersion forces increase, becoming relatively more effective than the dipole forces mentioned above. This is an important consideration in the study of aggregates of dye ions since dyes have fairly high molecular weights and in this connection Mukerjee and Ghosh[1058] describe the mutual association of Methylene Blue [(I), R = CH_3] with a number of heterocyclic compounds of varying size, obtaining a rough correlation between the free energy of mutual association and the molecular size of the heterocyclic compound. Giles et al.[517] studied the adsorption of a number of monoionic dyes (including Methylene Blue) on a variety of finely divided solids and calculated a coverage factor—the factor by which the adsorption exceeded that for a simple monolayer. These factors, which varied between two and ten, were found to increase with the cube of the ionic weight of the dye. They are assumed to represent the aggregation number of adsorbed stack-form ionic micelles of the dyes. In agreement with this assumption is the fact that the figure for Methylene Blue is

~2.0, equal to its aggregation number in solution found by other methods. Both these observations suggest that dispersion-type forces are important in the association of planar dye molecules in water and also adsorbed on surfaces.* The very rapid falloff of these forces with separation underlines the importance (in the self-association of dye ions) of planarity to which we have already referred. A number of early authors[802,1314] had suggested that dyes aggregate in aqueous solution as a result of additive forces of the van der Waals type, a view shared by Rabinowitch and Epstein[1207] in their study of Methylene Blue [(I), R = CH$_3$] and West and Pearce[1578] in their work on the cyanine dyes. The former authors, in stating that aggregation is a universal property of dyes in aqueous solution, interpreted the process semiquantitatively on the basis of London's theory of intermolecular forces, and, because of the mutual repulsion of two similarly charged dye ions, postulated an activation energy of dimerization.

West and Pearce[1578] investigated a series of thiacyanine dyes [(IV), $n = 0, 1, 2, 3$] and found that the tendency to form dimers in water as measured by the free energy of dimerization increased with chain length by an approximately constant amount per additional vinylene group. This result strongly suggests that the additive dispersion forces are important in the aggregation process.

5.2.3. *Hydrophobic Interactions*

This term is frequently used to describe the tendency of hydrophobic groups such as alkyl chains to associate and escape from an aqueous environment, and has been fully described in Chapter 1. A number of workers[131,965,1056,1064,1271,1505,1591,1628,1629] have considered the association of dyes in water in terms of the hydrophobic interaction and the results indicate that solvent-induced bonding of this type contributes significantly to the formation of dye aggregates. Much of this evidence comes from thermodynamic considerations which suggest that the process of dimerization ideally requires a negative enthalpy change and a positive entropy change. While the former condition is satisfied for dimerization, the estimated entropy change $\Delta S°$ is usually negative and of moderate magnitude. Analogous behavior is found in the thermodynamic changes involved in the transfer of hydrocarbons to aqueous solution, when the positive standard free energy change (representing low solubility) results primarily from a

---

* The structures of adsorbed cyanine dye aggregates and their spectral characteristics have recently been fully discussed in a series of papers from the Polaroid Corporation[564,1103,1552]

relatively large and negative entropy change. These anomalous effects have been explained in terms of the "iceberg" layer structure of water surrounding dissolved hydrocarbon molecules or groups suggested by Frank and Evans.[477] The association of two or more such molecules or groups will result in a breaking up of the surrounding water molecules resulting in a positive entropy change and this has been suggested as the thermodynamic cause of the hydrophobic interaction. Thus the unfavorable entropy change associated with the "ordering" of dye molecules during the aggregation process in water is more than compensated for by the positive contribution resulting from the water structure effects outlined above. The entropy changes involved in dimerization processes are usually derived from optical studies from which is calculated the equilibrium constant K of the reaction

$$2D^+ \rightleftharpoons (D_2)^{2+}$$

The entropy change involved in the dimerization of Methylene Blue [(I), $R = CH_3$] has been found to be -18.8 J mol^{-1} K^{-1}. In a study of this system Mukerjee and Ghosh[1058] have corrected this value to obtain the unitary entropy change.[759] This quantity is calculated by expressing it in mole fraction units and is characteristic of the interactions associated with dimerization. It is concluded that the unitary entropy change includes a number of contributions, and since these are in the main negative (e.g., arising from rotational, vibrational, and electronical effects), there must be a considerable hidden positive contribution. The most likely source is believed to be the effect due to the disordering of the structured water.

Further support for the postulate that the hydrophobic interaction is important in dye aggregation has come from a number of studies[15,57,773, 965,1056] in which the action of urea as a disaggregating agent has been examined. Urea, although capable of participating in hydrogen bond formation, appears to act mainly by virtue of its ability to reduce the intermolecular order in water.[447,478] Thus, according to Coates,[263] it interferes with the ordered water structure around a dye monomer, thereby reducing further the overall negative entropy change on dimerization. Mukerjee and Ghosh,[1056] in an optical study of Methylene Blue [(I), $R = CH_3$], investigated the contribution of "water structure" to dye aggregation, using urea as a probe. Its effect on the absorption spectrum of the dye in water is to enhance the monomer (M) peak at the expense of the dimer (D) peak. The authors give two reasons for concluding that solvent structure-induced hydrophobic interactions are important in dye aggregation, (i) the effect of urea is a general one for dyes, including the anionic[15,57,773] and cyanine[1056] classes; and (ii) dye aggregation is seldom found in nonaqueous media. Such evidence for hydrophobic effects involving the partial randomization of structured water may not, however, be conclusive and has been criticized by Holtzer and Emerson[676] and Valko.[1513] More direct evidence for hydrophobic interaction is indicated by Zollinger's spectral absorption measurements[1628] of a series of anionic dyes [(XIII), R = H, p-CH$_3$, m-Cl, p-Cl, m-NO$_2$, p-NO$_2$, p-CN]:

$$R \overset{}{\diagdown} \diagdown \diagup -N{=}N- \diagdown \diagup \diagdown \diagup -OH$$

$$NaO_3S$$

(XIII)

The dimerization constants for the series were obtained from spectral data and these constants were plotted against the corresponding Hammett σ value. The results are shown in Fig. 10, and are interpreted by Zollinger as indicating that the decrease in aggregation from the highly polar dyes [(XIII), R = CN and NO$_2$] to the unsubstituted member [(XIII), R = H] is due to contributions from van der Waals forces of the polar type. The introduction of a methyl group in the para position in the benzene ring results in increased aggregation for which hydrophobic effects may be responsible.

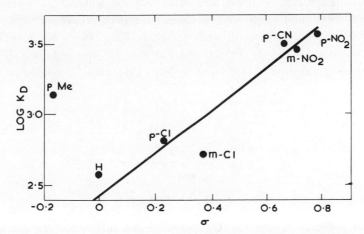

Fig. 10. Relation between dimerization of substituted 4-phenylazo-1-naphthol-6-sulfonic acids [dye (XIII)] and Hammett's σ constants. (Reproduced with permission from Zollinger.[1628])

It must be concluded that in spite of a considerable amount of investigation, the inferences as to the nature of the forces of attraction in the self-association of dye ions still remain to some extent speculative. This might be expected in view of the complex and diverse nature of these forces. The possession of structural features such as planarity and the ability to effect maximum charge separation on overlap with neighboring molecules predisposes certain dyes to aggregation in aqueous solution. The process is reversible with respect to temperature and concentration, and is also reversed by the addition of organic solvents and urea.*

6. THE ROLE OF WATER

The unique properties of water are well known, and in the present context it alone, of all solvents, readily induces the aggregation of similarly charged dye ions when dissolved in it. This has been interpreted, as we have seen in the last section, in terms of the existence of ordered water structures at solid/water interfaces. These so-called "vicinal water" structures have been widely studied, and in a review of the subject, Drost-Hansen[365] concludes from the experimental evidence that the structure of water near a water/solid interface frequently appears different from the bulk structure and that the vicinal water structures may extend considerably into the bulk liquid. Mention has been made (Section 5.2) of the disaggregating properties of urea and the explanation of this phenomenon in terms of structured water. Alcohols and other organic solvents have the same effect[196,965,1207] and relatively few cases have been reported[878,1035,1365] of the presence of aggregates in such solvents. That this is not a dielectric effect has been shown by Arvan and Zaitseva,[54] who obtained different degrees of aggregation in two water–solvent mixtures of the same dielectric

(XIV)

---

* In this connection, a recent detailed thermodynamic study[1253] of the cationic dye Acridine Orange [(V), R = CH$_3$] indicates that both short-range stacking and long-range electrostatic interactions are important in determining the tendency of this dye to aggregate.

constant. In addition, little, if any, aggregation occurs in a solvent such as formamide, the dielectric constant of which is higher than that of water. On the other hand, however, Coates,[263] studying the anionic dye (XIV), found that the molar absorption coefficient ε is the same for a series of seven water–solvent mixtures all of the same dielectric constant. It has been pointed out by Valko[1513] that urea, while having a similar disaggregating effect as alcohols, raises the dielectric constant. An alternative explanation of the action of urea and alcohols on aqueous dye aggregation can be found in postulates by Rath[1224,1225]; he includes these substances in a general group of hydrotropic substances. According to this hypothesis, these compounds form addition products with the solute, through hydrogen bonding, dipole attraction, or dispersion forces. The orientation of the hydrophilic radical of the hydrotropic substance into the water brings about solution. In support of this, Bubser and Eichmanns[196] found that the hydrotropic effect of a homologous series of alcohols, on aqueous solutions of (XII), increased with lengthening of the hydrophobic radical.

The aggregating power of water compared with that of dimethylformamide is clearly shown in Fig. 11, in which the anionic dye Congo Red [(III), R = H] has been examined over a 250- fold concentration range in the two solvents and the results expressed in the form of a composite Beer's law plot.

While, therefore, the disaggregating effect of urea and organic solvents is undisputed, the mechanism by which this occurs and the role played by water are as yet somewhat obscure.*

The participation of the water molecule in the structure of the dye aggregate has been proposed by a number of authors.[130,881,1376] As we have seen, the condition of minimum Coulombic repulsion between two cations in a dimer requires the delocalized charge to be alternately at each end of the planar ions; accordingly, Sheppard[1376] has suggested that the transitional position of maximum repulsion (i.e., the position where the charges on the two ions have minimum separation) could be buffered by the interposition of a water molecule, coordinated between the cations. It had previously been shown[870] that the presence of water is essential for the production of J-band structure and the mesophase of cyanine dyes, even when adsorbed on silver halide. Sheppard extended his concept of the intermediary role of water to the polymeric mesophase, proposing an arrangement whereby two water molecules are situated between the nitrogen atoms of

---

* The hypothesis that 'structured water' is important in the dimerization of dyes has been questioned by Padday.[1132]

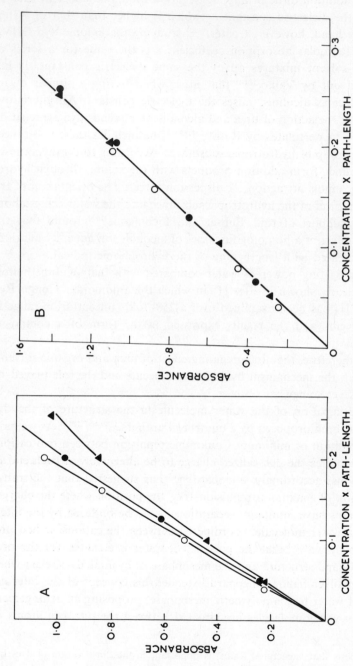

Fig. 11. Plots of absorbance vs. (concentration × path length) for dye (III) (R = H) in (A) water, λ_{max} 530 nm, and (B) dimethylformamide, λ_{max} 500 nm. (○) Concentration range up to 0.05 g liter^{-1}, path length 40 mm. (●) Concentration range 0.05–0.25 g liter^{-1}, path length 1 mm. (△) Concentration range 0.5–2.5 g liter^{-1}, path length 0.1 mm. (Reproduced with permission from Duff and Giles.[368])

each pair of opposite planar cations. Unsulfonated azo dye molecules in monolayers on water appear to hold water between each pair of contiguous azo groups.[518] These ideas are more fully developed elsewhere,[1375,1376] but there remains doubt as to the presence of water in cyanine dye aggregates since an IR study[307] of adsorbed H and J aggregates failed to reveal any evidence of –OH bands.

7. CONCLUSIONS

The self-association of dye ions in aqueous solution is well established; in addition to its academic interest in the field of stacking interactions, the phenomenon is of technological importance in the industries concerned respectively with dyes, photography, and, recently, lasers.[1156] The theme of the large number of publications devoted to the subject has tended to be concentrated in one or other of the three following specialized areas: (a) the (mainly) anionic textile dyes; (b) classical dyes such as Methylene Blue [(I), R = CH$_3$], Acridine Orange [(V), R = CH$_3$], or the Xanthene class, most of which are cationic; (c) the large class of photographic cyanine dyes.

We have tried in this review to deal with dyes from all classes: more detailed analyses of particular systems can be found elsewhere. The observations we have discussed seem sufficient to justify the following general conclusions.

(a) The presence of water is essential for aggregation to occur, and dyes aggregate only slightly, if at all, in other solvents.

(b) Aggregation is enhanced with dyes having planar molecular structures, particularly if this feature is accompanied by a maximum separation of the ionic charges when the dye ions overlap.

(c) The nature of the attractive forces responsible for the stability of the aggregates may vary from one dye system to another and may include contributions from hydrogen bonding and the various forms of van der Waals forces, i.e., dipole forces, dispersion forces, hydrophobic effects, and possibly π-electron interactions. The balance of the evidence, however, is in favor of the importance, in most systems, of the van der Waals forces.

CHAPTER 4

Lipids

Helmut Hauser

Biosciences Division
Unilever Research Laboratory Colworth/Welwyn
The Frythe, Welwyn, Herts, England

1. INTRODUCTION

In a volume dealing with the solution and aggregation of macromolecules in water, lipids must have their place. Lipids (Greek: lipos = fat) consist of a wide and quickly expanding range of amphipathic or amphiphilic (Greek: amphi = dual; pathi = sympathy; phile = liking) and surface-active compounds which contain a polar and a relatively large nonpolar portion. Although the single lipid molecule is small compared with proteins and other macromolecules, lipids, as a result of their amphiphatic character, associate to large aggregates, both in the solid state as well as in the presence of water. The size of lipid aggregates, which ranges from that of small micelles (50–100 Å) to macroscopic structures (>10 μm), depends on the nature of the lipid and the experimental conditions. In this review the accent will be placed on the aggregation of lipids and on the way in which water influences the behavior of these aggregated systems. It is to be hoped that the study of lipid systems and of the effects of water on these systems may also contribute to a better understanding of water itself.

The classification and nomenclature of lipids are still not uniform and far from satisfactory. Some improvement has been achieved recently by the replacement of ambiguous terms such as lecithin, cephalin, etc. used for phospholipid mixtures by systematic ones such as phosphatidylcholine and phosphatidylethanolamine, etc. (cf. Ref. 346 and 1587). This progress is the result of new preparative and analytical methods which permit the preparation, separation, and characterization of pure lipids.

The following compounds can be summarized under the term lipid[346]: (I) simple lipids comprising uncharged compounds, such as (a) neutral fats (glycerol esters or glycerides of fatty acids), (b) waxes (fatty acid esters of monohydric long-chain alcohols); (II) compound or conjugate lipids comprising charged lipids, such as (a) phospholipids or phosphatides (lipids with a phosphate group), (b) cerebrosides and gangliosides (lipids with a carbohydrate group), and (c) sulfatides (lipids with a sulfate group); and (III) derived lipids, comprising compounds derived from I and II, e.g., fatty acids, long-chain alcohols, sterols, hydrocarbons like carotenoids, Vitamins A, D, E, and K, sphingosine, etc.

On the basis of the interaction with water (i.e., the swelling behavior) three kinds of lipids may be distinguished according to Lawrence[858,859] (see Fig. 1). The compounds in the top row (Fig. 1) show only very limited swelling (less than 0.03 g H_2O/g lipid for hydrocarbon chains with 12 C atoms, to 0.01 g H_2O/g lipid for those with 18 C atoms).[859] The compounds of the second row swell significantly, up to some 0.8 g H_2O/g lipid and this behavior is further enhanced in the case of charged lipids[61] (bottom row). Soaps and detergents are discussed in Chapter 2 of this volume; other lipids showing significant swelling with water are the monoglycerides and the compound lipids (see Fig. 1). The emphasis of the discussion will therefore be on these classes of lipids, with particular reference to phospholipids.

Vauquelin's[1520] discovery of the phospholipids dates to 1811. The chemical nature of simple neutral lipids was established by Chevreul[248] as early as 1823 and despite the advantage of an early start which lipids had compared with proteins or carbohydrates, there was not much advance in lipid chemistry from about 1905 to the late 1930's. The reason for this was probably the lack of adequate techniques for separation, purification, and chemical analysis. It required the stimulating discovery of Gorter and Grendel,[551] Danielli,[312] and Danielli and Davson[313] that phospholipid bilayers are fundamental structural elements of biological membranes and the pioneering work of Schmitt and co-workers[101,1134,1328] to arouse new interest in the biochemistry and biophysics of phospholipids. In more recent years lipid chemistry, particularly that of phospholipids and glycolipids, has become one of the most active research areas. The reason for this is the development of new methods of separation and purification and the availability of new, powerful physical techniques. Some of the more recent advances in the field are clearly the result of a collaborative, interdiscipli-

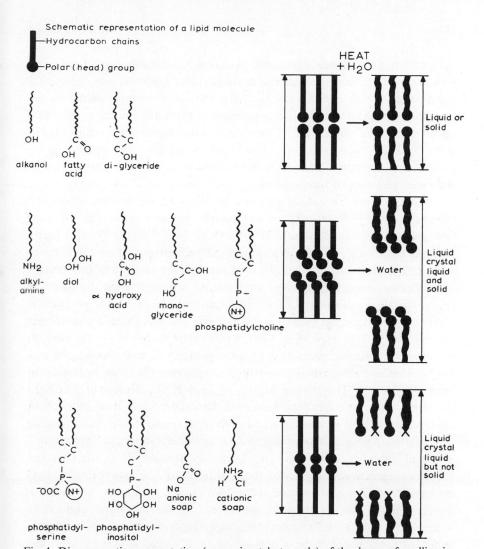

Fig. 1. Diagrammatic representation (approximately to scale) of the degree of swelling in water of lipids. The swelling behavior in the first row is characteristic of all monofunctional neutral lipids of class I and class III, i.e., lipids containing only one type of polar group, with the exception of the monofunctional alkylamines. This group of compounds incorporates significantly more water between the polar groups, as is characteristic for compounds in the second row. These comprise bifunctional lipids of class I and III and neutral (isoelectric) lipids such as phosphatidylcholine of class II (compound lipids). Charged lipids such as phosphatidylserine or long-chain cationic or anionic soaps can incorporate even larger quantities of water between their polar groups (see third row), whereby the actual thickness of the water layer depends on the charge density.[61] Schematic representations of various lipid structures are given on the left. The penetration of water occurs only above a minimum temperature which is the transition temperature from the crystal to the liquid crystalline state. The wavy lines on the right represent the thermal motion of the hydrocarbon chains above that temperature. (Adapted from Ref. 858.)

nary effort. Another reason is that lipids, particularly phospholipids and glycolipids, have been recognized as important structural and functional constituents of cell membranes, of many enzyme systems attached to cell membranes, and various other lipoproteins of plant and animal origin. Last but not least, the recognition of lipids as an important component of food-stuffs, and the wide range of technological applications of lipids, e.g., as emulsifiers, lubricants, cosmetics, medicinals, etc., has also contributed to the rapid advances in recent years.

Many reviews have dealt very adequately with the nomenclature and the chemistry[40,327,599,1277,1516] as well as the physical properties of lipids and particularly phospholipids[79,238,239,244,340,931,1161,1389,1430,1587] and there-fore the discussion of the physicochemical properties will be confined to more recent discoveries and those important in the context of this review. The information on lipid–water interactions which will be discussed in greater detail is still rather limited and scattered. As far as the reviewer is aware of, no concerted effort has been made so far to tackle this subject systematically. Two aspects of lipid–water interaction will be considered; first, the macroscopic interaction of water with lipids, and second, the mo-lecular aspect of this interaction, which will be referred to as hydration or binding of water. The former aspect of lipid–H_2O interactions has been described under the heading lyotropic mesomorphism in a number of comprehensive reviews and the discussion of lyotropic phases will therefore be restricted to points contributing to a better understanding of "hydration." The expression hydration has often been used too loosely in the past and requires an exact definition not only in terms of the quantities of H_2O associated with lipids but also in terms of motional parameters characteriz-ing the strength of association. Vague terms such as free, bound, and icelike water, etc. ought to be supplemented by motional parameters whenever possible. Lipid–water interactions result in dynamic structures and one principal aim of this chapter is to describe dynamic aspects in lipid systems and changes brought about by the interaction of lipids with water and aqueous solvents containing ions.

The material of this review is presented in three parts: first, there is a description of some of the fundamental physical properties of lipids which will help with an understanding of the subsequent discussion of the inter-action of lipids with water. The second part is devoted to the phase behavior of lipid–water systems (lyotropic mesomorphism), while the third part describes the interaction of lipids with water ("hydration").

2. PHYSICAL PROPERTIES OF LIPIDS

2.1. Amphipathic Character

The behavior of lipids both in the solid state (crystal) and in various lipid–H_2O phases is a consequence of the peculiar structure of this class of compounds. Lipids are amphipathic, elongated molecules, much smaller in weight and size than polymers (macromolecules) but sufficiently large to have two distinct regions of greatly differing polarity. One is the polar (hydrophilic) region at one end of the molecule, which favors the interaction with water, and the other is the apolar (hydrophobic) region consisting of hydrocarbon chains (Fig. 1). The amphipathic nature of the molecule is responsible for the molecular association phenomena characteristic of lipids in both crystals and lipid–H_2O systems. The intermolecular forces of cohesion differ according to the nature of the two regions. Van der Waals bonds are operative between the hydrocarbon chains of neighboring molecules ($\sim$1.8 kcal mol^{-1} of CH_3 and $\sim$1 kcal mol^{-1} of CH_2 [238]), while in the polar region hydrogen bonding and dipole–dipole interactions dominate, with interaction energies of the order of 1–10 kcal mol^{-1}. In the case of lipids with charged groups the principal interaction will be electrostatic, involving energies of about 100–200 kcal mol^{-1}. On the basis of energetic considerations it is expected that like regions interact with each other, i.e., that polar groups associate with each other and likewise apolar regions of the molecule cluster together when lipid molecules aggregate. This is indeed the principle governing the association of lipids in crystals as well as in aqueous systems. In addition to the interaction energies mentioned above, the association of the hydrocarbon chains is stabilized by a positive $T \Delta S$ term (ΔS is the entropy change) usually referred to as hydrophobic interaction.[477,759]

2.2. Surface Properties

A consequence of the amphiphilic character or the regional separation of hydrophilic and hydrophobic groups is the surface activity of lipids. When lipids are spread at interfaces, monomolecular layers of well-aligned lipid molecules are formed. Lipids with the correct hydrophobic/hydrophilic balance orient at the air/water or oil/water interface so that the hydrophilic group is anchored at the interface and in contact with water and the hydrocarbon chains project out of the water and associate with each other (see also Chapter 2). This alignment of lipids according to the polarity of different parts of the molecule leads to a significant decrease (up to about 40–50 mN m^{-1}) of the surface tension of pure water. Monolayer studies[340, 505,719,1023,1161,1369] are interesting from a historical point of view because they were the first to provide information about the dimensions of lipid molecules and about the molecular association at interfaces. Much has been learned about the nature of the molecular association of lipid molecules from monolayer studies, particularly in the area of lipid–lipid, lipid–ion, and, more recently, lipid–protein interaction. Recently, evidence has accumulated showing a good correlation between lipid monolayers and the bilayers present in the three-dimensional solids or liquids.[1161] As will be discussed later, the crystal lattice of lipids consists of pairs of monomolecular layers sandwiched together (= lipid bilayer) in which the principle of the molecular association between the molecules in any given layer is the same as in a monolayer. X ray-diffraction has shown that under certain experimental conditions water penetrates between the polar groups and the resemblance between individual lattice layers and monolayers is then even greater. This correlation between monolayers and bilayers, tacitly assumed by many workers in the past, may justify the use of monolayers as models for bilayers or membranes, provided certain limitations are borne in mind. The limitations of monolayers in their role as half bilayers have been discussed before.[1157] The reader interested in monolayer techniques and their application is referred to the book by Gaines.[505]

Recently Smith and Tanford[1408] measured the critical micelle concentration (CMC) of dipalmitoyl phosphatidylcholine at 20°C in water and obtained the extremely low value of $4.6 \pm 0.5 \times 10^{-10}$ M. From this value a unitary free energy of -15 kcal mol^{-1} for the transfer of one phospholipid molecule from water to the micelle was derived. This demonstrates the overwhelming preference of the lipid molecule for the micellar state. Hence the equilibrium between single molecules and the micellar state has been treated merely as a concept and terms such as phase or aggregate have been preferred to micelle, implying that the concentration of single molecules in solution is negligible.[621] An indication of the low concentration of monomers is the observation that aqueous phospholipid dispersions do not foam. At equilibrium the chemical potentials in both the bulk and the surface phase is the same. Using Tanford's value of the CMC and assuming a finite thickness of 20–30 Å for the interface, it can be calculated that the concentration of dipalmitoyl phosphatidylcholine in equilibrium with bilayers (aggregates) is not likely to have any significant effect on the surface tension of H_2O. The well-known fact that lipids with hydrocarbon chains exceeding 16 C atoms usually form insoluble monolayers is not at variance with the concept of a finite, however small, value of the CMC.

2.3. Physical Properties of the Solid State

2.3.1. *Polymorphism*

It is important to realize that long-chain amphypathic molecules arrange themselves in the solid state (crystals) in a fashion which is not drastically different from the arrangement of these molecules in the presence of water. The principle guiding this arrangement is the maximum interaction energy (minimum free energy) between the various portions of the amphipathic molecule. This requirement results generally in a packing in which the hydrocarbon chains form layers or zones bordered by layers of polar head groups. This packing allows for maximum van der Waals and hydrophobic interaction energy in the hydrocarbon chains and maximum dipole–dipole and electrostatic interaction energy in the polar head group region. The lamellar arrangement of hydrocarbon chain layers sandwiched between hydrophilic layers is shown in Fig. 2 and is characteristic of all long-chain lipids regardless of the nature of the polar head group. Also characteristic for compounds containing long hydrocarbon chains is their occurrence in more than one crystalline modification. While polymorphic modifications differ from one another in one or more physical properties, they give identical liquids and vapors. This phenomenon referred to as polymorphism is particularly prevalent with lipids. It was first and most widely studied with fatty acids and soaps and extensive reviews now exist on the polymorphism of these compounds,[239,931] phospholipids,[238,244,931,1161,1587] and glycerides.[237,929,969] Polymorphic forms may differ in one or more properties characterizing the packing of the individual lipid molecule in the crystal. First the hydrocarbon chains, usually in the extended trans configuration, when saturated may be either vertical or tilted with respect to the plane given by the polar head group region. Several different polymorphic forms will result, depending on the degree of tilting. Second, another source of polymorphism is the difference in packing density of the hydrocarbon chains, and third, lipids containing complex polar head groups may exhibit polymorphism due to differences in the conformation of these polar head groups. Which of the polymorphic forms, the lattice free energy of which may differ only marginally, is present will depend on a number of factors. First, intrinsic properties such as the nature and composition (length, saturation, branching) of the hydrocarbon chains, the

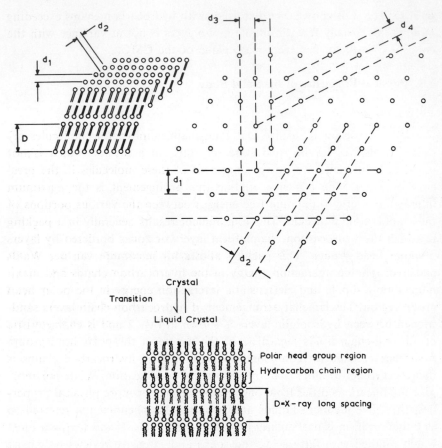

Fig. 2. Schematic representation of the crystal lattice of lipids. The X-ray long spacing D corresponds to the thickness of two molecular layers; the short spacings $d_1, d_2, \ldots, d_n$ represent a series of periodic planes (dashed lines) containing the long-chain axes of the lipid molecules. The drawing on the right is a top view of the crystal lattice. A schematic representation of a lipid bilayer in the lamellar liquid crystalline state (at temperatures above T_c, the crystal to liquid crystal transition temperature) is given below (Ref. 340).

nature and composition of the polar head group, and the interrelation between the hydrophobic and hydrophilic regions of the molecule. The latter, never described in quantitative terms, is usually discussed qualitatively as hydrophobic/hydrophilic balance. Second, extrinsic factors such as experimental conditions, the solvent and method of crystallization, and the presence or absence of impurities seem to have remarkable effects on polymorphic forms. For the nomenclature of the polymorphic forms and of the crystallographic subcells describing the symmetry relations of various hydrocarbon chain packing modes the reader is referred to the article by Williams and Chapman.[1587]

2.3.2. Crystal Structure

The complete crystal structure and the arrangement of the molecules in the unit cell has been reported for lipids and other long-chain molecules such as fatty acids, soaps, glycerides, and paraffins, but no such work has yet been carried out successfully with phospholipids. Due to the fundamental work on fatty acid single crystals by Müller[1066] and Vand *et al.*[1515] the crystal structures of fatty acids are best understood. The lack of phospholipid crystal structures is mainly due to the problem of preparing single crystals suitable for X-ray analysis.

2.3.2a. *Powder Studies of Phospholipids.* Some information concerning the packing of phospholipid molecules can be derived from X-ray patterns, which consist of certain periodic long spacings and short spacings. The interpretation of these powder patterns is greatly aided by crystallographic studies of fatty acids and paraffins and by the comparative studies of members of a homologous series (see Fig. 3). The basic feature derived from the crystal lattice of fatty acids is that both the hydrocarbon chains and the polar head group occur in parallel and equidistant planes. X-ray powder patterns have contributed significantly to our understanding of the polymorphism and mesomorphism (see below) of phospholipids because they provide information on (1) the hydrocarbon chain packing, (2) the angle of tilt of the hydrocarbon chain axis with respect to the bilayer plane, and (3) the orientation of the polar head group from the long spacings observed in a homologous series of phospholipids. For a detailed discussion of X-ray powder pattern of phospholipids the reader is referred to the review articles by Williams and Chapman[1587] and Luzzati.[931]

A better understanding of the interaction of water with the polar head group requires knowledge of the conformation of the polar group. The following brief discussion of X-ray powder patterns is addressed to this question. Figures 3a and 3b show the variation of X-ray long spacings D with number n of carbon atoms per hydrocarbon chain of 1,2-diacyl-L-phosphatidylcholine and 1,2-diacyl-DL-phosphatidylethanolamine. The main points of Fig. 3 are as follows.

· 1. The linear increase of the long spacing per CH_2 group of the hydrocarbon chain. This applies to all crystal forms and liquid crystalline phases, both anhydrous and at maximum hydration and is due to the general type of multilayered structure. The slopes of the straight lines (equal to the in-

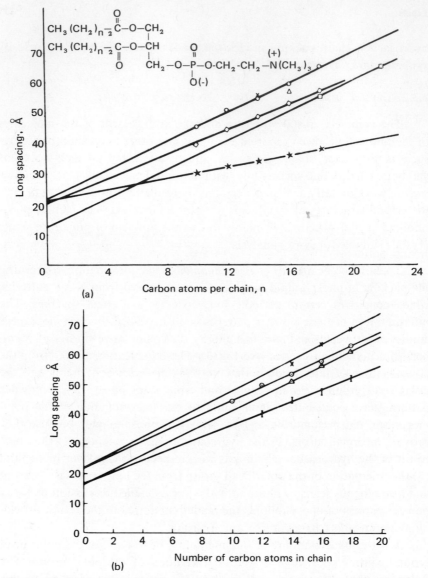

Fig. 3 (a) Variation of X-ray long spacings D with the number of carbon atoms per hydrocarbon chain for saturated 1,2-diacyl-L-phosphatidylcholines. (○) α_1 monohydrate; (△) α_2 monohydrate; (×) β' monohydrate; (◇) lamellar liquid crystalline at temperatures just above T_c; (★) liquid crystalline at 175°; (□) crystalline, anhydrous β form (from Ref. 1587). (b) Variation of X-ray long spacings D with number of carbon atoms per hydrocarbon chain for saturated 1,2-diacyl-phosphatidylcholines and phosphatidylethanolamines. (×) phosphatidylcholines at maximum hydration at 25°; (⊙) phosphatidylcholines, α_1 monohydrates at 23° (a least squares fit to data from Ref. 245); (△) anhydrous β-phosphatidylethanolamines at room temperature; (I) least squares fit to long spacings measured 3° above the liquid crystal transition temperature T_c of phosphatidylethanolamines at maximum hydration. This last curve is based on only the first-order diffraction of the lamellar repeat, so the accuracy is probably no better than ±1 Å. (from Ref. 1163).

crease in the thickness of the double layer per CH_2 group) vary with the crystal form or liquid crystalline phase depending on the angle of tilt of the hydrocarbon chain axes with the plane of the bilayer. Similar straight line relationships were obtained with crystals and liquid crystals of 1,2-diacyl-L-phosphatidyl-L-serines, 1,2- and 1,3-diacyl-DL-glycero-phosphatidic acid, fatty acids, diglycerides, and triglycerides. Straight line relationships mean that the tilt of the chains is independent of chain length when $10 \leq n \leq 20$.

2. The straight lines usually do not extrapolate to the origin but cut the "long spacing" ordinate at a positive value. These intercepts are summarized in Table I and are in general independent of the slope and hence the tilt of the hydrocarbon chain axis. To draw any conclusion on the dimension and conformation of the polar head group, a few assumptions have to be made. First, that the conformation of the polar head group is independent of chain length; second, that any disorder toward the terminal methyl group of the hydrocarbon chain is also chain length independent and thus will not affect the extrapolation procedure; and third, that the terminal groups of the chain pack like CH_2 groups. Different crystal forms and liquid crystals at different temperatures and degrees of hydration give the same intercepts at $n = 0$ indicating that phase transitions or addition of water need not alter the conformation of the polar head group. Changes in the X-ray powder pattern corresponding to phase transitions are therefore accounted for by changes in the hydrocarbon chain region with only minor changes in the polar head group region. This fact that different crystalline and liquid crystalline forms of phosphatidylcholines extrapolate to the same value at $n \to 0$ (Table I) is good support for the validity of the above assumptions.

All polymorphic forms of phosphatidylcholine, with quite different long spacings and slopes (Fig. 3), extrapolate to 20–22 Å. The same extrapolation procedure applied to the phosphatidylethanolamine data gives 15–16 Å. An alternative way of evaluating the plots in Fig. 3 is to extrapolate the straight lines to zero long spacing. The values for the intercepts on the abscissa are included in Table I. Except for the negative sign, these values may be viewed as the dimension of the head group expressed in carbon atoms perpendicular to the bilayer plane. The conformation of the head group of phosphatidylcholine is thus almost fully extended, corresponding to an equivalent of ten C atoms and an X-ray long spacing of about 11 Å per head group (= 22 Å per bilayer). Figure 3 also shows that this long spacing is unchanged by the addition of about ten water molecules,[1163] indicating that the H_2O molecules are fitted into the lattice of

TABLE I

Lipid	Polymorph[a]	Long spacing at zero number of CH_2 groups,[b] Å	Number of C atoms when the long spacing was extrapolated to zero[b]
Fatty acid	—	0.5–1.5	0–0.5
Diglycerides and triglycerides	—	4–8.5	2–3.5
1,2-Diacyl glycerophosphatidic acid	α form	12.6	7
	β form	7.8	6
1,3-Diacyl glycerophosphatidic acid	α form	8.9	6
	β form	10	7
1,2-Diacyl phosphatidylethanolamine	β form	16	6
	β' form	15	6
	Liquid crystalline 116 < T < 135°	16	—
	Liquid crystalline >135°	15	8.3
	Liquid crystalline at maximum hydration	16	
1,3-Diacyl phosphatidylethanolamine	Anhydrous	13	5
	β₂ form	14.8	8
1,2-Diacyl-L-phosphatidylcholines	β anhydrous	12	5
	α₁ monohydrate	22	10
	β' monohydrate	22	10
1,2-Diacyl phosphatidylcholines	Liquid crystalline (lamellar) just above T_c	20	10
	Liquid crystalline at 175°	22	
	At 25° and maximum hydration	22	9
1,2-Diacyl-L-phosphatidyl-L-serine	α₁ form	15	8
	α₂ form	10	4
	Liquid crystalline phase (hexagonal)	19	(20)

[a] The nomenclature is based on Larsson's classification of glyceride crystals (cf. Ref. 1587).
[b] This is a measure of the dimension of the polar head group in the direction perpendicular to the plane of the bilayer.

the polar head group. Measurements with molecular models show that the contribution of a fully extended glycerylphosphorylcholine group parallel to the hydrocarbon chains and with the OC–CN bond in the gauche conformation is 12 Å. Hence the head group must be slightly tilted with respect to the plane of the bilayer.[1163] Application of the extrapolation method to phosphatidylcholine above the liquid crystal transition temperature T_c at maximum hydration gives a long spacing consisting of two contributions, the long spacing due to the lipid polar group and that due to the interlamellar H_2O. Since the separation of these two contributions is not possible, no conclusions with respect to the conformation of the phosphorylcholine group above T_c and at maximum hydration can be drawn. In this context it is worth noting (see Fig. 3b) that the long spacing of dimyristoylphosphatidylcholine is 1.5 Å ($\sim$0.8 Å per polar head group), less than expected from a linear interpolation using the linear relationship of Fig. 3b. The experimental temperature at which dimyristoyl-lecithin was studied was close to the liquid crystalline transition temperature T_c, and consequently the polar head group was already undergoing rotational and oscillatory motions characteristic of the liquid crystalline state.[841] This suggests that there is no gross change in the time-average polar group conformation during the gel → liquid crystal transition.[1163] The only exception is the anhydrous crystal of phosphatidylcholine, which gives an extrapolation to a significantly lower value of about 12 Å, equivalent to about five C atoms. This suggests that in the anhydrous state the head group conformation is different and possibly folded back to an almost coplanar orientation with respect to the bilayer plane. This is consistent with nmr studies of Salsbury et al.[1300] and X-ray studies of Levine et al.[877] which show that the phosphatidylcholine head group unfolds into an extended conformation on addition of water.

The long spacings at $n \to 0$ obtained with phosphatidylethanolamines are significantly smaller than those of the phosphatidylcholines and are equivalent to six to eight C atoms (Table I). This and the similarity in dimensions between phosphatidylethanolamine and phosphatidic acid (Table I) suggest that the zwitterion of phosphatidylethanolamine is approximately parallel to the plane of the bilayer or alternatively it may be perpendicular to the plane of the bilayer and interdigitated with opposing groups of the neighbor bilayer.[1163] In both cases the net contribution to the long spacing would be 8 Å per head group or 16 Å per bilayer as derived from three-dimensional molecular models. There is no change in the long spacing $n \to 0$ and possibly in the conformation when phosphatidylethanolamines are fully hydrated at 3° above the liquid crystal transition temperature T_c. This shows that the presence of water does not affect this residual long spacing, indicating that this water is packed within the space of the polar head group. Phillips *et al*. suggested that it is oriented near the glycerol backbone.[1163]

Greater variations in the long spacing of the polar head groups are observed with phosphatidic acids and phosphatidylserines. With phosphatidylserines not only the long spacing and the slope (cf. Ref. 1587) but also the extrapolated long spacing ($n \to 0$) depends on the polymorphic form studied (Table I), the liquid crystalline hexagonal phase giving the largest long spacing (at $n \to 0$). The data of Table I suggest that the conformation of the glycerylphosphorylserine group in phosphatidylserine crystals is similar to that of phosphatidylethanolamine.

For comparison, the values of the long spacings for crystals of fatty acids, diglycerides, and triglycerides have been incorporated in Table I (cf. Ref. 340). The three crystalline forms of fatty acids, though differing in the tilt of the hydrocarbon chain axis and thus in the slope, extrapolate to similar values, which are close to zero and equivalent to about one C atom. Diglycerides and triglycerides give larger variations and the addition of the glycerol group to the hydrocarbon chain causes an extension of the molecule equivalent to 2–3.5 C atom (Table I).

2.3.2b. *X-Ray Crystallography of Lipid Constituents*. An alternative approach to the question of the polar head group conformation has been taken by Sundaralingam,[1447] who has recently reviewed the crystal structures of the following lipid constituents: glycerylphosphorylcholine,[1] the cadmium chloride complex of glycerylphosphorylcholine,[1448] ethanolaminephosphate,[814] glycerylphosphorylethanolamine,[345] the L isomer of serinephosphate,[1449] and the DL-racemic mixture of serinephosphate.[1204] From this and available crystal structures of triglycerides Sundaralingam[1447] proposed a number of possible phospholipid conformations. Despite certain restrictions imposed by structural requirements, the number of possible conformations is still very large. Which of these conformations is preferred in the absence and presence of water only future experimental work can show. Table II summarizes the preferred conformation and torsion angles of various groups derived from this approach. The most striking feature is the preference of the torsion angle α_5 of the $\overset{(+)}{N}$—C—C—O group in the $\pm$ gauche conformation (Fig. 4).* In all compounds listed by Sundaralingam[1447] the torsion angle α_5 is $\pm 60° \pm 15°$. In this conformation the posi-

---

* The notation for torsion angles is given in Fig. 4.

TABLE II. Preferred Conformation and Torsion Angles in Phospholipid Constituents[a]

Group	Torsion angle[b]	Definition[b]	Preferred conformation (range),[c] deg
Phosphodiester	α_1	C2–C1–O11–P	trans (171 $\pm$ 6; −174 $\pm$ 3)
	α_2	C1–O11–P–O12	gauche (65 $\pm$ 3; −75 $\pm$ 7)
	α_3	O11–P–O12–C11	gauche (68 $\pm$ 4; −65 $\pm$ 6)
	α_4	P–O12–C11–C12	trans (160 $\pm$ 20; −160 $\pm$ 20)
Choline	α_5	O12–C11–C12–N11	gauche (60 $\pm$ 15; −60 $\pm$ 15)
Glycerol	θ_1	O11–C1–C2–C3	trans
	θ_2	O11–C1–C2–O21	gauche
	θ_3	C1–C2–C3–O31	gauche
	θ_4	O21–C2–C3–O31	trans

[a] Collected from X-ray crystallography of phospholipid constituents: glycerylphosphorylcholine-CdCl$_2$·3H$_2$O,[1448] glycerylphosphorylcholine-1 and -2,[1] glycerylphosphorylethanolamine,[814] 2-aminoethanolphosphate, DL-serinephosphate, DL-serine, glycerol, di-Na-glycerophosphate·5H$_2$O, diglycerides and triglycerides, triacetylsphingosines, fatty acids, and hydrocarbons.

[b] The numbering of the atoms and the notations for the torsion angles in phosphatidylcholine are given in Fig. 4.

[c] For definition of torsion angles see Fig. 4c; positive angles are defined as 0 to 180° (clockwise rotation), negative angles are from 0 to −180° (counterclockwise rotation).

tively charged ammonium group and the anionic oxygen of the phosphate group are close to each other and on the same side with respect to the rest of the molecule. The intramolecular $\overset{(+)}{N}$–$O^{(-)}$ distances are close to the van der Waals distances and electrostatic interaction is probably the major contribution to the stabilization of the gauche conformation.[1445] The phosphodiester group (comprising the torsion angles α_2 and α_3) is always found in the gauche–gauche conformation, i.e., both torsion angles α_2 and α_3 are either in the positive or negative gauche conformation, similar to that in polynucleotides.[1446] It is interesting to note that in the few cases investigated so far the crystal structure seems to be little if any affected by the formation of metal complexes. The gauche conformation of the $\overset{(+)}{N}$—C—C—O group and the preferred gauche–gauche conformation of the phosphodiester group are unchanged when the CdCl$_2$ complex of glycerylphosphorylcholine is formed. In the crystal structure of the choline phosphate-CaCl$_2$ complex* the gauche conformation of the choline group is also unaffected. Furthermore, the finding of Sundaralingam and Jenson[1448] that in the glyceryl-

* J. McAlister, D. Fries, and M. Sundaralingam, private communication.

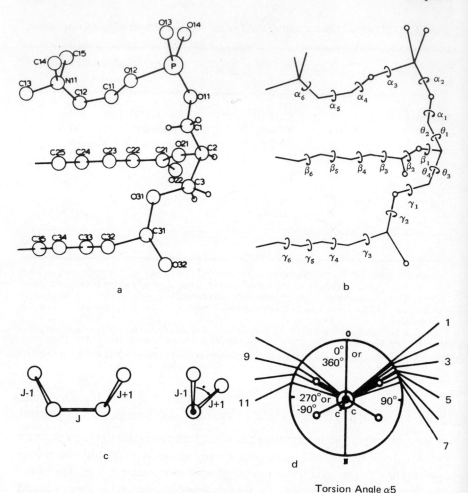

Torsion Angle α5

Fig. 4. (a) The numbering of the atoms and (b) the notation of the torsion angles in phosphatidylcholine. The torsion angles are positive for a right-handed rotation, i.e., the far bond $j + 1$ (c) rotates clockwise relative to the near bond $j - 1$ with the j bond being the axis of the rotation. Positive angles from 0 to 180° define a clockwise rotation negative angles from 0 to 180° define a counterclockwise rotation. In this notation the cis planar configuration of the $j + 1$ and $j - 1$ bonds is defined by a 0° torsion angle, while the trans planar configuration is defined by a torsion angle of 180°. For instance the torsion angle α_5 is defined as O12–C11–C12–N11 (Table II), where O12–C11 is the near bond, C11–C12 the bond of rotation, and C12–N11 the far bond. (d) The values of α_5 found in crystal structures of the following compounds: (1) glycerylphosphorylcholine-$CdCl_2$, 50°; (2) 2-aminoethanolphosphate, 52°; (3) glycerylphosphorylethanolamine, 55°; (4) DL-serine, 60°; (5) L-serinephosphate, 61°; (6) DL-serinephosphate, 66°; (7) glycerylphosphorylcholine-1, 72°; (8) 2-aminoethanolphosphate, −52°; (9) DL-serine, −60°; (10) DL-serinephosphate, −66°; (11) glycerylphosphorylcholine-2, −75°. (From Ref. 1447.)

phosphorylcholine-$CdCl_2$ complex the Cd^{2+} bridges two neighboring phosphate groups may be relevant to phospholipid–ion interactions observed in bilayers and biological membranes. In this context it is interesting to note that such a bridging may also be mediated by a single water molecule, e.g., in the glycerylphosphorylcholine-$CdCl_3 \cdot 3H_2O$ complex, one H_2O molecule was found[1448] to link two different phosphate groups in adjacent unit cells by hydrogen bonding. Water molecules hydrogen-bonded to two adjacent phospholipid molecules could contribute significantly to the stabilization of the bilayer, thereby "forming an infinite phosphate–water hydrogen bonded ribbon."[1447]

In the crystal structures of ethanolaminephosphate[814] and serine phosphate[1204,1449] hydrogen bonding has been observed between the phosphate group and the ammonium group. If this type of hydrogen bonding is maintained in the corresponding phospholipid, the $\overset{(-)}{P}$–$\overset{(+)}{N}$ zwitterion would be coplanar with the bilayer, as suggested by Phillips et al.[1163]

A survey of available X-ray crystal structures of glycerylphosphorylcholine shows that all have different conformations.[1447] This was ascribed to differences in the hydrogen bonding and crystal packing forces. The presence of the two hydrocarbon chains in phospholipids, however, imposes certain restrictions on the head group conformation. For optimum intramolecular interaction and packing the chains must be aligned parallel. It can be readily seen from three-dimensional models that in order to achieve this the torsion angles θ_2 (O11–C1–C2–O21) and θ_4 (O21–C2–C3–O31) of the β and γ oxygen atoms, respectively, must be in the $\pm$ gauche range. The trans conformation of θ_4 as found in glycerylphosphorylcholine and glycerylphosphorylethanolamine does not fulfil this requirement. In addition, maximum interaction also requires one of the acyl chains to deviate from the trans conformation about the first two C–C bonds adjacent to the carbonyl group. How relevant these results obtained by X-ray crystallography are to the question of the conformation of phospholipids in crystals or liquid crystals, in bilayers at maximum hydration, and in bilayers present in biological membranes remains to be seen. It is likely that the $\overset{(+)}{N}$—C—C—O group in phospholipids is also in the gauche conformation. Dufourcq and Lussan[369] investigated by proton magnetic resonance the conformation of phosphatidylcholine and of various compounds constituting the polar head group of phospholipids. They found that the $\overset{(+)}{N}$C–CO group in L-dimyristoyl, D,L-dipalmitoyl, and egg phosphatidylcholine dissolved in $C^2H_3O^2H$ is in $\pm$ gauche conformation. The same is true for the $\overset{(+)}{N}$C–CO group in choline phosphate and glycerylphosphorylcholine dissolved in ^2H_2O, while with ethanolaminephosphate in ^2H_2O no preferred conformation was detected, indicating that free rotation occurs about the C–C bond of the ethanolamine group. Proton magnetic resonance studies showed that the $\overset{(+)}{N}C$–CO group of acetylcholine chloride in aqueous solution is also in the gauche conformation,[304] consistent with the conformation of this compound in the solid state.[226]

2.3.3. Thermotropic Mesomorphism

Generally the physical states of a substance are adequately described by the following equilibria:

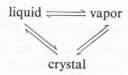

$$\text{liquid} \rightleftharpoons \text{vapor}$$

$$\text{crystal}$$

On heating, lipids do not pass directly from the crystalline form to an isotropic liquid. A number of intermediate states between these two extremes can exist; these have variously been referred to as mesomorphic (Greek: mesos = middle; morphic = form), liquid crystalline, or anisotropic liquids. Mesomorphism requires an additional fourth state:

$$\text{crystal} \overset{T_c}{\rightleftharpoons} \text{mesomorphic states (liquid crystal)} \overset{T_m}{\rightleftharpoons} \text{liquid} \rightleftharpoons \text{vapor}$$

where T_c is the transition temperature (Krafft point, see Section 3.1). The physical properties in these mesomorphic phases approach the properties of two extremes. For instance, certain properties of liquid crystals resemble those of crystals, e.g., the anisotropy which manifests itself in the positive birefringence, the rotation of the plane of polarized light, and in X-ray diffraction, while others resemble those of liquids, e.g., in certain directions liquid crystals show flow properties typical of liquids.

The characterization of lipid mesomorphic phases, of which undoubtedly those of soaps are best investigated, has been reviewed by Dervichian[340] and more recently by Luzzati,[931] who summarized all known mesomorphic phases of soaps and phospholipids. The structures present in phases may show periodicity in one, two, or three dimensions and will be discussed in Section 3.2. When lipid crystals are heated an endothermic transition is observed, the temperature of which is well below the capillary melting point. At this transition temperature T_c (Krafft point) a change of state from the crystalline to the liquid crystalline or mesomorphic state takes place.

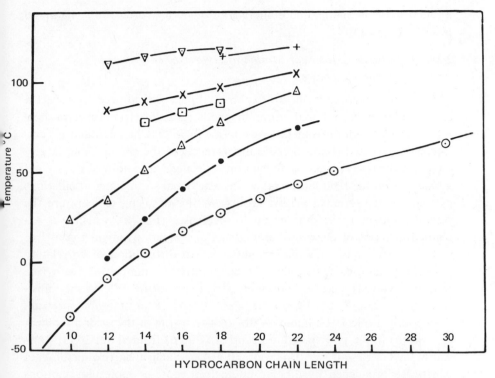

Fig. 5. Chain-melting transition temperatures for various phospholipids. The melting points of the *n*-alkanes (○) with the same hydrocarbon chain length are included for comparison. 1,2-diacyl-DL-phosphatidylethanolamines: (▽) anhydrous; (⊡) in excess water. 1,2-diacyl-L-phosphatidylcholines: (+) anhydrous (β form); (×) anhydrous (α ? form) obtained from β form on heating and from monohydrate on drying; (△) mono-hydrate (α_1 form); (●) in excess water (from Ref. 841).

The heat absorbed causes melting of the crystalline hydrocarbon chains. Figure 5 shows the transition temperature for various phospholipids as a function of hydrocarbon chain length. T_c increases with increasing chain length and parallels, more or less, the capillary melting point of the corresponding fatty acids. Thus T_c is highest with fully saturated hydrocarbon chains in all-trans configuration. Introduction of double bonds or branched fatty acids lowers T_c, cis double bonds being more effective than trans double bonds. Not only does T_c vary within a lipid class with chain length (Fig. 5), but T_c for phospholipids with different head groups and the same hydrocarbon chains also varies significantly. For example, T_c of 1,2-dimyristoyl-DL- phosphatidylethanolamine is significantly higher than that of 1,2-dimyristoyl-L-phosphatidylcholine. This clearly indicates that the T_c is a function of the nature and composition of both hydrocarbon chain and polar head group.

2.3.4. *Packing and Molecular Motion of Hydrocarbon Chains in Mesomorphic Phases*

At liquid nitrogen temperature phospholipids give infrared (IR) spectra characteristic of crystals and proton magnetic resonance (pmr) spectra show broad lines of width of 14.6 G (gauss).[240,242,243,1299] Chapman and his colleagues interpreted these observations in terms of the chains being largely frozen in the trans planar conformation. The slight deviation of the second moment from the rigid lattice value was attributed to torsional oscillations of some of the terminal methyl groups. With increasing temperature the second moment (linewidth) gradually decreases (Fig. 6), which was explained in terms of increasing oscillations and rotations about the hydrocarbon C–C bonds. This motion has a broad distribution of correlation frequencies and its amplitude and cooperativity (number of methylene protons involved) increases with increasing temperature.[1299] At the transition temperature T_c the IR spectra change from a crystal-type spectrum to one more typical of a liquid. A drastic narrowing of the second moment (linewidth) in the pmr spectrum occurred (Fig. 6) from about 10 G^2 (linewidth 6.7 G) to $\sim$0.3 G^2 (linewidth 0.11 G). Such a narrowing of lines which are broadened by dipolar interaction requires molecular motions with correlation frequencies comparable to, or greater than, the linewidth in frequency units ($\simeq$60 kHz). This indicates that above T_c the methylene groups of the hydrocarbon chains undergo rapid oscillatory, rotational, and translational motion. This motion is still not isotropic because the second moment of 0.11 G^2 is significantly greater than that expected for isotropic liquids: 10^{-4}–10^{-5} G^2.

At higher temperatures the intensity and the value of the X-ray long spacings become progressively smaller and at the transition temperature long spacings invariably decrease significantly with a concomitant change in the characteristic short spacings. The discrete short spacings degenerate to a diffuse band of about 4.6 Å which is almost identical to that of liquid paraffins. A series of distinct short spacings indicates a regular packing with discrete lateral distances between the hydrocarbon chains (Fig. 2) and the degeneration of these short spacings is an indication of a change in the hydrocarbon chain packing giving rise to a liquidlike structure. Consistent with this is the decrease in the long spacing. The "melted" chains with greatly increased, but still anisotropic, molecular motion, are partially coiled and folded back with reduced end-to-end distance. This leads to an increase in cross-sectional area as manifested in the lateral expansion of the crystal lattice. "Thinning" of the phospholipid bilayer occurs simultaneously and a reduction in the long spacing is observed. The ability of mesomorphic phases to dissolve steroids and other lipophilic substances,

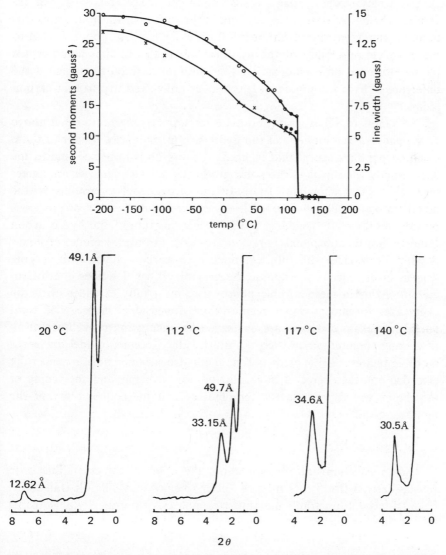

Fig. 6. (a) Second moment ($\times$) and linewidth ($\bigcirc$) of 2,3-dimyristoyl-DL-phosphatidylethanolamine (from Ref. 242). (b) Variation of X-ray long spacings of 2,3-dimyristoyl-DL-phosphatidylethanolamine with the diffraction angle 2θ and temperature (from Ref. 240).

often in large quantities, lends further support to the postulate of a liquid-like state of the hydrocarbon chains. Conversely, crystallization of the hydrocarbon chains causes "clustering," i.e., the lipophilic substance dissolved in the liquid crystalline phase becomes immiscible with the lipid and separates out in clusters. The liquidlike nature of the hydrocarbon chains in mesomorphic phases is also apparent in lipid mixtures. For example, mixtures of soaps and phospholipids with hydrocarbon chains differing in chain length and saturation form a single phase above T_c due to the complete miscibility of the hydrocarbon chains in the mesomorphic phase. However, on crystallization solid solutions are only retained if the difference in hydrocarbon chain length does not exceed two to four carbon atoms.[864,1165,1166]

In summary, all experimental evidence supports the notion that above T_c the packing and motions of the hydrocarbon chains are similar to those of liquid paraffins except that in lipids the motion is restricted due to the anchoring of the chains to the polar group. Up to this point general agreement exists. Unfortunately, details of the structure of n-paraffins in the liquid state are unknown; hence it is not surprising that the picture concerning details of the packing and molecular motion of the hydrocarbon chains is not unambiguous. It ranges from the disordered model of liquid paraffins to models with fully extended hydrocarbon chains with varying degrees of disorder. For example, Segerman[1362] has described the hydrocarbon chains in mesomorphic phases as almost fully extended with the chain axes forming a rather regular three-dimensional lattice. The local lateral interchain distances are envisaged as undergoing rapid fluctuation of varying amplitude. However, the most widely accepted model for mesomorphic phases[1161,1167] predicts that the hydrocarbon chains are not fully extended and the degree of molecular motion, twisting, and shortening of the chains will depend on the temperature and the cooperativity of the motion.

2.3.5. Melting of Lipids

Some structural information about the lipids in the crystalline state can be derived from their melting (point) behavior. While all transitions from the crystal to various mesomorphic phases yield optically anisotropic phases, the melting point* is the temperature at which the optically iso-

---

* Detection of the melting points of phospholipids can be complicated by decomposition; for instance, Ladbrooke and Chapman[841] reported gradual decomposition of phosphatidylcholines when heating in an atmosphere of oxygen-free nitrogen.

tropic melt is formed. With respect to the melting temperatures, lipids can be grouped into two classes. First, uncharged long-chain compounds such as alcohols, fatty acids and their esters, mono-, di-, and triglycerides, and amines have melting points rarely exceeding 70°. Second, charged compounds like the soaps or halides of long-chain amines have high melting points, usually ~300°. These two classes of compounds differ greatly in the forces of cohesion. Besides the relatively weak van der Waals forces which are operative between the hydrocarbon chains and thus common to both classes, the cohesion in the polar head group region is dominated by hydrogen bonding and dipole–dipole interaction in class 1 and electrostatic interaction in class 2. Table III summarizes the melting behavior of phospholipids,[340] giving average melting points for the stearoyl derivatives. Inspection of Table III reveals the following points:

1. All phospholipids except the phosphatidic acids fall into class 2, electrostatic forces being mainly responsible for cohesion.

2. The nature of the polar head group determines the melting point. The stronger the base, the larger the electrostatic interaction energy and the higher the melting point: phosphatidylcholines ≈ sphingomyelins

TABLE III. Melting Points of Phospholipids[a]

Phospholipid	Melting point, °C	Dependence on chain length (CL)
1,2-Distearoyl phosphatidylcholine	230	Independent of CL
1,2-Distearoyl phosphatidylethanolamine	196	Independent of CL
1,3-Distearoyl phosphatidylethanolamine	198	Independent of CL
1,2-Distearoyl phosphatidylserine	160	Decreases with decreasing CL
1,2-Distearoyl phosphatidylinositol	136	—
1,2-Distearoyl phosphatidylglycerol	67	—
1,2-Distearoyl phosphatidic acid	71	Decreases with decreasing CL
1,3-Distearoyl phosphatidic acid	70	Decreases with decreasing CL
Sphingomyelin	210	—
Stearoyl lysophosphatidylcholine	240	—
Stearoyl lysophosphatidylethanolamine	205	—

[a] Adapted from Williams and Chapman[1587]. For a more detailed discussion of the melting points of phospholipids see Refs. 340 and 1587.

> phosphatidylethanolamines > phosphatidylserines. The phosphatidic acids are exceptional. Their melting points are similar to those of the corresponding fatty acids, implying that the interaction energy in the crystal is significantly weaker and mainly due to hydrogen bonding.

3. Contrary to compounds of the first class, phospholipids have melting points independent of the nature and length of the hydrocarbon chain, nor does the number of hydrocarbon chains affect the melting points, as is evident from the melting points of lysophospholipids. This is another indication that the major factor controlling the capillary melting point of phospholipids is the electrostatic interaction between the polar groups. The melting point of sodium stearate is $\sim$300° and still higher than that of phosphatidylcholines. This indicates that there is greater ionic character in the polar head group of soaps than in the zwitterionic head groups of phospholipids.

3. LIPID–WATER PHASE BEHAVIOR

3.1. Lyotropic Mesomorphism

Similar to the phenomenon of thermotropic mesomorphism, whereby on heating, lipids do not pass directly from the crystal to the melt, lipids generally also do not pass directly from the crystalline state to the solution when interacting with water. Depending on the water content, different hydrated phases exist until an ideal solution of single lipid molecules occurs with the approach to infinite dilution. Such behavior is called lyotropic mesomorphism. The lyotropic phases also show thermotropic mesomorphism, i.e., the particular phase observed will depend on both the water content and the temperature. Lyotropic mesomorphism is a consequence of the amphipathic character of lipid, and in addition to the water content and the temperature, the nature of the phase will be determined by the hydrophilic–hydrophobic balance of the lipid molecule.[341,1640]* To illustrate this point, the concept of the hydrophilic–hydrophobic balance is considered qualitatively for three classes of lipids.

(a) Triglycerides, such as triolein, with three long hydrocarbon chains and with three ester groups representing the hydrophilic portion, are water insoluble and water does not penetrate their crystal lattices. Therefore no swelling is observed.

* The prediction of the phase behavior of lipids on the basis of the structural and physicochemical properties of the lipid is not yet possible.

(b) On the other hand, soaps or lysophosphatidylcholine have only one hydrocarbon chain and the hydrophilic portion is either an ionized carboxyl group or a zwitterion attached to the glycerol backbone. These compounds not only swell but generally dissolve to form micellar solutions. Up to a certain water content (about 63% in the case of egg yolk lyso-phosphatidylcholine[1240]) lyotropic/thermotropic phases are obtained. If more water is added, isotropic micellar solutions are formed.

(c) The phospholipids fall between these two extremes. The hydrophobic portion consists of two hydrocarbon chains which are counterbalanced by the polar head group comprising the glycerol backbone, ionized and/or sugar groups. This class of compounds usually swells when in contact with water above a certain threshold temperature T_c. As yet there is no report of isotropic micellar solutions with diacylphospholipid–H_2O systems. Ohki and Aono,[1106] in a theoretical paper, discussed the phospholipid bilayer–micelle transformation. These authors concluded that with diacylphospholipids in aqueous solution of 0.1 M NaCl the bilayer is the most stable state when the net charge per phospholipid molecule is $(0–1.25)e$ and the hexagonal phase is the most stable state in the range of $(1.25–2.0)e$. Spherical micelles are unlikely to be stable unless the charge is $>2e$. Consistent with this prediction, Hauser and Phillips found that with sodium phosphatidylserine, which has one net negative charge, no micelles are formed and the bilayer is the most stable state.[624]

In addition to the parameters discussed so far, the lyotropic phase observed will also depend on the size, shape, and stereochemistry of the lipid molecule. Phosphatidylcholines in the presence of H_2O ($>5\%$) have a preference for lamellar phases. This is not surprising considering that the isoelectric phosphatidylcholine molecule is roughly cylindrical and for steric reasons favors side-by-side stacking. However, phosphatidylcholines with shorter hydrocarbon chains (less than ten C atoms) do not form stable bimolecular lamellae, but break up to form micellar solutions.[619,1167] This becomes clear when the interaction energy of molecules in the lipid bilayer is considered. The heat of sublimation of the crystal can be taken as a measure of the lattice energy. With phospholipids the heat of fusion amounts to roughly one-half of the heat of sublimation and thus of the interaction energy in the crystal lattice (cf. Ref. 1167). Phillips *et al.*[1167] showed that the heat of fusion of 1,2-diacyl-L-phosphatidylcholines is linearly related to the number of C atoms per hydrocarbon chain. Extrapolation indicates that phosphatidylcholines with less than eight C atoms have zero interaction energy in the gel state (cf. Section 3.2). The minimum interaction energy for a stable molecular association in the bilayer is at least $10kT$, which is only realized with 12 or more C atoms[1167] per saturated phosphatidylcholine molecule. Hauser and Barratt[619] confirmed the instability of dicapryl and dilauroyl phosphatidylcholine bilayers and showed that bilayers of these compounds, though readily formed, tend to clump and/or fuse at room temperature in the course of Brownian collisions.

Besides the hydrocarbon chain length, the shape and ionic charge of the lipid molecule determines the phase behavior and influences the stability of bimolecular lamellae. Charged lipids or lipids whose shape differs significantly from that of a cylinder have a tendency to form curved lipid–water interfaces. This was found to be true for phosphatidylserine,[61,624] which has a net negative charge.[622] In dilute aqueous dispersions the curvature of phosphatidylserine bilayers was significantly greater than that of isoelectric phosphatidylcholines of similar hydrocarbon chain length. This was attributed to the electrostatic repulsion in the polar head group region since it could be shown that raising the ionic strength and shielding of the ionic charge favors less curved bilayers.[624] On the other hand, in lyso-phosphatidylcholine the polar head group has a significantly larger cross-sectional area than the single hydrocarbon chain attached to it. This wedge-shaped molecule attains maximum interaction energy by forming curved interfaces either in hexagonal phases at water contents up to 63% or in a micellar solution above this water content.[1240] Phosphatidylethanolamines have a smaller polar head group than phosphatidylcholines and this reduction in size may, among other reasons, contribute to the formation of a hexagonal II phase[1240] (for a description of the phases see below).

Water will only penetrate into the polar lattice of lipids if the temperature is above T_c. Water can then spontaneously penetrate ("dissolve") into the ionic lattice,[341] causing the crystal to swell and "myelin figures" to form.[863] This rule was first established by Dervichian,[341,342] who, using lipid mixtures, found a direct correlation between the ability of lipids to swell in the presence of water and the ability to form liquid expanded monolayers. The latter property was taken as a criterion for "melted" hydrocarbon chains. In the same way as the chain melting at T_c diminishes the cohesion in the hydrophobic part of the molecule, the penetration of water loosens the electrostatic association in the polar head group. Thus one would expect that the two phenomena, chain melting and penetration of H_2O into the ionic lattice, are interdependent. This is indeed observed, for, as the water content increases, so the transition temperature steadily decreases to a limiting value T_c^*. For instance, with 1,2-dipalmitoylcholine the limiting value T_c^* is reached at 0.25 g H_2O/g lipid. This is the minimum temperature at which water penetrates into the ionic lattice (see Fig. 17c).

Given a certain lipid class, these limiting transition temperatures T_c^* in a homologous series parallel the melting points of the analogous fatty acids. Thus we may conclude that in addition to the nature of the hydrocarbon chains and the polar head group, T_c also depends on the H_2O content and on all solutes present. The extremely important interdependence between the packing of the hydrocarbon chains and electrostatic interactions in the polar head group is clearly demonstrated by the effect of ions on the transition temperature T_c. Phospholipids are known to interact stoichiometrically with metal ions[620,622] involving the negatively charged phosphate group.[625] Chapman[239] showed that binding of UO_2^{2+} to dipalmitoyl phosphatidylcholine shifted the transition temperature T_c of the phospholipid from 41.5 to 46°. Similarly in a phosphatidylserine:Ca complex (mol ratio 1:1) the T_c was increased by 5 to 22°. Träuble and Eibl[1493] investigated the effect of pH and the presence of ions on the gel-to-liquid crystalline phase transition of several negatively charged phospholipids. Ionization caused loosening in the structure of the polar head group due to electrostatic repulsion and the transition temperature decreased linearly with increasing charge density. In contrast, the divalent cations Ca^{2+}, Mg^{2+}, and Ba^{2+} were found to increase the transition temperature by charge neutralization. This probably leads to a tighter packing in the polar head group which is propagated to the hydrocarbon chain region. Addition of Na^+ or Ca^{2+} to lamellar mesophases of phosphatidylserine led to a condensation and reduction in molecular motion in the glycerol backbone and the first two hydrocarbon methylene groups next to the polar backbone.[624] From combined proton magnetic resonance and electron spin resonance spin label experiments Hauser and Phillips[624] concluded that in the presence of Ca^{2+} the molecular motion of the hydrocarbon chain protons becomes more anisotropic. Similar results were reported for the interaction of Ca^{2+} with a lamellar, liquid crystalline phase of cardiolipin.[637] In the Ca^{2+}-cardiolipin complex the motional restriction and anisotropy of a fatty acid spin label was generally increased, the effect being more pronounced near the glycerol backbone. These observations are important because they clearly demonstrate that the transition temperature and hence the degree of hydrocarbon chain fluidity and related properties such as bilayer permeability and stability are related to the hydration and/or ion binding properties of the polar head group.

3.2. Phospholipid–Water Phases

A diagrammatic representation of possible lyotropic mesophases is given in Fig. 7. When the lamellar phase is cooled below T_c the hydrocarbon

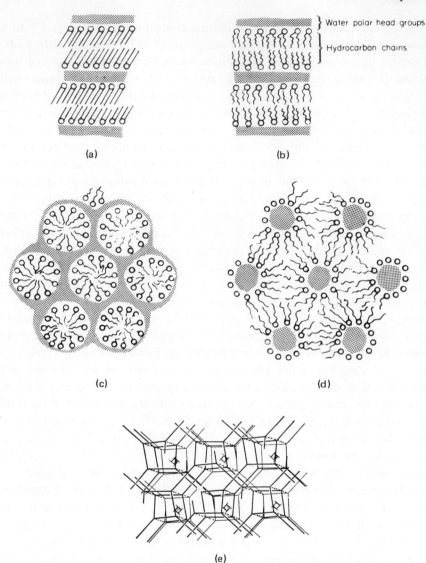

Fig. 7. Diagrammatic representation of the structures of some lyotropic mesophases. (a) gel; (b) lamellar liquid crystalline phase; (c) hexagonal phase I; (d) hexagonal phase II (from Ref. 1587); (e) body-centered cubic phase which is optically isotropic (e.g., that of phosphatidylcholines in the high-temperature ($\sim 80°$), high-lipid region (0.03 g H_2O/g lipid) of the phase diagram (cf. Section 3.2). The structure is similar to that of body-centered cubic phases of calcium and strontium soaps. Identical, rodlike elements of finite length are linked three by three into two distinct three-dimensional networks which are unconnected but mutually interwoven. The length of the rods does not exceed the diameter (~ 22 Å). The rods contain the polar groups and water, and the space between the rods is filled with liquidlike disordered hydrocarbon chains (from Ref. 931).

chains crystallize, but the water may be retained between the bilayers, so that a new lyotropic phase rather than the original crystal is formed. These so-called "gel phases" are frequently metastable and can be converted to the stable coagel state: microcrystals of the amphiphile in excess water. The hexagonal phases consist of a two-dimensional array of infinite cylinders. In the case of hexagonal I phases the interior of the cylinders is composed of lipid hydrocarbon chains and the outside is covered by the polar head groups, which are in contact with water, which also fills the space separating the cylinders. The hexagonal II phase is the inverse phase, where the cylinders are water channels lined by the polar head groups. The cubic phase first discovered for soaps has only been described for the phosphatidylcholine–water systems in the high-temperature, high-lipid-concentration range. It was originally thought to consist of a body- or face-centered lattice of water spheres surrounded by the polar head groups of the phospholipids whose hydrocarbon chains occupy the space between the spheres.[245,1240, 1404] Luzzati *et al.*[932] concluded from X-ray diffraction studies that the cubic phase is a more complex structure (Fig. 7). It consists of finite rods made up of water and the polar head groups embedded in a liquidlike hydrocarbon region. The rods are linked three by three and form an interwoven three-dimensional network. The isotropic, highly symmetric cubic phase has a high viscosity and has been reported as giving a high-resolution proton magnetic resonance spectrum.[245,1150] This is somewhat surprising, considering that the structure is both lipid and water continuous and highly viscous. It was suggested that the high-resolution spectrum is due to the rapid lateral diffusion of the lipid molecule in the cubic phase.[1150]

Most of the work on the phase behavior of phospholipids has been performed on synthetic and natural phosphatidylcholines. Chapman *et al.*[245] and Luzzati and co-workers[932] investigated the lyotropic phase behavior of a homologous series of 1,2-diacyl-L-phosphatidylcholines. The phase diagram of 1,2-dipalmitoyl-L-phosphatidylcholine is shown in Fig. 8. It is typical of all saturated 1,2-diacyl phosphatidylcholines, with T_c temperatures being disposed along the temperature axis dependent on the different hydrocarbon chains. Luzzati *et al.*[932] have studied the high-temperature, high-lipid region (<0.05 g H_2O/g lipid) and found a complex behavior under these experimental conditions. In addition to lamellar phases, body-centered cubic and two-dimensional hexagonal phases have been reported with less well-defined transition regions. Above the T_c line (Fig. 8) the phosphatidylcholine–water mixtures form liquid crystalline lamellar phases up to about 0.7 g H_2O/g lipid water. Addition of more water gives a two-phase system: The lamellar liquid crystalline phase at

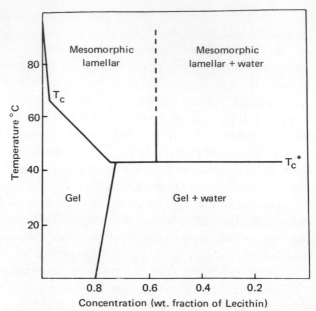

Fig. 8. Phase diagram of the 1,2-dipalmitoyl-L-phosphatidylcholine–water system (from Ref. 245).

maximum hydration and excess water. This lamellar phase consists of phospholipid bilayers separated by a few layers of H_2O. The X-ray long spacing D at 25° of 1,2-dipalmitoyl-L-phospatidylcholine increases steadily with increasing water content from $D = 56$ Å in the anhydrous state to $D = 64$ Å at 0.2 g H_2O/g lipid. Adding more water has no effect on the X-ray long spacing. At 25° a single, sharp short spacing of 4.19 Å is observed, indicative of the gel phase with highly ordered hydrocarbon chains packed in a hexagonal subcell (α-gel). The thickness of the lipid bilayer in the gel phase was obtained by plotting the long spacing as a function of $(1 - C)/C$, where C is the phospholipid concentration in g lipid/g lipid + g H_2O. A straight line relationship was obtained, the intercept of which at $(1 - C)/C = 0$ is the thickness $d' = 46$ Å of the lipid layer. The area S occupied per polar head group was obtained from $S = 2M/d'\varrho N_0 = 48$ Å², where M is the molecular weight and ϱ the density of phosphatidylcholine. From this the angle between the hydrocarbon chain axes and the phospholipid–water interface was calculated to be 58°. At 50°C the X-ray long spacings increased steadily up to 0.18 g H_2O/g lipid, where a sharp drop of 5 Å occurred. This drop in D is accompanied by a change in the short spacing from a sharp diffraction line at 4.19 Å to a diffuse one at 4.6 Å, indicative of the transition to the liquid crystalline phase. As the water content increases, so the long spacing also increases until at about 0.7 g H_2O/g lipid a limiting value of $D = 63.6$ Å is reached.

Small[1404] has investigated the phase behavior of egg phosphatidylcholine, which is similar to that of the synthetic, saturated diacyl phosphatidylcholines. Addition of water reduces the transition temperature T_c, which, with natural phospholipids, is a temperature range rather than a clearly defined point. The T_c line decreases from about 36–40° in the anhydrous state to below 15° at 0.14 g H_2O/g lipid (Fig. 9). The T_c line given by Small[1404] decreases continuously with increasing water content and differs from that obtained from the results of Reiss-Husson.[1240] The latter author showed that T_c reaches a constant value at ∼0.25 g H_2O/g lipid. At 24° there is a phase change from the crystalline to the liquid crystalline phase at about 0.09 g H_2O/g lipid. Between ∼0.18 and 0.80 g H_2O/g lipid (or 0.67 g H_2O/g lipid, in Ref. 1240) the X-ray long spacing (Fig. 9d) increases linearly from 51 to 64 Å, representing a single, homogeneous, lamellar liquid crystalline phase. Above ∼0.8 g H_2O/g lipid, where a two-phase system exists, the limiting value $D = 64$ Å remains constant. The maximum amount of H_2O of 0.7–0.8 g H_2O/g lipid[1240,1404] incorporated between phospholipid lamellar is similar to that observed with the synthetic phosphatidylcholines.[245] The increase in the X-ray long spacing from about 51 to 64 Å between 0.18 and 0.8 g H_2O/g lipid, respectively, represents the net result of changes in the lipid bilayer and water layer thickness due to the incorporation of increasing amounts of H_2O (Fig. 9b). The solid line at ∼0.8 g H_2O/g lipid in Fig. 9b represents the onset of a two-phase system, the lamellar liquid crystalline phase dispersed in excess free water. Small[1404] delineated in his egg phosphatidylcholine–water diagram the boundaries of a viscous isotropic (cubic) phase which can contain up to 0.18 g H_2O/g lipid. However, this range of existence is difficult to reconcile with that given by Luzzati *et al.*[932]

From the density of anhydrous egg phosphatidylcholine, glycerylphosphorylcholine, and propanediol, and assuming that the phosphorylcholine group is fully extended (∼8 Å long), Small[1404] and Reiss-Husson[1240] calculated the thickness of the hydrocarbon layer and the area per egg phosphatidylcholine molecule (Fig. 9c). The values for the hydrocarbon chain thickness calculated by Small are consistently smaller by 5 Å compared with those of Ref. 1240 and this constant difference is probably due to the assumptions involved in the calculations. As the long spacing increases linearly between 0.18 and 0.8 g H_2O/g lipid the thickness of the hydrocarbon layer decreases from ∼36 to ∼30 Å and the area per mole-

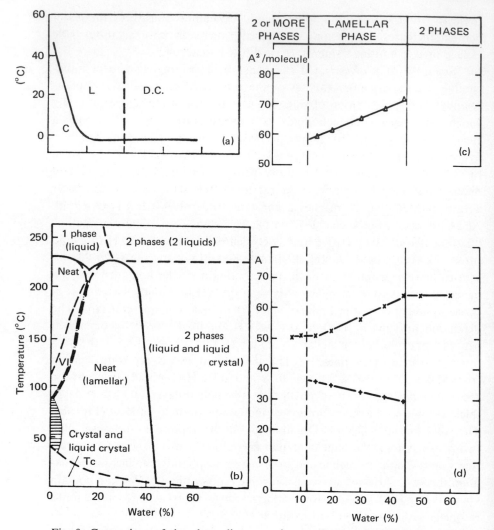

Fig. 9. Comparison of the phase diagrams of egg yolk phosphatidylcholine in water obtained by (a) Reiss-Husson (Ref. 1240) and (b) Small (Ref. 1404). (c) average surface area (Å²/molecule) of egg yolk phosphatidylcholine in the lamellar liquid crystalline phase at 24° (from Ref. 1404). (d) (×) X-ray long spacing D as a function of the water content for the same phase as in (c); (+) calculated thickness of the lipid layer. The dashed line at 12% water (=0.14 g H_2O/g lipid) represents the phase transition from the crystalline to the lamellar liquid crystalline phase. To the right of the line at 45% water (=0.82 g H_2O/g lipid) extends the region of the two-phase system, the lamellar liquid crystalline phase in excess water. The lamellar liquid crystalline phase at maximum hydration has a repeat distance of about 65 Å (from Ref. 1404).

cule increases from about 59 to $\sim$72 Å^2 (cf. Refs. 697 and 931). The surface area at 0.18 g H$_2$O/g lipid is close to that of egg phosphatidylcholine compressed to its collapse pressure,[1162] while at 0.8 g H$_2$O/g lipid the area per molecule has expanded to 72 Å^2, corresponding to a pressure of $\sim$20 mN m^{-1}. With the same assumptions mentioned above, Small[1404] also attempted to calculate the thickness d_W of the (water + phosphorylcholine) layer. Below about 0.18 g H$_2$O/g lipid ($\sim$8 mol H$_2$O/mol lipid, cf. Table IV) there is no "free" water and the phosphorylcholine groups of opposing bilayers are in contact or overlap. Above that water content a layer of "free" water appears which has a vapor pressure very close to that of pure water.[403] Between 0.18 and 0.8 g H$_2$O/g lipid the thickness of the free water zone steadily increases up to $\sim$19 Å and the sum of the thickness of the free water layer and the phosphorylcholine group increases from about 15 to 34.5 Å. Deducing the thickness of the free water layer from the total thickness of 34.5 Å gives 16 Å, which suggests that the extended conformation of the phosphorylcholine group does not change significantly as more water is incorporated between the phospholipid lamellae.[1163] At 0.18 g H$_2$O/g lipid, eight water molecules are associated with the phosphorylcholine group and this figure increases to 28–34 as the water content increases to 0.8 g H$_2$O/g lipid.

The assumptions underlying Small's calculation are, however, subject to criticism. It has been shown[1494] that the density of the hydrocarbon chain region and thus the density of phosphatidylcholine changes with the packing of the molecules.[1494] These changes in densities will affect the calculation of the thickness of the lipid and water layers. Using a simple geometric model and a constant polar group long spacing, it can be shown that the lateral expansion of the bilayer from about 50 to 70 Å^2/molecule at maximum hydration generates sufficient space around the phosphorylcholine groups within a given monolayer to accommodate seven more water molecules, thereby bringing the number of water molecules present within the polar head group lattice to a total of $\sim$15 mol H$_2$O/mol lipid. To accommodate the total number of bound and trapped molecules, which is 23 mol H$_2$O/mol lipid (cf. Section 4), an increase in the thickness of the (phosphorylcholine + water) layer by about 3.5 Å or an increase in the total bilayer thickness of about 7 Å suffices. This increase in the thickness of the water layer is significantly smaller than that calculated by Small[1404] and would indicate that opposing N(CH$_3$)$_3$ groups of adjacent bilayers are separated by $\leq$3 layers of water molecules. The observed increase in the X-ray long spacing must then contain a contribution due to changes in the packing of the hydrocarbon chains. For example, a change in tilt of the hydrocarbon chains in crystalline dipalmitoyl phosphatidylcholine bilayers of about 15° could account for an increase in the X-ray long spacing of about 7 Å. This change in tilt from, for instance, 60° to 75° (cf. Fig. 11) could result from a small conformational change (movement of about 2 Å) of the glycerol backbone as one possible consequence of the interaction of water with the polar head group. Thus, in contrast to Small's model, where the water in excess of about 10 mol H_2O/mol lipid occurs between adjacent bilayers as a zone of "free" water, in our model* a large proportion of the excess water is accommodated around the phosphorylcholine groups as the result of the lateral expansion of the bilayer; the amount of water inter-calated between bilayers corresponds to a thickness equivalent to <3 molecular layers of water. Thus, the observed increase in X-ray long spacing is the sum of the increase in the thickness of the water layer between adjacent bilayers and an increase in the thickness of the hydrocarbon chains. This observation is consistent with the general conclusion that changes in the polar group region of a phospholipid bilayer affect the hydrocarbon chain packing and vice versa (cf. effect of ions on phosphatidylserine bilayers[61,624]).

A complete phase diagram of synthetic phosphatidylethanolamine–H_2O systems is not yet available. Reiss-Husson[1240] has studied the phase behavior of the egg yolk phosphatidylethanolamine–H_2O system. In contrast to phosphatidylcholine–H_2O mixtures, which lack the hexagonal liquid crystalline phase, there are two liquid crystalline phases, a lamellar one and a hexagonal II and there is also a region where the two phases can coexist. Cullen et al.[303a] investigated the monolayer and bulk phase behavior of phosphatidylethanolamines extracted from *Pseudomonas fluorescens* grown at different temperatures. They found that as the growth temperature decreased, the relative amount of unsaturated and cyclopropane-containing fatty acids increased. In the presence of more than 0.25 g H_2O/g lipid a mixed-phase system was observed at room temperature; this consisted of a lamellar and a hexagonal phase which had limiting X-ray long spacings of 52.3 and 94 Å, respectively. At lower water content a lamellar phase only was present and at temperatures $>40°$ the hexagonal phase of the fully hydrated mixed-phase system grew at the expense of the lamellar phase until at 65° the hexagonal phase alone was present. The phase behavior of phosphatidylethanolamines extracted from organisms grown at 5° and 22° was identical, indicating that the physical properties of the phospholipids are controlled and remain constant over this temperature range.

* Phillips, Finer, and Hauser, unpublished results.

Recently Rand *et al.*[1216] have reported on the phase behavior of naturally occurring phosphatidylethanolamines extracted from pig erythrocytes and from blow fly larvae. Both phosphatidylethanolamines occur in lamellar and hexagonal phases and have regions where both phases coexist. In these regions of coexistence raising the temperature always shifted the equilibrium toward the hexagonal phase. Compared with phosphatidylcholine, it was found that less water was incorporated between the phosphatidylethanolamine lamellae and two-phase systems of lamellar liquid crystalline phases in excess water were formed at concentrations $C <$ 0.85–0.70 g lipid/g phase.

In comparison with phosphatidylcholines, phosphatidylethanolamines do not disperse as readily in water. The transition temperatures T_c for the gel $\rightarrow$ liquid crystalline transition of synthetic phosphatidylethanolamine–water systems are considerably higher than those of phosphatidylcholines with similar hydrocarbon chains.[841] This finding correlates well with monolayer experiments; high transition temperatures are consistent with condensed monolayers and low temperatures with expanded monolayers, and the force–area curves of synthetic, saturated 1,2-diacyl-phosphatidylcholines are more expanded than those of the corresponding phosphatidylethanolamines containing the same acyl groups.[1162] Phillips and Hauser[1164] have also pointed out that phosphatidylcholines spread at the air–water interface from the crystal at lower temperatures than do the equivalent phosphatidylethanolamine homologs. This was explained in terms of a lower crystal lattice energy of phosphatidylcholine due to a different polar head group orientation.[1163]

The complete phase behavior of pure acidic phospholipids has not yet been reported (cf. Ref. 1389). Both synthetic 1,2-dilauroyl- and 1,2-dipalmitoyl-L-phosphatidyl-L-serines and natural phosphatidylserine in the free acid form do not interact with water.[841] This conclusion is based on the finding that T_c of these phosphatidylserines is unaffected by the presence of water. When a mixture of the above phosphatidylserine with water is cooled the observed exothermic enthalpy change accounts for the freezing of all the water present in the mixture. Ladbrooke and Chapman[841] reported for the monopotassium salt of 1,2-dilauroyl-phosphatidylserine a single transition at a T_c which decreased steadily as more water was added. With dipalmitoyl phosphatidylserine, which in the absence of water gave two transitions, the lower-temperature transition increased in endothermicity at the expense of the transition at the higher temperature. With both phospholipids eutectic behavior was observed which was interpreted as an indication of the interaction of water with the ionic head group.

It has been reported that with nerve lipid extracts[101] a lamellar lipid crystalline phase is obtained which has an almost unlimited capacity of incorporating water between the lipid bilayers. As the water content increases, more water is incorporated, as evident from a steady increase in the X-ray long spacing.[101] Similar results were reported with 1,2-dipalmitoyl-DL-phosphatidylcholine containing ionic impurities.[245] The large water uptake was shown to be due to the net negative charge on the phospholipid head group. Addition of salt reduced the long spacing and the thickness of the water layer, and a two-phase system was formed: a lamellar liquid crystalline phase at maximum hydration and excess free water. This reduction of the interlamellar water content was a function of the screening effect of the electrolyte and depended on the concentration, charge, and hydration properties of the electrolyte added. It is worth noting that Ca^{2+} was much more effective than the monovalent ions K^+ or Na^+ (cf. Ref. 626). Recently Atkinson et al.[61] reported a similar increase in the X-ray long spacing of the monosodium salt of phosphatidylserine (extracted from bovine spinal cord) as the water content was increased (cf. Ref. 1389). Only at a water content exceeding 3 g H_2O/g lipid was a two-phase system observed at room temperature, consisting of a lamellar liquid crystalline phase and excess H_2O. The addition of electrolytes reduced the interlamellar water content significantly and the X-ray long spacing decreased from about 130 to 60 Å.[61] The effect of ions on the swelling behavior of phosphatidylserine was reversible, indicating that the continuous swelling of these lipids in the absence of ions is due to the electrostatic repulsion of the charged lipid bilayers.[61]

Sporadic information on the phase behavior of other pure phospholipids such as sphingomyelins, plasmalogens, and cardiolipins as well as mixed biological lipids has been presented and for a discussion of these data the reader is referred to the review articles by Luzzati,[931] Williams and Chapman,[1587] and Shipley.[1389] More recently the phase behavior of bovine brain cerebrosides and human brain sulfatides has been investigated by Abrahamsson et al.[1632] The most important finding with the cerebrosides was that the lamellar liquid crystalline phase was only observed at temperatures exceeding 70° (long spacing $D = 47.5$ Å, area/mol = 55.8 Å²) and a two-phase liquid crystalline dispersion was formed at a water content greater than 0.67 g H_2O/g lipid (cf. Ref. 1240). Sulfatides with a fatty acid composition comparable to that of the cerebrosides also gave a lamellar liquid crystalline phase the range of which extended to lower temperatures compared with the behavior of cerebrosides. For instance, over the concentration range 0.18–0.82 g H_2O/g lipid the lamellar liquid crystalline phase was present at temperatures higher than about $40°$; with excess water (>2.5 g H_2O/g lipid) the sulfatides formed micellar solutions and at an intermediate water content a cubic phase with "liquidlike" hydrocarbon chains was observed. The phase behavior of these glycolipids and of chloroplast monogalactosyl diglycerides and digalactosyl diglyceride, which contain a high proportion of polyunsaturated fatty acids and form hexagonal and lamellar phases, respectively, has been discussed in a recent review article by Shipley.[1389]

3.3. Conformation and Molecular Motion of the Hydrocarbon Chains in Lyotropic, Liquid Crystalline Phases

As with anhydrous liquid crystals, our understanding of the configurational freedom and molecular motion of hydrocarbon chains in liquid crystalline phases at maximum hydration is at best semiquantitative. Three possible model for the conformation of liquid paraffins are considered[1167]: (a) hydrocarbon chains fully extended and a high degree of order in the packing of the chains,[1363,1517] (b) hindered rotation about the C–C bonds, allowing many possible conformations, but still maintaining the approximately parallel orientation, and (c) free rotation about C–C bonds with the hydrocarbon chains in a chaotic, coiled conformation. A great deal of experimental data on this subject has accumulated, particularly more recently, when spectroscopic techniques have been applied to the question of hydrocarbon chain conformation and motion. A brief summary of the experimental evidence is given which should enable us to decide which of the three models may be the most credible.

Some quantitative information pertinent to the question of hydrocarbon chain fluidity can be derived from the thermodynamics of the crystal → liquid crystal transition. Unfortunately, suitable thermodynamic data for phospholipids are scarce and only available for 1,2-diacyl-L-phosphatidylcholines.[1167] Phillips et al.[1167] investigated the first-order crystal → liquid crystal transitions of 1,2-diacyl-L-phosphatidylcholines; the heat absorbed at T_c is equal to the transition enthalpy: $\Delta H = T_c \Delta S$. Inserting the experimentally determined values of ΔH and T_c in the above equation, entropy changes ΔS for the transition were calculated. Figure 10 summarizes ΔS at the transition temperature as a function of the alkyl chain length. The total entropy change comprises the configurational, positional, and orientational entropy increase, but only the configurational contribution will be chain length dependent. For all lipids listed straight line relationships were obtained, where $11 < n < 20$ (Fig. 10). From the slope

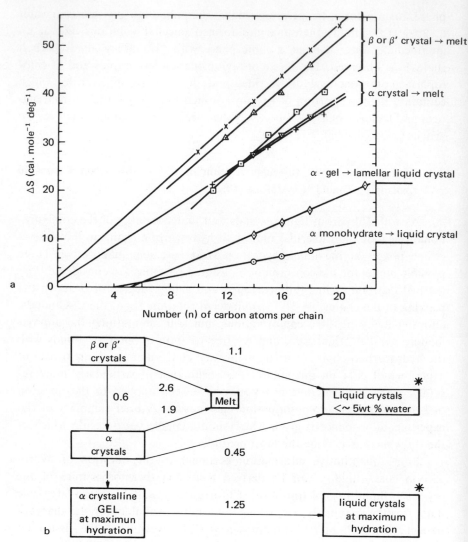

Fig. 10. (a) Dependence of the entropy change ΔS occurring at the chain-melting transition upon the number n of carbon atoms per hydrocarbon chain. ($\times$) Even n-alkanes (β form); ($+$) odd n-alkanes (α form); ($\triangle$) triglycerides (β form); ($\triangledown$) triglycerides (α form); ($\boxdot$) odd fatty acids (β form); $\diamond$ lecithin-water gel (α form); ($\bigcirc$) lecithin monohydrate (α form). (b) Configurational entropies per methylene group (cal deg^{-1} mole^{-1}) at the transition temperatures for the chain melting of lipids where $14 \leq n \leq 22$. For boxes marked with an asterisk particular liquid crystalline phases have not been specified because the entropy changes per methylene group at lamellar-to-hexagonal, lamellar-to-cubic, and cubic-to-hexagonal liquid crystalline phase boundaries are $<5\%$ of those of the crystal-to-liquid crystal transitions (from Ref. 1161).

the configurational entropies per methylene group for the chain melting of lipids at the transition temperature were obtained and are summarized in Fig. 10b. From this it is clear that the hydrocarbon chain fluidity in the anhydrous liquid crystals is about half that found in liquid n-alkanes at the transition temperature.[1167] This probably arises from an inhibition of rotation about C–C bonds. Phillips et al.[1167] pointed out that the hydrocarbon chain motions in anhydrous liquid crystals of different lipids are similar and hence depend more upon the physical state than on the molecular structure of the lipid. The reduction in chain motions in the liquid crystalline phases as compared with that in the melt is therefore due to the retention of the bilayer and the effect of the ordered structure in the polar head group region.

Extrapolating the linear relationship of ΔS for the α-gel $\rightarrow$ liquid crystal transition to $n = 0$ gives $\Delta S = -6$ cal deg^{-1} mol^{-1} (Fig. 10a). This is characteristic for the crystal $\rightarrow$ liquid crystal transition of phosphatidylcholines because for the other lipids (except for fatty acids with an odd number of carbon atoms) the intercepts on the ΔS axis as $n \rightarrow 0$ are positive and close to zero. In this case the ΔS increment per methylene group (end group effects are ignored) is constant over $0 \leq n \leq 20$, indicating that with n-alkanes and triglycerides the configurational disorder introduced in the hydrocarbon chains at the melting point is equal for all positions along the chain. This apparently is not so with fully hydrated phosphatidylcholine bilayers at the crystal $\rightarrow$ liquid crystal transition. As discussed by Phillips,[1161] the negative intercept indicates that there is a loss of about 6 cal deg^{-1} mol^{-1} over the whole chain because of the ordering effect of the polar group lattice on the closest methylene groups up the hydrocarbon chains. If it is assumed that the loss of entropy along the chain is all or nothing,[1161] then $\Delta S = 0$ for the first five methylene groups.[1276] The same extrapolation procedure applied to the plot for the α-monohydrate $\rightarrow$ liquid crystal transition gives $\Delta S_{(n \rightarrow 0)} \simeq -2$ cal deg^{-1} mol^{-1}, indicating that the entropy loss over the whole hydrocarbon chain is about 4 cal deg^{-1} mol^{-1} less in this case. Following the above assumption, the ordering effect of the polar head group appears to be confined to four methylene groups (i.e., $n = 4$ as $\Delta S \rightarrow 0$). The reason for the larger entropy loss observed with the fully hydrated liquid crystals is probably to be found in the interaction of the glycerylphosphate group with water molecules, interfering with the molecular motion of the first few methylenes near that group (cf. Finer and Phillips[448]). Furthermore, there may be a steric contribution in the sense that the more widely spaced, lateral packing in the fully hydrated liquid crystal (area per phosphatidylcholine molecule is 70 Å^2 as compared to 48 Å in the monohydrate) allows for a larger gradient in the motional freedom along the hydrocarbon chains anchored at one end to the polar backbone.[1276]

Figure 10b shows that the chain motions in the phosphatidylcholine monohydrate at the transition temperature are similar to those in the anhydrous liquid crystal. In both cases the configurational entropy increase per methylene group is about 1.5 cal deg^{-1} mol^{-1} (1.9–0.45 and 2.6–1.1, respectively), less than that for liquid n-alkanes at the melting point. In fully hydrated phosphatidylcholines the configurational entropy increase per methylene group for the α-gel $\rightarrow$ lamellar liquid crystal transition is 1.25 cal deg^{-1} mol^{-1}. This is 0.8 cal deg^{-1} mol^{-1} greater than that for the α-crystal $\rightarrow$ liquid crystal transition or 0.7 cal deg^{-1} mol^{-1} less than that of liquid n-alkanes. Provided the hydrocarbon chain structure is similar in both α-crystal and α-gel (cf. Ref. 245), the difference of about 0.8 cal deg^{-1} mol^{-1} must mean a significantly greater hydrocarbon chain mobility in the fully hydrated liquid crystalline state. X-ray diffraction data[245] support this suggestion, showing that with dipalmitoyl phosphatidylcholine the molecular area increases from 60 to $\sim$70 Å^2 as the water content increases from 0.25 to 0.70 g H$_2$O/g lipid. Aranow et al.[47] reported a configurational entropy increase of 2.2 cal deg^{-1} mol^{-1} per CH$_2$ group for the melting of n-alkanes. This figure has to be compared with 1.25 cal deg^{-1} mol^{-1}, or, since at present the loss in entropy cannot be described quantitatively as a function of the position in the chain, with an increment in ΔS per methylene group averaged over the whole hydrocarbon chain: 0.9 cal deg^{-1} mol^{-1}. Such a comparison gives an indication of the minimum inhibition in the motional freedom compared to that of liquid n-alkanes. To summarize, entropy measurements clearly show that the description of the hydrocarbon chain fluidity in the liquid crystalline state as liquidlike is an oversimplification. The fluidity, which is approximately half of that in liquid n-alkanes, is nonuniform along the hydrocarbon chain, such that it is significantly reduced in the vicinity of the glycerol backbone (cf. Ref. 668).

Träuble and Haynes,[1494] measuring volume increases at the gel $\rightarrow$ liquid crystalline transition, also concluded that the configurational freedom and molecular motion are not evenly distributed along the hydrocarbon chains, but increase toward the terminal methyl group. They compared the melting dilatation of lipids containing long hydrocarbon chains such as palmitic acid or tripalmitin with the dilatation of dipalmitoyl phosphatidylcholine at the gel $\rightarrow$ liquid crystalline transition. If the total volume increase is attributed to the hydrocarbon chains, it is $\Delta V/V = 0.014$ and still six to ten times smaller than the melting dilatation of analogous hydrocarbon compounds, indicating that the hydrocarbon chains of the phospholipid in

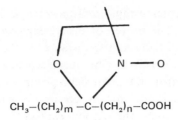

Fig. 11. Fatty acid spin labels. m is the number of CH_2 groups between the terminal CH_3 group and the carbon atom to which the nitroxide free radical is attached. n is the number of CH_2 groups between the latter carbon atom and the carboxyl group of the fatty acid. Using this notation, the N-oxyl-4,4′-dimethyloxazolidine derivative of the 5-keto-stearic acid is written as spin label (12,3), where $m = 12$ and $n = 3$.

the liquid crystalline state are still in a relatively ordered state. Volume increases at the gel → liquid crystalline transition of several saturated 1,2-diacyl-phosphatidylcholines were also reported by Melchior and Morowitz,[1638] but $\Delta V/V \simeq 0.025$ was higher than that reported in Ref. 1494.

More direct information about the packing and molecular motion in the hydrocarbon chains of phospholipids has been derived recently from spectroscopic studies, particularly spin-label electron spin resonance (esr) and nuclear magnetic resonance (nmr) studies. Hubbell and McConnell incorporated the spin-label 2,2,6,6-tetramethylpiperidine-1-oxyl (TEMPO)[689] as well as steroid and fatty acid (m, n) spin labels[690,691] (Fig. 11) in asolectin (a mixture of phospholipids extracted from soy beans). They were able to show that the hydrocarbon chains of the phospholipids present in aqueous dispersions are "fluid." The spin-labeled steroids and fatty acids were found to be oriented with their long axis preferentially perpendicular to the plane of the phospholipid bilayers and were shown to undergo rapid but anisotropic motion about the long axis and relatively restricted motion perpendicular to the long axis.[690] By comparing the motional freedom of the fatty acid spin labels $(m, n) = (17, 3)$, $(12, 3)$, and $(5, 10)$, Hubbell and McConnell[691] observed that the molecular motion of the (5, 10) label was far freer than that of the (17, 3) or (12, 3) labels. They concluded that the central region of the hydrocarbon chains is more "fluid" than the region close to the glycerol backbone.

Seelig[1358] investigated the motion of steroid and stearic acid spin labels $(m = 13, 12, 9, 5)^*$ incorporated in a lamellar liquid crystalline system consisting of mixtures of sodium decanoate, n-decyl alcohol, and water. He arrived at similar conclusions. The hydrocarbon chains are parallel and

* $n = 15 - m$.

rather extended, undergoing rapid and anisotropic motion with rotational frequencies $>10^8$ Hz. The degree of order, expressed as an order parameter, decreased exponentially as the distance of the nitroxide group from the polar head group increased. This increased disorder along the hydrocarbon chain starting from the polar head group was explained in terms of the Porod–Kratky[1636] model for hydrocarbon chain conformation. This assumes free rotation about the carbon–carbon bonds with the angle between adjacent carbon–carbon bonds being adjustable and with the hydrocarbon chains being "anchored" at the ester links, but, because of the first assumption, its usefulness as a realistic model is questionable.

Hubbell and McConnell,[692] using similar fatty acid (m, n) spin labels and phosphatidylcholine (m, n) spin labels (with the nitroxide group in the β-hydrocarbon chain) in egg phosphatidylcholine dispersions, derived a more detailed semiquantitative picture of the molecular motion in phospholipid alkyl chains. The rapid anisotropic motion about the long chain axis in terms of the order parameter[692] was determined as a function of the distance between the nitroxide group and the carboxyl carbon atom of the polar head group. The order parameter depended strongly on n and an exponential relationship was observed with the fatty acid spin labels but not with the phosphatidylcholine spin labels.* From a theoretical analysis of these relationships between order parameter and n, a model was derived which describes the possible conformation in the hydrocarbon chains anchored at one end in terms of rapid trans–gauche isomerization about various C–C bonds.

Libertini et al.[890] and Jost et al.[744] were the first to study by esr the molecular motion and orientation in hydrocarbon chains of oriented egg phosphatidylcholine multilayers containing a series of spin-labeled stearic acids with $(m, n) = (12, 3), (10, 5), (5, 10),$ and $(1, 14)$. In qualitative agreement with Hubbell and McConnell,[692] the anisotropy of the motion of the spin label was found to decrease as the position of the label was moved further away from the carboxyl end. This was interpreted to mean that the hydrocarbon chains are anchored to the polar head group and that the molecular motion of the spin probe progressively increases as it moves along the hydrocarbon chains. The same authors also concluded that their measurements are representative for the molecular motion and the fluidity of the phospholipid hydrocarbon chains in which the spin label is embedded. The effects of hydration and temperature were also studied. Increasing the

---

* For this type of spin label the relationship between order parameter s and number of carbon atoms n was linear.

degree of hydration enhanced the molecular motion and the effect was more pronounced the closer the label was to the carboxyl carbon atom. The motional characteristics along the hydrocarbon chains were maintained over rather a wide temperature range. The effect of hydration was also noticed by Hsia *et al.*[685] Exposing oriented multilayers of mixtures of egg phosphatidylcholine and cholesterol, labeled with a cholestane spin label, to water considerably increased the motional freedom about the long axis of the spin label.

McFarland and McConnell,[948] using phosphatidylcholine spin labels $(m, n) = (1, 14)$, $(5, 10)$, $(7, 6)$, and $(10, 3)$ incorporated in multilamellar arrays of egg phosphatidylcholine bilayers, showed that the CH_3-terminal half of the spin-labeled hydrocarbon chains is rather flexible and on the average oriented perpendicular to the bilayer plane. However, the half of the hydrocarbon chain anchored to the polar head group is comparatively rigid and tilted at an angle $\sim 60°$ relative to the bilayer plane (Fig. 12). The bent conformation of the hydrocarbon chain next to the polar head group is long-lived, with a lifetime $> 10^{-8}$ sec, and produces a carbon atom density that is $\sim 12\%$ higher than the carbon atom density near the terminal methyl group. The authors point out that this is about the difference in density between liquid and solid *n*-alkanes, indicating that the motional freedom toward the center of the bilayer approaches that of an isotropic liquid alkane, while in the same bilayer the motional freedom near the polar head group approaches that of a crystalline alkane.[1276] The authors suggest that although their conclusions were derived from experiments with spin-labeled phosphatidylcholine, the same average bending is likely to occur in the hydrocarbon chains of unlabeled phosphatidylcholine molecules.

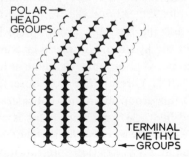

POLAR →
HEAD
GROUPS

TERMINAL
METHYL
← GROUPS

Fig. 12. Schematic, idealized representation of the packing of the hydrocarbon chains in a small section of one-half of a planar phosphatidylcholine bilayer. This packing allows greater motional freedom at the end of the methylene chains near the terminal methyl groups (from Ref. 948).

The picture of the molecular motion and orientation of the hydrocarbon chains derived from nmr studies is qualitatively consistent with that obtained from esr experiments. Proton magnetic resonance linewidth and relaxation measurements on egg phosphatidylcholine bilayers in the presence of excess water have shown that at room temperature the motion of the hydrocarbon chain protons is too slow to give complete averaging of the dipole–dipole interaction.[235,444,446,1112,1150] Thus multilamellar arrays of egg phosphatidylcholine bilayers both below and above the gel → liquid crystal transition do not give high-resolution nmr spectra. Nmr linewidth and relaxation measurements support the conclusion derived from esr measurements that the mobility in the hydrocarbon chains increases toward the terminal CH_3 group and that the packing of the bilayer is tightest in the region of the glycerol backbone, affecting the first two or three chain methylene groups.[318,445] Chan and his collaborators[235,236] concluded from relaxation measurements carried out on aqueous phosphatidylcholine dispersion in the lamellar liquid crystalline phase that there is an abrupt but not dramatic increase in the molecular motion close to the terminal CH_3 group of the hydrocarbon chains. This conclusion has been supported since by more detailed relaxation studies of Metcalfe et al.[867,1005] and Horwitz et al.[683] Metcalfe et al.[1005] reported spin–lattice relaxation (T_1) measurements for ^{13}C nuclei in lamellar liquid crystalline systems of egg phosphatidylcholine and ^2H_2O. The shortest T_1 in the hydrocarbon chain was observed with the carbon atoms close to the polar head group. The T_1 values for carbon atom 3 and carbon atoms 4–13 were similar and about five times longer, while the remaining three carbon atoms (14–16) showed a further sharp increase in the spin–lattice relaxation time. Spin–lattice (T_1) proton relaxation measurements on lamellar liquid crystalline phases of egg and dipalmitoyl phosphatidylcholine-^2H_2O mixtures[867] are consistent with the ^{13}C nmr results.[1005] Above the transition temperature T_1 values of the terminal CH_3 and the $N(CH_3)_3$ groups were longer than the single T_1 value measured for the hydrocarbon methylene groups. It is interesting to note that with dipalmitoyl phosphatidylcholine at temperatures ($T_c - 10$) $< T < T_c$ only the $N(CH_3)_3$ resonance was observed, indicating that in the presence of ^2H_2O this group maintains considerable motional freedom in spite of crystallized hydrocarbon chains. This is consistent with the observation of a small heat uptake by phosphatidylcholine gels prior to the main endothermic transition at T_c;[668,841] this "pretransition" was interpreted as being due to the onset of rotational motion in the polar head group.[841,1523]

Horwitz et al.,[683] measuring proton spin–lattice and spin–spin relaxation times of sonicated aqueous egg phosphatidylcholine above the transition temperature T_c, obtained similar results. The differences in T_1 values among various clearly discernible proton resonances were large enough to rule out spin diffusion as a dominating relaxation mechanism. The T_2 showed even greater variability. The shortest T_1 and T_2 were measured for the CH_2 group next to the carbonyl carbon, indicating that this part of the hydrocarbon chain is rather immobile. In agreement with Metcalfe et al.,[1005] the relaxation times for most of the CH_2 groups of the hydrocarbon chains were longer and characterized by a single T_1 and a composite T_2. The terminal CH_3 group relaxed considerably more slowly than the rest of the hydrocarbon chains. Somewhat inconsistent with the above ^{13}C and proton nmr relaxation studies are the results of Birdsall et al.[143] Fluorine-19 mono-fluoro stearic acids (substituted at the carbon atoms 3, 6, or 11) were incorporated in egg phosphatidylcholine bilayers and from linewidth measurement these authors showed a progressive and approximately linear increase in molecular motion of the hydrocarbon methylenes, starting from the polar head group toward the center of the bilayer.

With the exception of the last work,[143] the above relaxation data suggest that a rather large portion of the alkyl chain methylenes undergo relatively uniform molecular motion with a correlation time longer than that of the terminal methyl group. Levine et al.[1637] attempted to interpret qualitatively ^{13}C spin–lattice relaxation (T_1) measurements carried out with sonicated dimyristoyl and dipalmitoyl phosphatidylcholine vesicles in ^2H_2O in terms of effective correlation times for molecular motion. Three types of motion were considered: (a) the isotropic tumbling of the whole vesicle with a correlation time of about 10^{-6} sec, (b) motion of the dipalmitoyl phosphatidylcholine molecule as a whole about the long axis, which has a correlation time of 10^{-8}–10^{-9} sec as evident from esr measurements, and (c) motion about individual C–C bonds with correlation times $\sim 10^{-10}$ sec. The best fit to the experimental T_1 data was obtained for a model which involved a small decrease in correlation time along the hydrocarbon chain in that part close to the glycerol backbone and a sharp decrease toward the terminal methyl. The point in the hydrocarbon chain at which this occurred could not be defined. By this theoretical treatment of the T_1 data it was, however, possible to rule out a number of simple models, e.g., models involving equal motion about all C–C bonds or a linear or exponential increase in the motion from C_1 (carbon atom next to the ester group) to the terminal methyl. The authors concluded that the motion that dominates the spin–lattice relaxation is oscillatory (correlation time $\sim 10^{-10}$ sec), i.e., within a given conformation rather than rotational involving conformational changes.

To summarize the information on the motion and packing of the hydrocarbon chains in fully hydrated liquid crystalline phases, it should be stressed that the application of a range of physical techniques provides us with a picture which is as yet qualitative and only consistent in its crude outline. Sufficient evidence is available to conclude that the motional freedom and packing are not uniform along the alkyl chains. The motion and packing change from an immobile, highly ordered, probably quasicrystalline state to a flexible, more loosely packed state at the center of the bilayer. This motional and configuration freedom of the alkyl chains in the liquid crystalline state at the transition temperature is characterized by an average entropy increment per methylene group of 0.9 cal deg^{-1} mol^{-1}, which is about half that obtained for the β-crystal $\rightarrow$ melt transition. The details of the molecular motion and packing in the hydrocarbon chains are still a matter of debate. Somewhat conflicting conclusions have been derived from nmr and esr measurements. The discrepancies between the two techniques may either be due to a perturbation in the alkyl chains introduced by the nitroxide spin label[236,1633] or to differences in the time scale of the motions involved. We set out at the beginning of this section to test the validity of three feasible models describing the configurational and motional freedom of the hydrocarbon chains. It is clear from the experimental evidence discussed above that the description of the hydrocarbon chains as liquidlike is an oversimplification. Thus none of the three models describes the configurational and motional freedom of the hydrocarbon chains adequately, but they may be at least partially correct in the sense that they describe correctly the motion in a certain portion of the hydrocarbon chain.

4. HYDRATION OF PHOSPHOLIPIDS

It has been mentioned in Section 3 that only a limited number of lipids with an appropriate hydrophobic–hydrophilic balance will swell when in contact with water to form lyotropic mesophases. With few exceptions lipids that interact with water are confined to class II of compound or conjugate lipids (see Section 1) comprising phospholipids, cerebrosides, gangliosides, and sulfatides. Of the remainder only the neutral monoglycerides of class I and soaps and detergents of class III would be of interest in this context. Since soaps and detergents are dealt with in Chapter 2, this discussion will be restricted to compound or conjugate lipids, again with particular reference to phospholipids. As far as phospholipids are concerned, most of the hydration studies were carried out with phosphatidylcholines that were either synthetic, containing a single type of fatty

Various Physical Techniques As Discussed in the Text

Hydration number		Description of hydration	Technique	Ref.
mol H₂O/mol lipid	g H₂O/g lipid			
6	0.14	The first two hydration layers	Water vapor adsorption	403
18	0.44	Total hydration	Water vapor adsorption	403
13.6	0.33	Total hydration	Water distribution in two-phase system	335, 405
12	0.30	Total hydration	Water distribution in two-phase system	638
12	0.30	Total hydration	Water distribution in two-phase system	1551
4	0.10	Tightly bound	Tracer technique using gel filtration	1024
12–16	0.29–0.39	Total hydration	Hydrodynamic measurements	618, 686, 687
11–15	0.25–0.37	Unfreezable water	Differential scanning calorimetry	245
12–15	0.29–0.37	Total hydration	Differential scanning calorimetry, maximum reduction of T_c at 25°	245, 1299, 1523
12	0.29	Water in phosphorylcholine zone at maximum hydration	X-ray diffraction	1404
28–34	0.67–0.79	Bound + trapped[a] water	X-ray diffraction	245, 1404
4–5	0.11	First hydration layer	pmr	1523
10–12	0.24–0.29	First + second hydration layer	pmr	1523
6	0.14	Bound water	pmr	552
21	0.52	Bound + trapped[a] water	dmr	1523
4–10	0.1–0.24	Bound water	dmr	1301
21	0.52	Bound + trapped[a] water	dmr	1301
1	0.025	Tightly bound	dmr	443
12	0.29	Bound water	dmr	443
23	0.55	Bound + trapped[a] water	dmr	443
11	0.25	Bound water	Diffusion of ³HHO	1247
21	0.51	Bound and trapped[a] water	Diffusion of ³HHO	1247

[a] Trapped: intercalated between phospholipid lamellae.

acid, or extracted from natural sources and containing a mixture of fatty acids.

Table IV summarizes the results discussed in this section, which deals with quantitative aspects of the interaction of water (i.e., hydration) with phosphatidylcholine and other phospholipids of biological interest. A summary of the hydration numbers (Table IV) derived from various physical techniques is followed by a brief discussion of molecular details so far available.

4.1. Water Vapor Adsorption to Phosphatidylcholine

Demchenko[335] studied the uptake of water vapor by egg yolk phosphatidylcholine dissolved in various organic solvents (benzene, toluene, and xylene). He showed that the uptake of water by the organic phase is due to the dissolved phospholipid and is independent of the organic solvent used. The maximum quantity of water adsorbed was 0.33 g H_2O/g phosphatidylcholine. This value was confirmed by Elworthy and McIntosh,[405] who equilibrated benzene solutions of egg phosphatidylcholine with water by either shaking them with water or exposing them to water vapor. The same value was derived from viscosity measurements of phosphatidylcholine solutions in benzene using eqn. (1) (see below). The excellent agreement between values derived independently from different methods was interpreted as an indication that all the water was associated with phosphatidylcholine. In benzene this phospholipid is known to form spherical, inverted micelles and the water is thought to be associated with the polar head group forming a cavity of about 40 Å diameter in the center of the inverted micelles.[405]

Misiorowski and Wells[1024] used a radiotracer technique to show that water is bound to dioctanoyl and egg yolk phosphatidylcholine dissolved in diethyl ether/methanol (95:5 by volume) containing 4 μl/ml tritium-labeled water. By gel filtration of the phospholipids on Sephadex and other gels[1024] which were equilibrated with the above solvent these authors derived from the radioactivity associated with the phosphatidylcholine peak a value of approximately 1 mol of H_2O bound per mole of phosphatidylcholine. In the same solvent the phosphatidylcholines were also able to solubilize calcium and there was a competition of Ca^{2+} and water for the same binding sites. For instance, the interaction of Ca^{2+} with dioctanoyl phosphatidylcholine in methanol decreased as more water was added and was completely inhibited when 4 mol of H_2O was bound per mole of phospholipid.

By studying the adsorption isotherm of water vapor on anhydrous phosphatidylcholine and lysophosphatidylcholine at 25°, Elworthy[403] obtained the following saturation values: 18 mol H_2O/mol phosphatidylcholine (0.44 g H_2O/g lipid) and 14 mol H_2O/mol lysophosphatidylcholine (0.48 g/g lipid). This saturation value increased with increasing temperature and was $\sim$20 mol H_2O/mol phosphatidylcholine (0.48 g H_2O/g lipid) at 40°. Elworthy concluded from the changes in the slope of the adsorption isotherm that there are several types of water adsorption sites within the polar head group differing in their affinity for water. The first "hydration layer" amounts to 2.5 mol H_2O/mol lipid ($\sim$0.06 g/g) and this increases in the second hydration layer to about 6 mol H_2O/mol lipid (0.14 g H_2O/g lipid).

The strong adsorption of water vapor by anhydrous phosphatidylcholines was confirmed qualitatively by Chapman et al.,[245] who reported that anhydrous saturated 1,2-diacyl-phosphatidylcholines are highly hygroscopic and adsorb water vapor even when stored in a desiccator over silica gel or phosphorus pentoxide. However, single crystals of anhydrous 1,2-distearoyl-L-phosphatidylcholine only become hygroscopic after being heated to temperatures above the crystal $\rightarrow$ liquid crystal transition temperature. On the other hand, the hydrates of saturated phosphatidylcholines lose their hydration when dried under vacuum at 90° for several hours.

4.2. Hydrodynamic Measurements

With spherical particles hydration numbers can be derived from hydrodynamic quantities derived from viscosity, sedimentation, and diffusion measurements. When dispersions of phosphatidylcholine in excess water and at temperatures greater than T_c are sonicated, small, spherical vesicles of 200–300 Å diameter are obtained which are surrounded by a single bilayer.[618,623,686,1306] Such a system lends itself well to hydrodynamic measurements and the hydration numbers derived from these measurements are discussed in this section. The results are summarized in Table V.

Saunders et al.[1307] and Huang[686] carried out viscosity measurements on sonicated egg yolk phosphatidylcholine and from the intrinsic viscosity $[\eta]$ and the equation

$$100[\eta] = 2.5(\bar{V} + \bar{V}_1\delta_1) \tag{1}$$

where δ_1 is the quantity of water in grams associated with 1 g of lipid and $\bar{V}$ and $\bar{V}_1$ are the partial specific volumes of phosphatidylcholine and water, respectively, derived hydration numbers of $\delta_1 = 0.50$ g H_2O/g lipid

TABLE V. Hydration of Phosphatidylcholine Derived from Hydrodynamic
Measurements

Experimental method	Hydration number of phospholipid (mol H_2O/mol lipid)[a]		Ref.
	Egg phosphatidyl-choline	Egg lysophos-phatidylcholine	
Water vapor adsorption[b]	18 (0.44)[b]	14 (0.48)[b]	403
Viscosity		16 (0.58)	403
Light scattering + diffusion		11 (0.39)	403
Viscosity	21 (0.50)	—	1307
Viscosity	27 (0.64)	—	686
Light scattering + viscosity	7 (0.16)	—	65
Ultracentrifugation in H_2O/^2H_2O mixtures of varying density	21 (0.51)	—	687
Ultracentrifugation (sedimentation coefficient + MW by Archibald method)	24 (0.57)	—	c
Diffusion + ultracentrifugation (MW by Archibold method)	17 (0.42)	—	c
Diffusion + ultracentrifugation (MW from Svedberg's equation)	30 (0.72)	—	c

[a] Values in g H_2O/g lipid are given in parentheses.
[b] Included in the table for comparison.
[c] H. Hauser, unpublished results.

and $\delta_1 = 0.64$ g H_2O/g lipid, respectively (Table V). On, the other hand,
Attwood and Saunders,[65] comparing light scattering and viscosity measurements on sonicated egg phosphatidylcholine dispersions, concluded
that there is only a small amount of water associated with the phospholipid,
δ_1 amounting to 0.16 g H_2O/g phosphatidylcholine. Elworthy[403] carried
out similar viscosity measurements on egg lysophosphatidylcholine micelles
in water and obtained a hydration number of 16 mol H_2O/mol lipid (0.58 g
H_2O/g lipid). This value was compared with a hydration number derived
from light scattering and diffusion measurements. Light scattering studies
by Robinson[1639] showed that the micelles were almost spherical and gave
a micellar weight of 97,000. Assuming that this micellar weight corresponds
to an unhydrated sphere, the diffusion coefficient D_0 of such a sphere can
be calculated from the Stokes–Einstein equation if the partial specific vol-

ume $\bar{V}$ of lysophosphatidylcholine is known ($\bar{V} = 0.9821$ ml/g[1153]):

$$D_0 = RT/6N\pi\eta r_0 = 7.32 \text{ cm}^2 \text{ sec}^{-1} \tag{2}$$

where r_0 is the Stokes radius of the unhydrated sphere and η is the viscosity of water at the temperature of the molecular weight determination. From the ratio $D_0/D = 1.118$, where D is the experimentally determined diffusion coefficient, and the equations

$$f/f_{\min} = D_0/D \tag{3}$$

$$f/f_{\min} = (f/f_0)[(\bar{V} + \delta_1\bar{V}_1)/\bar{V}]^{1/3} \tag{4}$$

a hydration number $\delta_1 = 0.39$ g H_2O/g lysophosphatidylcholine was obtained (Table V). In eqns. (3) and (4) $f_{\min}$ is the frictional coefficient of an anhydrous sphere of micellar weight $M = 97,000$. f/f_0 represents the effect of shape alone on the frictional coefficient and is one for spherical particles such as lysophosphatidylcholine micelles. The figures for lysophosphatidylcholine derived from hydrodynamic measurements compare well with the hydration number determined by water vapor adsorption (see Table V).

Using a related approach, δ_1 for sonicated egg yolk phosphatidylcholine can be obtained from diffusion measurements, using eqns. (3) and (4) (see Table V). The D_0 of the unhydrated sphere is calculated from the molecular weight obtained independently from ultracentrifugation.[623] The hydration numbers based on diffusion measurements are then 17.5 mol H_2O/mol phosphatidylcholine (0.42 g H_2O/g lipid) and 30 mol H_2O/mol lipid (0.72 d H_2O/g lipid), respectively, depending on the molecular weight used in the calculation of D_0. In the first case the molecular weight M was obtained from the Archibald method and in the second from the sedimentation coefficients and the diffusion coefficient D using Svedberg's equation (Table V):

$$M = RTs/D(1 - \bar{V}\varrho) \tag{5}$$

Huang and Charlton[687] determined partial specific volumes and apparent specific volumes of phosphatidylcholine vesicles from sedimentation velocity measurements in H_2O–^2H_2O mixtures of varying, known densities. From the relationship between apparent specific volume ϕ and the partial specific volume $\bar{V}$

$$\phi = \bar{V} + \delta_1[\bar{V}_1 - (1/\varrho)] \tag{6}$$

$\delta_1 = 0.51$ g H_2O/g phosphatidylcholine was obtained. Hauser calculated the hydration of egg phosphatidylcholine from the average vesicle weight

M_{av} and average sedimentation coefficient s_{av} using eqn. (4) and

$$f/f_{min} = M_{av}^{2/3}(1 - \bar{V}\varrho)/3.72\pi\eta N^{2/3}s_{av}\bar{V}^{1/3} \qquad (7)$$

Ultracentrifugation and electron microscopy[623,686] showed that the vesicles present in sonicated aqueous dispersions are spherical, in which case $f/f_0 \rightarrow 1$. The hydration number thus obtained was 0.57 g H_2O/g lipid (Table V). It is subject to large errors (cf. Ref. 618); in particular, f/f_{min} is very sensitive to errors in the partial specific volume. For instance, a change in $\bar{V}$ of 0.1% causes a 30% change in the hydration number. Another complication of this approach is the instability of the vesicles under the high pressure of the centrifugal field. Aggregation and/or fusion of phospholipid vesicles was observed during sedimentation* giving rise to larger s values. This is probably also the reason that the vesicle weight depends on the method of determination and on the rotor speed used.

The hydration number averaged from all the hydrodynamic measurements summarized in Table V, except the extreme value of Ref. 65, is about 25 mol H_2O/mol lipid (0.6 g/g lipid). When comparing this figure with the value derived from the adsorption of water vapor we have to remember that all hydrodynamic measurements were carried out with single bilayer vesicles containing water in the internal cavity which will be included in the hydration numbers. The only exception is the value of 21 mol H_2O/mol lipid (0.51 g H_2O/g phosphatidylcholine[687]), which does allow for this contribution of the enclosed water and which is directly comparable to Elworthy's value of 18 mol H_2O/mol lipid (0.44 g H_2O/g egg phosphatidylcholine). All other hydration numbers of Table V have to be corrected for water contained in the internal cavity before they can be compared with other hydration data (see Table IV). Table VI gives the order of magnitude of the correction required. The quantities of water bound to the phosphatidyl choline bilayer or trapped in the internal cavity of the vesicle were calculated from experimentally determined parameters as a function of the vesicle diameter (Table VI). The average value of 12 mol H_2O bound/mol of phosphatidylcholine was taken from Table IV and used to calculate the amount of water bound to the lipid bilayer (columns 3 and 4 of Table VI). The average vesicle diameter is about 250 Å[618,623,686] and the calculated value for the total hydration is 25 mol H_2O/mol lipid (0.61 g H_2O/g lipid, column 5, Table VI), which is in good agreement with the average hydration number derived from hydrodynamic measurements (Table V). Allowing for the free water in the vesicle cavities, the total hydration number of

* H. Hauser, unpublished observation.

TABLE VI. Hydration of Spherical Egg Phosphatidylcholine Vesicles Surrounded by a Single Bilayer[a]

Outer vesicle diameter, Å	Total amount of water enclosed[b]	Water bound on inside of bilayer[c]	Water bound on outside of bilayer[c]	Total hydration[d]
230	0.28	0.068	0.22	0.50 (21)
260	0.40	0.077	0.21	0.61 (25)
300	0.56	0.087	0.20	0.76 (32)

[a] Values in g H_2O/g vesicle.
[b] This column gives the water enclosed in the vesicle cavity plus the water bound to the inner layer of the bilayer. The amount of water enclosed was calculated for spherical vesicles with outer diameters given in the first column, assuming a bilayer thickness of 51 Å[1404] and an area per phosphatidylcholine molecule of 70 Å.[1404] The same assumptions were used in the calculation of the values in colums 3–5. The molecular weight of egg phosphatidylcholine was taken as 750 (obtained from the fatty acid composition).[556]
[c] The values in this column were obtained assuming that 12 water molecules are bound to each phospholipid molecule.[443]
[d] The values in this column are the sum of the corresponding values in columns 2 and 4. The hydration numbers in mol H_2O/mol lipid are given in parentheses.

25 mol H_2O/mol lipid (0.6 g/g lipid) reduces to 12 mol H_2O/mol lipid (0.28 g/g lipid) for a vesicle diameter of 260 Å, in good agreement with the hydration values of Table IV. The agreement with the hydration numbers of Table IV derived from more direct measurements, such as the adsorption of water vapor to anhydrous phosphatidylcholine or from nmr techniques, is encouraging. It should be stressed at this point that the hydration numbers of Table V are bulk quantities and the equations from which they were derived do not allow any conclusion about the nature and mechanism of water binding.

4.3. Thermal Analysis

Thermal analysis of dipalmitoyl phosphatidylcholine–water systems[245] showed that the transition temperature of the endothermic gel → liquid crystal transition decreased steadily as the water content increased to about 0.25 g H_2O/g lipid (10–11 mol H_2O/mol lipid), yet up to this concentration no freezing of this water was observed, even on cooling to −100°. Chapman et al.[245] suggested that the nonfreezable water is due to the formation of a hydrate with the polar head group. Similarly, with egg phosphatidylcholine, water up to 0.35 g/g lipid (15 mol H_2O/mol lipid) and with egg phosphatidylethanolamine, water up to 0.15 g/g lipid (6 mol H_2O/mol lipid) was reported not to freeze at $0°.$[239,841] Chapman and co-workers designated this water as "bound" and argued that it may represent the limiting amount of water required to maintain the organization of biological membranes. As such it may also play an important role in the interaction of ions, drugs such as anaesthetics, and other small molecules with biological membranes. It can be argued that unfreezable water may not necessarily be identical with bound water. Indeed, the determination of unfreezable water can be misleading unless carried out carefully because water may be encapsulated or trapped within "molecular" crevices and may be prevented from freezing for sterical reasons. However, inspection of Table IV shows that with phosphatidylcholines the quantity of unfreezable water is in good agreement with that of bound water.

4.4. X-Ray Diffraction

Small,[1404] studying the phase diagram of egg phosphatidylcholine, observed that at $24°$ the lamellar repeat distance was constant at about 51 Å up to 0.18 g H_2O/g lipid (8 mol H_2O/mol lipid). Bewteen 0.18 g H_2O/g lipid and 0.80 g H_2O/g lipid (34 mol H_2O/mol lipid) the lamellar repeat distance increased linearly to a limiting value of about 65 Å (Fig. 9). Small's analysis shows that there is a simultaneous lateral expansion of the bilayer, the area per phosphatidylcholine molecule increasing from about 60 Å^2 (corresponding to a monolayer surface pressure $\pi = 43$ mN/m) to about 72 Å^2 (20 mN/m). As a result of this lateral expansion the space available between the phosphorylcholine groups increases and the amount of water incorporated in this region also increases from 0.18 to 0.3 g H_2O/g lipid (12.4 mol H_2O/mol egg phosphatidylcholine). It has been discussed (Section 3.2) that this quantity of water may amount even to 15 mol H_2O/mol lipid. On the basis of this analysis Small[1404] distinguished between water present within the fully extended phosphorylcholine groups (up to 0.3 g H_2O/g lipid) and free water sandwiched between adjacent bilayers (between 0.3 and 0.8 g H_2O/g lipid).

Chapman et al.[245] measured the lamellar repeat distance of 1,2-dipalmitoyl-L-phosphatidylcholine at maximum hydration below and above T_c. At $25°$ the lamellar repeat distance increased continuously with increasing water content from about 56 Å to a limiting value of about 64 Å at 0.37 g H_2O/g lipid (15 mol H_2O/mol lipid). Above T_c (at $50°$) the long spacing increases as the water content increases up to about 0.18 g H_2O/g lipid, where a sharp drop of some 5 Å occurs due to the gel → liquid crystalline transition. This is followed by a further increase in the lamellar repeat distance until at 0.67 g H_2O/g lipid (28 mol H_2O/mol lipid) the limiting long spacing of about 63.4 Å is reached. It is clear from these X-ray data that at maximum hydration the amount of water associated with various phosphatidylcholines is related to the packing of the molecules in the bilayer. The packing density decreases in the order dipalmitoyl phosphatidylcholine at 25° > dipalmitoyl phosphatidylcholine at 50° > egg phosphatidylcholine at 25°, and as the packing density decreases so the amount of water associated with each phosphatidylcholine molecule at maximum hydration increases (from 15 to 27 mol H_2O/mol dipalmitoyl phosphatidylcholine, to 34 mol H_2O/mol egg phosphatidylcholine, respectively). The work of Phillips et al.[1166] supports this hypothesis. These authors measured, by differential scanning calorimetry, the total heat uptake at the melting point of ice in various phosphatidylcholine–water mixtures. It was found that not all the water was frozen below 0° and the difference between the total water content and the amount of frozen water was designated as "bound" water. This quantity of bound water amounted to about 10 mol H_2O/mol of dipalmitoyl phosphatidylcholine and was increased to 12 mol H_2O/mol lipid with the less tightly packed dioleoyl phosphatidylcholine.

A comparison of the hydration numbers derived from X-ray diffraction with those derived from Elworthy's[403] adsorption isotherm reveals that the amount of about 8 mol H_2O/mol lipid is in good agreement with the amount of water adsorbed in the first two hydration layers (see Table IV). The values obtained at maximum hydration are in good agreement with values derived from other techniques, while the value for the sum of bound and trapped water is somewhat higher than those obtained by nmr methods.

4.5. Diffusion Coefficient of Water in Lamellar Liquid Crystalline Phosphatidylcholine–Water Systems

Rigaud et al..[1247] designed a method which enabled them to measure diffusion coefficients of ^3H-labeled water in lamellar liquid crystalline phases of egg phosphatidylcholine. For the experimental details of the method the reader is referred to the original paper. Figure 13 shows that the diffusion coefficient of water increases with water content and reaches a plateau at 11 mol H_2O/mol lipid. There is a second rapid increase in diffusion above 21 mol H_2O/mol of lipid.

The diffusion coefficient D_{H_2O} measured at the plateau in Fig. 13, $D_{H_2O} = 3 \times 10^{-6}$ cm^2 sec^{-1}, is only a fraction of that measured in water at 22°, D(bulk H_2O) $= 2.3 \times 10^{-5}$ cm^2 sec^{-1}. This method allowed measure-

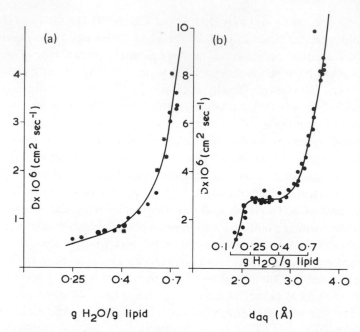

Fig. 13. (a) Diffusion coefficient D of ^{3}HHO at $22°$ in lamellar liquid crystalline phases of phosphatidylcholine–water mixtures as a function of the water content. Different symbols represent different experimental methods of preparing phospholipid—water mixtures (for details see Ref. 1247). (b) Diffusion coefficient D [as in (a)] as a function of water content and as a function of the thickness d_{aq} of the aqueous layer.

ment of diffusion coefficients up to a water content of about 0.77 g H$_2$O/g lipid. Above this water content the measurements were irreproducible due to the onset of a two-phase system (cf. phase diagram of egg phosphatidylcholine in Fig. 9). However, the domain of the lamellar liquid crystalline phase was extended up to a water content of $\sim$2.6 g H$_2$O/g lipid by incorporating 10% of the negatively charged phospholipid phosphatidic acid. The diffusion coefficient measured in these phases reached a plateau D_{H_2O} $\simeq 9.0 \times 10^{-6}$ cm^2 sec^{-1} at a water content of about 2 g H$_2$O/g lipid.

The above values of 11 and 21 mol H$_2$O/mol lipid compare well with the average values of Table IV for water bound and for the total amount of water (bound + trapped) intercalated between lipid lamellae. This agreement indicates that diffusion measurements monitor the state of water between the lipid lamellae and can distinguish readily between motionally restricted and free water. The diffusion coefficients of many nonelectrolytes show similar discontinuities as the curve in Fig. 13 when plotted against water content. These variations of the diffusion coefficient have been explained[1247] in terms of phospholipid–water interactions and in terms of steric hindrance of the water diffusion. The plateau value of the diffusion coefficient (Fig. 13) is probably characteristic of the diffusion within the region of the phospholipid head groups. As long as the water content is smaller than 10–11 mol H_2O/mol lipid, the water molecule will diffuse between the network of phosphorylcholine groups. X-ray diffraction shows[1404] that up to this water content the thickness of the aqueous layer is 20 Å, which is consistent with two fully extended phosphorylcholine groups.[1163] As more water is added the thickness of the water layer exceeds 20 Å and water is trapped between the lamellae. The finding of a constant diffusion coefficient as the water content increases from 10 to 21 mol H_2O/mol lipid may be explained as follows: Finer and Darke[443] have shown (see below) that there is fast exchange between water molecules hydrating the phosphorylcholine group and those trapped between lipid lamellae. Hence, if the added water is incorporated between opposing lamellae and resembles free water in its motional freedom, then an increase in the diffusion coefficient is to be expected. The plateau in the diffusion coefficient would suggest that for hydration between 11 mol H_2O/mol lipid (0.25 g H_2O/g lipid) and 21 mol H_2O/mol lipid ($\sim$0.5 g H_2O/g lipid) the water is mainly incorporated among the phosphorylcholine groups and that the motion of this water will be determined by that of the phosphorylcholine group. These suggestions are inconsistent with results derived from X-ray diffraction[1404] indicating a significant increase in the thickness of the water layer over that concentration range (cf. Section 3.2).

Another important finding of Rigaud et al.[1247] is that not only does the diffusion coefficient of H_2O and hydrophilic solutes, exhibit a stepwise increase with increasing water content, but so does that of hydrophobic molecules such as benzene. The diffusion of the latter type of molecule is confined to the hydrophobic interior of the lipid bilayer. An attempt was made to correlate the diffusion coefficient of benzene with hydration effects and the motional freedom in the hydrocarbon chains as determined by esr spin label experiment.[1248] Figure 14 shows the diffusion coefficient D of benzene and the rotational correlation time τ_c of the hydrophobic spin probe

as a function of the water content in phosphatidylcholine–water mixtures. Increasing motional freedom (decreasing τ_c) is accompanied by an increase in D (Fig. 14). Furthermore, the relationships of both D and τ_c show changes

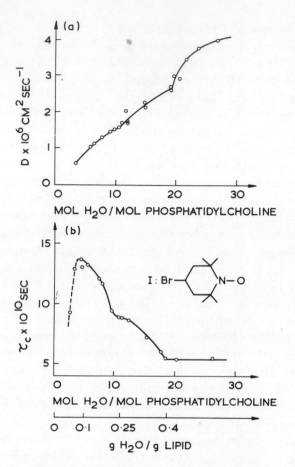

Fig. 14. (a) Diffusion coefficient D of benzene in lamellar liquid crystalline phases of phos-phatidylcholine–water mixtures as a function of hydration. (b) Rotational correlation time τ_c of spin-labeled probe I as a function of the hydration of phosphatidylcholine (from Ref. 1248).

in slope at the same water content, namely at about 10 and 18 mol H_2O/mol lipid. These values are again in good agreement with the hydration numbers of Table IV for bound water and maximum hydration, respectively. This suggests that the hydration of the polar head group determines the packing and motional freedom in the hydrophobic region of the bilayer. In order to support this suggestion, the motional freedom in the hydrophobic region of the bilayer has been monitored as a function of water content. For this purpose fatty acid spin probes (see Fig. 15) were incorporated in phos-phatidylcholine bilayers present in the lamellar liquid crystalline phase. The hyperfine splitting $\tilde{T}_{\parallel}$, which is a parameter related to the motional freedom, and the order parameter of the spin probe[692] decreased with increasing hydration, indicating that with all three spin probes tested (Fig. 15) the motional freedom increased with increasing water content. The effect is larger the closer the nitroxide radical is to the polar head group. The plots in Fig. 15 show three regions separated by changes in the slopes similar to those in Fig. 14. Up to 9 mol H_2O/mol lipid, $\tilde{T}_{\parallel}$ decreases rapidly, between 9 and 18 mol H_2O/mol lipid the slope is reduced, indicating a smaller in-

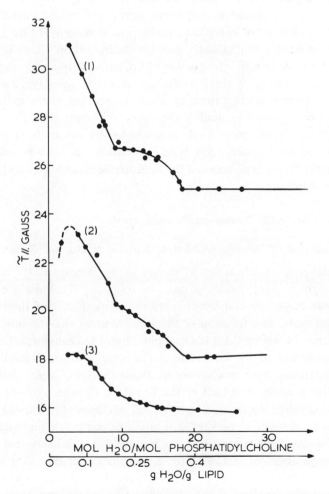

Fig. 15. Hyperfine splitting $\tilde{T}_{\parallel}$ of spin-labeled stearic acids (12,3) (curve 1), (5,10) (curve 2), and (1,14) (curve 3), incorporated in lamellar liquid crystalline phases of phosphatidylcholine–water mixtures, as a function of the water content (from Ref. 1248).

crease in motional freedom or disorder in this hydration range. At >18 mol H_2O/mol lipid no further change in the hyperfine splitting is observed. Figure 15 shows that the fluidity and disorder of the hydrocarbon chains both increase with water content and with distance from the polar backbone. The conclusion from these experiments is that the two regions of the bilayers, the polar head group and the hydrophobic region, are coupled to each other. Factors, such as hydration or the interaction with ions, that influence the packing and molecular motion in the polar head group will affect the packing and motion in the hydrophobic interior of the bilayer. Conversely, it is expected that changes introduced in the packing of the hydrocarbon chains will affect the packing and consequently the hydration in the polar head group. Changes in the hydrocarbon chain length, the number of double bonds, introduction of branched fatty acids, and changes in the packing caused by the introduction of other lipids (such as cholesterol) or proteins that are known to affect the packing of the hydrocarbon chains can be expected to modify the "state" (hydration and packing) of the polar head group region. Unfortunately, hardly any study of the hydration of mixed lipid systems has been reported so far which could throw light upon the "coupling" between the hydrophilic exterior and hydrophobic interior of a bilayer.

4.6. Proton Magnetic Resonance Measurements

4.6.1. *Hydration of Phosphatidylcholine Micelles in Organic Solvents*

Henrikson[638] studied the hydration of egg phosphatidylcholine micelles in carbon tetrachloride by pmr. By following the linewidth changes of the water resonance and the changes in chemical shift of the trimethylammonium signal as a function of the water content she was able to show that addition of about 0.1 g H_2O/g lipid (4 mol H_2O/mol lipid) caused a dramatic decrease in the linewidth of the water protons. Henrikson observed a dramatic drop in viscosity at about the same water content and suggested that some, if not all of the linewidth changes are due to a decrease in viscosity. However, Elworthy and McIntosh[405] showed that the first addition of water to a micellar solution of egg phosphatidylcholine in benzene increased the viscosity. In the light of this it is likely that the linewidth change observed at this water content is rather due to a loosening of the tight packing in the polar region of the lipid molecules allowing for greater motional freedom of the phosphorylcholine groups. Henrikson[638] concluded that the first few molecules of water are tightly bound. As progressively more water is added the changes observed in both the water and the trimethylammonium resonances approach a limit at a value of 0.3 g H_2O/g egg phosphatidylcholine ($\sim$12 mol H_2O/mol lipid). This quantity of water, giving the maximum line narrowing of the water signal and the largest upfield shift of the trimethylammonium group signal, was suggested to represent the total hydrogen-bonded hydration shell of the phosphoryl-choline group. This figure for the maximum hydration is in good agreement with the values of Table IV.

Walter and Hayes[1551] used a similar system except that carbon tetra-chloride was replaced by benzene as the organic solvent. They confirmed that addition and ultrasonic dispersion of water in the organic solvent caused line narrowing of the phosphorylcholine resonances, indicating in-creased motional freedom of this group. Figure 16 shows the linewidth dependence of the water and —$N(CH_3)_3$— resonances as a function of water content. With both resonances this treatment resulted in a stepwise decrease of linewidth as compared with the continuous decrease reported by Henrikson.[638] Both curves in Fig. 16 show three discrete regions with different slopes. The first two, A and B in each plot, reflect the completion of different hydration layers. The first layer would thus be completed at about 2.7 mol H_2O/mol egg phosphatidylcholine (0.06 g H_2O/g lipid) and the second one at 12 mol H_2O/mol lipid (0.28 g water/g lipid). It is important to note that both graphs in Fig. 16 show the same features, namely two dis-continuities in the relationships between linewidth and water content. These authors also interpret the initial steep slope as an indication that "water in the first hydration layer is much more strongly bound than in the second one." The first few water molecules induce a cooperative change in the packing of the lipid molecules. Particularly the polar head group region is affected, as indicated from the increased molecular motion in the phos-phorylcholine group. The hydrocarbon chain region is also affected, where a simultaneous loss in the chain motion was observed. As a possible ex-planation for the increase in the mobility of the phosphorylcholine group, the authors suggest that hydrogen bonding of water molecules to the phos-phodiester group replaces the electrostatic interaction between that group and the $N(CH_3)_3$ group. An alternative explanation would be that introduc-tion of water molecules between adjacent phosphate groups loosens the packing, leading to lateral expansion and increased motion in the polar head group region. The fact that only one water and one $N(CH_3)_3$ signal are observed is taken as an indication of fast exchange of H_2O molecules between different environments. These experiments clearly demonstrate that binding of H_2O is coupled to changes in the packing and molecular motion of the phospholipid molecules.

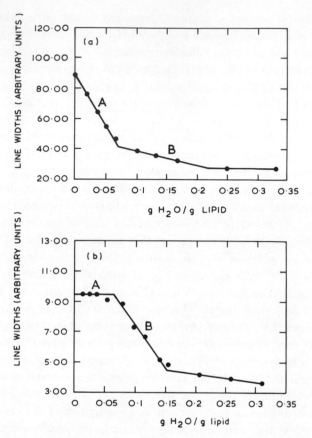

Fig. 16. Egg phosphatidylcholine was dissolved in benzene in which the phospholipid is known to form micellar solutions. Increasing amounts of water were added and dispersed by ultrasonic irradiation. Proton resonance spectra were taken from this system. Linewidth $\Delta\nu_{1/2}$ of (a) the $\overset{+}{N}(CH_3)_3$ and (b) the water signal, respectively, as a function of the water content of the system (from Ref. 1551).

4.6.2. Hydration of Phospholipids Dispersed in Deuterium Oxide

Cerbon,[233] using pmr, detected "immobilized" water at phospholipid–water interfaces by measuring linewidth and intensity of the water signal as a function of lipid and salt concentration. He used diluted, sonicated dispersions of a phospholipid mixture consisting mainly of phosphatidylcholine ($\sim$80%) and phosphatidylethanolamine and some lyso derivatives of these phospholipids. From changes in linewidth and intensity of the water signal Cerbon estimated the amount of immobilized water as about 190 mol H_2O/mol of phospholipids, which is considerably larger than the hydration numbers reported by Finer and Darke.[443] As pointed out by the author,[233] these measurements are subject to large errors, limiting the accuracy of the above hydration number. However, an important result of Cerbon's work is the increase in water signal intensity when $CaCl_2$ was added to the lipid dispersion. This indicates that the interaction of Ca^{2+} with the phosphate groups of the phospholipid reduces the hydration layer of the phospholipid by liberating "bound" water molecules. This "extrusion" of water from the lipid polar groups will in turn change the packing of the phospholipid molecules in the bilayer. From what was said above, a tighter packing in the polar head group resulting from charge neutralization and water loss is likely to be propagated to the hydrophobic region of the bilayer. A tighter packing of the hydrocarbon chains will in turn affect the conductance of the bilayer. Since lipid bilayers are now accepted as an integral part of biological membranes, this is a possible mechanism by which changes in ionic strength and environment may affect the ion permeability of membranes either directly or indirectly by influencing the interaction of the phospholipid with other membrane components, e.g., proteins.

Chapman and co-workers demonstrated by wide line nmr techniques[242,1299,1523] that introduction of water into phospholipid systems causes line narrowing and a reduction in the transition temperature. Thus, Chapman and Salsbury[242] observed line narrowing in the wide line spectra of dimyristoyl-phosphatidylethanolamine below T_c when one equivalent of ^2H_2O was added and that at the same time T_c was reduced from 116 to about 80°. The authors suggested that addition of ^2H_2O probably affects the packing of the polar groups, causing a loosening in that region of the crystal and hence facilitating the "melting" of the hydrocarbon chains. Similar phenomena were reported with other synthetic 1,2-diacyl-phosphatidylcholines.[1299] Increasing the H_2O content reduced the linewidth (second moment) of some broad components of dipalmitoyl and distearoyl phosphatidylcholine significantly and modified the shape of the broad lines at temperatures lower than T_c. At the same time the transition temperature of 1,2-dipalmitoyl-L- phosphatidylcholine was reduced from 93° to a limiting value of about 41° at 0.33 g H_2O/g lipid (12 mol H_2O/mol lipid).

A more detailed investigation of 1,2-dipalmitoyl-phosphatidylcholine by wide line nmr spectroscopy[1523] allowed further conclusions to be drawn about the hydration properties of the phosphatidylcholine polar head group. Figure 17a shows a series of wide line pmr spectra of dipalmitoyl phosphatidylcholine as a function of ^2H_2O content below T_c. All the spectra consist of three lines, the widths of which decrease with increasing ^2H_2O con-

(a)

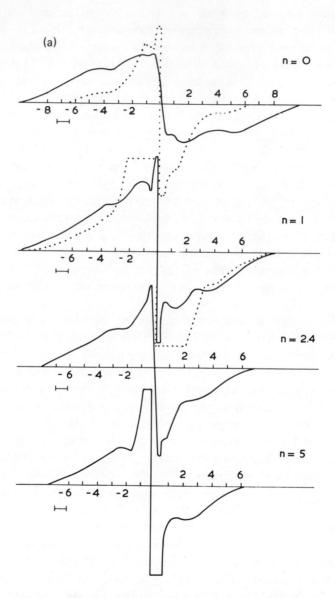

Fig. 17. (a) First derivative of absorption deuteron magnetic resonance spectra of 1,2-dipalmitoyl-L-phosphatidylcholine (solid lines) and of the perdeuterated analog (dotted lines) in the gel state (24°) as a function of water content ($n = $ mol H_2O/mol phospholipid). The amplitude of modulation is indicated and the abscissas are calibrated in G (gauss).

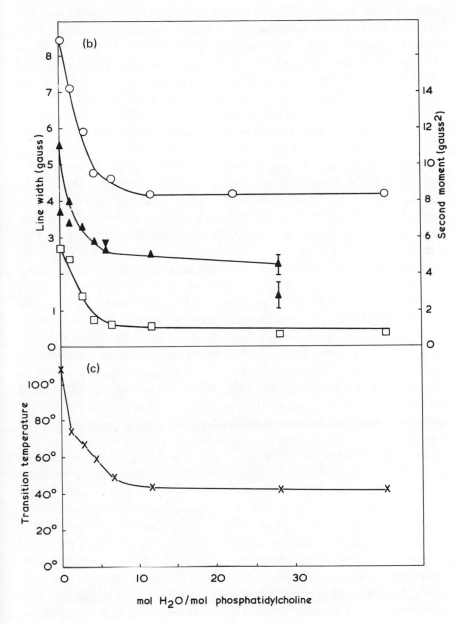

Fig. 17 (b) and (c) Variation of linewidth and second moment as a function of the water content; (○) broad component; (□) intermediate component; (△) second moment. (c) Variation of the transition temperature T_c as a function of water content (from Ref. 1523).

tent: the three components are a broad line of width ΔH between 9 and 4 G depending on the water content, a medium component of width 2.7–0.4 G, and a narrow, weak line of approximately 10^{-2} G. Figure 17b shows the variation of linewidths of the two broad components as a function of ^2H_2O content and temperature. Increasing amounts of ^2H_2O up to about 10 mol/mol phosphatidylcholine reduce both linewidth (second moment) and T_c to its minimum value of about 41° (Fig. 17c). The central, narrow component decreases only slightly with increasing ^2H_2O content. The rapid decrease in linewidths (second moments) of both broad and intermediate components to about 50% and 22% of the original values, respectively, by the addition of 4–5 mol ^2H_2O/mol phosphatidylcholine (Fig. 17b) and the considerably smaller effect of the next 5–6 mol ^2H_2O/mol lipid were interpreted by Veksli et al.[1523] as evidence for a primary and secondary hydration layer. In the first layer 4–5 mol ^2H_2O/mol of lipid (0.11 g/g lipid) is tightly bound and the second layer comprises a further 5–6 mol ^2H_2O/mol lipid more loosely associated with the lipid; thus the hydration totals about 10 mol ^2H_2O/mol lipid. Addition of ^2H_2O in excess of the first 5 mol of ^2H_2O has a much smaller effect on linewidth (Fig. 17b) and molecular motion and ^2H_2O in excess of 10 mol/mol lipid produces no observable effect.

By a comparison of the wide line spectra of dipalmitoyl phosphatidylcholine and of a dipalmitoyl phosphatidylcholine containing perdeuterated hydrocarbon chains, the authors were able to assign the three components observed in the wide line spectra. The broad component was assigned to the hydrocarbon chain protons, the two CH_2 groups of the choline residue, and some protons of the glycerol group. The intermediate line was identified with the protons of the $N(CH_3)_3$ group, and the narrow component of width $\sim 10^{-2}$ G with water protons. With these assignments changes in molecular motion in different regions of the phosphatidylcholine molecule could be monitored and related to the uptake of ^2H_2O. The dominant effect of the first five molecules of ^2H_2O is on the motion of the phosphorylcholine group, as is evident from the fivefold reduction in linewidth of the $N(CH_3)_3$ signal. This effect was ascribed to the penetration of the ^2H_2O molecules between the phosphorylcholine groups of the crystal lattice, forming hydrogen bonds to the oxygen atoms of this group, thereby weakening the ionic interaction between the phosphodiester PO_4^- and the $\overset{(+)}{N}(CH_3)_3$ groups. This ionic interaction represents a major contribution to the free energy in the crystal state and, as a consequence of the interaction with water, the phosphorylcholine group takes up a more extended conformation (as evident from X-ray diffraction results[1163]). However, it is likely that the phosphatidylcholine used by these authors, despite the thorough drying, was at least partially present as the monohydrate. In the monohydrate the head group is in the extended conformation[1163] and the increase in molecular motion as more water is added would in this case be due to the lateral expansion of the polar head group region. As mentioned above, the effects of the ^2H_2O penetration into the phosphorylcholine group, i.e., loosening of the packing in the polar head group, is propagated into the hydrocarbon chain region, resulting in a significant narrowing of the broad line assigned to the methylene chain protons.

Gottlieb et al.[552] measured the proton relaxation times T_1 and T_2 of water in mixtures with egg phoshatidylcholine. Knowing T_1 of bulk water $(T_1)_f$ and of bound water $(T_1)_b$, which was obtained from crystalline phosphatidylcholine hydrates, the mole fraction P_B of bound water was calculated from

$$\frac{1}{T_1} = \frac{1}{(T_1)_f} + \frac{P_B}{t_B + (T_1)_b} \tag{8}$$

where t_B is the lifetime of bound water $[t_B \ll (T_1)_b]$. In mixtures containing between 0.1 and 0.45 g/g lipid, 6 mol of water was found to be bound per mole of phosphatidylcholine. This is also the concentration at which the plot of water proton linewidth $\Delta\nu_{1/2}$ as a function of water content exhibits a distinct change in slope.[552] The use of eqn. (8) is based on the assumption of fast exchange of H_2O molecules between only two different states, the bound and the free states. Fast exchange between water molecules in different environments has also been observed in deuteron resonance studies of ^2H_2O–phospholipid mixtures (cf. Ref. 443 and 1301). However, the second assumption, of only two different types of water, eliminates the possible existence of intermediate states, as observed in deuteron resonance experiments (see below). This elimination of intermediate states is possibly the reason for the difference in the quantities of bound water derived by Gottlieb et al.[552] and those from deuteron magnetic resonance studies.[443,1301]

4.7. Deuteron Magnetic Resonance Measurements

Deuteron magnetic resonance (dmr) was originally applied in hydration studies of surfactants such as dimethyl dodecylamineoxide[861] and potassium laurate.[246] The first study of the hydration of phospholipids using this technique was carried out by Salsbury et al.[1301] and this was followed by a more detailed, comparative study by Finer and Darke[443] of the hydration of three natural phospholipids, egg phosphatidylcholine, egg phosphatidylethanolamine, and ox brain or spinal cord phosphatidylserine.

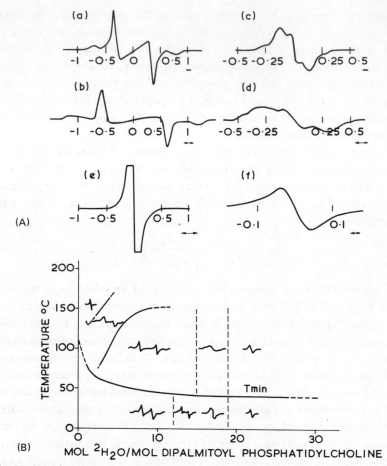

Fig. 18. (A) First derivative of absorption deuteron magnetic resonance spectra of dipalmitoyl phosphatidylcholine–^2H_2O mixtures as a function of water content for 8.7 mol ^2H_2O/mol lipid at (a) 23° and (b) 100°; for 17.3 mol ^2H_2O/mol lipid at (c) 23° and (d) 100°; and for 24 mol ^2H_2O/mol lipid at (e) 23° and (f) 100°. (B) The temperature T_{min} at which the dmr doublet splitting was a minimum, as a function of water content.

Figure 18 shows dmr spectra of dipalmitoyl phosphatidylcholine–^2H_2O mixtures at different temperatures. From a line shape analysis it can be shown (cf. Ref. 443) that the "doublet" spectra (e.g., Fig. 18a) are typical powder spectra characteristic of a nuclear spin $I = 1$ and the presence of an electric field gradient; furthermore, this electric field gradient is axially symmetric with respect to the ^2H_2O molecule, i.e., it is defined by an asymmetry parameter $\eta \simeq 0$.[249] Thus, the doublet splitting is $\Delta\nu = 3e^2qQ/4h$, i.e., three-quarters of the quadrupole coupling constant,[266] which is determined by the quadrupole moment eQ/h of the deuteron and by the electric field gradient eq at the deuteron nucleus.

From the splittings $\Delta \nu$ information about the motion of the ^2H_2O molecules is obtained, particularly about the degree of anisotropy of this motion. The maximum electric field gradient is along the O–D bond and the observation of a splitting $\Delta \nu$ or a nonzero quadrupole coupling constant indicates that the motion of the ^2H_2O molecule is insufficient to average out to zero the electric field gradient, or in other words, that the O–D bond is oriented with respect to the applied magnetic field (laboratory frame) over the time scale of the nmr experiment (~ 1 μsec). The splitting $\Delta \nu$ depends both on the ^2H_2O content and on the temperature. At low ^2H_2O content ($n \leq 9$ mol ^2H_2O/mol lipid) only doublets were observed, independent of temperature. The splitting $\Delta \nu$ of these doublets is of the order of 1–2 kHz and is small compared with that of crystalline hydrates or polycrystalline heavy ice (200 kHz). Salsbury et al.[1301] interpreted this as an indication that at low water content the bound ^2H_2O is undergoing rapid but anisotropic motion and that the rate and mode of reorientation are insufficient to average out the electric field gradient to zero. At a given ^2H_2O content the splitting $\Delta \nu$ decreased with increasing temperature and reached a minimum at T_{min}, which is close to, but consistently a few degrees lower than, the thermal transition temperature T_c of the same phospholipid–^2H_2O mixture (see line labeled T_{min} in Fig. 18B). At $T > T_{min}$, $\Delta \nu$ increases again, which may be explained in terms of the lateral expansion of lipid bilayers creating more space and openings for ^2H_2O molecules to penetrate between the polar head groups. Figure 18B shows an approximate version of the phase diagram of 1,2-dipalmitoyl phosphatidylcholine in the presence of ^2H_2O, together with the dmr spectra characteristic of various regions (cf. Fig. 8). The temperature T_{min} of the minimum dmr splitting decreases with increasing ^2H_2O, reaching a constant value at $n \sim 10$ mol ^2H_2O/mol phospholipid (Fig. 18B). At $n \gtrsim 12$ mol H_2O/mol lipid (see Figs. 18 and 20), the dmr spectra show a doublet together with a central singlet and at $n \gtrsim 20$, singlets only were observed at all temperatures. From these changes in the dmr spectra with ^2H_2O content three types of hydration were distinguished. Above the gel $\rightarrow$ liquid crystal transition temperature the innermost layer consists of 4 mol ^2H_2O/mol lipid; this water was considered as adsorbed (bound) and partially oriented in the sense that the O–D bond largely maintains its orientation; second, there is free water above $n \gtrsim 20$ mol ^2H_2O/mol lipid, characterized by rapid, isotropic motion; and third, there is water at $4 < n < 20$ in an intermediate state. These results agree in outline, although not in detail, with the more recent dmr study by Finer and Darke.[443] Salsbury *et al.*[1301] also compared the hydration of dipalmitoyl phosphatidylcholine with that of egg phosphatidylcholine; they found that with egg phosphatidylcholine the dmr splittings $\Delta\nu$ were somewhat smaller but in general the hydration behavior was similar to that of the saturated phosphatidylcholine.

Finer and Darke[443] studied the hydration of phosphatidylcholine, phosphatidylethanolamine, and phosphatidylserine as a function of the ^2H_2O content, temperature, and time. When the doublet splitting $\Delta\nu$ is

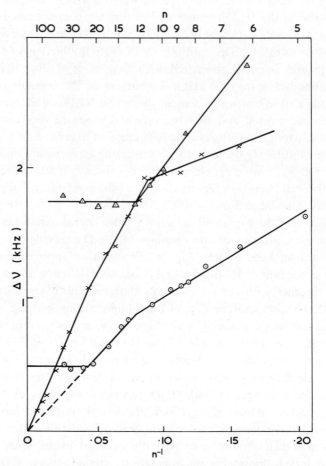

Fig. 19. Variation of the doublet splitting $\Delta\nu$ (in kHz) observed in deuteron magnetic resonance spectra of phospholipid–^2H_2O mixtures with $1/n$, where n = mol ^2H_2O/mol lipid. ($\triangle$) Phosphatidylethanolamine; ($\times$) monosodium salt of phosphatidylserine; ($\bigcirc$) egg phosphatidylcholine (from Ref. 443).

plotted as a function of $1/n$ ($n = $ mol H_2O/mol lipid), a series of linear regions with different slopes is obtained (Fig. 19). Now Δv is related to n by

$$\Delta v = (1/n) \sum_{i=1}^{j-1} n_i^0 (\Delta v_i - \Delta v_j) + \Delta v_j \qquad (9)$$

where n_i^0 is the maximum occupancy of the ith hydration shell and Δv_i is the doublet splitting characteristic of the ith shell.[442] A number of assumptions are involved in the derivation of eqn. (9); nevertheless the application of this treatment to the experimental data yields reasonable results, which are summarized in Table VII. The splittings Δv_i characteristic for various hydration states are obtained from the intercepts of the linear regions of the Δv axis (Fig. 19) and the maximum occupancy n_i^0 of various shells is obtained from the slopes of the straight lines. For the characteristic splitting Δv_1 of the innermost shell to be derived, an occupancy number n_i^0 has to be assumed. This was taken as $n_i^0 = 1$ and assumed to apply to all three phospholipids investigated.[442]

The results for phosphatidylcholine are in good agreement with the average hydration number of phosphatidylcholine (cf. Table IV). The results of Table VII are also in agreement with the conclusions derived from the time dependence of the dmr spectra (see below). The first two hydration shells, the inner and the main hydration shell (see Table VII), consist of 12 mol ^2H_2O/mol lipid in the case of phosphatidylcholine and ethanolamine and 11 mol ^2H_2O/mol lipid with phosphatidylserine. The three phospholipids differ significantly in the quantities of weakly bound and trapped water. Weakly bound water was only found with phosphatidylserine and this ^2H_2O was suggested to be associated with the sodium counterion. It would still undergo isotropic motion due to random motion of the Na^+ ion; however, the small residual splitting $\Delta v_3 = 0.23$ kHz (Table VII) indicates a weak association of Na^+ with the lipid polar head group.[622] While no water trapped between the lipid lamellae was detected with phosphatidylethanolamine, the value for phosphatidylserine of $n_4^0 \sim 120$ mol ^2H_2O/mol lipid was much larger than that for phosphatidylcholine, $n_4^0 = 11$ mol ^2H_2O/mol lipid. Atkinson et al.[61] studied the phase behavior of ox brain phosphatidylserine and observed a single phase up to a water content of 3 g H_2O/g lipid, consistent with about 140 mol H_2O/mol phosphatidylserine (cf. Table VII). The large quantities of water trapped between phosphatidylserine bilayers result from the repulsion between electric double layers at the interface of the negatively charged phosphatidylserine layers.

TABLE VII. Hydration Shells of Phospholipids with the Characteristic Doublet Splitting $\Delta \nu_n{}^a$

Hydration shell	Phosphatidylcholine	Phosphatidylethanolamine	Monosodium salt of phosphatidylserine
1. Inner hydration shell (assumed $n_1^0 = 1$)	$\Delta \nu_1 = 6.9$ kHz	$\Delta \nu_1 = 13.6$ kHz	$\Delta \nu_1 = 4.2$ kHz
2. Main hydration shell	$n_2^0 = 11$ $\Delta \nu_2 = 0.37$ kHz	$n_2^0 = 11$ $\Delta \nu_2 = 0.65$ kHz	$n_2^0 = 10$ $\Delta \nu_2 = 1.56$ kHz
3. Weakly bound water	None	None	$n_3^0 = 12$; $\Delta \nu_3 = 0.23$ kHz
4. Trapped water ($\Delta \nu_4 = 0$)	$n_4^0 = 11$	None	$n_4^0 \sim 120$
5. Condition for presence of bulk water	$n > 23$	$n > 12$	$n \gtrsim 140$

^{a} For details see text and Ref. 443.

The temperature dependence of the doublet splitting was similar to that reported for dipalmitoyl phosphatidylcholine–^2H_2O mixtures[(1301)] with a minimum near the transition temperature. The explanation given for this behavior is that the bilayer expands laterally at the transition temperature and the increased space available for each molecule in the bilayer allows a redistribution of bound water.[(443)]

The dmr spectra of both phosphatidylcholine and phosphatidylserine vary with time. Figure 20 shows the time dependence of phosphatidylcholine–^2H_2O spectra corresponding to four different concentrations. At low ^2H_2O content (up to 12 and 11 mol ^2H_2O/mol lipid for phosphatidylcholine and phosphatidylserine, respectively) no time dependence was observed. However, at a ^2H_2O content of 12–23 mol ^2H_2O/mol lipid for phosphatidylcholine and 11–140 mol ^2H_2O/mol lipid for sodium phosphatidylserine the initial, broad singlet in the spectrum changed progressively to a powder pattern and after equilibration the singlet completely disappeared. The time scales of these changes are of the order of a week and a month for phosphatidylcholine and phosphatidylserine, respectively. At water contents exceeding $n \sim 23$ and 140 for phosphatidylcholine and phosphatidylserine, respectively, the final, time-independent spectra still contained a singlet in addition to the doublet. These time-dependent changes were taken as evidence that all the ^2H_2O molecules are bound up to about 12 mol ^2H_2O/mol

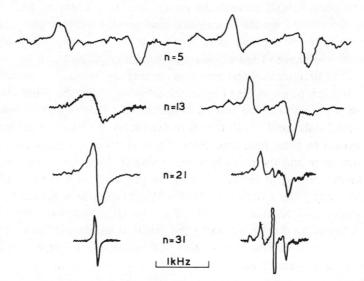

Fig. 20. First derivative of absorption deuteron magnetic resonance spectra of egg phosphatidylcholine–^2H_2O mixtures as a function of water content (n = mol ^2H_2O/mol lipid) and storage time of the sample (from Ref. 443).

lipid and that the ^2H_2O molecules, betwéen $n = 12$ and 23 for phosphatidylcholine and $n = 23$ and 140 for sodium phosphatidylserine, are trapped between the bilayers. While the bound water exhibits anisotropic motion characterized by doublet splittings $\Delta \nu^n$ (Table VII), the motion of the trapped water is rapid and isotropic, characterized by a singlet of very small intrinsic linewidth. The appearance of either broad singlets or of doublet spectra after equilibration was explained in terms of rapid exchange ($>10^4$/sec) between "bound" and trapped water, giving rise to averaged nmr spectra. Free water is present when $n > 23$ for phosphatidylcholine and $n > 140$ for sodium phosphatidylserine, and this is represented by the sharp singlet (Fig. 20 for $n = 31$). The presence of a sharp singlet superimposed on the doublet spectrum was interpreted as being due to slow exchange ($<10^2$/sec) between trapped and free water. The time-dependent changes from a broad singlet to two superimposed dmr spectra of a sharp singlet and a doublet were explained in terms of a time-dependent ordering process in the phospholipid lamellae. In the course of this ordering, multilamellar regions with strictly parallel bilayers which initially, after sample preparation, are locally confined and randomly oriented with respect to each other, fuse into more extensive regions. Similar time-dependent changes in phosphatidylserine dispersions were observed by the application of X-ray diffraction and electron microscopy.[61] The ^2H_2O is associated with the phospholipid polar head group and its orientation and motion will be determined by the orientation and motion of this group. In the initial environment the lipid molecules taken, for instance, over a distance of 0.5 μm (= length of the diffusion path of a ^2H_2O molecule in 10^{-3} sec, which is the time scale of the dmr experiment) are oriented randomly with respect to each other. A ^2H_2O molecule diffusing along the water channels between the lipid lamellae will exchange rapidly between the bound and the trapped states and it will therefore experience a variety of orientations when bound to these lipid molecules. There is no preferred orientation in the initial state and the result is a broad singlet. However, after equilibration the dimensions of the lamellae are of the order of 1 μm and all lipid molecules along the diffusion path of a ^2H_2O molecule will then have the same orientation because they are all in the same lamellae. Whenever a ^2H_2O is bound in this environment the orientation information is retained and the axis of its anisotropic motion will maintain a constant angle with respect to the magnetic field. In this case a doublet is observed in the dmr spectrum.

Finer[442] described the motional states of the ^2H_2O molecules in various hydration shells of phosphatidylcholine by calculating rotational correlation times τ_c. The values obtained are $\tau_c = 10^{-7}$ sec for the innermost shell of bound water, $\tau_c \lesssim 8 \times 10^{-10}$ sec for the main hydration shell of 11 ^2H_2O molecules, and $\tau_c \lesssim 3 \times 10^{-10}$ sec for trapped water. The characteristic splittings $\Delta\nu_2$ of the main hydration shells (Table VII) were shown[443] to be determined by the motions of the phospholipid polar groups to which the ^2H_2O molecules are bound. Inspection of Table VII shows that the anisotropy of motion of the binding sites, i.e., the polar head groups, increases in the order phosphatidylcholine < phosphatidylethanolamine < sodium phosphatidylserine. The conclusion that the phosphorylcholine head group has greater motional freedom than the phosphorylethanolamine is consistent with the larger area occupied per phosphatidylcholine molecule and with its different head group conformation.[1163] When the spectra of lipid N^2H_3 groups are analyzed similar conclusions regarding the motions of the polar groups[1163] are derived.[443] Observation of separate nmr signals for ^2H_2O and the N^2H_3 groups indicates that deuteron exchange is slow ($<10^4$ sec^{-1}). This may indicate that the principal hydration shell is centered around the phosphate group, rather than the nitrogen, which would also explain the similar occupancy of the principal hydration shell obtained with all three phospholipids.

The characteristic splitting $\Delta\nu_1$ of the inner hydration shell (Table VII) was shown[442] to be partly determined by the binding energy of ^2H_2O. The $\Delta\nu_1$ for phosphatidylcholine is smaller than that for phosphatidylethanolamine, consistent with the less anisotropic motion of the phosphorylcholine group. Comparison of $\Delta\nu_1$ values (Table VII) shows that the binding energy is in the order phosphatidylserine < phosphatidylcholine < phosphatidylethanolamine. That ^2H_2O in the inner hydration shell is less tightly bound to sodium phosphatidylserine was suggested to be due to adjacent phosphate groups being kept further apart by electrostatic repulsion. At acidic pH phosphatidylserine does not take up water at all and hence cannot be dispersed. In the acidic form the carboxyl groups of phosphatidylserine are undissociated and may form a hydrogen-bonded network both within and across lamellae.

Contrary to the findings of Finer and Darke,[443] Glasel,[526] from studies of T_1 of ^2H_2O in dilute solutions of macromolecules, concluded that only groups with proton donors were hydrated.* Ionized groups, lacking exchangeable protons, showed little or no effect on T_1. The explanation for this discrepancy may be that T_1 is less sensitive than the linewidth and

---

* These measurements, as they reflect the hydration of macromolecules, are also discussed in Chapters 5 and 7.

is not affected significantly by water molecules "bound" and undergoing rapid but anisotropic motion.

Rand et al.,[1216] studying the phase behavior of naturally occurring phosphatidylethanolamines, reported restricted swelling of this phospholipid, in agreement with Finer and Darke. Using the X-ray data of the former authors,[1216] it can be calculated that a two-phase system (lamellar liquid crystalline in excess water) is formed at water contents greater than 12 mol H_2O/mol lipid. This value is similar to that of the maximum hydration of phosphatidylethanolamine (Table VII). The difference in swelling properties between phosphatidylcholines and phosphatidylethanolamines has also been observed in calorimetric studies[841] which suggest that 10 mol H_2O/mol lipid is unfreezable in water–phosphatidylcholine mixtures but only 7–8 mol H_2O/mol lipid in water–phosphatidylethanolamine mixtures.

4.8. Mixed Lipid Systems

4.8.1. *Phosphatidylcholine–Cholesterol Mixtures*

Phosphatidylcholine and cholesterol are major constituents of membrane lipids. Hence, their physicochemical properties in aqueous phosphatidylcholine–cholesterol dispersions have been the subject of intensive investigation.[158,241,317,318,842,865,1161] The results up to 1972 have been summarized and reviewed in detail recently by Phillips.[1161] Bourgès et al.[158] have determined the complete phase diagram of the ternary system egg phosphatidylcholine–cholesterol–water at 25°C, i.e., somewhat above the transition temperature T_c for the phospholipid. With all mixtures containing more than 0.05 g H_2O/g lipid water, lamellar phases were observed and the maximum cholesterol content is one molecule of cholesterol per molecule of phosphatidylcholine. Lecuyer and Dervichian[865] pointed out that above the transition temperature T_c incorporation of cholesterol on the one hand and the addition of water on the other produced opposite effects on the phase behavior. Addition of cholesterol up to 0.25 g/g lipid (molar ratio phospholipid to cholesterol 2:1) causes a condensation of the bilayer (increasing the bilayer thickness), while addition of water loosens the packing (decreasing the bilayer thickness). The condensing effect of cholesterol above T_c was demonstrated by showing that both the mean molecular area (mean area of phospholipid plus cholesterol) as well as the partial molecular area of the phospholipid decreased continuously with increasing cholesterol content. This was detected in the lamellar liquid crystalline phase using X-ray diffraction and nmr spectroscopy[318,1111] as well as in mono-layers.[241] The effects of incorporating cholesterol into phosphatidylcho-line bilayers on the motional freedom of the methylene groups along the hydrocarbon chains was studied by proton[317,318] and deuteron magnetic resonance[1111] as well as electron spin resonance.[692] The first eight[692] to ten[318] methylene groups in the fatty acid chains are severely restricted in their motion when cholesterol is present, while the terminal methyl ends of the chains are rather free, with rapidly increasing probability of trans–gauche isomerization. Despite the tighter packing of the hydrocarbon chains close to the glycerol backbone, this latter group was shown[318] to have greater freedom of motion than it possesses in either the gel or the liquid crystalline phase of pure phosphatidylcholine bilayers.

Addition of cholesterol caused first an increase in the X-ray long spacing, the effect being more pronounced with dipalmitoyl phosphatidyl-choline[842] than with egg phosphatidylcholine,[865] and above a molar ratio of phospholipid/cholesterol of 9:1 and 2:1, respectively, the X-ray long spacing decreased, reaching a limiting value at equimolar ratio. The initial increase in the X-ray long spacing and the bilayer thickness was explained by Lecuyer and Dervichian[865] in terms of the condensation effect choles-terol has on the lipid hydrocarbon chains. The closer packing leads to a change in the angle of tilt of the hydrocarbon chains, which attained an almost upright orientation in the presence of cholesterol (with the chain axis perpendicular to the plane of the bilayer). This accounts partially for the observed increase in the X-ray long spacing. In order to account for the total increase, an increase in the thickness of the water layer was pos-tulated. The closer packing in the hydrocarbon chains would also lead to a closer packing in the head group region, squeezing out some of the water present around the phosphorylcholine groups in each layer. This would in turn lead to an increase in the thickness of the water layer associated with the polar head group. This condensing effect cholesterol has on the polar head group was first described by Chapman et al.,[241] who incorporated cholesterol into monolayers of distearoyl phosphatidylcholine, the hydro-carbon chains of which were already in the condensed state. This effect was further discussed by Finer and Phillips[448] and the explanation given is consistent with the experimental finding that the amount of bound water is not changed significantly; if anything, it decreases slightly when choles-terol is first added. The decrease in the long spacing as more cholesterol was added was explained as follows. As more cholesterol (molar ratio phospholipid:cholesterol < 2:1) is inserted, the condensing effect on the lipid hydrocarbon chains levels off and is compensated for by an increased spacing out of the phospholipid molecules. This steric effect allows water to penetrate between the phosphorylcholine groups, thereby restoring the polar head group hydration to its maximum level. This process will lead to a decrease in the thickness of the water layer in a system of fixed phospholipid–water composition as the cholesterol content is increased. While with pure egg phosphatidylcholine the total amount of water (bound plus trapped) amounts to 0.8 g H_2O/g lipid, in an egg phosphatidylcholine–cholesterol system (molar ratio 1:1) the maximum amount of water intercalated between bilayers is 0.55 g H_2O/g lipid. Assuming that the cholesterol is not hydrated and all the water is associated with the phospholipid molecules, 0.55 g H_2O/g lipid is the same number of water molecules/phospholipid molecule as 0.8 g H_2O/g lipid in the pure phospholipid system. In summary it may be concluded that addition of cholesterol initially causes a decrease in the amount of water associated with the phosphorylcholine groups due to the condensing effect of cholesterol on the hydrocarbon chains which is propagated to the polar head group. Tighter packing in the polar head groups causes water to be squeezed out from the space around the phosphorylcholine groups in each monolayer. This effect is outweighed at higher cholesterol concentration by spacing out of the phosphorylcholine groups so that the hydration of egg phosphatidylcholine in bilayers containing an equimolar mixture of phospholipid and cholesterol is similar to that of pure phosphatidylcholine at temperatures above T_c.

Ladbrooke et al.[842] studied the effect of cholesterol on phosphatidylcholine–water systems under quite different conditions. These authors used synthetic 1,2-dipalmitoyl-L-phosphatidylcholine and all their measurements were carried out at 25°C, well below the gel-to-lipid crystalline transition. Ladbrooke et al.[842] noted that equimolar mixtures of phospholipid and cholesterol are dispersible over a much wider temperature range than the phospholipid alone. The endothermic heat absorbed decreased with increasing cholesterol content and in equimolar mixtures no transition was observed. The X-ray repeat distance D increased first up to a cholesterol content of about 10 mol % from about $D = 64$ Å to $D = 80$ Å. This increase in D of 16 Å was ascribed partly to a change in the angle of tilt of the hydrocarbon chains which accounts for an increase in D of about 7 Å. In order to account for the total increase in the X-ray long spacing, an increase of 9 Å in the thickness of the water layer was postulated. The explanation for the large increase in the thickness of the (water + head group) layer given by Ladbrooke et al.[842] involves the reorientation of the phosphorylcholine group as the result of the perpendicular alignment of the hydrocarbon chains. Addition of more cholesterol up to a molar ratio of 1:1 had the following effects. The endothermic heat absorbed decreased; so did the X-ray long spacing, reaching a limiting value of about 65 Å. At the same time the proportion of water bound increased at the expense of free water intercalated between bilayers. This was deduced from the reduction in the heat absorbed at the melting point of ice in a frozen phosphatidylcholine–cholesterol–water system. This heat is directly proportional to the amount of free water and its reduction, i.e., the heat loss, was linearly related to the cholesterol content. This finding supports the explanation given above that the spacing out of phosphatidylcholine molecules as more and more cholesterol is inserted allows for maximum hydration of the phosphorylcholine groups. The increase in water binding and the increased disorder in the hydrocarbon chains introduced by interspersing cholesterol molecules accounts for the decrease in the thickness of both hydrophobic and hydrophilic regions and thus in the X-ray repeat distance. The maximum amount of water incorporated between dipalmitoyl phosphatidylcholine/cholesterol bilayers (molar ratio 1:1) was about 0.60 g H_2O/g lipid, similar to that found by Lecuyer and Dervichian.[865] The work[842,865] discussed above clearly shows that, depending on the packing of the phospholipid or the amount of cholesterol inserted, changes in the relative quantities of bound and trapped water can occur. Unfortunately, experimental data describing the hydration of phospholipid–cholesterol mixtures on a molecular level are not yet available.

4.8.2. *Glyceride–Phospholipid Mixtures*

Finer and Phillips[448] showed that the stability of colloidal particles containing phosphatidylcholine and diglycerides or triglycerides decreased and aggregation occurred as the phospholipid content decreased. The stability of phosphatidylcholine–diglyceride or -triglyceride emulsions in water decreased in the order dipalmitin > dilaurin > tripalmitin, while the molecular areas of the glycerides increase in this order. This explains why on a molar basis the triglyceride particles are least stable. From what was said about the hydration of phosphatidylcholine and from the fact that diglycerides and triglycerides are insoluble and do not swell in water it seems reasonable to assume that particles with a complete coverage of their surface with phosphatidylcholine are of lowest free energy. As the phospholipid content decreases, more glycerides will be exposed at the surface; the total hydration of the particle decreases until at a certain threshold value no further reduction in the hydration of the particle can be tolerated and flocculation commences. These dispersions are examples of making use of the special hydration properties of phosphatidylcholine in emulsifying water-insoluble compounds. Studies of the hydration of phospholipids on a molecular level such as described above have greatly aided our understanding of the mechanism of emulsification, which is of fundamental importance in food technology.

4.9. Conclusions

The agreement between the hydration numbers in Table IV is remarkable considering the different physical techniques used and considering that phosphatidylcholines with different hydrocarbon chains were used in different systems: as unsonicated lamellar liquid crystalline phases, as sonicated single bilayer vesicles, and as micellar solutions in organic solvents. Although macroscopic methods such as adsorption or hydrodynamic measurements can only give information about the quantities of water bound, without specifying the strength of the interaction, some of these methods[403, 638] nevertheless provide indirect evidence of the existence of different hydration states. The "molecular" methods, such as nmr spectroscopy, allow a differentiation to be made between various types of water in terms of strength of binding. The values for the total amount of water bound to phosphatidylcholine (Table IV) average at 12.5 mol H_2O/mol phosphatidylcholine. The average figure for the total amount incorporated between phosphatidylcholine bilayers is about 24 mol H_2O/mol lipid. The values for tightly bound water and the "inner" hydration shells given in Table IV show a much larger spread depending on the physical method used. The values vary between 1 and 6 mol H_2O/mol lipid. The first few molecules of water bound to the phospholipid probably differ significantly in their strength of binding and the range of one to six molecules probably reflects differences in the sensitivity of the physical techniques used. One criticism which can be leveled against the use of dmr for investigating the interaction of water with macromolecules is that the isotopic replacement of H_2O by ^2H_2O may produce significant changes in the hydration interactions. Walter and Hayes[1551] reported that the phosphorylcholine group in phosphatidylcholine was less effectively "hydrated" by ^2H_2O and the first and second hydration layers were decreased by one molecule of ^2H_2O per molecule of phosphatidylcholine. Furthermore, the linewidth of the $N(CH_3)_3$ group decreased with increasing quantities of ^2H_2O, but the limiting linewidth was significantly broader than that observed with H_2O. From these results and a comparison of the hydration numbers derived by proton and deuteron nmr (Table IV) we may conclude that there are differences in the interaction of H_2O and ^2H_2O with phospholipids, but the differences are small and insignificant, considering the accuracy of the measurements in Table IV.

5. HYDRATION OF GLYCERIDES

Glycerides are fatty acid esters of glycerol. Their uncharged polar head groups consist of ester groups and free OH groups according to $n_1 + n_2 = 3$, where the number of free OH groups n_1 is zero, one, or two and the number of ester groups n_2 is one, two, or three. A comparison of the hydration of phospholipids with that of simple glycerides should throw some light on the specific hydration properties of each polar head group.

The triglycerides ($n_1 = 0$, $n_2 = 3$) cannot be grouped in either of the three classes of amphiphathic molecules that Lawrence[858,859] distinguished on the basis of the interaction with water (see Fig. 1). With triglycerides the hydrophilic–hydrophobic balance is such that there is no significant interaction with water. Elworthy[403] reported that the adsorption of water vapor by tristearin and triolein was negligible compared with that of egg phosphatidylcholine. Triolein adsorbed slightly more water (8 mg H_2O/g lipid) than tristearin (5.6 mg H_2O/g lipid). A well-known consequence of the fact that triglycerides are practically immiscible with water is the requirement of pancreatic lipases for intestinal absorption. These enzymes hydrolyze the triglyceride components of foodstuffs to lipids, which interact with water and aqueous bile salt to form absorbable dispersions.

With the monoglycerides of class I (see Section 1) the hydrophilic character is greatly increased by the introduction of additional polar groups and as a consequence these compounds form lyotropic liquid crystalline phases. Monoglycerides are insoluble in water but at saturation take up some 0.8 g H_2O/g lipid, forming liquid crystalline phases over extensive areas of their phase diagrams.[859] The phase behavior of monoglycerides has been studied extensively by several groups.[400,855,860,930,970] Besides several isotropic phases the structures of which are less well characterized (see Ref. 856), lamellar (neat) and hexagonal I or II (middle) liquid crystalline phases, similar to those found in soap–water and phospholipid–water mixtures, have been observed with monoglycerides. The behavior of monoglycerides resembles in many respects that of phospholipids and similar arguments can be put forward to explain this observation .

Larsson[855,857] investigated the phase behavior of various 1-monoglycerides, from 1-monohexanoin up to monobehenin, and characterized the structural parameters of different phases from X-ray diffraction measurements. He also compared the phase behavior of 2-monoglycerides with that of the corresponding 1-monoglycerides and found that both isomers formed similar liquid crystalline phases except that the area per 2-monoglyceride molecule is about 3 Å² larger than that of the corresponding 1-monoglyceride. Similar to the lamellar liquid crystalline phases of phospholipids, he observed a decrease in the thickness of the bilayer and a concomitant increase in the area per monoglyceride molecules as the temperature and/or the water content increased. Since the area/molecule (or the area/alkyl chain) can be taken as a measure of the degree of motional freedom and/or orientation of the chains, this lateral expansion of the bilayer was interpreted as an increase in the motional freedom of the monoglyceride molecules. The values for the area/alkyl chain which Larsson derived from his X-ray analysis are very similar to those reported for the lamellar liquid crystalline phases of soaps[698] and detergents and are half of those characteristic of the lamellar liquid crystalline phases of phospholipids.[931] This fairly constant area/alkyl chain indicates that, within the experimental error, the orientation and packing of the hydrocarbon chains in lamellar liquid crystalline phases are similar, irrespective of the nature of the polar head group. Similar characteristic values for the area/alkyl chain were obtained for hexagonal and cubic liquid crystalline phases of monoglycerides, where the cross section varies from one end of the molecule to the other.

    Larsson and co-workers[855,857] also investigated the gel state of monoglyceride–water mixtures. Similar to the results with phospholipids, liquid crystalline phases of monoglyceride–water mixtures form, when cooled below the transition temperature, either coagels (mixtures of crystals and water) or metastable gels in which the lipid bilayers are still separated by layers of water of variable thickness (upper limit 15–20 Å). The fundamental arrangement in the gel is similar to that of the lamellar liquid crystalline phase and hence it is not surprising that the rheological properties of the gel are similar to those of the lamellar phase. The stability of metastable gel phases depends on the nature of the alkyl chains; for instance, the gel of 1-monopalmitin can exist for up to a few days at 25° before it is converted into the coagel (β-crystals + water). Longer chains and mixed alkyl chains were found to increase the gel stability and allow cooling to lower temperatures ("supercooling").[857] The dependence of the gel stability on the nature of the alkyl chains is yet another example of the interrelationship between alkyl chains and hydrophilic head groups which has been referred to repeatedly in the previous sections and seems to evolve as a general principle governing the behavior of long-chain, amphipathic molecules.

Larrson and co-workers[855,857] also described the packing of the alkyl chains in monoglyceride gels and what was said about phospholipid gels is also applicable to monoglycerides.

Lutton[930] investigated the phase behavior of both saturated and unsaturated 1-monoglyceride–water systems. He found that monoglycerides dissolve only to a fraction of a percent in water, but can incorporate up to 50% of water above or close to their melting points. Water initially lowers the melting point of the crystalline compounds before a lamellar liquid crystalline phase is formed. This is the prevailing liquid crystalline phase in the homologous series of saturated monoglycerides. As, however, the chain length increases, the proportion of the lamellar liquid crystalline phase gradually decreases,[930] being replaced by isotropic phases. With monoglycerides having sufficiently long hydrocarbon chains, such as mono-arachidin and monobehenin, hexagonal liquid crystalline phases were observed. The phase diagrams of unsaturated monoglycerides resemble those of saturated ones, with corresponding phases stable at lower temperatures, e.g., the lamellar phase of the trans unsaturated monoelaidin is shifted by about 30° to lower temperatures compared to monostearin and with the cis unsaturated monoolein the lamellar regions extend to still lower temperatures. With increasing chain length and unsaturation the tendency to form hexagonal liquid crystalline phases increases.[930]

Lawrence and collaborators[858–860] studied the phase behavior of the 1-monolaurin–water system. Up to the first few percent of water a lowering of the melting point of the β form of monolaurin from 61° to a eutectic minimum of 41° was observed[860] (Fig. 21). This is similar to the reduction, with increasing water content, of the gel → liquid crystalline transition temperature observed with phospholipids. As more water was added, the melting point rose to a maximum at 1 mol H_2O/mol of monolaurin. After that the melting point fell and leveled off at the saturation value of about 13 mol H_2O/mol monoglyceride. Pmr linewidth and second moment measurements[860] in monolaurin–H_2O mixtures below the crystal → liquid crystal transition temperature also support the formation of a monolaurin monohydrate. The linewidth and second moment increase significantly when 1 mol H_2O is added per mole of monolaurin (cf. Ref. 855). Further addition of water has no effect on the second moment. An indication for the formation of a monohydrate was also obtained with monoolein. Lawrence[858] observed that in the lamellar liquid crystalline phase of monoolein the vapor pressure of water reached that of pure water at a content slightly above that of the monohydrate.

The phase diagram of monolaurin is dominated by the extended region of the lamellar liquid crystalline phase (Fig. 21). As observed by Lutton with both saturated and unsaturated monoglycerides, the transition temperature at which the lamellar liquid crystalline phase melts increases

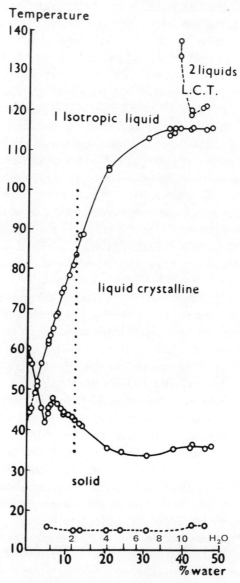

Fig. 21. Phase diagram of the monolaurin–water system (from Ref. 860).

almost linearly with increasing water content and reaches a maximum at the water saturation value. Similar to monolaurin, the lamellar liquid crystalline phase of monoolein also extends over a wide range of composition, showing a maximum of 55° in the liquid crystal → melt transition temperature at a composition of 2 mol H_2O/mol monoolein.[858] This suggests that a dihydrate may be an important complex in the structure of this liquid crystalline phase. Saturation is reached at about 13 mol H_2O/mol of monoolein, a value close to that observed with monolaurin–H_2O mixtures. With the latter system Lawrence[859] stressed the point that at maximum hydration the transition temperature from the lamellar liquid crystalline phase to the melt is 116°, i.e., 55° above the melting point of the anhydrous substance. This situation is strikingly different from that of phospholipids. With the latter class of compounds the transition temperatures between various liquid crystalline phases and from liquid crystal to melt are always lower than the melting points of the anhydrous material. Lawrence[859] regarded the unusual monoglyceride phase behavior as "a striking reminder that this water which flows out of taps so readily can also act as a very good cement between molecules to which it is hydrogen bonded." The difference in the behavior of monoglycerides and phospholipids is probably due to the absence of ionic interactions in the polar head group of monoglycerides. The decrease in the gel → liquid crystal transition temperature T_c and the simultaneous increase in the liquid crystal → melt transition temperature (see dotted line of Fig. 21) with increasing water content are difficult to reconcile. It may be explained in terms of Lawrence's interpretation[859] of an increased interaction energy in the polar group region. The interaction of the monoglyceride with water (i.e., the hydration) may allow for more favorable hydrogen bonding in this region. This hydrogen bonding may either result from water forming hydrogen-bonded bridges between monoglyceride molecules or from a conformational change in the polar head group which is induced by the interaction with water and which allows for more effective hydrogen bonding between neighboring lipid molecules and/or lipid–water molecules. Furthermore, when water interacts with the glycerol groups a lateral expansion of the lipid bilayer is observed which is due to the spacing out of the lipid molecules by water molecules accommodated between adjacent glycerol groups. This would account for the decrease in T_c due to the reduced van der Waals term in the interaction energy. In addition to hydrogen bonding, there is a hydrophobic term (entropy term) contributing to an increase in the interaction energy compared with that of the anhydrous state. This hydrophobic contribution would stabilize the hydrated bilayer against rupture which would lead to the unfavorable exposure of hydrocarbon residues to water. As a result of both hydrogen bonding and hydrophobic interaction the interaction energy of the monoglyceride molecule in the hydrated bilayer exceeds that in the anhydrous bilayer and hence bilayers of hydrated liquid crystals are stable up to temperatures exceeding the melting point of the anhydrous crystal. However, there may be a further contribution to the interaction energy of the hydrated bilayer which was not considered by Lawrence and his collaborators[859]; this invokes a simple geometric model. The spacing out of monoglyceride molecules by water means that the volume available to each hydrocarbon chain increases and this allows for greater molecular motion (twisting and bending, i.e., shortening) of the hydrocarbon chains as the temperature is raised. The cone swept out by each hydrocarbon chain increases with increasing temperature and with hydrated monoglycerides the maximum volume available will depend on the maximum hydration of the monoglyceride. This increase in motional freedom and thus in configurational entropy with increasing temperature may also contribute to the stabilization of the bilayer in the hydrated liquid crystalline phase. This is supported by experimental evidence as follows. Since the values for maximum hydration are similar, the maximum temperature up to which the lamellar liquid crystalline phase can exist is also expected to be similar. This was indeed observed by Lutton in a homologous series of long-chain monoglycerides which exhibited lamellar liquid crystalline phases up to about $100–120°$. Unfortunately, the existing experimental data are insufficient to allow an estimate of the relative contributions to the interaction energy of monoglycerides present in bilayers of hydrated liquid crystalline phases. A further complication is that the structure of the isotropic region (melt) of the phase diagram is not well characterized. Larsson[856] suggested that with long-chain monoglycerides, both anhydrous and in the presence of water (up to a 1:1 weight ratio), the structure is lamellar, similar to that of the lamellar liquid crystalline phases.

Ellis et al.[400] studied the phase diagram of monooctanoin–^2H_2O mixtures and, by applying nmr techniques, obtained information about the motional freedom in different regions of the bilayer. The phase diagram was similar to that of 1-monooctanoin–water mixtures reported by Larsson.[855] The extended region of the lamellar liquid crystalline phase showed a maximum in the liquid crystal → melt transition temperature at a composition of 4 mol ^2H_2O/mol monooctanoin, indicating that such a complex may be present in the lamellar liquid crystalline phase. However, deuteron magnetic resonance measurements could not provide any evidence for the existence of such a complex. Pmr spectra below the crystal → liquid crystal transition temperature consisted of two components, a broad one and a narrower component which sharpened further and grew in intensity as water was added or the temperature raised. In samples containing ^2H_2O in excess of a 1:1 molar ratio the broad line disappeared at the transition to the lamellar liquid crystalline phase and was replaced by a line of width of about 1 G. Similar results were obtained with monolaurin–water systems, except that in the lamellar liquid crystalline phase in addition to the very narrow and 1-G lines, two additional, broad components were observed whose linewidths varied between 2 and 4 G, depending on the water content. However, the lamellar liquid crystalline phase at temperatures closer to the melting point gave a spectrum consistent with that of the liquid crystalline phase of monooctanoin, consisting of a very narrow line and a line of width 1 G. This residual line was tentatively assigned to the glycerol group, suggesting that this group is most restricted in its molecular motion due to the hydrogen bonding between the glycerol residue and water.

The lamellar liquid crystalline phase of the 1-monooctanoin–water system gives a "powder"-type dmr spectrum, similar to the lamellar liquid crystalline phases of phospholipids.[400] As regards the doublet splittings $\Delta \nu$ observed, similar considerations apply as already discussed for the phospholipids. The overall line shape of the spectra suggests that the electric field gradient is axially symmetric characterized by the asymmetry parameter $\eta = 0$. Under these conditions the quadrupole coupling constant e^2qQ/h is obtained from the splitting as mentioned before. Its value is similar to that of the quadrupole coupling constants obtained by Lawson and Flautt[861] in liquid crystalline soaps and those reported for phospholipids (see Section 4.7). The spectra were analyzed in terms of rapid exchange between two types of deuterons, bound to the monoglyceride and free. The experimental values for the splitting ν_{exp} followed the equation $\nu_{exp} = X_M \nu_b$, where X_M is the mole fraction of monolaurin and ν_b is the splitting observed with bound deuterium. The value of ν_b was obtained by extrapolation of the linear relationship ν_{exp} vs. X_M to $X_M \rightarrow 1$, giving $\nu_b = 8.4$ kHz. This low value of ν_b (or the quadrupole coupling constant) indicates that despite appreciable ordering in the bimolecular structures, the molecules themselves possess considerable motional freedom. This is also consistent with the narrow lines in proton nmr spectra observed with the lamellar liquid crystalline phases of monoglycerides.

By evaluating the experimental data according to the method of Finer and Darke[443] quantitative information about the binding of water to the monoglyceride can be obtained. Thus, plotting the splitting ν_{exp} as a function of $1/n$ (n is the number of moles of ^2H_2O per mole of monooctanoin) suggests the existence of several hydration shells; the inner one is presumed to consist of one strongly bound water molecule ($v_1 = 4.4$ kHz) (cf. Ref. 443), the second hydration shell consists of a further one to two water molecules (characterized by $v_2 = 1.8$ kHz), and there are about ten molecules of trapped water ($v_3 = 0$).

Similar conclusions were derived from proton wide line nmr measurements carried out with macroscopically ordered samples of lamellar liquid crystalline phases of mono-octanoin–water mixtures.[946] The spectra consisted of two lines each of which showed angular dependent splitting. The narrow line was 3 ppm downfield of the broader one and was assigned to hydroxyl (water) protons, while the broad line was assigned to the alkyl chain protons (cf. Ref. 946). Maximum splitting occurred for each line when the angle θ between the direction of the magnetic field H_0 and the normal to the plane of the bilayers was zero. With such an orientation of the bilayers the splittings Δv_{max} were 0.225 and 3.77 G for the water (hydroxyl) and methylene signals, respectively. The splittings of both doublets varied with θ according to $\Delta v = K(3 \cos^2 \theta - 1)$, where K is Δv for $\theta = 90°$. As predicted from the above equation, both doublet collapsed to a singlet at $\theta = 55°$. The dipolar splittings (doublets) observed with both the water and the methylene protons indicate that in the lamellar liquid crystalline phase of the monoglyceride there is a nonspherical distribution of H–H vectors. The results were interpreted in terms of a small degree of ordering of the H–H vectors along the direction of the alkyl chain long axis. The small value of 1.23 G for the splitting of the water signal, which is about 1% of the rigid lattice value, is consistent with the results of dmr measurements. It indicates that the water molecules in these liquid crystalline phases have considerable rotational and diffusional freedom but at the same time experience some anisotropic orientation, characteristic of lamellar liquid crystalline phases.

To summarize, experimental evidence suggests that with monoglycerides 1 mol of H_2O is strongly bound. There is circumstantial evidence suggesting that there is a second layer of hydration consisting of 1–3 mol H_2O/mol of monoglyceride. The amount of trapped water is about 10 mol/mol monoglyceride, bringing the amount at saturation to about 13 mol H_2O/mol monoglyceride. The total amount of water intercalated between monoglyceride bilayers is significantly less than that present between phosphatidylcholine or phosphatidylserine bilayers, but is similar to that of phosphatidylethanolamine. This agreement with phosphatidylethanolamine may be coincidence, or it may be a consequence of the intermolecular (within the bilayer) charge neutralization of phosphatidylethanolamine resulting from the orientation of the $\overset{(-)\;(+)}{\text{P-N}}$ dipole parallel to the bilayer plane.[1163] An important feature of the hydration properties of monoglycerides is the extended region of the lamellar liquid crystalline phase up to temperatures exceeding the melting point of the anhydrous crystal. As the water content increases, the liquid crystal → melt transition temperature increases and simultaneously the gel → liquid crystal transition temperature decreases (Fig. 21). A possible explanation for this somewhat contradictory behavior is that the interaction with water increases the total interaction energy of the monoglyceride bilayers, but reduces the van der Waals interaction energy between the hydrocarbon chains. The latter reduction would account for the decrease in gel → liquid crystal transition temperature. Possible contributions to the net increase in interaction energy were discussed: First, in the presence of water there is a contribution due to hydrophobic interaction; second, there may be increased hydrogen bonding in the polar group region; and third, there may be an entropic contribution due to the spacing out of monoglyceride molecules by water molecules which allows for greater motional freedom in the hydrocarbon chains.

6. SUMMARY

1. Lipid molecules are amphipathic and tend to aggregate both in the dry as well as in the wet state. The principle governing the association of lipid molecules is maximum interaction energy (minimum free energy) and the structures of lipid aggregates in the presence of water are not fundamentally different from those of lipids in the anhydrous state. However, there are differences in the details of the structures of these two states.

2. The principle of the association (packing) of all long-chain lipids is that like regions interact with each other, i.e., that the polar residues associate with each other and likewise apolar residues associate to form hydrocarbon layers or zones bordered by layers of polar residues. This packing allows for maximum interaction energy (minimum free energy), i.e., the van der Waals and hydrophobic interaction energy in the hydrocarbon chains and the hydrogen bonding, the dipole–dipole and electrostatic interaction energy in the polar regions are maximized.

3. According to their melting points, two classes of lipids can be differentiated: uncharged lipids, which have melting points rarely exceeding 70°, and lipids containing charged groups, which have melting points of 200–300°. While with the former class of lipids the interaction energy is

due to relatively weak forces, e.g., van der Waals and dipole–dipole forces, hydrogen bonding, and hydrophobic interaction, the interaction energy with the latter class of lipids is dominated by electrostatic interaction.

4. With long-chain lipids the phenomenon of polymorphism is prevalent; in the absence of water these compounds exhibit thermotropic mesomorphism and in the presence of water lyotropic mesomorphism. The physical properties of mesomorphic states are somewhere in between those given by the two extreme states, the solid state and the isotropic liquid. Each mesomorphic state (phase) is determined by intrinsic factors, i.e., the chemical composition of the lipid, and extrinsic factors, i.e., experimental conditions such as water and ion content, temperature, pressure, the presence and absence of impurities, etc. The prediction of mesomorphic phases on the grounds of chemical composition and experimental conditions is not yet possible.

5. The crystal (gel) → liquid crystal transition temperature T_c for lipids forming liquid crystalline phases depends, as pointed out in the preceding paragraph, on the chemical composition of the lipid, e.g., on the nature and composition of the hydrocarbon chains as well as the nature and composition of the polar group, and the experimental conditions, e.g., the water content and the presence of ions and other small molecules (impurities).

6. The packing and molecular motion of lipid hydrocarbon chains both below and above T_c and in the presence and absence of water have been described in semiquantitative terms. The increase in molecular motion (decrease in the density of packing) at the crystal (gel) → liquid crystal transition has been demonstrated qualitatively with a whole range of physical techniques. Some information pertinent to the question of hydrocarbon chain motion at the transition temperature T_c can be obtained from the thermodynamics and the volume increase at the crystal (gel) → liquid crystal transition. The entropy increase observed at the transition temperature indicates that the motional freedom of hydrocarbon chains in the fully hydrated liquid crystalline state (at T_c) is about half of that in liquid n-alkanes and is not uniform along the length of the hydrocarbon chain, but is significantly reduced in the proximity of the glycerol group.

Qualitatively consistent with this conclusion is the volume increase measured at the transition temperature, which is only a fraction of that observed at the melting point of pure hydrocarbons, indicating that the hydrocarbon chains in the liquid crystalline state are in a relatively ordered state. A more detailed picture about the nature of the gradient describing the configurational and motional freedom along the entire length of the hydrocarbon chains is obtained from the use of spectroscopic techniques, particularly esr and nmr methods. The conclusions derived from esr and nmr measurements agree in outline, although not in detail, and may be summarized as follows. The configurational and motional freedom changes from an immobile, highly ordered, probably quasicrystalline state to a flexible, more loosely packed state at the center of the bilayer. The description of the hydrocarbon chains in liquid crystalline phases as liquidlike is an oversimplification. Models used to describe the configurational and motional freedom of hydrocarbon chains in liquid crystalline phases at maximum hydration range from (a) one with the hydrocarbon chains fully extended and a high degree of order in the packing, to (b) one with hindered rotation about C–C bonds, to (c) one with free rotation about C–C bonds leading to statistical bending of the hydrocarbon chains. Experimental evidence, particularly spectroscopic experiments, indicate that probably all these models are applicable, but only in a certain portion of the hydrocarbon chain. A motional profile along the length of a phospholipid molecule in a bilayer at maximum hydration and above the T_c can be given in qualitative terms; the configurational and motional freedom decreases continuously starting at the center of the bilayer and progressing toward the glycerol group, where it reaches a minimum. The molecular motion increases again along the polar group.

7. As water is added to and interacts with the polar group of phospholipid bilayers, a lateral expansion of the bilayer occurs and the area per lipid molecule increases. Consequently the van der Waals interaction energy in the hydrocarbon chains decreases, which manifests itself in a decrease in the crystal (gel) $\rightarrow$ liquid crystal transition temperature T_c. As the water content increases, the transition temperature decreases continuously until the phospholipid polar head group has attained its maximum quantum of water.

8. There is an interdependence between the packing density of the hydrocarbon chain and the packing density of the polar group. The interaction of water with the polar head group of anhydrous phospholipids loosens the packing in that region due to the weakening of the electrostatic interaction and/or the accommodation of water molecules between adjacent phosphate groups. This manifests itself in an increase in the motion of the phosphorylcholine group. Simultaneously, the motional freedom in the hydrocarbon chains increases, as evident from a decrease in T_c. There is a good correlation between the motional freedom of CH_2 groups in the hydrocarbon chains and the degree of hydration (or packing) in the polar group. For instance, divalent ions such as Ca^{2+}, Mg^{2+}, Ba^{2+}, UO_2^{2+}, which

have been shown to interact stoichiometrically with the negatively charged phosphodiester group, displace water molecules hydrogen-bonded to this group. The charge neutralization due to the formation of a metal complex and extrusion of water lead to a tighter packing in the polar group which is propagated to the hydrocarbon chains. This interdependence between the packing of polar groups and hydrocarbon chains may have important implications in the behavior of biological membranes. These are now known to contain lipid bilayers and changes in the packing of the bilayer induced by ionic interactions in the polar group will lead to changes in the bilayer (membrane) permeability and stability.

$$\text{phospholipid bilayer} + H_2O \rightarrow \text{looser packing}$$
$$(= \text{increase in bilayer permeability})$$

$$\text{phospholipid bilayer} + Me^{2+} \rightarrow \text{tighter packing}$$
$$(= \text{decrease in bilayer permeability})$$

Another example of the coupling between the packing and motion of the polar and hydrophobic bilayer region is the incorporation of cholesterol into phospholipid bilayers. The incorporation of cholesterol leads to a condensation of the hydrocarbon chains and as a consequence to the extrusion of water from the polar group region.

9. The hydration of phospholipids has been studied with a variety of physical methods, which are summarized in Tables IV and V. The term hydration comprises two aspects, first the macroscopic interaction between lipid and water, which is generally described as mass of water associated per unit mass of lipid and which does not define the association of water in molecular terms, and second, the molecular aspect of the lipid–water interactions, which defines the water associated with lipid in terms of an association constant and/or motional parameters such as exchange rates and relaxation or correlation times. Since lipid–water interactions result in dynamic structures, the latter aspect is particularly important in describing hydration properties.

Most of the hydration studies on phospholipids were carried out with phosphatidylcholines, both synthetic and naturally occurring. The results up to the end of 1973 are summarized in Tables IV, V, and VII. The agreement between the hydration numbers is remarkable, particularly the consistency between macroscopic methods such as water vapor adsorption or hydrodynamic measurements and molecular methods, such as nuclear magnetic resonance spectroscopy. The agreement between the hydration numbers derived from different physical techniques is also encouraging considering that the phosphatidylcholines used were present in grossly different structures ranging from the anhydrous crystalline state, to small spherical micelles present in organic solvents, to lamellar liquid crystalline phases consisting of microscopic (>1 μm) arrays of parallel bilayers, to sonicated liquid crystalline systems consisting of hollow, spherical particles of 200—600 Å diameter surrounded by a single bilayer. Results obtained with phospholipids do not confirm the observation made with other biopolymers, e.g., proteins, that the hydration numbers derived from nmr techniques are significantly smaller than those derived from macroscopic techniques.

10. The macroscopic methods (Tables IV and V) give the total amount of water associated with phosphatidylcholine and the average value from these tables is 23.5 ± 1.4 mol H_2O/mol lipid. This compares well with the average value for the total amount of water derived from nmr methods, which is 22 ± 0.6 mol H_2O/mol phosphatidylcholine. However, nmr spectroscopy allows us to differentiate among various types of water. The average value for the total amount of water bound to phosphatidylcholine is 12.5 ± 0.7 mol H_2O/mol phosphatidylcholine. The difference between the total amount of water and that of bound water constitutes the "trapped" water (between phospholipid bilayers), which in its motional characteristics approaches free water and which is in rapid exchange ($>10^4$/sec) with the bound water. The water in excess of 24 mol H_2O/mol lipid is free water which is outside the phospholipid bilayers, exchanging only slowly ($<10^2$/sec) with the water (bound + trapped) incorporated within the hydrated bilayers. Nmr methods also allow us to define quantitatively different hydration shells in terms of motional parameters. The innermost hydration shell of phosphatidylcholine consists of one tightly bound water molecule, as derived from deuteron magnetic resonance, with a rotational correlation time of $\tau_c = 10^{-7}$ sec. The main hydration shell consists of 11–12 water molecules with a rotational correlation time of $\tau_c \leqq 8 \times 10^{-10}$ sec, while the trapped water amounts to 11–13 mol H_2O/mol lipid with a rotational correlation time of $\tau_c \leqq 3 \times 10^{-10}$ sec. The values for tightly bound water in the "inner" hydration shell vary between 1 and 6 mol H_2O/mol lipid, depending on the physical method used (Table IV). The motional characteristics of the main hydration shell are determined by those of the phospholipid polar group. The interaction with the first or the first few water molecules induces a cooperative conformational change in the polar group of phosphatidylcholine such that the $\overset{(-)}{P}O_4-\overset{(+)}{N}$ dipole of the phosphorylcholine group changes from a conformation coplanar with the bilayer plane to one in which this dipole is more nearly perpendicular

to the bilayer plane. This conformational change in the polar group is coupled to changes in the packing and molecular motion of the hydrocarbon chains. Nmr measurements and the competition of water and Ca^{2+} for the same binding sites suggest that the main hydration shell is centered around the phosphate group. Similar results as with phosphatidylcholine were obtained with other phospholipids, although the comparison is limited to only a few studies.

11. Phospholipids with a net negative charge, such as the monosodium salt of phosphatidylserine, give similar results as regards the inner and the main hydration shells, but differ with respect to the large quantity of water trapped between adjacent bilayers with motional characteristics of free water: 120 mol H_2O/mol phosphatidylserine. A two-phase system (lamellar liquid crystals in excess water) is only observed at a water content greater than 3 g H_2O/g lipid or 140 mol H_2O/mol lipid. The ability of charged phospholipid bilayers to incorporate large quantities of water and to swell up to a bilayer thickness of about 140 Å (three times the original thickness) is ascribed to electrostatic repulsion between the bilayers.

12. The hydration numbers derived from the interaction of deuterium oxide ^2H_2O with phosphatidylcholine are, within experimental error, consistent with those derived from water–phosphatidylcholine systems. We may conclude that with phosphatidylcholine the isotopic replacement of H_2O by ^2H_2O does not produce significant changes in the hydration of the phospholipid; if anything, with ^2H_2O the hydration is less effective, i.e., the main hydration shell is reduced by 1–2 mol ^2H_2O/mol lipid.

13. The amount of bound water (average value 12.5 mol H_2O/mol phosphatidylcholine) is consistent with the quantity of unfreezable water derived from calorimetric measurements. We can conclude that the amount of water that does not freeze when fully hydrated liquid crystalline phases are cooled below 0° is identical with bound water present in the main hydration shell of the phospholipid.

14. The hydration behavior of the uncharged monoglycerides is in many respects similar to that of isoelectric phospholipids such as phosphatidylethanolamine. They dissolve in water only to a fraction of a percent, but similar to phospholipids, incorporate up to about 50% of water between bilayers close or above the crystal → liquid crystal transition temperature. The phase behavior of a homologous series of monoglycerides with the hydrocarbon chain length varying between six and 18 carbon atoms shows that lamellar liquid crystalline phases dominate, although with long-chain monoglycerides the proportion of hexagonal and isotropic cubic phases increases at the expense of the lamellar phase. The latter is present up to

temperatures exceeding the melting point of the anhydrous crystal. As with phospholipids, the interaction with water decreases the crystal (gel) → liquid crystal transition temperature T_c. At the same time, as the water content increases, the liquid crystal → melt transition temperature increases. This is explained in terms of a reduction in van der Waals interaction energy in the hydrocarbon chains which accounts for the decrease in T_c and increased motional freedom; however, the total interaction energy of the monoglyceride molecule in the bilayer increases at the same time. The increase in the total interaction energy is due to hydrophobic interaction, a possible increase in hydrogen bonding in the glycerol group, and a $T\Delta S$ term due to the spacing out of monoglyceride molecules by water molecules which gives the hydrocarbon chain greater configurational and motional freedom.

15. Deuteron nmr measurements show that with monoglycerides 1 mol of H_2O is strongly bound in the innermost hydration shell; there is some evidence of a second, strongly bound hydration layer consisting of 1–3 mol of H_2O, while the amount of "trapped" water is about 10 mol H_2O/mol monoglyceride, for a total at saturation of about 13 mol H_2O/mol monoglyceride. The total amount is significantly less than that obtained with charged phospholipids such as phosphatidylserine or with the isoelectric phosphatidylcholine, but is similar to that of phosphatidylethanolamine. A possible explanation for the agreement with phosphatidylethanolamine is that the conformation of the $\overset{(-)}{P}O_4 - \overset{(+)}{N}$ dipole in phosphatidylethanolamine in the presence of water is coplanar with the bilayer plane, enabling effective charge neutralization between adjacent phosphodiester and ammonium groups, leading to hydration properties similar to those of uncharged lipids.

ACKNOWLEDGMENTS

I am grateful to Drs. M. C. Phillips and E. G. Finer for many helfpul discussions, for valuable comments, suggestions, and criticism, and for reading the manuscript.

CHAPTER 5

Nucleic Acids, Peptides, and Proteins

D. Eagland

School of Studies in Chemistry
University of Bradford
Bradford, Yorkshire, England

1. INTRODUCTION

The roles played by the solvent and small solute–solvent interactions in aqueous solutions of proteins, polypeptides, and nucleic acids are in many aspects far from clear; the total effort which has so far been expended in advancing our knowledge of the phenomena has been quite considerable and yet many problems and anomalies remain to be explained. It is unfortunately not possible within the space limitations of a review of this kind to encompass the whole of the work performed within this field over the past few years. The best that one can hope to accomplish is to describe, in outline, some of the models proposed by several outstanding workers in the field which are helping to clarify a complex situation; additionally, we may summarize the evidence supporting such interpretations and that which points to the limitations of our knowledge at the present time. This, in turn, may serve as a warning against the dangers inherent in overenthusiastic interpretation of limited experimental data in complex situations.

Considerable evidence is available from hydrodynamic,[1471] high-resolution proton nmr,[828,828a,829,830] and density[832,833] measurements and dielectric studies[510,559,1349,1373] that biopolymers are hydrated to a considerable degree; this hydration appears to be associated with the molecule irrespective of whether it is in a predominantly native (e.g., helix) conformation or in a denatured (e.g., random coil) conformation.* Since certain aspects[828,828a,829,830] of the evidence suggest that at least some of this hydration water does not freeze at $0°C$, nor even at temperatures as low as $-60°C$, it is unlikely that it can be considered as normal water; exactly what kind of water it might be is, however, a problem to which convincing answers are as yet lacking. We shall return to this particular aspect of the subject in subsequent sections.

A different aspect of the hydration of proteins, polypeptides, and nucleic acids, is the role played by the solvent in stabilizing certain conformations of the macromolecule in preference to the many others statistically available.

The conformation adopted by a macromolecule such as a protein or polypeptide in an aqueous environment is the equilibrium result of the competing effects of the solvent, the component peptide units of the macromolecular chain, and the sequence of the units within the chain. The number of conformations which a macromolecule such as a protein, containing perhaps hundreds of peptide units, may take up is obviously extremely large; nevertheless, it is possible to divide them, in a very broad sense, into two main groups; one group contains those conformations in which interactions between peptide units predominates over interactions between the peptide unit and the solvent for the majority of the units; the second group contains those conformations for which peptide unit–solvent interactions predominate. The generally accepted viewpoint at the present time is that the first class, which in the limiting case would correspond to entirely peptide unit–peptide unit interactions, corresponds to a "native" helical configuration for the macromolecule, whereas the second class, in the limiting case consisting only of peptide unit–solvent interactions, is thought of as a "denatured" random coil configuration.

Such a division into two states is obviously an oversimplification of a complex situation but nevertheless it does form a very useful basis for thinking of the macromolecule as being either *predominantly* in its native (helix) conformation or *predominantly* in its denatured (random coil) conformation.

The equilibrium between the helix and coil configurations forms the basis of the two-state approach to protein denaturation proposed by

---

* There is a voluminous literature devoted to conformational transitions in biopolymer systems and to the exact meaning to be attached to terms such as native, denatured, renatured, unfolded, helical, random coil, etc. Within the context of this chapter expressions such as denaturation, helix → coil transition, and unfolding can be taken to be synonymous, indicating only a transition from a compact, ordered to a less ordered polymer configuration. *Editor's note.*

Brandts[172] in which an unspecified number of different conformations are possible each of which is characterized by a particular value of an experimentally observable parameter θ which is used to examine the system. In general only an average value of all the different parameters can be observed, $\bar{\theta}$, which will vary with temperature, pressure, and solvent composition since the probability of each conformational state is dependent upon the nature of the solvent, other effects such as the sequential composition of the polypeptide chain being unchanged.

If the assumption is made that $\bar{\theta}$ may be divided according to the two groups discussed previously, one corresponding to the helix conformation and the other corresponding to the random coil configuration, the following equation results:

$$\bar{\theta} = (1/P_H)\bar{\theta}_H + (1/P_C)\bar{\theta}_C$$

where P_H and P_C are the total sums of the probabilities of the configurations corresponding to helix and coil conformations, respectively. The equilibrium constant for the reaction

$$\text{helix} \rightleftharpoons \text{random coil}$$

can be formulated

$$K = (\bar{\theta} - \bar{\theta}_H)/(\bar{\theta}_C - \bar{\theta})$$

and it follows that $K = P_C/P_H = X_C/X_H$, where X_C and X_H are the mole fractions of the random coil and helix configurations, respectively.

The major problem which arises in the determination of the equilibrium constant for the two-state equilibrium is that of obtaining $\bar{\theta}$, $\bar{\theta}_H$, and $\bar{\theta}_C$ under the same conditions of temperature, pressure, and solvent composition. Only $\bar{\theta}$ is an experimentally obtainable quantity; $\bar{\theta}_H$ and $\bar{\theta}_C$ must be obtained by other means which usually involve extrapolation of the experimental data from the pretransition region (helix) and the post-transition region (random coil), so that a value of K for the midpoint of the region can be obtained.

If intermediate folded states occur over the transition region which do not occur either in the helical or the random coil domains, then the equilibrium does not fall within the classification of a two-state process.

It is therefore clearly possible on the basis of one experimental determination of the equilibrium constant to ascribe wrongly two-state behavior to a protein unfolding equilibrium. In order to avoid such an error, the multiple variable test has been proposed;[925,1187] if a transition is treated as a two-state process and different values of K, ΔH, ΔS, etc., are obtained when the transition is examined by different analytical techniques, then such a treatment is probably incorrect, since the relationships of the various techniques to the thermodynamic characteristics of the process are likely to be widely different when intermediate folded conformations arise.

An additional test of importance in deciding whether a transition has been incorrectly treated as a two-state process is the relationship between the enthalpy change obtained by a van't Hoff technique (ΔH_{vH}) and that determined by a calorimetric method (ΔH^*); incorrect treatment leads to lack of agreement between the two values. In addition, Lumry et al.[925] have pointed out that the loss of cooperativity introduced by the presence of intermediate states is likely to be reflected in the ratio $\Delta H^*/\Delta H_{vH}$ being in excess of unity.

A two-state denaturation reaction is, by virtue of the conditions proposed, a clear indication of the important role of the solvent medium in the process; a more detailed and quantitative analysis of the factors involved in a denaturation equilibrium will therefore enable us to examine this role more closely.

In the denatured state, say of a random coil conformation, a polypeptide chain will have a much greater degree of rotational flexibility than in the native, tightly folded (helical) conformation. An average value for the change in conformational entropy per peptide unit of the polypeptide chain of the order of 12–24 J mol^{-1} deg^{-1} is obtained from published data.[759,1020,1243,1318] The large number of such units in a polypeptide chain therefore provides a considerable destabilizing contribution to the Gibbs free energy change associated with the helix–random coil equilibrium.

In many instances hydrogen bonding has been suggested as a factor in stabilizing the native conformation of a protein; therefore it might be expected to provide a stabilizing contribution to the Gibbs free energy change associated with the process. Several authors[641,759,1020,1243] have cited values for the enthalpy change arising from the rupture of a peptide–peptide hydrogen bond ranging from 0 to 8 kJ mol^{-1}. However, Klotz[785] suggests that, on the basis of the changes in the near-infrared spectrum of N-methylacetamide,[788] the greater the hydrogen bonding ability of the solvent, the weaker the C=O$\cdots$H—N bond becomes. In water the enthalpy of formation of this bond is zero; the suggestion is therefore that the peptide–peptide hydrogen bond cannot provide a stabilizing contribution to the helical conformation in an aqueous solution. In contradiction to this viewpoint Bello[108] points out that the formation of one hydrogen bond between two water molecules is said to cause both molecules to become more polar[475] and hence additional hydrogen bonds with other water molecules are favored; thus similar reasoning applied to hydrogen bonding of a protein means that such stronger hydrogen bonds are formed between successive $C=O \cdots H-N$ bonds than is the case in N-methylacetamide dimerization.

Both proteins and nucleic acids are good examples of natural polyelectrolytes; it is therefore to be expected that a macromolecule of this kind should be sensitive to the hydrogen ion concentration of the aqueous environment, particularly in respect of its conformational stability. Brandts[172] has suggested that the electrostatic effect may be subdivided into two contributions: first a long-range electrostatic interaction which, according to Tanford,[1461] may be regarded as a Gouy–Chapman type of double layer effect. A change from a native to a denatured conformation will result in changes in size, shape, and charge of the macromolecule with a resultant contribution to ΔG associated with the overall process. The second contribution defined by Brandts arises from the presence of ionizable groups on the macromolecule that are capable of titration. If the ionizable group is buried within the interior of the macromolecule in the native helical conformation, then it will have a pK value several units removed from the value normally associated with such a group. If these groups revert to normal pK behavior during a denaturation process from the helix conformation to the random coil, then proton binding must change, resulting in an additional contribution to ΔG for the process.

An additional factor which may contribute to the overall ΔG for the denaturation process is the increased possibility of proton exchange between groups such as $-NH$ and the aqueous solvent medium when the macromolecule is in the denatured random coil conformation; since rapid exchange is only possible for hydrogen atoms exposed to the solvent, a proportion of the hydrogen atoms of the macromolecule exchange more slowly and the exchange rate will depend upon the degree of unfolding of the macromolecule. Experimentally[700,701] a wide variety of exchange rates are observed among buried peptide group hydrogens, in apparent contradiction of the two-state model, which predicts a single degree of unfolding in addition to a single rate of unfolding and hence a unique exchange rate characteristic of all buried hydrogens.

This problem is one illustration of several aspects of protein denaturation behavior apparently not compatible with the two-state theory. Hermans *et al.*[643] have suggested a solution to this problem by treating the macromolecule as a one-dimensional Ising lattice of finite extent[714] in which lattice sites exist in one of two conformations, the random coil and helix, with stabilization from nearest-neighbor contacts of sites in the helix conformation. This model predicts all-or-none unfolding transitions but a distribution of proton exchange rates.

This type of analysis will not invalidate the important basic assumption of the two-state theory relating to the predominant role of solvent–peptide interactions in the random coil configuration and we must now look in greater detail at this particular factor and its involvement in determining the stability of a particular conformation for the macromolecule.

The process helix→coil in general involves the removal of nonpolar side-chain peptide groups from the interior of the native conformation and their exposure to the aqueous environment in the random coil conformation. Such groups present to the solvent a surface consisting largely of hydrocarbon, hence the description of this state as *hydrophobic hydration* and the converse process of contract between nonpolar side chains with subsequent elimination of hydrated surface as *hydrophobic bonding*[1093] (see Chapter 1). Formation of the native conformation of the macromolecule from the random coil conformation can be expected to produce bonding of this kind. Clearly polar and ionizable side chains whose degree of solvent exposure is affected as a result of a conformational transition will also be subject to changes in hydration interactions and it is the sum total of the various solvent interactions which determines the stability of a given conformational state (see also Chapter 6).

The Gibbs free energy change associated with the transition that produces a hydrophobic bond has been described by Nemethy[1089] as a sum of several contributions. First, the free energy change arising from the structural change that occurs in the water no longer involved in hydrophobic hydration of the previously exposed nonpolar side group; such a change is markedly temperature dependent, and at 25°C has a value of approximately -500 J (mol CH_2)$^{-1}$; second, the van der Waals attraction between a pair of nonpolar side chains, which is of the order of -630 J mol^{-1} for aliphatic and -2.1 kJ mol^{-1} for aromatic groups; third, the interaction between exposed nonpolar groups and the water molecules formerly involved in hydrophobic hydration (-130 J mol^{-1} for aliphatic and -650 J mol^{-1} for aromatic groups). The ΔG arising from the partial loss in flexibility of the nonpolar side groups when the hydrophobic bond is formed is estimated to lie in the region $0–1.25$ kJ mol^{-1}. The number of water molecules involved in a particular hydrophobic bond is dependent upon the sizes of the nonpolar groups that constitute the bond but typical estimates are four in the case of an alanine–alanine bond and 12 for a phenylalanine–phenylalanine bond; somewhat surprisingly, no indication is given by Nemethy that these values might be temperature dependent, since it is accepted that structural change in water is temperature dependent, and unless this particular structure is considerably more stable than normal water structure, some temperature dependence should be apparent.

Brandts[171] has proposed a power series type equation to represent the overall temperature dependence of $\Delta G_{H\phi}$ associated with formation of a hydrophobic bond,

$$\Delta G_{H\phi} = AT + BT^2 + CT^3 \qquad (1)$$

Values of A, B, and C have been estimated for different side chains.

Estimates of the hydrophobic bond strength by the Nemethy and Scheraga treatment and by the method based upon eqn. (1) in general give good agreement at 25°C for $\Delta H_{H\phi}$ and $\Delta S_{H\phi}$ but unexplained differences, which are quite large, occur for particular side chains such as isoleucine and phenylalanine.

An experimental verification of the calculated ΔG associated with the transfer of a nonpolar peptide side chain into the interior of a protein has not to date been obtained but an analysis similar to that outlined above has been applied to the transfer of model compounds bearing a resemblance to protein side chain groups from water into a hydrocarbon medium; hopefully the process might be regarded as a credible model for the folding of a nonpolar side chain in contact with an aqueous environment into the interior of the native helical conformation.

Details of such transfer processes are given in Chapter 1 and do not require repetition here. However, one precaution with respect to their application to proteins, as pointed out by Nemethy,[1089] should be noted: The thermodynamic parameters for the transfer of a nonpolar side chain from one medium to another must always be smaller in magnitude than for transfer of a hydrocarbon molecule containing the same number of carbon atoms, since the bulk of the polypeptide chain presumably reduces the number of water molecules in contact with the exposed side chain compared to the free hydrocarbon molecule in contact with water.

Values have been calculated by Nemethy and Scheraga for the standard thermodynamic parameters associated with transfer of various nonpolar side chains from a hydrocarbon environment to one that is completely aqueous (Table I). In every case $\Delta G_t^{\ominus}$ is small and positive (in the range 5.5–8.4 kJ mol^{-1}). The value of $\Delta H_t^{\ominus}$ is negative and of the order of -8 kJ mol^{-1}; thus the transfer process is dominated by the negative $\Delta S_t^{\ominus}$, presumably due to the pronounced water "structuring" caused by the hydrophobic hydration that arises on transfer of the side chain from the hydrocarbon medium to the water.

TABLE I. Calculated Thermodynamic Parameters of Transfer of Amino Acid Side Chains from a Hydrocarbon Medium into Water at 25°C[a]

Amino acid side chain	$\Delta G_t^{\ominus}$, kJ mol^{-1}	$\Delta H_t^{\ominus}$, kJ mol^{-1}	$\Delta S_t^{\ominus}$, J mol^{-1} deg^{-1}
Alanine	$+5.5$	-6.30	-39.5
Valine	$+8.0$	-9.2	-57.5
Leucine	$+8.0$	-10.1	-60.1
Isoleucine	$+8.0$	-10.1	-60.9
Methionine	$+8.4$	-11.7	-67.2
Proline	$+8.4$	-9.2	-58.8
Phenylalanine	$+1.3^b$	-11.7	-42.4
	$+7.6^c$	-4.2	-39.9

[a] Data from Nemethy.[1089]
[b] For transfer from an aliphatic environment.
[c] For transfer from an aromatic environment.

Earlier discussion has indicated that several of the various contributions to hydrophobic bonding can be expected to be very sensitive to the effects of temperature; calculations by Brandts[172] on the temperature dependence of the transfer of a valyl side chain from a protein interior to an aqueous environment shows that $\Delta H_t^{\ominus}$ changes from -10.5 kJ mol^{-1} at 0°C to 2.1 kJ mol^{-1} at 70°C and $\Delta S_t^{\ominus}$ changes from -67 to -17 J mol^{-1} deg^{-1} over the same temperature range; a large and positive value for $\Delta C_{p_t}^{\ominus}$, the heat capacity change on transfer, is therefore observed for this process, changing from 126 to 3360 J mol^{-1} deg^{-1} over this temperature range.

Other contributions to the heat capacity change on denaturation may arise from changes in the internal degrees of freedom of the polymer. The work of Hill[665] and Gucker and Ayres[574] suggests that ΔC_p per side chain from this source is unlikely to exceed 4 J mol^{-1} deg^{-1}, which is considerably smaller than the changes attributable to hydrophobic effects.

If, then, a denaturation process is found to obey the two-state theory and to be associated with a large change in heat capacity, these observations can be taken as strong indications for the participation of hydrophobic interactions in the transfer process, since other contributory factors are not likely to be of comparable magnitude.

Brandts points out that caution must however be exercised because most of the earlier $\Delta H_t^{\ominus}$ data were obtained by van't Hoff techniques and are in fact *isosteric* heats, whereas the more recent data have been obtained by direct calorimetric methods. At infinite dilution the results should agree but earlier data obtained at higher concentrations show a concentration dependence down to quite low concentrations (see Chapter 1).

A somewhat different viewpoint of the role played by water in the denaturation process from helix to random coil has been taken by Sinanoglu and Abdulner.[1398] Three basic questions are posed: (1) What crucial physical properties of water, if any, are responsible for its unique role? (2) Do any other solvents possess properties which can give rise to "solvophobic" interactions and, if so, what are their relative strengths? and (3) What are the thermodynamic properties of the water layer surrounding the side chains of the macromolecule and is hydrophobic bonding still an operative factor if the side chains are not entirely apolar but have sizable dipoles, as in the case of the nucleotide bases of DNA?

These authors take the double-stranded DNA molecule as their basic model and consider the equilibrium process

$$\text{helix}_1 \rightleftharpoons \text{coil}_1 + \text{coil}_2$$

Intermediate folded states are not regarded as significant, i.e., a two-state mechanism is assumed.

The total ΔG for the process helix $\rightarrow$ coils in a given solvent is regarded as consisting of the standard Gibbs free energy change $\Delta G^{\ominus}$ for the process, i.e., not including any contribution arising from the partial molal entropy of mixing of helix with solvent, and a similar term for the two single strands.

The contributions to $\Delta G^{\ominus}$ arising from the solvent are characterized as follows: $\Delta G_c{}^{\ominus}$, the difference in free energy of creating a cylindrical cavity for the helix and spherical cavities for the bases in the coils; $\Delta G_{bs}^{\ominus}$, the differences in free energy of interaction of the bases in the helix and the bases in the coil with the solvent which surrounds them; $\Delta G_R{}^{\ominus}$ the free energy change arising from the reduction in base–base and phosphate–phosphate interactions in the helix due to the presence of the surrounding solvent.

The calculation of $\Delta G_c{}^{\ominus}$ is achieved by equating the free energy required to create the given surface area of the bubble or cylindrical cavity with the appropriate value of surface tension. $\Delta G_{bc}^{\ominus}$ is calculated by considering a base interacting with the surrounding liquid due to the polarization induced by the electric dipole of the base,[1115,1545] and $\Delta G_R{}^{\ominus}$ is obtained by using the theory of effective dielectric constants for electrostatic interactions[772] and for dispersion forces.[1379]

For a polynucleotide with the base sequence*

and water as solvent, $\Delta G_{bs}^{\ominus} = +187$ kJ per mol base pair, $\Delta G_R^{\ominus} = +17$ kJ mol^{-1}, and $\Delta G_c^{\ominus} = -158$ kJ mol^{-1}, giving a total $\Delta G^{\ominus}$ of $+46$ kJ per mol base pair. Similar calculations have been carried out for solvents ranging from formamide through a range of short-chain monohydric alcohols; $\Delta G_{bs}^{\ominus}$ and $\Delta G_R^{\ominus}$ in all cases show only small variations from the values obtained with water and only $\Delta G_c^{\ominus}$ shows a marked change with solvent. The conclusion can therefore be drawn that the unique stabilization of the helix by water arises from the amount of energy required to create the cavity capable of accommodating the helix, which is less than that required to accommodate the random coils, because of the smaller total surface area required. The helix conformation is therefore the thermodynamically preferred form and the association of nonpolar side chains in water, i.e., hydrophobic bonding, thus occurs because of the liberation of water molecules arising from the decrease in water surface area as the separate strands combine to form the helix.

The contribution to ΔG for the helix $\rightarrow$ coil transition arising from the entropy of mixing of the macromolecule and solvent is calculated on the basis of two components; first, the mixing of the DNA units, regarded as spheres, with the surrounding solvent, defined as external mixing; and second, the mixing of segments of the DNA macromolecule inside each sphere with solvent inside the sphere, termed internal mixing. Both internal and external terms are calculated using the Flory treatment of dilute polymer solutions.[456] A total value for the entropy of mixing of $+46$ kJ mol^{-1} deg^{-1} of base pair, with water as solvent, is obtained and little variation in the magnitude of this value is observed for other solvents. The conclusion already reached is therefore not invalidated, i.e., the energy required to produce the helix cavity is the factor which governs the stability of the helix conformation.

The structural restrictions placed on the water molecules which make up the surface of the bubble cavity are equated to the restrictions involved in the formation of clathrate hydrate cages around small nonpolar solutes

---

* Throughout this chapter the commonly accepted abbreviations for the nucleotide bases are used, i.e. A = adenine, C = cytosine, G = guanidine, T = thymine. *Editor's note.*

such as methane and ethane. The formation of such a clathrate can be regarded as a two-step process; first, the formation of an empty cavity within the water, and second, the filling of the cavity with the nonpolar solute. If ideal mixing is assumed, then the total entropy and enthalpy changes for the formation of the clathrate each consist of contributions from these two steps. The entropy and enthalpy contributions from the second step can be calculated by reference to the intermolecular potentials derived from the dissociation pressures of the solid hydrate. The differences between the experimentally determined total values of the entropy and enthalpy of formation of the clathrate and the contributions from the second step result in values for the formation of the empty cavity. The entropy change for this step, taking the cavity to have a radius of 39.5 nm and comprising 20 water molecules, is -57 ± 2 kJ mol^{-1} deg^{-1} for a variety of nonpolar solutes, yielding a value of -82.7 J mol^{-1} dcg^{-1} for each mole of cage water.

Experimental and theoretical considerations of the total enthalpy change for the formation of the clathrate[531,1518] indicate the value to be approximately zero (see also Volume 2, Chapter 3).

The structural restriction of water molecules surrounding a large macromolecule such as DNA must additionally produce an interfacial entropy term arising from the water in contact with the nonpolar surface. An approximate estimate of the magnitude of this contribution can be obtained from the entropy of formation of an interface between water and nonpolar organic liquids. This indicates a decrease of approximately -3.0 J mol^{-1} deg^{-1} of surface water, assuming a radius of 14 nm for a water molecule. The value obtained by Sinanoglu and Abdulner, based on the assumption that the entropy decrease arises only from the aqueous side of the interface, thus agrees well with that obtained from the gas hydrate model for small solutes, namely -2.7 J mol^{-1} deg^{-1}.

Lewin[886] has proposed a similar model for the water surface energy contribution to the stability of a protein configuration and additionally points out that the denaturation of proteins arising from the addition of alcohols or surfactants such as sodium dodecylsulfate may be linked to the known decrease in the surface tension of water which occurs upon the addition of these solutes.

A further indicator for the involvement of the solvent in the process helix → random coil has recently been proposed by Lumry and Rajender,[927] who suggest that the enthalpy–entropy compensation phenomenon which is observed for many processes involving proteins[116,127,364,420,681,1194] is a consequence of the properties of liquid water, irrespective of the solute or the particular process under investigation, and may be used as a diagnostic test for the participation of water in protein reactions.

The particular variety of thermodynamic compensation discussed by these authors is that in which the enthalpy change associated with an isothermal process is linearly related to the entropy change. The slope of the straight line obtained from a compensation plot of standard enthalpy change versus standard entropy change or of standard enthalpy of activation against standard entropy of activation for a rate process is the basis of comparison for differing examples of compensation behavior and has the dimension of temperature; the slope of the line has thus been termed the compensation temperature T_C. Compensation processes giving rise to similar values for T_C may be taken to reflect the same property of the solvent; values of T_C lying in the range 270–300°K are said to be a strong indication of the involvement of water in the compensation process.

In small solute–solvent systems the necessary range of values of ΔH and ΔS to produce a compensation plot can be obtained by two general methods: (a) variation of the structure of the solute, such as ascending a homologous series, or (b) variation in the composition of the solvent medium; in the case of water this is usually achieved by the addition of small amounts of short-chain monohydric alcohols. Both methods are equally effective with protein systems, but in addition, suitable variations in the ΔH and ΔS may be obtained by changing the pH of the solvent; the reasons for the pH effects will be similar to those outlined by Brandts[172] in proposing the two-state theory of protein denaturation.

Lumry and Rajender go on to discuss a large number of processes involving small solutes in water for which the evidence of linear compensation is regarded as satisfactory[52,119,190,718] and on this basis consider it justifiable to extend the linear compensation concept to reactions of proteins in an aqueous environment and particularly to the denaturation process.

The authors point to the data of Brandts and Hunt[173] on the thermal denaturation of ribonuclease A at acid pH values in water and various mixtures of water and ethanol as a particularly good example which illustrates how the various pointers to the involvement of water in the denaturation process relate to each other. Brandts has shown the denaturation process to be of a two-state nature, exhibiting a large increase in ΔC_p in passing from the native to the unfolded state; Shiao and Pool* calculated the compensation plots shown in Fig. 1 from the data of Brandts and Hunt; the symbol $\Delta\Delta H$ denotes the change in the standard enthalpy of denatur-

* D. F. Shiao and G. L. Pool, data published in Ref. 927.

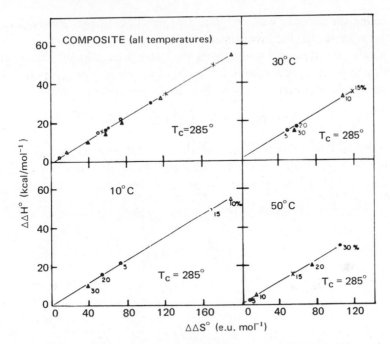

Fig. 1. Effect of ethanol on the reversible thermal unfolding transition of ribonuclease A at various temperatures. Representative points calculated by Shiao and Pool (data published in Ref. 927) from data of Brandts and Hunt.[173] The symbol $\Delta\Delta H°$ denotes the change in the standard enthalpy of unfolding relative to the pure water solution produced by adding ethanol to the percentage marked at each point.

ation (relative to that in pure water) observed for the addition of ethanol to the given concentration. The compensation temperature is found to be 285°K regardless of the experimental temperature.

The kinetic data of Pohl[1183,1184] for the corresponding transition in trypsin show the same compensation behavior for the addition of ethanol; variation of the alcohol concentration produces only a small entropy factor contributing to the rate of folding. Almost all the effect of the alcohol on the equilibrium appears to be concentrated in the rate constant for the denaturing or unfolding step. The dependence of ΔC_p upon the alcohol concentration is, within the limits of experimental error, in agreement with the values obtained by Brandts and Hunt for the ribonuclease transition. It is interesting to note that Brandts and Hunt found that at low temperatures, low concentrations of ethanol had an increased capacity for stabilizing the native conformation rather than the random coil; the correlation of this effect with the known effects of relatively low concentrations of ethanol in promoting the order in water[486] (see Chapter 1) is further strong circumstantial evidence for the involvement of water in the transition.

Lumry and Rajender suggest that the states predominantly involved in a two-state protein denaturation are in fact two states of water rather than two states of the protein; thus for the equilibrium process

$$\text{helix (h)} + nW_1 \rightleftharpoons \text{coil (c)} + nW_2$$

where W_1 and W_2 represent water in the two different states and n is the number of water units involved in the transition, the standard enthalpy and entropy changes for the process are given by

$$\Delta H^{\ominus} = \Delta H^{\ominus}_{\text{h} \to \text{c}} + n\, \Delta H^{\ominus}_{W_1 \to W_2}$$

and

$$\Delta S^{\ominus} = \Delta S^{\ominus}_{\text{h} \to \text{c}} + n\, \Delta S^{\ominus}_{W_1 \to W_2}$$

$$\Delta H^{\ominus} = \Delta H^{\ominus}_{\text{h} \to \text{c}} - \left(\frac{\Delta H^{\ominus}_{W_1 \to W_2}}{\Delta S^{\ominus}_{W_1 \to W_2}} \right) \Delta S^{\ominus}_{\text{h} \to \text{c}} + \left(\frac{\Delta H^{\ominus}_{W_1 \to W_2}}{\Delta S^{\ominus}_{W_1 \to W_2}} \right) \Delta S^{\ominus}$$

and

$$T_C = \Delta H^{\ominus}_{W_1 \to W_2} / \Delta S^{\ominus}_{W_1 \to W_2}$$

$$\Delta G^{\ominus}_{W_1 \to W_2}(T = T_C) = 0$$

Thus T_C is described as a property of the water rather than the protein.

Enthalpy–entropy compensation, although perhaps a good diagnostic tool for the detection of the involvement of water in a protein process,[41, 106,267,543,893,1211] does not, in itself, provide any information on how the solvent is involved in the process; Lumry and Rajender, by applying the Gibbs–Duhem relationship to the protein–water system for a single protein solute, deduce the equation

$$A\, d\bar{\gamma} = (V_W\, d\beta_W / \beta_W{}^2) + (V_p\, d\beta_p / \beta_p / \beta_p{}^2)$$

where V_W is the total free volume within the water, V_p is the total protein volume, β_W is the compressibility of water with respect to its change in free volume, and β_p is the compressibility of the macromolecule; $\bar{\gamma}$ is the (partial) specific interfacial free energy and A is the area of the interface between the protein and the water.

It is suggested that a good reference state for $\bar{\gamma}$ is the surface energy of water, or more precisely, the free energy required to make a hole in liquid water with a given surface area A. Such a definition of the reference state should result in the equivalence of this value with $\Delta G_C{}^{\ominus}$ for the Sinanoglu and Abdulner model discussed on p. 313. This model suggests that the overriding importance of the surface energy contribution to the stability of the helical conformation arises to a great extent from an interfacial energy contribution due to restriction placed upon the solvent molecules in the protein–solvent interface. The Lumry and Rajender model, on the other hand, indicates that the surface energy contribution may be attributed to changes in the free volumes of the water and the protein. Unfortunately, at the present time knowledge of variations in such factors as the interfacial area between water and protein, protein volume, and free volume of water are too imprecise to allow realistic quantitative assessments of their relative importance to be made, but the directions in which research into protein–water interactions may well develop are becoming clearer.

2. EXPERIMENTAL TECHNIQUES

Changes in the conformation of a biopolymer may be accompanied by changes in the length or shape of the macromolecule, perhaps by the exposure of groups previously buried within the interior of the molecule, or in the degree of ionization of ionizable side chains, and certainly by a marked change in the degree of order (e.g., helix content) of the polymer. In addition to changes in the macromolecule itself, such a change in conformation may be accompanied by changes in the counterion distribution and/or the extent of ordering of the surrounding solvent medium. Experimental techniques which are capable of measuring any of these changes are therefore of possible use in detecting changes in the conformation of the biopolymer; this section will briefly review the more important techniques which have been used to this end.

2.1. Optical Rotation, Rotary Dispersion, and Circular Dichroism

One of the most important conformational changes which many biopolymers exhibit, and that of most concern in this review, is the change in the helix content of the macromolecule. Such changes can be observed by monitoring changes in the optical activity of the biopolymer which is manifested in two related phenomena; circular dichroism and optical rotation.

If a beam of *circularly* polarized light is passed through a solution of an optically active substance, absorption of the left- and right-handed circularly polarized components of the beam is dependent upon the wavelength of the beam. Circular dichroism at a particular wavelength λ is defined as $\Delta\varepsilon = \varepsilon_L - \varepsilon_R$, the difference in extinction coefficients for the two components. Thus a particular absorption band is characterized by its rotational strength, i.e., the integrated extinction coefficient difference over the wavelength difference of the band

$$R = \frac{2.303 \times 3000hc}{32\pi^3 N} \int \frac{\Delta\varepsilon}{\lambda}\, d\lambda$$

where h is Planck's constant and c is the velocity of light. Circular dichroism, it must be noted, is only observed in the wavelength bands in which the substance *absorbs* light.

In the case of optical rotation, however, a *plane* polarized beam of radiation of *any* wavelength when passed through a solution of an optically active substance will be subject to a rotation of the plane of polarization. The angle of rotation α is dependent on the nature of the optically active material, its concentration c (in g cm^{-3}), and the thickness of the sample d (in dm). Thus

$$\alpha = [\alpha]dc$$

where $[\alpha]$ is termed the specific rotation and ideally is dependent only upon the nature of the optically active material and is independent of its concentration; this is very often not in fact the case, and a marked concentration dependence of $[\alpha]$ may often be observed. Since rotation of the plane of polarization of the beam may be right or left (looking down the beam toward the source), the first is termed dextrorotation and given a positive sign, while the latter is termed levorotation and given a negative sign. Optical rotation occurs at all wavelengths and is not therefore an absorption but a scattering phenomenon.

The relationship of circular dichroism to optical rotation can be made clearer by considering the interaction of an electromagnetic wave with the electrons of the active species; at wavelengths far from the resonance frequencies (absorption bands) of the species only induced oscillations of the electrons occur, with subsequent scattering of the wave and, in the forward scattering direction, a refraction effect. At wavelengths close to the resonance frequency absorption of energy by the species will also occur. If an optically active material shows a differing absorptive capacity for left and right circularly polarized light, then there will be at all wavelengths a difference in refractive index for the two components; both components will be retarded by refraction but one more so than the other. Since a plane polarized beam can be represented as the sum of two circularly polarized beams of opposite sense and equal amplitude, the resulting effect will be a rotation of the plane of polarization.

At wavelengths that correspond to circular dichroism (CD) absorption bands it is clearly the case that, in addition to rotation of the plane of polarization of a beam of plane polarized light, differences in absorption of the left and right circularly polarized beams which constitute the plane polarized radiation will occur. The resultant component vectors of the emergent radiation will therefore be of unequal amplitude and the beam will no longer be plane polarized but converted to elliptically polarized radiation. The extent of ellipticity is defined in terms of a molar ellipticity $[\theta]$ which is related to circular dichroism by the relationship

$$[\theta] = 3300 \, \Delta\varepsilon$$

If the dimensions of $\Delta\varepsilon$ are $l \, cm^{-1} \, mol^{-1}$, then $[\theta]$ is expressed in units of $deg \, cm^2 \, dmol^{-1}$.

It has previously been noted that rotation of a plane polarized beam by an optically active species will occur at all wavelengths, but that the sense of the rotation may change with changes in the wavelength of the polarized beam, giving rise to optical rotatory dispersion (ORD). We have already observed that a refractive effect is a result of electromagnetic radiation inducing dipole oscillation in the optically active species; variation of the frequency of the incident electromagnetic radiation forces the driven oscillator to behave in a characteristic manner. At low driving frequencies (long wavelengths) the oscillator is in phase with the driving force; as the frequency of the radiation approaches a resonance frequency of the oscillator, the response of the oscillator becomes very large and the initial in-phase relationship disappears. At high frequencies the oscillator lags 180° out of phase with the incident radiation; this results in a phase shift of the forward-scattered radiation by 180° and in consequence the transmitted beam leads rather than follows the incident beam. The effect on the optical rotation is a change in sign of the rotation; if the rotation is positive in character at long wavelength, it passes through zero at a given wavelength which corresponds to the CD absorption band, becoming negative at high frequencies. Such a change in sign is called a Cotton effect.

The extent of rotation arising in ORD is defined by the molar rotation $[m^1]$, where[1508]

$$[m^1] = [\alpha] \, \frac{3}{n^2 + 2} \, \frac{M}{100}$$

with n the refractive index of the medium and M the molecular weight of the solute; often in the case of biopolymers M is an average residue weight and $[m^1]$ is then called the mean residue rotation.

The variation of $[m^1]$ with wavelength in regions unaffected by an absorption band is described by the Drude equation

$$[m^1] = \frac{96\pi N}{hc} \frac{R\lambda_0^2}{\lambda^2 - \lambda_0^2}$$

where c is the velocity of light, R is the rotational strength, and λ_0 is the wavelength of the absorption band. At wavelengths close to λ_0 the Drude equation becomes a poor approximation since at λ_0, $[m^1]$ is predicted to become infinite, which is typical of the response of an undamped oscillator driven by a periodic field; inclusion of a damping term would yield a curve similar to the observed Cotton effect curve.

The interrelationship of circular dichroism and optical rotatory dispersion is quantified by the Kronig–Kramer relationship

$$[m^1]_\lambda = 2.303 \frac{9000}{\pi^2} \int_0^\infty \Delta\varepsilon_{\lambda_0} \frac{\lambda_0}{\lambda^2 - \lambda_0^2} \, d\lambda_0$$

which enables the ORD value at a particular value of λ to be calculated from the CD spectrum.

From a practical viewpoint it may be wondered why, since CD and ORD are so closely interrelated, techniques have been developed for separate measurement of the phenomena. Higher resolution of CD curves can be obtained since each peak corresponds to a particular absorption frequency, whereas the ORD curve which corresponds to each transition is spread over a much wider spectral region; thus resolution of a complex series of optically active bands can be extremely difficult, although McMillen and co-workers[954] have shown that it is possible to identify the minimum number of independent spectral components contained within the data by the method of matrix rank analysis. The ORD measurements are, however, advantageous when the optically active transitions lie at inaccessible wavelengths since the ORD curve extends far beyond the absorption band, enabling, for example, the rotational transitions of the far ultraviolet to be studied by examination of the ORD in the near-ultraviolet region of the spectrum.

The helical content of a biopolymer may be estimated from the Moffit–Yang equation, which arises from summation of the Drude equation for a number of transitions; if a number of transitions at wavelengths λ_{i0} and rotational strengths R_i contribute to the optical rotation at a particular wavelength, then

$$[m^1] = \frac{96\pi N}{hc} \sum_i \frac{R_i \lambda_{i0}^2}{\lambda^2 - \lambda_{i0}^2}$$

when λ is far from any transition wavelength λ_{i0}. Expansion of $\lambda_{i0}^2/\lambda^2 - \lambda_{i0}^2$ as a Taylor series results in the expression

$$[m^1] = \frac{a_0 \lambda_0^2}{\lambda^2 - \lambda_0^2} + \frac{b_0 \lambda_0^4}{(\lambda^2 - \lambda_0^2)^2} + \cdots$$

where a_0 and b_0 are constants.

If the value of λ_0 is known (usually taken as 212 nm), $[m^1](\lambda^2 - \lambda_0^2)$ versus $(\lambda^2 - \lambda_0^2)^{-1}$ will give a straight line with an intercept $a_0\lambda_0^2$ and slope $b_0\lambda_0^4$.

2.2. Spectrophotometry

Spectrophotometry, in the context of biopolymer systems, in general means determination of the absorption spectrum of the molecule. Emission spectra are usually obtained at temperatures which are too high to preserve the complex structure of biological molecules intact; exceptions to this rule are, however, provided by fluorescence and phosphorescence emission spectra, which are also employed in investigations of biopolymer systems.

Absorption spectra may arise from rotational and vibrational transitions of the whole or part of the biopolymer molecule. In addition, absorption spectra arise from electronic transitions within the molecule.

Pure rotational spectra, which arise in the far-infrared region of the spectrum, are rarely observed for substances of biological interest; vibrational transitions, however, which are observed in the near-infrared regions of the spectrum, give rise to extremely complex spectra. This is due to the fact that a complex molecule containing n atoms can have $3n - 6$ normal modes of vibration; each normal mode involves displacement of all, or nearly all, the atoms in the molecule, but while in certain modes all atoms may be displaced to approximately the same extent, in others displacements of small particular groups may be much more marked than those of the remainder. Thus normal modes of vibration can be divided into two groups: skeletal vibrations, which involve all the atoms to some extent, and the characteristic group vibrations involving only a particular part of the molecule.

Skeletal modes arise essentially, as the name implies, from the backbone structure of the molecule and, although very complex and incapable of resolution in total, are highly typical of that particular molecular structure. Such spectra act as "fingerprints" enabling the macromolecule to be positively identified.

Group frequencies are independent of the structure of the molecule as a whole, but do undergo small shifts when involved in processes such as hydrogen bonding; thus groups which are of particular interest for biopolymer studies, —OH, —NH$_2$, —C$=$O, have characteristic frequencies of 3600, 3400, and 1700 cm^{-1}, respectively, but these show reductions to approximately 3500, 3200, and 1600 cm^{-1}, respectively, when the groups are involved in hydrogen bonding. Thus a polypeptide in an α-helical conformation produces a —C$=$O vibration frequency of 1650 cm^{-1}, which is close to that observed in crystalline amides, where X-ray diffraction shows hydrogen bonding to occur.

A major restriction on the use of infrared analysis for the examination of biopolymers arises from the extremely strong absorption bands of water in the regions around 3400 and 1600 cm^{-1}, which makes examination of hydrogen bonding of biopolymers in aqueous solution an extremely difficult proposition. Although modern double-beam instruments and extremely thin cells of adjustable thickness make it possible in principle to minimize the intensity of the water bands and to balance out the water contributions, it is in practice a difficult and tedious procedure. To some extent this difficulty can be overcome by working in D$_2$O solutions, when the solvent absorption bands are moved to 1200 and 2500 cm^{-1}; the drawback to this procedure is that one must assume that no subtle changes in structure and interaction arise from the change in solvent environment.

One of the most recent advances in the techniques of infrared spectroscopy is the application of Fourier transform spectroscopy, which is now becoming available in the "ordinary" infrared region, 4000–400 cm^{-1}. Fourier transform spectroscopy has two major advantages: The first is its speed; thus the whole spectrum is contained in an interferogram which is computer recorded within 1 sec. Second, the greater the storage capacity of the computer, the greater the number of storage points of the spectrum which are obtained and hence the greater the resolution of the spectrum obtained. In addition, several scans of the same sample can be added to the computer store in order to improve the signal-to-noise ratio.

Absorption spectra in the visible region of the spectrum arise from transitions between outer shell electrons, i.e., relatively low-energy transitions. Biopolymers with transitions in this range include compounds containing metal ions, e.g., metalloproteins and large aromatic structures and conjugated double bond systems. Proteins such as hemoglobin owe their absorption mainly to the conjugated heme group rather than the metal ion, with very marked absorption occurring in the 405–410 nm Soret region.

Small conjugated ring systems, both aromatic and heterocyclic, exhibit strong absorption bands in the near-ultraviolet region of the spectrum; thus amino acids such as tyrosine, phenylalanine, and tryptophan absorb quite strongly in the 200–400 nm region, with the peak in the absorption band near 280 nm. Use is made of this in denaturation studies by observing the change in absorption at 280 nm as the degree of exposure of the amino acid group to its solvent environment changes with conformation.[525] Nucleotides, nucleosides, and nucleic acids, however, produce a very strong band in the region of 260 nm; absorption studies on nucleic acids are therefore more usually undertaken at this wavelength.

The general spectroscopic technique utilized in the visible and near-ultraviolet regions of the spectrum is that of *difference spectroscopy*. The optical density difference is measured between the biopolymer in a known reference state under constant experimental conditions and the material under differing experimental conditions, perhaps as a function of time, in the case of a conformational transition in the macromolecule; this technique is much facilitated by the use of a double beam spectrophotometer.

The quantitative measure of difference is defined as hyperchromicity H, the normalized absorbance change after treatment of the sample A^* compared to the absorbance of the reference sample A^0,

$$H = (A^* - A^0)/A^0$$

At an earlier point in the discussion it was mentioned that emission spectra of biopolymers are not usually available, with the exception of fluorescence spectra; such spectra can be a very sensitive tool for the examination of changes in conformation undergone by proteins and polypeptides; it will not therefore be out of place to include a brief description of the basic aspects of the techniques of *steady-state* polarization and *fluorescence* lifetime measurements. When a large molecule such as a biopolymer is in its lowest vibrational level of the electronic ground state, energy absorption with subsequent excitement of the electrons to a higher energy level can occur without large displacements of nuclei; the excited state may subsequently lose its excitation energy in a nonradiative manner or, alternatively, lose part of its energy as heat to reach the lowest vibrational level of the excited state, followed by a further energy loss as emission of radiation of lower frequency than the exciting radiation in fluorescence. Since the excited molecules generally reemit from their lowest vibrational levels, the fluorescence spectrum is completely independent of the wavelength of the exciting light; the fluorescence spectrum can thus, like the skeletal absorption spectrum, act as clear means of identifying the compound.

Absorption of polarized light by a long macromolecule such as a biopolymer depends upon the relative orientation of the absorbing chromophoric group and the direction of the polarization of the light; thus in the case of fluorescence when the exciting light is plane polarized, absorption is most likely for chromophores oriented with their transition moments parallel to the direction of polarization. If the macromolecule did not rotate in the short time interval between absorption and emission of light, then the fluorescence would be polarized to the same extent. Perrin[1149] and Weber[1567] have shown, however, that some depolarization occurs because of rotational Brownian movement; the treatment was developed for rigid spheres or ellipsoids and based upon the following assumptions: (1) equipartition of energy pertains; (2) the solvent is a continuum; in particular, the microviscosity of the solution is the same as the measured solvent viscosity; (3) the chromophores are randomly placed on the molecule; (4) there is no internal rotation in the molecule; and (5) changes in the orientation of the chromophores can yield all possible directions. If the material is irradiated continuously with plane polarized light, then the polarization P of the fluorescence is defined by

$$P = (I_\perp - I_\parallel)/(I_\perp + I_\parallel)$$

where $I_\perp$ is the intensity of the fluorescence perpendicular to the plane of polarization of the incident light, and $I_\parallel$ is the intensity in the parallel direction. The polarization P will depend upon the ratio of the average excited-state lifetime τ to the rotational relaxation time ϱ. The rotational relaxation time is usually approximated by

$$\varrho \simeq 3\eta V/RT$$

where η is the solvent viscosity and V is an effective molar volume; the following equation results:

$$\frac{1}{P} + \frac{1}{3} = \left(\frac{1}{P_0} + \frac{1}{3}\right)\left(1 + \frac{RT}{V\eta}\tau\right)$$

where P_0 is the polarization of the fluorescence that would be observed if no molecular rotation occurred. Since P_0 is not usually known, measurements of fluorescence are made under conditions of varying viscosity and/or temperature when a plot of $(1/P) + \frac{1}{3}$ versus T/η will be linear. Since ϱ is dependent upon the dimensions and the conformation of the macromolecule, a technique is thus available for monitoring changes in conformation by changes in the fluorescence spectrum. This relationship has been shown for several proteins[1569,1613] and synthetic polypeptides[522] to depart from linearity at high values of T/η. From fluorescence studies of protein-dye complexes using 1-dimethylaminonaphthalene-5-sulfonyl chloride Omenn and Gill[1113] established that the departure of the polarization from the Perrin equation was due to the onset of accelerated internal rotation and could be used as a sensitive measure of the internal structure of the molecule.

Techniques are now available[1414,1548] which allow direct measurement of the rotational relaxation time of the macromolecule, for example, by the use of light pulses of extremely short duration (~ 1 nsec). The fluorescence emission following each pulse is measured by detectors fitted with polarizers and the results expressed as emission anisotropy A defined as

$$A = (I_{\parallel} - I_{\perp})/(I_{\parallel} + 2I_{\perp})$$

The anisotropy decays exponentially with time according to the relationship

$$A(t) = A_0 e^{-3t/\varrho}$$

Very short relaxation times (~ 1.0 nsec) may be measured by this technique, which is thus a very powerful probe for changes in molecular structure.

2.3. High-Resolution Nuclear Magnetic Resonance

A spinning charged particle such as a nucleus behaves as a magnet and possesses a magnetic dipole moment; according to the laws of quantum mechanics, the moment of the nucleus may have only certain allowed values, measured in terms of the spin quantum number I, which can have half-integral or integral values. Nuclei that have been investigated include $^1H(\frac{1}{2})$, $^2H(1)$, $^{13}C(\frac{1}{2})$, $^{14}N(1)$, $^{17}O(\frac{5}{2})$, and $^{31}P(\frac{1}{2})$, the values in parentheses being the nuclear spin quantum number; by far the largest amount of work, particularly from a biopolymer viewpoint, has involved investigation of the proton nuclear magnetic resonance (pmr).

When a nucleus has a spin value other than zero, it will interact with an externally applied magnetic field, i.e., resonate, taking certain allowed orientations with respect to the applied field, corresponding to different energy levels for the nucleus. The resonance condition for a proton in a magnetic field of strength H is given by

$$\omega = \gamma H$$

where $\omega = 2\pi\nu$, with ν the frequency of the transition, H is the strength of the magnetic field, and γ is the ratio of the magnetic moment of the nucleus to its angular momentum. The magnetic field experienced by a particular proton depends not only on the externally applied field, but also on the fields produced by electrons and other nuclei in the molecule; thus protons in different chemical groups resonate at slightly different frequencies and the position of the resonance peak with respect to the position of a reference peak from an external reference compound is known as the *chemical shift*. These chemical shifts can be measured with high precision and provide a sensitive probe of the nuclear environment, for example, alterations caused by conformational changes.

The allowed energy levels for the proton are somewhat closely spaced, so that transitions between them correspond to radiation in the very far infrared, the microwave region; in addition, the spacing depends upon the external applied field. Thus it is possible to keep the microwave radiation at a constant frequency and sweep the applied field over a range until absorption by the different protons is successively observed.

Since the spacing of the allowed energy levels is proportional to the strength of the applied field, much better resolution of the spectra of complex molecules such as the globular proteins will be obtained at higher frequencies of the microwave radiation; much of the published work has involved the use of instruments capable of frequencies of 60 or 100 MHz, but more recently spectrometers operating at 220 and 300 MHz have become commercially available, which has resulted in dramatic improvements in the resolution of the protein spectra.[945]

We have already observed that the allowed energy levels for the proton nucleus are very closely spaced; the separation of these levels is given by

$$\Delta E = \gamma(h/2\pi)H$$

Since the energy separation of the levels is very small, the differences in population are also small; however, increasing the strength of the applied magnetic field, according to the above equation, produces a large difference in populations, and increased sensitivity is obtained. At resonance all the nuclei do not immediately occupy the appropriate energy level since a sharing of energy between nuclei occurs. The mechanism of sharing is referred to as a relaxation process; in fact, two differing relaxation processes occur: in the first the excess spin energy equilibrates with the surroundings (the lattice) by spin–lattice relaxation giving a "spin–lattice," "longitudinal," or "thermal" relaxation time T_1. Sharing of the excess spin energy between nuclei via spin–spin or transverse relaxation results in a "spin–spin" relaxation time T_2.

The times T_1 and T_2 have a marked effect upon the shapes of the resonance signals since they are dependent upon the rates of intramolecular and intermolecular motion in the system; the relationship between T_2 and the linewidth of the resonance peak is given by

$$T_2 = 1/\pi \nu_{1/2}$$

where $\nu_{1/2}$ is the width (in Hz) of the peak at half peak height. It is therefore clear that a molecular process that results in a long T_1 and T_2 (short correlation times) produces a narrow resonance peak, whereas a short T_2 (long correlation time) will in turn produce a broad peak. Increasing rigidity in the macromolecule produces a broadening of the nmr spectrum. Thus, in general, since the helical form of a protein involves increased rigidity of the molecular backbone, compared with the random coil conformation, a broadening of the nmr spectrum is observed. Since the advent of the higher-resolution spectrometers, it is now easier to observe the effects of denaturation upon the proton resonance of particular residue side chains of the protein,[945] with consequent greater insight into the detail of the conformation changes that occur on denaturation.

Additional sensitivity in terms of resolving power in nmr spectroscopy has been achieved by the technique known as computer averaging of transients (CAT); here a spectrum can be broken up into 2000 sampling points and the output from each sampling point stored in a 2000-channel computer; by repeated scanning over N scans a net gain in signal-to-noise ratio of $N^{1/2}$ is obtained. More recently the Fourier transform technique has been extended to nmr spectroscopy; here the sample is irradiated with pulses of "white" energy of a few microseconds duration, separated by intervals of approximately 1 sec to allow for the decay to the equilibrium energy distribution. The resultant emitted energy is a composite of the resonance frequencies of all the nuclei in the sample and the spectrum is obtained by Fourier transformation; as in the case of Fourier transform infrared spectroscopy, the great advantage is speed, only some 20 sec being required for a complete spectrum. A further improvement in sensitivity can be obtained by computer averaging of the transients.

Sciter and co-workers[1364] have recently reported a modification of the standard Fourier transform technique for the analysis of spectra containing both broad and narrow components. The broad resonances are filtered out from the spectrum by introducing a delay time between the end of the energy pulse and the start of data collection. Each component of the transient signal decays to $\exp(-\Delta t/T_2)$ of its initial value before recording of the transient signal begins; Δt is the length of pulse time plus the chosen decay time. The spectrum so obtained contains all the components of the continuous wave spectrum but with each component weighted by its own factor $\exp(-\Delta t/T_2)$. The effect of this weighting process is to enhance the narrow components of the spectrum compared with the broad resonance bands.

2.4. Calorimetry

The upper limit of cooperativity in a conformational transition is that exhibited in a two-state process. If θ represents the degree of conversion from helix to random coil in such a two-state process, then the equilibrium constant for the process is given by $K = \theta/(1 - \theta)$ and the van't Hoff equation, which represents the variation of K with temperature, can be written as

$$d\theta/dT = \theta(1 - \theta)\,\Delta H_{\mathrm{vH}}^{\ominus}/Rt^2$$

where $\Delta H_{\mathrm{vH}}^{\ominus}$ is the standard enthalpy change accompanying the two-state process. In general, unfolding reactions involving biopolymers exhibit enthalpy and entropy changes which are extremely temperature dependent due to the large heat capacity difference between the two structural forms. This heat capacity difference results in a nonlinearity of the van't Hoff plot which in itself has been the subject of much discussion as to its value as a criterion for the occurrence of a particular process, for example, changes in hydration of proteins.

Studies of thermodynamic changes during such processes as protein helix–coil transitions are potentially capable of yielding much useful information regarding the nature of the change. It is therefore important that van't Hoff studies of conformational transitions should be supplemented by direct calorimetric measurements of the enthalpy and heat capacity changes which occur during the process, since, for example, if the transition is not of a two-state type, then the van't Hoff and the calorimetric heats will not be in agreement. Further, since heat capacity changes during a conformational change must be determined from the second derivative of the experimental data, it is essential that such data be of sufficiently high precision for reliable second derivatives to be evaluated.

For any process which occurs in a system, according to the first law of thermodynamics,

$$\Delta U = q + w$$

where ΔU is the change in internal energy in the system, q is the heat absorbed from the surroundings, and w is the mechanical and any other work done on the system by the surroundings. Thus, for a process which occurs at constant volume, $w = 0$ and $\Delta U = q$; for a process which occurs at constant pressure, $\Delta U = q - p \Delta v$, i.e., $q = \Delta H$. Calorimetry is, however, not the science of measurement of q; in most types of calorimeter the extent of heat exchange with the surroundings is kept to the minimum. Calorimeters can be divided into two major classes. In the first, the constant-temperature environment type, the calorimeter is insulated to a certain extent from the surroundings, which are at a constant temperature; if the temperature of the calorimeter varies, heat exchange between the calorimeter and surroundings will occur; a correction must then be made for the net heat flow during the process. In the second class the amount of heat exchanged with the surroundings is maintained at zero, often by means of a vacuum jacket, hence the classification of adiabatic. This type of calorimeter has been the most widely used in measurements of conformation changes and hydration of biopolymers.

In an adiabatic process, since ΔU for the whole process must be zero, a change in temperature within the calorimeter must occur, rising for an exothermic process or falling for an endothermic process. If the temperature of the calorimeter is kept constant, e.g., for an endothermic process by the simultaneous addition of electrical energy which exactly compensates for the cooling effect of the chemical process, then an isothermal adiabatic system results. In the case of an exothermic process the amount of electrical energy necessary to duplicate exactly the heating effect of the experimental process is measured.

For such a system a precise comparison must be obtained between an electrical calibration process and the experimental process; in order that the required degree of precision can be obtained, the temperature control of the surroundings must be extremely good (better than $0.001\,^\circ\text{C}$); timing and temperature measurements must also meet an extremely high standard. There are major problems in experimental thermochemistry which lie outside the limits of this review and the interested reader is referred to the original literature (see, for example, Refs. 9, 314, 1036, 1547).

In the systems discussed previously, the apparatus consists essentially of only one calorimeter and separate calibration and experimental runs are unavoidable. This has led to the development of twin adiabatic calorimeters —differential calorimetry. Privalov et al.[1196] have developed a system which involves continuous electrical heating of the reaction calorimeter and an identical calorimeter containing only solvent. The temperature difference between the two calorimeters arising from the reaction produces a proportional voltage difference which is amplified and, by a feedback process, additional heat is supplied to the reaction calorimeter in order to maintain a very small temperature difference between the two calorimeters. Recording of the power feedback enables the heat imput to be obtained by graphical integration.

Danforth and co-workers[310] have reported a slightly different method in which the electric power needed to balance the effect of the chemical process is automatically integrated throughout the reaction range; this results in the temperature difference of less than ± 25 μdeg between the calorimeters.

One extremely useful development for studies of biopolymers from the increased calorimeter sensitivity now available commercially is that only small quantities of valuable materials need be used—hence the term microcalorimetry—and because of the extreme sensitivity, the interesting range of low solute concentrations can be studied.

2.5. Other Techniques

One of the most useful supplementary techniques for the investigation of aqueous solutions of biopolymers involves measuring viscosity. The contribution of a macromolecular solute to the viscosity of a solution depends in large measure upon the effective volume occupied by the macromolecule; since the effective volume occupied by the polymer is dependent upon the conformation of the macromolecule and, to an undefined extent, upon the hydration of the macromolecule, information on changes in conformation and/or changes in hydration are in principle accessible from viscosity studies.

Einstein showed that for spherical noninteracting particles the viscosity of the solution was described by the equation

$$\eta/\eta_0 = 1 + a\phi$$

where η is the measured viscosity of the solution, η_0 is the viscosity of the solvent, ϕ is the volume fraction of solute present, and a is a numerical factor, shown by Einstein to be 2.5 for this limiting case.

For solutions other than extremely dilute ones and for solutes of shape differing from spherical, this equation is inadequate and, as a result, a power series type of equation has evolved:

$$\eta/\eta_0 = \eta_r = 1 + a\phi + b\phi^2 + \cdots$$

where η_r is termed the relative viscosity of the solution, and a and b are constants.

In studies of biopolymer solutions, since at least the average molecular weight of the solute is known, ϕ can be replaced by cV, where c is the concentration in g cm^{-3} and V is the specific volume of the solute. Then

$$\eta_r^{-1}/c = \eta_{sp}/c = aV + bcV^2 + \cdots$$

where η_{sp} is termed the specific viscosity of the solution and η_{sp}/c, at the limiting value when $c \rightarrow 0$, is known as the intrinsic viscosity $[\eta]$.

In the limit of infinite dilution it is implicit in the above approach that the effects of interaction are eliminated and $[\eta]$ is dependent upon the shape of the macromolecule and its specific volume, which must include, to an undefined extent, its hydration shell. For many proteins the "anhydrous" specific volume is approximately 0.75 cm^3 g^{-1}, so that for the limiting case of a spherical protein with no water of hydration $[\eta]$ would be of the order of 2 cm^3 g^{-1}. Some globular proteins have values ranging from three to eight, suggesting a near spherical shape, but rodlike molecules such as DNA and myosin may yield values of $[\eta]$ larger by a factor of 10^2.

The coefficient of viscosity η is by definition the ratio of the shear stress developed by the application of a shear rate (in units of sec^{-1}) to the solution; the upper practical limiting rate of applied shear in straightforward viscosity measurements is of the order of 10^4 sec^{-1} but if higher rates of shear ($\sim 10^6$ sec^{-1}) were available the shear rate would be of comparable magnitude to the frequencies associated with molecular relaxation processes. This is impossible in normal viscosity measurement, but is possible by the adiabatic propagation on an ultrasonic wave of such a frequency through the solution; the wave motion will perturb molecular equilibrium at rates which depend upon the frequency, and in consequence molecular energy level populations are perturbed at the expense of acoustic wave energy and absorption occurs. Kessler and Dunn[765] have used the ultrasonic technique in an investigation of the conformational changes of bovine serum albumin in aqueous solution under the influence of changes in pH. For a plane-progressive sinusoidal wave propagated through a homogeneous, infinite medium the amplitude of the wave decays exponentially according to the equation

$$P(x, t) = P_0 \exp(-ax) \exp[i(\omega t - kx)]$$

where P is the instantaneous value of the acoustic pressure amplitude as a function of distance x and time t, a is the absorption coefficient, ω is the angular frequency, and k is the wave number. The authors employed the pulse technique, in which pulses of acoustic energy were transmitted covering the frequency range 0.3–163 MHz; over the range of 9–163 MHz two matched 3-MHz fundamental frequency X-cut quartz transducers were employed, one as generator and the other as receiver of the acoustic pulses. The acoustic path length was varied by displacing one transducer relative to the other at constant velocity. The amplitude of the received pulse varied according to the above equation.

At frequencies below 9 MHz, because of diffraction effects which arise when the wavelength of the sound approaches the dimensions of the transducer, the authors used the comparison technique of Carstensen[229] in which the acoustic properties of the protein solution were determined relative to those of water.

Data from such an investigation are usually presented in terms of the excess frequency-free absorption per unit concentration

$$A = \Delta a / c f^2$$

where c is the concentration (g cm^{-3}) of solute and Δa is the difference in the absorption coefficients between the solution and the solvent.

Urick[1507] has shown that for a solution that can be considered homogeneous to the sound wave the velocity of sound C in the solution is given by

$$C = C_0 \left(\frac{1}{(1 + \phi \sigma_K)(1 + \phi \sigma_\varrho)} \right)^{1/2}$$

where

$$\sigma_K = (K_0 - K_1)/K_0, \qquad \sigma_\varrho = (\varrho_1 - \varrho_0)/\varrho_0$$

In this equation the subscripts zero and one refer to solvent and solute respectively, K is the bulk modulus, ϱ is the density, and ϕ is the volume fraction of the solute in the solution; it is therefore possible to correlate changes in velocity of the sound wave with changes in volume of the macromolecule in solution.

A good deal of interest is currently being expressed in the technique of buoyant density measurements as a means of measuring the "hydration" of biopolymers in aqueous solution. A great deal of the initial work was conducted by Hearst and Vinograd[635]; essentially the technique is based on the sedimentation of the macromolecule through a salt solution which forms a density gradient, until the equilibrium position is obtained where the density of the hydrated macromolecule is equal to that of the salt solution. Since the macromolecule is in fact a colloidal system, perforce high gravitational fields are required, the equilibrium sedimentation is therefore conducted in an ultracentrifuge rotating at speeds up to ∼50,000 rpm.

The buoyant density is determined from the relationship

$$\varrho_0 = \varrho_e + (dp/dr)(r - r_e)$$

where ϱ_0 is the buoyant density, ϱ_e is the density of the cell contents, determined after centrifugation, r is the equilibrium distance of the macromolecular band from the center of rotation, and r_e is the root mean square position of the cell from the center of rotation.

The buoyant density determined experimentally is that of the solvated complex and is given by

$$1/\varrho_0 = (M_3\bar{v}_3 + \Gamma_n M_1 \bar{v}_1)/(M_3 + \Gamma_n M_1)$$

where the subscripts one and three refer to water and the anhydrous biopolymer, respectively. M is the molecular weight, $\bar{v}$ is the partial specific volume, and Γ_n is the hydration expressed as the number of moles of water per mole of polymer.

3. THE IMPORTANCE OF WATER AND HYDRATION INTERACTIONS IN DETERMINING THE CONFORMATIONAL STABILITY OF BIOMACROMOLECULES IN AQUEOUS SOLUTION

In Section 1 the outlines of several approaches to the interactions of proteins, polypeptides, and nucleic acids with their aqueous environment were summarized; here we discuss recent published evidence which illustrates the involvement of hydration in the conformational stability of natural macromolecules.

3.1. Conformational Transitions in Nonfibrous Proteins and Polypeptides

3.1.1. *Ribonuclease and the Chymotrypsinogen Family of Proteins*

Jackson and Brandts[720] in a recent high-precision calorimetric study of the thermal denaturation of chymotrypsinogen at low protein concentrations (0.21–0.26%) and low pH values (1–3) found $\Delta H_{\mathrm{cal}}^{\ominus}$ values ranging from $+433$ kJ mol^{-1} at pH 1.85 with a transition temperature T_m of 40.6°C, to $+588$ kJ mol^{-1} at pH 3.0 and a T_m value of 54°C. The change in the heat capacity associated with the change from the native conformation to the denatured conformation over the temperature range 20–70°C was from 17.64 kJ mol^{-1} deg^{-1} to 11.76 kJ mol^{-1} deg^{-1}, i.e., a change of -5.88 kJ mol^{-1} deg^{-1}. Enthalpy values for the denaturation process obtained from a van't Hoff analysis of spectrophotometric data[170] ($\Delta H_{\mathrm{vH}}^{\ominus}$) over similar ranges of temperature and pH lie between $+385$ and $+567$ kJ mol^{-1}. The good agreement between the $\Delta H^{\ominus}$ values obtained by the different techniques is strong evidence that the denaturation is in fact a two-state process; the large value of ΔC_p obtained is also indicative of the involvement of hydrophobic hydration in the denaturation process, probably due to the nonpolar side chains, which are buried within the interior of the protein in its native conformation, coming into contact with the solvent when exposed in the random coil conformation.

A comparison experiment using a higher protein concentration (1.85%) over the temperature range 5–65° was also conducted to obtain a more accurate value of the absolute specific heat of the native conformation of the protein; a value of 1.68 J g^{-1} deg^{-1} obtained at 20°C is in good agreement with the value of 1.68 J g^{-1} deg^{-1} reported by Bull and Breeze,[201] but somewhat lower than a value of 1.93 J g^{-1} deg^{-1} obtained earlier by Kresheck and Benjamin.[816] The data of Jackson and Brandts reveal that the heat capacity of native chymotrypsinogen in solution has a strongly positive temperature coefficient (C_p increases by 0.7% deg^{-1}), which is similar to that of the anhydrous protein,[699] although there is a difference of 30% in the absolute values. Jackson and Brandts suggest that the positive temperature coefficient of C_p is caused by thermal excitement of the internal degrees of freedom of the protein since the temperature dependence appears to be independent of the anhydrous or solvated state of the protein. Bull and Breeze find similar differences in C_p between the solvated native conformation of ovalbumin and the anhydrous form of the protein, and have suggested that the explanation may be that the solvated native conformation has a greater structural lability than the anhydrous protein, or that the water involved in hydration of the macromolecule has an anomalously high C_p due to contact with the exposed hydrophobic surface of the native protein.

Biltonen and Hollis* have suggested, on the basis of nmr studies, that chymotrypsinogen in the native conformation contains approximately 20% of the nonpolar side chains in an unfolded state, whereas α-chymotrypsin, which nmr data suggest to be fully folded in the native conformation, shows the same heat capacity for the native state in solution and the anhydrous solid. It therefore seems likely that differences in values of the absolute C_p between the native conformation in solution and the anhydrous solid are due to hydrophobic hydration of the unfolded side chains rather than to increased lability of the solvated protein over that of the anhydrous state.

* R. Biltonen and D. Hollis, unpublished results.

The average value of ΔC_p for the denaturation of chymotrypsinogen over the temperature range 20–70°C obtained by Jackson and Brandts is 14.3 kJ mol^{-1} deg^{-1}, in quite close agreement with 13.4 kJ mol^{-1} deg^{-1} reported by Shiao et al.,[1380] and obtained by a van't Hoff treatment of spectrophotometric data. Jackson and Brandts point out that the negative temperature dependence of ΔC_p for the denaturation process (approximately 200–400 J mol^{-1} deg^{-1}) is also in agreement with the interpretation that denaturation of chymotrypsinogen results primarily from exposure of non-polar side chains to be solvent.

Biltonen and Lumry[138] followed the thermal denaturation of chymotrypsinogen (CGN), α-chymotrypsin (CT), monomethionine sulfoxide-α-chymotrypsin (MMSCT), and dimethionine sulfoxide-α-chymotrypsin (DMSCT) by measurement of the differential extinction coefficient of the solutions at 293 nm, ΔE, which is defined as the difference in extinction between the protein at a given pH and temperature and the protein in its native conformation at the reference condition of pH 4.0 and 25°C. These reference conditions were chosen since the four proteins were found to exist completely in the native conformation under these conditions.[136,137]

In their analysis, which can be compared with that used by Brandts,[172] the authors summarize the various contributions to $\Delta G^\ominus$ for the denaturation reaction as follows:

$$\Delta G^\ominus = p(N\,\Delta h_{\mathrm{HB}}^\ominus - TN\,\Delta S_c) + AT + BT^2 + CT^3$$

where N is the total number of peptide units in the chain, p is the fraction of these units experiencing environmental change in the denaturing process, $\Delta h_{\mathrm{HB}}^\ominus$ is the standard enthalpy change due to peptide–peptide hydrogen bond rupture, and ΔS_c is the conformational entropy change per peptide unit of the macromolecule chain. The thermodynamic parameters for the denaturation of the proteins, which is shown to be a two-state process (agreement of the van't Hoff ΔH_{vH} with earlier calorimetric data[925]), are shown in Table II for CGN, CT, and DMSCT. According to the Brandts model, only the fraction p might be expected to change for any given protein, and the data show that this is indeed the case.

The authors point out that differing p values may mean different conformational states for the native proteins, the denatured proteins, or perhaps both. Sedimentation data[135] and ORD data[139] suggest that it is in fact the native states of the proteins that differ, while the conformational states of the denatured proteins are identical; it follows therefore that the number of exposed side chains in the denatured proteins should be identical for all the proteins and hence they should have identical heat capacities.

TABLE II. Thermodynamic Parameters for Denaturation at pH 3.0 and 0.01 mol dm^{-3} Cl$^-$ [a]

Protein	p	$\Delta h_h^{\ominus}$, kJ mol^{-1}	$\Delta S_c^{\ominus}$, J mol^{-1} deg^{-1}
CGN	0.63	3.36	21.4
CT	1.0 ± 0.02	3.28	21.4
DMSCT	0.59 ± 0.04	3.32	21.8

[a] Data from Ref. 138. Parameters calculated using $A = -22{,}702$, $B = 18.122$, and $C = -0.02792$.

Direct calorimetric measurements of the enthalpies of denaturation of CGN, CT, and DMSCT have been reported[140] for a protein concentration range of 0.20–1.0% in 0.01 M KCl adjusted to two reference pH states, 2.00 and 4.00, by the addition of 0.1 M HCl or NaOH. Values of ΔH obtained from direct calorimetric measurement and by a van't Hoff analysis of ORD data are shown in Table III. The good agreement confirms that the denaturation process is in fact two-state in character. Table IV contains the data of the apparent molar heat capacities ϕC_p of CGN, CT, and DMSCT in the native and denatured states; at 25° the heat capacities of all three proteins are very similar but the strong temperature dependence of ϕC_p appears to be different for each protein, which suggests different native structures, supporting the views of Biltonen and Lumry[138] and Biltonen and Hollis.*

TABLE III. Enthalpy of Unfolding[a]

Protein	Temp., °C	$\Delta H_{\text{cal}}^{\ominus}$	$\Delta H_{\text{vH}}^{\ominus}$
CGN	50	516	512
CT	40	452	504
DNSCT	25	134	151
	40	307	302

[a] Data from Ref. 140. Values in kJ mol^{-1}.

* R. Biltonen and D. Hollis, unpublished results.

TABLE IV. Apparent Molar Heat Capacities[a]

Protein	Native state pH 4.00			Denatured state pH 2.00		
	25°	40°	50°	25°	40°	50°
CGN	39.5	—	44.5	—	—	57.1
CT	40.3	42.8	—	—	56.7	—
DMSCT	39.5	48.3	—	57.5	56.3	—

[a] Data from Ref. 140. Values in kJ mol^{-1} deg^{-1}.

The ϕC_p values of the denatured proteins are identical within experimental error, supporting the view[138] that the conformations of the three proteins in the denatured state are identical. The heat capacities of the proteins in the denatured state show no temperature dependence, while the heat capacities of the native conformations show a positive temperature dependence. Therefore ΔC_p for the denaturation process shows a negative temperature dependence, in agreement with the data of Jackson and Brandts[720] for CGN and lending further support to the role of hydrophobic hydration of the nonpolar side groups in the denaturation process.

Privalov and co-workers[1195] have reported differential scanning calorimetric data on the thermal denaturation of α-chymotrypsinogen, ribonuclease, and myoglobin at concentrations of 0.1–1.0% in aqueous solution buffered to a maximum of pH 5. The authors suggest that the energy absorption plot as a function of temperature can in each case be divided into two distinct regions, giving first an initial slope, beginning in the case of chymotrypsinogen at 28° irrespective of the pH of the solution; this is subsequently followed by a sharp increase to a peak at a temperature which increases with increasing pH from 45° at pH 2.3 to 62°C for pH 4.0.

Supporting evidence is reported from optical density data at 282 nm for a protein concentration of 0.02%; the change in optical density is fully reversible with temperature at pH values up to 2.8 for chymotrypsinogen; van't Hoff plots derived from the optical density data are shown to be linear if the initial slope, corresponding to the small heat absorption region in the calorimetric data, is deducted from the total value.

The van't Hoff ΔH for chymotrypsinogen at pH 4 and 61°C of +600 kJ mol^{-1} is in good agreement with the calorimetric value corrected for the predenaturation thermal effect ($+588$ kJ mol^{-1}), which gives further support for characterization of the denaturation as a two-state process.

The calorimetric and optical density data of Privalov and co-workers for solutions of ribonuclease yield a ΔH_{cal} (corrected in a similar manner to that for chymotrypsinogen) of $+218$ kJ mol^{-1} and a ΔH_{vH} of $+256$ kJ mol^{-1} at the transition temperature of 36°C. Although agreement between the two values may be regarded as reasonable, it should be noted that considerable discrepancies exist between these values and those of Shiao[1380] ($+313$ kJ mol^{-1}) and Brandts[170] ($+315$ kJ mol^{-1}).

The authors suggest that the major portion of the heat capacity change associated with the total denaturation process occurs in the early predenaturation stage and that less than 4 kJ mol^{-1} can be attributed to the denaturation step—they consider that the predenaturation stage is due to change in the extent of contact of hydrophobic side chains of the native conformation with the solvent since unfolding is said to be insignificant in this stage.[673] This interpretation is apparently in agreement with the suggestion* that in the native conformation chymotrypsinogen contains a proportion of unfolded side chains. α-Chymotrypsin, on the other hand, is thought to be tightly folded in the native conformation and should, on the basis of this model, show no predenaturation stage, leaving only the denaturation step, and consequently, the heat capacity change for the process should show very little temperature dependence.

The work of Brandts and Jackson[720] has resulted in a value lying between -200 and -400 J deg^{-1} mol^{-1} for the temperature dependence of ΔC_p for the denaturation of chymotrypsinogen; however, the results of Shiao et al.[1380] for the chymotrypsinogen family of proteins suggest that ΔC_p is invariant with temperature for both chymotrypsinogen and the diphenyl carbonyl derivative of α-chymotrypsin. A contradiction seems to exist, but the discrepancy may be apparent rather than real, since Shiao et al. admit that the error limits on their data for this second temperature derivative of ΔH are considerable, and the value reported by Jackson and Brandts appears to lie inside these error limits. It is thus not possible at this stage to be certain about the temperature dependence of the heat capacity change associated with the denaturation of α-chymotrypsin and to draw inferences with regard to the interactions of the protein with water in the predenaturation stage.

Crescenzi and Delben,[299] using a differential scanning calorimetry and ORD, found for chymotrypsinogen (protein concentration 5.92%,

* R. Biltonen and D. Hollis, unpublished results.

pH 3.0) a ΔH_{cal} of unfolding of $+647$ kJ mol^{-1} and a ΔH_{vH} of $+567$ kJ mol^{-1}; these values are in good agreement with those already discussed. An apparent energy of activation for the denaturation process of $+336$ kJ mol^{-1} was obtained by the method of Beech,[104] and this agrees well with a value of $+342$ kJ mol^{-1} at the same pH, reported by Kim.[767]

Similar data have been reported for bovine ribonuclease,[331] $\Delta H_{cal} = +420$ kJ mol^{-1} and $\Delta H_{vH} = +430$ kJ mol^{-1}. No direct comparison is available with other reported data on the transition since this work was carried out at a relatively high pH of 9.24, although the ΔH values obtained were found to be independent of protein concentration over the concentration range studied (3.41–7.22%), implying no aggregation even at these high pH values and concentrations. However, ΔH in agreement with the extrapolated value obtained from Privalov's data, which range from $+218$ kJ mol^{-1} at pH 2.4 to $+378$ kJ mol^{-1} at pH 6.0.

The evidence strongly suggests, therefore, that the denaturation of ribonuclease is a two-step process over a pH range 2.0–9.0. The corresponding ΔC_p obtained by Delben et al. ($\Delta C_p = 8.4$ kJ mol^{-1} deg^{-1}) also supports this view. Tsong et al.,[1499] however, from differential scanning calorimetry on ribonuclease at a concentration of 0.5% over a pH range of 0.4–7.8 and a range of protein concentrations (0.1–2.7%) at pH 2.8, suggest that the thermal denaturation is a two-state transition at pH <2.0 but a more complicated pattern is involved at higher pH values. The calorimetric ΔH at pH 2 reported by Tsong et al. ($+275$ kJ mol^{-1}) is in good agreement with $+281$ kJ mol^{-1} obtained from a van't Hoff analysis of the calorimetric data; the value is somewhat greater than that obtained by Privalov ($+218$ kJ mol^{-1}) at pH 2.4 and rather lower than that of Brandts ($+315$ kJ mol^{-1}) at pH 3. It is in good agreement with the result of Shiao et al. ($+273$ kJ mol^{-1}) at pH 2.0. At pH >2.0 discrepancies are, however, apparent between the calorimetric and van't Hoff values; for example, in 0.2N NaCl solution at pH 7, $\Delta H_{cal} = +706$ kJ mol^{-1} and $\Delta H_{vH} = +407$ kJ mol^{-1}; Tsong et al. therefore propose that two transitions may occur in the denaturation process (cf. the proposals of Privalov and of Jackson and Brandts), but whereas both Privalov and Jackson and Brandts suggest that the heat absorption that occurs prior to the DSC absorption peak corresponding to the denaturation is related to a change in the heat capacity of the protein in its native conformation, possibly via solute–solvent interactions, Tsong et al. propose that a small, weakly cooperative transition occurs prior to the denaturation transition. An interesting point resulting from the analysis of Tsong is the apparently constant ΔC_p over the complete range of temperature and pH. It is suggested that both the native and denatured forms of the protein have apparent heat capacities which are independent of temperature. The calorimetric data of Biltonen et al.[140] have shown that the heat capacity of the denatured form of the protein is independent of temperature but that the heat capacity of the native form of the protein can be expected to have a positive temperature dependence. It seems likely therefore that the denaturation process incorporates a predenaturation change in heat capacity, involving solute–solvent interactions, followed by a two-stage transition.

In a recently published report, Tsong et al.[1498] reported kinetic studies of the reversible slow folding and unfolding of ribonuclease A by measurement of the change in optical absorbance at 287 nm associated with a temperature jump technique at neutral pH values. The data follow single exponential time decays better at neutral than at low pH values, and since the authors consider that the kinetic studies are a more sensitive test of two-state behavior than equilibrium studies, a discrepancy arises between these data and those presented in the previous paper. The authors suggest that the paradox may be resolved if the two-state approach is replaced by a simple sequential model; this consists of a nucleation step and $(n - 1)$ folding steps, reducing in the limit to only the nucleation step, i.e., two-state behavior. The nucleation step is regarded as the first step in the sequence on the basis of data of Pohl[1183–1185] concerning the activation energies for folding and unfolding of proteins; Pohl observed that the activation energy for unfolding is large and temperature independent whereas that for folding is small and strongly temperature dependent. This behavior is suggested as characteristic of nucleation dependent sequential reactions in which nucleation occurs at an early stage of the reaction sequence. The characteristics of the simple sequential model are that nucleation limits the rate of folding, and that folding and unfolding follow definite pathways and are cooperative reactions dependent upon the equilibrium constant for nucleation and the number of steps in the reaction; apparently the unfolding approximates to a two-state process even with appreciable concentrations of partly folded intermediates. The most important prediction of the model is the division of the unfolding kinetics into two steps, a rapid, transient nucleation step followed by a slow, steady-state unfolding with a *single exponential time course*; on this basis it is proposed that the reaction being observed in the ribonuclease transition is the slow, steady-state unfolding and that the initial nucleation reaction is faster by several orders of magnitude.

Tsong et al.[1497] report absorbance values at 287 nm obtained by a rapid temperature jump technique on ribonuclease A in aqueous solution at pH 1.3. The data confirm the existence of a rapid, transient nucleation step with a time constant in the millisecond range which is strongly temperature dependent and thus strongly implicates hydrophobic hydration in the nucleation step of the sequential model.

Pohl[1183] has studied the kinetics of the reversible denaturation of α-chymotrypsin over the pH range 2.0–2.5 (protein concentration range 0.01–1.0%) by optical density and optical rotation measurements based on a temperature jump technique. A value of $+344$ kJ mol^{-1} was reported for the apparent activation energy of unfolding. In a subsequent publication,[1184] reporting the kinetics of denaturation of trypsin in water and water–ethanol mixtures, it is pointed out that the apparent activation energy for denaturation E_f^* ($+271$ kJ mol^{-1} for trypsin) shows no temperature dependence; thus the established dependence on temperature of the enthalpy of denaturation ΔH, where

$$\Delta H = E_f^* - E_b^*$$

must be due to the folding step or to the denatured state of the protein as reflected in E_b^*, the apparent activation energy for folding. This is interpreted in terms of water being excluded from the immediate region of the hydrophobic side chains in the activated state, and the thermodynamics of hydrophobic bond formation is said to be responsible for E_b^* but not for E_f^*.

3.1.2. Lysozyme

Calorimetric and optical rotation (589 nm) measurements of the enthalpy of denaturation of lysozyme have been reported[1121] at a protein concentration of 2.5% and pH 1.0. The value of ΔH_{cal} was $+235$ kJ mol^{-1}, in fair agreement with ΔH_{vH} ($+265$ kJ mol^{-1}), and in agreement with a value of $+277$ kJ mol^{-1} reported earlier.[1141] The value of ΔC_p associated with the denaturation is apparently on the high side, at 29 kJ mol^{-1} deg^{-1}, and is said to be constant over a narrow temperature range. The evidence strongly suggests that the denaturation reaction is a two-state process and the large value of ΔC_p argues for considerable involvement of hydrophobic hydration in the unfolding process.

McDonald et al.[945] obtained high-resolution proton magnetic resonance measurements at 220 MHz of the thermal denaturation of lysozyme in water at pH 3.3 and in D_2O at pD 3.3 over the temperature range 35–80°; they report a ΔH of $+307$ kJ mol^{-1}. The authors designate the process as two-state since the pmr data indicate that the process is completely reversible and no evidence is observed of intermediate folded states. The pmr spectrum also indicates that the unfolded protein is in an extended conformation in the solvent and that all the peptide side chains of the protein are in an equivalent solvent environment.

3.1.3. *Myoglobin*

Appel and Duane Brown[42] have carried out optical rotation (233 nm) and absorption (407 nm) measurements on the denaturation of sperm whale myoglobin in H_2O and D_2O over the pH and pD range 2.6–8.0 (protein concentration 0.017%). The equilibrium curves of optical rotation or absorbance with D_2O as solvent are slightly displaced to the alkaline side compared to the values in H_2O (50% transition at pH 4.4 in H_2O and pD 4.9 in D_2O); this effect has also been observed with synthetic polypeptides[43] and is attributed to the weaker acidity of ionizable groups in D_2O.

The shapes of the optical rotation and absorption curves for the respective solvents are exactly superimposable and the transition curve is extremely steep; the authors therefore suggest that the denaturation is a two-state process and the steepness of the transition curve is due to the exposure to the solvent during transition of the large number of nonpolar groups buried in the interior of the molecule in its native conformation.

Privalov and his co-workers,[1195] in addition to their work on chymotrypsinogen and ribonuclease, also report calorimetric and optical density data over a pH range of 11.00–12.25 for cyanometmyoglobin at a concentration of 0.50%. It is shown that the change in partial heat capacity commences at 20°C irrespective of the pH of the solution, but unlike in the cases of chymotrypsinogen and ribonuclease, a distinct change of partial heat capacity also takes place in the denaturation stage of the process. The magnitude of the change is virtually independent of pH or temperature and is 11.3 kJ mol^{-1} deg^{-1}. The value of ΔH_{cal} ranges from $+560$ kJ mol^{-1} at pH 11.0 (ΔH_{vH} 567 kJ mol^{-1}) to $+307$ kJ mol^{-1} at pH 12.25 (ΔH_{vH} 330 kJ mol^{-1}). The authors point out that in conjunction with the marked heat capacity change on denaturation of myoglobin being considerably larger than the corresponding values for chymotrypsinogen and ribonuclease, the solubility of this protein in the denatured state is much lower than the solubility of the other proteins; this correlates well with the observation[1193] that the solubility of denatured egg albumin is even lower and the change in heat capacity more marked than is the case for myoglobin.

It is suggested that the denaturation enthalpies of the three proteins chymotrypsinogen, ribonuclease, and cyanometmyoglobin are in close agreement at corresponding temperatures and, further, that the enthalpy values depend in very similar ways upon the denaturation temperatures. The values of the enthalpy and entropy of denaturation thus appear to be interrelated via the denaturation temperature and hence lend support to the enthalpy/entropy compensation model.

Brunori et al.[193] have reported ORD measurements in the range 360–480 nm, absorption spectra in the range 330–650 nm, and fluorescence emission spectra in the range 270–420 nm for Aplysia myoglobin in the concentration range 5–$9.7 \times 10^{-6} M$ and the pH range 5.5–9.5. The transition equilibrium exhibits similar curves for the different methods used and linear van't Hoff plots are obtained. The agreement between the data obtained by the different techniques is a strong indication of two-state behavior and the value obtained for ΔH_{vH} ($+390$ kJ mol^{-1}) is similar to other values reported for two-state denaturation. In contrast, however, to the data of Privalov et al., it is suggested that ΔC_p is small ($\lesssim 4$ kJ mol^{-1} deg ^{1}); no comment is made upon this result, which seems to be somewhat at variance with two-state behavior and the probable importance of hydrophobic hydration for the process.

3.1.4. Polyamino Acids

Several investigations have probed the importance of solvent–solute interactions for the conformation of protein-type macromolecules, based on "tailor-made" polyamino acids in which the various effects of hydration can be deliberately emphasized and their relative importance more carefully assessed than is often the case in a large protein molecule with many differing peptide residues. Scheraga and co-workers have contributed an interesting and continuing series of reports covering this important aspect of protein and polypeptide chemistry based upon the α-helical form of the series of homopolyamino acids, polyglycine, poly-L-alanine, poly-L-leucine, and poly-L-valine and related compounds. The data obtained from such a series might be expected to be particularly informative with respect to the effects of hydrophobic bonding upon formation of the α-helix because of the increase in size of the nonpolar side chains with the ascending homologous series; however, as the size of the nonpolar side chain increases, the solubility of the polyamino acid decreases. Solubility can, however, be conferred on the polyamino acid by sandwiching the nonpolar group region between two blocks of a soluble polyamino acid. Since the investigation is concerned with the degree of helicity of the particular polyamino acid contained in the middle portion of the block copolymer, it is also advantageous for the end blocks to be of a random coil configuration under the experimental conditions of the investigation. Two such "ideal" partners in the block copolymer, having good aqueous solubility and a suitable random coil conformation, are poly-D,L-lysine and poly-D,L-glutamic acid.

Ingwall *et al.* reported one of the earlier studies of such a nature, involving poly-L-alanine[709] sandwiched between two blocks of poly-D,L-lysine. A series of block copolymers was investigated, containing 10, 160, 450, and 1000 alanyl residues, respectively (three typical examples are $lys_{15}ala_{40}lys_{15}$, $lys_{30}ala_{60}lys_{30}$, $lys_{45}ala_{120}lys_{45}$). The helical contents of the copolymers were obtained at pH 5–6 by ORD measurements in the range 190–500 nm and application of the Moffit–Yang equation[1028]; the helix–coil transition was observed over the temperature range 5–80°C. The copolymer containing only ten alanyl groups showed no helicity; application of the Lifson–Roig theory[892] to the data obtained from the other copolymers gave for the coil to helix transition per residue a ΔH of -800 J mol^{-1} and a ΔS of -2.3 J mol^{-1} deg^{-1}, which compares well with -546 J mol^{-1} and -1.9 J mol^{-1} deg^{-1}, respectively, reported[916] for the transition of poly-N^5-(3-hydroxypropyl-L-glutamine) in aqueous solution.

The enthalpy change for the transition contains contributions from all the interactions involving the side chains, such as dipole interactions and amide–water and amide–amide hydrogen bonds; Scheraga's estimate[1318] of the ΔH of formation of an amide–amide hydrogen bond is approximately -4.2 kJ mol^{-1}. The contribution from dipole–dipole interactions is also negative[169] and since the total enthalpy change for the coil-to-helix transition is only -800 J mol^{-1} for poly-L-alanine in water, Ingwall and co-workers suggest this is strongly indicative of an additional positive contribution to ΔH from hydrophobic bonding of nonpolar side chains in the helix conformation; theoretical estimates of the enthalpy change in hydrophobic bond formation in alanine–alanine range from $+1960$ to $+2940$ J mol^{-1}. The authors consider that neighbor–neighbor hydrophobic bonds in the random coil, as discussed by Poland and Scheraga,[1186] are not important in this system, perhaps due to the influence of electrostatic repulsions from the lysine residues on the random coil conformation.

Interpretation by Ingwall *et al.* of ΔS for the coil-to-helix transition is much more speculative, consisting, as it does, of contributions from the backbone of the molecule, hydrogen and hydrophobic bond formation, and any additional interactions involved in forming the helix conformation. The entropy contribution from the loss of rotational freedom, according to the Lifson–Roig theory, is taken to be -37 J mol^{-1} deg^{-1}, other estimates by Scheraga[1318] being in the range -13.4 to -30 J mol^{-1} deg^{-1}, and the contribution from hydrophobic bonding in the helix is estimated to lie in the range 8.8–19.7 J mol^{-1} deg^{-1}.[1318] The proposed hydrophobic bonding in the helical conformation is therefore considered the most likely explanation of the small, negative value for the entropy change associated with the coil-to-helix transition.

Joubert *et al.* have reported a high-resolution nmr investigation of polymers in the group poly-N^5-(ω-hydroxyalkyl-L-glutamines), which have the structure $[NHCH(CO)CH_2CH_2CONH(CH_2)_mOH]_n$.[745] In aqueous solution the helix content depends upon the number of methylene groups in the side chain; for example, at 90° the ethyl polymer PHEtG ($m = 2$; (degree of polymerization DP 200), the *n*-propyl PHPrG ($m = 3$; DP 750), and the *n*-butyl PHBuG ($m = 4$; DP 500) have 0, 40, and 80% helix content, respectively, as determined by ORD measurements. As the authors point out, the combination of ORD and nmr measurements is particularly informative since ORD discriminates between different backbone conformations such as the helix and random coil and the extent of helix formation but gives no information about the conformations and mobilities of side chains attached to either conformation, while nmr can detect differences in conformation and mobility in both backbone and side chains. The ORD and nmr measurements cover the temperature range 0–70°C and polymer concentrations for the nmr study was 2% w/v in D_2O.

Over the temperature range investigated it was shown that PHEtG (DP 200) exists in the random coil conformation and well-developed resonance peaks in the nmr spectra are observed[916]; the resolution of the splitting patterns improves as the temperature is increased to 70°. PHPrG (DP 270) has a helix content of 40% at 0°C, decreasing to 5% at 70°,[928] and a marked change in the spectrum occurs, particularly in the resonance peaks relating to the α-CH and the protons of the inner side chain. PHBuG (DP 650), which is 80% helical at 0°C, shows similar but even more marked changes in the resonance peaks due to these protons.

In general, therefore, the randomly coiled polymer (PHEtG) shows well-developed resonance peaks, but polymers with considerable helix content show a broadening of the peaks for the inner side-chain protons, those due to the outermost side-chain protons, however, remaining relatively sharp. That this phenomenon of broadening of the inner side-chain proton peaks is a function of helix content alone can be shown from a comparison of spectra at 10° of PHBuG samples having DP's of 650, 70, and 20 and helix contents of 75, 50, and 0% respectively. Despite the large differences in molecular weight and helix content, only the inner side-chain proton peaks exhibit any appreciable change. Earlier studies[916] of the helix–coil transition of PHPrG by ORD measurements gave values for ΔH and ΔS per residue of -546 J mol^{-1} and 1.9 J mol^{-1} deg^{-1}, respectively. These results are closely analogous to those obtained for the helix–coil transition of poly-L-alanine,[709] which illustrated the importance of hydrophobic bonding in the transition process. Joubert and co-workers suggest that their data show that an increasing degree of hydrophobic bonding occurs on passing from PHEtG to PHBuG; this is associated with increasing restrictions on the outermost groups of protons in the polymer as the size of the nonpolar residue increases, providing additional stability to the helical conformation in aqueous solution.

Ostroy and co-workers[1126] have examined the conformational behavior of block copolymers of poly-L-leucine in aqueous solution by ORD measurements over the wavelength range 190–250 nm. Full sandwich-type copolymers of the type $(D,L\text{-lys})_m(L\text{-leu})_x(D,L\text{-lys})_n(\text{gly})$ and half-sandwich polymers of the type $(D,L\text{-lys})_m(L\text{-leu})_x(\text{gly})$ were investigated. The ORD results for the half-sandwich copolymers $(D,L\text{-lys})_{16}(L\text{-leu})_{11}(\text{gly})$ and $(D,L\text{-lys})_{50}(L\text{-leu})_{20}(\text{gly})$ show the 11-mer to have a helix content of 20% at 25° and the 20-mer a helix content of 48% at 25°C; for the full-sandwich copolymers $(D,L\text{-lys})_{29}(L\text{-leu})_{11}(D,L\text{-lys})_{37}(\text{gly})$, $(D,L\text{-lys})_{74}(L\text{-leu})_{21}(D,L\text{-lys})_{37}(\text{gly})$ and $(D,L\text{-lys})_{60-n}(L\text{-leu})_{56}(D,L\text{-lys})_n(\text{gly})$, ORD results indicated helix contents of $\sim$10, 20, and 70% respectively. The ΔH and ΔS values for the transition from coil to helix as given by the Lifson–Roig theory[892] were $+420$ J mol^{-1} and $+2.9$–4.2 J mol^{-1} deg^{-1}, respectively. Molecular weights for both the half-sandwich and full-sandwich copolymers were obtained by calculation and by sedimentation analysis; the results indicated that the full-sandwich copolymers were free of aggregation in solution but the half-sandwich copolymers were considerably aggregated, which, the authors suggest, accounts for the enhanced helical content indicated by the ORD measurements.

Comparison of the data of Ingwall et al. on poly-L-alanine with those of Ostroy on poly-L-leucine shows that for comparable short chain lengths poly-L-leucine has a higher helix content, but a startling difference arises in the behavior of the degree of helicity θ_n as a function of temperature. This is illustrated in Fig. 2, where theoretical $\theta_n(T)$ curves computed for both polymers at various DP values are plotted; θ_n decreases rapidly with increasing temperature for poly-L-alanine but rises very slightly for poly-L-leucine. The values of the thermodynamic parameters for the coil $\rightarrow$ helix transition in aqueous solution of several polyamino acids are listed in Table V together with the enthalpy and entropy contributions to the transition, which, according to Ostroy, are due to interactions between backbone and side chain atoms and between atoms of different side chains, which can be approximated as the net contribution from hydrophobic bonding in

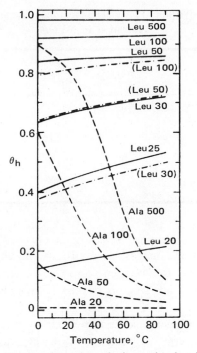

Fig. 2. Theoretical θ_h vs. T curves for poly-L-alanine and poly-L-leucine of various DP's (see Ref. 1126).

TABLE V. Thermodynamic Parameters for Coil → Helix Transition for Several Polyamino Acids in Aqueous Solutions[a]

Polyamino acids	Solvent	ΔH, J mol^{-1}	ΔS, J mol^{-1} deg^{-1}	ΔH_{sc}, J mol^{-1}	ΔS_{sc}, J mol^{-1} deg^{-1}
Poly(L-glutamic acid)[b]	0.1 M KCl	−4704	−14.4	−2654	−11.8
Poly-L-lysine[b]	0.1 M KCl	−3717	−11.3	−1657	+1.5
Poly-L-lysine[c]	0.1 M KCl	−5074	—	—	—
Poly-L-lysine[d]	H$_2$O	−3738	−8.4	—	—
Poly-L-alanine	H$_2$O	−800	−2.3	+1250	+10.9
Poly-N³-(3-hydroxy-propyl)-L-glutamine	H$_2$O	−651	−2.1	+1436	+11.13
Poly-L-leucine	H$_2$O	+420	+2.9 to +4.2	+2460	+16.2 to +17.4

[a] Data from Ostroy et al.[1126]
[b] Data of Hermans[641].
[c] Data of Chou and Scheraga.[253]
[d] Data of Davidson and Fasman.[321]

TABLE VI. Thermodynamic Parameters for the Homopolymer Coil → Helix Transition in Water[a]

Polyamino acid	ΔH, J mol^{-1}	ΔS, J mol^{-1} deg^{-1}
Polyglycine	−2050	−13.2
Poly-L-alanine	−800	−2.3
Poly-L-leucine	+420	+2.9–4.2
Poly-L-valine	+890	+6.1

[a] Data from Ref. 1126.

the helix and the coil; the increasing importance of hydrophobic bonding on passing down the table can be seen from the increasingly positive contributions to both ΔH_{sc} and ΔS_{sc} on passing from poly(L-glutamic acid) to poly-L-leucine.

The authors neatly summarize the effects of hydrophobic bonding in the polyamino acid series by utilizing the assumption that the observed enthalpy and entropy changes associated with the transition consist of backbone and hydrophobic bonding contributions; hence they calculate theoretical values for ΔH and ΔS for polyglycine and poly-L-valine. The thermodynamic parameters for nonpolar homopolyamino acid coil → helix transitions per residue in water are given in Table VI and it can be clearly seen that ΔH and ΔS become increasingly positive as the increasing size of the nonpolar side chain produces increasing hydrophobic stabilization of the helix.

The work of Ostroy et al.[1126] strongly suggests that hydrophobic bonding plays a minimal role in promoting helical stability in polyglycine; it is therefore interesting to consider a recent report[704] concerning the helical content of block copolymers containing varying ratios of glycine to alanine. As is the case with other polyamino acids, solubility is achieved by block polymerization between blocks of soluble poly(D,L-glutamic acid); the particular copolymers which were the subject of the investigation were (D,L-glu)$_{23}$(ala$_2$gly)$_{10}$(D,L-glu)$_9$–ala, (D,L-glu)$_{19}$(ala$_2$gly)$_6$(D,L-glu)$_9$–ala, and (D,L-glu)(ala$_3$gly)$_7$(D,L-glu)$_{11}$–ala. The ORD measurements were obtained over the range 250–450 nm at pH 6.0–7.0 in water. Infrared spectra confirmed that the polymers formed only α-helix or random coil conformations and the helix content of each polymer was determined from the Moffit–

Yang equation.[1028] A block copolymer $(D,L\text{-gly})_n(ala)(D,L\text{-gly})_n$ was also prepared in which the ratio of alanine to D,L-glycine was 38:1 and the α-helix content of the four copolymers was shown to decrease in the order $(D,L\text{-gly})_n(ala)(D,L\text{-gly})_n > (D,L\text{-glu})_{23}(ala_2gly)_{10}(D,L\text{-glu})_9\text{–ala} > (D,L\text{-glu})_{19}$ $(ala_2gly)_6(D,L\text{-glu})_9\text{–ala} > (D,L\text{-glu})(ala_3gly)_7(D,L\text{-glu})_{11}\text{–ala}$. Thus, although the $(ala_3gly)_7$ copolymer has a relatively high alanine content, its degree of helicity is rather low, in contrast to polyalanine, which shows the highest degree of helicity; the stability of the α-helix does not therefore coincide with the glycine/alanine ratio, though the data of Fraser et al.[494] and Block and Kay[152] confirm the results of Ostroy et al.[1126] in showing that the glycyl residue markedly reduces the stability of the helix.

The author suggests that the distribution of glycyl (G) and L-alanyl (A) residues on the surface of the right-handed α-helix is such that if any residue is designated as 0, then successive residues are 1, 2, etc. in the —CO—C—NH— direction and -1, -2, etc. in the opposite direction; thus for poly-L-alanyl–L-alanyl–glycine, if G is located at 0, then successive G's occupy positions ±3, ±6, etc. In poly-L-alanyl–L-alanyl–L-alanyl–glycine positions 0, ±4, ±8, etc. are occupied by G's, which leads to the conclusion that glycyl residues at positions 0 and ±4 decrease the stability of the helix compared to the situation when glycyl residues occupy positions 0, ±3, etc. Iio suggests therefore that the stability of the α-helix is enhanced when hydrophobic bonding is possible between the CH_3 group of the ith L-alanyl residue on the surface and the CH_2 group of the $(i + 3)$th L-alanyl residue, rather than interaction involving the ith L-alanyl and the $(i + 3)$th glycyl residues.

The homopolymer of L-lysine provides an interesting molecule from the viewpoint of hydration effects, since not only is it easily soluble in aqueous solutions without requiring block copolymerization to another polymer to render it soluble, but suitable adjustment of the pH of the solution produces the random coil or the α-helix conformation, a behavior which parallels that of the proteins themselves. At pH < 8, when the NH_2 groups are almost completely protonated, the polymer in aqueous solution is in the random coil conformation and at pH ≥ 12, when the NH_2 groups are virtually uncharged, the transition to the helix is almost complete.

Davidson and Fasman[321] have investigated the thermal transition of poly-L-lysine hydrochloride at pH 11.6 from the α-helix conformation to the random coil conformation in the temperature range 10–40°C. ORD spectra were obtained for the helix–coil transition of the homopolymer in water, 0.2 mol dm^{-3} NaCl, 0.06 mol dm^{-3} LiBr, 15% ethylene glycol solution, and various combinations of these solvents; the transitions were all of a similar form and did not show a sharp change as a function of temperature.

The standard enthalpy change for the helix–coil transition in H_2O was obtained from a van't Hoff plot, the slope of which is $-\Delta H^{\ominus}/R\sigma^{1/2}$.[44] The ratio $K = $ (random coil)/(α-helix) is obtained from the ORD data by the method of Moffit and Yang[1028] and σ is the helical conformational partition function,[1623] assumed to be 3×10^{-3} from the data of Rifkind et al.[1246] and Snipp et al.[1409]; the value of $\Delta H^{\ominus}$ obtained was $+3738$ J mol^{-1} and the corresponding values of $\Delta G^{\ominus}$ and $\Delta S^{\ominus}$ for the transition were $+6.3$ kJ mol^{-1} and -8.4 mol^{-1} deg^{-1}, respectively. The authors suggest that the entropy decrease can be due either to too low a value of σ being used or to increased ordering of water around the lysyl side chains in the solvated random conformation.

The value of $\Delta H^{\ominus}$ is in good agreement with $+3717$ J mol^{-1} reported by Hermans[641] from a pH titration of poly-L-lysine; however, $\Delta G^{\ominus}$ is higher by approximately 4.2 kJ mol^{-1} than the value obtained by Hermans and Miller and Nylund,[1020] who report a positive $\Delta S^{\ominus}$ associated with the helix–coil transition.

The authors suggest that the thermal transition is too broad to be a completely cooperative process but Rifkind and Applequist[1246] and Miller and Nylund[1020] agree that the pH-monitored helix–coil transition appears to be cooperative for poly-L-lysine and for poly-L-glutamic acid and Fasman et al.[426] have shown that the thermal transition for poly-L-glutamic acid is also very broad. Observed broadening is believed to be due to hydrophobic interactions between the nonpolar side chains leading to increased stability of the α-helix conformation. The summarized data on the thermodynamic parameters for the helix–coil transition of the series of polyamino acids reported by Ostroy[1126] confirm the interpretation of Davidson and Fasman.

Parker et al.[1139] have reported ultrasonic sound absorption measurements of poly-L-lysine in aqueous solution; velocities and attenuations were determined over a frequency range of 3–100 MHz as a function of temperature and pH at constant ionic strength (0.6 mol dm^{-3}) for poly-L-lysine (DP 675) in aqueous solution. No significant dispersion was observed; at 35.6° and pH 9.22 the velocity of sound in 0.156 mol dm^{-3} of poly-L-lysine in 0.6 mol dm^{-3} of NaCl is 1560.6 m sec^{-1} at 3 MHz, 1561.0 m sec^{-1} at 11. 5 MHz, and 1561.2 m sec^{-1} at 50 MHz. The frequency dependence of the observed excess acoustic absorption α/ν^2 (α is the ultrasonic attenuation and ν is the frequency) is clearly described in terms of a process involving only a single relaxation time τ of 20–40 nsec. Optical rotation measurements at 300 nm were used to determine the helix content as a function of pH and correlated with intrinsic viscosity measurements over a similar pH range.

Data were obtained over the temperature range 20–36°C from all three experimental techniques and the general trend showed that at constant pH the helix content of the copolymer decreased with increasing temperature; a more detailed analysis of the data at 20.3°C, however, shows that the optical rotation undergoes a rapid change over the pH range 9.5–11.0 with the inflection point at pH 10.2. The intrinsic viscosity reveals a sharp minimum at the same pH value and the ultrasonic absorption rises rapidly on approaching this value, higher pH values being unobtainable due to the onset of aggregation. This striking agreement of different experimental techniques strongly suggests that the helix–coil transition is a two-state process; this is further supported by the single relaxation time for the process, and is in agreement with the work of Rifkind and Applequist[1246] and Miller and Nylund.[1020]

The authors consider that the sound absorption of poly-L-lysine occurs at too low a frequency and is too large at low values of pH to be due to a proton transfer equilibrium; they also consider that the large, pH-dependent attenuation observed is not related to any variation in the shear viscosity and, since at all values of pH the high-frequency absorption approaches a common limit close to that obtained for pure water,[1179] the result is not consistent with the existence of a large, pH-dependent bulk viscosity; therefore the observed absorption is not due to viscous relaxation.

Burke et al.[214] have attributed the sound absorption of poly-L-glutamic acid in water–dioxan mixtures to the perturbation of a solvation equilibrium, but Parker et al., from a consideration of diglycine and triglycine ultrasonic absorption in water, suggest that at the high pH values involved there is no evidence of a large absorption due to interaction of water with individual peptide groups. The latter authors suggest that, if the absorption in solutions of polylysine is to be attributed only to solvation, then either interaction of water with the polypeptide backbone must occur, which is unlikely, or a cooperative effect requiring the simultaneous proximity of a number of neutral side chains must be involved. This is not unlikely since the data obtained by the authors seem to point quite strongly to a two-state process for the helix–coil transition, a transition which, by its very nature, must be highly cooperative. This reasoning seems to confirm the interpretation of Parker et al. that the observed α/ν^2 is due to perturbation of the helix–coil transition. Also, the magnitude of τ for the process suggests that the change in conformation of a uniform polypeptide is an extremely fast process; an interesting correlation may exist therefore between these data for a uniform homopolymer and those of Tsong *et al.*[1497] indicating a rapid nucleation step for the helix–coil transition of the protein ribonuclease.

Chou and Scheraga[253] have recently reported one of the few calorimetric determinations of a thermally induced helix–coil transition for a polyamino acid. Calorimetric and ORD data were reported on the transition of poly-L-lysine hydrochloride (DP 670) in a 0.1 mol dm^{-3} solution of KCl. The $\Delta H^{\ominus}$ was determined by changing the pH of the solution from a value at which the configuration is helical to pH 7, when the configuration is a random coil; the technique used was to break a glass vial containing 1.2 ml of 0.55 mol dm^{-3} HCl in 40 ml of a 0.25% w/v solution of the polypeptide and measure the heat evolved. The observed ΔH_{cal} was corrected for the heat of breaking the sample vial, the heat of dilution of HCl, the heat of neutralization of OH$^-$ ion, and the heat of ionization of the NH$_2$ groups in the random coil conformation. The value obtained for $\Delta H_{cal}^{\ominus}$ of transition at both 15 and 25° was -5040 J mol^{-1}, which differs from the ΔH_{vH} value of -885 J mol^{-1} reported by Hermans[641] and Pederson *et al.*[1144] and a value of -890 J mol^{-1} reported by Davidson and Fasman,[321] based on an analysis of potentiometric titration data. The value of Chou and Scheraga is much closer to the only other calorimetric measurement of ΔH for a polyamino acid, -4620 J mol^{-1} for polyglutamic acid in 0.1 mol dm^{-3} KCl.[1243]

ORD experiments were conducted under similar pH conditions to the calorimetric experiments and the helix content at various pH values obtained from the Moffit–Yang theory. The results confirmed that poly-L-lysine was completely in the α-helix form at pH 11 and 15° and essentially 95% in the helix form at 25°; at pH 7 the polymer was completely in the random coil conformation at both temperatures. The pH at the transition point agreed very closely with the value of 10.2 obtained by Parker *et al.*,[1139] the transition again being very sharp, in agreement with several other workers[1020, 1139,1246] and lending further support for the importance of hydrophobic bonding in contributing to the helix–coil transition of poly-L-lysine.

3.2. The Helix–Coil Transition in Nucleic Acids, Nucleosides, and Nucleotides

In the previous discussion of the stability of proteins and polypeptides the central theme was the helix–coil transition; while a broadly similar conformation change occurs in aqueous solutions of nucleic acids, nucleosides, and nucleotides, it is more generally referred to as a stacking interaction or layering of base pairs upon each other. The stacking interaction is the dominant stabilizing force for the helical conformation of poly-nucleosides and nucleic acids in aqueous solution, and is observed even in simple nucleosides and dinucleotides without base pairing by hydrogen bonding, electrostatic effects, or cooperative phenomena. The thermodynamic contributions to the stacking interaction from solvent–solute and solute–solute interactions are not clearly understood, although the reduction in free energy obtained by the formation of a cavity in the solvent for a helix conformation in preference to a random coil conformation[1398] and hydrophobic bonding[1089] undoubtedly provide contributions to the overall energetic changes associated with stacking. It must not be overlooked, however, that the nucleic acids contain base groups that are highly polar in addition to having charged phosphate groups; it is therefore very possible that *hydrophilic bonding* by water molecule bridges between strategically placed charged centers in the helix[885] plays a much greater role in stabilizing the helix conformation than will be the case for the corresponding conformation in proteins and polypeptides.

The two-state model for the helix–coil transition has been notably successful in the interpretation of the interactions involving proteins and polypeptides, but problems arise in the application of this model to the base stacking reaction of nucleotides; success with the proteins is due to the fact that transitions occur within the experimentally accessible temperature range. In the case of nucleotides, however, the transition range can often exceed the working range[918]; and since experimental verification of the model requires experimental data attributable to the two extreme conformations, it may thus be difficult to apply the model to the stacking of nucleotides. The transition equilibrium is usually interpreted as between a state when the bases are stacked and a random conformation when the bases are completely solvated. Even when it is not possible to obtain experimental verification of the two states, it is possible to determine whether the data fit a two-state model within defined limits; this can be done if the experimentally obtainable property can be represented as a linear combination of the experimental properties of the two extreme states.

McMullen *et al.*[954] applied a matrix analysis technique to ORD data on tobacco mosaic virus RNA over the range 230–350 nm at pH 7.5–8.2 and showed that the experimental data could be represented as a sum of two spectra representing the extreme states. A van't Hoff analysis of the data produced a value for the standard enthalpy of stacking of 84 kJ mol^{-1}. Previous estimates of the enthalpy of formation of a stacked base pair of the order of -28 kJ MBP^{-1} (mol base pair) led the authors to suggest that the helical regions each contain roughly three base pairs.

Broom *et al.*[184] have reported vapor pressure measurements and pmr results on a range of 14 purine nucleosides in aqueous and D_2O solution, in certain cases to concentrations as high as 0.2 M, and have shown that the compounds associate to a marked degree by a stacking mechanism and not by hydrogen bonding. The observed concentration dependences of the various proton chemical shifts showed that the free energies of stacking of the nucleosides do not correlate with the dipoles and monopoles of the corresponding bases. This is confirmed when the self-stacking of the bases in organic solvents of low dielectric constant is considered. If the base stacking is not dominated by solvent interactions but by electrostatic interactions, then stacking would be expected to be greater in an organic solvent; this is not the case and the bases in fact self-associate by hydrogen bonding. Also, the extent of stacking in aqueous solution is greater for the purine than it is for pyrimidine derivatives.[425,1410] It is also noted that the methylation of the base moiety results in an enhancement of the stacking association ($\Delta G^{\ominus}_{stacking}$ of purine is -1.8 kJ MBP^{-1} and that of 6-methyl purine is -4.70 kJ MBP^{-1}). The range of values for $\Delta G^{\ominus}$ at 25° is from $+1.3$ kJ MBP^{-1} for uridine to -8.0 kJ MBP^{-1} for N-6-dimethyladenosine. Farquhar and co-workers[425] show that for the pyrimidine derivatives cytidine, uridine, and thymidine, corresponding $\Delta S^{\ominus}$ values are -29.4, -42, and -42 J MBP^{-1} deg^{-1}, respectively; these estimates are within the range of 21–57 J MBP^{-1} deg^{-1} reported by Gill *et al.*[521] and van Holde and Rossetti.[1519]

Subsequent vapor pressure and pmr studies[1496] on 7-methylinosine (which is highly polar due to an electron deficiency in the five-membered ring and an electron excess in the six-membered ring) in aqueous and D_2O solutions confirm the lack of an energy contribution from electrostatic interaction of monopoles and dipoles to the nucleoside stacking process. It is pointed out that, although the overall entropy of stacking is negative, there are some indications of a positive entropy contribution to the association, e.g., the free energy of association of caffeine (-6.3 kJ MBP^{-1}) is much larger than that of purine (-1.8 kJ MBP^{-1}) or 6-methyl purine (-4.7 kJ MBP^{-1}), yet the enthalpy of stacking for purine (-17.6 kJ MBP^{-1}) or 6-methyl purine (-25.2 kJ MBP^{-1}) is considerably larger than that of caffeine (-14.3 kJ MBP^{-1}). The reason for self-association of caffeine is therefore due to a smaller negative entropy (-25.2 J MBP^{-1} deg^{-1}), as against -54.6 J MBP^{-1} deg^{-1} for purine, and similarly for purine nucleosides (-29.4 J MBP^{-1} deg^{-1}) versus pyrimidine nucleosides (-42 J MBP^{-1} deg^{-1}). The comparisons suggest the existence of a positive entropy contribution to association hidden in an overall negative entropy, which is most probably due to competition between solute–solvent interaction and solvent–

solvent interaction. Evidence to support this viewpoint is provided by the association of actinomycin and deoxyguanosine[302]; the likely form of association by analogy with the structure of DNA is that the deoxyguanosine purine ring forms a π complex with the actinomycin chromophore, possibly with a hydrogen bond from the actinomycin carboxamide group to the deoryribose ring oxygen. The degree of association has been measured by absorbance changes at 42.5 nm in a phosphate buffer (pH 6.95). The association process is believed to be an appropriate model system for the study of the importance of hydrophobic interactions in the stacking of polynucleotides. The change in standard free energy for the formation of the complex in phosphate buffer is -18.6 kJ mol^{-1}, which increases to -14.7 kJ mol^{-1} in a mixture of 40% methanol and phosphate buffer; corresponding values for the enthalpy change are -43.6 and -56.3 kJ mol^{-1} and for the entropy change are -93.6 kJ mol^{-1} deg^{-1} and -139.44 J MBP^{-1} deg^{-1}. The increased methanol content of the solvent results in an increased tendency to destabilization due to an increasingly negative entropy of association, partially compensated by an increasingly negative enthalpy of association. Such a result is consistent with the view that the solvent, probably by a hydrophobic bonding mechanism, is responsible for the association. A further point of interest arising from these data is that application of the Lumry–Rajender enthalpy–entropy compensation model[927] to the results produces an average compensation temperature of approximately 250°, which is within the range of temperatures considered as an indication of the importance of solvation interactions for the transition process.

Lubas et al.,[922] by use of a nmr spin-echo technique, observed the shortening of the longitudinal and transverse spin–lattice relaxation times T_1 and T_2 associated with the thermal transition of calf thymus DNA in water as compared to other solvents. The correlation time τ_1 of the denatured DNA was found to be lower than that for DNA in its native conformation, which the authors interpret as being due to a more limited freedom of movement. The DNA concentrations used were 2.90 mg/ml in 0.15 NaCl; comparison data were obtained from spectrophotometric absorbance curves over the range 25–95°C, the transition being complete at the upper temperature. Values of T_1 and T_2 were obtained over the same temperature range and the region of high slope in T_1 and T_2 corresponds to the thermal transition point in the 260-nm absorption curve. In a previous paper Lubas and Wilczok[919] suggest on the basis of nmr studies that protons in the hydration shells surrounding DNA are bound nonrotationally and this influences T_1 and T_2. The protons of the DNA have very little movement and hence a long relaxation time and do not influence T_1 and T_2. The observed decrease reported in this work is attributed to fast exchange of protons between free and nonrotationally bound water; hence changes in the degree of hydration of the DNA molecule occur before any destabilization arising from hydrogen bond rupture between base pairs.

Pmr measurements on yeast DNA in 0.015 M NaCl and 0.015 M sodium citrate buffer with varying ratios of H/D in the solvent may also suggest[1415] that the decrease in T_1 observed by Lubas et al. is not due to an appreciable alteration of water structure in the hydration shell of the DNA; instead, the explanation may lie in the presence of magnetic impurities in the sample material. It does seem unlikely, however, that if this were the case that such a close correlation would have been observed by Lubas et al. between the region of high slope of T_1 and T_2 over the temperature range of transition and the thermal transition temperature obtained from absorbance values.

Klump and Ackermann[795] obtained direct calorimetric and UV absorbance estimates of the denaturation process for a variety of DNA's at neutral pH values (DNA concentrations 5–6 mg/ml in 0.001 M KCl and 0.0015 M sodium citrate giving a $pH \sim 6$). The DNA from M. lysodeikticus observed over the temperature range 45 to 95° showed a marked change in heat absorption and hyperchromicity at the transition temperature of 90° with a small, earlier endotherm at 60–65°. DNA from Cl. perfringens revealed a marked change at 80° and an earlier endotherm at 50°. DNA obtained from salmon sperm showed a marked change at 60° but no earlier endotherm, in contrast to the other samples. By combining these data with those of Privalov[1196] and Sturtevant et al.[1438] the authors obtained a linear plot of stacking enthalpy vs. G–C base content. For DNA from Cl. perfringens (32% G–C content), $\Delta H^{\ominus}$ for denaturation is 32.5 kJ MBP^{-1}; for DNA from salmon sperm (41%), $\Delta H^{\ominus} = 33$ kJ MBP^{-1}; and for DNA from M. lysodeikticus (72%), $\Delta H^{\ominus} = 35.8$ kJ MBP^{-1}. The extrapolated value of $\Delta H^{\ominus}$ for pure d(AT) polymer is 30.2 kJ MBP^{-1} and for pure d(GC) is 37.8 kJ MBP^{-1}. The value for d(AT) is in fair agreement with 31.8 kJ MBP^{-1} for an alternating copolymer of deoxyadenylic acid and deoxythymidylic acid measured calorimetrically by Scheffer and Sturtevant.[1312] The authors suggest that the variation of the enthalpy of stacking with G–C content can be discussed from two viewpoints: If the stacking enthalpy is equal for G–C and A–T pairs, then the increase in ΔH can be attributed to G–C pairs forming one more H bond than A–T pairs, i.e., 7.6 kJ MBP^{-1} will be the net contribution of one additional H bond to the stabilizing enthalpy. The enthalpy for transition for pure A–T (30.2 kJ MBP^{-1}) is divided into two parts, 15.1 kJ MBP^{-1} for two hydrogen bonds and 30.2 kJ MBP^{-1} from stacking interactions. The transition enthalpy of pure G–C can thus be separated into 22.7 kJ MBP^{-1} for three hydrogen bonds and 15.1 kJ MBP^{-1} from stacking.

The alternative view discussed starts from the position that the postulate of different numbers of hydrogen bonds in base pairs cannot by itself account for the increase in ΔH with increasing G–C content, since evidence is available[386] that the interaction of two polymeric strands of nucleotides is not ascribable to hydrogen bonding between bases; even relatively high concentrations of nucleosides in water show no evidence of hydrogen-bonded pairing. Utilizing earlier data[303] which show that the change in free energy associated with the stacking of an already hydrogen-bonded base pair upon an existing base pair is approximately −28 kJ MBP^{-1}, it has been shown that the *total* $\Delta G^{\ominus}$ for the transition of *Cl. perfringens* DNA (−3.8 kJ MBP^{-1}) is in good agreement with −4.2 kJ MBP^{-1} obtained by Crothers and Zimm[303] and −5.0 kJ MBP^{-1} reported for phage T_2 DNA.[1199] Taking the calorimetrically determined ΔH for salmon sperm DNA, the contributions to the overall values from stacking interactions and other sources are calculated and shown in Table VII; a positive value for the stacking entropy of 47.5 J MBP^{-1} deg^{-1} is obtained, although the total entropy change for the transition is −99 J MBP^{-1} deg^{-1}.

This effective concealment of a positive entropy within an overall negative entropy is in agreement with the work of Tso and co-workers[1496] and Crothers and Ratner[302] and is a strong indication of the involvement of solvent–solute interactions, such as hydrophobic bonding, in the transition. It is interesting to note that the values for the denaturation entropies

TABLE VII. Contributions to the Thermodynamic Parameters for the DNA Transition from Stacking and Other Interactions[a]

	ΔG, kJ MBP^{-1}	ΔH, kJ MBP^{-1}	ΔS, J MBP^{-1} deg^{-1}
Calorimetric	−3.2	−32.9	−99.1
Stacking	−29.4	−15.1	47.5
Other contributions	+26.2	−17.8	−146.6

[a] Data from Ref. 795.

of the various DNA's (*Cl. perfringens*, 98.3 J MBP^{-1} deg^{-1}; *M. lysodeikticus*, 101.2 J MBP^{-1} deg^{-1}; salmon sperm, 98 J MBP^{-1} deg^{-1}; phage T_2, 98.3 J MBP^{-1} deg^{-1}) yield an average value of 340° for the compensation temperature based upon variation in G–C content; this is at the upper end of the range regarded by Lumry and Rajender as indicating the involvement of the solvent in the transition process. There seem therefore to be quite good grounds for suggesting that the increased stability associated with the increased G–C content of DNA is a reflection of the increased involvement of hydrophobic bonding in the stacking process.

Gennis and Cantor[509] have examined the circular dichroism of a number of different DNA's in 0.1 *M* NaCl, 0.001 *M* phosphate at *p*H 7 over the spectral range 220–310 nm. The following DNA's were used in the study: *S. maxima* (29% G–C),[924] *Cl. perfringens* (31% G–C), salmon sperm (41% G–C), *B. subtilis* (42% G–C), calf thymus (42% G–C), *S. miribilis* (42% G–C),[924] *E. coli* (50% G–C), *M. lysodeikticus* (72% G–C), and *S. chrysomallus* (72% G–C)[924]; in addition, poly d(A–T), poly d(G–C), and a series of d(A–T)$_n$ oligomers with $10 < n < 21$ were used. The authors were interested in the premelt transition behavior of DNA and remark that the largest change in the CD spectrum occurs with poly d(A–T), decreasing to the smallest change with poly d(G–C), but they are hesitant to relate the transition changes to the percentage G–C content of the DNA; such a relationship does, however, appear realistic.[795]

Analysis of the molar ellipticity of the oligomers of (A–T) as a function of temperature over the premelt region results in a family of essentially parallel curves; thus ΔH_{vH} for the structural transition, assuming two-state behavior, has a maximum value of 21 kJ MBP^{-1} which is independent of *n*. The transition also shows no dependence upon ionic strength but the authors point out that this may not be the case for lower salt concentrations.[920,1501] It is also suggested that the data support the viewpoint that a definite conformational transition occurs in the premelt region, possibly to the C-type DNA which has been related to a dehydrated form of DNA,[1554] but the possibility of an $n–\pi^*$ transition cannot be excluded. Results from oligo A and poly A nucleotides suggest that an $n–\pi^*$ transition is responsible for substantially increased ellipticity in the 280-nm region only when adenine is removed into the interior of the molecule with subsequent dehydration.[197] Gennis and Cantor note that the CD spectrum at 270 nm of salmon sperm DNA in 40% ethanol is very different from that in H_2O and that the same spectral region shows greatest change for calf thymus DNA upon changing temperature. On the basis of this evidence the authors suggest that hydration of DNA increases as the percentage content of A increases.

Davis and Tinoco[326] have reported ORD, hypochromism, and pmr studies on a number of dinucleoside phosphates as a function of temperature over the range 0–85° at pH 6.7 in a buffer solution of 0.15 M KCl and 0.01 M phosphate and pH 6.75 in a solution of 0.15 M NaClO$_4$ and 0.02 M phosphate; no differences in properties between the two buffer solutions was observed. The nucleic acid materials used in the study were as follows: A, C, G, Ap, pC, ApG, UpC, CpA, CpU, CpC, GpU, GpC, UpG, CpG, GpA, ApA, ApC, ApU, UpU, and UpA. The $\Delta H_{vH}^{\ominus}$ and $\Delta S_{vH}^{\ominus}$ values for the unstacking process were determined and led to several points of interest: The two-state model seems to be unsatisfactory in accounting for the stacking process; the evidence used to support this view is the variation of $\Delta H_{vH}^{\ominus}$ for unstacking of ApA, obtained from a variety of sources, as ranging from 22.3 kJ MNP^{-1} to 42 kJ MBP^{-1}. Consideration is given to a different model, the oscillating dimer.[529] The two bases of the dinucleoside phosphate are represented as circular disks arranged such that one disk is directly above the other and a harmonic torsional spring connecting the two bases defines their relative potential energy. The ORD from interaction between the two bases is considered to arise from interaction of the point transition dipole ($\sim$260 nm) at the center of one base with the analogous point transition dipole at the center of the other base. The rotation caused by the interaction of the two bases is related to the low-temperature limit of the optical rotation by the expression

$$\phi = \phi e^{-2kT/K}$$

where K corresponds to the force constant of the harmonic torsional spring.

An interesting correlation arises between the $\Delta H_{vH}^{\ominus}$ of unstacking and the apparent force constant of the spring (see Table VIII) when the dinucleoside phosphates are divided into the groups shown. This may give

TABLE VIII. Average Value of $\Delta H_{vH}^{\ominus}$ for Unstacking and K for the Dinucleosides[a]

	$\Delta H_{vH}^{\ominus}$, kJ MBP^{-1}	K, J (mol rad^2)$^{-1}$
Purine–purine	21.8	840
Pyrimidine–purine	24.4	840
Purine–pyrimidine	30.7	1092
Pyrimidine–pyrimidine	30.2	1092

[a] Data from Ref. 325.

an indication of the forces responsible for stabilizing the conformations; simple overlapping of the area of the bases is not believed to be responsible, since otherwise purine–purine pairs would be expected to exhibit the largest values. The pyrimidine derivatives are considerably more soluble than their purine counterparts, which can be attributed to strong hydration, particularly at the carbonyl and amino groups. The authors subsequently propose that if the low-temperature form of the dinucleoside phosphate starts a right-handed helix with both bases in antiorientation about the glycosidic bond (the most common situation in crystal and fiber structures), then the carbonyl and amino groups of a 3'-pyrimidine would be above the other base while the carbonyl and amino groups of a 5'-pyrimidine will be exposed to the solvent, thus acting as a possible nucleus for solvent ordering and contributing to a large $\Delta H_{vH}^{\ominus}$ value for unstacking. Davis and Tinoco are therefore in essence proposing hydration at specific sites as a contribution to the stability of the stacked form of the nucleoside. This interpretation of the data has considerable attraction and lends support to the proposals of Lewin for hydrophilic hydration as a stabilizing force in nucleic acid structure.[885] It is, however, worth considering another possible contributory factor to the energetics of the stacking process; Table IX lists the average values of $\Delta S^{\ominus}$ for unstacking of the four groups of nucleosides. The smallest entropy change is associated with the purine–purine group; this corresponds to the least negative entropy of stacking, and suggests that a larger, positive entropy term may be concealed in the overall value as in the cases discussed earlier.[184,302,1496] This would imply a hydrophobic bonding contribution to the stacking process which is predominant in the purine–purine group. It also seems to be the case, from the ΔH and ΔS data tabulated by Davis and Tinoco, that the most favorable ΔG for stack-

TABLE IX. Average Values of $\Delta S^{\ominus}$ for Unstacking of the Dinucleosides[a]

	$\Delta S^{\ominus}$, J MBP^{-1} deg^{-1}
Purine–purine	91.1
Pyrimidine–purine	93.7
Purine–pyrimidine	112.6
Pyrimidine–pyrimidine	105.0

[a] Data from Ref. 325.

ing is obtained for the purine–purine group, which would agree with other evidence[184,302,425,1410,1496] that the extent of stacking of purines in aqueous solution is greater than that of pyrimidine derivatives.

A study of solvent effects on the conformation of dinucleosides and mononucleotides[918] is of particular interest to this discussion. ApA, AMP, UMP, thymidine, UpU, and TpT were the substances studied over a pH range of 6.4–7.1 by absorption, ORD, and CD techniques. The effect of temperature on the conformation of ApA in water over the range 1–84°C was observed by means of ORD. Analysis of the spectra followed a simplified matrix analysis technique[954] and showed that the experimental results could be represented as a sum of two spectra representing the extreme conformational states. The data can therefore be interpreted by a two-state model and a van't Hoff analysis yielded a ΔS of stacking of -27.7 kJ MBP^{-1} deg^{-1}; the corresponding ΔH_{vH} of stacking was in good agreement with the value of -22.6 kJ MBP^{-1} obtained by Davis and Tinoco. Similar studies on TpT and UpU[1560] showed smaller hypochromic effects than ApA and smaller differences in the ORD spectra and this is taken as evidence for a weaker stacking interaction, in agreement with Tso[1495] and Solie and Schellman.[1410] The authors suggest that the effect of solvent structure changes, as implied by the application of the two-state model, is predominantly reflected in a decrease in the magnitude of the stacking ΔH rather than an entropic contribution which might be expected from the existence of hydrophobic bonding in the stacking process. This viewpoint suggests support for the effects of hydrophilic hydration in stabilizing the helical form of the nucleoside. It is almost certainly the case that hydrophilic hydration has a role to play in the stabilizing process but, as has been discussed above,[184,302,326,425,1496] it is dangerous to draw conclusions as to the role of hydrophobic contributions to the stacking interaction without making a careful analysis of data from as wide a range of purine and pyrimidine nucleosides and nucleotides as possible.

3.3. The Helix–Coil Transition of Native Collagen and Soluble Tropocollagen and the Development of the Collagen Fold by Gelatins in Aqueous Solution

3.3.1. The Helix–Coil Transition of Native Collagen and Soluble Tropocollagen

The parent protein of tropocollagen and the gelatins is collagen, the principal protein in mammals, accounting for approximately 60% of the total mammalian protein. Collagen is unique among proteins for its unusual amino acid composition, containing, as it does, large amounts of hydroxyproline, proline, and glycine; however, cysteine does not occur at all, and other sulfur-containing amino acids are present in only extremely small quantities. Collagens exist mainly in the form of fibers or fibrils, hence the term fibrous protein, as distinct from the globular proteins of earlier discussion. The basic unit of native collagen is tropocollagen, which is soluble in water and is thought to exist in solution at low temperatures as long, thin rods of about 300 mm length and 1.4 nm diameter. The basic molecular structure is that of a triple helix consisting of three polypeptide strands in which the sequence glycine–proline–hydroxyproline–glycine has been shown to occur to a considerable extent.[822,1346] Several models for the structure of tropocollagen have been proposed[1213,1245] but a common feature is that each demands the presence of glycine at every third peptide position in the polypeptide chain in order that intramolecular hydrogen bonds may occur between —NH groups on the backbone of one chain with a —C=O group on the backbone of another chain since the pyrrolidine residues proline and hydroxyproline cannot form peptide hydrogen bonds. In this context it should be noted that glycine accounts for approximately 33% of the amino acid residues in tropocollagen, and the pyrrolidine residues for an additional 25%. At temperatures in excess of 40° tropocollagen loses its ordered triple helical structure to adopt a random conformation which consists of a mixture of single, double, and triple strands called α-, β-, and γ-gelatins, respectively, which are all in random coil conformations with apparently very little helical structure remaining. Cooling of the gelatins initiates a reversion to a helical form—the so-called reversion to the collagen fold, but this is not a simple transition back to a complete tropocollagen unit, at least not in the case of the single strand α-gelatin; the situation is more uncertain with the double- and triple-strand gelatins and unfortunately almost completely confused in many systems which have been reported containing a mixture of α-, β-, and γ-gelatins.

It is significant that the large proportion of pyrrolidine ring residues in tropocollagen and the gelatins results in a considerable amount of nonpolar hydrocarbon surface in the molecule, which makes them interesting subjects to study from the viewpoint of possible hydrophobic interactions.

Privalov and Tiktopulo[1201] have reported a calorimetric study of the thermal transition from helix to coil of several tropocollagens of increasing imino acid content, (i.e., increasing proportion of the gly–pro–hydroxypro unit). If the tropocollagen denaturation is regarded as a phase transition of the melting type and the free energy change at the transition point is zero, then $\Delta G_d = \Delta H_d - T_d \Delta S_d = 0$, $T_d = \Delta H_d / \Delta S_d$, and the transition temperature is related directly to enthalpy and entropy of transition from the ordered helical state to the disordered random coil conformation. Several authors[213,743,984,1171,1174] have reported a linear dependence of T_d upon the imino acid content. As pointed out by Privalov, the increase in T_d can be due either to an increase in ΔH_d or to a decrease in ΔS_d with increasing imino acid content. If it is assumed that the tropocollagen structure is stabilized by intramolecular hydrogen bonds, as has been suggested,[611,613,616,1218] then the increase in T_d cannot be accounted for by the enthalpy factor since, as we have previously noted, imino acid residues cannot form intramolecular hydrogen bonds and an increase in their concentration should lead to a decrease in ΔH_d. It is therefore possible that ΔS_d is the principal factor in determining the stability of the tropocollagen triple helix, but if intramolecular hydrogen bonds are mainly responsible for the triple helix stability, then ΔH_d must in some manner decrease, without an accompanying appreciable increase in ΔS_d, with increasing imino acid content. Highly precise calorimetric measurements of the heat change associated with the transition are a particularly direct method, therefore, of gaining information about the processes involved in the transition. The acid-soluble white rat, pike, cod, and merlang skin tropocollagens investigated by Privalov and Tiktopulo were extracted and purified at 4°C; each tropocollagen was shown by light scattering to have a molecular weight close to the theoretical value (360,000 ± 20) for a manodisperse system with no aggregates. The transition process was observed over a pH range of 2.0–6.0 and the degree of denaturation was followed by ORD measurements over the range 210–600 nm, the actual value being found to be independent of wavelength within this range. Calorimetric measurements were performed with a differential scanning calorimeter and to prevent accidental thermal denaturation of the tropocollagens, all experimental work was conducted in a cold room at 4°C. The calorimetric heat of denaturation of all the tropocollagens was found to be independent of the protein concentration in the range studied (0.034–0.35%) and of pH in the range 2.0–4.0. The denaturation temperature T_d was found to be directly dependent upon the pH of the medium; in the case of rat skin tropocollagen it decreased from 42° at pH > 4.0 to 35° at pH < 2.2: all data were therefore normalized to pH 3.5. Table X lists the thermodynamic parameters associated with the denaturation of the various tropocollagens in water at pH 3.5 together with their related imino acid content per 1000 residues.

The authors conclude that the variation of ΔH_d with imino acid content demonstrates the importance of the enthalpy change for the transition but is in contradiction with the model which suggests the structure of the molecule to be stabilized by a net of intramolecular bonds. The sign of the

TABLE X. Thermodynamic Parameters for the Denaturation of Tropocollagens of Differing Imino Acid Contents in Aqueous Solution at pH 3.5[a]

Tropocollagen	Imino acid content per 1000 residues	T_d, °C	ΔH_d, J mol^{-1}	ΔS_d, J mol^{-1} deg^{-1}
Rat	226	40.8	+6426	+20.6
Pike	199	30.6	+5208	+17.2
Merlang	—	21.5	+3696	+12.6
Cod	155	20.0	+3150	+10.9

[a] Data from Ref. 1201.

enthalpy change also appears to mitigate against a possible contribution by hydrophobic bonds between the nonpolar pyrrolidine residues. Electrostatic interactions, judging from the small change in ΔH_d with variation of pH, also do not appear to play a significant role in the stabilization process. Privalov and Tiktopulo therefore suggest that the results support a model which envisages stability of the molecular structure resulting from the formation of water bridges in the vicinity of the imino acid groups of the protein, but that the stabilization mechanism includes several layers of water molecules adjacent to the tropocollagen structure and may include the formation of supramolecular structures in solution by mutual stabilization of the macromolecule and water. The strong dependence of the denaturation enthalpy of the imino acid content leads the authors to suggest, on the basis of the model of Loeb and Scheraga,[912] that the effective melting enthalpy of the cooperative region is only 5% of the total melting enthalpy of the whole molecule. This discrepancy is explained as arising from the existence in tropocollagen of independently melting cooperative regions, possibly due to the regions of differing content of polar and nonpolar residues which have been reported.[561,671,823]

Traub and Piez[1490] have recently criticized the data of Privalov and Tiktopulo referring to tropocollagens in salt-free solution, since in solutions containing 0.1 mol dm^{-3} of NaCl a two-stage heat absorption transition is obtained, the first stage of which was not correlated with any change in the ORD spectrum. No similar two-stage thermal transition was reported by McClain and Wiley[942] resulting from a calorimetric determination of ΔH_d for a range of collagens and tropocollagens. No difference in ΔH_d was observed for intact collagens obtained from a wide range of sources; the experimental ΔH_d of all the materials was $+4431$ J mol^{-1}. The authors are in agreement with Privalov and Tiktopulo that it is unlikely that hydrogen bonding can be the major factor responsible for the stability of the helical structure, and conclude from the constant value of ΔH_d for a variety of imino acid contents that the explanation lies in the relationship between ΔS_d and the imino acid content of the molecule. The authors calculate that the ΔS_d values for calf skin, bovine muscle, rat skin, and cod skin are 13.9, 14.3, 16.0, and 16.4 J mol^{-1} deg^{-1}, respectively, and conclude that the increasing entropy change per residue as the imino content decreases is compatible with the concept of Flory and Garrett[461] that the imino acid content might be related to the entropy term because the presence of a pyrrolidine ring in the polypeptide chain should decrease the configurational entropy of the random coil and hence decrease ΔS_d. The change in ΔS_d with imino acid content is in the opposite direction to that obtained by Privalov and Tiktopulo. This discrepancy poses problems in any attempt to correlate the data; however, the differences may in fact lie in the value attributed to T_d, the transition temperature. McClain and Wiley unfortunately did not obtain correlating data, such as ORD curves, for the systems they investigated; it is uncertain therefore whether the temperatures tabulated refer to the onset of the heat absorption step or to the temperature corresponding to the peak height. The data of Privalov and Tiktopulo do, however, show that the transition temperature estimated from ORD data correlates well with the calorimetric estimate of T_m. Values of ΔS_d from the data of McClain and Wiley, obtained by using the peak height temperature as the transition temperature, reveal very little change for the intact collagens (12.9–13.0 J mol^{-1} deg^{-1}) except for cod skin collagen (13.7 J mol^{-1} deg^{-1}), but the soluble tropocollagens (bovine semimembranosus and calf skin) have the same value of 13.7 J mol^{-1} deg^{-1} and rat skin gives 13.5 J mol^{-1} deg^{-1}. The denaturation entropy of the soluble tropocollagens thus shows a slight decrease from 13.7 to 13.5 J mol^{-1} deg^{-1} with decreasing imino acid content from 232 to 217 per 1000 residues, this trend being in agreement with that observed by Privalov and Tiktopulo but not to the same marked extent. It is interesting to note that an enthalpy–entropy compensation plot based upon the imino acid content of the various collagens and tropocollagens utilized by McClain and Wiley[942] and Privalov and Tiktopulo[1201] would give a compensation temperature of 335 and 301°K, respectively, both values being of the order that Lumry and Rajender consider as being indicative of the involvement of the solvent medium in the transition.

Cooper,[279] by utilizing an alternative approach to the thermal unfolding of tropocollagen to that discussed by Schellman[1317] and Harrington,[611] proposed that the mean enthalpy and entropy changes arising from changes other than those due to backbone flexibility can be written

$$\Delta H_r = (\Delta S_c)T_d^2 \, \partial p / \partial T_d$$

$$\Delta S_r = (\Delta S_c)[T_d - (\partial p / \partial T_d) + (p - 1)]$$

where p is the fractional content of pyrrolidine residues and ΔS_c is the mean configurational entropy change per mole of residues (excluding pyrrolidine residues, for which the entropy change is assumed to be zero[611]). Taking a value of 17.2 J mol^{-1} deg^{-1} for ΔS_c (taken from literature values on simple polymers and peptides), the computed variation with T_d of the entropy, enthalpy, and free energy of interaction per residue may be obtained. The values obtained are indicated in Fig. 3, together with the experimental results of Privalov and Tiktopulo.[1201] The agreement between theory and experiment for the data of Privalov and Tiktopulo[1201] reinforces the suggestion that hydrogen bonding is unlikely to be the major factor in stabilizing the helix. The effective compensation obtained between the enthalpy and entropy results in the average interaction free energy remaining constant at approximately 4 J mol^{-1} over a wide range of imino acid content, decreasing only at the highest values to an extrapolated value of approximately 2 J mol^{-1} for poly(gly–pro–pro) at $p = 0.67$ even though ΔH_r and ΔS_r have corresponding extrapolated values of 24.8 kJ mol^{-1} and 59 J mol^{-1} deg^{-1}, respectively. The extrapolated enthalpy value of Privalov and

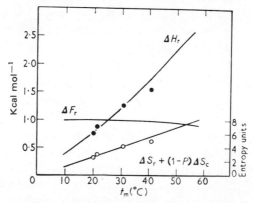

Fig. 3. Computed variation with denaturation temperature of the entropy, enthalpy, and free energy of interaction per residue. The experimental points are taken from the calorimetric data of Privalov and Tiktopulo.[1201] (See Ref. 279).

Tiktopulo[1201] to $p = 0.67$ is 23.1 kJ mol^{-1}, in good agreement with the extrapolated theoretical value. Cooper points out that the poly(gly–pro–pro) molecule would be much less stable than any of the collagens if it were not for the conformational restrictions imposed by the proline residues; the author tentatively concludes that some kind of structural organization of the water surrounding the macromolecule is the most likely explanation of the size and magnitude of the thermodynamic parameters.

The possibility of the tropocollagen molecule being surrounded by an extended region of ordered solvent is supported by the dilatometric data of Cassel and Christiansen,[230] based on the volume change occurring when tropocollagen units aggregate to form native collagen fibrils. A ΔV of 0.8×10^{-3} cm^3 g^{-1} of collagen in phosphate buffer at pH 7–7.5 was observed; this ΔV was independent of ionic strength over the range 0.20–0.50. The authors attribute the observed ΔV to hydrophobic bonding between the tropocollagen units arising from the loss of solvent structure initially surrounding the tropocollagen macromolecule.

Viscosity and light scattering studies, supplementing the earlier calorimetric investigation[1201] of the thermal denaturation of tropocollagen, have been reported by Privalov.[1200] The tropocollagen solutions were shown to exhibit non-Newtonian behavior at low rates of shear, the viscosity of the solution increasing with decreasing shear rate; the authors also observed that the viscosity changed continually with time for a concentration range of protein (0.0003–0.016%) and for each protein concentration a time interval was observed after which the viscometer rotor came to rest, this period being shorter the smaller the shear gradient applied to the material. The authors deduce that such behavior can be explained by the buildup at an optimum concentration of a spatial structure and not by the orientation of the elongated tropocollagen particles, the optimum concentration for structure formation in salt-free solutions being that which gives a distance of 300 nm between the centers of the macromolecule, comparable to the length of the tropocollagen molecule. The optimum concentration effect precludes the buildup of a spatial intermolecular network but is more suggestive of a liquid crystal type of organization in which the macromolecules are stabilized at a definite distance from each other. The temperature dependence of the intrinsic viscosity at a constant shear rate shows that in salt-free solution the supramolecular structure disappears in the temperature range of tropocollagen denaturation obtained from earlier calorimetric measurements.[1201]

Light scattering data[1200] on tropocollagen solution at 20°C give a Zimm plot leading to a molecular weight of 390,000 and a radius of gyration of 123 nm, in good agreement with values obtained by other workers. The

Zimm coordinates show the same curvature reported by other workers, but a plot of cK/R_0 against $\sin(\theta/2)$, where c is the concentration of protein in solution and R is the Raleigh scattering factor for a scattering angle of θ, results in a value for the mass per unit length of the tropocollagen molecule of 53 nm^{-1}, which is about half the value obtained by assuming that the tropocollagen molecule is rodlike; the ordinate intercept also turns out to be negative, which does not suggest a rodlike particle. At temperatures between 32 and 34° the observed anomalies in the light scattering data disappear, the mass per unit length increases to 115 nm^{-1}, and the scattering pattern corresponds to that expected of rodlike particles; at temperatures higher than 34° abrupt changes in the scattering pattern occur and the Zimm coordinates correspond to particles in a random coil conformation with a molecular weight of 140,000 and a radius of gyration of 38 nm.

The evidence discussed seems to suggest that the tropocollagen molecule is stabilized in aqueous solution by hydrophilic hydration, possibly of a type similar to that suggested by Lewin as a factor in the stability of DNA, but that interaction between tropocollagen units to form collagen fibrils may occur by a hydrophobic bond mechanism.

3.3.2. The Development of the Collagen Fold by the Gelatins in Aqueous Solution

Earlier studies of the reversion of the gelatins from a random coil to helical conformation—the so-called collagen fold—have been reviewed by von Hippel.[1536] Unfortunately, interpretations of experimental results are complicated by the heterogeneity of the protein material, in many cases all three gelatins α, β, and γ being present in the system. This problem is particularly troublesome for investigations which attempt to ascertain the extent of solvent participation in the reversion process. Additional problems arise from the method of preparation of the gelatins, and since it is not possible to discuss these problems in detail, suffice it to say that gelatins may be of the "acid precursor" type, with an isoelectric point at pH 7.0–9.0, or of "alkaline precursor" type, with an isoelectric point at pH 5.0–7.0. It is probable that the considerable variations in methods of preparation account for the differences which arise between the results of various workers. It has recently been shown[376] that for rat tail tendon α-gelatin, prepared by very mild procedures involving alcohol coacervation with exclusion of salts, the optical rotation and viscosity in solution varies considerably, in a nonlinear manner with concentration, in the presence of bromides, ranging from sodium to tetrabutylammonium bromide. An ex-

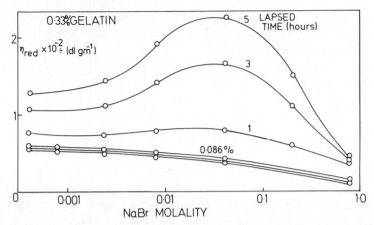

Fig. 4. The reduced viscosity of α-gelatin as a function of NaBr concentration and lapsed time after quenching from 40 to 15°C.

ample of this behavior is shown in Fig. 4, which illustrates the viscosity behavior of α-gelatin, prepared by a salt-free technique, as a function of added NaBr concentration.

It is also shown[376] that gel permeation chromatography of gelatin-type molecules, recommended by Piez and Sherman[1175] as an effective separation procedure in a solvent of 0.15 mol dm^{-3} sodium acetate, is impossible for salt-free gelatins in aqueous solution. The gelatin is immediately and firmly bound to the gel material, but can, however, be removed by washing the column with 0.15 mol dm^{-3} NaCl solution. This evidence is in direct conflict with the view proposed by von Hippel and Schleich[1538] that salt concentrations of this order have no serious effect upon the properties of gelatin in solution. This viewpoint is without justification since very little is known about salt-free collagens and gelatins, and as the data of Fig. 4 illustrate, the fact that small differences are observed at salt concentrations of the order of 0.15 mol dm^{-3} is due to the marked suppression of the magnitude of the effect being observed, whether it be optical rotation or viscosity. It is therefore doubly unfortunate that in the data published for α-gelatin considerable variations exist in the salt concentrations present in the solution.

Piez and Carrillo,[1173] in one of the earlier studies of well-characterized gelatins, reported optical rotation, viscosity, and light scattering results on single-chain α-gelatins and double-chain β-gelatin. The gelatins were prepared by salt and acid extraction of rat skin collagen and dissolved at pH 4.8, in 0.15 M potassium acetate buffer; the gelatin concentration range was 0.01–0.5% w/v. Two different α-gelatins were investigated in this study, α-1 and α-2, the former having an imino acid content of 226 residues per thousand and the latter only 196.

The α-1 and α-2 gelatins formed helix conformations by concentration-dependent reactions, the rate of helix formation being much slower for α-2 than α-1. The authors interpret the data as indicating stabilization of the helix by interchain association. β-Gelatin, however, was shown to form a helix by a concentration-independent reaction which the authors interpret as an intramolecular conformational change associated with helix formation in the double-strand molecules. Melting curves of 0.15% w/v solutions of the three gelatins were obtained by these authors by cooling the samples to 10° for 24 hr prior to melting. The melting process was observed by optical rotation measurements, which showed the melting to occur over a much broader temperature range than that expected for a collagen, the melting temperatures being 22.5° for α-2, 27° for α-1, and 26° for β. The extent of the optical rotation change showed that the amount of helix formation does not extend to more than two-thirds of that to be expected of the original collagen, although Engel[411] has shown that an annealing process of alternate cooling and warming of a gelatin solution sharpens the melting curve, making the transition appear more "collagenlike," but without increasing optical rotation, i.e., without increasing the helix content of the macromolecule. The authors envisage that, on cooling, the helix conformation is initiated by the pyrrolidine-rich regions of the molecule; propagation and stabilization of the fold conformation is by means of association with other chains, or parts of the same chain, to form a compound helix ultimately with bonds between the chains or parts of the same chain. With β-gelatins the process is considered more efficient since appropriate regions for interchain bonding are held in the correct orientation.

Drake and Veis[363] studied the recovery of optical rotation and viscosity as a function of temperature and time for an α- and γ-gelatin derived by a salt and alkali extraction procedure. The protein concentrations studied were in the range 0.009–0.2% w/v and the protein was dissolved in 0.1 mol dm^{-3} KCl and 0.1 mol dm^{-3} acetic acid or a 0.15 citrate solution at pH 2.8 or 3.6. The recovery of viscosity was first order for the γ-gelatin, but of intermediate character for the α-gelatin; recovery of optical rotation was found to be second order for both gelatins, but the rate constant for the γ-gelatin was approximately 30 times greater than that of α-gelatin at the same concentration. The apparent activation enthalpy for renaturation of γ-gelatin at 15° obtained from optical rotation measurements was -44.1 kJ mol^{-1} and from viscosity measurements -2.5 kJ mol^{-1}; the authors do not comment on this discrepancy. The corresponding values for the apparent free energy and entropy of activation for renaturation obtained by the different techniques were $+96.1$ and $+96.5$ kJ mol^{-1}, and -449 and -311 kJ mol^{-1} deg^{-1}, respectively. The properties of the transition state are clearly dominated by the very large negative entropy value; if $\Delta S^{\ddagger}$ for the complete transition is taken to be of the order of -20 J mol^{-1} deg^{-1},[1201] the large, negative $\Delta S^{\ddagger}$ requires a transition state that is considerably more ordered than the final folded conformation. Such a situation could be achieved if the transition state were to have associated with it considerable quantities of ordered solvent, a situation which seems very unlikely. The authors[363] interpret their data as indicating an intramolecular refolding mechanism for both gelatins, involving the interaction of the pyrrolidine-rich segments of the same molecule: The differences between the α- and γ-gelatin are considered as arising from differences in the degree of specific interaction and cooperativity.

In a study by Bensusan and Nielsen[125] purified calf skin tropocollagen was dissolved at a concentration of 0.15% in 0.067 m dm^{-3} citrate buffer at pH 3.2 with 0.8% tetramethylammonium chloride. After heating to 60° for 10 min and cooling to 14°C the transition was followed by optical rotation and viscometric techniques; aliquots of the reaction mixture were simultaneously lyophilized and redissolved in D$_2$O and the extent of exchange of peptide-group hydrogen atoms followed. All the peptide group hydrogen atoms in the gelatin at zero quench time were found to exchange rapidly with the same first-order rate constant of 0.16 min^{-1}, a value similar to that obtained with earlier exchange data for randomly coiled poly-D,L-alanine.[194] As the collagen fold develops the peptide-group hydrogens separate into two classes, one with an exchange rate constant of less than 1 day^{-1}. The rate of disappearance of fast exchange protons was observed by the authors to correlate with the gain in optical rotation.

The authors point out two possible reasons for the slow exchange rate of the peptide protons; the hydrogen atoms involved are removed from free contact with bulk solvent either by being involved in strong hydrogen bonding or by being forced into a largely hydrophobic region within the interior of the molecule. The first interpretation is favored, i.e., the results indicate the formation of interpeptide hydrogen bonds resulting in a collagen triple helix after the helical coiling of the individual peptide chains, in agreement with the mechanism proposed by Flory and Weaver.[463] The later calorimetric data of Privalov and Tiktopulo,[1201] McClain and Wiley,[942] and Cooper[279] show, however, that strong interpeptide hydrogen bonds of the type outlined by Bensusan and Nielsen[125] are unlikely, and it therefore now seems probable that the alternative proposal may provide a more likely explanation, i.e., that the exchangeable protons are buried within the hydrophobic regions of the molecule during the process of helix formation, possibly by some type of hydrophobic interaction, involving progressive "squeezing out" of a proportion of the water molecules from the interstices between the nonpolar residues in the molecule.

In a recent study Harrington and Rao[614] reported optical rotation data on a variety of single-chain gelatins of varying pyrrolidine contents, isolated by salt gradient chromatography.[1172] The optical rotation measurements, at several temperatures ranging from 10 to 30°C, were made on solutions of the gelatins which had first been denatured for 10 min at 50 or 60°C and then cooled to the required temperature. The concentrations of gelatins ranged from 0.0001 to 0.2% w/v in 0.15 mol dm^{-1} NaCl and 0.05 mol dm^{-1} sodium citrate buffer at pH 4.8. Initial rates of helix formation, determined from the optical rotation data extrapolated to zero time, showed that at low concentrations (<0.01% w/v) the kinetics of all the gelatins obeyed a first-order dependence, indicating an intramolecular generation of the folded regions in the molecule. Above the 0.01% w/v concentration the order of the rate constant was said to increase with increasing concentration, which the authors interpret in terms of the involvement of intermolecular processes in the initial stages of helix regeneration.

The initial rates of helix formation in the concentration-independent region for the various α-gelatins are shown to undergo a marked increase as the imino acid content of the chains is increased, with two exceptions, as shown in Table XI; earthworm cuticle and RCM-Ascaris show anomalous behavior. The authors point out that the residual specific rotation remaining above the helix-to-coil transition temperature, in excess of that calculated on the basis of amino acid composition and a completely random conformation, varies with the origin of the gelatin, being smallest for earthworm and largest for RCM-Ascaris.[614] A much closer correlation is therefore obtained between the initial rate of helix reformation and the calculated number of contiguous pyrrolidines per thousand triplets, the only exception now being the RCM-Ascaris which, as pointed out by other workers, often shows anomalous behavior by comparison with other gelatins.

The authors support the viewpoint that the first-order folding reaction in the low-concentration regions reflects a completely intramolecular folding to form a reverse folded structure; the rate-determining step in the process is envisaged as the formation of a triple helical nucleus stabilized by hydrogen bonding. After the initial hydrogen-bonded pattern is established, propagation of the structure proceeds in a stepwise manner as the chain

TABLE XI. Effect of Imino Acid Residues on Initial Renaturation Rate of α-Gelatins at $\Delta T = 20°$ [a]

Gelatin	Σ Pyrrolidines (per 1000 residues)	Predicted contiguous pyrrolidines (per 1000 triplets)[b]	Initial rate $-d[\alpha]_{215}/dt$, deg min^{-1}
Earthworm cuticle	169	10	34
α_1 Cod skin	155	49	92
Dog fish Sharkskin	166	57	94
α_2 Ichthyocol	187	72	88
α_1 Ichthyocol	199	78	124
α_2 Rat skin	197	78	144
α_1 Rat skin	226	102	200
α_1 Calf skin	232	110	234
Reduced carboxymethylated Ascaris	310	211	84

[a] Data from Ref. 613. Solvent is 0.15 mol dm^{-3} NaCl + 0.01 mol dm^{-3} acetate, pH 4.8. $\Delta T = 20°$ is the degree of undercooling of the gelatins below their respective melting temperatures T_m.
[b] Predicted frequency of neighboring pyrrolidine residues estimated according to Josse and Harrington.[743]

segments adjacent to the nucleus wrap themselves around to form the triple helix. At the higher concentrations the rate-determining step is considered to be the formation of a hydrogen-bonded nucleus by interaction of polypeptide segments from three separate chains.

Harrington and Rao thus consider that the critical step in the production of a helical conformation in the α-gelatins, whether at high or low concentrations of protein, is the initiation of a hydrogen-bonded nucleus by the imino acid regions of the molecule. The calorimetric data of Privalov and Tiktopulo[1201] and McClain and Wiley,[942] on the other hand, strongly suggest that hydrogen bonding is not responsible for the stability of the collagen-type helix and, as in the case of tropocollagen of varying imino acid content, the dominant factor in the formation of the helix is an entropy change associated with the formation of a hydration network surrounding the macromolecule. In a continuation of the previous report,[614] Harrington and Karr[612] report optical rotation and ORD studies on RCM-Ascaris and α rat skin gelatin in an ethylene glycol–water (1:1 v/v) mixture. The renaturation was followed by initial rate determinations over the temperature range from -20 to $+35°$; the protein concentration was 0.038% for all experiments and the solutions were prepared by dilution of samples in 0.30 mol dm^{-3} NaCl and 0.02 mol dm^{-3} sodium acetate at pH 4.6 either by water or anhydrous ethylene glycol immediately prior to each experiment.

The results show that the renaturation rate of RCM-Ascaris gelatin passes through a maximum about 46° below the thermal transition temperature of the native collagen. The authors interpret this in terms of competing reactions of opposing temperature dependence and analyze the overall mutarotation as the sum of two first-order processes; first, a nucleation stage which shows a marked increase in rate as the degree of cooling below T_m increases, and, second, a propagation stage, with a positive temperature dependence, reflecting the addition of random coil residues to the nucleus. Superimposed on the two major processes is a much slower reaction which is believed to result from the annealing of segments of the chain initially frozen into a less ordered conformation (cf. earlier work by Engel[410]).

The data so far discussed for the various collagens and gelatins illustrate the problems associated with studies of the mechanism of triple helix formation arising from the heterogeneity of available material. Piez and Sherman[1175] have attempted to simplify the picture by the splitting of α-1 rat skin gelatin into discrete peptides and subjecting one of them, a 36-residue peptide (α 1-CB2) to investigation by optical rotation, circular dichroism, molecular sieve chromatography, and ultracentrifugation. The residue, which was obtained by cyanogen bromide digestion, was found to undergo the reversible transition from helix to coil which is characteristic of collagen. The peptide was shown to have the sequence

pro-ser-gly-pro-arg-gly-leu-hyp-gly-pro-hyp-gly-ala-hyp-gly-pro-gln-gly-

phe-glu-gly-pro-hyp-gly-glu-hyp-gly-glu-hyp-gly-ala-ser-gly-pro-hse

Thus the peptide consists of triplets of the form gly–X–Y and contains 33% proline and hydroxyproline.

The similarity between the CD spectra of rat skin collagen and α 1-CB2 at 2°, illustrated in Fig. 5, clearly shows that the helical form of α 1-CB2 resembles that of collagen, the helix content of the peptide being estimated to be of the order of 95%. The helical form of the peptide is therefore considered to exist as an almost completely intact triple helix of the type characteristic of collagen and no evidence was found for the existence of dimers in the solution.

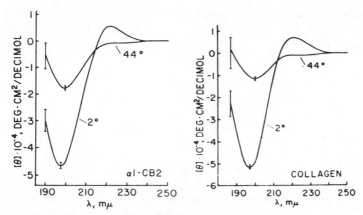

Fig. 5. (a) CD spectra of α 1-CB2 small peptide from collagen measured at 2 and 44° at a concentration of 0.162% w/v. The sample at 2° contained 93% helical form and 7% random coil. (b) CD spectra of native and denatured rat skin collagen at 2 and 44° at a concentration of 0.145% w/v. (See Ref. 1175.)

In the second paper of the series Piez and Sherman[1176] report optical rotation data at 313 nm in cells with path lengths as short as 5 mm. The small volume of the cells allowed temperature changes to be made very rapidly; thermistor measurements showed that for the largest temperature difference utilized, 26°, the temperature jump was 90% complete in 5 sec and to within 0.1° after 20 sec. The peptide was dissolved in a 0.15 mol dm^{-3} potassium phosphate buffer at pH 4.8. The authors reported that after heating or cooling to a given temperature the optical rotation approached the same equilibrium value, irrespective of the direction of the temperature shift. Figure 6 shows the molecular rotation $[\phi]$ in degrees, defined as $1000\,\alpha M/Lc$, where M is the molecular weight of the peptide, L is the path length in mm, c is the concentration in mg cm^{-3}, and α is the measured optical rotation, for two concentrations of the peptide between 1 and 25°. The reversibility of the change in $[\phi]$ during either melting or renaturing confirms that the equilibrium is attained at each temperature. The authors note that at 1° the concentration dependence of $[\phi]_{\text{equil}}$ indicates that transition to the helical form is not complete, but the values of $[\phi]$ above 25° show that only the random coil conformation is present. The equilibrium constant for the monomeric random coil-to-triple helix transition was obtained from the rotation data illustrated in Fig. 6; from a van't Hoff plot $\Delta H_{\text{vH}}^{\ominus}$ the standard enthalpy of formation of the helix was determined as -390.6 kJ mol^{-1}, and $\Delta S^{\ominus}$ was calculated to be -11.53 kJ deg^{-1} mol^{-1} of trimer formed.

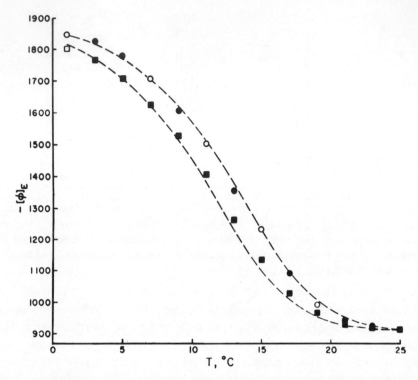

Fig. 6. Equilibrium curves for the helix–coil transition in α 1-CB2 peptide at several concentrations (○ and ●, 1.44 mM; □ and ■, 0.80 mM). Open symbols denote molecular
rotation at equilibrium after a downward temperature shift, closed symbols denote values
obtained after an upward temperature shift. Dashed curves were computed for a monomer–trimer equilibrium with $\Delta H^\circ = -93$ kcal mol^{-1} trimer at the above concentrations.
(See Ref. 1176.)

The values of optical rotation obtained during the period 0–500 sec
were found to fit a third-order polynomial and from this smoothed curve
the initial rate of reversion was obtained as a function of temperature between 1 and 17°. A good fit of experimental data to the theoretical line was
obtained in a plot of ln[initial rate] vs. $1/T$ when the equilibrium random
coil monomer ⇌ triple helix was assumed; a value of -75.6 kJ mol^{-1} of
trimer was obtained for $\Delta H^\ddagger$ of helix formation, giving a $\Delta H^\ddagger$ for melting
of $+215$ kJ mol^{-1} of trimer.

The pattern of a large, positive $\Delta H^\ddagger$ for melting and a small, negative
$\Delta H^\ddagger$ for helix formation is very similar to that observed by Pohl[1184] for
the reversible denaturation of α-chymotrypsin and trypsin. If the analogy
is pursued, it suggests that water is removed from the hydrophobic groups

in the activated state and it may be the thermodynamics of hydrophobic bond formation which are responsible for the apparent ΔH of folding but not for that of unfolding.

The authors[1176] determine $\Delta H_{vH}^{\ominus}$ and $\Delta S^{\ominus}$ per residue for helix formation as -4.03 kJ mol^{-1} and -12.2 J mol^{-1} deg^{-1}, respectively, and point out that if the enthalpy change for the transition is to be attributed to formation of interchain hydrogen bonds, of which, according to Traub's computer studies of possible theoretical models,[1492] there is a minimum of one per triplet, then the average enthalpy change per hydrogen bond formed is of the order of -12.2 kJ mol^{-1}. Such a value suggests either a particularly stable hydrogen bond in a collagen-type helix, which is contrary to the calorimetric evidence,[942,1201] or that hydrogen bonding involving the solvent must be invoked in the stabilization of the triple helix, a view which is supported by Privalov[1201] and Ramachandran and Chandrasckharan.[1212] Further support for the role of solvent in the stabilization process is given by the fact that the value of $\Delta H_{vH}^{\ominus}/\Delta S^{\ominus}$ is 335°, which is of the magnitude that Lumry and Rajender[927] suggest for the compensation temperature when water is involved in the transition process. A compensation temperature of 335° results in a melting temperature of T_m of 62° under standard conditions, which, as Piez and Sherman point out, is in good agreement with the predicted value of 57° for a collagen of corresponding composition obtained by extrapolation of the data of Josse and Harrington[743] on other collagens.

Piez and Sherman[1176] go on to point out that the linear behavior of $\ln K$ as a function of $1/T$ except at low temperatures means that ΔC_p between the two states is negligible and contrast this with the denaturation of globular proteins, which in general is associated with a considerable value for ΔC_p (up to 5 kJ mol^{-1} deg^{-1}). The authors therefore conclude that no change in the degree of exposure to the solvent of hydrophobic groups occurs during the unfolding process.

The work of Brunori et al.[193] on the transition of *Aplysia myoglobin* shows that the transition may occur with little apparent change in C_p, although other evidence[42] clearly implicates hydrophobic bonding in the transition. The dilatometric evidence of Cassel and Christensen[230] implicates hydrophobic bonding in the association of tropocollagen units to form collagen fibrils and, if the original single polypeptide chains have an appreciable hydration envelope associated with them, it is difficult to visualize their aggregation into a triple helix, even if the cross-linking is by water bridging, as suggested by Ramachandran and Chandrasckharan, without the loss of some hydration structure. It thus seems reasonable to propose that association between single chains, when it occurs, to form the triple helix of a collagen type is associated with a hydration envelope of hydrophilic hydration possible as envisaged by Lewin,[885] but during the association some loss of hydration occurs, i.e., weak hydrophobic bonding.

Recently Eagland and Pilling[376] observed the reversion of salt-free rat tail tendon α-gelatin by optical rotation (at 546 nm) and viscometric techniques over a protein concentration range of 0.1–0.5% w/v. The optical rotation measurements were carried out using a cell of 2.5 ml volume; a thermistor probe within the cell showed that the contents of the cell achieved the desired temperature of 15° within 2 min of being placed in the cell at an initial temperature of 40°.

The data on the initial rate of optical rotation were analyzed by the method of Harrington and Rao[614] relating the initial rate to the order of reaction by the equation

$$d[\alpha]_0/dt = kc^{n-1}$$

where $d[\alpha]_0/dt$ is the initial rate of mutarotation, k is the rate constant, c is the gelatin concentration, and n is the "order" of reaction. Harrington and Karr[612] used this argument to show that the order of reaction changes from one to higher orders at gelatin concentrations between 0.01 and 0.1%. The results of this work, shown in Fig. 7, present a similar picture of an apparent increase in order with increase in gelatin concentration. It should be noted with regard to an analysis of this kind, however, that a similar effect is produced if the order of reaction remains constant at one and the rate constant changes with gelatin concentration. Reactions of this type

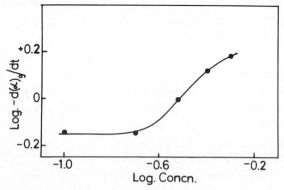

Fig. 7. Logarithmic plot of the initial rate of optical rotation versus α-gelatin concentration in aqueous solution at 15°C.

have been designated as pseudo-first-order reactions[527] since the reaction obeys the first-order kinetic law with respect to the concentration of gelatin but is affected by some other component in the reaction mixture. A complete kinetic analysis of the extended optical rotation data showed that at the gelatin concentrations studied the reversion reaction consisted of an initial "fast" process followed by a "slow" one, a situation first observed by von Hippel and Harrington[1536] from studies of collagenase digestion of ichthyoid gelatin. The rate of reversion obeys a first-order kinetic law in both the "fast" and "slow" regions, the rate constant k' of the slow reaction being independent of gelatin concentration over the range 0.034–0.538% at a value of 1.8×10^{-2} min^{-1}. The rate constant for the fast reaction was more difficult to determine with accuracy, particularly at the lower concentrations, and more results at time intervals shorter than 3 min are required to confirm the results, but the trend appears to be an increase in rate constant with decreasing gelatin concentration from 3×10^{-2} min^{-1} at 0.538% w/v gelatin concentration to 16×10^{-2} min^{-1} at 0.034%.

The reversion process of salt-free rat tail tendon α-gelatin to a collagen fold appears to be an intramolecular process within the concentration range 0.034–0.538% w/v; the initial fast rate, probably corresponding to the nucleation reaction of the imino acid rich regions, is pseudo first order, suggesting that the initiation reaction is an intramolecular process with respect to the gelatin molecules but involves some other species in the nucleation reaction. The evidence of Privalov et al.[1201] and Ramachandran et al.[1212] suggests that the species is water, the water molecule perhaps being involved in bridging between imino acid-rich regions of the same molecule brought into close correlation to each other by the gelatin molecule folding back on itself in the manner proposed by Harrington and Rao.[614]

The "slow" stage in the reversion process appears to correlate with the propagation stage defined by Harrington and Rao, the stepwise addition of chain segments adjacent to the nucleus to the helical region.

Viscosity measurements[376] performed in conjunction with the optical rotation studies already discussed showed on kinetic analysis that at gelatin concentrations between 0.138 and 0.462% w/v the rate of change of viscosity paralleled the change in optical rotation and a similar value for the rate constant of 1.0×10^{-2} min^{-1} was obtained. It was not possible to obtain viscosity data at time intervals less than 10 min, due to cooling problems associated with the larger volume of solution; thus only the "slow" stage of the reversion to the collagen fold could be observed by viscometric techniques.

Fourier transform nmr measurements of the coil-to-helix transition of salt-free rat tail tendon α-gelatin* at concentrations of 0.2 and 0.5% as function of time produced a similar rate constant for the transition of 1.5×10^{-2} min^{-1} for the "slow" reversion stage.

Molecular weight determination of the salt-free rat tail tendon α-gelatin by a light scattering technique as a function of time at 5, 10 and 15°[378] confirmed that aggregation of single α chains to form a triple helix does not occur, the molecular weight remaining constant at 110,000 within the experimental error of $\pm 10,000$. If the reversion had involved the formation of a triple-stranded helix in either the fast or slow reaction, the molecular weight obtained would have been nearer the value 380,000 reported for γ-gelatin.

The radius of gyration determined from a Zimm plot[942] was found to be 40 nm at all temperatures, which is somewhat larger than the values of approximately 17 and 30 nm reported by Baedtker and Doty[72] and Gouinlock et al.[553] for gelatins in the presence of high salt concentrations. Conversely, the second virial coefficient, which is related to the molecular weight of the gelatin, to the average effective volume occupied by the molecule, and to solute–solute interactions, is smaller than the values obtained by other workers, i.e., 1.5×10^{-4} ml mol g^{-2} at 5°, 1.0×10^{-4} ml mol g^{-2} at 10°, and approximately zero at 15°, compared with values of the order of 2.6×10^{-4} ml mol g^{-2} obtained by Baedtker and Doty[72] and Gouinlock et al.[553] Determination of the radius of gyration of the α-gelatin molecule from asymmetry measurements resulted in a value of 100 nm, considerably at variance with the Zimm plot value of 40 nm. A plot of Kc/R_θ vs. $\sin(\theta/2)$ as outlined by Privalov et al.[1201] produced the same marked discrepancies as those reported by Privalov et al. for the melting of tropocollagen; the value of mass per unit length obtained by this method was 2.7 nm^{-1}, compared to 8.3 nm^{-1} obtained from asymmetry measurements.

The pattern observed by Privalov and co-workers[1201] in the stabilization of salt-free tropocollagen in water seems to have correlations in the reversion behavior of salt-free α-gelatin in water. Thus, it was observed that 0.5% solutions of α-gelatin in salt-free water changed into a gel with time, without any change in the first-order kinetics, as monitored by optical rotation, viscosity, pmr, and dielectric relaxation.[376,377] Thus gelation apparently occurs without gelatin chain cross-linking. This is compatible with the postulate of a long-range network of structurally modified water.[1201]

---

* E. G. Finer, private communication.

3.3.3. *Conformational Transitions of Polyamino Acids Which May Serve As "Model" Fibrous Proteins*

One of the most interesting polypeptides in this group is poly-L-proline, the polymeric form of one of the major constituents of collagen. X-ray diffraction[1521] yields a pattern very similar to that of collagen. The polymer, however, may exist in either of two conformational forms, I and II, the structures of which have been determined, also from X-ray diffraction data[1491]; conformation I is a right-handed helix with all the peptide bonds in the cis configuration, whereas conformation II is a left-handed helix with all the peptide bonds apparently in the trans configuration, confirming the model of Cowan and McGavin.[294]

Poly-L-proline does not possess NH groups; therefore it is not possible for intrapeptide hydrogen bonding to contribute to the stability of the helices; predictions of the conformational energy of the two forms, solely on the basis of steric effects,[343] suggest that form I is slightly more stable than form II. Such calculations do not take into account the possibility of interactions between the polypeptide and the solvent and there is considerable evidence[159,550,1421,1434] that the nature of the solvent plays the predominant role in determining the stable conformation; pyridine, for example, stabilizes form I, whereas water, acetic and formic acids, trifluoroethanol, benzyl alcohol, and *n*-butanol have been shown to stabilize form II. Blout and co-workers[154] reported ORD measurements on poly-L-proline II and calf skin collagen in aqueous solution, and noted that both compounds produced marked absorption maxima and minima at 194 and 216 nm and 190 and 210 nm, respectively. This type of absorption behavior, contrary to that observed for most L-α-helical polypeptides, which form right-handed helices, thus confirms that both collagen and poly-L-proline appear to exist as left-handed helices in aqueous solutions.

Smith *et al.*[1406] have reported laser-excited Raman spectra of poly-L-proline in both the solid state and in aqueous solution. Two clear differences are observed between the two spectra; the amide I band at 1650 cm^{-1} in the solid state shifts to 1631 cm^{-1} in the aqueous solution and a weak band at 836 cm^{-1} in the solid spectrum disappears and is replaced by a new band at 826 cm^{-1} in the solution spectrum. The change in the amide I band, corresponding to the frequency of the C=O stretching mode, confirms infrared spectroscopic data.[1454] This would be expected if the carbonyl group is hydrogen-bonded to water. The 836-cm^{-1} band is suggested by the authors to be associated with the vibrational mode of the pyrrolidine ring, and the decrease in frequency of this band cannot therefore be due to a conformational change in the backbone, but very probably involves a solvent effect on the vibrational modes of the pyrrolidine ring.

A similar decrease in frequency from 1471 to 1460 cm^{-1} is observed for an absorption band associated with the bending mode of the pyrrolidine ring between the solid and the solution.

The authors conclude that there is no evidence that direct hydrogen bonding of water to the backbone contributes significantly to the conformational energetics of poly-L-proline II in solution, but solvent stabilization of the helix does occur by interaction with the pyrrolidine rings. Since in both α-gelatin and tropocollagen the poly-L-proline II conformation of the macromolecule leaves the pyrrolidine rings exposed to the solvent environment, this evidence suggests that a similar kind of solvent stabilization is to be expected in both collagen and the gelatins.

In a later publication Deveney et al.[347] report laser-excited Raman spectra of poly-L-hydroxyproline in both the solid state and aqueous solution over a pH range of 1–13. The spectrum of poly-L-hydroxyproline in solution is shown to be remarkably insensitive to the pH, only one band showing any change, from 1466 to 1470 cm^{-1} at pH 13. The spectra show that, as in the case of poly-L-proline, the conformation of the polypeptide in the solid and in aqueous solution is the same and the major differences in the spectra are associated with the influence of the solvent upon the pyrrolidine rings of the polymer in solution.

Isemura and co-workers[713,1107] have reported an interesting series of investigations involving infrared and far-ultraviolet absorption spectra and optical rotation measurements on poly-L-proline and the oligomers of tert-amyl oxycarbonyl-L-proline, ranging from the monomer to the octomer tert-amyl oxycarbonyl-L-(pro)$_7$proline. Infrared absorption bands of the pentamer, hexamer, and octamer were very similar to those of poly-L-proline II in the 1800–75 cm^{-1} region, from which the authors conclude that the molecular structure of the pentamer, hexamer, and octamer was based upon a left-handed helical structure on a threefold screw axis. They also tentatively concluded from the existence of an absorption band at about 400 cm^{-1} in the spectrum of the tetramer, which is characteristic of poly-L-proline II, that this molecule might also have a left-handed helix, probably of one turn.

Ultraviolet absorption spectra over the range 190–240 nm of 0.005% polymer solutions in water showed that the tetramer exhibits and absorption peak at 201 nm which increases to 205 nm being observed in the octamer. The ORD spectra of the oligomers over the range 200–240 nm revealed a minimum in the absorption curve for each polymer between 210 and 220 nm, but the depth of the minimum was much more marked from the trimer to the octamer and with a clear switch in the position of the minimum between the trimer and tetramer.

The authors conclude that the poly-L-proline II helix conformation commences at the tetramer and when the number of residues is ≥ 5 the helical structure of poly-L-proline II is essentially complete.

The amino acid glycine constitutes the major proportion of the amino acid content of collagen and the gelatins and, as we have seen, the sequence gly–X–Y, where either X or Y is L-proline or L-hydroxyproline, is associated with an increased triple helical structure in the macromolecule. Several authors have reported investigations into small polypeptides involving varying tripeptide sequences, in attempts to correlate the peptide sequence with generation of a poly-L-proline II conformation. Traub[1489] showed that the polymer (pro–gly–gly), although capable of intramolecular hydrogen bonding, does not form a collagenlike structure. In a subsequent paper Segal and Traub[1360] reported the synthesis and structural investigation of poly(ala–pro–gly) in aqueous solution and the solid state. The lyophilized polymer was found to be not readily soluble in water but could be dissolved in 35% acetic acid and the solution diluted without precipitation of the polypeptide. Optical rotation measurements utilizing the sodium D line, on solutions of the polymer in formic acid–water mixtures, gave an extrapolated value of specific rotation in pure water very similar to that of (gly-pro-ala)$_n$ obtained by Bloom et al.,[153] which has been shown to behave like a random coil in solution.[1122] Molecular weight determination of a particular fraction of the polymer indicated that aggregation of the polymer does not occur in aqueous solution to a significant degree.

Poly(ala–pro–gly) was found to exist in several different forms in the solid state but only that obtained by crystallization from aqueous solution gave oriented specimens suitable for X-ray diffraction. The unit cell was found to consist of two tripeptide units together with approximately two water molecules and the authors speculate that water bridging occurs between neighboring parallel chains.

Segal[1359] has reported the synthesis and investigation of four polyhexapeptides, (gly-ala-pro-gly-pro-pro)$_n$, (gly-pro-ala-gly-pro-pro)$_n$, (gly-ala-pro-gly-pro-ala)$_n$, and (gly-ala-ala-gly-pro-pro)$_n$. The specific optical rotations of all the polyhexapeptides show a collagenlike reversible temperature dependence, with the transition temperature T_m dependent both upon molecular weight and amino acid sequence. Figure 8 illustrates the general behavior pattern of all four polymers, a marked lessening of the sigmoidal shape of the curve with decreasing molecular weight being observed in all cases.

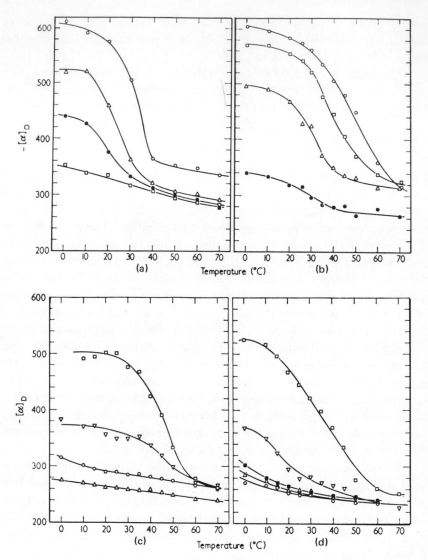

Fig. 8. (a) Thermal transitions of (gly-ala-pro-gly-pro-pro)$_n$ in water, concentration 0.2–0.3% w/v. (○) $M = 11,500$; (△) $M = 7200$; (●) $M = 3800$; (□) $M = 2600$. Molecular weights were determined at 20°C in 0.2 M NaCl. (b) Thermal transitions of (gly-pro-ala-gly-pro-pro)$_n$ in water, concentration 0.2–0.3% w/v. Molecular weights as for part (a). (c) Thermal transitions of (gly-ala-pro-gly-pro-ala)$_n$ in 2% acetic acid, concentration 0.2–0.3% w/v. (□) $M = 7900$; (▽) $M = 4300$; (○) $M = 3000$; (△) $M = 2300$. Molecular weights were determined at 20°C in 2% acetic acid. The fractions with molecular weights 7900 and 4300 were turbid below 40°C. (d) Thermal transitions of (gly-ala-ala-gly-pro-pro)$_n$ in water, concentration 0.2–0.3% w/v; (□) $M = 6300$; (▽) $M = 4150$; (■) $M = 3600$; (△) $M = 2600$; (○) $M = 1900$. Molecular weights were determined in 0.2 M NaCl. (See Ref. 1359.)

X-ray diffraction studies[1361] of the four polyhexapeptides in the solid state have shown that all the polymers exhibit typical collagenlike diffraction patterns which can be interpreted in terms of a structure containing one hydrogen bond between peptides. Segal[1359] suggests that the polyhexapeptides adopt similar conformations in aqueous solutions at low temperatures. Molecular weight determinations by an ultracentrifuge technique confirm that the molecular unit is in fact a triple helical structure at low temperatures, becoming single stranded at higher temperatures.

The stabilization of the collagen fold can be interpreted as arising from steric restrictions imposed by the high amino acid contents of the polypeptides and interchain hydrogen bonding; hydrogen exchange studies conducted on the polypeptides showed that in the helical conformation the polymers contained significant numbers of slowly exchanging protons, which the authors suggest as arising from their involvement in hydrogen bonding. The calorimetry data of Privalov et al.,[1201] McClain et al.,[942] and Cooper,[279] together with the hydrogen exchange studies of Bensusan and Nielsen[125] on tropocollagen, suggest that hydrogen bonding is unlikely to be a major factor in stabilizing the helical conformation; solvent interaction, involving both water bridging, according to the model of Ramachandran et al.,[1212] and solvent stabilization of the pyrrolidine groups [347,1406] seems to be a much more likely possibility.

Brown et al.[188] have reported the synthesis and investigation of two polytripeptides, poly(pro–ser–gly) and poly(pro–ala–gly), by far-ultraviolet circular dichroism. The CD studies were performed on aqueous solutions containing 0.15% of peptide, the peptide being easily soluble in cold water. Figure 9 compares the CD spectra of the tripeptides in water, and includes the spectrum of guinea pig skin collagen; (pro–ala–gly)$_n$ shows identical peak positions to those of the collagen at 221.5 and 198 nm, indicating a similar type of helix formation, but the low rotational strength of (pro–ser–gly) is interpreted by the authors as indicating a random coil conformation.

The CD spectra of (pro–ala–gly)$_n$ and (pro–ser–gly)$_n$ in 1,3-propanediol reveal much more marked maxima and minima, implying a more ordered structure even for (pro–ser–gly)$_n$. The authors conclude that water tends to randomize (pro–ser–gly)$_n$, possibly due to specific interaction of H_2O molecules with the side chain serine OH group. In an extended investigation of the CD spectra of the polymers in a variety of hydrogen bonding solvents (1,3-propanediol, 1,4-butanediol, diethylene glycol, ethylene glycol, glycerol, water, hexafluoroisopropanol, and trifluoroethanol), the authors clearly relate increased structuring to decreased hydrogen bond strength of the solvent; for the two strongest, hexafluoroisopropanol and trifluoroethanol,

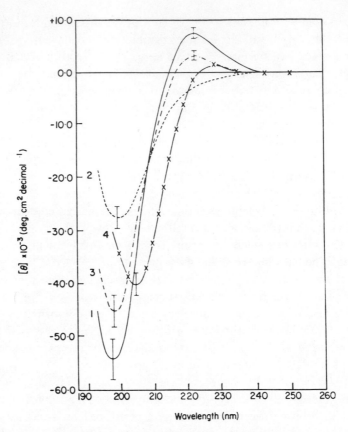

Fig. 9. CD spectra of (pro–ser–gly)$_n$, (pro–ala–gly)$_n$, and (pro–gly–pro)$_n$ compared with guinea pig skin collagen in water. Curve 1, guinea pig skin collagen; curve 2, (pro–ser–gly)$_n$, $M = 18{,}000$, fractionated; curve 3, (pro–ala–gly)$_n$, $M = 14{,}000$, fractionated; curve 4, (pro–gly–pro)$_n$, $M = 5500$. In all cases the solvent used was water, the concentration of the polypeptide was 1.25 mg/ml, and the temperature $+24°C$. (See Ref. 188.)

neither polymer assumes a helical conformation. The authors conclude from their data that the efficiency of the weaker hydrogen bonding solvents in promoting a collagenlike structure is due to a reduced tendency of the solvent to bind to the polypeptide chains, hence allowing easier formation of interchain hydrogen bonds, which they consider necessary for stabilization of the triple helix-type conformation. The data presented do, however, allow for an alternative model for stabilization of the collagen-type helix; only the dihydric alcohols and water promote the helical conformation and only in water are the polymers easily soluble; solution in the polyhydric alcohols requires heating to 80° for some time and several days elapse before the CD spectrum is completely developed. Significantly, the polymers are not soluble in monohydric alcohols; thus we have the situation that only polyhydric alcohols and water (which are all capable of producing extended hydrogen-bonded networks linked to the polymer) can dissolve the polymer *and* stabilize a collagen-type structure, and only water, with its unique hydrogen bonding capability, can dissolve the polymer easily. A model such as this envisages the possibility of a solvent network surrounding the polymer which has a close correlation with the model proposed by Privalov and Tiktopulo for tropocollagen in solution.[1201]

The authors clearly show by a gel filtration procedure that the molecular weights of the helical species correspond to a threefold aggregation of the monomer, thus confirming the close similarity to a collagen structure.

It is also pointed out that the sequence of the residues within the polypeptide is largely responsible for the particular conformation adopted; thus the CD spectrum of (ser–pro–gly)$_n$ differs markedly from those of collagen and poly-L-proline II.[187] Segal et al.[1360] have also shown that the tripeptide (ala–pro–gly)$_n$ has a random coil conformation in solution. Therefore not only is the proline content of the polypeptide important, but its sequential position within the tripeptide is of crucial importance.

Brown et al.[189] have compared the thermal stability in solution of (pro–ser–gly)$_n$, (pro–ala–gly)$_n$, and (pro–gly–pro)$_n$ by CD spectroscopy. The spectra were compared in 1,3-propanediol and water as solvents with guinea pig skin collagen as the control. (Pro–gly–pro)$_n$ was shown to produce a much more stable triple helix relative to (pro–ala–gly)$_n$ and (pro–ser–gly)$_n$. From an analysis of the melting temperatures of the polytripeptides a semiempirical equation based on a two-state model is derived, involving contributions from the primary residue sequence, the molecular weight and polydispersity of the polymer, and the extent of polymer–solvent interaction. The enthalpy and entropy parameters, excepting those involving the polymer–solvent interactions, are calculated by conformational analysis; the polymer–solvent contributions, unfortunately, due to lack of information, form the empirical part of the relationship, since to evaluate the polymer–solvent interaction energies, it is necessary to calibrate the equation for each solvent system.

Several interesting predictions can be made from the model: For instance, a comparison of the melting temperatures of pro–X–gly with those of X–pro–gly. The predicted transition temperature of ala–pro–gly is −40°, which is inconsistent with collagenlike structures being able to exist in water at room temperature. Pro–ala–gly, on the other hand, is predicted as forming a stable triple helix at room temperature, but this has not yet been verified experimentally. The calculated internal energy and entropy terms of the (pro–X–gly) sequence are lower than for (X–pro–gly), which is supposed to arise from the more pronounced steric restrictions in the former sequence. The polymer–solvent interaction of X–pro–gly is thought to be equal to or more destabilizing than pro–X–gly.

It seems permissible, on the basis of the foregoing evidence, to draw two main conclusions about collagen-type polypeptides; first, that steric restrictions involving the proline residues are very important in determining the stability of the triple helices, and additionally, solvent interactions, involving the pyrrolidine groups, also provide an important contribution to the stability of the triple helical poly-L-proline II conformation.

3.4. Hydration Structures Associated with Biopolymers

This section is devoted to attempts to quantify the amount of water that can be considered as hydrating the protein, rather than to evidence of the involvement of water in conformational transitions of the macromolecule. A primary difficulty arises, however, when one attempts to define "hydration" in an unambiguous manner; the difficulty stems from the fact that different experiments give different results for the "extent of hydration."

Techniques such as nmr, infrared spectroscopy, and dielectric dispersion appear to be capable of detecting only the most tightly bound water. Hydrodynamic measurements, on the other hand, appear to measure the average amount of water which migrates through the solution with the macromolecule. These water molecules may be indistinguishable from bulk water in terms of those properties that are monitored by techniques such as nmr, since rapid interchange of this water with the bulk water probably occurs. Techniques that might perhaps be best described as hydrostatic, as distinct from hydrodynamic, form a third group of methods that may be used to study the water of hydration of the macromolecule; these include density gradient ultracentrifugation, isopiestic, and calorimetric methods.

3.4.1. *Hydration of Proteins and Polypeptides*

Kuntz and co-workers have recently performed an interesting investigation of the hydration of a variety of proteins and their constituent amino acids.[828,828a,829,830] The initial report describes a proton nmr technique involving rapid freezing of the protein solution in liquid nitrogen, followed by equilibration at the temperature of the experiment ($-35°$). The high-resolution nmr signal shows a single, broad (0.2–2 kHz) Lorentzian signal and the area under the curve is a direct measure of the amount of unfrozen, "bound" water associated with the macromolecule; the area is taken to be the product of linewidth and peak height.

The bound water signal is shown to be directly proportional to the protein concentration and significant variations in both linewidth and relative hydration are observed. Ionic strength is also observed to alter the signal; in the absence of salt a relatively broad band is found, which narrows to approximately half the width at 0.1 mol dm^{-3}, then broadens again at salt concentrations in excess of 1 mol dm^{-3}. This effect is observed with KCl, NaCl, and $(NH_4)_2SO_4$.*

The effect of temperature is to increase the linewidth exponentially as the temperature falls, producing an Arrhenius activation energy $\Delta E^{\ddagger}$ of between 17 and 21 kJ mol^{-1} for all protein solutions, in good agreement with the values reported[181,503] from the measurement of spin–spin and spin–lattice relaxation times of water sorbed by solid proteins as a function of temperature. A $\Delta E^{\ddagger}$ value of this magnitude appears to be consistent with a single hydrogen bond barrier[181] by comparison to a $\Delta E^{\ddagger}$ of 50.4 J mol^{-1} in ice.[151] The authors speculate that the bound water can be of an adsorbed form, as a monolayer or perhaps multilayers, held by hydrogen bonds to suitable groups on the surface of the macromolecule, or perhaps by hydrophobic interactions. Alternatively, consideration is given to the water being held in interstices within the interior of the macromolecule in insufficient quantity to form a normal ice lattice, but, as the authors point out, this is unlikely because a conformational change, such as denaturation, would make such water accessible and it would then be expected to lose its "nonfreezing" capability; there is no evidence that this occurs.

The studies have been extended to a wide range of polypeptides in a variety of conformational modes to determine the extent of hydration of each peptide residue.[828] Table XII contains the details of the proposed peptide residue hydrations under standardized experimental conditions of −35° and pH 6–8, with some exceptions. All the solutions contained the polypeptide at a concentration of 5–10% in 0.01 mol dm^{-3} KCl with the pH adjusted with KOH or HCl.

All the amino acids with ionized side chains are found to be heavily hydrated and proline, hydroxyproline, and the uncharged forms of the

---

* It is curious that Eagland and Pilling, in their study of the reversion of α-gelatin to the collagen fold,[376] noted a similar effect on the extent of the reversion, which increased to a maximum as the salt concentration increased to approximately 0.1 mol dm^{-3} and decreased markedly with a further increase in salt concentration; the salts used in that investigation were NaBr, $(CH_4)_4NBr$, and $(C_4H_9)_4NBr$.

TABLE XII. Proposed Amino Acid Hydrations[a]

Amino acid	Hydration, mol H_2O/mol amino acid
Acidic groups	
Asp⁻	6.0
Glu⁻	7.5
Tyr (uncharged)	3
Asp (*p*H 4)	2
Glu (*p*H 4)	2
Tyr⁻ (*p*H 12)	7.5
Basic groups	
Arg⁺	3.0
His⁺	4
Lys⁺	4.5
Arg (*p*H 10)	3
Lys (*p*H 10–11)	4.5
Hydrophilic groups	
Asn	2.0
Glu	2
Pro–OH[b]	4
Pro	3
Ser, thr	2
Trp	2
Hydrophobic groups	
Ala	1.5
Lys, met	1
Gly	1
Ile, leu	1
Phe	0
Val	1

[a] Data from Ref. 828. At 35° and *p*H 6–8 unless otherwise stated.
[b] As pro plus one water per hydroxyl group.

TABLE XIII. Protein Hydration Expressed As g H_2O/g Protein[a]

Protein	Conformation	Model	Calc.	Obs.
Gelatin	N	FE	0.50	0.45
Myoglobin	N	FE	0.45	0.415
Bovine albumin	N	FE	0.445	0.40
Bovine albumin	D urea	FE	0.445	0.44
Bovine albumin	D pH3	FE,T	0.32	0.30
Hemoglobin	N	FE	0.415	0.42
Chrymotrypsinogen	N	FE	0.39	0.34
Lysozyme	N	FE	0.36	0.34
Lysozyme	N	FE,B	0.335	0.34
Ovalbumin	N	FE	0.37	0.33
Chymotrypsin	N	FE	0.36	0.37

[a] Data from Ref. 828. N, native protein, 0.01 M KCl; D, denatured protein; FE, all residues fully exposed; FE,T, residues fully exposed but all carboxylate ions assumed titrated; FE,B, corrected for buried residues.

basic amino acids also hold considerable amounts of water. The degree of hydration is calculated on the assumption that each residue in the polypeptide is fully exposed to the solvent and there are no buried residues.

Table XIII illustrates the hydration of a variety of proteins, as determined directly by this nmr technique, and also calculated from the data listed in Table XII. The author notes that the results on urea and acid denatured samples of bovine serum albumin appear to agree with predictions that urea exposes the hydrophobic groups that pick up a small amount of water; acid, on the other hand, by titrating the carboxyl groups, dehydrates the protein.

Since the linewidth of the water resonance peak is related to the spin–spin relaxation time of the water protons,[181,376,503] it can be expected to be sensitive to the type and conformation of the macromolecule.[828a] In fact, the linewidth of the water signal is sensitive to the nature of the polypeptide, the temperature, and, for ionic peptides, the hydrogen ion concentration of the aqueous environment.

The $\Delta E^{\ddagger}$ values for the polypeptides cover a wide range, from 12.2 J mol^{-1} for polyglycine to 75.60 J mol^{-1} for polyaspartine at pH 8.4, in contrast with the narrow range of $\Delta E^{\ddagger}$ of 17–21 J mol^{-1} exhibited by the globular proteins.

No general correlation is found of the linewidth with the nature of the ionic side chains of the polypeptide, but the polypeptides that are considered to exist in a random coil conformation at room temperature exhibit sharper water lines than the corresponding polypeptides in a helical conformation; the change in linewidth occurs over a pH range close to that expected from conventional solution studies.[151] Tentative proposals for the conformations of the nonionic polymers can be made; polyvaline, for example, which is suggested as being limited to a random coil form in solution due to steric restrictions,[1146] shows a very narrow water line. In contrast, poly-L-proline and polyglycine show considerable broadening of the water signal, presumably due to the structural conformations causing restriction in the freedom of movement of neighboring water molecules.

Glasel[526] has reported deuteron magnetic spin–lattice relaxation (T_1) experiments on D_2O solutions of poly-L-glutamic acid, poly-L-lysine, polymethacrylic acid, polymethylvinyloxazolidinone, and polyvinylpyrrolidone. The assumption is made that the lifetime of a D_2O molecule in association with the macromolecule is less than 0.1 sec and the two-phase equation for the observed relaxation rate is used in its fast exchange approximation[43]

$$(1/T_1)_{obs} = (1/T_1)_{free} + cWK\tau_r \tag{1}$$

where $(1/T_1)_{obs}$ is the measured relaxation rate for the solution, $(1/T_1)_{free}$ is the relaxation rate of the solvent, c is the macromolecular concentration in g/g of solvent, W is the time-averaged weight of solvent associated per g of macromolecule, K is the quadrupole coupling term, and τ_r is the rotational reorientation time of the solvent molecules associated with the macromolecule. Plots of $1/T_1$ and relative viscosity as a function of concentration

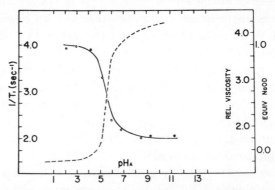

Fig. 10. Relaxation rate and relative viscosity vs. apparent pH of a 3% solution of polymethacrylic acid ($M = 300,000$ in D_2O): solid line, relaxation rate; dashed line, viscosity. (See Ref. 526.)

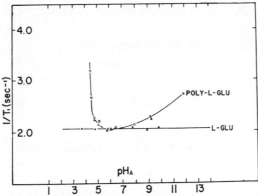

Fig. 11. Relaxation rate vs. apparent pH for 3% w/v solutions of poly-L-glutamic acid and L-glutamic acid; 0.1 N NaCl in D_2O: ($\bigcirc$) $M = 100,000$; ($\triangle$) $M = 3000$. (See Ref. 526).

for polymethylvinyloxazolidinone and polyvinylpyrrolidone show no correlation and the very small slope of relaxation rate vs. concentration suggests little or no interaction of water with the macromolecule. Figure 10 shows the relation of the relaxation rate and the relative viscosity for polymethacrylic acid as a function of pH. D_2O interacts strongly with the macromolecule in its un-ionized form but, in the ionized form at higher pH, the D_2O shows a much lower degree of interaction.

Figure 11 shows the relaxation rate as a function of pH of L-glutamic acid and poly-L-glutamic acid; the polymer is shown to interact strongly with the solvent when in the ionized state (negatively charged) but the monomer reveals very little interaction. The interaction of the macromolecule decreases with increasing degree of neutralization of the carboxylate group and reaches zero at pH 6, the middle of the helix–coil transition region; it then increases rapidly as the polymer becomes completely neutralized.

Glasel concludes that in order for strong interaction with water to occur, the macromolecule must have a proton donor group to form a hydrogen bond to a water molecule, and he sees no evidence for interaction of water with carbonyl groups on the macromolecule. The conclusion that macromolecules containing pyrrolidone groups do not interact to any significant extent with water is at variance with the evidence[828,1406] for the strong interaction of water with poly-L-proline and the suggestion that hydration of the carbonyl group does not occur is in conflict with the hydration model for solid collagen[120] which assumes water bridging via the carbonyl groups on the macromolecule.

In the case of polymethacrylic acid the mean τ_r of the associated D_2O molecules is found to be 1.6×10^{-8} sec, leading to an estimate of the energy of activation for the reorientation process of 29 kJ mol^{-1} compared to 9.7 in pure D_2O.

Glasel concludes from the evidence that

$$\diagdown \diagup C=O, \qquad \diagdown NH, \qquad \overset{\displaystyle O}{\underset{\displaystyle \|}{-C}}-O^-M^+, \qquad C-NH_3^+X^-$$

where M$^+$ is a counterion, do not interact strongly with water, but that the groups

$$\overset{\displaystyle O}{\underset{\displaystyle \|}{-C}}-OH, \qquad -C-N\diagup^H_\diagdown{}_H$$

take part in hydration interactions.

One of the most interesting points arising from Glasel's work is the observation that the interaction of poly-L-glutamic acid with water reduces to zero in the middle of the helix–coil transition; this appears to correlate with a decrease in the intrinsic viscosity of the polymer in solution[360] and with the dielectric relaxation time.[1459] Glasel points out that the structural relaxation should reach a maximum at this point and the relative amounts of helix and coil should change rapidly[166,926,1350,1426] with fluctuations whose lifetimes are 10^{-3}–10^{-2} sec, thus effectively destroying the interaction of water with the macromolecule in this region.*

---

* Glasel's findings are so unexpected that a closer scrutiny is in order. The use of eqn. (1) implies that the motion considered is isotropic *and* that $\tau_r \ll \omega_0^{-1} \ (= 2 \times 10^{-8}$ sec.). For D_2O bound to polymethacrylic acid τ_r is stated to be 1.6×10^{-8} sec, so that the second assumption is not valid. As regards the nature of the motion, detailed nuclear magnetic relaxation studies of D_2O in phospholipid systems[442] indicate that the motion of "bound" water attached to polar groups is anisotropic and has a "slow" component which can only be studied by spin–spin relaxation measurements, i.e., $T_2 \neq T_1$, whereas Glasel's treatment implies that $T_1 = T_2$.

As regards the observed pH dependence of $(T_1)^{-1}$ and the resulting conclusions, it can be argued that protonation of the ionized groups gives rise to a different type of mechanism, i.e., very slow motion of the –O–D group attached to the polymer, plus deuteron exchange with the solvent. The measured T_1 represents an average of the two processes. It is also quite possible that the quadrupole coupling term [K in eqn. (1)] for the two states is not the same.

The above considerations therefore suggest that T_1 measurements *alone* coupled with eqn. (1) cannot provide unambiguous information about the extent or nature of the biopolymer–water interaction in solution.

Ikegami[706] has investigated the hydration of polyacrylic acid, poly-methacrylic acid, and poly-L-glutamic acid by measuring the change in refractive index as the polymers were progressively neutralized with tetra-butylammonium hydroxide and sodium hydroxide. The results were inter-preted in terms of the change in apparent molar volume during the neutral-ization, given as

$$\Delta \bar{V}_\alpha = \bar{V}_{\text{RCOO}^-} + \bar{V}_{\text{H}_2\text{O}} - \bar{V}_{\text{RCOOH}} - \bar{V}_{\text{OH}} + \Delta \bar{V}_c$$

where $\Delta \bar{V}_\alpha$ is the change in apparent molar volume at a particular degree of neutralization α and $\Delta \bar{V}_c$ is the volume change accompanying any con-formational transition. The change in the partial molar volume associated with the neutralization process, $\bar{V}_{\text{H}_2\text{O}} - \bar{V}_{\text{H}^+} - \bar{V}_{\text{OH}^-}$, was found to be 28.2 ml mol^{-1}.[705] Assuming $\bar{V}_{\text{H}^+}$ to be equal to -5 ml mol^{-1}, the sum of the partial molar volumes associated with the hydration process of the ionizable groups at a given value of α, $\Delta \bar{V}_h (\equiv V_{\text{RCOO}^-} - V_{\text{RCOOH}})$, and the conformational transition from helix to coil, $\Delta \bar{V}_c$, can be obtained, $\Delta \bar{V}_c$ appearing only in the range of the transition. Figure 12 illustrates the behavior of $\Delta \bar{V}_h + \Delta \bar{V}_c$ as a function of the degree of neutralization for

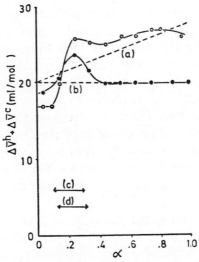

Fig. 12. Dependence of partial molar volume associated with the hydration process and conformational transition of poly-L-glutamic acid, $\Delta \bar{V}_h + \Delta \bar{V}_c$, on the degree of neutral-ization α: ($\bigcirc$) Neutralized by (Bu)$_4$NOH; ($\bullet$) neutralized by NaOH; (a) calculated value of $\Delta \bar{V}_h$ neutralized by (C$_4$H$_9$)$_4$NOH; (b) calculated value of $\Delta \bar{V}_h$ neutralized by NaOH; (c) the range of the helix–coil transition neutralized by (C$_4$H$_9$)$_4$NOH; (d) the range of the helix–coil transition neutralized by NaOH. (See Ref. 706.)

TABLE XIV. The Total Change in Volume Associated with the Hydration Process and Conformational Change for Polyacids[a]

Polyacid	Counterion	ΔV_h, ml mol^{-1}	ΔV_c, ml mol^{-1}	α
Polyacrylic acid	Bu$_4$N$^+$	−20.2	—	—
	Na$^+$	−14.6	—	—
Polymethacrylic acid	Bu$_4$N$^+$	−23.7	−0.8	0.20
	Na$^+$	−19.7	−0.7	0.22
Poly-L-glutamic acid	Bu$_4$N$^+$	−14.5	−2.1	0.40
	Na$^+$	−14.0	−1.7	0.39

[a] Data from Ref. 706. Here α is the degree of neutralization at the midpoint of the conformational structure.

poly-L-glutamic acid and Table XIV summarizes the values of ΔV_h and ΔV_c, the total changes in volume associated with the hydration process and the conformational change, respectively.

Ikegami suggests that two regions of hydration surround the macromolecule; in the primary region water molecules are considered as being oriented to ionized groups on the macromolecule without cooperative interaction between neighboring groups. This first hydration region is considered as a spherical region of ∼0.31 nm radius surrounding each charged group. The change in excess volume $\Delta \bar{V}$ as α increases noted with the polyacids neutralized with Bu$_4$NOH is attributed to cooperative interaction between charged groups; the interaction, it is suggested, only occurs when the distance is of the order of the diameter of the first hydration region.

In poly-L-glutamic acid the distance between nearest neighbors is larger than this limit, and no cooperative interaction should occur. Figure 12 indeed shows that the volume change, other than that due to the helix–coil transition, is virtually constant.

Figure 13, which shows the change in $\Delta \bar{V}_h$ associated with the neutralization for polyacrylic acid, illustrates the case where cooperative interactions can be expected to occur, in this particular case without a helix–coil transition arising.

In an earlier report Ikegami[705] suggests that the binding of Na$^+$ ions destroys the second cooperative region: The difference between the values of $\Delta \bar{V}_h$ when neutralized by Bu$_4$NOH and NaOH is attributed to this effect, the Bu$_4$N$^+$ ion being assumed to have no effect on the hydration region

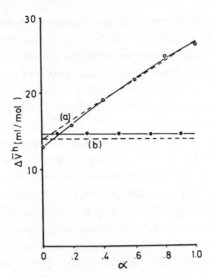

Fig. 13. Dependence of partial molar volume associated with the hydration process of polyacrylic acid, $\Delta \bar{V}_h$, on the degree of neutralization α: (○) Neutralized by $(C_4H_9)_4NOH$; (●) neutralized by NaOH; (a) calculated value of $\Delta \bar{V}_h$ neutralized by $(C_4H_9)_4NOH$; (b) calculated value of $\Delta \bar{V}_h$ neutralized by NaOH. (See Ref. 706.)

around the charged groups and the polymer skeleton. The larger initial $\Delta \bar{V}_h$ at $\alpha = 0$ for polymethylacrylic acid is said to be due to hydrophobic hydration associated with the methyl group.

By making assumptions as to the extent of the primary and hydrophobic hydration regions, values of $\Delta \bar{V}_h$ can be calculated which are shown as dashed lines in Figs. 12 and 13. The differences between the observed and calculated values of $\Delta \bar{V}_h$ at low values of α are attributed to $\Delta \bar{V}_c$; thus the conformational transition from helix to coil apparently produces a negative change in the apparent volume. This change appears to be in the opposite sense to that suggested by Sinanoglu and Abdulner,[1398] who predict that the coil → helix transition should be accompanied by a decrease in the volume of the macromolecule.

Several points arise from the above data which may contradict the reasoning of Ikegami; first, the steady increase in $\Delta \bar{V}_h$ for the case where cooperative interactions arise when the neutralization involves the Bu_4N^+ ion would suggest that, far from having no effect upon the hydration region, cooperative bonding involving the hydrophobic regions of the Bu_4N^+ ion occurs, giving rise to increased stability of the hydration shell. The data also show that at low values of α, $\Delta \bar{V}_h$ involving the Bu_4N^+ ion is smaller than for the Na^+ ion, only becoming larger as the neutralization proceeds.

This suggests that the presence of an increasing number of carboxylate groups is required to involve the Bu_4N^+ ion in the hydration shell, which then results in increased stabilization of the hydration shell.

A comparison of polymethacrylic acid and poly-L-glutamic acid based on the nmr[526] and volumetric[706] studies reveals some very marked differences in the interpretations presented by the respective authors. Glasel considers, for the case of polymethacrylic acid, that hydration of the macromolecule only occurs in the un-ionized form with a marked reduction when the polyacid is neutralized, whereas Ikegami appears to favor the explanation that primary hydration of the macromolecule arises from the presence of the carboxylate groups on the polymer. It should be noted, however, that Glasel fails to explain satisfactorily the evidence of the relaxation data that the ionized form of poly-L-glutamic acid appears to be appreciably hydrated, although not to the same extent as the un-ionized form, a situation which appears to correlate with Ikegami's data for the neutralization of the polymer.

Horne and co-workers[682] have attempted to gain insight into the problem of protein hydration by studying the effect of ionic solutes upon the cloud point (34°) of polyvinylmethyl ether, i.e., the temperature at which rapid collapse of the molecular hydration sheath is believed to occur, leading to aggregation of the polymer. The authors, by equating the cloud point to a denaturation process, found a somewhat anomalous correlation between the Jones–Dole viscosity B coefficient of the ionic solute[738] and its effectiveness in denaturation. The normal trend[1540] is that decreasing values of the B coefficient (believed to denote a decreasing tendency to structure water) are associated with a decreasing effectiveness in denaturation.

Horne *et al.* found that the effect of alkali metal cations is apparently the reverse of what might be expected, since even the structure-making ions decrease the cloud point, but to a lesser extent, than do the structure-breaking ions. The tetraalkylammonium ions, however, show different behavior to the alkali metal ions (Fig. 14); increasing values of the B coefficient occur with increasing stabilization of the polymer in solution.

The authors suggest that the electrostrictively hydrated alkali metal cations are preferentially excluded by the hydrophobic hydration surrounding the polymer; thus, the more positive their B coefficients, the more strongly they are excluded, and the less capable they are of disrupting the hydrophobic hydration regions. The hydration envelopes of the larger tetraalkylammonium ions, however, being of a hydrophobic hydration character, are assumed to stabilize further the hydrophobic hydration sphere

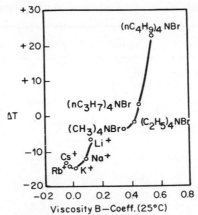

Fig. 14. The change in cloud point temperature of polyvinylmethyl ether as a function of the B coefficients of the tetraalkylammonium cations (0.75 M). (See Ref. 682.)

of the polymer. This interpretation would also appear to be in agreement with the explanation previously proposed for the anomalous data of Ikegami.[706]

The balance of hydrophobic and hydrophilic interactions of a macromolecule with its solvent environment can be expected to produce measurable dielectric effects. These effects can be utilized to measure the amount of "bound" water associated with a protein and changes in the extent of protein hydration which may accompany changes in the conformation of the protein.

Grant and co-workers[559] have investigated the dielectric dispersion behavior of bovine serum albumin as a function of protein concentration, temperature, and pH over the frequency range of 2×10^8 to 2×10^9 Hz. Particular attention was paid to the frequency region 10–1000 MHz, termed the δ region, which lies between the dispersion region attributed to the protein and that of the free water; it has been suggested[558,1348] that this dispersion originates from the relaxation of the water bound to the protein.

A dielectric mixture model of the form

$$\varepsilon' - \varepsilon_1 = kp(\varepsilon_2 - \varepsilon_1) \tag{2}$$

is assumed, where ε' is the measured dielectric constant, ε_1 and ε_2 refer to the solvent and solute, respectively, p is the volume concentration of the solute, and k is a constant dependent upon the shape of the protein molecule. For most proteins $1 < p < 1.8$. The hydration of bovine serum albumin over a range of pH values at three frequencies, calculated for two repre-

TABLE XV. Variation of the Hydration of Bovine Albumin with pH at $25°$ [a]

Frequency, MHz	Hydration, g H_2O/g dried protein							
	$k = 1.1$				$k = 1.3$			
	pH 5.07	pH 7.85	pH 8.8	pH 9.8	pH 5.07	pH 7.85	pH 8.8	pH 9.8
250	0.52	0.8	—	—	0.32	0.23	0.1	—
700	0.52	0.42	0.45	0.3	0.32	0.20	0.22	—
2000	0.52	0.64	0.25	0.4	0.32	0.44	0.1	0.21

[a] Data from Ref. 559.

sentative values of k, is shown in Table XV; the amount of bound water near the isoelectric point was found to vary from 0.18 to 0.64 g/g of protein, in agreement with other estimates.[450]

The dielectric dispersion of horse hemoglobin (concentration 107 gl⁻¹) has also been measured at a stabilized frequency of 9.57 GHz and a pH of 6.8 in H_2O and D_2O.[1534] The results indicate that if ε_2 is constant for a given frequency region ($\varepsilon_2 = 4$ for hemoglobin[1534]) and the dispersion of the dielectric constant of water ε as a function of frequency is given by

$$\varepsilon = \varepsilon_\infty + [(\varepsilon_0 - \varepsilon_\infty)/(1 + i\lambda_c/\lambda)]$$

where ε_0 is the static dielectric constant of water, ε_∞ is the far-infrared value, and λ_c and λ are the vacuum wavelengths corresponding to the relaxation time and the measuring frequency ($\lambda_c = 1.78$ cm for pure water at 20°),[199] then combination of this equation with eqn. (2) gives

$$\varepsilon' = \varepsilon_\infty(1 - kp) + \varepsilon_2 kp + \frac{\varepsilon_0(1 + kp) - \varepsilon_\infty kp}{1 + i(\lambda_c + \delta\lambda_c)/\lambda}$$

where the amount of hydration water per gram of protein is derived from p, and $\delta\lambda_c$ is the shift in the relaxation wavelength. The calculated values of the shift of the water relaxation wavelength and the amount of "bound" hydration water are presented in Table XVI; the hydration values are in quantitative agreement with other reports,[1337] but the calculated $\delta\lambda_c$ for horse methemoglobin differs from that calculated by Buchanan et al.[198] (0.06 cm).

TABLE XVI. Shift of Relaxation Wavelength λ_c and Bound Water of Horse Hemoglobin[a]

Temperature	Sample	Solvent	Shift $\delta\lambda_c$, cm	Hydration, g/g
20	Oxy. hb	H_2O	0.13 ± 0.03	0.24 ± 0.02
25	Met. hb	H_2O	0.15 ± 0.02	0.18 ± 0.02
25	Met. hb	D_2O	0.08 ± 0.02	0.15 ± 0.02

[a] Data from Ref. 559.

Masuzawa and Sterling[990] have measured the dielectric constant as a function of frequency for a variety of high-concentration gels, including gelatin. The value of ε' is much larger than that of water at the low frequencies, only becoming lower than that of water at high frequencies (Fig. 15).

At high frequencies ε' values of the gels do not follow a curve parallel to that of water, ice, or theoretically bound water[1348] and at low frequencies ε' of the gels is higher than that of liquid water, even in an uncharged starch gel. It is therefore suggested that polarized water molecules exist in the

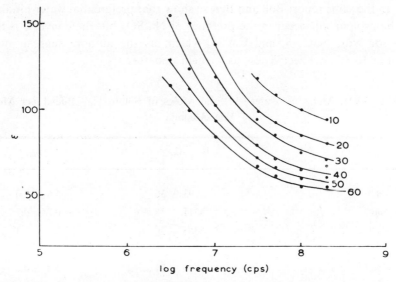

Fig. 15. Dielectric constant ε as a function of the logarithm of frequency and concentration (%) for gels of gelatin (unfractionated) (see Ref. 990).

ordered gel state; presumably, from Fig. 15, the degree of ordering of the water increases with decreasing solute concentration, the effect being most pronounced in a gelatin gel.

Bull and Breeze[202] have presented a theory of solute and water binding by proteins for which confirmatory experimental data have been obtained by an osmotically corrected isopiestic investigation of the binding of solutes and water to egg albumin in solutions of alkali metal chlorides (concentration 1–4 m); the results are presented in Table XVII. The data for CsCl are in good agreement with those of Hade and Tamford.[586]

The cations are arranged in a lyotropic series which is related to the hydrated ionic radii; thus it seems that the larger the hydrated radii of the cations, the stronger is the capacity of the ion to dehydrate the protein. The Li^+, Rb^+, and Cs^+ ions are apparently bound to the protein to the same degree, but Na^+ and K^+ show no significant binding. This appears to confirm the suggestion of Ikegami[706] that the Na^+ ion is unable to penetrate the primary hydration shell of the macromolecule.

In a previous report Bull and Breeze[586] found that at a relative humidity of 0.92 at 25° in the absence of electrolyte, 1 mol of egg albumin binds 740 mol of water; thus the addition of the alkali metal chlorides results in the loss of "bound" water, a fact, the authors point out, that is probably responsible, at least in part, for the wide divergence in reported values of protein hydration.

In the same report Bull and Breeze show that preferential water binding by the protein will occur in the presence of Na_2SO_4 but the converse is the case for NaI, NaCNS, and $CaCl_2$, which, as the authors suggest, may account for their effectiveness as protein denaturants.

TABLE XVII. Moles of Water Δn_1 and Moles of Solute Δn_2 Bound per Mole of Egg Albumin

Salt	Δn_1	Δn_2
LiCl	193 ± 36	5.6 ± 1.0
NaCl	211 ± 18	-0.2 ± 0.7
KCl	365 ± 36	2.4 ± 1.3
RbCl	409 ± 26	6.2 ± 1.0
CsCl	493 ± 21	5.2 ± 0.8

[a] Data from Ref. 202.

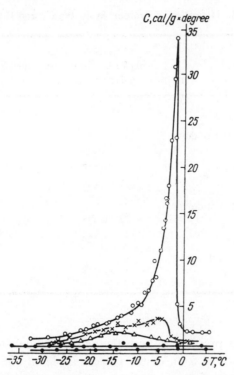

Fig. 16. Temperature dependence of the heat capacity of 1 g of tropocollagen in the presence of different amounts of water (0, 0.35, 0.64, 0.93, 1.26, and 2.00 g) (see Ref. 1198).

Few calorimetric studies of protein and polypeptide hydration have been reported. This is surprising, since the claimed accuracy for the method ($\sim$0.15%) is better by at least an order of magnitude than that claimed for dielectric dispersion and nmr techniques. Privalov and Mrevlishvili[1198] have reported calorimetric studies of several proteins in both the native and denatured form; the technique involved cooling of the solution to $-35°$ and heating to temperatures of the order of 5°. Figure 16 shows the temperature dependence of the heat capacity of tropocollagen in the presence of varying amounts of water; only in the presence of >30% by weight of water does a heat absorption peak appear, indicating the presence of a hydrogen-bonded, structured system, but even at a water level of 2.00 g/g of collagen the absorption peak does not correspond to that of ice.

The amount of "bound" water associated with the protein is determined from the area of the heat absorption peak; Table XVIII contains details of the hydration of the macromolecules in both the native and denatured state. It is interesting to note that tropocollagen is much more heavily hydrated

TABLE XVIII. Hydration of Proteins in the Native and Denatured States[a]

Protein	Hydration of native conformation, g H_2O/g protein	Hydration of denatured conformation, g H_2O/g protein
Tropocollagen + water	0.465	0.519
Tropocollagen + 0.15 mol dm^{-3} citrate buffer	0.485	0.492
Serum albumin + water	0.315	0.330
Egg albumin + water	0.323	0.332
Egg albumin + 0.15 mol dm^{-3} borate buffer pH 10.0	0.322	0.333
Hemoglobin + water	0.324	0.339
Hemoglobin + 0.15 mol dm^{-3} borate buffer pH 10.0	0.336	0.345

[a] Data from Ref. 1198.

than the globular proteins, in both the native and denatured form, but in general the denatured form of the protein is slightly more hydrated than the native conformation. A marked variation in hydration behavior is noticeable for tropocollagen, and to a lesser extent with the globular proteins, in water and in 0.15 mol dm^{-3} citrate buffer; this behavior parallels that observed by nmr,[828a] optical rotation, and viscosity[376] studies for a variety of proteins, i.e., considerable differences are observed between pure water and an ionic strength of $\sim$0.15.

Privalov and Mrevlishvili equate the hydration of 0.465 g of water/g of protein for tropocollagen to 2.4 H_2O per residue, which, if the sequence gly–pro–HO–pro is considered as representing a triple residue sequence for tropocollagen, is in very good agreement with the value obtained from Kuntz's calculated values of hydration per residue, 2.7 H_2O. Good agreement is found between the above hydration values for the globular proteins and those obtained by Fisher,[450] who assumed that hydration of the macromolecule was due only to the amino groups present on the surface of the macromolecule.

There seems to be considerable agreement between the work of Privalov and Mrevlishvili and other workers as to the extent of the primary hydration shell of the proteins, which appears to be mainly due to a mixture of ionic hydration and hydrogen bonding, but the increased extent of hydration on denaturation suggests a contribution from hydrophobic hydration; explanations are lacking, however, for the extended region, involving up to 50 water molecules, which shows evidence of slight restriction of the motional freedom of the water molecules due to the presence of the macromolecule.

An interesting recent report of infrared data[212] which support the results of Privalov[1198] and Kuntz[828a] covers the hydration of lysozyme and bovine serum albumin deposited from solution and dried to various moisture contents. Differential spectra, relative to a completely dry film, during drying displayed two intense bands at 3450 cm^{-1} and below 3300 cm^{-1} and a shoulder at 3060 cm^{-1}. The first band decreased rapidly during the early stages of drying and, on further prolonged drying, the second band and shoulder decreased markedly. The position of the shoulder indicates that the –NH band profile is affected by water, contrary to Glasel's proposals.[526] The authors suggest that the band in the 3300-cm^{-1} region results from the type of hydration observed by Kuntz[828] in his low-temperature nmr studies; this was confirmed by slowly condensing water vapor onto a lysozyme film held at near liquid nitrogen temperatures; a marked increase in the intensity of the 3300-cm^{-1} band resulted. Only at very high water coverages did the profile band of ice become apparent. Thus IR data suggest that –OH stretching at 3300 cm^{-1} is due to the hydration water bound to the protein molecule. It is also possible that the band at 3450 cm^{-1} may correspond to the outer region of hydration observed by Privalov and Mrevlishvili[1198] in their calorimetric experiments and to the region of oriented water dipoles proposed by Masuzawa and Sterling,[990] based on their dielectric constant measurements at low frequencies.

Difference spectra of aqueous solutions against water in the near-infrared region can distinguish the amount of water excluded by a hydrated solute, the hydrating water, and any absorption due to the solute.[939,940] Such hydration spectra have been obtained for poly-L-glutamic acid and poly-L-lysine[1439] as part of a study of changes in hydration during the helix–coil transition of the two polypeptides.

The difference spectra for poly-L-lysine display a positive peak at 6554 cm^{-1} and another at 7114 cm^{-1}. The 6554-cm^{-1} band, which is absent for polyglutamic acid, is attributed to the un-ionized ε-NH$_2$ group in the helical conformation; the negative absorption below 6554 cm^{-1} observed for the random coil form is attributed to the ionized NH$_3^+$ group. The peak at 7114 cm^{-1} is shown to be due to a water molecule hydrogen-bonded to NH$_2$ but hydration of –NH$_3^+$ groups is not observed.

The broad absorption band at approximately 6800 cm^{-1} in the difference spectrum of polyglutamic acid is attributed to the carboxylate group of the helical form, which interacts with water in a similar manner to the –COOH group of acetic acid.

The authors observe that CO and NH groups might be expected to be more hydrated in the random coil than in the helix conformation but this is not the case and no corresponding peptide hydration band is observed.

Several general conclusions can be drawn from the foregoing data about the nature of the hydration of proteins and polypeptides. Primarily, the role of salts in the hydration process is, at best, ambiguous. The evidence[200,376,828a,1198] indicates that hydration of the macromolecule increases in the presence of salt up to a maximum value at an ionic strength of 0.1–0.15 but marked dehydration occurs at high ionic strengths.

The higher degree of hydration associated with collagen compared to the globular proteins seems to be clearly confirmed from several different viewpoints; nmr data[828] show that the broadening of the water signal in poly-L-proline and polyglycine could be due to restriction of the movement of water molecules by their proximity to the macromolecular surface. Calorimetric data[1198] confirm extremely well the quantitative measure of this hydration but also indicate a much larger region around the molecule within which the water molecules are subject to some kind of restriction. This seems to be confirmed by the dielectric data,[990] which suggest that some kind of long-range ordering of water occurs around gelatin, which increases with decreasing concentration of the protein; since the lowest concentration studied was $\sim 10\%$, it seems clear that at lower concentrations a maximum value will be reached (the extended hydration of Privalov and Tiktopulo[1201]?).

The evidence for increased hydration of the denatured form of the protein compared to the native conformation is considerable and appears to give strong support for at least some contribution from hydrophobic hydration to the overall hydration structure.

It is surprising that such conflict is apparent as to the extent of hydration of particular residues within the protein or polypeptide; Kuntz[828a] indicates that ionized acidic groups are much more heavily hydrated than the un-ionized groups and is supported by the data of Ikegami[705] and others. Glasel,[526] however, suggests that the un-ionized form is the more heavily hydrated; this proposal receives little support but his associated contentions that CO and NH groups are not appreciably hydrated do receive support from the data of Subramanian and Fisher[1439] and Smith.[1406] Buontempo and co-workers,[212] on the other hand, suggest that the –NH group is affected by hydration.

There is general agreement that the basic amine group $-NH_2$ is heavily hydrated, but not, however, the $-NH_3^+$ form.[1439]

3.4.2. *Hydration of Nucleic Acids, Nucleotides, and Nucleosides*

In Section 3.2 the role of water in the stacking interactions of nucleic acid materials was outlined; it is certainly the case that the presence of water is essential for the preservation of the helical structures of such compounds. Falk and co-workers[421–423] have shown that the normal state of solid NaDNA (the B form) is stable only at high relative humidities (above 75–80%) and below 55% relative humidity a general loss of helical structure occurs. At relative humidity values above 92% the observed infrared bands are said to be characteristic of those of liquid water; at 80% relative humidity all the specific hydration sites are occupied and approximately 10 mol of water is bound per mol of nucleotide. In the range 80–55% relative humidity 4–5 mol H_2O/mol nucleotide is progressively removed, mainly from the grooves in the surface of the molecule. Falk *et al.* conclude that the phosphate groups bind water most strongly and that 2 mol H_2O/ phosphate group remains even at low humidities. In a recent study[424] of the OH and OD stretching bands of isotopically dilute HDO in hydrated and partially deuterated films of DNA as a function of relative humidity and temperature, it was confirmed that no regions of ordered water exist in the hydration shell of DNA. Consecutive layers of water surrounding the macromolecule have similar distributions of hydrogen bond strengths; even at 3% relative humidity DNA still holds one molecule of water per nucleotide and its hydrogen bonding does not differ appreciably from that of succeeding layers. The authors are careful to point out that though hydrogen bond energies associated with water–biopolymer and water–water interactions are similar, the mobility of water molecules near the biopolymer surface would be expected to be lower than in bulk water.

Infrared spectra at low temperatures and relatively low humidities[424] (nine water molecules per nucleotide) show a complete absence of crystallization of the water; at higher water contents (14 water molecules per nucleotide) partial freezing of the water suddenly occurs. The results suggest an inner hydration layer (ten molecules of water per nucleotide) which will not freeze, and a second layer (three molecules of water per nucleotide) which freezes with difficulty. Lewin[885] has made definite proposals as to the nature of the hydration of polar sites on the surface of the nucleotide. Beginning from the viewpoint that association with water is a prime requirement for the stability of the normal (B) form of DNA in the solid state, he considers that water bridging can occur on the surface by the following mechanisms: (1) by reduction in interstrand electrostatic repulsion by citing of a water molecule between phosphate groups, and (2) by binding of complementary strands via the ring nitrogen, free carbonyl, ribose oxygen, and amino groups. From studies with Courtauld atomic models Lewin considers it feasible for single water molecules to form bridges between the free carbonyl of thymine and the amino groups of guanine which lie exclusively in the narrow 100-nm groove on the surface of the nucleotide. The wide groove can be divided into an upper hydrophobic band, a central hydrophilic band of amino and carbonyl groups, together with ring nitrogens and a second lower hydrophobic band. Multiwater links are possible from the upper part of the hydrophilic belt across the hydrophobic region to polar sites in the adjacent narrow groove. It is particularly interesting that an increased proportion of G–C in the nucleotide is said to increase the possible number of water bridges, which, Lewin suggests, might be expected to favor increased stability.

It seems to be the case that hydration of DNA in solution produces a structure which is very close to the normal B form of the nucleotide in the solid state at high relative humidities, but it is also extremely difficult[424] to define exactly what is meant by hydration; is it to be restricted to the innermost layers of water which do not freeze on cooling? Or are the outer layers, which are also involved in some manner, to be included?

Nuclear magnetic resonance studies appear to see only those water molecules most firmly bound to the polynucleotide; Lubas and Wilczok,[919] using a spin–echo technique, have measured the spin–lattice and spin–spin relaxation times T_1 and T_2 for calf thymus DNA in a 0.15 mol dm^{-3} NaCl and 0.015 mol dm^{-3} sodium nitrate solution at 20°C (calculated CG content, 46%); two fractions of DNA, IV and V, were used, having molecular weights of 7.7×10^6 and 6.0×10^6, respectively.

The observed shortening of both relaxation times is interpreted by the authors as indicating that DNA IV in the native conformation is hydrated to the extent of 10.4% by weight, and DNA V by 8.9%; and DNA V in the denatured conformation by 6%. These results suggest that the tightly bound hydration of native DNA is quite small, corresponding[1501] to only about two molecules of water per base. This value is similar to that observed by Falk *et al.*[421] for DNA films at 3% relative humidity; thus it appears that nmr techniques can only measure the water molecules bound by the phosphate groups of the macromolecule.

As regards the effect of the thermal helix → coil transition of DNA upon the nonrotationally bound water associated with DNA, it is found that about 10° below the beginning of the transition the nonrotational hydration decreases, reaching values which do not permit further shortening of T_1.[922] At temperatures above the melting temperature T_m the relaxation times cease to increase; T_m appears to represent a state in which the water molecules are removed to a defined extent from the lattice of the macromolecule.

Lubas and Wilczok[920] have also reported the effect of ionic strength upon T_1 and T_2 of calf thymus DNA when subjected to thermal denaturation. The extent of nonrotational hydration of the native conformation[921] as a function of temperature up to temperatures immediately below T_m at ionic strengths of 0.02, 0.15, and 1.0 in NaCl solutions, together with the nonrotational hydration values of the denatured form w_a, are listed in Table XIX. At 20° the values of w_n for DNA in 0.15 and 1.0 mol dm^{-3} solutions of NaCl are identical at 8.85%, but for DNA in 0.02 mol dm^{-3} NaCl, w_n becomes 11.25%; in all cases an increase in temperature leads to dehydration. The denatured form of DNA on cooling shows an increasing tendency to hydration, but to a much smaller extent than the native conformation.

The authors interpret their data as evidence for the capacity of Na$^+$ ions, present in solution at the higher ionic strengths, to bind water molecules liberated from the DNA at elevated temperature. The ratio of w_d/w_n is apparently independent of both temperature and ionic strength, being constant at 0.42, a fact which the authors suggest is evidence for the involvement of the water molecules rotationally bound to DNA in the structural stabilization of secondary and tertiary structure of the macromolecule.

There are also reports of the rotational correlation times τ_c (defined as the mean time during which the rotating molecule changes its position over an angle of $\sqrt{\frac{2}{3}}$ rad, calculated from T_1 and T_2).[921] The molecular rotation of DNA in the native conformation at 20° and at the higher ionic strength of 0.15 and 1.0 mol dm^{-3} NaCl is described by $\tau_c = 1.0$–1.2×10^{-8} sec, *reducing* slightly with increasing temperature below T_m; for the denatured form at 20°, τ_c is approximately 1.5 times larger. The τ_c of native DNA in 0.02 mol dm^{-3} NaCl at 20° was, however, lower ($\tau_c = 2.23 \times 10^{-8}$ sec) but *increased* rapidly with temperature to a maximum at 65°, approximately 10° below T_m of DNA at this ionic strength. The denatured form showed a broadly similar pattern of behavior, with τ_c decreasing from 3.34×10^{-8} sec at 20° to 0.87×10^{-8} sec at 60°. Both the native and denatured forms of DNA then show marked increases in τ as the temperature approaches T_m; a marked slowing down in the rotation of the macromolecule as T_m is approached is shown at all ionic strengths investigated.

The authors point out that only in the 0.02 mol dm^{-3} NaCl solution is the theoretically expected decrease in τ_c observed at temperatures of

TABLE XIX. Results of Calculation of Nonrotational DNA Hydration Depending on the Temperature and NaCl Concentration[a]

Temperature, °C	DNA in 0.02 M NaCl			DNA in 0.15 M NaCl			DNA in 1.0 M NaCl		
	w_n, %	w_d, %	w_d/w_n	w_n, %	w_d, %	w_d/w_n	w_n, %	w_d, %	w_d/w_n
20	11.25	5.50	0.489	8.85	3.05	0.343	8.85	3.56	0.403
25	9.25	4.28	0.463	7.70	2.36	0.325	5.505	2.50	0.435
30	7.85	3.90	0.498	6.50	2.01	0.310	3.90	1.92	0.490
35	7.70	3.48	0.452	5.60	1.93	0.344	3.25	1.44	0.448
40	6.70	3.18	0.468	4.90	1.92	0.390	2.70	1.21	0.449
45	6.27	2.92	0.466	4.55	1.82	0.400	2.44	1.07	0.438
50	6.00	2.60	0.435	4.20	1.73	0.412	2.19	0.92	0.420
55	5.65	2.60	0.440	3.80	1.63	0.428	2.03	0.81	0.400
60	5.50	2.20	0.400	3.62	1.56	0.430	1.94	0.82	0.422
65	5.25	1.41	0.268	3.48	1.50	0.432	1.89	0.82	0.434
67.5	3.15	0.645	0.205	—	—	—	—	—	—
70	1.55	0.395	0.255	3.38	1.38	0.407	1.80	0.83	0.461
72.5	0.375	—	—	—	—	—	—	—	—
75.0	—	—	—	3.10	1.26	0.407	1.70	0.79	0.466
80	—	—	—	2.54	0.91	0.338	1.62	0.814	0.505
82	—	—	—	1.52	0.65	0.428	—	—	—
84	—	—	—	0.86	0.40	0.465	1.56	0.796	0.502
86	—	—	—	0.324	—	—	1.33	0.722	0.536
88	—	—	—	—	—	—	1.15	0.548	0.415
90	—	—	—	—	—	—	0.883	0.372	0.462
92	—	—	—	—	—	—	0.560	0.250	0.592
94	—	—	—	—	—	—	0.320	0.168	0.525
96	—	—	—	—	—	—	0.232	—	—
98	—	—	—	—	—	—	0.141	—	—

[a] w_n, hydration of native DNA; w_d hydration of denatured DNA. Data from Ref. 921.

20–65°, attributable to an increase in the rate of molecular rotation arising from an increasing thermal energy contribution. The reason is advanced that in solutions of higher ionic strength the thermal energy is used almost completely for dehydration of the DNA molecules and, in the 0.02 mol dm^{-3} NaCl solution, the strength of the water dipoles bound to DNA is relatively lower than at the higher salt concentrations, presumably because the water is bound by a different mechanism.

A similar ionic strength effect upon the size of viral DNA has been reported[849]; diffusion-controlled adsorption of DNA from bulk solution of known ionic strength onto a monolayer of surface denatured cytochrome C was used to characterize, by electron microscopy, the size and shape of DNA molecules from bacteriophage T3 and T1 and bovine papilloma virus. The salts used in the investigation were NaClO$_3$, NaCl, and NH$_4$OAc; in each case the contour length of the DNA molecule was constant over a range of ionic strength from 0.1 to 0.5, but below an ionic strength of 0.1 the contour length increased by up to 20% with decreasing ionic strength.

The increase in contour length has been attributed to a partial unwinding of the B conformation of the macromolecule, possibly as a preliminary step to the denaturation of DNA in salt-free water, said to have been observed by Inman and Jordan.[711] It is significant that similar ionic strength effects have been observed in proteins,[376,828a] the changes in size and degree of hydration of the macromolecule apparently occurring over the same range of ionic strength; the fact that this ionic strength lies close to the osmotic value of the vertebrate species is perhaps not without significance.

Hydrodynamic measurements of the extent of hydration associated with the macromolecule, in contrast with nmr measurements, appear to be much more sensitive to the outer layers of water molecules, which probably migrate with the macromolecule through the solution. Since the outer H$_2$O molecules of this region will exchange rapidly with the bulk water of the solution, it is unlikely that this water can be detected by techniques such as nmr.

Hearst and co-workers[563,634] have determined an effective hydrodynamic radius for the hydrated DNA macromolecule from the variation of the sedimentation coefficient with molecular weight, and obtained a value of 120–130 nm, which they attribute to a layer of water molecules surrounding the helix. Tunis and Hearst,[1501] however, subsequently pointed out that this is not a particularly satisfactory method for determination of hydration envelopes and, in essence, only the approximations used enable a reasonable agreement to be obtained with earlier self-diffusion coefficient measurements of H$_2$O in DNA solutions.[1555]

These latter are still regarded as among the most definitive results on the hydration of DNA, indicating about 0.35 g H_2O/g NaDNA, or approximately 7 mol of water per nucleotide, which is of the same order of magnitude as that suggested by Falk.[421]

Thermodynamic equilibrium measurements appear to have attractive possibilities for the determination of hydration values of macromolecules, and Hearst and co-workers have published a succession of reports on the hydration characteristics of DNA, based on density gradient ultracentrifugation to determine the buoyant density of the macromolecule in a variety of salt solutions. One of the earlier reports[635] deals with measurements of the degree of hydration of salmon sperm DNA in density gradients of various cesium salts and of mixtures of lithium bromide and cesium bromide. The net hydration of the DNA macromolecule was a monotonic function of the water activity and apparently independent of the nature of the cation.

A subsequent report[633] covers the effect of anions on salmon sperm DNA. In addition to NaCl and Na_2SO_4, the effect of $NaClO_4$ was investigated, since Hamaguchi and Geiduschek[594] had reported that DNA has a lower range of melting points in solutions of sodium perchlorate than in sodium chloride, attributable to the water-structure-breaking of the ClO_4^- ion. Figure 17 illustrates the results for all three salts, the solid curve being that obtained by Hearst and Vinograd[635] for different cesium salts. Hearst also determined the maximum contribution w_{max} to the observed value of hydration w' to be expected from electrostatic interactions, suggesting that w_{max}/w' is the fraction of w' that, at most, is attributable

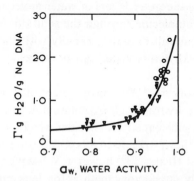

Fig. 17. Net hydration of sodium deoxyribonucleate in various salt solutions; ($\triangledown$) NaCl, ($\bigcirc$) Na_2SO_4, ($\bullet$) $NaClO_4$. The data points were determined by an isopiestic technique, the solid line was calculated from buoyant densities in density gradients of various cesium salts in the ultracentrifuge.

TABLE XX. Net Hydration of DNA and Estimates of the Relative Importance of the Electrostatic Contribution to the Net Hydration[a]

Salt	Water activity	Concentration of salt, mol dm^{-3}	w_{max}	w'	w_{max}/w'
NaCl	0.80	5.2	0.164	0.40	0.41
	0.85	4.1	0.238	0.47	0.50
	0.90	2.9	0.402	0.60	0.67
	0.95	1.5	0.904	1.20	0.75
NaClO$_4$	0.92	2.4	0.58	0.75	0.77
	0.95	1.5	0.95	1.10	0.86
	0.98	0.7	2.43	1.70	1.43
Na$_2$SO$_4$	0.95	1.6	0.95	1.50	0.63
	0.97	0.9	1.49	1.75	0.85
	0.98	0.5	2.26	2.40	0.94

[a] Data from Ref. 633.

to long-range electrostatic interactions. The data, tabulated in Table XX, show that a large fraction of the observed net hydration arises from short-range interaction between the DNA and the solvent and appears to be insensitive to the nature of the anion.

Tunis and Hearst[1501] later pointed out that, on the evidence of Liquori and co-workers,[56,276] the charge density on the surface of the DNA molecule decreases on denaturation, which should result in a decrease in hydration due to the long-range electrostatic interaction. The authors therefore investigated the buoyant density of a denatured and native bacteriophage T-7 DNA in a density gradient of potassium trifluoroacetate; this salt was chosen because T_m for DNA is much lower in this salt than in others used in density gradient experiments. Assuming that the partial specific volume of DNA does not change on denaturation, the change in hydration upon denaturation was obtained; in addition, the change in hydration with temperature and the heat of hydration for native and denatured DNA were determined by the method of Vinograd et al.[1533] Table XXI gives these values for native and denatured DNA in potassium trifluoroacetate adjusted to a water activity of 0.70.

By assuming that the denaturation of DNA can be considered as a two-state process, the authors applied a Clausius–Clapeyron equation to determine the enthalpy of denaturation of DNA in potassium trifluoroacetate, obtaining a value of 17.6 kJ MBP^{-1} at $a_w = 0.7$ and $T_m = 54.5°$,

TABLE XXI. Results of Temperature-Dependent Buoyant Density Studies on Native and Denatured DNA Potassium Trifluoroacetate Solution[a]

DNA	ΔH_{hyd}, kJ mol^{-1} H$_2$O	$(w'/T)_{a_w}$, mol H$_2$O/mol nucleotide °C	w', mol H$_2$O/mol nucleotide	
			25°	54.5°
Native	1.67	−0.024	6.9	6.2
Denatured	1.43	−0.015	5.3	4.9
Three DNA's in CsCl[b]	1.72	−0.028	6.6–7.4	—

[a] Data from Ref. 1501.
[b] Data of Vinograd et al.[1533]

the value at 25° being approximately the same. The magnitude of ΔH is approximately half of that determined by other workers,[209,1196] but under very different experimental conditions, since a water activity of 0.7 corresponds to a concentration of potassium trifluoroacetate of approximately 5 mol dm^{-3}.

On the basis of this evidence it is suggested that trifluoroacetate destabilizes the DNA helix by removal of water, leading to denaturation, but that the nature of the destabilization is not easily apparent; specific interaction between the TFA$^-$ ion and the bases in denatured DNA would lower ΔH, the measured hydration of the denatured DNA would be lowered, and the difference in hydration between the native and denatured form, w', would be greater. The capacity of the salt to inhibit recovery of secondary structure below T_m supports this interpretation, but the large value of $\Delta w'$ appears to be due to extensive hydration of the native form rather than marked loss of hydration of the denatured form; this is incompatible with the above line of reasoning.

It is well known that variations in the G–C content of a polynucleotide will influence the extent of base stacking in the macromolecule, and it is therefore of particular interest that Tunis and Hearst[1502] have observed a dependence of the net hydration of DNA upon the base composition of the macromolecule. They repeated the density gradient experiments of Hearst and Vinograd[635] in a series of cesium salt solutions and solutions containing mixtures of LiBr and CsBr, using *E. coli* DNA (50% G–C), *M. lysodeikticus* DNA (72% G–C), and *B. megatherium* DNA (38% G–C). The calculated degrees of hydration of the polymers are shown in Fig. 18.

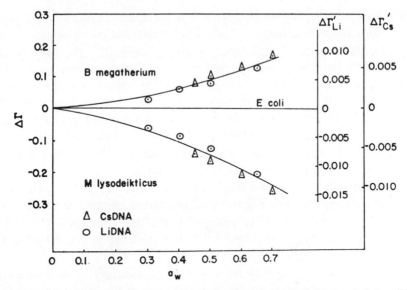

Fig. 18. Differential hydration of *B. megatherium* DNA (38% GC) *M. lysodeikticus* (72% GC) compared with *E. coli* DNA (50% GC) (see Ref. 1502).

The hydration differences are almost the same for LiDNA and CsDNA on the basis of mol of water per mol of DNA and the data show that an A–T pair binds on average two more water molecules than a G–C pair.

The authors are uncertain whether the differences in differential hydration of LiDNA and CsDNA are real, but this does not obscure the importance of the result that increasing G–C content results in a decrease in the extent of hydration of the macromolecule. This evidence provides strong support for proposals[302,1496] that hydrophobic bonding plays a considerable role in stabilizing the helical structure of the native helical conformation of DNA, but is in disagreement with the suggestion of Lewin[885] that an increased proportion of G–C groups increases the number of water bridges on the macromolecule and hence the stability of the helix conformation.

Density increments in solutions of calf thymus DNA in NaCl and CsCl over a wide range of concentration indicate that the partial specific volume of NaDNA at 25° is 0.500 ml g^{-1} in pure water and that of CsDNA is 0.440 ml g^{-1}.[265] Values of the net hydration of NaDNA and CsDNA as a function of water activity, derived from the density increments and partial specific volumes, are in good agreement with those reported by Hearst and Vinograd.[635]

Vijayendran and Vold[1528] have recently carried out buoyant density determinations of T-4 DNA by equilibrium sedimentation in a density

gradient of mixed solutions of cesium and magnesium chlorides and bromides. The authors selected the Mg^{2+} ion for detailed study because of its
acknowledged biological importance in cell replication,[635] the stabilization
of the double helix by phosphate binding ions,[385] the magnesium requirement for enzymatic synthesis of DNA,[801] and the important role attributed
to magnesium in holding the ribosomal nucleoprotein complex intact.[1480]
Mixed salt solutions were used in this work because attempts to produce
DNA bands in $MgCl_2$ and MgBr solutions were unsuccessful; the high
solution density required could not be obtained and the resulting density
gradient was too narrow to support stable bands.

The net hydration w', expressed as mol of water per mol of nucleotide,
is plotted against the water activity in the mixed electrolyte solution in
Fig. 19; w' is calculated on the assumption that the complex is $(Mg^{2+}/2)$-
DNA rather than Mg^{2+}-DNA.[362,936,1152] The solid line in Fig. 19 refers to
the data for uni-univalent electrolytes.[1501]

It is apparent that, as with the uni-univalent electrolytes, the hydration
of DNA in the presence of the divalent Mg^{2+} ions is dependent only on the
thermodynamic activity of the water in the solution, and is independent
of the nature of both cation and anion.

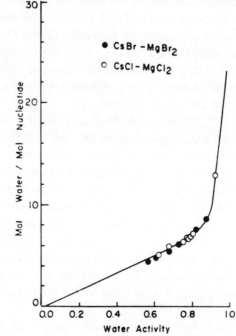

Fig. 19. Hydration of T-4 DNA as a function of water activity (see Ref. 1528.)

Further, the fact that the hydration of DNA at a given water activity is not sensitive to the ratio Cs^+/Mg^{2+} shows that the intrinsic solvation of the counterion is not important; possibly the observed buoyant density data may be explained on the basis that they relate to the solvated DNA plus the desolvated counterions in the Stern region of the electrical double layer surrounding the macromolecule.[1424]

Vijayendran and Vold have also reported sedimentation coefficients of calf thymus and T-7 DNA in solutions of $MgCl_2$, $MgBr_2$, and $MgSO_4$.[1529] A plot of the product of the relative viscosity and sedimentation coefficient against the density of the solution is found to be linear only in the case of $MgSO_4$; the zero intercept occurs at a density of 1.41 cm^{-3}, which corresponds to a water activity of 0.89. The net hydration determined from this value is 10.5 mol water/mol nucleotide, in good agreement with earlier data[1555] for uni-univalent electrolytes.

In the previous section the intrinsic advantages of calorimetric measurements for determining hydration values for macromolecules were outlined; Privalov and Mrevlishvili,[1198] as part of their extensive calorimetric investigations of the hydration of macromolecules, showed that native DNA in aqueous solution does not show any appreciable heat absorption on being heated from $-35°$ to $0°$ until water in excess of 0.5 g/g of dry DNA was present, and even at water contents as high as 2.0 g H_2O/g dry DNA the absorption peak does not occur at $0°$. The authors suggest that, as in the case of proteins, the heat absorption peak appears when the water contains modified structures that may be frozen and unfrozen by changes in temperature; the diffuse state of the peak and its displacement to temperatures below $0°C$ are interpreted as a strong influence on the macromolecule upon the water–water interactions, apparently extending in an attenuated manner to water contents as high as 2.0 g H_2O/g DNA.

It is clear that the shape of the fusion curve for water is markedly dependent upon the nature of the macromolecule. Figure 20 illustrates the dependence of the heat capacity of DNA on the amount of water present; it is clear that fusion begins at lower temperatures than for the proteins, even for tropocollagen. This effect is attributed to the concentration of charged groups, particularly phosphate groups, on the surface of the molecule, and the amount of bound water, in this case 0.61 g/g DNA, is determined from the observed area of the peak. An amount of 0.61 g of water corresponds to 11.6 mol/mol of macromolecule and to a hydration layer of 40 nm thickness.

It is interesting to note that a corresponding experiment for DNA dissolved in 0.15 mol dm^{-3} NaCl produced a lower hydration value of

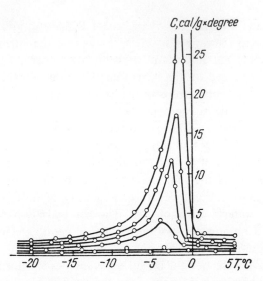

Fig. 20. Temperature dependence of the heat capacity of 1 g DNA in the presence of different amounts of water (0.05, 0.75, 1.00, and 2.00 g.) (See Ref. 1198).

0.526 g H_2O/g DNA, or 10.5 molecules of water/molecule of nucleotide, a trend also observed by Lubas and Wilczok[922] from nmr measurements. The apparent discrepancy in the degrees of magnitude of hydration suggests that calorimetric studies are capable of distinguishing between the first and subsequent hydration layers, whereas the nmr method observes only the innermost hydration region.

It has also been found that denatured DNA is hydrated to a slightly greater extent, 0.645 g H_2O/g DNA, and in 0.15 mol dm^{-3} NaCl, 0.584 g H_2O/g DNA; this is at variance with the results of Hearst and co-workers[563] and Lubas and Wilczok,[922] although it should be remembered that the density gradient data of Hearst refer to high salt concentrations ($\sim$5 mol dm^{-3}) and therefore the data are not easily compared.

Chattoraj and Bull[247] have recently conducted a calorimetric study of water–DNA interactions. Lithium, potassium, cesium, and magnesium salts of calf thymus DNA were added to water to form a gel which was then cooled to -3 or $-16°$, the amount of heat required to raise the temperature to $25.5°$ being subsequently measured; the water in the gels supercooled and no ice was produced down to a bath temperature of $-8°$. The samples were then frozen in a bath of dry ice and acetone before being allowed to equilibrate at $-3°$ or $-16°$, followed by heating to $25.5°$. The difference in heat supplied to the two samples represents the heat required to melt the ice in the samples and hence the amount of unfrozen or "bound"

TABLE XXII. Moles of Water Remaining Unfrozen in Association with Various Salts of DNA[a]

System	Bath temperature, °C	Mol H_2O unfrozen/mol nucleotide
Na DNA	−3	29.4 ± 1.3
Na DNA	−6	24.1 ± 0.9
Na DNA	−9	21.0 ± 1.2
Na DNA	−16	21.4 ± 1.6
Li DNA	−6	30.4
K DNA	−6	22.1
Cs DNA	−6	13.6
Mg DNA	−6	16.7

[a] Data from Ref. 247.

water associated with the DNA salt. Table XXII contains the values for the unfrozen water associated with Li-DNA, K-DNA, Cs-DNA, and Mg-DNA.

The authors also report that the hydration of the nucleotide is significantly increased if a concentrated solution of an inorganic salt of common cation is used instead of water.

The apparent heat capacity of the aqueous DNA gel increases rapidly with increasing water content, up to 50% by weight, the increase becoming less marked at higher water contents. The sharp rise at low water contents is attributable to strong interaction of water with dry DNA and the shallower rise at higher water contents is taken as indicating an interaction between water and the DNA gel, possibly via the metal cation.

On the evidence summarized in Table XXII it is suggested that the extent of hydration of the DNA is directly related to the nature of the cation associated with the DNA macromolecule, increasing with increasing cation hydration. This completely contradicts the observations of Hearst and co-workers,[376,563,635,919,1555] which are confirmed by Vijayendran and Vold,[1529] but the authors do not comment upon the discrepancy; in addition Chattoraj and Bull extrapolate their salt hydration data to zero salt concentration, when the DNA gel is theoretically free of cations, and find that 4.5, 12.5, 9.0, and 4.5 mol water/mol nucleotide remains unfrozen for Li-DNA, Na-DNA, K-DNA, and Cs-DNA, respectively. Thus the curious anomaly arises that the most heavily hydrated cation produces the least hydrated macromolecule, although at high salt contents the reverse is true.

In support of Chattoraj and Bull, it should be noted that their data indicate that at 11° approximately nine molecules of water interact strongly with Na-DNA, which closely parallels Falk's evidence[421] that a relative humidity of 75–80% is required for DNA to retain its native helical structure and, at approximately 80% relative humidity, all the specific hydration sites on the macromolecule are filled and approximately 10 mol of water bound per nucleotide. It is difficult, however, to find support for their suggestion that the hydration at higher water contents, e.g., when 50 molecules of water per molecule of nucleotide are present in the gel, (in agreement with Privalov and Mrevlishvili) involves the metal cation.

Although many pitfalls remain, it is perhaps now possible to sketch in fairly bold outline the regions of particular interest from the viewpoint of hydration interactions. First, it seems certain that an ionic strength of 0.1–0.15 in a solution of nucleotide or protein has particular significance for both structure and hydration of the macromolecule, and this value appears to correspond to the osmotic value of vertebrate species. Just why this is the case remains a matter for speculation; it is, however, not simply a straightforward ionic strength effect,[376] although it seems likely to arise from competing effects. It is clear, nevertheless, that at lower ionic strengths a large increase in the degree of hydration results.

The infrared[421–424] and nmr data[919] appear to be in agreement; at 3% relative humidity Falk suggests that approximately two water molecules are held per group. The fact that this decreases on denaturation (in contrast to other experimental evidence on denaturation) is in agreement with the proposals[56,276] that charge density decreases on denaturation and hence the hydration resulting from this cause should also show a decrease.

The density gradient and sedimentation coefficient data[563,633–635,1501, 1528,1529,1555] appear to be sensitive to the first hydration layer of approximately ten water molecules which are said to fill the specific hydration sites on the macromolecule at approximately 80% relative humidity[421–424] and are essential for the stability of the helix. This water is not held by the charged phosphate groups, but is subject to other interactions; it should not be forgotten that Hearst's data[1502] on the dependence of hydration on G–C content strongly implicate hydrophobic bonding in this hydration region, in contrast to Lewin's proposals[885] of increased water bridging.

Calorimetric data[247,1198] clearly distinguish a hydration region which corresponds to that observed by Falk, Hearst, and Vold and Wang, i.e., the first hydration shell of approximately ten water molecules; the technique appears, however, to be sensitive to a secondary, more extensive hydration region, which only the infrared technique seems capable of detecting, at least in part.

It seems logical, on the evidence of Hearst and Vold, that the first hydration shell, which is said to be of about 50 nm thickness, is involved in the Stern region[1424] of the electric double layer surrounding the molecule. The second hydration layer, on the basis of the infrared data, must contain water which is very similar to normal bulk water; this is potentially the most interesting hydration region, but it is also the most difficult to examine from an experimental point of view.

The evidence of Hearst *et al.* on the dependence of hydration of the macromolecule on the G–C content suggests that the increased hydration observed on denaturation may be associated with an increase in hydrophobic hydration; in this context it should be of considerable interest to study the temperature dependence of hydration as a function of G–C content of the macromolecule.

4. SOLVENT STRUCTURE EFFECTS ON MACROMOLECULAR STRUCTURE, CONFORMATIONAL STABILITY, AND TRANSITIONS IN THE PRESENCE OF NEUTRAL ELECTROLYTES

4.1. The Nature of the Neutral Electrolyte Effect on the Structure and Conformational Stability of the Macromolecule

The observed effect of a neutral electrolyte present in an aqueous solution of a macromolecule, such as a protein, upon the structure or conformational stability of the protein is the sum of several differing and perhaps competing interactions, such as protein–solvent, protein–electrolyte, and electrolyte–solvent interactions.

The effects can be arbitrarily divided into two major classes; the first is due to "direct interactions" of the ions with specific charged groups on the macromolecule. The literature dealing with this subject is somewhat confused and has led to a wide range of values for the degree of "direct interaction" or "site binding," but in essence the predominant factor is the magnitude of the charge on the ions. The second class of effects arises primarily from an indirect process whereby the ions affect the macromolecule through their modification of the solvent environment of the macromolecule.

The two effects outlined may be regarded as extremes of behavior because one is unlikely to achieve either type of interaction in isolation of the other; it should be permissible, however, to discuss an observed effect as being a combination with one or other of the types of interaction predominating.

The manner in which the electrolyte ions modify the aqueous environment of the macromolecule is markedly dependent upon the structure and constituents of the ions; for example, salts such as the alkali metal halides modify the environment mainly by polar forces, monopole–monopole and monopole–dipole. Salts such as the tetraalkylammonium halides can modify the aqueous environment by polar forces but they also interact with the solvent via the nonpolar hydrocarbon residues present on the ion, i.e., hydrophobic interactions. Salts such as guanidinium chloride, on the other hand, may interact with the solvent by both the forementioned effects and, in addition, by direct hydrogen bonding.

The observed result of the combination of such effects on the structure of the macromolecule, e.g., conformational transitions of a protein or nucleic acid, or aggregation of the species, has led to the construction of series in which ions are listed in order of efficiency in promoting the given result. The data of Bull and Breeze[202] on the extent of water binding by egg albumin in the presence of various ions is an example of such a series; the extent of hydration was found to increase in the order $Li^+ < Na^+ < K^+ < Rb^+ < Cs^+$. Similarly, Dix and Strauss[352] have found the stability of DNA to thermal denaturation to vary with the species of counterion present in the order Sr^{2+}, Ba^{2+}, $Mg^{2+} > Ca^{2+} >$ guanidinium$^+ > Li^+ > NH_4^+ > Cs^+ > Na^+ > K^+$. Ross and Scruggs[1275] have found from electrophoretic measurements that counterions bind to DNA in an order of decreasing effectiveness of $Li^+ > Na^+ > K^+ > (CH_3)_4N^+$, and von Hippel and Wong[1539] reported the effect of various ions on the formation and stability of the collagen fold in gelatin to be in the order $(CH_3)_4N^+ < NH_4^+ < Rb^+ < Na^+ < K^+ < Cs^+ < Li^+ < Ca^{2+}$, and a similar series involving anions is of the form $SO_4^- < CH_3COO^- < Cl^- < Br^- < NO_3^- < ClO_4^- < I^- < SCN^-$.

This ability of ions to act independently of each other in regard to their effect upon the stability and conformation of the macromolecule was first studied quantitatively by Hofmeister[672] in his investigation of the efficiency of various salts in precipitating or salting-out euglobins from aqueous solution. Series of this kind, characterizing the effectiveness of ions upon some property of the macromolecule, are now known as Hofmeister, or sometimes lyotropic, series.

It is pertinent at this juncture to examine more closely the effect of concentration of the neutral electrolyte upon the macromolecule; data reviewed in the previous section upon the extent of hydration of macromolecules[830] have shown that a sharp change occurs at electrolyte concentrations less than 0.1–0.15 mol dm^{-3}, compared to the extent at higher electrolyte concentrations. Similarly, the degree of conformation of proteins[376] and DNA[922] undergoes a marked change over this concentration region.

It is singularly unfortunate that this difference in behavior as a function of ionic strength at low electrolyte concentrations has passed unnoticed by some authors, to such an extent, in fact, that test solutions often contain 0.2–0.3 mol dm^{-3} of electrolyte, usually as part of a buffer system. This concentration of electrolyte has then been ignored as a factor in subsequent discussion of the data; thus, many workers have tended to regard such an electrolyte concentration as negligible and have considered the solutions as salt-free. The data reviewed here indicate that this is unlikely to be the case, and conclusions drawn about the behavior of macromolecules in such solutions should be regarded with some caution.

Several workers have referred to interactions involving electrolyte ions with a macromolecule as being "binding" processes with various degrees of binding, but the whole picture of electrolyte concentration and binding appears to be somewhat confused. The experimental facts that are now available should enable us to form a rather clearer picture of the situation. The facts are that a neutral uni-univalent electrolyte in solution will promote an observable change in properties of the macromolecule at a concentration of ∼1.0–1.5 mol dm^{-3} irrespective of whether the macromolecule is a protein of fairly low charge or a relatively highly charged molecule such as DNA. This effect is also irrespective of the nature of the counterion, providing it is univalent; divalent ions, such as Mg^{2+} ions, are found to be orders of magnitude more effective in promoting change than singly charged ions such as Na^{+}.[432] This large difference between univalent and divalent ions, together with the lack of differentiation between univalent ions, provides the clue to the probable origin of the effect.

Macromolecules such as proteins and nucleic acids are particles of colloidal size in their own right. Charge effects on colloidal particles give rise to an electric double layer which is largely responsible for the colloidal stability of the particle in the case of hydrophobic particles, but only partly so in the case of hydrophilic particles, such as proteins. A more detailed picture of the structure of the double layer will be discussed in Volume 5; it is sufficient at this stage to point out that in the case of hydrophobic particles, i.e., particles for which stability is in large measure a function of the thickness of the electric double layer, the difference in flocculating power between univalent and divalent ions upon the colloidal particle is in the ratio of the reciprocal of the valence raised to the sixth power. This means that a divalent ion is approximately 100 times more effective than a monovalent ion in promoting flocculation of a hydrophobic colloid; the electrophoretic studies of Ross and Scruggs[1275] indicate that a similar power law relationship is observed with hydrophilic colloids such as the proteins, leading not necessarily to flocculation but to more subtle changes in the stability of the system.

Flocculation, in the case of the hydrophobic colloids, occurs when the electric double layer is effectively suppressed to less than 1.0 nm by the concentration of neutral electrolyte; the thickness of the double layer is also a function of the valence of the counterion and is independent of the charge on the colloidal particle. For a neutral uni-univalent electrolyte the thickness of the particle double layer is found to be of the order of 1.0 nm at an electrolyte concentration of 0.1 mol dm^{-3}. Flocculation does not occur in the case of hydrophilic colloids such as the proteins and nucleic acids, because of the marked hydration of the particle and its subsequent stabilization. It does, however, mean that intercharge repulsion between charge centers on the macromolecule will be at a minimum and interstrand repulsion, in the case of duplex molecules such as DNA and triplex molecules such as collagen, will also be at a minimum.

Concentrations of neutral electrolyte lower than 0.1 mol dm^{-3} in the case of hydrophobic colloids, which tend to be compact in shape, result in a double layer of increased thickness and hence increased stability with regard to flocculation for the colloidal particle. Hydrophilic colloids, however, produce a situation which is much more complex; the macromolecule, unlike the hydrophobic particle, is long and threadlike and, as the helix–coil transition shows, is capable of marked changes in conformational order. Lowering of the electrolyte concentration leads to increased thickness of the double layer, as in the case of hydrophobic particles, but additionally, because of the internal mobility of the macromolecular chain, to repulsion between adjacent charged groups and resulting changes in conformation. Such changes in conformation, since they may lead to the exposure of previously buried hydrophobic or hydrophilic groups, may also lead to changes in hydration of the macromolecule.

Changes in the hydration and conformation of the macromolecule at neutral electrolyte concentrations in excess of 0.1 mol dm^{-3} do not involve changes due to variation of the thickness of the double layer, since this is effectively suppressed. Such changes are most likely to occur because of interactions of the ions with the solvent and exchange between ions in the solvent and ions more firmly held by the macromolecule.

4.2. The Salt Effect on Deoxyribonucleic Acid

The addition of neutral electrolyte tends generally to stabilize the helical form of DNA, within an upper electrolyte concentration limit of ~ 0.1 mol dm^{-3}. This effect might be expected since the molecule is highly charged with a large number of charge groups, i.e., the phosphate groups, being of the same sign and therefore likely to repel each other strongly, leading to destabilization of the helical form.

The effect of uni-univalent electrolytes on the helix–coil transition of a variety of DNA's and polynucleotides has been investigated by several workers.[362,482,1128,1324] Several empirical equations have been produced which demonstrate a dependence of the temperature of the midpoint in the helix–coil transition, i.e., melting temperature T_m, upon the concentration of NaCl in the solution and the base composition of the DNA. Figure 21, taken from the work of Frank-Kamenetskii,[482] illustrates the dependence of T_m upon the logarithm of the sodium ion concentration. The empirical equation corresponding to the straight line in Fig. 21 is of the form

$$T_m = 176 - (2.6 - x_0)(36 - 7.04 \log C_s)$$

where C_s is the concentration of uni-univalent electrolyte in mol dm^{-3} and x_0 is the fraction of G–C base pairs in the particular DNA. The upper concentration limit of this equation is of the order of 0.1 mol dm^{-3}; at higher concentration of neutral electrolyte a different type of dependence arises.

Several workers[808,1117,1199,1202,1324] have derived equations involving a dependence of T_m upon $\log C_s$ from a theoretical model; one of the most recent reports, which is shown to correlate well with those of previous workers, is that of Manning,[979] based upon an application of polyelectrolyte theory to the helix–coil transition of DNA in the low electrolyte region. The model considers the effect of electrolyte on melting of DNA within this region in terms of the effect of a change in the thickness of the electric double layer surrounding the macromolecule and the effect of ions that can be regarded as bound, but not necessarily site-bound, to the macromolecule. The distinction here is essentially between ions that are regarded as being in the diffuse or Gouy–Chapman region of the double layer and

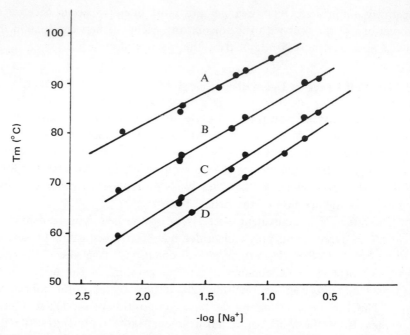

Fig. 21. Dependence of DNA melting temperature T_m on log[Na$^+$] for: (A) *M. lysodeikticus* ($x_0 = 0.72$); (B) *E. coli* ($x_0 = 0.50$); (C) *S. saprophyticus* ($x_0 = 0.33$); (D) *M. myaides var. capri* ($x_0 = 0.24$). Straight lines computed from the empirical equation of Frank-Kamenetskii.[482]

those in the fixed or Stern region of the electric double layer. An equation of the form

$$G_c = N_p \mu_c + (N_{AB} + \xi_c^{-1} N_p)\mu_{A+} + N_{AB}\mu_{B-} + N_{H_2O}\mu_{H_2O}^{\circ}$$

is proposed for the coiled form of DNA, where N_p is the total number of phosphate groups (or bases), μ_c is the chemical potential per base of the coil, N_{AB} is the number of salt "molecules" present, μ_{A+} and μ_{B-} are the chemical potentials of the particular ions, N_{H_2O} is the total number of water molecules present in the system, and $\mu_{H_2O}^{\circ}$ is the chemical potential of pure water. ξ is the linear charge density assuming that DNA is a linear polyelectrolyte modeled by a uniform and continuous line charge, given by the equation

$$\xi = e^2/\varepsilon k T b$$

where e is the electronic charge, ε is the bulk permittivity, k is the Boltzmann constant, T is the absolute temperature, and b is the average spacing of the projection of charged groups onto the axis of the fully extended polyion (0.17 nm for the helical form and 0.68 nm for the coil form of DNA).

A similar equation is derived for the helical conformation and the difference between the two yields ΔG, the total free energy change for the helix–coil transition,

$$\Delta G = \Delta\mu + \eta\,\Delta\mu_{A+}$$

where $\Delta\mu = \mu_c - \mu_h$ and $\eta = \xi^{-1} - \xi_h^{-1}$. An imaginary transition between uncharged forms of the coil and helix provides a reference state, so that

$$\Delta G = \Delta\mu^\circ(T) = \Delta\mu' + \eta\mu_{A+}$$

where $\Delta\mu = \Delta\mu_1' + \Delta\mu_2'$, with $\Delta\mu_1'$ the interaction of the polyion with its electric double layer and $\Delta\mu_2'$ any other contributions. $\Delta\mu_1$ is related to a parameter by the equation

$$\Delta\mu_1' = -\eta kT \ln \alpha$$

where α is described as the Debye screening parameter given by

$$\varkappa^2 = (4\pi e^2/\varepsilon kT)(2n_{A+})$$

where n_{A+} is the concentration of A^+ in molecules cm^{-3} and $1/\varkappa$ is the thickness of the double layer. Since μ_{A+} can be written in the form

$$\mu_{A+} = \mu_{A+}^*(T) + kT \ln(n_{A+}/n_{H_2O})$$

where μ_{A+}^* is independent of solute concentration and n_{H_2O} is the solvent concentration in molecules cm^{-3}, the following equation results:

$$\Delta G = \Delta\mu^\circ(T) + \Delta\mu_2'(T) + \eta\mu_{A+}^*(T) - \eta kT \ln \varkappa + \eta kT \ln(n_{A+}/n_{H_2O})$$

At the equilibrium point of the transition between the helix and coil, $\Delta G = 0$ and the temperature is the melting temperature T_m. Dividing throughout by T_m and differentiating with respect to T_m^{-1} yields an expression relating the enthalpy change on melting to the slope of the T_m vs. log(salt concentration) curve which, per mol, becomes

$$\Delta H^\circ(T_m) = 1.15\eta RT_m^2(dT_m/d \log m_{A+})^{-1}$$

where m_{A+} is the concentration of added salt. Since the change in T_m^{-2} $\Delta H^\circ(T_m)$ is small for a given change in T_m, the slope is essentially constant.

Manning shows that this equation relates to previously published expressions[811,1117,1199] and claims that experimental data[795] are in agreement with the equation within the electrolyte concentration range ≤ 0.1 mol dm^{-3} insofar as the slope of the T_m vs. log m_{A^+} graph is independent of G–C content. The recent work of Gruenwedel and co-workers,[572] however, shows that the slope of the T_m vs. log m_{A^+} graph decreases slightly with increasing G–C content and equally important, a marked difference occurs in the slope $dT_m/d(\log m_{A^+})$ as a function of the nature of the cation; the slopes in the case of Cs$^+$, K$^+$, and Na$^+$ are similar, but the slope of the Li$^+$ salt is very much steeper.

In a detailed analysis of the contributions of the different terms to ΔG, Manning suggests that, since the ln $\varkappa$ term appears with a negative sign, addition of salt destabilizes the helix, and Manning concludes that this is due to a higher proportion of unneutralized phosphate groups in the coil conformation, i.e., a higher charge density. This interpretation is contrary to the evidence of Liquori and co-workers,[276] who have shown that the coil conformation of DNA is expected to have a lower charge density than the helix, and are supported in this by the nmr data of Lubas and Wilczok.[921]

It is possible that this discrepancy is more apparent than real; however, Manning takes the free energy contribution from the interaction of the polyion with its electric double layer as

$$\Delta\mu' = -\eta kT \ln \varkappa$$

from an earlier paper [eqn. (17) in Ref. 978] and appears to have incorrectly omitted the distance parameter ϱ from the equation; thus $\Delta\mu'$ should be given by the equation

$$\Delta\mu' = -\eta kT \ln \varkappa\varrho$$

Thus the term $\varkappa\varrho$ becomes, in the limit, the ratio of the double layer thickness to the axial radius of the DNA molecule ϱ; a $\varkappa\varrho$ value of unity, because of the small magnitude of the axial radius of the DNA molecule, means the virtual suppression of the double layer, and $\varkappa\varrho < 1$ results in the $-\ln \varkappa\varrho$ term being positive and hence indicating a stabilizing influence for the helix.

The model, as formulated, takes no account of the variation in G–C content of the DNA nor of the degree of hydration associated in various ways with such groups. It seems probable that the variation of $dT_m/d(\log m_{A^+})$ noted for differing G–C contents of the DNA and for differing salts and the poor agreement of the model with experimental data for ΔH_m are at least partially accounted for by these omissions.

It now seems clear that the linear dependence of T_m upon the log(salt concentration) reaches an upper limit at approximately 0.1 mol dm^{-3} for a uni-univalent electrolyte, when the diffuse region of the electric double layer is effectively suppressed; thus, attempts to extrapolate to higher concentrations of electrolyte, on the assumption that the contributions to ΔG from electrostatic interactions continue at these higher electrolyte concentrations,[1538] are suspect and a different model must be developed to account for the behavior of DNA in this region of electrolyte concentration.

Few experimental investigations covering a sufficiently wide range of neutral electrolyte concentration have been reported, but Gruenwedel *et al.*[572] have recently published an extended investigation of T_m of a variety of DNA's of varying G–C content as a function of neutral electrolyte concentration, $\sim$0.003 to 1–3 mol dm^{-3} for LiCl, NaCl, KCl, CsCl, Li$_2$SO$_4$, and K$_2$SO$_4$. DNA's utilized in the study were synthetic poly d-AT (% G–C = 0), T-4 bacteriophage DNA (% G–C = 34), calf thymus DNA (% G–C = 42), *Escherichia coli* DNA (% G–C = 50), and *M. lysodeikticus* DNA (% G–C = 72). The heat denaturation was followed by using an optical absorbance technique at 260 nm, and T_m and the breadth of the transition σ were observed. T_m was found to increase linearly with the log(cation concentration) up to approximately 0.1 mol dm^{-3}; at higher concentrations T_m leveled off and, in some cases, fell slightly. The slope $-dT_m/d(\log m_{A+})$ fell slightly with increasing G–C content for all the DNA's and all the salts investigated; the slope was identical, within experimental error, at each G–C content for the Cs$^+$, K$^+$, and Na$^+$ ions but was markedly greater for Li$^+$, this being the case in both chloride and sulfate solutions. Dove and Davidson,[362] however, in their earlier work could not detect any dependence of $-dT_m/d(\log m_{A+})$ upon the base content of the DNA.

LiCl and Li$_2$SO$_4$ also appear to give rise to higher values of T_m with increasing salt concentration than the other salts, in addition to the higher values of $-dT_m/d(\log m_{A+})$. Zimmer and Venner[1624] have reported that T_m decreases linearly with increasing ionic radius of the counterions in the series Li$^+ <$ Na$^+ <$ K$^+ <$ Rb$^+ <$ Cs$^+$, which is in broad agreement (with the exception of Cs$^+$) with the work of Gruenwedel *et al.* It is unlikely, however, that the increasing ionic radius is the effective factor in depressing T_m (destabilizing the helical conformation) because this implies increasing ion binding to the macromolecule; electrophoretic data[1275] show that "binding" of the alkali metal ions to DNA is in the reverse order, Li$^+ >$ Na$^+ >$ K$^+ >$ Rb$^+ >$ Cs$^+$. The data are compatible if the decrease in T_m is associated with the decreasing degree of ionic hydration in the series

TABLE XXIII. T_m and σ Values (°C) of Bacteriophage T-4 DNA (34% G–C) at High Salt Concentrations[a]

pM	LiCl		CsCl		NaCl		KCl		LiSO$_4$	
	T_m	σ	T_m	σ	T_m	σ	T_m	σ	T_m	σ
0.75	85.7	1.6	83.4	1.3	82.2	1.4	81.9	1.6	85.3	1.7
0.50	88.0	1.5	85.8	1.3	85.2	1.3	85.4	1.3	88.2	1.2
0.25	90.1	1.1	88.5	1.2	88.2	1.3	88.5	1.2	90.3	1.2
0.00	92.0	1.3	90.0	1.1	90.7	1.3	90.8	1.3	92.9	1.1
−0.25	92.1	1.0	92.0	0.9	91.5	1.1	91.3	1.0	93.8	1.0

[a] Data from Ref. 572.

$Li^+ > Na^+ > K^+ > Rb^+ > Cs^+$, particularly since the Li^+ ion is markedly more hydrated than the other ions in the series and is shown to raise T_m by about 7–10°.

The authors also observed that the degree of cooperativity of the helix–coil transition showed a similar marked dependence upon the concentration of neutral electrolyte; the variation of T_m and the breadth of the transition σ for bacteriophage T-4 DNA at high salt concentrations are shown in Table XXIII and for *M. lysodeikticus* DNA in Table XXIV. The degree of

TABLE XXIV. T_m and σ Values (°C) of *M. lysodeikticus* DNA (72% G–C) at High Salt Concentrations[a]

pM	LiCl		CsCl		NaCl		KCl		LiSO$_4$	
	T_m	σ	T_m	σ	T_m	σ	T_m	σ	T_m	σ
0.75	99.9	1.9	96.7	2.1	96.3	2.3	97.4	2.2	98.2	1.7
0.50	101.1	1.7	98.9	1.7	98.7	1.8	99.8	1.8	100.0	1.7
0.25	102.6	1.7	101.0	1.8	100.9	1.7	101.8	2.0	101.8	1.5
0.00	103.3	1.8	102.0	1.5	102.3	1.4	103.0	1.7	102.9	1.4
−0.25	102.9	1.3	102.5	1.0	103.5	1.4	102.5	1.4	102.6	1.4
−0.50	101.9	1.3	102.4	1.0	102.4	1.3	102.1	1.1	103.0	1.3

[a] Data from Ref. 572.

cooperativity is shown to decrease as the electrolyte concentration falls below approximately 0.1 mol dm^{-3}, but increases markedly at higher salt concentrations, tending to a limiting value of $\sim 1.0°$, irrespective of the G–C content of the DNA; Dove and Davidson[362] have reported σ values for *E. coli* and calf thymus DNA, at an electrolyte concentration of 3×10^{-4} mol dm^{-3}, of 6.6 and 6.3°, respectively, compared to values here of the order of 1–2° at the high salt concentrations. It is interesting to note that ϱ reveals little dependence upon the nature of the electrolyte except in the DNA of higher G–C content, for example, at an electrolyte concentration of 0.6 mol dm^{-3}, *M. lysodeikticus* DNA yields $\sigma = 1.9$ and 1.7° in LiCl and Li$_2$SO$_4$, respectively, but $\sigma = 2.1$, 2.3, and 2.2° in CsCl, NaCl, and KCl—a behavior which closely correlates with the trend in $-dT_m/d(\log m_{A+})$.

Table XXV lists the slopes of the plot of T_m vs. percentage G–C of DNA (X_{G-C}) over a range of electrolyte concentrations; the data show that little or no change in dT_m/dX_{G-C} occurs with increasing electrolyte concentration until an upper limit of 0.1 mol dm^{-3} is reached.

Since, as earlier discussion has shown, stacking interactions are the major contributor to helix stability and an increasing G–C content results in greater stability of the helix (as indicated by T_m and enthalpies of stack-

TABLE XXV. Values of dT_m/dX_{G-C} As a Function of Salt Concentration and Salt Composition[a]

pM	dT_m/dX_{G-C}, °C							
	NaCl	KCl	LiCl	CsCl	Na$_2$SO$_4$	K$_2$SO$_4$	Li$_2$SO$_4$	Cs$_2$SO$_4$
2.22	45	44	41	44	44	41	39	42
2.00	44	44	41	43	44	41	37	42
1.50	43	44	41	40	44	40	36	40
1.00	40	44	41	38	41	39	33	36
0.50	35	38	35	35	36	36	29	33
0.25	34	35	34	33	34	35	28	31
0.00	33	33	31	30	30	31	26	25
−0.25	32	31	29	29	24	—	21	19
−0.50	30	26	29	26	18	—	18	12
−0.65	28	—	—	23	11	—	—	—

[a] Data from Ref. 572.

ing[795]), it is clear that for electrolyte concentration less than 0.1 mol dm^{-3} the presence of electrolyte has little effect upon the base stacking energy due to the G–C pairs. At the same time, however, increasing electrolyte concentration (up to 0.1 mol dm^{-3}) does produce a much greater degree of cooperativity in the helix since σ narrows significantly with increasing electrolyte concentration even at high G–C content; here Li$^+$ salts seem to be especially effective in producing a higher degree of cooperativity than the other alkali metal salts.

Generally speaking, an increasing G–C content of the polynucleotide is associated with a broadening of the transition, modified, as we see, by the effect of electrolyte, which makes the data of Inman and Baldwin[710] all the more curious; these workers found that poly-d(G–C) (100% G–C) at a low electrolyte content of 0.00534 mol dm^{-3} Na$^+$ ion gives $\sigma = 1.7°$, which undergoes no appreciable change with increasing electrolyte concentration up to 0.1 mol dm^{-3}, although the expected linear dependence of T_m upon the logarithm of the Na$^+$ ion concentration is observed. Inman and Baldwin also observed that σ for poly-d(A-T) was equal to 0.4° at 0.002 mol dm^{-3} Na$^+$ ion but broadened with increasing electrolyte concentration, which was attributed by the authors to extra tightening of the helical screw by the additional electrolyte. Thus, a situation arises in which we find the two homogeneous polymers behaving in an opposite manner to that attributed to them in naturally occurring DNA. Gruenwedel and co-workers[572] find that poly-d(A–T) melts 5° below the temperature expected from extrapolation of the T_m vs. X_{G-C} plot to $X = 0$; it therefore seems that great caution should be used in utilizing data from poly-d(G–C) and poly-d(A-T) in interpreting the behavior of DNA.

It may be recalled from a previous discussion that Lang and co-workers[849] have observed a shortening of the macromolecule, presumably tightening of the helical screw, in DNA from bacteriophage T-3 and T-4 and bovine papilloma virus to an extent of 20% as the electrolyte concentration was increased to 0.1 mol dm^{-3}; thus the effect observed by Inman and Baldwin also seems to occur in natural DNA's.

A possible explanation of the behavior pattern may result if, as seems very likely, the G–C pairs exert their effect upon helix conformation predominantly via stacking interactions involving some measure of hydrophobic bonding; this can be justified because the values of σ for all the DNA's so far observed at 0.1 mol dm^{-3} of neutral electrolyte approximate to that of poly-d(G–C) pairs; the greater the degree of hydration involved in the process, the more cooperative the process might be expected to become, providing that the tightening of the helical screw is subject to some restriction. The G–C pairs in the nucleotide will effectively prevent the development of a tighter helical screw which would arise with long, uninterrupted sequences of A–T pairs. G–C pairs are associated with water molecules and the greater effect of the Li^+ ion upon σ at higher G–C contents presumably involves the greater degree of hydration of the Li^+ ion and the hydration of the G–C pairs.

It has been clearly shown that in the presence of neutral electrolyte T_m is a linear function of $\log m_{A^+}$ up to an electrolyte concentration of 0.1 mol dm^{-3}. Manning[979] has attempted to determine the standard enthalpy of transition of DNA from this basis but the values obtained, 92 kJ (mol base pair)$^{-1}$, are some three times larger than experimentally determined values.[1533] Calorimetric data[1199] on bacteriophage T-2 DNA suggest that ΔH, the enthalpy of transition, changes by approximately 5%, while T_m changes by 30% over the electrolyte concentration range up to 0.1 mol dm^{-3}; this approximation is also supported by calorimetric data[811] on the helix–coil transitions of poly A–poly U complexes. Since at T_m, $\Delta G = 0$, therefore $\Delta H = T_m \Delta S$; for this equality to hold with increasing electrolyte concentration, ΔS, the transition entropy, must increase at a relatively smaller rate than ΔH; this implies that the total increase in disorder created during the transition tends to become smaller with increasing electrolyte concentration. The data of Lang et al.,[849] showing that the degree of helicity of DNA increases with increasing electrolyte concentration, would lead to the reverse of the observed change in ΔS. Since loss of hydration also occurs during the transition, it appears that increased electrolyte concentration produces a smaller loss of hydration on melting. Such a situation is explicable if the electrolyte has a tendency to dehydrate the biopolymer, leaving less water to be lost during the melting process.

The results of Vinograd et al.[1533] suggest that A–T pairs are more heavily hydrated than G–C pairs; thus if electrolyte withdraws water from the surface of the macromolecule, leaving less to be lost on melting, ΔT_m should increase with increasing A–T content of the DNA, which is shown to be the case by the data of Gruenwedel et al.[572] Since Li^+ is the most strongly hydrated ion in the alkali metal series, it might be expected that this ion will be the most effective in removing water from the environment of the macromolecule, and hence giving the highest values of T_m, which is again an observed fact.

Such a model requires that the degree of hydration of the macromolecule should decrease with increasing electrolyte concentration; unfortunately, no data appear yet to have been published upon the extent of hydration as a function of electrolyte concentration within the concentration region below ~ 0.1 mol dm^{-3}, but nmr[830,921] and calorimetric data indicate that at an electrolyte concentration of 0.1–0.15 mol dm^{-3} a marked dehydration of the macromolecule is observed compared with its degree of hydration at very low electrolyte concentrations. In addition, Bull and Breeze[204] observed in their protein hydration studies that lithium salts were the most effective of the alkali metal salts in removing water from the macromolecule.

The essence of this model is therefore that, within the electrolyte concentration region up to ~ 0.1 mol dm^{-3} of univalent counterion, the greater the degree of hydration of the counterion, the more effective it becomes in stabilizing the DNA helix against melting by partial dehydration of the macromolecule, probably the major portion of water coming from the A–T-rich regions.

This proposal is largely speculative, but it is worthwhile showing that the model currently thought by some workers to explain the effect of neutral electrolyte upon T_m is incorrect in this important respect. The ΔH_m and T_m data so far discussed clearly show that ΔS_m decreases with increase in the electrolyte concentration and increasing degree of hydration of the counterion. The model which relates the increase in DNA helix stability to the formation of a more ordered water structure around the macromolecule, induced by the counterion and becoming stronger with increased hydration of the counterion, is incompatible with the evidence. Such a model would require relatively larger regions of ordered water to disappear on melting, i.e., ΔS_m would tend to become relatively larger and hence T_m would have to *decrease* with increasing electrolyte concentration. Furthermore, T_m would decrease to the greatest extent when the counterion has the largest degree of hydration. This is the exact reverse of the observed state of affairs.

In this context it is interesting to note a recent report[352] on the stability of salmon sperm (42% G–C) and lambda phage (50% G–C) DNA to thermal denaturation in the presence of univalent and divalent counterions at an ionic strength of 0.01. The melting temperatures, which were obtained by Applequist's method[45] from optical density–temperature profiles at 260 and 280 nm, are outlined in Tables XXVI and XXVII. Unequal contributions from A–T and G–C pairs to DNA hyperchromicity at a given wavelength result in an optical density–temperature profile at 260 nm reflecting more strongly the melting of A–T-rich regions, while that at 280 nm reflects the melting of G–C rich regions. Figure 22 illustrates the behavior of the guanidinium salt, which is typical of the univalent cations (with the exception of the NH_4^+ ion), where no difference between the 260-

TABLE XXVI. Melting Temperatures of DNA in 0.01 Ionic Strength Solution of Monovalent Cations[a]

Cation	Salmon sperm DNA T_m (260 nm), °C	λ-phage DNA	
		T_m (260 nm), °C	T_m (280 nm), °C
Guanidinium$^+$	71.2	77.5	77.5
Li$^+$	67.0	74.0	74.1
NH$_4$$^+$	64.7	72.1	72.2
Cs$^+$	64.0	70.4	70.5
Na$^+$	62.6	69.6	69.8
K$^+$	62.3	69.1	69.2

[a] Data from Ref. 352.

and 280-nm profiles could be detected. At $T < T_m$ the data show that the remainder of the univalent cations promote earlier changes in the A–T-rich regions compared to the G–C-rich regions. The univalent cation sequence of T_m values is in agreement with that of Gruenwedel *et al.*[572] and the sensitivity of the A–T-rich region to melting lends support to the proposal that hydration loss from the A–T-rich regions of the DNA is an important factor in the helix–coil transition. The data pertaining to the

TABLE XXVII. Melting Temperatures of DNA in 0.01 Ionic Strength Solution of Divalent Cations[a]

Cation	Salmon sperm DNA T_m (260 nm), °C	λ-phage DNA	
		T_m (260 nm), °C	T_m (280 nm), °C
Sr^{2+}	82.8	87.3	86.9
Ba^{2+}	82.7	85.9	85.4
Mg^{2+}	82.3	precipitated	
Ca^{2+}	78.4	82.4	81.3

[a] Data from Ref. 352.

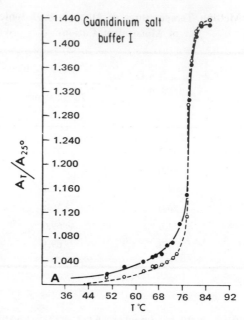

Fig. 22. Absorbance–temperature profile for the guanidinium salt of lambda phage DNA. Denaturation solvent was 0.01 ionic strength with 0.01 mol dm^{-3} concentration of guanidinium chloride in buffer I (5.7×10^{-4} mol dm^{-3} Tris-HCl, 5.0×10^{-4} mol dm^{-3} NaEDTA, pH 7.5). ($\bullet$) Measurement at 260 nm; ($\circ$) measurement at 280 nm. (See Ref. 352.)

NH$_4^+$ and guanidinium ions are particularly interesting in this respect; studies of the extent to which guanidine hydrochloride dehydrates the protein egg albumin indicate that only 40 molecules of water are bound to each molecule of protein.[203] The data relate to concentrated salt solution but they clearly indicate the strong dehydrating capability of the salt, which correlates particularly well with the evidence of Dix and Strauss for the marked capability of the guanidinium ion to raise T_m of the DNA. On the other hand, the NH$^+$ ion is not considered to significantly disturb water structure[667] and no appreciable difference is observed in the premelt behavior of the A–T- and G–C-rich regions.

   The correlation of T_m of DNA and the degree of hydration of the counterion may be subject to minor variation, as shown by the position of the Cs$^+$ ion in the relationship of T_m to the degree of cation hydration; Cs$^+$ might be expected to follow the K$^+$ ion in the sequence but, in fact, follows the NH$_4^+$ ion. This anomalous position may be due to the fact that the Cs$^+$ ion appears to be bound to the macromolecule to a much greater extent than are K$^+$ or Na$^+$ ions.[202]

The optical density–temperature profiles of λ-phage DNA in the presence of divalent cations reveal a different situation to that observed with the univalent ions. The melting temperature again shows the relationship to the degree of hydration of the cation; all the alkaline earth ions studied are much more heavily hydrated than the alkali metal ions and higher T_m values result. The Mg^{2+} rather dramatically illustrates the limiting effect to be expected if the counterion is particularly effective in promoting dehydration as well as charge neutralization; in fact, precipitation is observed. As with Cs^+ in the case of the univalent ions, an exact correlation of T_m with the degree of hydration is not obtained, probably due to variations in the extent of ion binding to the macromolecule. Binding of this kind by the divalent ions is to be expected because of the possibility of cation bridging between phosphate groups, which is not possible with the univalent ions.

A further difference between the effects of univalent and divalent ions upon the optical density–temperature profile of λ-phage DNA is illustrated in Fig. 23; this shows that in the presence of divalent ions A–T pairs are relatively more stable than G–C pairs. Dix and Strauss point out that this

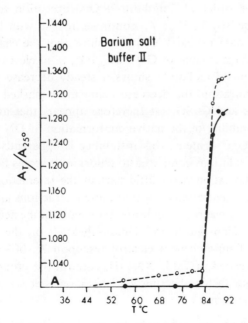

Fig. 23. Absorbance–temperature profile for the barium salt of lambda phage DNA. Denaturation solvent was ionic strength 0.01 with 0.0033 mol dm^{-3} of barium chloride in buffer II (0.001 mol dm^{-3} Tris-HCl, pH 7.9) ($\bullet$) Measurement at 260 nm; ($\bigcirc$) measurement at 280 nm. (See Ref. 352.)

behavior is unlikely to be due to a preferential binding affinity of the cation for A–T regions since the evidence of Shapiro *et al.*[1370] shows that Ca^{2+} ions appear to bind equally to A–T- and G–C-rich regions of DNA. The authors suggest that increased hydration of the A–T-rich regions may be the answer, but this is extremely unlikely since their own evidence and that of others[247,1528,1529] shows that the Mg^{2+} ion is particularly effective in dehydrating the macromolecule. Unfortunately, calorimetric and density gradient data, although able to provide information on the effects of univalent ions upon the extent of hydration of DNA, are lacking for divalent ions; until more quantitative information of this kind is available it is difficult to speculate further upon the details of the interaction.

The helix–coil transition of DNA in neutral electrolyte solution at electrolyte concentrations in excess of that corresponding to suppression of the electric double layer of the macromolecule would be expected to show a different concentration dependence than in the region below ~ 0.1 mol dm^{-3} (for a univalent electrolyte). The data of Gruenwedel *et al.*[572] show that at concentrations higher than this, T_m tends either to change very little or to fall slightly, even when the concentration of electrolyte rises to something of the order of 3 mol dm^{-3}. One exception to this behavior pattern is provided by Li_2SO_4; T_m continues to rise with higher salt concentrations. The data (Table XXV) also show that the change in T_m with change in percentage content of G–C, dT_m/dX_{G-C}, at electrolyte concentrations greater than 0.1 mol dm^{-3} shows a steady decrease with increasing electrolyte concentration, the decrease being most marked in solutions of Na_2SO_4, Li_2SO_4, and Cs_2SO_4. It therefore appears that at high salt concentrations the stability of the native conformation of DNA may decrease with increasing G–C content; this instability is apparently further accentuated in sulfate solutions compared to chloride. It seems fairly conclusive, therefore, that the cation plays little part in the interactions occurring at high electrolyte concentrations and that these interactions appear to involve the G–C pairs of the macromolecule and the anions of the neutral electrolyte.

The data of Hamaguchi and Guiduschek[594] on the effect of 4 mol dm^{-3} solutions of uni-univalent electrolytes upon T_m of sea urchin DNA (37% G–C), correlated in Table XXVIII, confirm the proposition that the anion is the major factor affecting T_m. A Hofmeister series can hence be constructed of the form

$$CCl_3COO^- \gg CF_3COO > CNS^- > ClO_4^- > Ac^- > Br^- > Cl^- > H \cdot COO^-$$

in which the anions show decreasing effectiveness in lowering the melting

TABLE XXVIII. Melting Temperature of Sea Urchin DNA in 4 mol dm^{-3} Aqueous Electrolyte Solutions at Neutral pH

Anion	Cation		
	Li$^+$	Na$^+$	K$^+$
CCl$_3$COO$^-$	—	25	—
CF$_3$COO$^-$	—	63	—
CNS$^-$	—	68	76
ClO$_4^-$	67.4	74.0	—
I$^-$	78.8	76.4	81.8
CH$_3$COO$^-$	—	83.5	86.7
Br$^-$	89.5	89.1	91.1
Cl$^-$	91.6	90.0	—
HCOO$^-$	—	92.6	—

[a] Data from Ref. 594.

temperature of DNA. Robinson and Grant[1259] noted a similar anionic series in the solubility of the bases thymine, adenine, cytosine, deoxyadenosine, and adenosine in concentrated solutions of the salts; from the derived activity coefficients of the bases as a function of salt concentration these authors suggest that the denaturation of DNA by these concentrated salt solutions is due to their effects upon the activity coefficients of the constituent bases and is attributed to the increased exposure of the bases to solvent in the random coil compared to the helix conformation.

In the earlier discussion of hydration of nucleic acids and polynucleotides we related the extent of biopolymer hydration in concentrated electrolyte solutions to the activity of the water alone and showed that it is unaffected by the nature of either the cation or anion, even for powerful denaturing agents such as NaClO$_4$. Hamaguchi and Guiduschek[594] showed that such a relationship does not exist between T_m and log a_w but they do show that the linear relationship of T_m vs. X_{G-C} is still observed in 7.2 mol dm^{-3} NaClO$_4$ and 6.5 mol dm^{-3} CF$_3$COONa, but with larger slopes than those observed by Gruenwedel et al.[572] Combination of the data of Gruenwedel et al. and Hamaguchi and Guiduschek on the variation of T_m vs. X_{G-C}

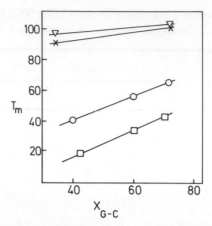

Fig. 24. Relationship of the melting temperature of various DNA's with the G–C content of the DNA in the presence of (○) 7.2 mol dm^{-3} NaClO$_4$, (□) mol dm^{-3} CF$_3$COONa, (×) 3.0 mol dm^{-3} NaCl, and (▽) 3.0 mol dm^{-3} Na$_2$SO$_4$. The data for NaClO$_4$ and CF$_3$ CF$_3$COONa were taken from the paper of Hamaguchi and Guiderschek,[594] those for NaCl and Na$_2$SO$_4$ from the paper of Gruenwedel et al.[572]

results in Fig. 24; the data of Gruenwedel et al. were obtained at a some-what lower salt concentration of 3 mol dm^{-3} but the trend of their data shows that the differences observed would be even more marked at the higher salt concentrations. Figure 24 shows that T_m decreases in the ex-pected manner with increasing efficiency of the anion in promoting de-naturation, but the *relative* efficiency of the denaturants decreases with increasing G–C content of the macromolecule.

Robinson and Grant[1259] have shown that the solubility of the bases in electrolyte solutions increases in the sequence A < T < C in a similar order as denaturation of DNA by these solutions. Unfortunately, no in-formation is available as to the solubility of guanine; Gruenwedel et al.[572] suggest that the solubility of guanine might be approximated to that of cytosine, resulting in G–C pairs being more soluble than A–T pairs; this in turn suggests that DNA's rich in G–C would be more easily denatured. This is clearly not the case, since T_m increases with increasing G–C content. T_m also increases most rapidly with increasing G–C content in solutions of the most effective denaturants, which again is the reverse of what is suggested.

The molecular structure of guanine suggests that its solubility might be expected to be closer to that of adenine than to cytosine, which may result in A–T pairs being more soluble than G–C pairs; there is some evidence for this from observations of solubility and stacking behavior of a wide range of dinucleoside phosphates in 0.15 mol dm^{-3} KCl.[326] It was found that the pyrimidine derivatives were considerably more soluble than their purine counterparts. This in turn would suggest that denaturation of DNA might be favored by an increased A–T content.

It is particularly difficult to draw definitive conclusions on the role of water in the denaturation process at these high electrolyte concentrations. The data of Hearst and co-workers reviewed in Section 3.4.2 seem to show quite clearly that the extent of hydration of the biopolymer at high electrolyte concentrations is a function of water activity alone and independent of the nature of the electrolyte. This would appear to rule out any differential changes in hydration of DNA by differing anions being responsible for variations in T_m; the data of Lubas and Wilczok[920] and Tunis and Hearst[1117] show, however, that denaturation in concentrated electrolyte solutions, as in dilute solutions, is accompanied by an overall decrease in hydration. Unfortunately, no information appears to exist about the extent of dehydration as a function of anion type.

Robinson and Grant[1259] suggest that the denaturation of DNA by concentrated electrolyte solutions is attributable to the greater exposure of the bases to the solvent in the denatured form of DNA than in the native form, the exposure arising from a direct interaction of anions with the polar groups of the bases. The data show, however, that the ratio of the solubilities of the bases in salt solutions and in water is only greater than unity for Br$^-$ < I$^-$ < ClO$_4^-$ < CNS$^-$ < Cl$_3$COO$^-$ and is, in fact, less than unity for the series SO$_4^{2-}$ < Cl$^-$. Since this latter *salting-out* effect is unlikely to be due to ion binding to the bases, the authors suggest that it may be due to changes in the internal pressure of the water[336] or in the free energy of cavity formation to allow the accommodation of the DNA molecule[1398] due to the raising of the surface energy of the salt solutions compared to that of water.

It is, however, worth noting that Bull and Breeze[202] in their isopiestic investigation of the hydration of egg albumin at high salt concentrations noted differential hydration of the biopolymer due to the anion; in particular, these authors found in terms of their model that the protein was *preferentially* hydrated in the presence of SO$_4^{2-}$, less so with Cl$^-$, and even less so with Br$^-$; CNS$^-$ and I$^-$, however, were shown to *dehydrate* the protein. Hearst and co-workers utilized both isopiestic and density gradient techniques in their investigations of the hydration of DNA and observed identical behavior. It is therefore an open question whether the different treatments by various authors of their respective data is responsible for the observed discrepancies, or whether a real difference exists between the effects of electrolyte on protein and nucleic acid hydration. It cannot be denied, however, that the sequence observed by Bull and Breeze[202] is very similar to that observed by Hamaguchi and Geiduschek[594] and Robinson and Grant,[1259] particularly with respect to the agreement between *salting-in* and *salting-out* effects and the denaturing behavior of the salts with respect to DNA and the preferential dehydration or hydration by these salts.

It should also be noted that Tunis and Hearst[1500] found from ORD studies on native calf thymus DNA that between 1 and 7 mol dm^{-3} salt the ORD spectrum changed as a function of the salt concentration. The authors suggest that this is caused by a gradual change in the structure of the DNA from the normal B form to the A form. Circular dichroism studies of calf thymus and *E. coli* DNA films as a function of relative humidity indicate that at low relative humidity the spectrum obtained is very similar to that obtained from DNA in the presence of high salt concentrations.[1504] Studdert *et al.*[1437] have recently observed the temperature dependence of the CD spectrum of calf thymus DNA and poly-d(A–T) in the helix conformations in solutions of LiCl, NaCl, KCl, NH$_4$F, and guanidinium chloride over the concentration range 0.5–8 mol dm^{-3}. These authors also noted changes in the secondary structure of DNA but are hesitant to attribute such changes to definite structures such as A, B, or C and consider that a large number of intermediate structures are possible. They also found that the CD spectra were strongly dependent upon the nature of the cation and only weakly dependent on the anion, in agreement with Alder and Fasman[7]; the guanidinium ion, however, behaved in an opposite manner to the other ions studied, a situation for which a satisfactory explanation appears to be lacking.

The authors observed systematic differences in the temperature dependence of the CD spectra of DNA, the largest variation occurring in LiCl solutions, decreasing through NaCl, KCl to NH$_4$F. The variation of ellipticity at 276 nm as a function of temperature at 0° and constant solvent composition is correlated with the heat of hydration of the cation in Table XXIX. Since poly-d(A–T) apparently behaves in a similar manner to calf thymus DNA (57% A–T), the authors conclude that differential interaction of the bases with their solvent environment is not important and that the changes in the structure of the molecule are due to interaction of the sugar phosphate backbone with its environment. The marked correlation of the temperature dependence of the CD spectra with ΔH of hydration of the cation suggests a similar correlation in the extent of hydration of the DNA molecule and the degree of hydration of the cation which is observed at electrolyte concentrations below $\sim$0.1 mol dm^{-3}.

TABLE XXIX. Correlation of CD Spectra Changes at 276 nm and Cation Heat of Hydration[a]

Cation	Heat of hydration, kJ mol^{-1}	Concentration, mol dm^{-3}	$d(\theta)_{276}/dt$ at 0°C
Li$^+$	520.8	1.81	64
		2.96	95
		3.50	96
		4.52	120
Na$^+$	407.4	1.79	57
		3.58	86
		4.48	85
K$^+$	323.4	1.80	48
		3.00	45
NH$_4^+$	306.6	5.00	40

[a] Data from Ref. 1437.

4.3. The Effect of Salts on the Helix–Coil Transition of Collagen and Related Polypeptides and the Reversion of Gelatins to a Collagen Fold Structure

Calorimetric studies[1198] have shown that tropocollagen behaves in a similar manner to the globular proteins in that the protein appears to be more extensively hydrated in 0.15 mol dm^{-3} KCl than in salt-free water. Kuntz and co-workers[829] observed changes in the bound water associated with gelatin with ionic strength comparable to those observed for the globular proteins; the bound water nmr signal decreased in width from that found with salt-free water, underwent a minimum at 0.1 mol dm^{-3}, and subsequently broadened again at higher salt concentrations.

Evidence of this kind suggests that an electric double layer effect is observed with collagen and the gelatins analogous to that found with the nucleic acids and the globular proteins. Confirmation for this suggestion is obtained from light scattering data[1522] on unfractionated bovine corium extract gelatin. Turbidities were obtained as a function of ionic strength from 0 to 1 mol dm^{-3} (Fig. 25); the data show that, although hydrolytic degradation occurs in all solutions as a function of time, maximum turbidity is consistently obtained in the 0.10 mol dm^{-3} KCl solution. In addition to changes in turbidity, the authors also observed complete renaturation, presumably by an annealing process,[411] with consequent precipitation of

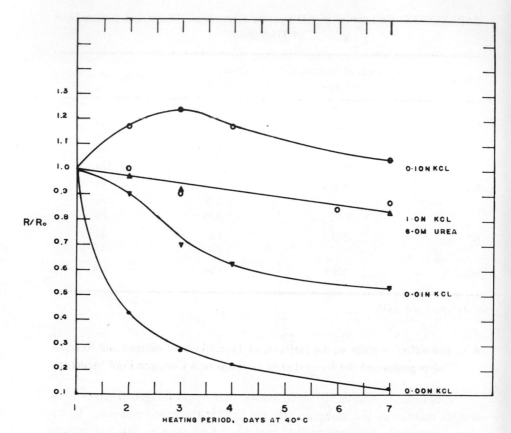

Fig. 25. Effect of salt concentration and time on collagen fold formation and fiber precipitation at 40° as measured by the relative turbidity of an unfractionated gelatin solution (see Ref. 1522.)

fibers; precipitation was very rapid at zero ionic strength but completely inhibited in 1 mol dm⁻³ KCl and 6 mol dm⁻³ urea. At low gelatin concentrations, in 0.01–0.1 ionic strength KCl, an electron microscope investigation of the fibers showed them to be very clean with sharply indicated cross-striations, whereas those obtained at very low salt concentrations appeared to be coated with amorphous material. It thus appears that the optimum ionic strength for the most effective renaturation of collagen is of the order of 0.1 mol dm⁻³.

Eagland and Pilling[376] have recently observed a similar ionic strength effect in their studies of the reversion of rat tail tendon α-gelatin to the collagen fold (as distinct from complete renaturation to collagen fibers). Viscosity and optical rotation at 546 nm and a temperature of 15° were

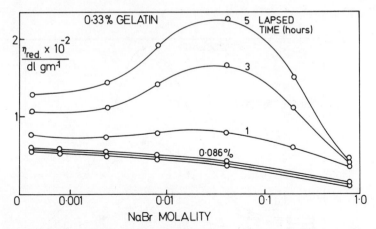

NaBr MOLALITY

Fig. 26. Reduced viscosity of α-gelatin solutions at a concentration of 0.086% w/v and 0.33% w/v; data were obtained at 15° as a function of added NaBr over periods up to 5 hr.

measured as a function of gelatin concentration over the range 0.034 to ~0.5% w/v and of salt concentration over the range 0–1 mol kg⁻¹ for NaBr, $(CH_3)_4NBr$, and $(C_4H_9)_4NBr$; data were also obtained as a function of time to enable a kinetic analysis to be undertaken. Figures 26–28 illustrate the viscosity data for the three salts at low and high gelatin concentrations.

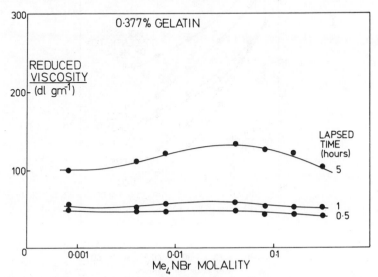

Me₄NBr MOLALITY

Fig. 27. Reduced viscosity of α-gelatin, concentration 0.377% w/v, as a function of added $(CH_3)_4NBr$ and lapsed time at 15°C.

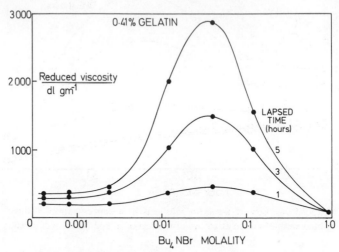

Fig. 28. Reduced viscosity of α-gelatin, concentration 0.41% w/v, as a function of added $(C_4H_9)_4NBr$ and lapsed time at 15°C.

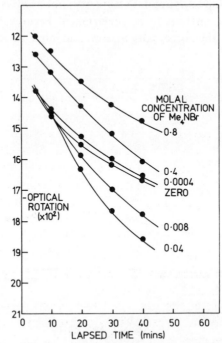

Fig. 29. Optical rotation of α-gelatin as a function of lapsed time and added NaBr concentration at 15°C. α-Gelatin concentration 0.45% w/v.

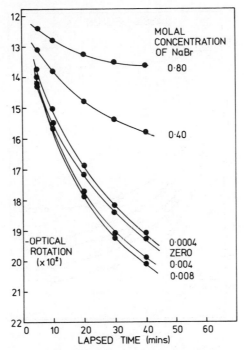

Fig. 30. Optical rotation of α-gelatin as a function of lapsed time and $(CH_3)_4NBr$ concentration. α-Gelatin concentration 0.44% w/v.

Figures 29–31 illustrate the comparable optical rotation data. The highest values for viscosity and mutarotation are obtained in the range of ionic strength for which other authors have observed maximum effects, suggesting that the maximum extent of folding of the protein occurs in this concentration region.

The data show that at concentrations of salt in excess of approximately 0.5 mol dm^{-3} a drastic reduction in the helicity is observed, to the extent that at salt concentrations of the order of 1 mol dm^{-3} the helix content of the macromolecule is less than that observed in salt-free water. Also, for similar gelatin and salt concentrations the extent of helix formation is a function of salt type; although the maximum amount of folding occurs at the same salt concentration for all three salts investigated, folding is somewhat greater in $(CH_3)_4NBr$ compared to NaBr, but is markedly greater than either in $(C_4H_9)_4NBr$.

Kinetic analysis of the viscosity data, commencing 10 min after the solutions, previously warmed to 40°, were cooled to 15°, and the optical rotation data, commencing 3 min after cooling to 15°, indicated that after the lapse of 10 min both the viscosity and optical rotation changes obeyed

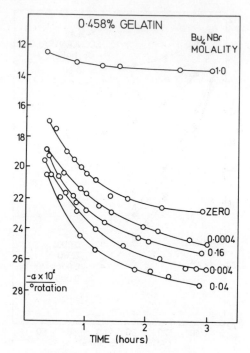

Fig. 31. Optical rotation of α-gelatin as a function of lapsed time and $(C_4H_9)_4NBr$ concentration et 15°C. α-Gelatin concentration 0.46% w/v.

a first-order rate law; rate constants determined for all gelatin and salt
concentrations were sensibly constant within the range $1–3 \times 10^{-2}$ min^{-1}.
The data reflect a folding process of an intramolecular character, presumably
involving a reverse folded structure, with no dimer or trimer formation and
unaffected by the nature or concentration of the salt in the solution.

The kinetic results cannot explain the remarkable maximizing effect
in fold formation observed in the region of 0.1 mol dm^{-3} added electrolyte.
Extrapolation of the optical rotation data to zero time gives initial rates
of reversion which exactly mirror the effect of salts observed in the optical
rotation data at longer time intervals; in addition, analysis of the optical
rotation data at time intervals between 3 and 10 min, although at the limits
of experimental accuracy in the case of the lower gelatin concentrations,
suggest that a pseudo-first-order process is occurring over this time region
with respect to the gelatin. The variation of the rate constants with salt
type and concentration strongly suggests that at shorter time intervals a
different process is occurring, with involvement of the salt in the folding
process; the initial rate data confirm that this is the case.

The above observations are in agreement with the proposal[1536] that the critical step in the production of helix formation in a gelatin is the initiation of a folded nucleus, presumably by the imino-acid-rich regions of the molecule. This nucleation region is considered to be hydrogen-bonded[614] but this is in conflict with the observation that the dominant factor in formation of the helix is an entropy change associated with a hydration network surrounding the macromolecule.[942,1201] The data of Eagland and Pilling appear to support such a view. Thus, although helix formation achieves a maximum value as the salt concentration approaches 0.1 mol dm^{-3}, the absolute values of helix formation appear closely related to the water structure enhancement of the cations.

In a report which is still regarded as one of the most definitive studies of the effects of salts on gelatin, von Hippel and Wong[1539] investigated the effect of a variety of salts over a wide concentration range upon the melting of refolded gelatin and the initial generation of the collagen fold at a single gelatin concentration. The data for CaCl$_2$, expressed as logarithm of the initial rate versus concentration of the salt, are shown in Fig. 32;

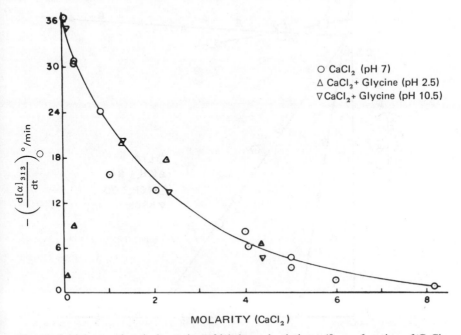

Fig. 32. Initial rate of optical rotation of ichthyocol gelatin at $4°$ as a function of CaCl$_2$ concentration at $4°$. Gelatin concentration 0.156% w/v. (⊙) CaCl$_2$ (*p*H 7); (△) CaCl$_2$ + 0.02 *M* glycine (*p*H 2.5); (▽) CaCl$_2$ + 0.02 *M* glycine (*p*H) 10.5. (See Ref. 1539.)

CaCl$_2$ at pH 7 and CaCl$_2$ + glycine at pH 5 apparently produce linear plots, but CaCl$_2$ + glycine buffered at pH 10.5 shows a maximum at a salt concentration of approximately 0.1 mol dm^{-3}; this very marked difference in behavior is, however, not commented on by the authors. The behavior of the gelatin in CaCl$_2$ solution buffered at pH 10.5 is very similar to that observed by Eagland and Pilling with other salts at neutral pH.[376] They found, however, that the behavior is also very dependent upon the gelatin concentration.

The linear plot observed by von Hippel and Wong[1540] at the higher salt concentrations for a variety of salts is shown in Fig. 33; similar behavior was observed by Eagland and Pilling at salt concentrations in excess of 0.1 mol dm^{-3} for NaBr, (CH$_3$)$_4$NBr, and (C$_4$H$_9$)$_4$NBr. It is interesting to note that the optical rotation data[376] indicate that the extent of collagen fold formation is very markedly suppressed as the concentration of (C$_4$H$_9$)$_4$NBr

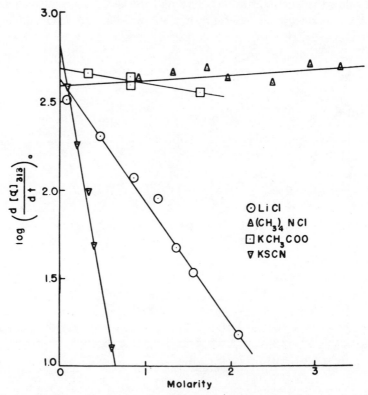

Fig. 33. Logarithm of initial rate of optical rotation of ichthyocol gelatin at 3° as a function of added salt; pH 7; protein concentration $\simeq 0.12\%$ w/v (see Ref. 1539.)

TABLE XXX. The Effect of Alkylammonium Ions on the Limiting ($t = 0$). Rate of Formation and the Stability of the Collagen Fold for Ichthyocol Gelatin[a]

Salt	$(d[\alpha]_{3/3}/dt)_0$, deg min^{-1}	T_m, °C
$(CH_3)_4NBr$	-8.2 (1 M)	18.4 (2 M)
$(C_2H_5)_4NBr$	-2.8 (1 M)	15.4 (2 M)
$(C_3H_7)_4NBr$	0 (1 M)	—
$(C_4H_9)_4NBr$	0 (1 M)	—

[a] Data from Ref. 1539.

approaches 1.0 mol dm^{-3}, which correlates with the suggestion by von Hippel and Wong[1540] that at 1 M concentrations of $(C_4H_9)_4NBr$ no collagen folding could be observed (Table XXX) in ichthyocol gelatin.

The kinetic data[376] show that the propagation or growth of the collagen fold in the gelatin chain, once initiated, is completely unaffected by salt at all concentrations investigated. In the initiation stage, however, at a salt concentration approaching 0.1 mol dm^{-3} the rate of initiation is greatest for $(C_4H_9)_4NBr$; thus the salt that at higher concentrations is the most effective in suppressing helix formation[1539] has the greatest effect at the low salt concentrations in promoting helix formation. The maximum helix formation observed in the concentration region of 0.1 mol dm^{-3}, since it is found with all three salts investigated, is obviously primarily caused by electrostatic interaction resulting in a reduction of the thickness of the electrical double layer to negligible proportions, but superimposed on this effect is an additional factor favoring helix formation, which appears to be dependent upon the hydration characteristics of the cation.

The evidence of von Hippel and Wong[1539] from data at higher salt concentrations shows that both anions and cations affect T_m of gelatin in much the same way as they affect T_m of globular proteins such as ribonuclease; for example, the sulfate and acetate ions are shown to raise T_m and the effect of the alkali metals can be classified in a Hofmeister series $Li^+ > Cs^+ > Na^+ > K^+ > Rb^+$. The evidence considered for salt effects on globular proteins indicates that the most effective depressants of T_m at high salt concentration operate by a complex mechanism which appears to involve changes in the internal pressure of the solvent by the ions and preferential concentration or exclusion of ions from the macromolecule–water interface.*

In the case of the tetraalkylammonium halides, however, the rules applicable to inorganic ions appear to break down; Deno and Spink[336] have shown that in the general theory of internal pressure due to Long and McDevit[915] the electrostriction effect of these salts is, in fact, negative and in consequence these salts have a salting-in rather than a salting-out effect. Desnoyers *et al.*[344] investigated the solubility of benzene in solutions of substituted quaternary ammonium bromides and found the solubility to increase with increasing size of the alkyl group. They therefore proposed, in agreement with Diamond,[349] that salting in is caused by formation of a complex between the tetraalkylammonium ion and the nonelectrolyte, induced by the "structure" of the water, the driving force for the association being van der Waals interaction between the large ion and the macromolecule.

The McDevit–Long theory, although predicting the correct order of salts in their capacity to salt out nonelectrolytes, is somewhat poor in quantitative agreement with experiment, and it seems that the scaled particle theory mentioned in earlier discussion has some advantage in this respect. The scaled particle theory as proposed by Shoor and Gubbins[1393] and modified by Masterton and Lee[988] expresses the solubility of the nonelectrolyte in terms of two contributions, the free energy change required to create a cavity large enough to hold the nonelectrolyte molecule and the free energy of interaction when the nonelectrolyte is introduced into the cavity. The first term involves the surface energy of the cavity interface and contributes to the salting out of the nonelectrolyte, whereas the second involves van der Waals forces across the interface between the nonelectrolyte and the solvent and contributes to the salting-in process. In the case of inorganic ions that are preferentially desorbed from such an interface the surface energy is increased and the salting-out contribution is dominant. In the case of the substituted quaternary ammonium salts, which are preferentially absorbed at the interface, the salting-out contribution is correspondingly reduced and the salting-in contribution enhanced; the observed result is a capacity to salt in nonelectrolytes, this capacity increasing with increasing size of the alkyl group. An association of this kind will have the effect of removing water molecules in contact with the nonelectrolyte, i.e., "dehydrating" the molecule.

---

* It is questionable whether the concept of internal pressure has any real meaning when applied to aqueous systems. Thus at 4° the internal pressure of water is equal to zero, and below that temperature it becomes negative!

The depressant effect observed by von Hippel and Wong[1539] of the tetraalkylammonium halides upon T_m of refolded gelatin would appear to correlate with this pattern; at the low electrolyte concentrations investigated by Eagland and Pilling, however, the situation is complicated by the electric double layer surrounding the molecule. Helix formation is clearly enhanced as the thickness of the double layer is reduced but is further favored by the presence of salts such as tetrabutylammonium bromide. It seems probable from the evidence of Masterton and Lee[988] that the capacity of the large ion to remove water from the macromolecule while electrostatic interaction between charge groups on the polymer is being simultaneously reduced favors increased folding of the imino acid groups into a collagen fold structure.

The model discussed earlier for tropocollagen[942,1201] supports this interpretation; the poly-L-proline II helix with intervening water molecules between the imino acid residues would be expected to occur more easily when electrostatic double layer repulsion is reduced and some of the water molecules are removed. At higher salt concentrations, however, when no electrostatic repulsion is experienced, water removal can be expected to become much more severe, effectively dehydrating the protein with subsequent destabilization of the helix conformation. Unfortunately, no evidence is yet available over the low range of electrolyte concentration on the effect of anions such as the sulfate ion upon the extent of helix formation; results for the sulfate ion should be particularly interesting in this respect because of its reported capability for favoring preferential hydration of proteins and salting-out capability with regard to nonpolar groups.

At salt concentrations in excess of 0.1 mol dm^{-3} von Hippel and Wong[1539] have shown that the effect of the salt on the melting temperature of refolded, unfractionated ichthyocol gelatin at a concentration of 0.15% w/v is a linear function of the salt concentration, the data following the equation

$$T_m = T_m{}^0 + Km$$

where T_m is the melting temperature at a salt concentration m and $T_m{}^0$ is the melting temperature in "salt-free" water (0.15 mol dm^{-3} KCl). The value of K is therefore a measure of the effectiveness of a particular salt in disrupting the helical conformation of the folded gelatin; Table XXXI contains values of K given by von Hippel and Wong and also obtained by these authors from the data of Bello et al.[108] on porcine gelatin gels at a rather higher gelatin concentration of 5%. The data in Table XXXI follow very similar trends to those observed for globular proteins and DNA and are clearly not governed by electric double layer effects. Absolute differences

TABLE XXXI. The Equilibrium Parameter K Describing the Effect of Neutral Electrolytes on the Stability of the Collagen Fold

Salt	K	
	Ref. 108	Ref. 1539
NaF	4.1	—
$(NH_4)_2SO_4$	—	3.8
$(CH_3)_4NCl$	—	2.2
$(CH_3)_4NBr$	—	−0.4
CH_3COOK	—	−0.8
KCl	—	−1.4
NaCl	−2.4	−1.6
CsCl	—	−1.6
LiCl	−5.0	−4.1
NaBr	−7.6	—
$MgCl_2$	−7.5	—
$NaNO_3$	−8.1	—
$CaCl_2$	−14.3	−8.8
KSCN	—	−10.0
NaSCN	−14.4	—
NaI	−14.4	—
LiI	−14.8	—

between the values of K for the same salt from the data of the two groups of workers are most likely to be due to differing imino acid contents of the two gelatins, since von Hippel and Wong show that the data for $CaCl_2$ with calf skin gelatin yields $K = -12.8$ compared to the -8.8 with ichthyocol gelatin.

The values of K obtained from the data of the two groups of workers suggest that the change in T_m is independent of gelatin concentration and von Hippel and Wong therefore postulate that the salts affect T_m by an intramolecular process in which the ionic constituents of the salt effects the collagen fold by competition for the available water with the macromolecule.

Such an interpretation is in agreement with the suggestion[202] that proteins are preferentially hydrated in the presence of the sulfate ion, which thus raises T_m, and dehydrated by Li^+ ion, which lowers T_m. However, the Li^+ ion can also be expected to bind preferentially to the protein,[202] in addition to dehydrating the macromolecule; but von Hippel and Wong specifically rule out this possibility. It is of particular interest therefore to consider the results of Harrington and Kurtz[837] from optical rotation and viscosity studies of the interactions of LiBr with the polypeptide poly-L-proline in both aqueous and nonaqueous solutions. It was observed that addition of LiBr to a solution of poly-L-proline II in anhydrous acetic acid caused the optical rotation and viscosity to fall immediately to values of the same order as those observed in concentrated aqueous LiBr solutions; the authors also noted that on dilution with water the aqueous lithium bromide–poly-L-proline system rapidly regenerated the poly-L-proline II structure. Dilution of the concentrated acetic acid–LiBr–poly-L-proline system with anhydrous acetic acid produced very little regeneration of the poly-L-proline II structure. The authors subsequently isolated a poly-L-proline–LiBr complex from anhydrous acetic acid and observed the rate of mutarotation of this complex in acid solvents of varying water content.

Harrington and Kurtz[837] propose that their results are consistent with binding of LiBr at the peptide linkages, possibly by binding of the lithium ion at the carbonyl oxygen,[107,351] a proposal which received further support from the work of Nandi and Robinson.[1081,1082] Such binding would lead to a lowering of the potential barrier to rotation at the peptide linkages and thus promote rapid destruction of the helical conformation of the poly-L-proline II structure.

Mattice and Mandel[995] have recently reported optical rotation, CD, and viscosity studies of the effect of concentrated $CaCl_2$ solutions upon the conformational properties of poly-L-proline solutions. The authors examined the hydrodynamic and optical properties of the polypeptide in dilute solution as a function of molecular weight, calcium chloride concentration, and temperature in an attempt to quantify the polymer conformation obtaining in high concentrations of this salt. By using Flory's[457] empirical equation

$$[\eta] = K^a M$$

where $[\eta]$ is the intrinsic viscosity, M is the molecular weight, and K and a are constants for a given solvent and temperature, the authors report that in water poly-L-proline exists in the II (left-handed) conformation as a rod of limited flexibility, but complete transition to random coil is completed at a $CaCl_2$ concentration of ~3 mol dm^{-3}. The authors did not find it possible to reproduce the CD spectra of the polymer in concentrated $CaCl_2$ solution by using a linear combination of the spectra for poly-L-proline I and II; it is thus not possible for the random state to consist of reasonably

long sequences of residues in any one form, implying that any such sequences must be extremely short. Since rotation of the peptide bonds in
the polymer would be expected to produce such a result, the data suggest,
but do not clearly prove, the existence of ion binding at the peptide group.

Swenson[1453] has recently attempted to clarify whether ion binding
occurs at the carbonyl oxygen of the peptide linkage in poly-L-proline.
He observed the infrared absorption spectra of poly-L-proline in concentrated solutions of LiBr, NaI, $CaCl_2$, NaSCN, KBr, NaBr, and LiCl. The
spectra of poly-L-proline in all the salt solutions studied showed two absorptions in the carbonyl C–O stretching region and two absorptions in the CH
stretching region, in contrast with the results in salt-free solution, which
give only one component in each spectral region. The data show that the
conformation of the peptide linkage, cis or trans, determines the frequency
of the carbonyl absorption and that the effect of salt on the band position
is relatively much smaller, suggesting that an interaction of the
—C—O⁻···Li⁺ type is unlikely. Swenson points out that, on the basis of
the results, it is not possible to distinguish whether the isomerized mixture
is of all-cis and all-trans chains or chains containing a mixture of both;
the author quotes nmr results[1487] which suggest that the latter is the case.
The two carbonyl absorptions observed are not form I cis or form II trans
but are absorptions of "cislike" and "translike" carbonyls in the disordered
structure.

Johnston and Krimm[733] also consider the infrared spectra of poly-L-
proline in aqueous $CaCl_2$ and, at the highest salt concentrations of 6 mol
dm^{-3}, obtain spectra for the —CO and CH absorption regions which are
identical with those of Swenson. Previous infrared studies[713] have shown
that the second band in the CO stretching region, observed in $CaCl_2$ solutions, is the same one which is observed in both poly-L-proline I and II in
the solid state; the authors thus conclude that it is impossible to distinguish
the cis from the trans imide bond by means of the C=O frequency and
suggest that the upward shift in the C=O frequency is due to replacement
of hydrogen bonding to water by a binding mechanism to the Ca^{2+} ion.
These authors also consider that the observed frequency differences between the CH_2 bending frequency in the solid state and in aqueous solution
support the alternative to a cis–trans isomerism, i.e., a change in the
C^α—C'=O rotation angle; thus the polymer chain adopts a range of local
conformations rendering its CD spectra similar to that of unordered polypeptides and proteins.

One of the most comprehensive investigations into neutral salt effects
upon the conformation of poly-L-proline in aqueous solution is that recently

reported by Schleich and von Hippel[1325]; it involves optical rotation studies at 546 nm of 0.1% w/v poly-L-proline II in salt solutions containing 0.1 mol dm KCl plus other salts at concentrations ranging from 0 to 8.0 mol dm^{-3}. The authors observed the same linear correlation of specific rotation with concentration of added salt as previously described,[615]

$$[\alpha] = [\alpha]_0 - BM$$

where $[\alpha]_0$ is the specific rotation of the solution containing 0.1 mol dm^{-3} KCl *only* and B is the slope of the plot of $[\alpha]$ versus concentration of neutral salt. The values of B obtained for the various salts are summarized in Table XXXII. The linear decrease in $[\alpha]$ with increasing salt concentration means that the poly-L-proline II helix does not undergo a cooperative transition of the type regarded as characteristic of proteins, polypeptides, and nucleic acids. In this sense, therefore, it is perhaps a poor model compound for studies of protein–solute–solvent interactions; nevertheless, since collagen and gelatins are known to adopt a similar structure to poly-L-proline II, while also being capable of undergoing cooperative transitions,

TABLE XXXII. B **Values of Various Salts for the Mutarotation of Poly-L-Proline II in Aqueous Solution**^{a}

Salt	B, deg mol^{-1}	Salt	B, deg mol^{-1}
Formamide	0.0	NaClO$_4$	−55.0
Guanidine HCl	0.0	LiClO$_4$	−55.0
LiCl	−21.0	LiSCN	−62.0
NaCl	−21.0	NaSCN	−62.0
KCl	−21.0	KSCN	−62.0
$\frac{1}{2}$CaCl$_2$	−27.8	Me$_4$NBr	−48.8
LiNO$_3$	−23.0	Et$_4$NBr	−70.0
NaNO$_3$	−23.0	Pr$_4$NBr	−96.5
LiBr	−39.0	Bu$_4$NBr	−149.0
NaBr	−39.0	LiCl^b	−23.0
KBr	−39.0	LiBr^b	−27.0
LiI	−51.8	—	—
NaI	−51.8	—	—
KI	−51.8	—	—

^{a} Data from Ref. 1537.
^{b} In formamide. 4% H$_2$O (v/v) present.

it is important to understand the effects of salts upon the stability of this left-handed helical structure.

Schleich and von Hippel also found, in agreement with Harrington and Sela,[615] that salting out of poly-L-proline occurs at room temperature in the presence of certain electrolytes but the precipitation is dependent upon the type and concentration of the salt and the time of standing. The lithium salts, with the significant exceptions of sulfate and acetate, did not precipitate the polymer at any of the salt concentrations studied and, for the overall capacity of the various ions to induce precipitation, Hofmeister series of the following order were constructed, with the concentration required for precipitation increasing from left to right: $SO_4^{2-} < Ac^- < Cl^- < Br^- < SCN^- < I^- < ClO_4^-$ and $K^+ < Na^+ \ll Li^+ \sim Ca^{2+}$. These series are strikingly similar to those observed for hydration changes of bovine albumin caused by salts[202] and for the salting-in and salting-out capabilities of these salts for the peptide groups and nonpolar side chains of polypeptides.[1082]

Evidence is available[259,973] that precipitation of the polymer takes place without change in the conformational structure of the poly-L-proline II helix; it seems very probable that the unfavorable contributions to the free energy of transfer of the nonpolar groups from water into salt solution are responsible for the salting out of the polypeptide, i.e., preferential hydration in the presence of sulfate, acetate, and chloride ions of the nonpolar groups of the macromolecule effectively promotes precipitation. Similarly, in the case of the cations, the Li^+ ion is the least effective of the alkali metal cations in salting out the nonpolar group, but the most effective in salting in the peptide group, hence requiring relatively larger amounts for precipitation than K^+ or Na^+ ions.

Plots of the extent of left-handed helix conformation in poly-L-proline as a function of salt concentration show a somewhat different dependence upon the salt type to that found for salting-out effects; for the univalent alkali metal cations it is clear from the B values listed in Table XXXII that the Li^+, Na^+, and K^+ ions have identical effects upon the conformation of the polymer; the B values of the anions, on the other hand, show a marked specificity, giving a series, in order of decreasing capacity for reducing the optical rotation of poly-L-proline II, as follows: $SCN^- > ClO_4^- > I^- > Br^- > NO_3^- > Cl^-$.

The guanidinium ion is less effective than the alkali metal cations in reducing the optical rotation of poly-L-proline; the tetraalkylammonium ions, however, show a marked effect upon the mutarotation, which increases with increasing size of the alkyl group.

The authors suggest, from inspection of molecular models, that randomization of the poly-L-proline II helix to a more disordered form involves no change in the extent of exposure of the polypeptide chain to the solvent; therefore the effects of the ions cannot be attributed to solvent structure but are to be expected if ion binding to the polypeptide chain occurs. Using the existence of LiBr–poly-L-proline complexing in non-aqueous solvents[615] as supporting evidence, Schleich and von Hippel suggest that local anion binding by the imine nitrogen brings about increased rotation of the $C^\alpha—C'{=}O$ linkage in a similar manner to that described by Johnston and Krimm,[733] with subsequent disruption of the poly-L-proline II conformation.

It is pertinent at this point to draw attention to studies of the effect of concentrated salt solutions on the activity coefficient of acetyltetraglycine ethyl ester.[1261] These indicate that two effects are operative; first, salting out, which is attributed to changes in the internal pressure of the solvent, as previously discussed. Second, salting in of the amide group is postulated, which may occur by direct association of the anion with the amide group, as suggested by Schleich and von Hippel; support for this view is given by a comparison of the binding of anions to anion exchange resins, which follows a pattern of order similar to the effectiveness of anions in salting in acetyl-tetraglycine ethyl ester and of anion binding to cationic sites on proteins,[1309] namely $SCN^- > ClO_4^- > I^- > NO_3^- > Br^- > Cl^- > CH_3COO^-$. This sequence is similar to that observed[1325] for the effect of anions upon the mutarotation of poly-L-proline. One additional common factor between the data of Robinson and Jencks[1261] and Schleich and von Hippel[1325] is the apparent insensitivity of both processes to the nature of the cation; because of this behavior the alternative explanation of the observed effects in terms of the scaled particle theory should be given serious consideration. Thus, the properties of the surface layer of water surrounding the polymer are believed to be markedly altered by the presence of ions within this layer. The distributions of cations and anions in this layer are completely different; inorganic cations, because of their hydration requirement, are almost completely excluded from a hydrocarbon water interface[792]; anions, because of their smaller degree of hydration, show wide variations in the extent of their exclusion. The cations Na^+, K^+, Rb^+, Cs^+, and NH_4^+ are all excluded equally and almost totally from the air–water and water–hexane interface[792]; the exclusion of anions from the interface, however, is, in decreasing order of exclusion,

$$F^- > OH^- > Cl^- > BrO_3^- > Br^- > NO_3^- > I^- > ClO_4^- > SCN^-$$

which is in close agreement with their salting-in capacity toward acetyl-triglycine ethyl ester and their binding to proteins[1309] and poly-L-proline II.[1325] This formalism therefore has the advantage of being able to account for the insensitivity of the various processes to the nature of the cation, and for the large but graded effectiveness of anions. It also has the advantage, discussed earlier, of accounting for the marked capability of substituted quaternary ammonium salts to suppress the poly-L-proline II helix con-formation.

Such an interpretation would also be in agreement with the data of Klotz and Shikama[792] on the reversible binding of methyl orange by bovine serum albumin and polyvinyl pyrrolidone; since the binding is suppressed by the addition of urea but can be recovered by removal of the urea by dialysis, it is unlikely that anion binding is a direct interaction but is more likely to involve the solvent in some manner, possibly as described above.

Threlkeld and co-workers[1217] have reported sedimentation studies on unfractionated pigskin gelatin and rat tail tendon collagen in a limited series of concentrated salt solutions. Results are given for KSCN, $(C_4H_9)_4NBr$, $(CH_3)_4NBr$, and KCl and are interpreted in terms of the number of moles of salt sedimenting with 1 mol of the macromolecule (Table XXXIII). Compared to the number of moles of salt "bound" per mole of egg albumin,[202] the magnitude of the binding seems remarkably small, although the data for KSCN and KCl are in satisfactory agreement with data previously obtained by a radioisotope comparison technique.[1479] In the light of this evidence of the binding of $(CH_3)_4NBr$ and $(C_4H_9)_4NBr$

TABLE XXXIII. Electrolyte Adsorption to Gelatin at 25° and the Effect on the Shrinkage Temperature of Collagen[a]

Salt (0.4 mol dm^{-3})	$\Delta C_s/C_p$, mol salt bound/mol macromolecule	ΔT_s,[b] (1 mol dm^{-3}), °C
KSCN	0.042	25
Bu$_4$NBr	0.040	17
Me$_4$NBr	0.023	12
KCl	0.012	0

[a] i.e., the temperature at which shrinkage of the fibers is observed. Data from Ref. 1479.
[b] Depression of the shrinkage temperature of rat tail tendon collagen in 1 mol dm^{-3} solution of the salt.

it is difficult to accept the explanation of Schleich and von Hippel that the depressive effect of the tetraalkylammonium ions on the mutarotation of poly-L-proline II is caused by the low dielectric constant of the medium surrounding the macromolecule; on a simple charge interaction basis, binding of tetramethylammonium ion, the most compact ion, with the highest surface charge density and the smallest degree of hydration, should certainly be stronger than that of the tetrabutylammonium ion. The reverse is the case and association appears to be greatest for the ion with the greater degree of hydration.

4.4. Conformational Transitions of Globular Proteins and Related Polypeptides

In the introductory remarks of Section 4.1 some stress was laid on the difficulties that may arise in the interpretation of data associated with the conformational transition of a macromolecule if the solvent medium surrounding the molecule contains added electrolyte at low concentrations. This problem is particularly relevant with regard to the helix–coil transition of proteins and polypeptides.

The work of Kuntz et al.[829] and Privalov et al.[1198] clearly indicates that marked changes in the conformational structure and degree of hydration of a wide variety of proteins occur when the concentration of uni-univalent electrolyte in the solvent medium increases from zero to approximately 0.1 mol dm^{-3}. Since this is the region of electrolyte concentration within which suppression of the electric double layer surrounding the macromolecule will occur, it is perhaps not unexpected that associated changes in conformation and hydration should accompany such a change. It is probable that neglect of this parameter has led to discrepancies between the reported data of different authors who claimed to observe the same conformational transition. Shiao and co-workers[1380] and Privalov et al.[1195] have reported enthalpy data for the helix–coil transition of ribonuclease under identical conditions of hydrogen ion concentration, pH 2.4; both groups of workers reported melting temperatures which were very similar, $35°$ and $36°$ respectively, but obtained values for the ΔH of transition which were markedly different, 313 and 218 kJ mol^{-1}, respectively. It is significant that the results of Shaio et al. were obtained at an electrolyte concentration of 0.01 mol dm^{-3} in Cl$^-$, unbuffered, whereas Privalov worked at an electrolyte concentration of 0.1 mol dm^{-3} in a buffered solution of glycine acetate/phosphate. Comparison of the data suggests that at the melting temperature, when the free energy change for the transition is zero, the close agreement in the value of T_m indicates that a smaller entropy change on transition is associated with the solvent of higher ionic strength.

It is also noteworthy that Privalov *et al.* observed a pretransitional melting at this higher ionic strength, not observed by other workers[140] but which has been reported for the helix–coil transition of DNA at higher ionic strengths.[795] Ginsburg and Carroll[525] have reported studies of specific ion effects on the conformation of ribonuclease at low ionic strength which appear to confirm that the above ΔH discrepancies are due to variations in ionic strength. These authors observed the thermal transition of ribonuclease at pH 2.1 and an ionic stength of 0.019 in mixtures of HCl–KCl, $H_2PO_4^-$–H_3PO_4, and SO_4^{2-}–HSO_4^-. The transition was observed by an optical absorption technique at 287 nm, optical rotation at 365 nm, and intrinsic viscosity measurements; the good agreement between the data obtained by the three techniques, shown in Table XXXIV, confirmed that the transition is a two-state process. The observations of the transition in the HCl–KCl mixture under conditions of pH and ionic strength very close to those adopted by Shaio *et al.* show excellent agreement in the ΔH of transition with the value obtained by the latter (214 and 218 kJ mol^{-1}, respectively).

The results of Ginsburg and Carroll also illustrate the marked differences between the effects of different anions upon the stability of the native conformation of the protein at this low pH value; the sulfate ion is clearly a better stabilizing entity than both the phosphate and chloride

TABLE XXXIV. Thermodynamic Parameters for the Reversible Denaturation of Ribonuclease at pH 2.1 and 0.019 Ionic Strength[a]

	KCl–KCl			$H_2PO_4^-$–H_3PO_4		SO_4^{2-}–HSO_4^-		
	$\Delta\varepsilon_{287}$	$\Delta(\alpha)_{365}$	$\Delta(\eta)$	$\Delta\varepsilon_{287}$	$\Delta(\alpha)_{365}$	$\Delta\varepsilon_{287}$	$\Delta(\alpha)_{365}$	$\Delta(\eta)$
Concentration mg ml^{-1}	2.34	12.55	0	2.18	13.36	2.68	12.60	12.60
Transition temperature T_m, °C	29.9	29.9	28.8	31.9	32.8	42.7	44.1	44.1
ΔH_{vH}° kJ mol^{-1}	197.4	214.2	193.2	235.2	247.8	319.2	289.8	319.2
ΔS°, J mol^{-1} K^{-1}	659	706	672	764	806	1003	916	1003
ΔT_m, °C	7,5	7.9	8.2	6.1	6.3	5.9	6.1	6.1

[a] Data from Ref. 525.

ions, since T_m in the sulfate buffer solution is 14° higher than in the chloride solution and approximately 12° higher than in the phosphate buffer. The transition temperature range over which unfolding of ribonuclease occurs is also markedly more narrow, at 6.0° compared to approximately 8.0° in the chloride solution; this behavior contrasts with that observed by von Hippel and Wong[1540] at salt concentrations in excess of 0.1 mol dm^{-3}; they found that for a wide variety of salts ΔT was constant at 4.7 $\pm$ 1°.

It is concluded that the sulfate ion, by binding to the macromolecule, induces a particularly ordered form of the helical conformation compared to that induced by binding of the chloride ion, and that after melting, a somewhat more ordered form of the macromolecule is still obtained in the sulfate buffer than is the case in the chloride solution. There are indications that binding of the sulfate ion to the macromolecule is unlikely to be the cause of the more ordered structure but preferential hydration of the biopolymer in the presence of the sulfate ion is to be expected, and this may be the cause of such behavior.[202,1259]

The protein molecule after denaturation, unlike the DNA molecule, is at least as hydrated as in the native state and in general appears to be slightly more so[1198]; thus the largest entropy change on denaturation, observed in the presence of the sulfate ion, would be expected if hydration of the macromolecule played a major role in the denaturation process.

The proposal of Ginsburg and Carroll that anion binding is the major factor in affecting the thermal transition of the protein is also in disagreement with results on the binding of anions by bovine serum albumin.[792] It has been reported that the binding of Methyl Orange is suppressed by the addition of urea, but the inhibition is reversible, binding being restored on removal of the urea by dialysis or dilution. Also the binding of Methyl Orange to polyvinylpyrrolidone which is in a random configuration and swollen in water is similarly suppressed by the addition of urea. Since urea is shown to behave in a similar manner in aqueous solutions of many natural and synthetic polymers of very different chemical structures, it seems unlikely that the anion binding effect is by direct interaction with the macromolecule, but is much more likely to be via hydration interactions which are markedly affected by the addition of urea.

Peggion and co-workers[1145] have recently reported circular dichroism data on the effect of sodium perchlorate on the coil–helix transition of poly-L-lysine and copolymers of L-lysine and L-phenylalanine over a range of electrolyte concentration from zero to approximately 1 mol dm^{-3}. The results are summarized in Fig. 34, and show that apparently the (L-lysine)

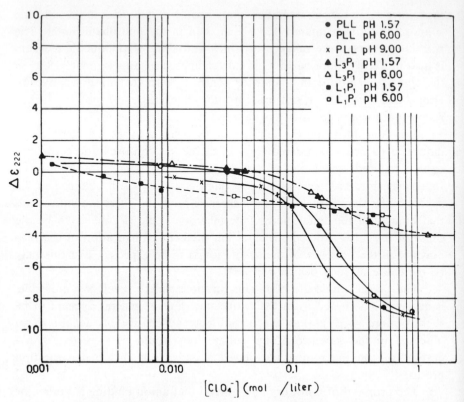

Fig. 34. Semilogarithmic plot of the $\Delta\varepsilon$ values at 222 nm of PLL and of the L_3P_1 and L_1P_1 copolymers versus the molarity of ClO_4^- (see Ref. 1145).

(L-phenylalanine) (L_1P_1) copolymer does not undergo a cooperative conformational change, but poly-L-lysine (PLL) and (L-lysine)$_3$(L-phenylalanine) (L_3P_1), irrespective of the degree of protonation of the side chains, undergo a cooperative transition from coil to helix in the salt concentration range 0.1–0.2 mol dm^{-3}. The authors consider that the explanation lies in specific site binding of ClO_4^- ions at the peptide groups. Evidence of other workers (see later discussion) indicates that the ClO_4^- ion does not bind to the biopolymer and acts as a denaturant, not a stabilizer, by virtue of its interactions with the solvent. It is therefore clear that the primary role of the salt at the concentration of $\sim$0.1 mol dm^{-3} is in suppressing the electric double layer surrounding the macromolecule, thus allowing helix formation to occur more easily, although the degree of ionization of the side chains has a slight effect, possibly through hydration interactions, in displacing the transition point to higher or lower salt concentrations.

The earlier work by Peggion and co-workers[1146] appears to support this viewpoint; the copolymer (L-lysine)(L-phenylalanine) assumes a random coil conformation in salt-free water at pH 4.36, it assumes a partial β-sheet conformation at the same pH and 0.1 mol dm^{-3} KCl, and it assumes an α-helical conformation at higher salt concentrations. The data thus tend to show that the primary effect of the neutral electrolyte at 0.1 mol dm^{-3} is suppression of the electric double layer in conjunction with hydration interactions dependent on the nature of the ions present in the solvent, which, from an energetic point of view, favor a particular conformation of the biopolymer.

Nmr studies indicate[920] that the degree of hydration of DNA at an ionic strength of 0.02 mol dm^{-3} is markedly greater than that at an ionic strength of 0.15 mol dm^{-3}, whereas the data of Privalov et al.[1198] for the extent of hydration of proteins show that in general hydration of proteins at an ionic strength of 0.15 mol dm^{-3} tends to be greater than at very low ionic strength. It is therefore particularly disappointing that only a limited number of systematic studies appear to have been undertaken upon the effect of electrolyte concentration, within region 0–0.1 mol dm^{-3}, on the conformational stability, degree of hydration, or enthalpy of transition of proteins, and no calorimetric work comparable to that of Privalov et al.[1199] on the effect of ionic strength on the enthalpy of transition of DNA has been reported. This is clearly a province of protein behavior for which very little factual evidence is available, and it is to be hoped that further research within this region will be initiated in the near future.

The effect of electrolytes at concentrations in excess of $\sim$0.1 mol dm^{-3} upon the helix–coil transition of proteins and related compounds has been the subject of much more detailed investigation than the low-concentration region. One of the most comprehensive studies is that of von Hippel and Wong[1540] on the effects of various electrolytes as a function of concentration, upon the thermal transition of ribonuclease. All the solutions, in addition to containing the various electrolytes, were buffered to neutral pH and contained 0.15 mol dm^{-3} KCl and 0.013 mol dm^{-3} sodium cacodylate; the thermal transition of the protein was followed by optical rotation measurements at 366 nm.

The changes in the transition temperature of ribonuclease so obtained are summarized in Figs. 35–37; it is immediately apparent that for a common cation a Hofmeister series of anions in order of decreasing effectiveness in denaturing ribonuclease is obtained, very similar to that observed by Hamaguchi and Geiduschek[594] for the denaturation of DNA. It is also apparent that for a common anion a cationic Hofmeister series can be constructed which for the alkali metal salts shows that the degree of effectiveness in denaturation is paralleled by the extent of hydration of the cation; this correlation is also observed for the family of tetraalkylammonium halides, although the nature of hydration of the tetraalkylammonium ion is largely hydrophobic, unlike the alkali metal ions.

The authors point out that the effect of the salts upon T_m is apparently additive, which strongly suggests that the interactions of the macromolecule with cations and anions are independent of each other. Since hydration interactions are strongly implicated for both series of ions, it appears

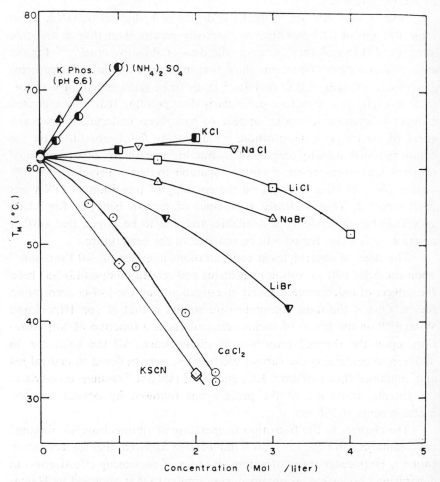

Fig. 35. Transition temperature of ribonuclease as a function of the concentration of various added salts. All the solutions were adjusted to pH 7.0 and laso contained 0.15 mol dm^{-3} sodium cacodylate; ribonuclease concentration $\simeq 0.5\%$ w/v (see Ref. 1540).

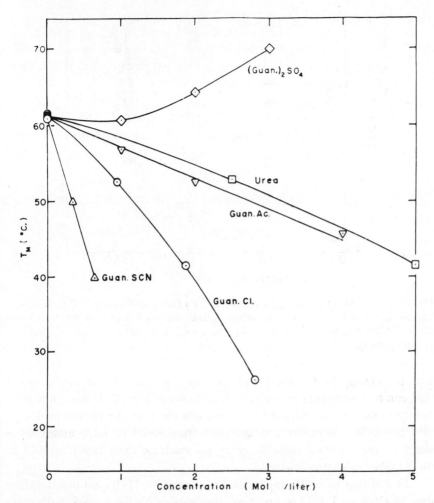

Fig. 36. Transition temperature of ribonuclease as a function of concentration of urea and various guanidinium salts. All solutions were adjusted to pH 7.0 and also contained 0.15 mol dm^{-3} KCl and 0.013 mol dm^{-3} sodium cacodylate; ribonuclease concentration $\simeq 0.5\%$ w/v (see Ref. 1540).

that the regions of hydration due to each type of ion are largely independent.*

Von Hippel and Wong have also reported the effect of the salts upon the width of the melting transition $\Delta[\alpha]_{366}$ of ribonuclease; the data are summarized in Fig. 38; the authors suggest that this effect is not correlated

* See Chapters 1 and 6, Volume 3.

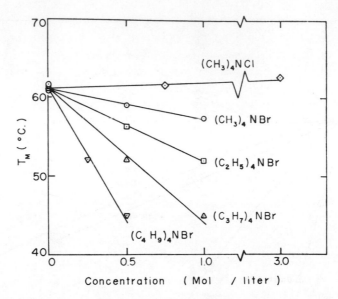

Fig. 37. Transition temperature of ribonuclease as a function of concentration of various tetraalkylammonium salts. All the solutions also contained 0.15 mol dm^{-3} KCl and 0.013 mol dm^{-3} sodium cacodylate, pH 7.0; ribonuclease concentration $\simeq 0.435$ w/v (see Ref. 1540).

with the change in T_m induced by the salt and that all the uni-univalent salts can be represented by the solid line drawn in Fig. 38. It does, however, appear to be the case that LiCl is responsible for a smaller change in $\varDelta[\alpha]_{366}$ at a particular electrolyte concentration than NaCl or KCl, and LiCl is clearly a more potent destabilizer of ribonuclease than NaCl or KCl at these higher salt concentrations.

In addition to producing a smaller change in $\varDelta[\alpha]_{366}$ on denaturation than NaCl or KCl, LiCl is, however, also responsible for a marked decrease in the optical rotation of the native form with increasing salt concentrations, as shown in Fig. 39, taken from the data of Mandelkern and Roberts,[974] which contrasts with the behavior of $CaCl_2$ as shown in Fig. 40.[1540] Although both salts lower T_m with increasing salt concentration and, in addition, show a decrease in $\varDelta[\alpha]_{366}$, LiCl produces a decrease in the optical rotation of the native form whereas $CaCl_2$ produces a marked increase. This implies that Li$^+$ salts promote a form of the protein which is more ordered than the native state; this correlates extremely well with the observation of a similar effect of the Li$^+$ ion on the extent of rotation of native DNA.[1503] Studdert *et al.*[1437] observed a similar effect and noted a marked correlation of the temperature dependence of the CD spectra with the enthalpy of hydration of the cation. These authors also found, in agreement with Alder and Fasman,[7] that the guanidinium ion behaves in an opposite manner to the other ions studied, which also appears to correlate well with the data of von Hippel and Wong,[1540] who show that the size of the thermal transition of ribonuclease increases with increasing concentration of the guanidinium ion. Noelken[1097] has also reported a similar effect of reduction of $\Delta(\alpha)$ of a protein when lithium salts are part of the solvent. This author studied the unfolding of S-cysteinyl bovine serum albumin as a function of concentration of added lithium salts. Noelken considers, however, that the effect of the lithium salts is a general nonspecific solvent effect rather than

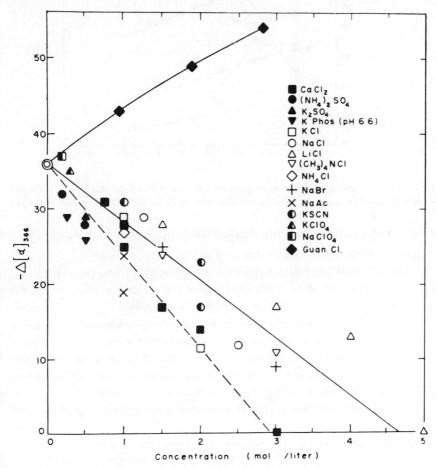

Fig. 38. Size of the thermal ribonuclease transition at 366 nm in various salts; ribonuclease concentration $\simeq 0.5$ w/v (see Ref. 1540.)

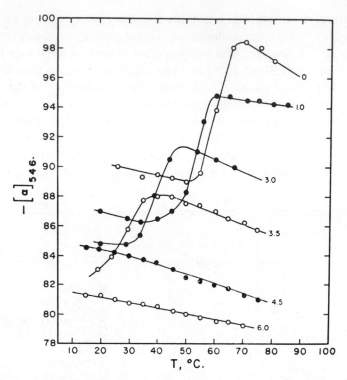

Fig. 39. Plot of specific rotation at 546 nm as a function of temperature for a 1.4% w/v ribonuclease solution containing various concentrations of lithium bromide (see Ref. 974).

a conformational change in the protein, in contradiction to the evidence of Tunis and Hearst[1503] and Studdert et al.[1437]

The effects of the tetraalkylammonium bromides upon the value of T_m for ribonuclease[1540] show quite clearly that the larger the alkyl group, the more effectively the tetraalkylammonium ion depresses T_m. Since there is now a large body of evidence indicating the strong tendency of the tetraalkylammonium ions to stabilize water structure, this tendency increasing with increasing size of the alkyl group, it is clear that an increased degree of hydrophobic hydration of the cation is associated with an increasing effectiveness in depressing T_m.*

Von Hippel and Wong also show that the denaturing capacity of the guanidinium ion, as measured by T_m, is markedly affected by the nature of the associated anion; thus at concentrations above 1 mol dm^{-3} guani-

---

* For up-to-date accounts of the physical chemistry of aqueous solutions of tetraalkylammonium halides see Ref. 1570 and Volumes 2 and 3 of this treatise.

dinium sulfate is in fact a stabilizer of the native conformation. Urea appears to behave in a manner similar to guanidinium acetate, which, since the acetate ion appears to have very little effect on T_m of a variety of biopolymers, might be regarded as being due to the guanidinium ion alone.

Bull and Breese[204] have shown that in the alkali metal ion family the effectiveness in removing water from the protein decreases from Li^+ to Cs^+. Hence restricting discussion to the effect of cations upon the protein confirms the trend shown by von Hippel and Wong[1540] that the greater the degree of hydration of the cation, irrespective of the nature of the hydration (i.e., whether it is by hydrophobic hydration, hydrogen bonding, or Coulombic interaction), the more effective the ion becomes in removing water from the environs of the macromolecule. This is in agreement with the model proposed previously for the effect of cations at low ionic strength upon the extent of hydration of DNA and the comparable effect of these ions upon the structure of the native conformation of both DNA and

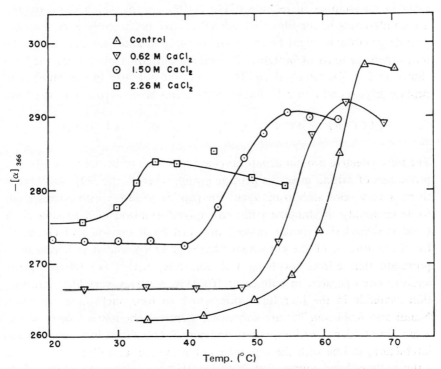

Fig. 40. Plot of specific rotation at 366 nm as a function of temperature for a 0.5% w/v ribonuclease solution in various concentrations of $CaCl_2$ at pH 7.0. All the solutions also contained 0.15 mol dm^{-3} KCl and 0.013 mol dm^{-3} sodium cacodylate. (See Ref. 1539.)

ribonuclease. The behavior of T_m as a function of cation concentration for the two types of biopolymer is, however, somewhat different; in the case of DNA, T_m falls only slightly with increasing cation concentration[472] and appears relatively insensitive to the nature of the cation, whereas in the case of proteins T_m falls appreciably with increasing concentration of the cation and is insensitive to the nature of the cation.

This difference in behavior with respect to T_m would appear to be associated with the differing extent of hydration of the biopolymer after denaturation, the DNA molecule being appreciably less hydrated, whereas the protein molecule is at least as much hydrated in the denatured conformation as in the native form, and probably more so. This difference in behavior may reasonably be attributed to the cation in the light of the work of Nandi and Robinson,[1081] who observed the effects of several salts on the activity coefficients of the N-acetyl ethyl esters of glycine, diglycine, and triglycine, and on formamide, acetamide, and N-methyl acetamide at salt concentrations up to 2.0 mol dm^{-3}. These compounds form a series which differ in the number of peptide (CH_2CONH) groups which they contain. Thus differences in the observed salt effects are presumably effects on the peptide group and might be expected to apply in a similar manner to the repeat peptide units of proteins. The salting-out constants so obtained are shown in Fig. 41, which shows that the effect is linear in the number of peptide groups and that a Hofmeister series can be constructed of the form

$$Cl_3CCOO^- < SCN^- < ClO_4^- < I^- < Br^- < Cl^- < SO_4^{2-}$$

The most effective protein denaturants are thus seen to be the most effective promoters of salting in of the peptide group, whereas the SO_4^{2-} ion is seen to be a very poor stabilizing agent for peptide group, which corresponds to its capability of stabilizing the native conformation of a protein. This trend is remarkably similar to that obtained by Robinson and Grant[352] for the salting in of the constituent bases of DNA and it is tempting to postulate that effects involving CO and NH, which are independently additive, are operative in both cases. There is, however, very little information available in the literature upon which to base such an assumption. Nandi and Robinson[1081] are somewhat sceptical about the role of water, if any, in the denaturation process and postulate a direct ion–amide group interaction, in line with the arguments of Robinson and Grant.[1259] These latter authors have pointed out, however, that the salting-out effects of the sulfate and chloride ions are unlikely to be due to ion binding. In addition, the hydration studies on egg albumin have clearly shown that the protein

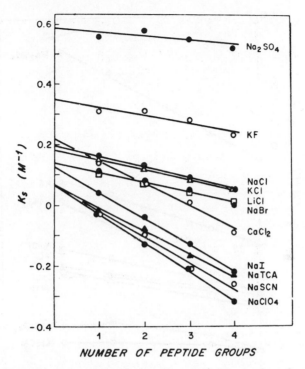

Fig. 41. Salting-out constants of glycine peptides at 25°. The number of peptide groups equals the number of glycyl groups. NaTCA is sodium trichloroacetate. (See Ref. 1081.)

is preferentially hydrated by the SO_4^{2-} ion, less so by the Cl^- ion, whereas Br^-, SCN^-, and I^- dehydrate the protein[202]; there is thus a clear relation between the denaturing capacity of the various anions, their capacity to dehydrate the protein, and their ability to stabilize or salt in the peptide group. It is an additional point of interest that the data of Nandi and Robinson[1081] also show that for the alkali metal chlorides the lithium salt is the most effective salting-in agent, which corresponds to its effective role, previously discussed, in promoting denaturation of the protein.

In a subsequent publication Nandi and Robinson[1082] have shown that the effects of salts upon the peptide groups and nonpolar groups in proteins are independently additive. The authors obtained activity coefficients of N-acetylamino acid ethyl esters with apolar side chains, of the general formula

$$\underset{\substack{\| \\ CH_3-CNH-CHCOC_2H_5}}{\overset{\substack{O \quad\quad R \;\; O \\ \| \quad\quad | \;\; \|}}{}}$$

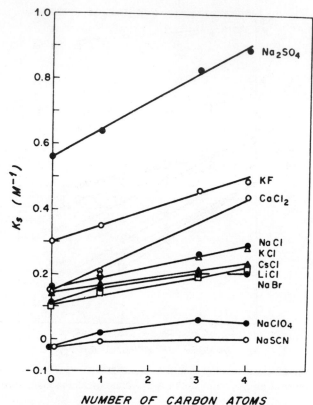

Fig. 42. Salting-out constants of acetylamino acid ethyl esters as a function of the number of carbon atoms in their unbranched aliphatic side chains at 25°C. The compounds corresponding to zero, one, three, and four carbon atoms are acetylglycine ethyl ester, N-acetyl-L-alanine ethyl ester, N-acetyl-L-norvaline ethyl ester, and N-acetyl-L-norleucine ethyl ester. (See Ref. 1082.)

(where R represents alanine, β-alanine, valine, norvaline, leucine, norleucine, or phenylalanine), at salt concentrations up to 2 mol dm^{-3}. The salt effects on acetylglycine ethyl ester are subtracted from effects on the other compounds to give the contributions due to the apolar chains. The results in Fig. 42 show that the side chains are salted out by all the salts investigated and that the extent of salting out is generally an additive function of the number of carbon atoms in the aliphatic side chain; branching of the side chains had no significant effect.

    In general the ions most effective in salting in the peptide group are found to be the least effective in salting out the alkyl side chain, whereas those ions, such as the sulfate ion, that salt out the peptide group are shown to strongly salt out the alkyl side chain. This effect is emphasized since the data pertaining to the sulfate ion were in fact obtained at a salt concentration of 1.0 mol dm^{-3} and would be even more marked at 2.0 mol dm^{-3}. The observed sequence Na$_2$SO$_4$ > CaCl$_2$ > KF > NaBr, NaCl, KCl, LiCl > NaClO$_4$ > CsCl > NaSCN is in close agreement with that of Long and McDevit[915] for the salting-out effects of these salts on benzene.

The evidence presented here lends strong support to the view that the salt effects upon the peptide and nonpolar groups in proteins are additive; the effect of CaCl$_2$ is a particularly good example for detailed consideration. This salt has one of the strongest salting-in effects on the peptide group, but as the number of glycine residues decreases in the acetylglycine esters the effect of CaCl$_2$ changes from salting in to salting out. Increasing the number of carbon atoms in the aliphatic side chain of the acetylamino acid ethyl ester results in an increasingly large salting-out effect by CaCl$_2$, which corresponds with its effects on nonpolar compounds.[8]

The similarity between the effects of salts on apolar side chains of models for proteins and their effects on small nonpolar molecules suggests that the same mechanism operates in both cases and one can conclude that the answer may lie in the change of internal pressure of the water caused by the addition of salts; Long and McDevit[915] have shown that the effect of salts on the activity coefficients on nonpolar compounds can be related to their effects on the internal pressure of water which arises from the electrostriction caused by the salt. An alternative explanation also pointed out by the authors makes use of the scaled particle theory[8]; in this case the effect of the salt is related to the increased work required to create a cavity in the solvent capable of accommodating a nonpolar solute, due to the increased surface energy of the solvent caused by the presence of the salt. This effect is partially compensated by a van der Waals effect based upon the fact that the interaction between a solute and solvent across the interface of the cavity may be stronger when salts are present in the solution. With regard to this latter possibility, however, it should be noted that, although the theory correctly predicts the order of salt effects on a nonpolar solute such as benzene, an arbitrary factor is required to give good agreement between experiment and theory.[1393]*

Nandi and Robinson[1082] have attempted approximate calculations of the effects of four different salts upon the free energy of denaturation of ribonuclease and the results are shown in Table XXXV; only contributions

---

* The application of the scaled particle theory to the treatment of aqueous solutions of apolar compounds is discussed in Section 3.4 of Chapter 1.

TABLE XXXV. Approximate Effects of Salts of Common Cation on the Free Energy of Nonpolar and Peptide Groups in Unfolded Ribonuclease at 25°C [a]

	Na_2SO_4 1.0 mol dm^{-3}	NaCl 2.0 mol dm^{-3}	NaSCN 2.0 mol dm^{-3}	$CaCl_2$ 2.0 mol dm^{-3}
Nonpolar groups	+58.4	+46.6	+12.6	+94.5
Peptide groups	−8.0	−31.1	−87.8	−98.3
Total	+50.4	+15.5	−75.2	−3.8
ΔT_m,[b] deg	+12[c]	+1	−28	−24

[a] Data from Ref. 1082. Free energies in kJ mol^{-1}.
[b] ΔT_m is the change in the transition temperature, taken from Ref. 1540.
[c] $(NH_4)_2SO_4$.

from peptide groups[472] and nonpolar side chains are considered, all other contributions being neglected; it is also assumed for the purposes of the calculation that all these groups are fully exposed to the solvent.

Agreement between the changes in free energy and ΔT_m is good, with the exception of $CaCl_2$; the salting-in effect of $CaCl_2$ appears to be small, which contrasts with the strong denaturing effect of the salt. The authors suggest, in agreement with Tanford,[1466] that in this solvent the preferred conformation of the protein is not that of a random coil but one in which the nonpolar groups are partially shielded from the solvent.

Similar approximate calculations performed for the alkali metal halides, shown in Table XXXVI, also show good agreement between the changes in

TABLE XXXVI. Approximate Effects of Salt of Common Anion on the Free Energy of Nonpolar and Peptide Groups in Unfolded Ribonuclease[a]

	KCl	NaCl	CsCl	LiCl
Nonpolar groups	+46.2	+46.6	+39.0	+35.0
Peptide groups	−28.5	−31.1	−28.5	−41.5
Total	+17.7	+15.5	+10.5	−6.5
ΔT_m, deg	+1.5	+1	—	−1

[a] Data from Ref. 1082. Free energies in kJ mol^{-1}. Salt concentrations, 2.0 mol dm^{-3}.

free energy and ΔT_m, except for 2.0 mol dm^{-3} CsCl, for which information on ΔT_m is lacking. It is interesting, however, that the Hofmeister series is of the order K$^+$, Na$^+$ < Cs$^+$ < Li$^+$, which agrees with the sequence observed by von Hippel and Wong[1539] for the effect of various ions upon the stability of the collagen fold in gelatin and by Dix and Straus[352] on the transition temperature of DNA; perhaps most important, it is also in good agreement with the work of Bull and Breese[202] (Table XVII) on the extent of ion binding by the alkali metal ions to egg albumin. Thus the Li$^+$ ion, the most effective denaturant of ribonuclease, has the greatest salting-in effect upon the peptide group and the smallest salting-out effect upon the nonpolar group. The Cs$^+$ ion has a much smaller salting-in capacity for the peptide group, comparable to that of the K$^+$ and Na$^+$ ions, but has a much smaller salting-out capacity for the nonpolar group compared to K$^+$ and Na$^+$ ions, which are of comparable strength. This correlates with the binding of Li, Na, K, and Cs chlorides to egg albumin, being 5.6, 0.2, 2.4, and 5.2 mol/mol, respectively.

The evidence of Nandi and Robinson[1081,1082,1259] indicates convincingly that the effects due to the peptide and nonpolar groups of the macro-molecule are additive in their effect upon T_m. Von Hippel and Wong[1540] have also suggested that the effects of salts upon T_m are additive in character and the binding data of Bull and Breese[202] suggest that hydration effects due to the cation may be predominant in the case of the peptide groups, while those due to the anion may predominate in the nonpolar group interactions. If similar information to that of Nandi and Robinson[1082] and von Hippel and Wong[1540] were available for other alkali metal salts, such as the nitrates, it would be feasible to conduct a Kohlrausch type analysis to determine if the additivity relationship holds for the contributions of individual ions to the effect upon T_m.

Schleich and von Hippel[1326] have recently reported studies on the temperature dependence of the water-proton nmr chemical shift in a variety of concentrated electrolyte solutions in an attempt to extend to proteins the studies of Gordon et al.,[547] who showed that in a variety of concentrated sodium perchlorate solutions calf thymus DNA is denatured at a constant value of the (sodium-ion-corrected) water-proton chemical shift. The variation of the sodium-ion-corrected sodium perchlorate-induced chemical shift of water protons with temperature and sodium perchlorate concentration, and the transition temperature of ribonuclease in these same solutions, are shown in Fig. 43. An excellent linear relationship of the kind observed by Gordon et al. is obtained, but other salts (NaBr, NaCl, NaSCN) do not give similar horizontal lines (Fig. 44).

Gordon *et al.*[547] interpreted their data as indicating that perchlorate affects the stability of DNA through its power to disorganize water structure and not by direct interaction with the macromolecule. Schleich and von Hippel suggest that their data confirm this view of the effect of perchlorate but in the cases of the other salts examined, different interactions, involving perhaps binding of the ion to the macromolecule, destroy this linear relationship.

Noelken[988] in a report on the interaction of *S*-cysteinyl bovine serum albumin with lithium salts and guanidinium hydrochloride, based upon equilibrium analysis and differential refractometry, has shown (Table XXXVII) that preferential solvation of this protein by the sulfate ion is accompanied by an increased extent of helix formation. The lithium salts and guanidinium hydrochloride are shown to denature the helix conforma-

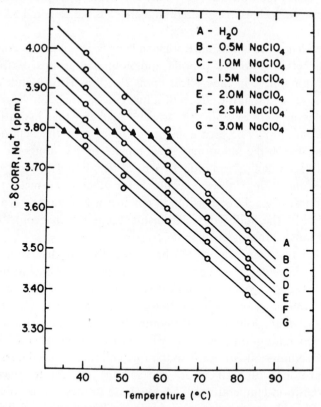

Fig. 43. Variation of the water proton chemical shift (sodium ion corrected) with temperature for a variety of sodium chlorate solutions, versus the transition temperature of ribonuclease in the same solutions (see Ref. 1326).

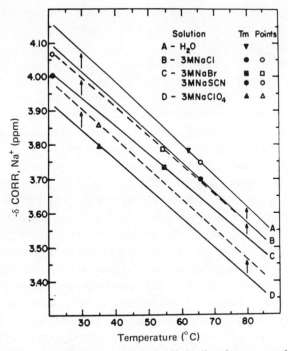

Fig. 44. Variation of the water proton chemical shift (sodium ion corrected) with temperature for a variety of sodium chlorate solutions, versus the transition temperature of ribonuclease in the same solutions (see Ref. 1326).

TABLE XXXVII. Optical Rotatory Dispersion Parameter b_0 and Preferential Solvation of S-Cysteinyl Bovine Serum Albumin in Aqueous Salt Solutions[a]

Salt	Concentration, mol dm^{-3}	$-b_0$	Mol salt/mol protein	Mol water/mol protein
LiSO$_4$	1.7	290	−60	1160
LiCl	6.6	135	65	—
LiNO$_3$	6.4	140	0	—
LiBr	7	65	120	—
LiI	7	55	160	—
LiSCN	7.2	65	160	—
GaHCl	6	40	100	590
NaCl	4.6	320	−55	—
KCl	3.2	310	−65	870

[a] Data from Ref. 1098.

tion preferentially by what the author describes as "a thermodynamically reversible process that is accompanied by relatively nonspecific protein–denaturant interactions."

The evidence presented in this discussion of the effects of salts would seem to show that at high salt concentrations an additivity relationship exists for a wide variety of proteins. Similarly, an additivity relationship involving the contributions of nonpolar and peptide groups to the stability of the native conformation of proteins appears very likely. The relative involvement of cation and anion in the salting in or salting out of the non-polar and peptide groups is less clear-cut; it does appear, however, to be the case that the major stabilizing effect of the SO_4^{2-} ion upon the native conformation of a protein, by promoting extensive hydration around the protein, is through the prevention of the salting in of the nonpolar group leading to denaturation, rather than by involvement with the peptide group to any major extent. The Cl^- ion is less effective in this context and the SCN^- ion, which has been shown to dehydrate the protein, appears to be effective mainly by a salting in of the peptide groups rather than a salting out of the nonpolar groups. This is perhaps to be expected, since a reduction in the hydration envelope surrounding the protein should make it relatively easier to salt in the nonpolar group, which appears to be the case. In the case of the alkali metal ions the Li^+ ion, which has been shown to dehydrate the protein molecule, has the most pronounced tendency to salt in the peptide groups and the lowest tendency to salt out the nonpolar groups, which is again to be expected, as in the case of the SCN^- ion. It is therefore not surprising that LiSCN is the most effective of all the lithium salts in causing dehydration of the protein.[1098]

5. SOLVENT EFFECTS OF AQUEOUS NONELECTROLYTES ON THE CONFORMATIONAL STABILITY AND STRUCTURE OF NUCLEIC ACIDS, PROTEINS, AND PEPTIDES

5.1. Urea and Related Compounds As Probes for the Investigation of Solvent Effects on the Stability and Structure of the Biopolymer

The nature of the effect of urea and substituted ureas on the stability of the native conformations of proteins has been of continuing interest; at low to moderate concentrations (2–4 mol dm^{-3}) there is some evidence[1284] that urea has a stabilizing influence on the native conformation, but at higher concentrations urea and its related compounds act as denaturants.

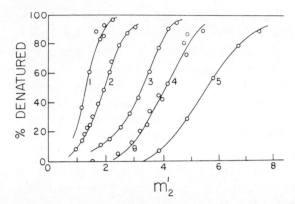

Fig. 45. Percentage denaturation of egg albumin as a function of the molal concentration of the denaturants in the reference solutions. Curve 1, guanidine HCl; curve 2, NaCNS; curve 3, NaI; curve 4, urea; curve 5, acetamide. One week at 25°C (See Ref. 203.)

The mechanism of urea denaturation of proteins has not been unequivocally defined but at various times particular explanations have been favored.[1467]

One of the important aspects of the capacity of a salt or nonelectrolyte to act as a denaturant for proteins is the capability for preferentially hydrating or dehydrating the protein; the data of Bull and Breese[203] on the degree of solvent and solute binding of urea and guanidine hydrochloride to egg albumin are therefore of particular interest. These authors, on the basis of isopiestic data, determined the extent of denaturation as a function of concentration of the denaturant, shown in Fig. 45, and the number of moles of solute and solvent bound by the protein at denaturant concentrations in excess of 5 mol dm^{-3} (Table XXXVIII). The data in Fig. 45 indicate that denaturation by guanidine hydrochloride occurs over a cosolute concentration range of 1–2 mol dm^{-3}, but for urea, much higher concentrations and a

TABLE XXXVIII. Number of Moles of Water Δn_1 and Solute Δn_2 Bound per Mole of Egg Albumin^a

Solute	Δn_1	Δn_2
Urea	88.9 ± 14.6	32.5 ± 2.5
Guanidine HCl	40.7 ± 10.0	32.8 ± 1.3

^{a} Data from Ref. 203.

wider concentration range are required (2–5 mol dm^{-3}). The data in Table XXXVIII show that above concentrations of 5 mol dm^{-3}, for both denaturants, the protein binds the same amount of urea and guanidine hydrochloride, but urea is not as effective a dehydrating agent, the protein retaining approximately 90 mol water/mol protein, whereas in the presence of guanidine hydrochloride only 41 mol of water is retained. It appears from the data of Bull and Breese that the unfolding in the presence of urea is much less cooperative in character than is the case with guanidinium hydrochloride.

Kuntz and Brassfield,[829] using the rapid freezing technique described earlier,[830] measured the pmr spectrum of water remaining unfrozen when a protein solution is cooled to −25°; data are reported for a 10% solution of bovine serum albumin in the presence of urea, methyl urea, ethyl urea, and guanidine hydrochloride (Table XXXIX) and show that the linewidth of the pmr signal associated with bound water broadens by a similar amount for all the denaturants, indicating a lack of specificity. On the assumption that pmr signal is primarily due to bound water and that the linewidth of the signal is directly proportional to a correlation time the authors estimated activation entropies (Table XL). Effects of ureas can be ascribed to denaturation of the protein and the ureas do not contribute directly to the pmr signal; the ureas do not therefore act by a direct, large-scale replacement of bound water. This appears to contradict the evidence of Bull and Breese,[203] but it must be remembered that pmr measurements only account for the most firmly "bound" water molecules, whereas the isopiestic data of Bull and Breese may include a much larger region of "hydration water."

TABLE XXXIX. Denaturant Effects on Water Line Width^a

Solute	Line width ± 10%, Hz
None	550
0.5 mol dm^{-3} urea	1000^b
1 mol dm^{-3} urea	1400
8 mol dm^{-3} urea	1600
1 mol dm^{-3} methyl urea	1500
1 mol dm^{-3} ethyl urea	1400
0.5 mol dm^{-3} guanidine HCl	1100^b

^{a} 10% bovine serum albumin 0.01 mol dm^{-3} KCl, −25°. Data from Ref. 829.
^{b} Denaturation probably not complete.

TABLE XL. Activation Parameters for Water, Ice, and "Bound" Water[a]

	Bulk water		Ice $-25°C^c$	"Bound" water, $-25°C$	
	$+5°C^b$	$-25°C^c$		Native protein	Urea-treated protein
Correlation time, sec	2×10^{-11}	7×10^{-11}	5×10^{-4}	10^{-8}	3×10^{-8}
Activation enthalpy, kJ mol^{-1}	$+18.9$	$+21$	$+54.6$	$+21$	$+37.8$
Activation entropy, J mol^{-1} K^{-1}	$+31.1$	$+42$	$+42$	-42	$+16.8$
Amount of bound water, g H$_2$O/g protein	—	—	—	0.40	0.45

[a] Data from Ref. 829.
[b] Dielectric relaxation data of Collie et al., taken from Eisenberg and Kauzmann.[387]
[c] Extrapolated from the data of b.

It is, however, interesting to note that the relative increase in "bound" water observed by Kuntz and Brassfield when the protein is denatured by urea is similar to the increase in hydration observed by Privalov and Mrevlishvili[1198] from direct calorimetric observation of the thermal denaturation of bovine serum albumin.

The small number of urea molecules required to produce 50% denaturation of bovine serum albumin ($\sim$6.5 ureas/protein) led Kuntz and Brassfield to conclude that the most plausible explanation for the denaturation of the protein at low urea concentrations in frozen solutions is that the low temperature tends to destabilize the protein and that the ice lattice, by excluding urea, concentrates the denaturant in the neighborhood of the protein. However, it is also possible that the increased linewidth (as well as the increase in the energy of activation of the bound water) is related to an increased ordering of the water as hydrophobic residues are exposed during denaturation and/or a restriction of proton mobility due to interaction of water with the binding sites of the denatured protein.

In the context of evidence for or against the binding of urea to biopolymers, the previously reported evidence of Klotz and Shikama[792] should be recalled; binding of Methyl Orange by bovine serum albumin is suppressed by urea, as is also the case with a random coil synthetic polymer polyvinyl pyrrolidone. The mechanism of the action of urea is thus not limited to disruption of N—H$\cdots$O=C bonds or breaking of hydrogen bonds between the anion and the polymer, since these do not exist in polyvinyl pyrrolidone. Viscosity evidence[791] shows that in 6 mol dm^{-3} urea solution polyvinyl pyrrolidone shows no change in its intrinsic viscosity; thus urea produces no change in the unordered conformation of this water-swollen polymer. The effect of urea, on the basis of this evidence, involving very different natural and synthetic polymers, is unlikely to be by direct combination with the polymer, thus implying that interactions involving the solvent are the cause.

The data of Robinson and Jencks,[1260,1261] however, appear to contradict this view of the role of urea; these workers reported studies of the effect of urea, several substituted ureas, guanidine hydrochloride, and its substituted derivatives on the activity coefficients of acetyltetraglycine ethyl ester

$$\underset{\displaystyle CH_3CNHCH_2CNHCH_2CNHCH_2CNHCH_2COC_2H_5}{\overset{\displaystyle O \qquad O \qquad O \qquad O \qquad O}{\| \qquad \| \qquad \| \qquad \| \qquad \|}}$$

in aqueous solution. The data, shown in Table XLI, indicate that increased alkyl group substitution of urea or guanidine hydrochloride results in a lowering of the solubility of acetyltetraglycine ethyl ester, whereas the converse is the case for compounds that contain nonpolar groups. The data thus show the "nonhydrophobic" nature of the effect of urea on acetyltetraglycine ethyl ester, confirmed by the effect of temperature on the activity coefficient effect of urea on the peptide decreasing with increasing

TABLE XLI. Solubility and Free Energy of Transfer of Acetyltetraglycine Ethyl Ester at 25°C in Various Urea Solutions[a]

Solvent[b]	Concentration, mol dm^{-3}	S/S_0,[c]	ΔG_t, J mol^{-1}
Urea	3	1.85	−1722
Ethylurea	3	1.68	—
Ethyleneurea	3	1.46	—
1,3-Dimethylurea	3	1.41	−1323
Tetramethylurea	3	0.74	—
Guanidine hydrochloride	3	3.5	−6216
1,1,3,3-Tetramethyl guanidine hydrochloride	3	0.63	+113

[a] Data from Refs. 1260, 1261.
[b] All solutes are in 3 mol dm^{-3} solution.
[c] S_0 = solubility in water; S = solubility in solvent.

temperature over the range 0–40°; the reverse effect is observed with hydrocarbons, increasing with increasing temperature between 5 and 50°.[1580]

The capability of reducing the solubility of the peptide shown by the alkyl-substituted ureas and guanidinium hydrochloride closely matches their denaturing effect on bovine serum albumin.[548] The authors suggest that a major part of the denaturing capability of urea and guanidine for some proteins is accounted for by a specific, nonhydrophobic interaction of the solvent with the peptide and amide groups of the protein. This appears to be confirmed by the solubility data of carbobenzoxydiglycineamide in a corresponding series of solvents to those used with acetyltetraglycine ethyl ester; this molecule contains a large hydrophobic group in addition to several amide groups. The solubility is increased by alkyl-substituted ureas and guanidine hydrochlorides; a similar solubility dependence on denaturant concentration is observed with small molecules, such as toluene. The effectiveness of the various reagents in influencing the solubility of carbobenzoxydiglycineamide can be regarded as a combination of the effects on the hydrophobic and amide parts of the molecule, which corresponds well to the denaturation of proteins such as ovalbumin.

The solubility of acetyltetraglycine ethyl ester is shown to increase linearly with increasing urea or guanidine hydrochloride concentration (Fig. 46), which implies complex formation between the peptide and co-

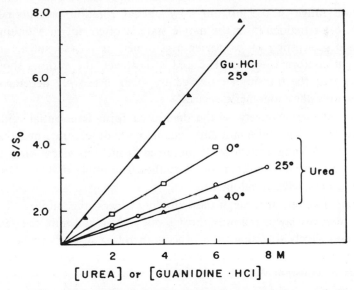

Fig. 46. Solubility of acetyltetraglycine ethyl ester in urea solutions at 0, 25, and 40° and in guanidine hydrochloride solutions at 25° (see Ref. 1260).

solute since nonspecific solvent effects are often observed to follow a logarithmic increase with increasing denaturant concentration.[915] The greater solubilizing effect of guanidine hydrochloride over that of ureas on polytetraglycine ethyl ester can suggest that binding occurs by the two mechanisms

with structure II having a higher stability than structure I.

Pace and Tanford[1129] have examined the thermodynamics of unfolding of the protein β-lactoglobulin A in aqueous urea solutions at pH 2.5–3.5; the denaturation was followed at several temperatures over a urea concentration range of 0–9 mol dm^{-3} by optical rotation measurements at 365 nm. β-Lactoglobulin in the native state is often not in a unique conformation, consisting of monomer and dimer; it is also subject to pH-dependent conformations at pH 5 and 8; however, the authors show conclusively that the transition observed by them is not pH dependent and involves only the monomeric species.

The intrinsic viscosity of the denatured form is essentially identical in 8 mol dm^{-3} urea and 6 mol dm^{-3} guanidine hydrochloride and the slope of the Moffit–Yang equation[1028] obtained from optical rotation data is close to zero in both solvents; the data therefore imply that the denatured form is a completely random coil.

From a kinetic analysis of the unfolding process the authors show that a first-order rate law is followed throughout and, in addition, the denaturation is completely reversible; the process therefore appears to be of a reversible two-state nature and the data can be expressed in terms of an equilibrium constant K

fraction in native conformation $\rightleftharpoons$ fraction in random conformation

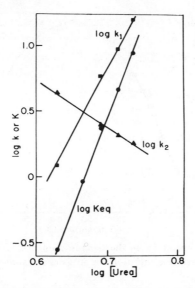

Fig. 47. Comparison of the effects of urea concentration on the equilibrium constant and on the rate constants for the forward and reverse reactions at pH 2.8 and 15°C for the unfolding of β-lactoglobulin A (see Ref. 1129).

The authors found K to be a linear function of the logarithm of the urea concentration and Fig. 47 illustrates the behavior of log K vs. log[urea], together with the values of k_1 and k_2, the rate constants for the denaturation and renaturation processes; k_1 is much more dependent upon the concentration of urea than is k_2, which is the opposite behavior to that observed for the unfolding of lysozyme by guanidine hydrochloride.[1472] The effect of temperature on the denaturation at a pH of 3.0 is shown in Fig. 48, which illustrates the relationship between the urea concentration for 50% denaturation [urea]$_{1/2}$ and the temperature; the thermodynamic parameters for the denaturation process are summarized in Table XLII. No systematic trend is observed, indicating that ΔH and ΔC_p are constant within experimental error and unaffected by the type or concentration of the denaturant; the large value of ΔC_p is characteristic of the unfolding process of globular proteins, as discussed in Section 4.4.

Since log K is a linear function of the logarithm of the urea concentration over the larger part of the transition region, the reaction is described by the equation

$$K = K_0 C^n$$

where C is the concentration of urea in mol dm^{-3} and K_0 is a constant without

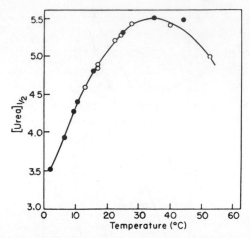

Fig. 48. The effect of temperature on the denaturation of β-lactoglobulin A at pH 3.0. [urea]$_{1/2}$ represents the urea concentration for 50% denaturation of the protein. (See Ref. 1129.)

physical significance; at the midpoint of the transition $K = 1$ and $C = C_{1/2}$; the urea concentration at the midpoint of the transition equals $(1/K_0)^{1/n}$.

For a two-state transition Tanford[1463] has shown that the value of n is related to the standard free energy of unfolding $\Delta G_n^{\ominus}$ by the equation

$$RTn = -[\partial(\Delta G_n^{\ominus})/\partial \ln C]_{C=C_{1/2}}$$

irrespective of the mechanism of the urea effect upon denaturation.

TABLE XLII. Enthalpy and Heat Capacity Changes Accompanying the Denaturation of β-Lactoglobulin A[a]

Urea concentration mol dm^{-3}	pH	ΔH at 25°, kJ mol^{-1}	ΔC_p, J mol^{-1} K^{-1}
4.42	2.50	−75.6	8946
4.48	2.50	−79.8	9072
5.01	2.71	−88.2	8526
5.03	3.00	−88.2	9030
5.09	2.57	−96.6	9786
5.27	2.78	−88.2	8148
5.53	2.55	−84.0	7812
Guanidine hydrochloride			
3.37	3.20	−88.2	8820

[a] Data from Ref. 1129.

The derived value of n for the rate constant k_1 is much larger than that for k_2, which leads the authors to suggest that the activated complex for the denaturation of β-lactoglobulin is similar to the random coil conformation in its degree of unfolding, in contrast to lysozyme,[1472] where the activated complex appears to be quite highly folded. The authors suggest that the value of n for the overall denaturation process[1463] confirms a previous proposal[1463] that the action of urea involves local solvent effects.

The magnitude of ΔC_p for the transition is very similar to that obtained with other proteins in the absence of urea or guanidine hydrochloride, indicating the involvement of hydrophobic hydration in the process as previously buried nonpolar side chains are exposed to the solvent during denaturation.

Hopkins and Spikes[677,678] have monitored the denaturation of chymotrypsin, chymostrypsinogen, trypsin, trypsinogen, and some of their derivatives by fluorescence measurements using 290-nm UV radiation; the data were expressed as a function of urea concentration at neutral pH. No shift in fluorescence wavelength was observed with any of the proteins at a urea concentration of 5 mol dm^{-3}, but the rate of shift increased rapidly as the urea concentration was increased from 6 to 8 mol dm^{-3}. The denaturation reactions for all of the biopolymers studied were found to be first order, with the rates for chymotrypsin and trypsin being very similar, of the order of 1.0 min^{-1}; the rates for trypsinogen, diisopropyl-fluorophosphate-treated trypsin, and 1-chloro-2-tosylamido-7-amino-2-heptone-treated trypsin become progressively larger.

Pace and Tanford[1129] also observed first-order kinetics in the urea denaturation of β-lactoglobulin and, in addition, showed the denaturation to be completely reversible. Quantitative data on the number of urea molecules "bound" per protein molecule indicates quite small numbers, while at the same time requiring a denaturant concentration of the order of 8 mol dm^{-3} for denaturation to occur.

The evidence thus far reviewed would seem to indicate that "binding" of urea by the macromolecule is unlikely to play a major role in the denaturation of the protein; interactions via the solvent appear to be a much more likely possibility.

Hermans and co-workers[644] have reported the effect of urea and ethanol on the denaturation of sperm whale myoglobin as a function of pH, the solutions containing 0.01% of protein, 0.1 mol dm^{-3} of KCl, and 0.01 mol dm^{-3} of citrate. Optical rotation studies at 233 nm and optical density measurements at 410 nm were used to follow the change in conformation as a function of pH at each solute concentration. The authors report that denatured myoglobin retains a large degree of helicity and suggest that the addition of urea or ethanol to the denatured form of the macromolecule induces a change in conformation of the protein and that this is not a cooperative transition of the kind associated with denaturation. The authors conclude that the interpretation of large values of ΔC_p of denaturation as evidence of hydrophobic interactions should be treated with caution because of an undetermined contribution to ΔC_p from changes in the conformation of the denatured species with temperature. This may possibly also be the case with myoglobin, but the work of Tanford and Pace[1129] has shown that no residual helicity remains in proteins such as β-lactoglobulin, large values of ΔC_p still being observed which are unaffected by the concentration of denaturant.

Hermans et al. determined the dependence of the free energy of denaturation of myoglobin at neutral pH upon the concentration of denaturant, and the results are presented in Table XLIII. The authors suggest that the decrease in $\Delta G_{\text{denat}}^{\ominus}$ upon addition of ethanol can only be obtained if relatively more hydrophobic groups than peptide bonds become exposed to the solvent during denaturation; presumably a similar reasoning must also apply in the case of urea addition.

Herskovits and co-workers[649] have studied the effects of urea and various alkyl-substituted ureas on the denaturation of sperm whale myoglobin, horse heart cytochrome C, and α-chymotrypsinogen by ORD, spectral, and difference spectral methods; protein concentrations were in

TABLE XLIII. Free Energy Changes for the Denaturation of Myoglobin at Neutral pH Values in Various Solvents[a]

Solvent	$\Delta G_{\text{denat}}^{\ominus}$, kJ mol^{-1}
Water	78.1
2 mol dm³ Urea	68.5
4 mol dm^{-3} Urea	49.1
6 mol dm^{-3} Urea	30.7
10% EtOH	74.3
20% EtOH	58.8
30% EtOH[b]	47.5

[a] Data from Ref. 644. All solutions contained 0.1 mol dm^{-3} KCl.
[b] An interpolated value.

TABLE XLIV. Concentration of Various Denaturants Required to Obtain the Midpoint of the Denaturation Transition for Sperm Whale Myoglobin, Cytochrome c, and α-Chymotrypsinogen[a]

Denaturant	Myoglobin, mol dm^{-3}	Cytochrome, mol dm^{-3}	α-Chymotrypsinogen, mol dm^{-3}
Urea	6.6	6.6	6.9
Methylurea	5.6	5.6	6.5
1,3-Dimethylurea	5.4	4.9	6.5
Tetramethylurea	3.6	3.5	—
Ethylurea	3.8	4.7	4.7
1,1-Diethylurea	2.1	2.8	—
1,3-Diethylurea	1.9	3.1	—
Propylurea	1.8	2.3	3.1
Butylurea	0.6	0.7	—

[a] Data from Ref. 649.

the range 0.007–0.14% and all solutions contained 0.1 mol dm^{-3} (pH 5.7) acetate buffer. The concentrations of denaturant producing 50% denaturation of the proteins are given in Table XLIV.

Herskovits *et al.* also determined the solubilities of the model compounds N-acetyl-L-tryptophan ethyl ester and indole in the various ureas. The results, expressed as S/S_0, where S and S_0 represent the solubility of the compound in water and that in the mixed solvent, respectively, are listed in Table XLV, together with the free energy of transfer of the compound from water to the various urea solutions at a concentration of 1 mol dm^{-3}.

The denaturing effectiveness of the ureas increases with increasing alkyl substitution; similarly, the solubilities of indole and N-acetyl-L-tryptophan ethyl ester increase with increasing alkyl content of the substituted ureas and the free energy of transfer becomes more negative. Behavior of this kind is characteristic of hydrophobic interactions involving the protein.

The data of Robinson and Jencks[1260,1261] (Table XLI) show that for the peptide acetyltetraglycine ethyl ester the converse situation holds; increasing alkyl substitution of the urea results in a less favorable free energy of transfer of the peptide from water to the urea solution. Thus,

TABLE XLV. Solubility and Free Energy of Transfer of Indole and *N*-Acetyl-L-Tryptophan Ethyl Ester in 1.0 mol dm^{-3} Urea Solutions at 25°C[a]

Solvent	pH[b]	S/S_0	ΔG_t, J mol^{-1}
N-acetyl-L-tryptophan ethyl ester			
Water	6.5	1.00	—
Urea	5.2	1.24	−533
Methylurea	5.3	1.45	−924
1,3-Dimethylurea	5.0	2.00	−1726
Tetramethylurea	5.0	4.11	−3532
Ethylurea	5.0	1.80	−1462
1,1-Diethylurea	5.0	3.09	−2806
1,3-Diethylurea	5.0	2.73	−2499
Propylurea	5.0	2.33	−2104
Butylurea	5.1[c]	1.84[c]	−304[c]
Indole			
Water		1.00	—
Urea	—	1.24	−533
Methylurea	—	1.47	−958
Ethylurea	—	1.64	−1235
Propylurea	—	1.86	−1546
Butylurea	—	1.60	−2344

[a] Data from Ref. 649.
[b] pH of unbuffered 1.0 dm^{-3} aqueous solutions.
[c] Data given are obtained on 0.5 mol dm^{-3} butylurea solutions and ΔG_t is calculated for 1.0 mol dm^{-3} solution assuming that the log(S/S_0) function is linear with increasing butylurea concentration.

as pointed out by Herskovits and co-workers for the proteins examined, the hydrophobic contributions involving the nonpolar parts of the molecule outweigh the urea–peptide bond interactions in their effects upon conformational stability.

From the theories of Peller[1148] and Flory[458] the following equation was derived by Herskovits *et al.* to account for the effect of salts and denaturants on the protein transition:

$$\theta_m = - \left[\frac{\Delta T_m \, \Delta H_m}{2.303 R T_m T_m{}^0 \nu} \right] \frac{1}{K_s}$$

where θ_m is the solvent denaturation midpoint, T_m and $T_m{}^0$ are the denaturation temperatures in the presence and absence of denaturant, $\Delta T_m = T_m{}^0 - T_m$, ΔH_m is the enthalpy change of unfolding per peptide unit, and $\bar{\nu}$ is the effective number of denaturant molecules bound per peptide unit in the unfolded conformation of the macromolecule. The constant K_s is related to the free energy of transfer thus

$$\Delta G_{\mathrm{tr}} = -RT \ln(\theta/\theta_0) = 2.303 RTK_s C_D$$

where C_D is the concentration of the denaturant in mol dm^{-3}.

Table XLVI presents a comparison of experimentally determined midpoints and calculated values of θ_m for myoglobin; $T_m{}^0$ and ΔH values of $355°K$ and 2.1 kJ mol^{-1}, respectively, were taken from the data of Hermans and Acampora[642] and a value of 0.1 is assumed for $\bar{\nu}$. Absolute agreement between the results is not particularly good but the same trend in the data is clearly present, and if θ_m is calculated solely on the basis of a free energy contribution from binding of the alkyl ureas, values of 4.7, 1.8, and 0.7 mol dm^{-3} for ethyl, propyl, and butyl urea are obtained, compared with the experimental values of 3.8, 1.8, and 0.6 mol dm^{-3} for myoglobin, a trend which the authors suggest strongly implies an increasing importance of hydrophobic interactions.

Very little attention recently has been paid to the effects of urea on the fibrous proteins such as collagen; earlier reports[1521] have indicated that at concentrations as high as 8 mol dm^{-3} urea denatures the natural triple helix form of collagen. Schleich and von Hippel[1325] have reported that urea has

TABLE XLVI. Comparison of Calculated and Experimental Solvent Denaturation Midpoints, θ_m of Sperm Whale Myoglobin Obtained with Various Solvents at pH 5.7 and 25°C[a]

Denaturant	θ_m (calc)	θ_m (esp)
Urea	6.3	6.6
Methylurea	3.7	5.6
1,3-Dimethylurea	2.0	5.4
Tetramethylurea	0.9	3.6
Ethylurea	2.3	3.8
1,1-Diethylurea	1.2	2.1
1,3-Diethylurea	1.3	1.9
Propylurea	1.6	1.8
Butylurea	1.1	0.6

[a] Data from Ref. 649.

a stabilizing effect upon the poly-L-proline II form, the extent of mutarotation increasing apparently linearly over the range 0–8.0 mol dm^{-3}. Thus
urea alone of all the compounds studied by these authors gave a positive
value for B ($+27°$ mol^{-1}) in the linear equation

$$[\alpha] = [\alpha]_0 - BM$$

where $[\alpha]_0$ is the specific levorotation of the "salt-free" system (actually
0.1 mol dm^{-3} KCl) and B is the slope of the plot of $[\alpha]$ versus the concentration M of the additive; the authors make very little comment upon
this exceptional behavior with regard to urea. Since the poly-L-proline II
structure does not have the normal protein type of peptide bond, it would
appear that "nonhydrophobic" effects of the type described by Robinson
and Jencks[1260] cannot operate in this case. The linear increase in the degree
of mutarotation of poly-L-proline with increasing urea concentration indicates that the process is noncooperative in nature, and suggests a continuous
tightening of the left-handed helix as the urea concentration increases. It is
not possible on the evidence available to distinguish what type of interaction
is responsible for the increased extent of folding of the peptide helix; whether
this involves changes in the configurational interactions of the proline rings
or changes in the extent of hydrophobic interactions is an open question at
the present time; it should be particularly interesting to study the effect of
alkyl substitution of urea on its efficiency in promoting formation of the
poly-L-proline II type helix.

The Russel and Cooper[1284] have reported that urea has a stabilizing effect
upon the collagen fold in gelatin solutions at moderate concentration
(0–2 mol dm^{-3} urea). Renaturation kinetics were observed by these authors
for acid-soluble calf skin collagen in 0.15 mol dm^{-3} potassium acetate buffer
at a pH of 4.8 after heating to 45° for 15 min prior to cooling to the temperature at which renaturation was studied; the protein concentration used
in these experiments was 0.086% w/v.

The initial rate of reversion to the collagen fold, observed by optical
rotation at 365 nm and a temperature of 15°, showed a progressive decrease
with increasing urea concentration, but the rates of reversion after 2 hr
were greater than in the case of the urea-free solution. Reduced viscosity
studies confirmed the optical rotation data and further showed that the
optimum urea concentration for recovery of the collagen fold is of the
order of 1 mol dm^{-3}.

The authors suggest that the decrease in the initial rate of reversion
to the collagen fold with increasing urea concentration is consistent with
the von Hippel and Harrington[1536] model of a nucleation step followed

by a propagation step for generation of the collagen fold; it is suggested that the presence of urea retards the nucleation of the poly-L-proline II type helix by local configurational changes in the imino acid regions of the peptide chain by decreasing the rotational restriction about the α-carbon-to-carbonyl carbon bond in these regions. This picture is, however, at variance with the data of Schleich and von Hippel,[1325] who show that urea is capable of stabilizing this particular conformation. Russell and Cooper[1284] suggest that the enhancement of the propagation step by the presence of urea can be attributed to competitive hydrogen bond formation by the amide at the carbonyl and imino groups on the exposed chains of the denatured gelatin; however, this proposition appears in conflict with the evidence of Privalov and Tiktopulo,[1201] who have shown that water bridging between the imino acid groups of the molecule plays a considerable role in the stability of the helix conformation.

Veis[1521] has reported that at urea concentrations of the order of 2–4 mol dm^{-3} the poly-L-proline I conformation is preferentially converted into the II conformation. It seems to be the case that the poly-L-proline II molecule, and to a certain extent the poly-L-proline II type structure of collagen, is preferentially stabilized by urea; for lack of evidence any further suggestions as to the role of urea must be highly speculative; however, it seems quite possible that the stabilizing effect of urea may be due to its replacing the bridging water molecules between the imino acid groups and performing the bridging function itself. Its denaturing effect on collagen and gelatin at higher concentrations appears to be analogous to its effect upon globular proteins and presumably functions by similar interactions.

5.2. The Influence of Alcohols and Glycols on the Protein Denaturation–Renaturation Equilibrium

As part of their wider study of the effects of various electrolytes and nonelectrolytes on the conformational stability of globular proteins, von Hippel and Wong[1540] reported on the effects of various monohydric alcohols and ethylene glycol on the thermal transition of ribonuclease. They found, in agreement with earlier work,[1345] that the alcohols had a destabilizing effect upon the native conformation of the protein, T_m being progressively lowered by an increase in the alcohol concentration. Also, the denaturing capability of the alcohol increased with increasing alkyl group size, the dihydric alcohol, ethylene glycol, being the least effective denaturant of the group studied. Figure 49 shows T_m of ribonuclease as a function of concentration of various alcohols.

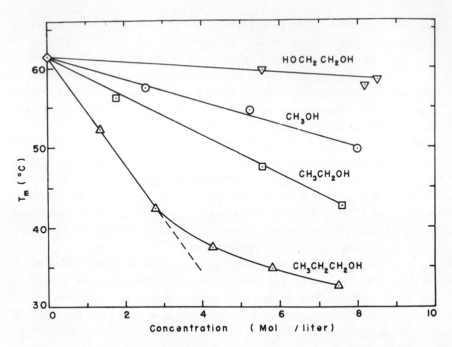

Fig. 49. Melting temperature of ribonuclease as a function of concentration of various added alcohols. All the solutions contained 0.15 mol dm^{-3} KCl and 0.013 mol dm^{-3} sodium cacodylate adjusted to pH 7.0 before adding the organic component; ribonuclease concentration $\simeq$ 0.5% w/v. (See Ref. 1540.)

The nature of the alcohol does not apparently affect the size of the ribonuclease transition. Table XLVII reveals that $-\varDelta[\alpha]_{366}^{T_m}$, the change in optical rotation at 366 nm which accompanies melting, is the same for different alcohols over the concentration range 0–8 mol dm^{-3}; this is true for ethylene glycol, where the transition is only just discernible. The mechanism of alcohol denaturation is thus very different from that caused by either neutral electrolytes or compounds such as urea.

Brandts and Hunt[173] have also examined the effect of ethanol upon the thermal transition of ribonuclease, but under acid conditions (pH 3.20)· The change in extinction coefficient at 287 nm was used to monitor the extent of denaturation as the temperature increased; all solutions were buffered with 0.04 mol dm^{-3} glycine. Measurements were taken over an alcohol concentration range of 0–28.1% w/w and T_m decreased from approximately 44° in pure water to 36° in 28.1% ethanol. The calculated values of the changes in free energy, enthalpy, and heat capacity as a function of temperature and ethanol concentration are shown in Figs. 50 and 51. The

TABLE XLVII. Effect of Alcohols on the Magnitude of the Thermal Ribonuclease Transition[a]

Alcohol	Concentration, mol dm^{-3}	$-\Delta[\alpha]_{366}^{Tm}$, deg
Control	—	36
MeOH	2.5	32
	5.2	37
	8.0	36
EtOH	1.8	37
	5.6	37
	7.6	37
1-PrOH	1.4	38
	2.8	39
	4.3	34
	5.8	33
	7.5	28
Ethylene glycol	5.6	30
	8.2	26
	8.5	28
	10.7	19

[a] Data from Ref. 1540. All solutions contained 0.15 mol dm^{-3} KCl and 0.013 mol dm^{-3} sodium cacodylate and were adjusted to pH 7.0. Ribonuclease concentration, 0.5% w/v.

free energy change accompanying the denaturation passes through a clearly defined minimum at low temperatures (10°) but this becomes less pronounced at higher temperatures; the minimum disappears at 30°, but the curve remains convex with respect to the alcohol concentration axis; at 50° the curve becomes concave. The enthalpy (van't Hoff) $\Delta H^{\ominus}$ shows a maximum at the same alcohol concentration; a corresponding maximum will also arise in the entropy change accompanying denaturation. A minimum value in ΔC_p at similar alcohol concentrations accompanies the changes in $\Delta H^{\ominus}$, $\Delta G^{\ominus}$, and $\Delta S^{\ominus}$.

At low temperatures (<25°) ethanol has a marked stabilizing effect on the native conformation of the protein, the maximum stabilizing capability arising over the ethanol concentration range of 8–15% w/w, a destabilizing capability only appearing at higher alcohol concentrations.

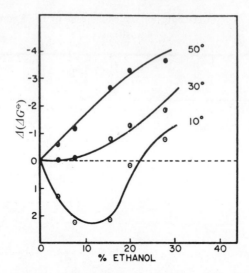

Fig. 50. The dependence of the free energy of denaturation of ribonuclease on the alcohol concentration at 10, 30, and 50°. The ordinate $\Delta(\Delta G^{\ominus})$ corresponds to the difference in ΔG° in aqueous ethanol and in pure water, the values being compared at the same temperature and pH. Free energy values are in k cal mol^{-1}. (See Ref. 173.)

The authors point out that the effects of ethanol on the thermodynamics of solution of nonpolar compounds[171,759,1462] are very similar to the effects observed with ribonuclease. From this evidence it is suggested that ethanol replaces water molecules, presumed to be in a clathrate-type structure, around exposed hydrophobic groups and subsequently interferes with the strongly cooperative nature of the clathrate structure, i.e., that both solvent components are involved in the solvent sheath immediately surrounding the nonpolar side chains of the protein. The authors do not, however, make it clear why, at low temperatures, maximum stability arises at an ethanol concentration of 8–15% w/w, except by drawing a parallel between this observation and the solubility of nonpolar model compounds in alcohol–water mixtures.

The discussions of Franks et al.[483,486,492] on the nature of alcohol–water mixtures can be employed to propose a rather more quantitative explanation of what is occurring in the process (see also Chapter 5, Volume 2). Negative deviations occur in the partial molar volume of ethanol in aqueous solution over a similar concentration range to that for which maximum stabilizing effects upon the native conformation of ribonuclease are observed; the magnitude of the deviation in the partial molar volume of ethanol decreases with increasing temperature and becomes much less pronounced at higher temperatures. Deviations of this kind have been attributed to incipient clathrate formation around the alkyl group of the molecule, which may be further stabilized by the capacity of the OH group of the alcohol to hydrogen bond with water; at higher alcohol concentrations hydrogen bonding between alcohols probably arises, which is incompatible with the tetrahedral type of hydrogen bonding occurring in water, and rapid collapse of the clathrate structures ensue. At low temperatures therefore it is extremely unlikely that contact hydrophobic bonding between the nonpolar part of the alcohol molecule and the exposed non-

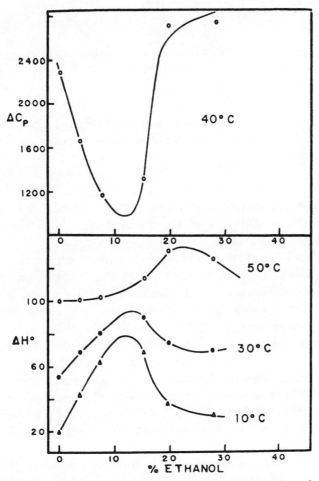

Fig. 51. Dependence of the enthalpy and heat capacity of denaturation of ribonuclease in aqueous ethanol upon the alcohol concentration (see Ref. 173).

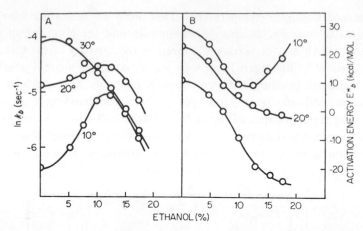

Fig. 52. Dependence of (A) the rate constant K_B and (B) the activation energy of the renaturation E_b^* of trypsin upon the ethanol concentration at 10, 20, and 30° (see Ref. 1184.)

polar group on the protein occurs; more probably the alcohol produces a stabilizing effect upon the clathrate cage of the exposed nonpolar region by the interaction of several clathratelike hydration systems. At higher temperatures disruption of the clathrate cages around both the exposed protein hydrophobic group and the alcohol can be expected; this in turn would give the opportunity for contact hydrophobic bonding to occur between the nonpolar parts of the two molecules, with subsequent denaturation of the protein.*

The data of Pohl[1184] on the kinetics of the reversible denaturation of trypsin in water–ethanol mixtures lend support to the above model. The dependence of the rate constant and the activation energy of the renaturation reaction on the ethanol content of the system, studied by absorption at 293 nm, is shown in Fig. 52. Pohl observed the renaturation at similar temperatures and alcohol concentrations to those used by Brandts and Hunt;[173] the data show that at 10° maximum stability of the native form of the protein occurs over the same concentration range of ethanol found by Brandts and Hunt to stabilize the native conformation of ribonuclease. The dependence of the rate constant for the renaturation reaction on ethanol concentration shows a very similar trend to the thermodynamic parameters associated with denaturation of ribonuclease; at 10° the rate of renaturation is fastest at the alcohol concentrations where structuring of the water by

* For current ideas on hydrophobic hydration and solute interactions see Chapter 1.

the alcohol is thought to be at its most pronounced. At 30°, when structuring of the water by the alcohol is much reduced, preferential stabilization at the same alcohol concentration is rapidly disappearing. The activation energy data of Pohl at 10° show a similar pattern to the plot of log(rate constant) for renaturation as a function of alcohol concentration; at higher temperatures, however, the activation energy for renaturation becomes negative; behavior of this kind is often an indication that a complex process is involved. In this context it would be particularly interesting to have information on the associated changes in the entropy of activation of protein renaturation which accompany changes in the ethanol concentration.

Herskovits and co-workers[647] have examined the effect of alcohols and glycols on the structural stability of α-chymotrypsinogen, cytochrome c, and sperm whale myoglobin; the alcohol denaturation of the proteins was followed by means of mean residue rotation at 233 nm, difference spectra at 292 nm, optical rotatory dispersion at 233 nm, and, in the case of sperm whale myoglobin, by changes in the heme absorption at 409 nm. The data shown in Figs. 53–55 confirm the sequence observed by von Hippel and Wong[1540] that increasing alkyl group size of the alcohol increases its denaturing capability and that glycols are much less effective denaturants than are the monohydric alcohols. The data also illustrate the fact that

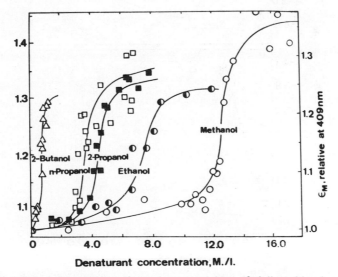

Fig. 53. The alcohol denaturation of cytochrome c at 25 ± 1°, followed by changes in the optical absorption at 409 nm. All the solutions were buffered with 0.1 mol dm^{-3} pH 5.7 acetate buffer. (See Ref. 648.)

chain branching in the alcohol reduces its capability for denaturing the native protein. Some discrepancies exist between the different techniques as to a particular alcohol concentration responsible for 50% denaturation, the most obvious being the effect of methanol upon the denaturation of α-chymotrypsinogen, where the transition midpoint occurs at approximately 7.5 mol dm⁻³ methanol according to difference spectra at 292 nm, but at approximately 12.0 mol dm⁻³ by ORD measurements at 233 nm. The authors prefer the data obtained by difference spectra and construct a table of alcohol concentrations required to produce 50% denaturation. If, however, their ORD data are utilized instead, a somewhat different sequence obtains (Table XLVIII) and close correlation is seen to occur between the alcohol concentrations required for 50% denaturation of the various proteins, with the possible exception of ethanol and cytochrome c.

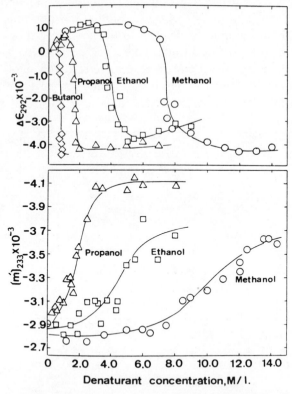

Fig. 54. The alcohol denaturation of α-chymotrypsinogen, pH 2.8, 0.01 mol dm⁻³ Cl⁻, followed by changes in the difference spectra at 292 nm ($\Delta\varepsilon_{292}$) and the optical rotatory dispersion at 233 nm ($[m']_{233}$) (see Ref. 648).

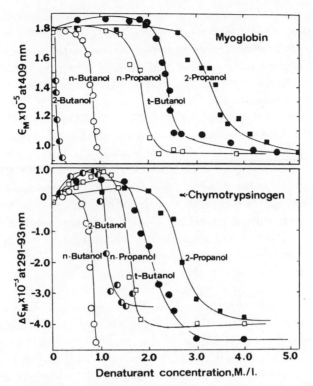

Fig. 55. The effects of alcohol branching on the denaturation of sperm whale myoglobin in 0.1 mol dm^{-3} acetate (pH 5.7) and α-chymotrypsinogen in 0.01 mol dm^{-3} pH 2.8 chloride at $25 \pm 1°$. The myoglobin denaturation was followed by changes in the heme absorption at 409 nm, while that of α-chymotrypsinogen was followed by the difference spectral changes at 292 nm. (See Ref. 648.)

TABLE XLVIII. Alcohol Concentrations Required for 50% Denaturation of Various Proteins[a]

Denaturant	α-Chymotrypsinogen, mol dm^{-3}	Myoglobin, mol dm^{-3}	Cytochrome c, mol dm^{-3}
Methanol	12	12.4	12.5
Ethanol	5	5.3	7.4
Propanol	2	2	1.8
Butanol	1	1	1

[a] Data from Ref. 649.

The data of Herskovits *et al.* on the effect of alcohols on myoglobin are in good agreement with the results of Brunori *et al.*[193] if the data are brought to a common concentration scale; the latter workers examined the effects of methanol, ethanol, and butanol upon the denaturation of *Aphlysia* myoglobin by the heme absorption spectrum, protein fluorescence, and optical activity studies.

The widely accepted viewpoint at the present time[1464] is that alcohols denature the native conformation of proteins by "dissolving out" the hydrocarbon interior of the macromolecule by a hydrophobic bonding mechanism, the denaturing capability of the alcohol being increased by increased alkyl residue content. The fact that corresponding branched-chain alcohols are less effective in promoting denaturation has been attributed to steric effects, in that the alcohol molecule experiences difficulty in approaching sufficiently close to the macromolecule for hydrophobic bonding to occur with exposed apolar groups.

The evidence presented here suggests that, at least in the case of *isothermal* denaturation of the protein by increasing alcohol concentration, an alternative and perhaps more realistic explanation is possible. The data of Brandts and Hunt[173] and Pohl,[1184] previously discussed, show that *maximum* stability of the native conformation of a protein is observed at the ethanol concentration which Franks[483] suggests corresponds to maximum structuring in the solvent caused by the presence of the alcohol. At higher alcohol concentrations rapid structural disruption in the solvent mixture occurs and denaturation of the protein follows. The data of Herskovits *et al.*[647] and Brunori *et al.*[193] confirm that denaturation of the myoglobins and of α-chymotrypsinogen is observed at the ethanol concentration where disruption of structure begins to occur, as evidenced, e.g., by the attenuation of the negative deviation of the partial molar volume of ethanol in water, but also by many other physical properties. The correlation is not definite in the case of cytochrome c and ethanol, but for all other proteins and alcohols investigated the correlation between the alcohol concentration at which solvent structure is believed to collapse and protein denaturation occurs is extremely close.

The available evidence therefore suggests that denaturation of the protein arises from entropic effects due to the loss of solvent structure surrounding the exposed nonpolar residues of the protein. The data of Herskovits *et al.*[647] on the effects of chain branching upon the denaturing capability of alcohols confirm this interpretation; chain branching reduces the denaturing capability. This is currently explained as due to a decreased capability of the alcohol to form a hydrophobic bond with an exposed nonpolar side chain on the macromolecule. However, the data on the effects of chain branching among the butanols on their denaturing capability also show a close correlation between the alcohol concentration required to produce protein denaturation and the alcohol concentration at which structure disruption in the solvent occurs (Fig. 56). It has been shown that of the butanols, tert-butanol has the greatest structure-stabilizing effect on water;[492] Herskovits' data show that at butanol concentrations lower than that required to produce protein denaturation, i.e., at alcohol concentrations that correspond to maximum solvent structure stabilization, tert-butanol actually enhances helix formation in the protein.

The observed pattern thus suggests that at the alcohol concentration that promotes the maximum extent of order in the solvent the ordered form of protein is stabilized; in the case of alcohols such as tert-butanol, which are particularly strong solvent order promoters, actual *enhancement* of "native" structures may occur. At the higher alcohol concentrations which correspond to collapse of solvent structure, denaturation of the protein occurs, presumably due to collapse of solvent structure around the apolar groups of the protein. Once denaturation has occurred it is quite possible that contact hydrophobic bonding between the "free" alcohol and the apolar residues, as visualized by Tanford,[1464] may occur, but this will be a *secondary* factor favoring denaturation and not the *primary* cause of the disruption.

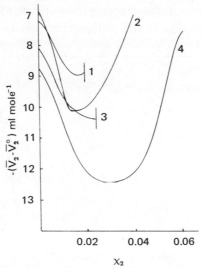

Fig. 56. Partial molar volumes at $20°$ of aqueous solutions of the isomeric butanols: (1) n-BuOH, (2) sec-BuOH, (3) iso-BuOH, and (4) tert-BuOH. The vertical lines denote miscibility limits. (See Ref. 483.)

In the case of *thermal* denaturation of a protein in the presence of the alcohol the mechanism of the process cannot be explained in such a relatively straightforward manner; Fig. 49, taken from the data of von Hippel and Wong,[1540] shows that T_m of ribonuclease decreases with increasing alcohol concentration, even at alcohol concentrations at which maximum stability of solvent structure occurs. The depression of T_m is also greater, the larger the alkyl chain of the alcohol; such a situation has led to the proposal of hydrophobic bonding being responsible for unfolding.[1464] There are, however, anomalies which are difficult to explain by this mechanism, but their discussion is postponed for later consideration.

Conio and co-workers[273] have investigated the coil–helix transition of the polypeptide poly-L-ornithine in the presence of a variety of alcohols; from measurements of the mean residual rotation at 233 nm the transition at 25° was observed as a function of alcohol concentration for methanol, ethanol, 1-propanol, 2-propanol, 1-butanol, 2-butanol, and tert-butanol; the data are presented in Fig. 57. Increasing alcohol concentration favors the formation of the helix and the larger the alkyl group of the alcohol, the more effective the alcohol becomes at stabilizing the helix conformation.

The alcohol concentrations at which Conio *et al.* find maximum helix promotion are very similar to the concentrations at which Herskovits *et al.*[647] found these same alcohols to be effective disrupting agents for the native conformation of the proteins α-chymotrypsinogen, cytochrome c, and myoglobin. To repeat, the concentration of each alcohol that promotes maximum helix formation in the polypeptide poly-L-ornithine is close to

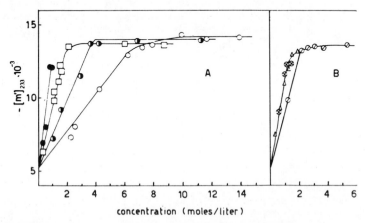

Fig. 57. Variation of the mean residue rotation at 233 nm with alcohol content for poly-L-ornithine at 25°. A: (⊙) Methanol; (◑) ethanol; (□) 1-propanol; (●) 1-butanol, B: (⊘) 2-Propanol; (△) tert-butanol: (⊗) 2-butanol. (See Ref. 273.)

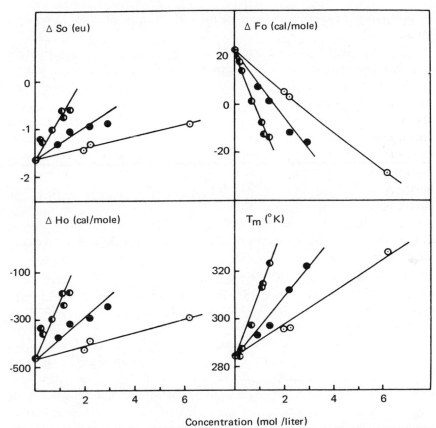

Fig. 58. Values of the enthalpy, entropy, and free energy of formation of the uncharged helix, $\Delta H°$, $\Delta S°$, $\Delta G°$, respectively, and melting temperature T_m for poly-L-ornithine as a function of alcohol concentration. (⊙) Methanol; (●) ethanol; (◑) 1-propanol (see Ref. 273).

the concentration that has been found to be effective in denaturing proteins. The common factor here is the pertinent alcohol concentration, which, as has been discussed earlier, is that at which collapse of alcohol/water structuring is thought to occur.[486]

The thermodynamic functions for the formation of the uncharged poly-L-ornithine helix, determined by Conio et al.,[273] are shown in Fig. 58; the operative factor in the increasingly negative free energy of formation with increasing alcohol concentration is seen to be the increasingly positive entropy of helix formation, since the change in enthalpy with increasing alcohol concentration is unfavorable. The observed result arising from the ΔH–ΔS balance is that for each alcohol the melting temperature of poly-L-ornithine is seen to increase with increasing alcohol concentration.

Conio *et al.*[273] also determined the free energy of helix formation of poly-L-glutamic acid as a function of alcohol concentration, as shown in Fig. 58; the ethanol data agree well with their own earlier data[272] and with those of Hermans.[640] The free energy of helix formation of the uncharged poly-L-glutamic acid decreases in a similar manner with increasing alcohol content as that for poly-L-ornithine. Conio and co-workers therefore suggest that a common mechanism operates in polypeptides as dissimilar as poly-L-ornithine and poly-L-glutamic acid, the former not having a marked capacity for helix formation, while the latter has a very strong tendency for helix formation.

Conio *et al.*, using the results of Franks,[483] speculate that the crucial factor involved in the stabilization of the helix arises from a decrease in the entropy of mixing of the peptide group with the solvent, resulting from an increasing degree of structuring of the latter in the presence of the alcohol. Unfortunately, such a supposition fails to explain why a maximum degree of helicity occurs at the alcohol concentration where collapse of solvent structure is assuming major proportions.

The model for helix stabilization, as previously outlined, based upon the data of Franks[483] is successful up to a point in reconciling the behavior of proteins and polypeptides in the presence of alcohols. The native conformation of the protein, the normal conformation in aqueous solution at low and room temperatures, is further stabilized, at least under isothermal conditions, by increasing the alcohol content to that concentration at which disruption of the solvent structure becomes serious; beyond this concentration denaturation of the protein ensues. In the case of the polypeptides the helix conformation is also stabilized by increasing alcohol concentration, achieving a maximum at the alcohol concentration at which maximum solvent structure is obtained; however, a further increase in alcohol concentration, with disruption of solvent structure, does not result in denaturation of the helix conformation, which continues to be the stable form, and the polymer may in certain cases precipitate from solution in the helix conformation at high alcohol concentrations.[272] Associated with this differing behavior of the proteins and polypeptides at higher alcohol concentrations is the differing behavior of T_m; in the case of the proteins T_m falls with increasing alcohol concentration, whereas in the case of the polypeptides it increases with increasing alcohol concentration.

For the polypeptides both the enthalpy and entropy of helix formation become less negative with increasing alcohol concentration; however, the relative decrease of ΔS is smaller than for ΔH, leading to an increase in T_m with increasing alcohol concentration. Unfortunately, no data appear to be available to indicate how the relative parameters for the melting of the helix change with increasing alcohol concentration to produce a decrease in T_m with increasing alcohol concentration; it would appear reasonable, however, to suggest that ΔS changes at a relatively greater rate than ΔH, producing a decrease in T_m.

Klotz et al.[787,788,818] have investigated the stability of the amide hydrogen bond in water and a nonpolar solvent, showing that the amide group has a greater ability to hydrogen bond in a nonpolar solvent than in water; the evidence points to enthalpic rather than entropic factors determining the stability of the bond. The opposite appears to be the case for both proteins and polypeptides; thus the helix conformation owes its stability to entropic rather than enthalpic factors.

Similar alcohol effects are observed in denaturation behavior of fibrous protein, such as collagen. Schnell[1333] has studied the influence of various alcohols on the thermal denaturation of acid-soluble calf skin collagen and on the gelatin–collagen refolding process of the unfractionated gelatin obtained from the calf skin collagen: the thermal transition was observed at a protein concentration of 0.12% w/v by optical rotation measurements at 405 nm. The effects of the different alcohols on T_m and related properties of the protein at a constant alcohol concentration of 2.53 mol dm^{-3} are compared in Table XLIX; as with the globular proteins, the capacity for

TABLE XLIX. The Effect of Different Alcohols on T_m and Related Parameters of Acid-Soluble Calf Skin Collagen[a]

Alcohol	Concentration, mol dm^{-3}	T_m, °C	$\Delta[\alpha]_{405}^{T_m}$ [b]
None	—	35.3	−643
Methanol	2.53	34.5	−660
Ethanol	2.53	33.6	−654
Propanol	2.53	28.8	−644
2-Propanol	2.53	31.8	−650

[a] Data from Ref. 1333.
[b] $\Delta[\alpha]_{405}^{T_m}$ is defined as the difference in optical rotation at 405 nm between the native form of the collagen molecule at 20° and its denatured state at 40°—taken as a measure of the helix content of the macromolecule.

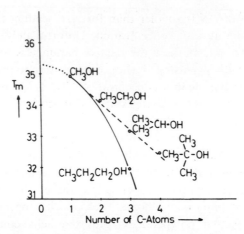

Fig. 59. The effect of different alcohols at 1 mol dm⁻³ concentration on T_m of acid-soluble collagen (see Ref. 1333).

protein denaturation increases with increasing size of the alkyl group in the alcohol. Schnell's data also show that $\Delta[\alpha]_{405}^{T_m}$ for calf skin collagen does not change with the character of the alcohol, which resembles the effect of alcohols on the denaturation of ribonuclease.[1540] Figure 59, which illustrates the effect of the alcohols, including branched-chain derivatives, on T_m of the collagen, also shows that the branched-chain alcohols are less effective in promoting denaturation of the protein than the corresponding straight-chain alcohol; this behavior again resembles that observed with globular proteins.

Similar results have also been reported by Harrap[610] for the effect of monohydric alcohols and glycols upon the thermal denaturation of rat tail tendon collagen at pH 3.0, as studied by optical rotation at 365 nm at a protein concentration of 0.08%. At low alcohol concentrations ($<$2 mol dm⁻³) the decrease in T_m with alcohol concentration was observed to be linear (Fig. 60), in parallel with the effects observed[1540] for ribonuclease; at higher concentrations the lines become positively curved, the onset of curvature varying with the alcohol, commencing first with tert-butanol and last with methanol. The author suggests that hydrophobic bonding of the kind outlined by Tanford[1464] is responsible for the behavior pattern observed.

At a concentration characteristic of a given alcohol, positive deviations of the linear depression of the melting temperature of collagen occur; it is of particular interest that the deviation in the case of tert-butanol occurs at the same alcohol concentration and in the same direction as that observed by von Hippel and Wong[1540] for the effect of tert-butanol on T_m of ribonuclease. Similar effects therefore appear to be operating in the case of both globular and fibrous proteins.

It was pointed out earlier that hydrophobic bonding between the exposed apolar residues on the protein and the alkyl group of the alcohol has been suggested as the driving force for alcohol denaturation of proteins. This mechanism is unsatisfactory as an explanation of the isothermal denaturation of proteins by increasing alcohol concentration and the data of von Hippel and Wong,[1540] Conio et al.,[272,273] and Harrap[610] indicate

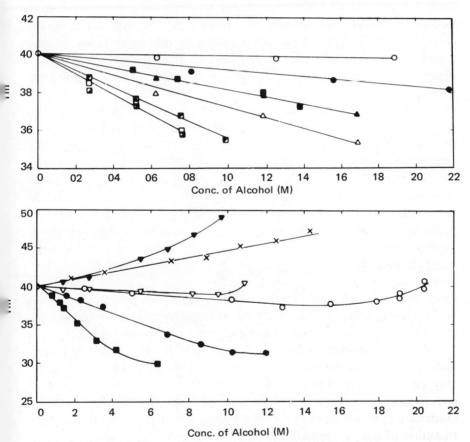

Fig. 60. (a) Plots of T_m for tropocollagen dissolved in monohydric alcohol–water mixtures (pH 3.0) at $C_M < 2.5$ mol dm^{-3}: (○) methyl; (●) ethyl: (△) n-propyl; (▲) isopropyl; (□) n-butyl; (□) iso-butyl; (□) sec-butyl; (■) tert-butyl. (b) Plots of T_m for tropocollagen dissolved in alcohol–water mixtures at pH 3.0 over a wider concentration range: (○) methyl; (●) ethyl; (■) tert-butyl; (×) ethylene glycol; (▽) propane-1:2-diol; (▼) propane-1:3-diol. (See Ref. 610.)

the shortcomings of this type of explanation for the thermal denaturation of proteins in the presence of alcohols.

The experimental data show that the linear depression of T_m with alcohol concentration ceases with each alcohol at that concentration at which disruption of solvent structure is becoming a dominant factor. If hydrophobic bonding between the alcohol and the protein were the cause of denaturation, it is difficult to see why the melting temperature of the protein should begin to increase over the alcohol concentration region where more "free alcohol," i.e., alcohol unencumbered by solvent interactions, becomes available for hydrophobic bonding and subsequent disruption of the native conformation of the proteins. The evidence of Klotz et al.[787,788,818] precludes hydrogen bonding as the major factor in this complicated pattern of events, which requires that the various sequences observed find an explanation in the entropy changes that arise during the helix–coil transition.

A further complicating factor may be the behavior of the polypeptide poly-L-proline II, for which the degree of specific mutarotation decreases linearly with increasing alcohol concentration,[1325] the rate of decrease becoming greater with increasing alkyl chain length of the alcohol. Ethylene glycol apparently equals ethanol in its denaturing capability. The experimental data do not give any indication of the upper alcohol concentration limit, so that no information is available to show whether deviations in the linear decrease of the specific mutarotation occur at the appropriate alcohol concentrations where solvent structure disruption is expected to occur. It does, however, appear to be the case that, at least at the lower alcohol concentrations, increasing concentration of alcohol leads to a disruption of the left-handed helix conformation of this polypeptide, which is the reverse effect to that observed with polypeptides such as poly-L-glutamic acid and poly-L-ornithine, which form right-handed α-helices.

The poly-L-proline molecule is incapable of forming amide hydrogen bonds and therefore the effect of alcohols upon the extent of helicity presumably involves some form of hydrophobic interaction of the polypeptide and its solvent medium; unfortunately, no information appears to be available on the enthalpy and entropy changes associated with the helix–coil transition of this polypeptide.

The evidence discussed indicates the problems involved in explaining the effects of alcohols upon the stability of protein and polypeptide conformations. A consistent thread seems to run through all the data considered, a change in the property under investigation, be it degree of helicity, or melting temperature of the helix, which occurs when the alcohol concentration reaches the point at which disruption of the solvent structure becomes of major importance. It has been shown how such an explanation fits the isothermal helix–coil transition data[647] for different proteins; in addition it is seen that the thermal transition results show marked changes at the same alcohol concentrations, T_m showing a linear decrease with increasing alcohol concentration up to this point. Information about ΔH_m and ΔS_m of transition of proteins in alcohol–water mixtures is lacking, but the above lines of apparently conflicting evidence can be reconciled if, as seems likely, the decrease in T_m is due to ΔS_m increasing relative to ΔH_m as the alcohol concentration increases, i.e., a relatively larger entropy change occurs on transition, which would be expected if a greater degree of solvent structuring is involved. At the higher alcohol concentrations where T_m is observed to increase, this again is the expected result, if ΔS_m is becoming relatively smaller, due to loss of solvent structure about the nonpolar groups of the protein.

In essence this interpretation requires that over the region of alcohol concentration where solvent structure is the dominating factor the helix is the preferred conformational structure; however, because of the greater degree of solvent structure involved, this does not stabilize the helix conformation against thermal denaturation. Thus, the greater the extent of structuring in the total system, the more sensitive it becomes to the effects of temperature, i.e., T_m decreases. At the higher alcohol concentrations the helix conformation is not the preferred form but, because of the marked decrease in the extent of solvent structure, ΔS_m becomes smaller and T_m increases, i.e., the helix conformation is stabilized against thermal denaturation.

Such an interpretation can explain the data on the stability of the poly-L-proline II helix but difficulties arise when attempts are made to extend the model to other polypeptides. In the case of polyornithine and polyglutamic acid the helix conformation is certainly stabilized at the lower alcohol concentrations, but it continues to be the preferred form at higher alcohol concentrations. In addition, T_m *increases* at these lower alcohol concentrations, although no experimental evidence appears to be available as to the behavior of T_m for these polypeptides at the higher alcohol concentrations.

The model as presented cannot, therefore, on the limited data available, explain the behavior of the polypeptides, but it must be reiterated that the hydrophobic bonding model previously discussed is even more severely handicapped in accounting for the experimental data, the most telling fact being that the marked deviations observed in conformational behavior at given alcohol concentrations cannot be explained by this model.

5.3. The Solubility of Hydrocarbons in Solutions of Proteins

The importance of hydrophobic interactions in the complex series of interactions which are responsible for the stability of a particular protein conformation has become clear from the work reviewed in the preceding pages; it would therefore be inappropriate to complete this review without considering the interactions that arise between proteins and hydrocarbon solutes, since such interactions should exhibit all the characteristics of hydrophobic interactions.

Mohammadzadeh-K *et al.*[1029] have examined the solubilities of pentane, heptane, nonane, and mixtures of pentane and nonane in solutions of bovine serum albumin, β-lactoglobulin, chicken ovomucoid, and chicken ovalbumin; the extent of binding of the hydrocarbons to the proteins was determined by a gas–liquid chromatography technique. The equilibrium solubilities of the hydrocarbons in protein solutions were determined in the presence of 0.1 mol dm^{-3} sodium phosphate buffer (pH of 6.8). Table L presents the solubilities of the alkanes in solutions of bovine serum albumin and β-lactoglobulin.

TABLE L. Solubilities of Pentane, Heptane, and Nonane in Solutions of Bovine Serum Albumin and β-Lactoglobulin[a]

Alkane	Solubility	
	mol alkane/ mol protein	mol alkane/ 10^4 g protein
Bovine serum albumin		
C_5H_{12}	3.80	0.551
C_7H_{16}	2.80	0.406
C_9H_{20}	1.42	0.206
β-Lactoglobulin		
C_5H_{12}	1.95	0.542
C_7H_{16}	1.47	0.408
C_9H_{20}	0.95	0.264
Ovomucoid		
C_5H_{12}	1.65	0.590
Ovalbumin		
C_5H_{12}	1.09	0.237

[a] Data from Ref. 1029.

The binding of the alkanes to bovine serum albumin, β-lactoglobulin, and ovomucoid is very similar and shown by the authors to be primarily a function of the total number of amino acid residues in the protein and the number of methylene groups in the alkane molecule. The lower degree of binding to ovalbumin is suggested as being due to the higher proportion of polar amino groups compared to the other proteins. The number of moles of pentane bound by β-lactoglobulin is in good agreement with earlier data.[1590]

Wishnia,[1589] from solubility studies, has calculated the thermodynamic functions for the transfer of butane, pentane, and neopentane to water, from ideal solutions, dodecyl sulfate micelles, ferrimyoglobin, deoxyhemo-globin, β-lactoglobulin at pH 2.0, β-lactoglobulin dimer at pH 5.3, and apomyoglobulin. The data, presented in Table LI, particularly the large values of ΔC_{p_t}, argue very strongly for hydrophobic interactions of the kind discussed previously; the binding of pentane and butane to β-lacto-globulin and ferrihemoglobin is particularly strong, arising from an exceptionally large ΔH_t. Wishnia suggests that the "excess" part of this

TABLE LI. Thermodynamic Functions of Alkane Binding to a Variety of Molecular Species at 0 and 25°C[a]

Species	Pentane			Neopentane			Butane		
	ΔH_t	ΔS_t	ΔC_{p_t}	ΔH_t	ΔS_t	ΔC_{p_t}	ΔH_t	ΔS_t	ΔC_{p_t}
Ideal	-12.2	-139	307	-10.5	-126	139	-13.0	-126	256
	-5.0	-113	269	-5.5	-109	269	-6.3	-105	302
Dodecyl sulfate	-10.9	-122	386	-10.1	-113	344	-8.0	-97	113
	-1.7	-88	433	0	-80	437	-2.9	-80	273
Ferrimyoglobin	-7.1	-118	386	—	—	—	-5.5	-92	185
	-3.8	-80	479	—	—	—	-0.4	-76	273
Ferrihemoglobin	4.2	-84	356	0.4	-80	818	-3.4	-97	269
	14.3	-50	445	6.7	-59	294	4.2	-71	328
β-Lactoglobulin (pH 2.0)	-1.3	-109	433	—	—	—	3.8	-80	256
	12.6	-63	588	—	—	—	10.1	-59	269
β-Lactoglobulin (pH 5.3)	0.8	-105	433	-9.2	-109	302	5.9	-71	197
	13.9	-63	588	-1.3	-80	357	12.2	-50	298

[a] Data from Ref. 1592. ΔH_t in kJ mol^{-1}, ΔS_t and ΔC_{p_t} in J mol^{-1} deg^{-1}. The upper and lower sets of values refer to 0 and 25°C, respectively.

large ΔH_t cannot arise from alkane–water interactions since these are expected to be similar to those of the dodecyl sulfate micelles, nor from alkane binding with water displacement from the binding site[1335,1336] or particularly strong alkane–protein interactions in the complex.[1476,1596] Instead this "excess" ΔH_t is said to be due to a "strain" in the *unoccupied* binding site which is released by the admission of a molecule of favorable geometry, i.e., pentane and butane, but *not* neopentane, and it is this relaxation which produces the extra free energy of binding.

On the evidence presented, Wishnia distinguishes between three kinds of hydrophobic regions in the protein molecule: (1) The typical region, which produces good van der Waals contacts, but cannot expand to include the alkane molecule without energetically upsetting the rest of the protein molecule. Such sites are weak binders of alkanes and constitute considerable regions in some proteins, e.g., ribonuclease.[1598] (2) The second kind also shows good van der Waals binding contacts but, in a like manner to a dodecyl sulfate micelle, can solubilize the included alkane molecule by expanding without strain; enthalpies of binding of alkanes are therefore thought to be similar to those encountered with dodecyl sulfate micelles. (3) In the third kind of site "superbinding" occurs with a heat of dissociation of 12–16 kJ mol^{-1} higher than that of dodecyl sulfate micelles: This heat of dissociation arises from release of strain in the previously unoccupied sites.

CHAPTER 6

Polysaccharides

A. Suggett

Biosciences Division
Unilever Research Laboratory Colworth/Welwyn
Colworth House, Sharnbrook, Bedford, England

1. INTRODUCTION

In contrast to proteins, polypeptides, and nucleic acids, polysaccharides are an often neglected class of biopolymer. Despite the fact that the first observation of a helical biopolymer was of a polysaccharide,[1282] in general the solution conformations of polysaccharides tend to be less ordered and less understood than their protein, polypeptide, or nucleic acid counterparts; and of course *less ordered* to many scientists implies *less interesting*. However, polysaccharides are of universal importance in living matter, and recent studies of their primary, secondary, tertiary, and even quaternary structures,[328] and the factors, including solvent, which control the adoption of particular structures are gradually awakening a more general interest in these substances.

Nevertheless at this time it appears that polysaccharides still lag behind certain other biopolymers in terms of our understanding of their solution properties, and nowhere is this better illustrated than in a consideration of the interactions between solute and solvent, and the role of the solvent in influencing the conformation. Direct observations of solute–water interactions in polysaccharide systems are as yet few in number, and, as we shall see, some of the more striking examples of the influence of the solvent have emerged from studies of the related lower-molecular-weight (but better characterized) oligosaccharide systems.

2. STRUCTURAL FEATURES OF POLYSACCHARIDES

Before any consideration of solute–water interactions it is pertinent to describe some of the structural features of polysaccharides. This is included first because the monosaccharide "building blocks" and the ways in which these are arranged in the polymers are not generally so familiar as their counterparts in, for example, polypeptides. But more importantly, in any discussion of polysaccharide–water interactions and of the ways in which the solute and solvent structures are mutually modified, some picture of the molecular structure of the polysaccharide is a necessary prerequisite. In this section the basic ideas of polysaccharide structure will be outlined as much as possible, using as examples molecules which will figure prominently in subsequent sections.

2.1. The Monosaccharide Building Blocks

The term carbohydrate originally derives from the early observations that these compounds have the empirical formula $(C \cdot H_2O)_n$ in which $n \geq 3$. The most important simple monosaccharides have $n = 5$ (pentose sugars) or $n = 6$ (hexose sugars). However, as a result of the high proportion of asymmetric (chiral) carbon atoms in monosaccharides and the cyclic nature of the molecules, the formulas $C_5H_{10}O_5$ and $C_6H_{12}O_6$ represent a large number of possible structures. Let us consider a few of the relevant factors.

1. We must first distinguish between aldoses and ketoses, i.e., monosaccharides that in the straight-chain form are formally terminated by an aldehyde or ketone function, respectively. Related aldohexoses and ketohexoses, for example, are glucose and fructose, and the linear representations and Haworth perspective formulas of the D-sugars are given below:

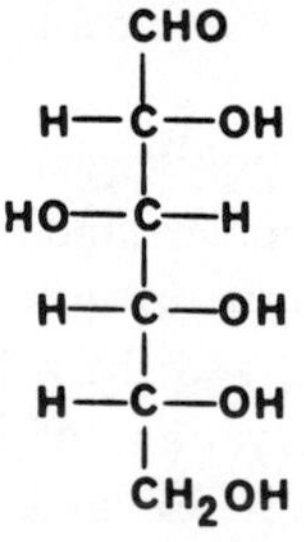

(I) 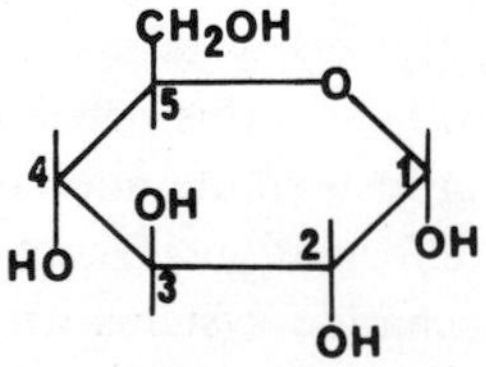   (II)

α-D-glucose

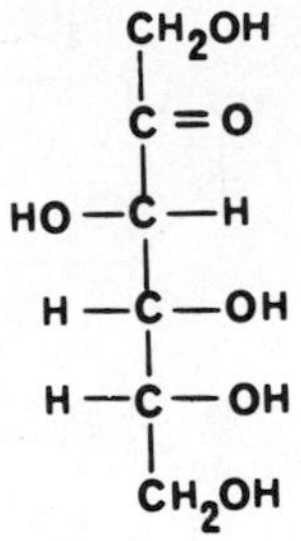

(III) 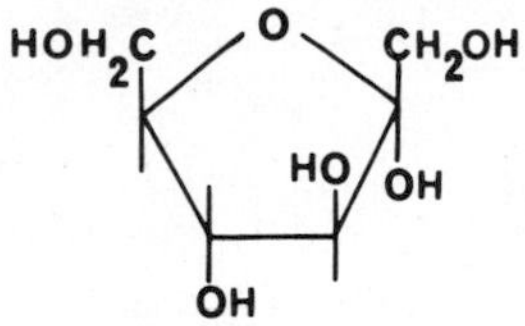   (IV)

D-fructose

The mirror image forms, the L-sugars, are only rarely found in nature.

2. Although the linear structures of aldoses (such as D-glucose above) do not exist to any significant extent, the equilibria between these and the corresponding cyclic forms are relatively mobile and lead to the interconversion of *anomeric* α and β forms of the monosaccharide (the so-called mutarotation equilibria). Thus D-glucose, for example, can exist in two forms (structures II and V) which differ only in the orientation of the H, OH at carbon atom number one (C-1), and at equilibrium in aqueous solution at ambient temperatures the $\alpha:\beta$ ratio is about 9:16.

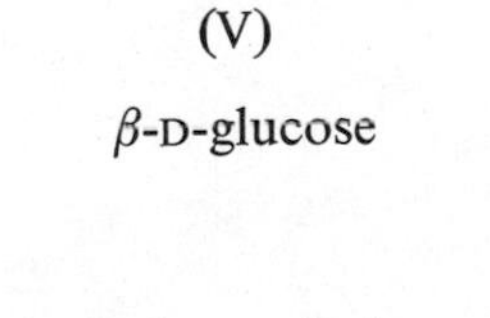

(V)

β-D-glucose

3. By changing the orientations of the groups at the chiral carbon atoms 2, 3, and 4 we find a total of eight pairs (α and β forms) of aldohexoses. Two of the more common in nature are mannose and galactose:

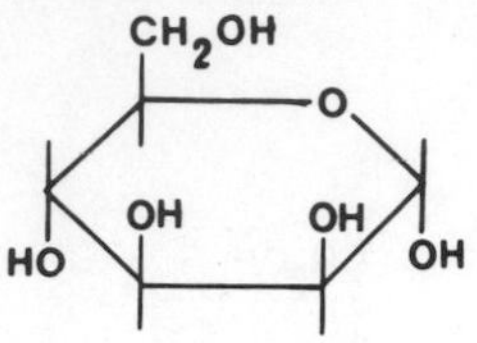

CH$_2$OH

HO

OH   OH

HO   OH

(VI)

α-D-mannose

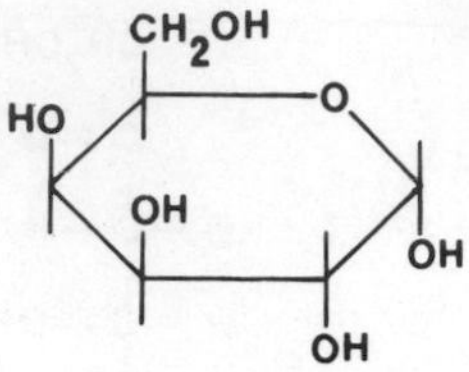

CH$_2$OH

HO

OH

OH

OH

(VII)

α-D-galactose

which differ from glucose only in the orientations of the hydroxyls at C-2 and C-4, respectively.

4. For an individual sugar there also exists in principle the possibility of five-membered ring (furanose) forms or six-membered ring (pyranose) forms; for example, α-D-glucocopyranose (II) and α-D-glucofuranose (VIII):

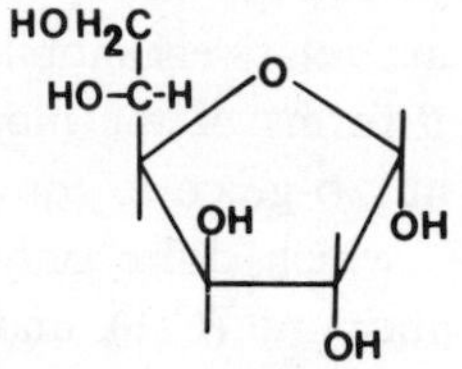

HOH$_2$C

HO–C–H   O

OH   OH

OH

(VIII)

α-D-glucofuranose

or β-D-ribopyranose and β-D-ribofuranose

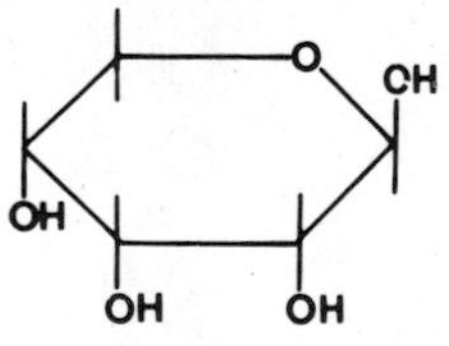

O   OH

OH

OH   OH

(IX)

β-D-ribopyranose

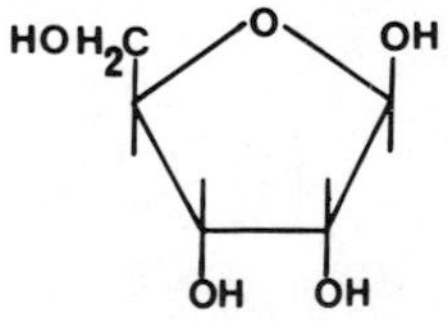

HOH$_2$C   O   OH

OH   OH

(X)

β-D-ribofuranose

Proton magnetic resonance spectroscopy of solutions of sugars in D_2O[27, 28,872,1280] in fact cannot detect ($<1\%$) furanose forms of a number of monosaccharides (including glucose and mannose), but for some aldoses, particularly altrose, idose, talose, and ribose, the proportion of furanose forms may be 25% or more.

5. The Haworth formulas presented in the preceding discussion, useful though they are, do not tell us very much about the relative spatial orientations of the various groups on the ring. The ring cannot of course be planar for tetrahedral or near-tetrahedral coordination of the carbon atoms, and a pyranose monosaccharide could in principle adopt any one of a number of chair, boat, or other alternative forms. A great many observations have, however, indicated that in fact chair forms greatly predominate, with the Reeves C1 conformation (Fig. 1) usually preferred for the D-sugars[27,872, 1280] (>99% for all the D-aldohexopyranoses except α-D-idose and α-D-altrose). For the aldopentopyranoses the alternative 1C conformation may predominate (as in α-D-arabinose) or occur in equilibrium with C1 (as in α- and β-D-ribose).

With respect to the monosaccharides, two major points must be reiterated. First, the simple formulas $C_5H_{10}O_5$ and $C_6H_{12}O_6$ comprise a large number of alternative monomer units for polysaccharide structures. Second, simple monosaccharides are unlikely to exist in aqueous solution in a single, well-defined conformation. The complexity of the situation for D-ribose is illustrated in Fig. 2, in which mutarotation, pyranose–furanose, and other conformational equilibria are superimposed. The ways in which the solvent can modify these conformational equilibria will be considered in Section 3.1.4.

A number of derivatives of the simple monosaccharides are also of exceptional biological importance as units of polysaccharides.

(a) *Uronic acids*, in which the $-CH_2OH$ group on C-5 has been oxidized to $-CO_2H$:

(XI)

β-gluco(pyran)uronic acid

Also important are the corresponding methyl esters.

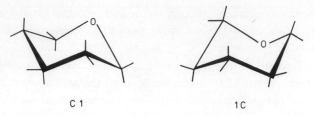

C 1   1 C

Fig. 1. The Reeves C1 and 1C conformations for aldohexopyranoses.

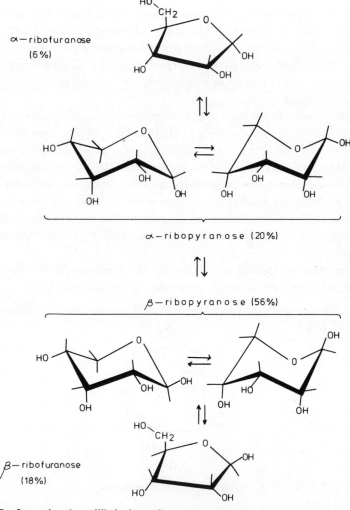

Fig. 2. Conformational equilibria in D-ribose. Percentages of the various components refer to a temperature of 35°.[1168]

(b) *Amino sugars*, e.g., 2-D-glucosamine and 2-D-galactosamine very commonly occur in nature, usually as their *N*-acetyl derivatives;

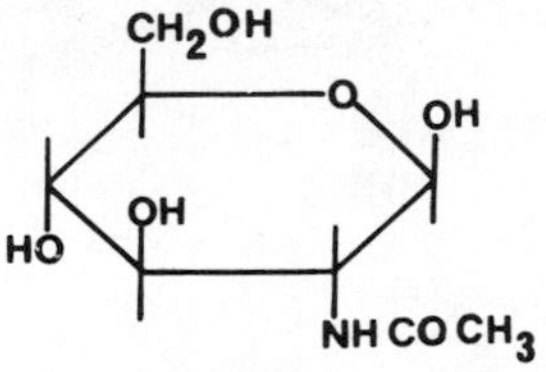

(XII)

2-acetamido-2-deoxy-D-glucose
(*N*-acetyl glucosamine)

(c) *Methylpentoses*, e.g., rhamnose and fucose, which as the L-forms are widely distributed in plant materials.

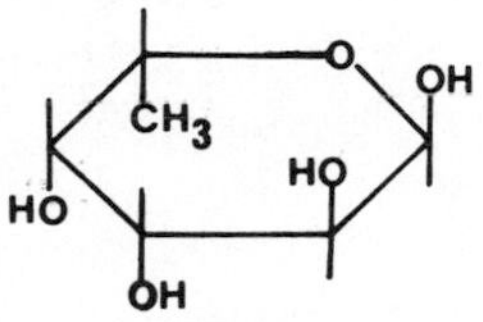

(XIII)

α-L-fucopyranose

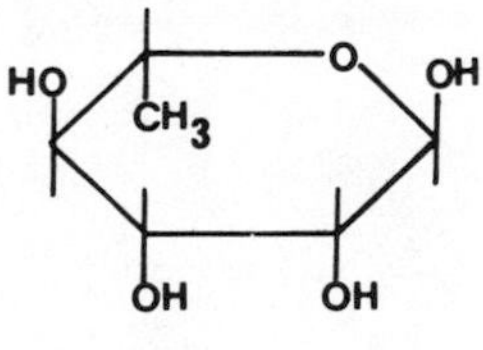

(XIV)

α-L-rhamnopyranose

(d) Finally, one or more hydroxyl groups in the simple monosaccharides may be replaced by OSO_3^-.

2.2. Disaccharides and Oligosaccharides

Having now considered the variability in the monomeric units of polysaccharides, the next stage is to examine the alternative ways of joining

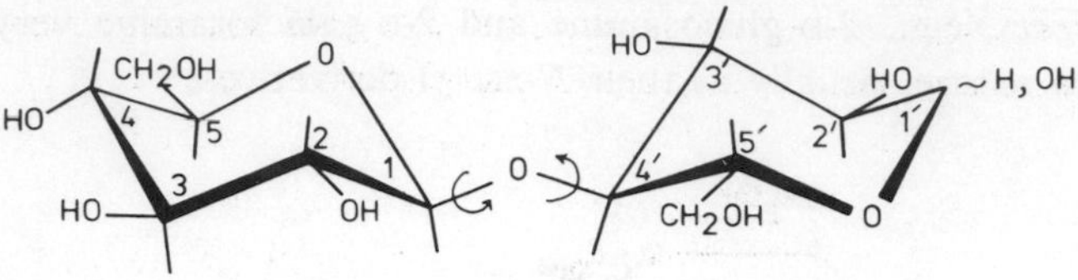

a) CELLOBIOSE
(β, 1→4 linkage)

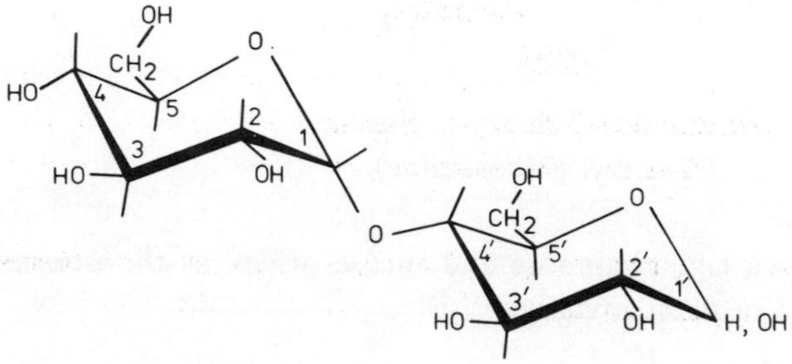

b) MALTOSE
(α 1→4 linkage)

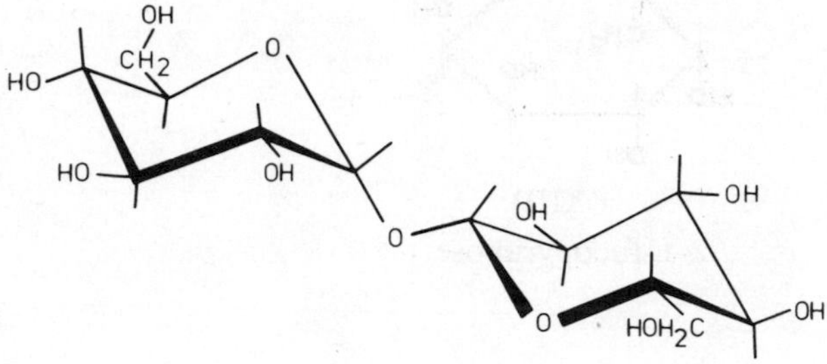

c) α, α TREHALOSE
(α, 1→1 linkage)

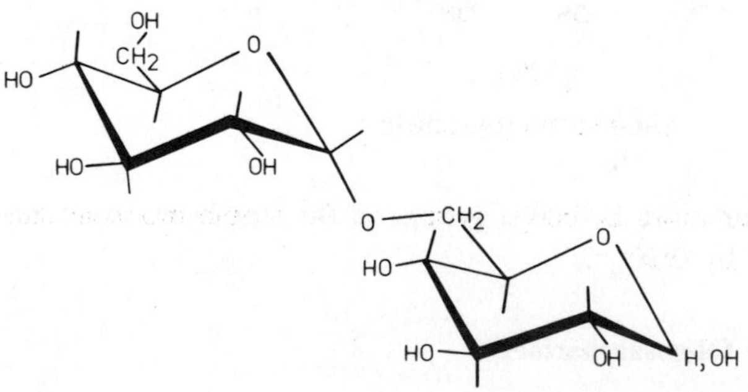

d) ISOMALTOSE
(α, 1→ 6 linkage)

Fig. 3. Diglucoses with different glycosidic linkages.

these monomeric units together. The glycosidic linkage joins carbon atom C-1 on one monosaccharide ring via a hemiacetal oxygen to any one of C-1, C-2, C-3, C-4, or C-6 on a second ring. This lingage can be either in the α or β orientation with respect to C-1 on the first ring. Figure 3 shows four well-known disaccharides each of which consists of two D-glucopyranose rings but have different glycosidic linkages. Compared to the monosaccharide case, extra uncertainties in the molecular conformation are also now introduced by the possibility of rotation about the glycosidic bonds (Fig. 3), although in certain cases these rotations may be restricted by inter-ring hydrogen bonding. For example in the crystal[185,256] and in DMSO solution[231] the cellobiose conformation is stabilized by hydrogen bonding between O-5 (the ring oxygen) and O-3'. Maltose in the crystal,[255] on the other hand, appears to contain a hydrogen bond between O-2 and O-3' (although, as we shall see in due course, the crystal conformation is unlikely in this case to make a significant contribution to the aqueous solution conformation).

Because of the difficulty in obtaining pure monodisperse samples of polysaccharides and in some cases also because of limited solubility of the polysaccharide, considerable attention has been directed toward oligomeric alternatives. For example, the extension of the maltose structure (Fig. 3) i.e., $1\alpha4$-linked glucose units, gives maltotriose, maltotetraose, etc., which are collectively known as the maltodextrins. The same monosaccharide and the same glycosidic linkage also occurs in the cyclodextrins (alternatively

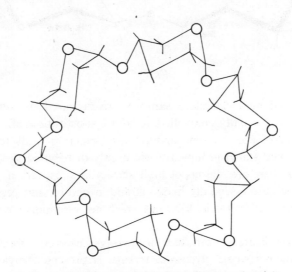

Fig. 4. Cyclohexaamylose (or Schardinger α dextrin).

known as the cycloamyloses or Schardinger dextrins), a family of fully cyclic oligomers (6–12 glucose units) of which the first member, cyclohexaamylose (Schardinger α dextrin), is shown in Fig. 4.

2.3. Polysaccharides

The first major division in classifying polysaccharide structures is between *homopolysaccharides* based upon one kind of sugar unit only, e.g., glucans, mannans, and fructans, and *heteropolysaccharides* based upon two or more sugar units, e.g., arabinoxylans and galactomannans. The simplest types of homopolysaccharides are *linear* macromolecules with the same glycosidic linkage between all the sugar units. The most frequently encountered examples of this type are the glucans cellulose ($1\beta4$ linkages) and amylose ($1\alpha4$ linkages), which correspond to the extensions of the disaccharides cellobiose (Fig. 3a) and maltose (Fig. 3b). Cellulose itself is virtually insoluble in water, but aqueous solubility can be conferred by partial methylation or carboxymethylation. Allowing, for example, the methylation reaction to proceed to completion would generate chains of the following type:

(XV)

Commercial methylcellulose samples are prepared by heterogeneous reaction which is usually controlled to allow substitution of, on average, about one-half of the hydroxyl groups. This reaction leads to a product in which the methylated groups are not evenly distributed throughout the chains; rather there are regions of high density of substitution (as in structure XV) which are essentially hydrophobic in nature, and regions of low density of substitution (as in Fig. 3a) which are essentially hydrophilic in nature.[1308]

More complicated, branched homopolysaccharide structures result when more than a single glycosidic linkage is present. Amylopectin and glycogen are $1\alpha4$-linked glucans which are branched by $1\alpha6$ linkages.

a)
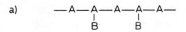

b)   —A—B—A—B—A—B—

c)   —A—A—A—A—A—
          —B—B—B—B—B—
          —A—B—A—B—A—

Fig. 5. Ways in which two different monosaccharide units can be incorporated into a heteropolysaccharide structure.

Dextrans are predominantly $1\alpha6$-linked glucose units with $1\alpha3$ or $1\alpha4$ branching linkages.

The term *heteropolysaccharide* covers a multitude of structures and so let us consider first those polymers containing *two sugars only*. The simplest types are those in which one sugar is present in the main chain and the other in the side chain (Fig. 5a), or when the two sugars are arranged consecutively in a single linear chain (Fig. 5b). Of the former type, the galactomannans serve as a good example; these are $1\beta4$-linked D-mannans with occasional residues substituted in the C-6 position by α-D-galactose stubs. The galactose stubs have a tendency to be grouped together[291,292] to generate so-called[328] "smooth" and "hairy" regions. Alternating *A–B–A–B* sequences are of great importance in both the animal and plant kingdoms, with glucose, galactose, and their derivatives the most common monosaccharide constituents. Three examples of gelling polysaccharides having this structural pattern are shown in Fig. 6. Hyaluronic acid is universally present in connective tissues of animals and in the vitreous and synovial fluids, and has β-D-glucopyranuronic acid and 2-acetamido-2-deoxy-β-D-glucopyranose as constituents linked $\beta1,3$ and $\beta1,4$ as shown in Fig. 6. Agarose and $\varkappa$-carrageenan are seaweed polysaccharides commercially important for their gelling properties. Agarose contains 4-linked 3,6-anhydro-α-L-galactopyranose and 3-linked β-D-galactopyranose, while $\varkappa$-carrageenan contains mainly 3-linked β-D-galactopyranose-4-sulfate and 4-linked 3,6-anhydro-α-D-galactopyranose.*

* In these alternating structures such absolute regularity is unusual; indeed the observed irregularities, i.e., insertion of a different A or B, are thought to be of importance in determining gelling properties.[1233]

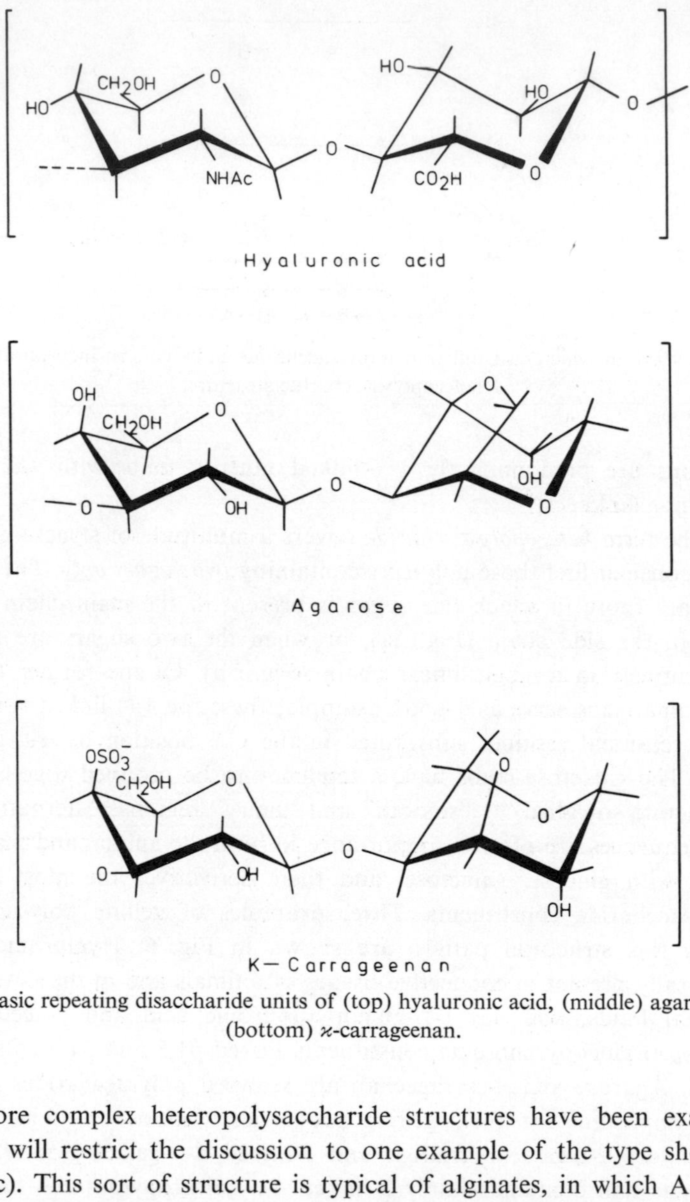

Fig. 6. Basic repeating disaccharide units of (top) hyaluronic acid, (middle) agarose, and (bottom) ×-carrageenan.

More complex heteropolysaccharide structures have been examined but we will restrict the discussion to one example of the type shown in Fig. 5(c). This sort of structure is typical of alginates, in which A and B are the uronic acid derivatives of the monosaccharides mannose and gulose; more specifically β-D-mannopyranosiduronate and α-L-gulopyranosiduronate both joined through position 4. Alginates from different sources vary in the relative proportions of the A blocks, B blocks, and AB blocks.

2.4. Potential Sites for Interaction with Water

Many of the chemical groupings which occur commonly in carbo-hydrates, e.g.,

$$-OH, \quad -NHCOCH_3, \quad \diagup\overset{O}{\diagdown}, \quad -CO_2H, \quad CO_2^-, -OSO_3^-, \quad -CH_3$$

can of course be considered (in isolation from neighboring groups) with regard to their interactions with water. A detailed coverage of the inter-actions between water and small "hydrophobic" and "ionic" solutes is given in Volumes 2 and 3 of this treatise, and the above groups could be classified under similar headings. However, it is apparent that carbohy-drate–water interactions cannot simply be thought of in terms of the inter-actions of the individual chemical groups with water. For example, the single most important mode of interaction of carbohydrates with water is by means of hydrogen bonding from the sugar hydroxyls, and this would be expected to be highly orientation dependent. Thus the precise spatial arrangement of the many hydroxyls in the solute might have a pronounced influence on the nature and extent of the solute–solvent interactions. This of course suggests that studies of small-molecule solutes possessing a single hydroxyl group (e.g., monohydric alcohols) will not necessarily tell us very much about the hydration of carbohydrates. One question we shall there-fore attempt to answer is, "Are the differences in conformation between chemically similar carbohydrates reflected in differences in their interactions with solvent?" (Or put in another way, "Are certain spatial arrangements of hydroxyl groups more favorable for hydration than others?") The im-plicit emphasis here lies in the modification of the properties of the solvent. However, any discussion of the interactions between carbohydrate and water must, by definition, include a consideration of the possible mutual influences of solvent and solute. We shall therefore also examine the pos-sibility that the conformation of the carbohydrate may be controlled, at least in part, by the *structure* of the solvent, in a way which maximizes the extent of hydrogen-bonded interactions between the components.

Although the hydration of polysaccharides is probably in general dominated by hydrophilic interactions (hydrogen-bonded, electrostatic), it would be unwise to totally neglect the influence of hydrophobic groups. In molecules such as methylcellulose the hydrophobic nature of certain parts of its structure (structure XV) is obvious, and is reflected in its rather unusual aqueous solution and gel properties. However, it must also be remembered that monosaccharide rings contain C–H as well as C–OH groups, and the adoption of certain polysaccharide conformations can generate areas of the polysaccharide surface which are essentially hydrophobic—capable, for example, of complexing nonpolar molecules.

3. AQUEOUS SOLUTIONS OF SMALL CARBOHYDRATES

3.1. Monosaccharides

3.1.1. Bulk Solution Measurements of Hydration

There has been a number of attempts[382,703,1085,1215,1381] in recent years to determine the extents of hydration (i.e., hydration numbers) of carbohydrates form a range of bulk solution measurements. Each of these incorporates its own particular set of assumptions and is not easy to judge which, if any, is the most quantitatively reliable. Probably of greater significance than the absolute magnitude of the hydration are the trends obtained as a function of temperature or as a function of carbohydrate structure; for example, the hydrations of D-glucose and tetrahydropyran-2-carbinol (structure XVI) have been determined as a function of temperature from

(XVI)

measurements of their intrinsic viscosities.[703] Figure 7 shows the values obtained for the polyfunctional glucose and the monofunctional THPA. The hydration of THPA is observed to "melt off" more quickly with increasing temperature than that of the glucose. This is consistent with the hydration of the glucose being of a specific hydrogen-bonded nature but that of the THPA being dominated by *hydrophobic hydration* characteristic of monofunctional solutes, for example, simple alcohols, amines, and ethers.[293] Another method of determining hydration, popular particularly in Japanese schools, is to measure the adiabatic compressibility of the solutions and make some simplifying assumptions, usually that the solute is incompressible and the "hydration water" has the compressibility of ice.[508,1381,1382] Although it is possible that these assumptions may be reasonable for a particular sugar, it is perhaps unlikely that the compressibility

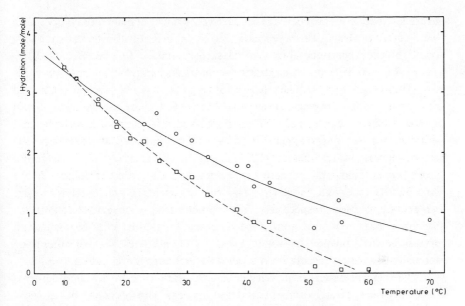

Fig. 7. Viscometric estimation of the hydration of glucose (○) and tetrahydropyran-2-carbinol (□) as a function of temperature (from Ref. 703).

of the hydration water should be similar for all sugars (particularly in view of the data to be presented later concerning differences in the *nature* of the hydration of various sugars).

3.1.2. *Thermodynamic Studies and Development of Simple Hydration Model*

Taylor and Rowlinson[1473] observed in 1955 that the thermodynamic behavior of solutions of glucose and sucrose resembled hydrogen peroxide–water mixtures rather than alcohol–water mixtures, and concluded that the hydrogen bonding between the sugars and water is stronger or more extensive than that between water molecules themselves. Stokes and Robinson,[1431] in fact, using the concept of the semiideal solution (i.e., the solution is an ideal one provided due allowance is made for hydration), have suggested that the observed concentration dependence of the thermodynamic quantities is due to solute–water interactions and can be expressed in terms of hydration equilibria:

$$S_{i-1} + H_2O \xrightleftharpoons{K_{i-1}} S_i, \qquad i = 1, 2, \ldots, n$$

where n is the number of potential sites for hydration. Assuming all hydration sites (and therefore all equilibrium constants K_{i-1}) are equivalent, an average hydration number can be derived which depends only upon the water activity. Using this approach, Stokes and Robinson found that they could fit the experimental activity data for glucose ($n = 6$, $K_{i-1} = K_i = \cdots = K_n = 0.789$) up to saturation conditions, and for sucrose ($n = 11$, $K = 0.994$) up to 6 M. The hydration numbers which can be extracted from their data correspond to near 2 mol H_2O/mol glucose and 5 mol H_2O/mol sucrose, both at 25°. This simple hydration approach is consistent with the Raman observations[1550] of the effect of added sucrose upon the intermolecular water band at 152–175 cm^{-1}. The ratio I_2C_1/I_1C_2 is found to be greater than unity, where I represents the integrated intensities of the band and C the water molarity, and the subscripts 1 and 2 refer to pure water and solution, respectively. The absence of any observable change in the band shape on addition of sucrose argues against contribution from intramolecular sucrose hydrogen bonds, and indicates that the increased intensity ratio must result from solute–water hydrogen bonding.

The interpretation of activity coefficient data in terms of simple hydration processes, after closer examination, appears, however, to be an oversimplification. Data will be presented, for example, in this section to demonstrate the nonequivalence of hydration sites.

It has also been suggested[488] that for carbohydrate–water systems free energies, as a result of enthalpy–entropy compensation effects, may be fairly insensitive to the hydration model chosen, whereas temperature and pressure derivatives of the free energy should be more discriminating. In this context it has been reported[1341] that $\partial(\ln \gamma)/\partial T$ and $\partial(\ln \gamma)/\partial P$ can no longer be fitted by the simple hydration model of Stokes and Robinson.

3.1.3. *Effect of Solute Conformation on Hydration Properties*

We will now consider what evidence exists to support the premise advanced in Section 2.4 that the nature and extent of hydration of sugars depend to some degree upon the stereochemistry of the sugar molecule. First, the effects of the sugar conformation on solvent participation in the mutarotation process (Section 2.1) have been clearly demonstrated in studies of the thermodynamics of the mutarotation of a number of carbohydrates in aqueous solution.[748,749] Whereas the standard free energies of mutarotation are all similar, the corresponding enthalpies and entropies are markedly different. For example, the standard entropies of mutarotation ($\alpha \rightarrow \beta$) range from -0.52 e.u. for xylose to $+0.70$ e.u. for maltose. These data was interpreted in terms of the equatorial hydroxyls being more strongly hydrated than axial hydroxyls. It was pointed out that a β-D-

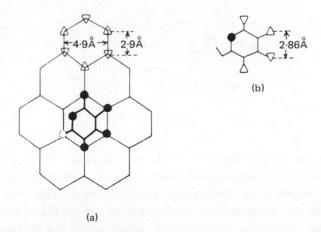

(b)

(a)

Fig. 8. Possible model for the hydration of monosaccharides. (a) Tridymite water structure at 25°; (b) β-D-Glucose. The orientation of the triangles indicate whether an oxygen atom is above or below the plane of the ring.

glucose molecule could replace almost exactly a "chair conformation of water molecules" in a tridymite ice lattice whose dimensions had been expanded to ambient temperatures, with glucose–water hydrogen bonds replacing the water–water hydrogen bonds (Fig. 8). In this context it should be recalled that Warner[1557,1558] later reinforced this sort of correlation between oxygen–oxygen spacings of about 4.9 Å in many biological molecules and the oxygen next-nearest neighbor distance for water as obtained from the X-ray radial distribution function.[311,1042]

   Second, influences due to solute conformation were also recognized in the thermodynamic measurements of Franks et al.[488] Whereas, for example, the apparent molal volumes of monosaccharide solutions are independent of solute concentration over the concentration range studied (0–0.05 mol kg^{-1}), indicating the absence of solute–solute interactions, subtle differences are observed between sugars differing only in the configuration at a single carbon atom. The limiting apparent molal compressibilities $\phi_K{}^\circ$ likewise show variations; introduction of successive OH groups into a cyclic ether ring renders $\phi_K{}^\circ$ more negative, but again small but significant differences between axial and equatorial substituents are apparent. The effect produced by myo-inositol (structure XVII), which has five equatorial hydroxyls, is particularly striking, the $\phi_K{}^\circ$ being of the same magnitude as those of electrolytes which derive their very negative ϕ_K from electrostrictive hydration (as shown, for example, by their low limiting apparent molar volumes[492]).

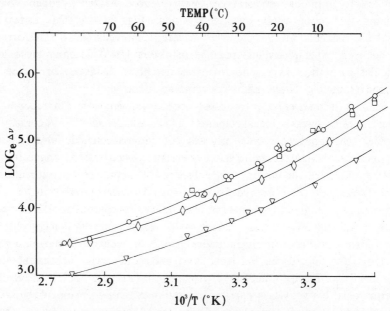

(XVII)

Finally, an attempt has been made by Tait *et al.*[1458] to investigate the hydration of monosaccharides, using relaxation methods. Oxygen-17 nmr measurements of the transverse relaxation time T_2 were used to generate information about the time-averaged properties of the aqueous component, while time domain spectroscopic measurements (for details of these novel techniques see Refs. 430, 431, 913, 1440, 1441, 1444) provided the dielectric spectrum which was resolvable in terms of the various relaxing components. A plot of the ^{17}O linewidth ($\Delta\nu = 1/\pi T_2$) against reciprocal temperature is shown in Fig. 9 for water and equimolar solutions of four monosaccharides: D-glucose, D-galactose, D-mannose, and D-ribose. At

Fig. 9. The ^{17}O linewidths corrected for inhomogeneity as a function of temperature. ($\bigtriangledown$) Water; ($\Diamond$) 2.69 *m* D-ribose; ($\triangle$) 2.79 *m* D-galactose; ($\bigcirc$) 2.79 *m* D-glucose; ($\square$) 2.79 m D-mannose.

lower temperatures there is a clear distinction between the hydration properties of the three hexose sugars (five hydroxyls) and ribose (four hydroxyls). However, it is clear that the hydration properties are not simply related to the numbers of OH groups, since the temperature dependences of the two sets are quite different. At temperatures in excess of $80°$, in fact, the T_2 values for all the monosaccharide solutions examined are indistinguishable. A "frequency domain" representation of the dielectric spectrum of a D-glucose solution is shown in Fig. 10, illustrating the large deviations from the dielectric spectrum of pure water. For a simple dielectric relaxation process the complex plane diagram in which the dielectric loss ε'' is plotted against the permittivity ε' (Cole–Cole plot) takes the form of a semicircle with its center on the ε' axis. The corresponding diagram for the data of Fig. 10(a) (Fig. 10b) clearly indicates the complex nature of the relaxation in glucose solutions. Analysis of both the frequency-dependent and time-dependent permittivities led Tait et al.[1458] to the conclusion that there were probably three separate relaxing components. The highest frequency (shortest time) process exhibited a relaxation time not very different from pure liquid water and a relaxation amplitude which decreased as the solute concentration increased, and was therefore assigned to the bulk water component. A second process at somewhat lower frequencies appears to be due to the reorientation of the solute, in view of the fact that the relaxation time is increased by the substitution of a disaccharide for the glucose.[1443] A third, minor process at yet lower frequency was assigned to a modified water component. Examination of the amplitudes of these relaxation processes[489] led to the estimation of the amount of water whose mobility has been significantly modified by the solute. For example, D-glucose from these calculations appears to be hydrated to the extent of about six molecules of water per molecule of sugar, at $5°$, whereas D-ribose is much less hydrated (about 2–3 mol H_2O/mol sugar, at $5°$).* These values are consistent with the qualitative interpretation of the nmr data of Fig. 9. The information generated by dielectric and nuclear magnetic relaxation is also quantitatively consistent in other ways; for example, they both indicate that the average correlation time for the water molecules in 2.8 m aqueous glucose at $5°$ is lengthened by a factor 2.5–3 over that for pure water at the same temperature.

All in all, therefore, despite the scarcity of experimental information,

---

* The activity coefficient of ribose solutions up to 3 m has been reported (H. Uedaira, personal communication to editor) to be very close to unity. According to the semi-ideal treatment, this would imply negligible hydration of ribose.

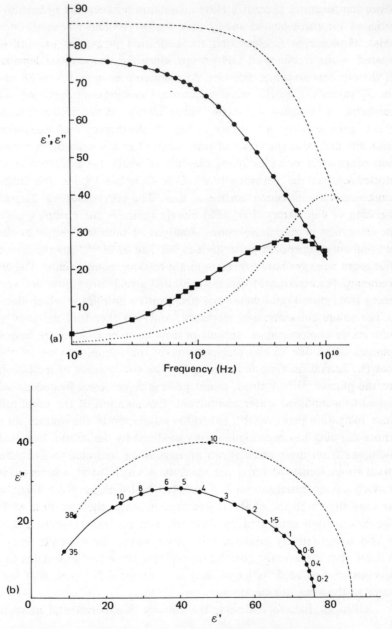

Fig. 10. Dielectric spectrum of 2.8 m D-glucose relative to that of pure water, at 5°. (a) Frequency spectrum: (●) ε', glucose; (■) ε'', glucose; broken lines, pure water spectrum (from Ref. 913). (b) Representation on "Cole–Cole" diagram: (●) glucose; broken line, pure water. Frequencies indicated at the measurement points are in GHz.

there is now good reason to believe that the conformation of simple sugars does have an influence on their hydration properties. An explanation for this has been put forward[489,1458] in terms of the compatibility between the oxygen–oxygen spacings in liquid water and those in the sugar molecule. This specific hydration model is based upon the earlier ideas of Kabayama and Patterson[748] (Fig. 8). D-Glucose, D-galactose, and D-mannose, for example, have pyranose C1 conformations[27] in which the majority of the hydroxyls are in equatorial configurations. The distance between the hydroxyl oxygens on the same side of the ring, e.g., those attached to carbon atoms 1 and 3, or 2 and 4, of β-D-glucose, is 4.86 Å,[1557] and similarly in water at ambient temperatures, X-ray diffraction studies[311] have indicated a high concentration of water molecules 4.9 Å apart. This agrees with a calculated second-nearest-neighbor distance for oxygen atoms linked in an icelike lattice in which the next-nearest oxygen atoms are coplanar. The OH groups on opposite sides of the glucose ring can thus, it is suggested, hydrogen-bond to two layers of water molecules, one above and one below the plane of the ring. Whereas β-D-glucose has all its hydroxyls in equatorial configurations and therefore fits very well into a tetrahedral arrangement of water molecules, the conformation of D-ribose is by no means so ideal. It is known that ribose adopts both α and β furanose and α and β pyranose forms (see Section 2.1, Fig. 2) and is not able to fit so well into such a water structural arrangement. As a result, it is argued, ribose is hydrated to a much smaller extent. As the temperature is increased, we have seen (Fig. 9) that the difference between the hydration properties of, for example, glucose and ribose gradually disappears. Likewise the X-ray radial distribution function for water shows a decreasing probability of the 4.5–5.3 Å spacing with increasing temperature until at 75° this peak has virtually disappeared.[1083]

If such a tetrahedral arrangement of water molecules is equated with the "bulky" (or "structured" or "lattice") component of a two-state mixture model for water structure,

$$H_2O(\text{bulky}) \rightleftharpoons H_2O(\text{dense})$$

then the incompatibility of the ribose stereochemistry with this component might tend to suggest[1458] that ribose might interact preferentially with the less structured component and, like urea,[478] act as a "statistical structure breaker." As far as the author, is aware there is no direct evidence as yet for such an influence of ribose upon water, but such considerations must eventually lead us to consider whether in fact the *nature* of the interactions between various sugars and water are fundamentally different, and potentially of major biological significance.*

Recent observations, then, are fully consistent with a specific hydration model in which the compatibility between the sugar stereochemistry and the intermolecular order in liquid water is of great significance. Warner[1557–1559] has discussed the correlation between the stereochemistry of sugars and sugarlike molecules and their hydration behavior, with reference, for example, to biological specificity and cryoprotective action. The cyclitol *scyllo*-inositol has a structure in which each of its six (equatorial) oxygens are ideally spaced to fit into the "expanded ice lattice." Warner[1559] relates the observation[959] that *myo*-inositol, but not *scyllo*-inositol, is oxidized by *Acetobacter suboxydans* to their respective hydration structures. The single axial hydroxyl substituent of *myo*-inositol has the only oxygen atom that does not fit into the expanded ice lattice, and it is this axial substituent which is specifically oxidized. The keto oxygen of the reaction product is, however, fairly close to an oxygen position in this ice lattice. Similar reasoning has been applied to the stabilization by sugars and related compounds of biological systems at low temperatures and under dehydrating conditions. Warner[1559] again has made the point that the positions of the hydroxyls in the sugar molecules correspond best with an ice lattice which has been slightly expanded to agree with observed water oxygen–oxygen distances at physiological temperatures. In "real ice," therefore, substances like glucose or glycerol (which can adopt a conformation in which the oxygen atoms on carbon atoms 1 and 3 are spaced at 4.85 Å) cannot now hydrogen-bond into the lattice, and their hydration structure may inhibit the growth of large ice crystals and hence minimize freezing damage to the biological specimen. It is also interesting to note that the loss of organized structure produced by dehydrating DNA from *E. coli* may be prevented by addition of *myo*-inositol.[1566] This observation has also been interpreted[1559] in terms of the close resemblence of the hydroxyl spacings in the inositol with the spacings in liquid water, and the potential ability therefore of *myo*-inositol to replace the hydration water requirements of DNA.

3.1.4. *Effect of Solvent upon Solute Conformational Properties*

In order to predict the relative stabilities of monosaccharide conformations in vacuum and in various solvents, we ideally require a knowledge of

---

* For example, the evolution of nucleic acids containing ribose and 2-deoxy ribose (rather than other sugars) might arguably be a consequence of the rather poor hydrophilic hydration properties of these sugars, bearing in mind the proximity of the predominantly hydrophobic bases.

the relative free energy in these states. Although a number of "vacuum calculations" have been performed, [1221,1407,1450,1527] it is difficult to judge their value in this present context since one of the criteria of success of such calculations appears to have been how well the results correlate with experimental observation (i.e., in solution). An empirical approach was adopted by Angyal, [27] who obtained approximations to the relative free energies of, for example, the C1 and 1C pyranose forms of various monosaccharides, by summing estimates of the energies associated with nonbonded interactions with ligands, and making allowances for electronic interaction and entropy differences. Although the interaction between the sugar and water is not specifically allowed for in these calculations, some contribution from this source may be implicitly included since the "interaction energies" were in fact obtained from studies of the equilibria of cyclitols with their borate complexes *in aqueous solution*, and of the anomeric equilibria of pyranose sugars *in aqueous solution*. Thus in our attempt to consider what, if any, are the effects of the solute–water interactions upon the conformation of the *solute*, we cannot at this time be at all certain as to what the relative stabilities of the various possible conformations would be in the absence of solvent. The arguments here cannot therefore be in any way conclusive, and we will attempt only to relate a few pieces of recent information which perhaps point the way to further study in this area.

Tait *et al.*[1458] pointed out a possible correlation between hydration and conformation in the case of D-ribose. From measurements of the relative amounts of the α and β furanose forms (Fig. 2) in aqueous solution as a function of temperature[1168] it was noticed[1458] that, as the temperature was decreased below 80°, the form most likely to be hydrated according to the specific hydration model[1458] (namely β-ribopyranose) increased in concentration, whereas the form least likely to be hydrated (α-furanose) decreased in concentration. The basis for these observations was a collection[1168] of proton magnetic resonance spectra of D-ribose in D_2O. The spectra were obtained and analyzed in three different laboratories[872,1168,1280] (two temperatures only in each laboratory) and showed considerable scatter in terms of the relative proportions of the various forms. A recent study[894] has clarified the situation by obtaining the relative proportions (on the same sample and on the same instrument) over the temperature range 0–85°. The results confirm that the major effct of increasing temperature is to markedly decrease the β-pyranose concentration. The C1 form of the β-pyranose is the one which is most compatible (in terms of hydrogen bond formation) with a tetrahedral arrangement of water molecules. It may be, therefore, that the greatly enhanced stability of this form at the lower temperatures (Fig. 11) reflects its more extensive interaction with the solvent as a result of the increased intermolecular order in the water at these temperatures. There are, however, a number of alternative explanations for these phenomena (they may arise, for example, from modification of the electronic interaction induced by the changes in the bulk dielectric constant) and the case for subtle influences of water struc-

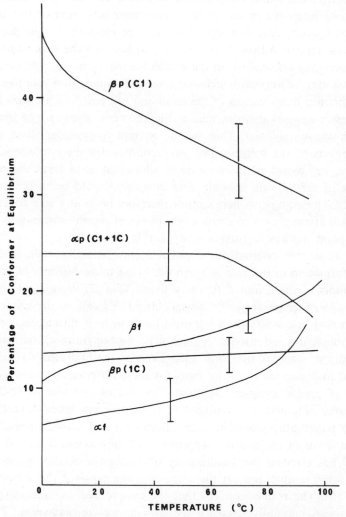

Fig. 11. Temperature dependence of the various conformations of D-ribose. For structured identification, refer to Figs. 1 and 2. βp(Cl), for example, represents the β-pyran form in the Cl conformation, etc. Error bars represent the total maximum uncertainties, over the whole temperature range, from integration of NMR peaks and interpretation of coupling constants in terms of Cl and lC forms (from Ref. 894).

ture upon ribose conformation must await further examination of the alternative hypotheses.

The effects of gross changes in solvent properties are more obvious. Optical rotation measurements,[601,1475] for example, have indicated that the conformation adopted by many sugars in aqueous solution is different to that in nonaqueous solvents such as dioxane or dimethylsulfoxide. Furthermore, as the temperature is increased, in general the optical rotatory properties of the aqueous sugars tends toward the nonaqueous values.[1475] These changes may reflect a distortion of the chair conformation. In water–dioxane mixtures[601] the optical rotation varies in a complex manner with solvent composition, but again suggests a pronounced influence of the solvent on the solute conformation. Another recent demonstration of a solvent-dependent conformation comes from nmr pseudo-contact shift measurements on transition metal complexes of adenosine monophosphate (AMP).[99] It has been clearly shown that in aqueous solution the ribose ring takes up a well defined orientation with respect to the adenine residue which is quite different to the orientation in dimethylsulfoxide solution.

3.2. Disaccharides and Oligosaccharides

The studies of monosaccharide–water systems outlined above have indicated something of the influences of the solute–solvent interactions on the "structures" of both components. The same general features could be reiterated for disaccharides and oligosaccharides. However, since we are considering these small carbohydrates as model systems for the polymers, we shall here discuss only the "extra contributions" that studies of disaccharides and oligosaccharides can make to the understanding of polysaccharide–water systems.

Compared to the monosaccharide–water systems, we now must also consider the effects of the conformational variability introduced because of the possibility of rotation about the glycosidic bonds (Fig. 3), and also the possibility of intramolecular interactions between the different monosaccharide residues. The presence of interresidue hydrogen bonding has been demonstrated in dimethylsulfoxide solution using nmr and IR spectroscopy.[231] For the $1\beta4$-linked cellobiose (Fig. 3), for example, an $O_3'-O_5$ hydrogen bond is indicated, while $1\alpha4$-linked polyglucoses show evidence of an $O_3'-O_2$ hydrogen bond. In the latter case the tendency for hydrogen bond formation increases in the series maltose < maltodextrins < amylose < cyclodextrins.

However, aqueous solutions of oligosaccharides cannot be subjected to the above type of proton magnetic resonance study, and we must look elsewhere, at least at present, in order to derive some information about the influence of the aqueous solvent in these systems. Optical rotation has been shown[1232] to be highly sensitive to the conformation at the glycosidic linkage, and some recent optical rotation measurements on oligosaccharide–water systems[1475] have provided an unexpected source of information about the effects of the solvent upon solute conformation. The linkage conformation of a disaccharide is given by the two dihedral angles ϕ and ψ in structure XVIII, and Rees[1232] has shown that these angles can be related to a param-

(XVIII)

eter known as the "linkage rotation" $[\Lambda]_D$. This in turn is obtained experimentally from the relationship

$$[\Lambda] = [M_{\mathrm{NR}}] - \{[M_{\mathrm{MeN}}] + [M_{\mathrm{R}}]\}$$

where $[M_{\mathrm{NR}}]$ is the molecular rotation for a given disaccharide which contains a nonreducing (N) and a reducing (R) residue, $[M_{\mathrm{MeN}}]$ is the molecular rotation of the methyl glycoside of N having the same anomeric configuration as the disaccharide, and $[M_{\mathrm{R}}]$ is the rotation of the reducing sugar.

Rees[1232] found that using conformational details from crystal structures and the optical rotations of the component monosaccharide derivatives, he was able to predict the molecular rotations of β-cellobiose and α-lactose derivatives in water and of cyclohexaamylose (Fig. 4) in dimethylsulfoxide to within a few degrees. The molecular rotation of cyclohexaamylose in water differs by about 30° per residue from the predicted value based upon the crystal structure, but this anomaly disappears when the inclusion complex is formed with saturated hydrocarbon derivatives. The distorsion that this cyclic oligomer undergoes in aqueous solution will be discussed later in this section.

Very recently the above approach has been applied[1475] to a range of disaccharides and oligosaccharides in dimethylsulfoxide, dioxane, and water. Let us first of all consider the results for the disaccharides, cellobiose, maltose, and trehalose, whose structures are shown in Fig. 3. (For the first two measurements were made on the β-methyl glycosides to eliminate any effects due to differences in the mutarotation equilibria.) From the crystal coordinates for β-cellobiose,[185,256] which appears to be stabilized by an $O_3'-O_5$ hydrogen bond, a linkage rotation $[A_{calc}]_D$ of $-62°$ was calculated for methyl β-cellobioside. This is very close to the observed $[A]_D$ in both dimethylsulfoxide and water at 25°C, namely $-59°$ and $-60°$, respectively, and it is reasonable to assume that a conformation very close to that in the crystal is adopted by this disaccharide in both solvents. For methyl β-maltoside and α,α-trehalose, however, the behavior is markedly solvent dependent, as indicated by Fig. 12. The dioxane and dimethylsulfoxide solutions behave similarly, but the aqueous solution, both in the magnitude of $[A]_D$ and the direction of its temperature dependence, is very different.

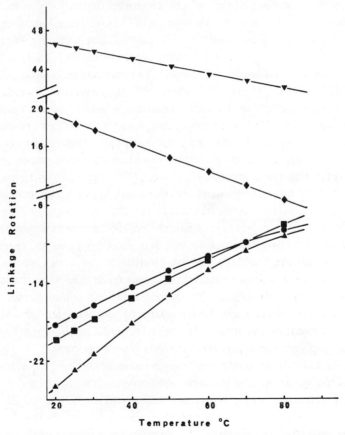

Fig. 12. Linkage rotation $[A]_D$ versus temperature for methyl β-maltoside in water (▼), dimethylsulfoxide (■), and dioxane (▲), and for α,α-trehalose in water (♦) and dimethylsulfoxide (●) (from Ref. 1475).

TABLE I. Observed and Calculated Linkage Rotations for β-Methyl Maltoside

	Observed[1475]				Calculated	
	In DMSO	In dioxane	In H_2O	From crystal conformation[255]	van der Waals minimum[1220] (I)	van der Waals minimum[146] (II)
$[\Lambda]_D$, deg	-19	-23	$+46$	-109	-24	$+85$

Table I indicates the observed $[\Lambda]_D$ values at $25°$ in the three solvents together with $[\Lambda_{calc}]_D$ values based upon a number of different conformations, and it is obvious from this that, unlike the cellobioside, methyl β-maltoside does not exist in solution in the crystal conformation. Conformational energy calculations[1220] indicate that the van der Waals minimum is at $\phi = -30°$, $\psi = -10°$, which corresponds to $[\Lambda_{calc}]_D = -24°$. However, other calculations[146] suggest an additional minimum corresponding[1475] to $[\Lambda_{calc}]_D = +85°$. Rees[1232] suggested that methyl β-maltoside in dimethylsulfoxide might be a mixture of both these conformations which minimize van der Waals repulsions with the hydrogen-bonded conformation. This would then be consistent with the observations of proton magnetic resonance spectroscopy[231] discussed earlier. In aqueous solution the linkage rotation measurements of Thom[1475] suggest that the molecule must spend a much larger proportion of its time in the conformation with $[\Lambda_{calc}]_D = 85°$. Such a conformation has in fact been suggested independently[1086] to account for the apparent specific expansibility of maltose (Fig. 13). Here the two glucose residues have folded together to screen their hydrophobic surfaces from the solvent. Further evidence in favor of some sort of folded conformation for maltose comes from the ^{13}C nmr measurements of Dorman and Roberts,[358] who found that, although the spectra for cellobiose and lactose can be reconstructed by a simple summation of their monosaccharide spectra, this is not so for maltose, showing that steric or proximity effects occur between the two glucose residues.

The folded maltose conformation is an example of how the interactions between a polyhydroxylic solute and water may be influenced by the partial *hydrophobic* character of the solute molecules. Neal and Goring,[1086] in fact refer to maltose folding via an "intramolecular hydrophobic bond" (in an analogous way to that suggested for proteins by Kauzmann[759] and discussed more fully in Chapter 1). Alternatively it may be argued[1475] that the folding presents a hydrophilic surface which is more compatible with a

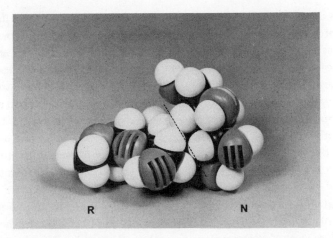

Fig. 13. CPK space filling model of methyl β-maltoside in the folded conformation. The hydrogen atoms H-4 and 2H-6 on the R residue are in contact with the H-3 and H-5 on the N residue (from Ref. 1475).

tetrahedral arrangement of water molecules (the sort of argument presented for monosaccharides in Section 3.1.3). Whatever the true explanation, there is a clear message from this work that the nature of the aqueous solvent can have a pronounced influence upon the conformation of disaccharides, particularly those with α-glycosidic linkages.

An extension of the above approach to $1\alpha4$-linked oligoglucoses, both linear and cyclic, has led to an increased awareness of the possible dependence of the conformations of the related polysaccharides upon the nature of the solvent. The observed linkage rotation of cyclohexaamylose (Fig. 4) in dimethylsulfoxide and water and calculated values based on a number of different conformations are shown in Table II.

The structure of cyclohexaamylose is such that the surface of the cavity is essentially hydrophobic in character and Rees[1232] has suggested that

TABLE II. Observed and Calculated Linkage Rotations for Cyclohexaamylose

	Observed[1475]		Calculated[1475]			
			Potassium acetate complex[702]	Alkanoic acid complex[1327]	Hexahydrate[980]	Hexahydrate plus one maltose fold
	DMSO	Water				
$[\Lambda]_D$, deg	−99	−65	−105	−90	−85	−66

whereas dimethylsulfoxide can solvate the cavity, in water the undistorted cavity probably represents a source of high free energy. The formation of an inclusion complex, it is further suggested,[1232] would relieve the conformational distortion. Recently the X-ray structure of cyclohexaamylose hexahydrate has been determined,[980] and in the crystal the strain is apparently relieved by the rotation of a single residue rather than by a symmetric puckering of the whole molecule. As shown in Table II, if in solution the hexahydrate structure is retained with one disaccharide unit in the folded conformation, as previously suggested for β-methyl maltoside, then a $[\Lambda]_D$ value can be calculated which is in excellent agreement with the observed aqueous solution value. The conclusion must therefore be drawn [1475] that in aqueous solutions of cyclohexaamylose the unfavorable interactions of the hydrophobic cavity with water are overcome by the molecule forming a sort of "intramolecular inclusion complex," with one maltose unit in the folded conformation filling what would have been the hole. As the degree of polymerization increases from cyclohexaamylose to cycloundecaamylose the observed linkage rotations are consistent with a larger contribution from the folded conformation, as the glycosidic linkages gain more freedom to relieve torsion and van der Waals strain and to make themselves more compatible with the solvent structure.

The cycloamyloses are ideal models for a consideration of the interactions between water and helical or potentially helical 1α4-linked polysaccharides (such as amylose). The information gained from these model systems will later be brought to bear upon the considerations of the possibility of helical amylose conformations in the presence and absence of hydrophobic complexing agents.

For the linear 1α4 oligomeric series based upon maltose (the maltodextrins) $[\Lambda]_D$ rapidly converges on the amylose value, indicating that in the polymer there is some interaction present which reinforces the O_2–O_3' hydrogen bond between neighboring residues. Rees[1232] has suggested that consecutive hydrogen bonds may be favored by the consequent parallel alignment of dipoles. Table III shows the trends in linkage rotation for the 1α4-linked oligoglucoses together with the value for amylose.

Movement left to right across this table represents a decreasing tendency to adopt the folded linkage conformation, and it is interesting to note that the linkage conformation in the maltodextrins of DP $\geq$ 7, the cycloamyloses of DP $\geq$ 11, and amylose are all essentially the same—a fact that will be recalled in Section 4.2.1.

So far in our considerations of disaccharide– and oligosaccharide–water systems the emphasis has been placed firmly on the side of the in-

TABLE III. Linkage Rotations of 1α4-Linked Oligosaccharides of Glucose

Degree of polymerization (glucose units)	$[A]_D$, deg
Maltodextrins[1475]	
2	69
3	46.4
4	37.2
5	38.6
6	33.2
7	17
(Amylose[1232])	
—	16
Cycloamyloses[1232]	
12	17
11	16
10	11
9	1
8	−21
7	−48
6	−65

fluence that the solvent may exert upon the conformation of the solute. In the $1\alpha4$ glucose series which we have taken as our example the obviously desirable parallel studies upon the state of the water are as yet barely in progress. Gekko and Noguchi[508] have, however, estimated the "bound water" in a series of dextran fractions ($1\alpha6$-linked oligoglucoses) whose number-average molecular weights varied from 410 to 200,000. Using the adiabatic compressibility approach of Shiio,[1381,1382] which has been discussed in Section 3.1.1, they found that the "bound water" decreased rapidly as the average degree of polymerization increased from about 2.4 to 8, but for DP > 10 was independent of molecular weight. In terms of mol bound water/mol OH group the data indicate that in the polymer a hydroxyl group is only about half as likely to be hydrated as in the disaccharide or trisaccharide.

4. POLYSACCHARIDE SOLS AND GELS

Any attempt to categorize polysaccharides into those that form sols and those that can gel becomes more than a little meaningless since most polysaccharides can be induced to gel under some conditions (it is also exceedingly difficult to define clearly what is or is not a gel). Moreover, it was felt that it is unsatisfactory to separate the properties of a particular polysaccharide–water system in its sol and gel states. Because of the uncertainty in any artificial compartmentalization of polysaccharide systems, we shall consider individually a number of systems, particularly from the point of view of the effect of solute–water interactions. It should be stressed that no attempt has been made to make this account in any way comprehensive. However, before considering these examples in detail it is pertinent to consider the mechanisms by which it is believed that solutions of polysaccharides can form a gel.

4.1. Gelling Mechanisms

We will consider here only gels that can form reversibly. No account will be given of covalently cross-linked polysaccharide gels, such as those formed by the dextrans in gel filtration media. Some sort of classification of reversibly formed polysaccharide gels can be made, based upon macroscopic observations, such as: gels formed on (a) cooling; (b) heating; (c) addition of certain cations; (d) cooling in sucrose solutions; (e) addition of other polysaccharides.

However, it is more revealing to attempt to classify according to the different molecular mechanisms. For a solution of a polymer to form a gel, some sort of cross-linked network must develop. We can therefore conveniently differentiate between the mechanisms from the point of view of the nature of the cross-linking. The cross-links are sometimes called "junction zones."

In recent years the subject of gelling polysaccharides, particularly the molecular mechanisms via which the gels form, has become synonymous with the name of D. A. Rees, and the bulk of the information presented in this section comes from the publications of Rees and co-workers. More detailed accounts of polysaccharide gelation mechanisms than will be given here can be found in a number of other review articles.[1231,1233,1234]

4.1.1. *Helical Junctions*

Gels of this type are formed, for example, by agarose and the family of carrageenans (the basic repeating structures of agarose and x-carrageenan are shown in Fig. 6). Cross-links are considered[1231,1233,1234] be to of the form shown in Fig. 14, although, particularly for agarose and the less sulfated carrageenans, it is believed[1234] that the individual double helices quickly aggregate to form "superjunctions," imparting an extra rigidity to the gel. Gels of this type set on cooling and melt on heating, often with a pronounced hysteresis loop, as exemplified in Fig. 15 for the x-carrageenan case by optical rotation measurements.[1236]

The regular A–B–A–B structure for these polysaccharides indicated in Fig. 6 is not strictly observed. Occasional residues are replaced in a "masked repeating structure"[25,1231] by residues having an inverted ring conformation. This causes a sudden change in the direction of propagation of the polymer chain. According to Rees,[1233] it is likely that these "kinks" have

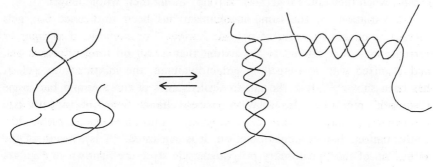

Fig. 14. Gel network formation by double helical junctions (from Ref. 1231).

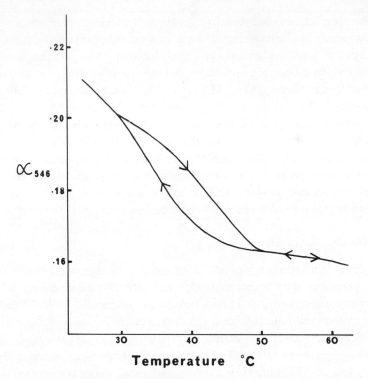

Fig. 15. Optical rotation α_{546} changes with temperature in a $\varkappa$-carrageenan system showing hysteresis loop (from Ref. 1236).

both a physical and biological function in the sense that they cause each chain to form double-helical structures with more than a single partner, thus facilitating the formation of a three-dimensional network. This is in contrast to other, more familiar double-helical macromolecules such as DNA, in which the regularity of the structures ensure that two individual chains, when they come together, "zip up" along their whole length.

A variation on the same mechanism has been implicated for gels formed by certain mixtures of polysaccharides. For example, if an agar or carrageenan sol is diluted to the extent that it can no longer form a gel, and is mixed with a nongelling galactomannan, the mixture may gel. It has been shown[328] that the unsubstituted parts of the mannan backbone ("smooth" region) in this gelation process change from the random coil form in which they normally exist in solution into an extended, ribbonlike conformation. This synergistic action, it is suggested,[328] is a result of the formation of the "quaternary polysaccharide structure" shown in Fig. 16, or an "aggregated version" of this structure.

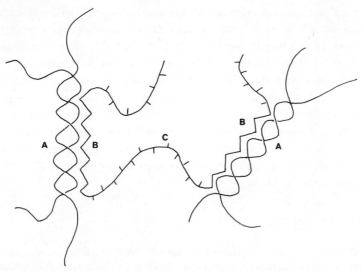

Fig. 16. Model for quaternary polysaccharide structure in mixed carrageenan-galactomannan system leading to gel formation (from Ref. 1233). (A) Carrageenan double helices; (B) galactomannan "smooth" regions; (C) galactomannan "hairy" regions.

4.1.2. Stacked Junctions

In this category we are primarily considering gels formed by the addition of salts of a suitable cation, e.g., Ca^{2+}, and examples are the gels formed by alginates and pectins.

The basic structure of alginates has been described in Section 2.3 and Fig. 5(c), and gels can be formed by diffusion of Ca^{2+} ions into a solution of the soluble sodium alginate. It appears[560,1043] that the cross-links are formed preferentially by chains of α-L-gulopyranosiduronate residues which stack with Ca^{2+} ions between the chains, as schematically shown in Fig. 17.

Pectins resemble alginates in that they have structures based upon linear chains of 1,4-linked uronic acid residues, although in this case the

Fig. 17. Model for junction zones in alginate gels. Each pair of guluronate residues is represented by a kink with calcium ions (•) stacked between the chain (from Ref. 1234).

uronic acid is D-galacturonic acid and it occurs predominantly either as the anion or the methyl ester. Pectin chains also contain residues of L-rhammose (XIV) which cause "chain kinking."[1234] Commercial pectins with a high methoxyl ester content can form gels on cooling a slightly acidic solution containing high concentrations of sucrose. Other solutes such as glycerol and ammonium sulfate can be used instead of the sucrose and it is therefore likely that its role is simply related to a lowering of the water activity. Low methoxy pectins, on the other hand, can form gels at low pH and in the presence of Ca^{2+} ions in a similar manner to the alginates. It is believed[1231] that the mechanisms for both types of pectic gelation involve chain stacking.

4.1.3. Micelle Junctions

This type of mechanism, in contrast to the first two, is a consequence of a type of solute–solvent interaction peculiar to aqueous systems (hydrophobic hydration) and involves the concept of solute–solute hydrophobic interactions ("hydrophobic bonding") which is discussed more fully in Chapter 1. Examples of this category are the gels formed by the cellulose derivatives, e.g., methylcellulose (XV), and hydroxypropylcellulose. They show unusual behavior in that solutions gel on heating and melt on cooling (this fact alone suggests a very different type of mechanism). The tendency for methylcelluloses to gel has been shown[1308] to be enhanced by increasing the proportion of di- and tri-O-methyl-D-glucose residues.

It is significant that methylcelluloses, as discussed in Section 2.3, are thought to consist of cellulose chains having zones of high density of methyl substitution (less water soluble) and other zones with a low density of substitution. The behavior of the hydrophobic parts can be compared to the aqueous solution behavior of simple, essentially apolar, molecules such as tert-butanol, the solubility of which decreases with increasing temperature to the point at which lower critical demixing is observed. Likewise, it is then expected that the hydrophobic parts of the methylcellulose, from different chains, can come together with increasing temperature to form separate "microphases," leaving the less substituted segments exposed to solvent. These "microphases" then serve as junction points for the gel network.

Such junctions have been described[1234] as micelles by analogy with detergent micelles, and unlike the other types of junctions considered earlier, in this case there is no requirement for a precise stereochemical compatibility between the polysaccharide chains in the junctions. As pointed out by Rees,[1234] the cross-links in methylcellulose gels are formed because parts of the chains are expelled from contact with water to cluster together without any necessary fit. The properties of the gels depend rather upon the hydrophilic–hydrophobic balance along the length of each chain.

4.2. Starch

Although many volumes have been written over the years on the chemistry of starches, it is often difficult to separate scientific observations from folklore. Nevertheless, some information is available about its inter-actions with water, particularly relating to the linear fraction of starch, amylose.

Starch granules consist in the main of linear (amylose) and branched (amylopectin) molecules associated, presumably by hydrogen bonding, into a structural arrangement that has only a limited capacity for water sorption. The granules are insoluble in water, despite the fact that they are composed of highly hydrophilic sugar residues. If, however, an aqueous suspension of a starch is subjected to the action of heat or various chemicals, the granules can swell, leading to the process called *gelatinization*. It is perhaps pertinent, before concentrating our attention on the amylose component of starch, to relate some observations on the influence of the nature of the aqueous solvent upon the starch gelatinization process.

If a starch suspension is heated, the granules do not change in ap-pearance until a certain critical temperature is reached. This temperature depends upon the origin of the starch, and can be altered by the addition of certain simple solutes. For example, the addition of sodium sulfate represses gelatinization (higher gelatinization temperature), whereas so-dium nitrate and urea have the opposite effect.[862] An alternative approach to the study of the solvent dependence of gelatinization is to determine the concentration of various reagents required to induce gelatinization at a fixed temperature. Mangels and Bailey[977] observed many years ago that a number of reagents can induce gelatinization at 30°; in order of decreas-ing effectiveness they are sodium hydroxide, potassium hydroxide, sodium salicylate, sodium, ammonium, or potassium thiocyanate, sodium or po-tassium iodide, urea, calcium chloride, sodium or potassium bromide. For the sodium salts the relative gelatinizing powers correspond to the Hof-meister series, familiar in protein chemistry for the salting-in/salting-out effects of various reagents. In this context, the relative effects of sodium sulfate, sodium nitrate, and urea discussed earlier also correlate with their positions in the Hofmeister series.

A possible explanation as to why the effect of aqueous anions on the stability of native structures of many biopolymers (proteins, nucleic acids, and, as we shall see, polysaccharides) follows a common sequence has been advanced[784,1538] (see also Chapter 4, Volume 2). This suggests that the common factor is the modification of the "water structure" by the anions. On this basis, therefore, one might similarly suggest that the effectiveness of various small solutes in facilitating starch gelatinization is associated with their capacity to break down the intermolecular order in water.

Amylose, the linear fraction of starch, is almost insoluble in water at ambient temperatures, although some degree of solution can be attained if the amylose is converted into a helical form (V-amylose) by complexation with certain hydrophobic molecules such as 1-butanol. On the other hand, amylose does dissolve in aqueous alkali (for example, 1 m NaOH), and upon neutralization metastable solutions are obtained. After some time (which depends critically upon the source of the amylose, its molecular weight, pH, temperature, and solvent composition), the amylose component begins to aggregate and ultimately precipitate—this is the phenomenon known in starch chemistry as *retrogradation*. The amylose precipitate is known to possess an ordered fibrillar structure with a B-type X-ray diffraction pattern similar to that of the original starch granule,[496] hence the term *retrogradation*, implying a return to the native state.

Three areas of amylose chemistry will now be noted in which a solvent role may be identified.

4.2.1. *Amylose Conformation in Solution*

The conformation of amylose in aqueous solution has been a matter of great conjecture over the years. It has been suggested, for example,[1222, 1455] that amylose behaves as a stiff chain with helical segments. On the other hand, it has been repeatedly demonstrated that, hydrodynamically, amylose behaves as a flexible coil.[80–83] Furthermore, with the glucose units in the C1 conformation, a hypothetical amylose helix would have all its hydroxyl groups on its external surface, the internal surface being lined with C–H groups and glycosidic oxygen atoms.[83] In this state it is suggested[83] that water cannot maintain its usual structure within the helix, and to avoid an energetically unfavorable situation amylose therefore retains a random conformation. Support for this view is taken from the optical rotation studies of Rees,[1232] which indicate that cyclohexaamylose (Fig. 4) adopts a strained conformation in water which is relaxed on addition of a complexing agent.

Optical rotation studies, however, while supporting the lack of any pronounced helical character, do tentatively suggest that the amylose coil is anything but random and is determined in part by solvent-dependent effects. If we recall, for example, the studies on amylose model systems described in Section 3.1.4, it is seen that the linkage rotations for malto-dextrins with $DP \geq 7$, cycloamyloses of $DP \geq 11$, and amylose are all very similar (Table III).

The conclusion has been drawn[1475] that for the case of amylose in neutral aqueous solution for each glycosidic linkage there is a dynamic equilibrium among (a) the hydrogen-bonded conformation (which, if it were the only contribution, would cause amylose to zip up into a helix), (b) the van der Waals minimum energy conformation, and (c) the "folded conformation." The third contribution is probably a major influence and is probably stabilized by hydrophobic interactions as discussed earlier. Since such interactions are characteristic of aqueous systems, one might expect the linkage rotation to be significantly different in nonaqueous solvents. The $[\Lambda_{\text{obs}}]$ of $-52°$ for amylose in dimethylsulfoxide, compared to the value of $+16°$ in water,[1232] is therefore consistent with the above interpretation, and indicates a much larger contribution from the hydrogen-bonded conformation. A significant contribution from this conformation in DMSO is also indicated from nmr studies.[231]

4.2.2. Retrogradation and Gelation of Amylose

The structure of retrograded amylose is thought to consist of aligned helices, which are held together by strong secondary forces, presumably hydrogen bonds. The act of retrogradation is thought[467] to involve inter-action between neighboring molecules, mutual alignment, expulsion of water, and formation of the new intermolecular forces.

Retrograded amylose gives the B-type X-ray diffraction pattern with a repeat distance of 10.4 Å. Blackwell et al.[101] interpret their diffraction patterns in terms of a sixfold single helix structure. It has been suggested elsewhere,[407,1235] however, that a double helical structure in which the hydrophobic regions of each chain are juxtaposed in the center of the helix is a credible alternative. We will not be concerned here with a detailed consideration of the kinetics of the retrogradation process itself; the discussion will be restricted to cover only the possible role of the solvent.

Loewus and Briggs[914] found that the presence of monovalent cations and anions retarded the retrogradation rate, with $F^- < Cl^- < Br^- < I^-$, and $Li^+ < Na^+ < K^+$, in order of increasing retardation. The anions exhibited a wider spread of effects than the cations. Other relevant observations were the strong retarding influence of NO_3^- (about as effective as I^-) and the less strong influence of urea (about as effective as NaCl). These observations regarding the 1:1 electrolytes have been confirmed by later studies.[71,1442] Recent measurements[1442] have in fact extended the investigations of the dependence of the amylose retrogradation rate upon added small solutes. The most striking effects are demonstrated by certain larger monovalent anions, with

$$Cl^- < Br^- < HCO_3^-, \ NO_3^- < I^- < benzoate < salicylate$$

in terms of increasing retardation on a molar basis. Also of interest are the different effects of various monosaccharides; for example, while ribose and glucose both cause a retardation of the retrogradation process, the effect produced by ribose is much more dramatic. This may be a significant observation in view of the suggested (see Section 3.1.3) differences between the hydration properties of glucose and ribose.

The similarity between the anionic series given above and the Hofmeister series for the stability of native protein conformations is surely not coincidental. Other protein denaturants such as urea, sodium dodecyl sulfate,[1442] and OH^- (and high temperature) also have been demonstrated to stabilize amylose in solution. Thus the concept of the (helical) retrograded amylose as the "native" form and of the flexible coil form in solution as the "denatured" form is easy to visualize. Now, therefore, we have indications of a common sequence for the effects of salts and certain nonelectrolytes upon the stability of the native conformations of certain proteins, nucleics acid, *and polysaccharides*. In view of the considerably different structures of these biopolymers, these observations must add further weight to the arguments that the salts, etc., are acting primarily by means of their influence on the solvent structure.

The gelation of amylose is a process closely related to retrogradation. If, for example, a fairly concentrated hot amylose solution is cooled slowly, then precipitation via the retrogradation process will usually occur. On the other hand, rapid cooling can lead to a rigid, irreversible gel.[1334] Alternatively, a controlled neutralization of, for example, a 2% amylose solution in alkali leads to a metastable solution which after a period of hours retrogrades and precipitates, whereas a rapid neutralization causes gel formation[1442] (Fig. 18). Thus, gel formation can be regarded as a disorganized and hindered precipitation. The dependence of the retrogradation process upon the addition of water structure-perturbing agents discussed above

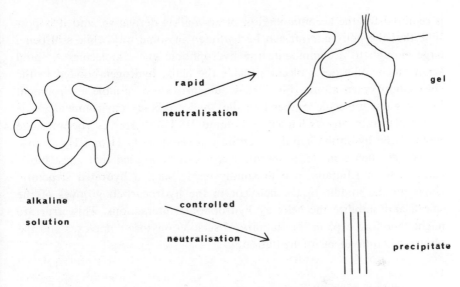

Fig. 18. Retrogradation and gelation of amylose (idealized picture).

therefore refers also to gel formation. Agents that, for example, markedly retard the retrogradation process also hinder gel formation.[1442] The nature of the junction zones in amylose gels in terms of the classification of Section 4.1 must remain in doubt until the structure of retrograded amylose is more fully established.

4.2.3. Possible Role of Water in V-Amylose Complexes

V-Amyloses can be prepared by the addition of a variety of complexing agents, such as hydrocarbons and derivatives, alcohols, esters, ketones, fatty acids, and iodine. It is believed [498,1281] that the complexing agent normally enters the hole inside the helix, which can adjust to six, seven, or eight glucose residues per turn depending upon the size of guest molecule and sometimes the availability of water for hydrate formation.[205] The sixfold V-amylose helix is thought to have hydrogen bonds between O-2 and O-6 of adjacent residues on the helix surface.[495,702] Removal of the complexing agent and humidification converts the V structure to the "B-form" which is shown by some natural starches and retrograded starch.

In the V-amylose structure the interior of the helix must be essentially nonpolar; of all the oxygens in the molecule only the oxygens of the glycosidic linkage point inward. An examination of space-filling molecular models, however, reveals that the hole inside the helix is considerably larger than is required for the accommodation of an *n*-alkyl derivative, and it is possible that the helix interior can be hydrated in some way while still being large enough to accommodate the hydrophobic guest molecule. A spiral arrangement of water molecules inside the helix, hydrogen-bonded to the glycosidic oxygen atoms (Fig. 19), would provide a suitable environment for the guest molecule (rather like the "hydrophobic" environment inside an ice clathrate cage (Chapter 3, Volume 2). The "grooves" between the turns of the hydration spiral are entirely covered by C–H or CH$_2$ groups.

While such a model is speculative, it can be argued that guest molecules such as 1-butanol can fit snugly inside such a hydrated structure, either up the middle of the helix or in the hydrocarbon grooves of the spiral, and stabilize the helix by hydrophobic interactions. This structure might therefore explain the stability of such V-amylose complexes and the observed requirement for hydration.

Fig. 19. "Internal hydration" of V-amylose helix, CPK space filling model of a portion of helix in which the dots indicate the spiral of water molecules hydrogen-bonded to the glycosidic oxygens of the amylose.

4.3. Agarose and Related Polysaccharides

The gelation behavior of agar (major component agarose; see Fig. 6) and the state of the water in agar gels have attracted considerable attention. Gels can be formed at agar concentrations of a fraction of 1%, and have been used as model systems for the ordering of water in more complex macromolecular systems such as muscle. Studies of the properties of the water in agar gels have been dominated by nmr measurements, and the history of these agarose measurements provides a prime example of the power of the nmr method for examining the state of water in macromolecular systems, but also of its difficulty in providing an unequivocal explanation of the observations.*

First of all, however, let us recall how these polysaccharides are thought to form gels, and discuss some of the differences within the agar–carrageenan family of polysaccharides. This family shows a spectrum of varying composition, from agarose (which is essentially sulfate-free) through furcelleran (about 40% of one sulfate residue per disaccharide unit) and $\varkappa$-carrageenan (one sulfate residue per disaccharide unit) to ι-carrageenan (two sulfate residue per disaccharide unit). All are thought to gel via double helical junctions, although the gels can become more tightly cross-linked and more rigid as helices associate to form aggregates or "superjunctions." Rees[1234] has discussed how changing the level of sulfate alters the gelling properties. The more highly sulfated chains have a greater tendency to enter into aqueous solution, because of contact between the ionic substituents and the water, and a reduced tendency to associate. Thus the most highly sulfated member, ι-carrageenan, requires much higher concentrations to gel, does not appear to form "superjunctions," and gives clear, elastic gels. In contrast, agarose gels at very low concentrations, aggregates extensively to form "superjunctions," and gives turbid and rigid gels. The properties of furcelleran and $\varkappa$-carrageenan are intermediate between those of agarose and ι-carrageenan, consistent with their sulfate contents.

Early studies of the state of water in agar gels appeared to indicate considerable restriction of the motion of the water molecules as compared to pure liquid water. Hechter et al.[636] for example, using high-resolution proton magnetic resonance spectroscopy, observed that the water proton signal is significantly broadened in the gel state and suggested that the water in agar gels is in a modified state intermediate in mobility between liquid

---

* For details of the problems associated with the interpretation of such nmr data see Berendsen (Chapter 6, Volume 5).

water and ice. Later Sterling and Masuzawa[1423] found that if the agar concentration was increased, then the linewidth of the pmr signal also increased, and they interpreted the changes in signal area in terms of "solid," "bound," and "free" water molecules. Hazlewood *et al.*[632] found similar high-resolution line broadening and used this as evidence for "ordered" water in macromolecular systems.

The microwave dielectric relaxation properties of water in agar gels had in the meantime been investigated by the Loor and Meijboom.[333] Although this work is not generally well known, it shows that the dielectric relaxation time of the great majority of the water molecules in a 4% agar gel is essentially the same as for pure liquid water. They also calculated, from the difference between ε_0 in Fig. 20 and that anticipated for the particular volume fraction of water, that a further fraction of the water ($\sim$0.8 g/g agar) did not contribute to the microwave spectrum and must be assumed to be "bound" to the agar. The dielectric measurements do not therefore support any hypothesis based upon the ordering or restriction of the bulk of the aqueous component.

Nmr studies of water in agar gels were put on a much firmer basis by the work of Child *et al.*[250] (using agarose) and Woessner *et al.*[1592,1593] (using agar). These studies were carried out independently and in terms of the experimental data are in very good agreement. Figure 21 shows the temperature dependences of the two proton nuclear relaxation times T_1 and T_2, clearly showing the hysteresis effect (gels set about 40° but require heating to about 90° before melting occurs). The T_2 minimum with temperature observed (for both deuteron and proton signals) near the sol–gel

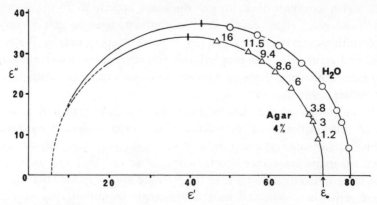

Fig. 20. Complex plane representation of the dielectric relaxation of water (○) and 4% agar gel (△) at 20°C. Frequencies indicated at the measurement points are in GHz. (From Ref. 333).

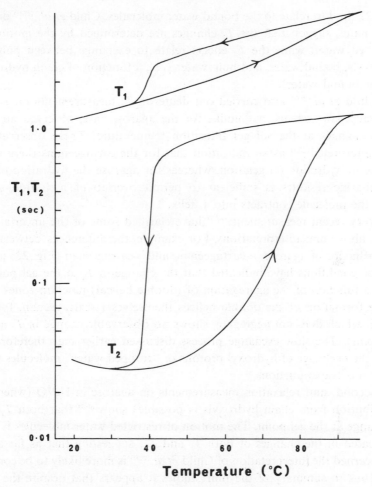

Fig. 21. Temperature dependence of T_1 and T_2 for 5% agarose (from Ref. 895).

transition point is characteristic of an exchange phenomenon, and the recognition of this minimum is crucial to the arguments presented by Child et al.[250] and Woessner et al.[1592,1593] Both groups conclude that the motions of the great majority of the water molecules are unaffected by the presence of the agarose, but thereafter the interpretations differ. Woessner suggests that the fraction of bound water must be less than 1% in a 10% agar gel, and that there must be two distinct nuclear correlation times, differing by at least two orders of magnitude. The shorter correlation time determines T_1, and is assigned to the motion of the protons on the agarose hydroxyl groups, while the longer correlation time is considered to determine T_2, and to relate to the bound water molecules. Child *et al.*,[250] on the other hand, suggest that the T_1 changes are determined by the motion of restricted water, while the T_2 effects relate to exchange between polymer hydroxyls, bound water, and bulk water (i.e., a function of chain hydroxyls and/or bound water).

Child *et al.*[250] also carried out deuteron T_1 measurements on ι- and $\varkappa$-carrageenan systems, but unlike for the agarose case, observed no dramatic changes at the sol–gel transition temperature. These observations were interpreted[250] as an indication that for the carrageenans there is no change in hydration on gelation, whereas for agarose the hydration of the disaccharide residues is sufficient to permit overlap of hydration shells when the molecule contracts into a helix.

Very recent measurements[895] have clarified some of the uncertainties of the above nmr interpretations. For example, the differences between the nmr behavior of agarose, $\varkappa$-carrageenan, and ι-carrageenan (Fig. 22) under various conditions have indicated that the change in T_2 at the gel point is more a function of the aggregation of (double helical) junction zones than of the formation of the double helices themselves (ι-carrageenan, for example, which does not aggregate, shows no observable change in T_2 at the gel point). The slow exchange process discussed earlier may therefore involve the exchange of hydroxyl protons or "trapped water" molecules from inside the "superjunctions."

Second, nmr relaxation measurements on agarose in $H_2^{17}O$ (where no contribution from chain hydroxyls is possible) show[895] that both T_1 and T_2 change at the gel point. The motion of restricted water molecules is thus implicated in the changes of both T_1 and T_2, suggesting that as far as T_1 is concerned the interpretation of Child *et al.*[250] is more likely to be correct.

Thus in summary of the nmr studies it appears that despite the pronounced changes in the linewidths (i.e., in T_2) of ^1H, ^2H, and ^{17}O signals accompanying the gelling of agarose, no really significant change in the hydration of the agarose occurs at the gel point, and no significant ordering of the bulk solvent occurs. The amount of water with a restricted motion is very small (although a quantitative estimate of this water is still required). The most reasonable explanation of the T_2 changes at the gel point is that they are a consequence of the slowing down of the rate of exchange of water molecules (and possibly also chain hydroxyl protons) when the agarose molecules associate into aggregates of double helices.

Very recent dielectric relaxation studies* of agarose in both the sol

---

* A. Suggett, unpublished work.

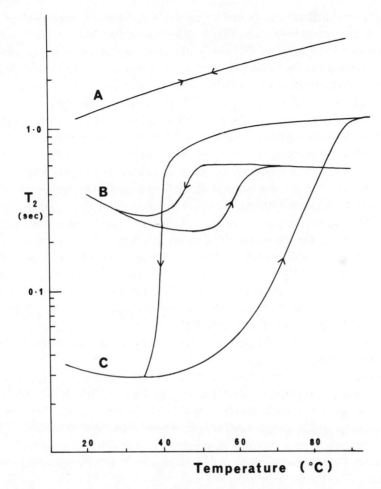

Fig. 22. Temperature dependence of T_2 for 5% sols/gels of (A) ι-carrageenan; (B) $\varkappa$-carrageenan; (C) agarose (from Ref. 895).

and gel states, in contrast to the earlier work of de Loor and Meijboom, indicate only a very small fraction of the water not contributing to the bulk water relaxation process. No significant change in the hydration of the gel point is observed.

4.4. Water in Polysaccharide Gels

Views expressed in this area have been rapidly changing in recent years, and, although the situation appears somewhat clearer now, there is still a great deal of unexplored territory. Furthermore, the sophisticated experimental methods which must be used to probe inside the bulk system and focus on the properties of the water are more prone to misinterpretation. However, on the whole, the use of relaxation techniques (for example, nmr and dielectric) appear to offer the most promising prospects for future studies of the role of water in polysaccharide solutions and gels.

Until very recently it was generally considered that for 99.5 parts of water to be "held together" in a gel by 0.5 parts of a polysaccharide, some considerable structuring of the water must have occurred. We now think we know better. For a number of polysaccharides that gel via helical junctions (Section 4.1.1) there is no observable change in the properties of the bulk water on passing through the sol–gel transition—they remain indistinguishable from normal liquid water. Likewise, there is no marked change in the extent of the short-range hydration as observed by relaxation methods. The gelation mechanisms, in fact, could be visualized to occur in nonaqueous systems. However, these polysaccharides have not been observed to gel in nonaqueous media, and water structure-breaking solutes can prevent gelation.[260] So it is likely that water plays some part in gel formation, although it is clear that the "solidlike" nature of these polysaccharide gels is largely a result of polymer–polymer interactions and not large-scale ordering of the water.

Gels classified under Section 4.1.2 (micelle junctions) result from a mechanism which is peculiar to aqueous system and which has been discussed earlier. For such polysaccharides the polymer–water interactions are critically important in determining whether the system under a given set of conditions is a mixture of precipitate and solvent, a sol, or a gel. But again, the gel network is mainly a result of polymer–polymer cross-linking.

Water, it seems, is essential for the formation of polysaccharide gels, and yet after gelation its properties are scarcely changed. This statement can perhaps lead us to a convenient description of the role of water in biological systems in general: *Water—the original catalyst*—intimately involved in all biological processes and yet, in the end, unchanged.

5. SUMMARY

This is not so much a review of all the work carried out on carbohydrate–water systems that relates to possible influences of solute–water interactions, as a generalized account of the *ways* in which these interactions can play a role in determining the physical properties of such systems. In the area of polysaccharide chemistry the effects of the aqueous solvent have, to a much greater degree than in protein chemistry, been largely ignored, and this chapter has therefore deliberately focused on the role(s) of the water. Examples have been selected which confirm that *in those cases* the role of the water is a very significant one, but should the reader feel that the author has to some extent been playing "devil's advocate" and presenting only one side of the total picture, then I refer him to the large number of polysaccharide reviews in which water is treated implicitly as the universal inert filler.

ACKNOWLEDGMENTS

The author would like to express his thanks to Prof. D. A. Rees, Dr. D. Thom, and Dr. P. J. Lillford for permission to report work in advance of publication, and for helpful discussion.

CHAPTER 7

Synthetic Polymers

Philip Molyneux

Pharmacy Department
Chelsea College (University of London)
Manresa Road, London, England

1. INTRODUCTION

In recent years increasing use has been made of synthetic polymers in a wide variety of applications; because of the prevalence of water, and of aqueous systems in general, in the environment, the topic of the interactions between water and synthetic polymers has thus many directly practical aspects, as well as being of some academic and theoretical interest, particularly to scientists in the "water community" and to those in the "polymer community." In addition, the study of these interactions casts light upon the parallel interactions between water and the naturally occurring polymers in their native environment.

Mainly because of the technical and industrial developments in this field, the science and the technology of polymers now comprises an extremely large and rapidly increasing volume of information; as with so many other related areas of knowledge, it is becoming more and more difficult to keep abreast even with the most significant advances in this field. Accordingly, despite the claim of the volumes in this treatise to be "comprehensive," the most that can be done within the scope of this chapter is to provide a framework within which to outline the main aspects of the interactions between water and the synthetic polymers; emphasis will be placed upon the general principles involved, illustrating these with some examples but not considering particular polymers in great detail.

Similarly, although the aim has been to include as much recent work as possible, older work from the literature has also been discussed, particularly where the theoretical and practical consequences of the results obtained do not seem to have been absorbed into current thought.

It is useful at this early stage to define two extremely general aspects of these interactions in relation to the interests of the scientific community as a whole. First, there is the manner in which the behavior of water toward *polymers* compares with that of water toward *small-molecule substances*. This might be expected to be a topic of some interest to the "water community," for whom, however, the subject of "polymers" is apparently viewed as overly complex, perhaps simply because the molecular weights involved are around 1000 times those of the more familiar small-molecule substances. In fact, most polymers (particularly synthetic polymers) have quite simple chemical structures for their monomer units, and the main novelty arises from the linking together of large numbers of such units into long chains. An opportunity is therefore taken in the early parts of this chapter to restate some of the key features of polymers, and also to emphasize the important role that the mode of its preparation plays in the properties of the polymer.

The second general aspect is the manner in which the behavior of polymers toward *water* compares with that of polymers toward other *small-molecule liquids*. Here again, this might be expected to be a topic of some interest to the "polymer community," for whom the subject of "water" is one they would rather avoid because of the complexity of the liquid. Indeed the author, himself a polymer scientist, must confess to being more confused than enlightened by much of the current conflicting theories of the structure of liquid water; for the present purposes this structure will be taken to involve an almost completely four-coordinated, hydrogen-bonded system which is steadily reduced in its degree of bonding as the temperature is raised.

It is intended, therefore, that the present chapter will serve as a bridge linking together and providing common ground for the interests of these two rather distinct communities.

1.1. Sources of Information on Polymers

In most cases specific references will be given in the text to the literature where more details may be found on the topic discussed; where no specific reference is given, further information may generally be found in one or other of the following sources.

For both reference purposes and for general background information on the theoretical principles of polymer science (particularly as they apply to synthetic polymers) the monograph by Flory[457] is still a very useful source, and by no means outdated despite its impending coming of age. More recently Tanford[1461] has dealt with the subject as it applies both to synthetic and to natural polymers, while Morawetz[1038] has covered rather similar ground but with more emphasis on the behavior of polymers in solution and on the correlation with the behavior of small-molecule systems. The technical aspects of polymers and their applications have been covered admirably by Rodriguez,[1268] who places especial emphasis on their role as components in complex systems. Two very useful handbooks[168,1270] provide collected data on the properties of the commercially most important polymers, especially the synthetic and the semisynthetic (modified natural) ones. Finally, where its high cost is no obstacle, a multivolume encyclopedia of polymer science and technology[982] serves as an additional source of information in this field, while its very bulk illustrates the mountain of information that is piling up on the subject.

1.2. General Nomenclature and Molecular Structural Features of Polymers

The term "polymer" denotes a substance whose molecules are composed of a large number of similar (if not identical) units covalently linked. The units so linked are termed "monomer units," or more simply "mer units."* If the mer units are essentially or nominally identical, then they form a "homopolymer"; with mer units of two or more types a "copolymer" results.

The number of mer units in a particular polymer molecule is termed the degree of polymerization, which is often abbreviated DP; in the present case this quantity will be symbolized by x. There is, of course, no sharp demarcation between polymeric and nonpolymeric ("small-molecule") substances; generally speaking, however, a true high polymer is normally taken to be one having a molar mass (molecular weight) of at least 10^4, corresponding (for a common average monomer molar mass of 100) to a value of about 100 for x. Shorter chain homologs are termed "oligomers";

---

* One mole of these units of the polymer chain is termed a "basemole" or "monomole," and this is used as a convenient measure of the amount of polymer in situations where it is the reactivity of these individual units which is the important factor, as in, for example, the titration behavior of a polyacid or a polybase.

these are of interest both in themselves, and in forming a link between small-molecule and truly polymeric substances.

Although each particular polymer molecule will have a definite mass, in the great majority of samples of polymers encountered in practice there will be a range of lengths of polymer chains, i.e., such samples are normally "polydisperse" rather than "monodisperse." Where a single-value result is obtained from a molecular weight determination (e.g., by osmometry, viscometry, or light scattering measurements) this will be some particular type of average which depends upon the specific method used (i.e., number-average, $\bar{M}_n$, viscosity-average $\bar{M}_v$, and weight-average $\bar{M}_w$, for the three respective techniques). In the present context we will not consider this polydispersity further, except to emphasize that its effects must be kept in mind especially when interpreting phase separation and polymer solubility data.

Although the term polymer is generally taken to imply an organic substance, in fact there are quite a number of inorganic substances (particularly, the silicates and the phosphates, as well as more recent synthetic materials such as the phosphonitrilic chlorides) which fall logically within this category.[305,524] In general, however, these inorganic polymers are essentially insoluble in common solvents so that they are difficult to study by the normal techniques of polymer science; even when they are, in particular, water soluble, then frequently their behavior in aqueous solution is made complex by hydrolytic reactions. In the present case, therefore, attention will be focused almost entirely upon organic polymers.

Considering the structural features of polymer molecules, the simplest type of structure is that of a homopolymer composed of difunctional mer units M linked together in a linear chain; unless the chain is cyclic it must be terminated by end groups (say, A and Z) chemically distinct from M, so that the structure becomes

$$A \left(M \right)_{\bar{x}} Z$$

So long as x is fairly large, then A and Z will make a negligible contribution to the chemical composition and properties of the chain, so that for most purposes the structure may be more simply represented as

$$\left(M \right)_{\bar{x}}$$

If a trifunctional unit Y is incorporated in the chain, then this can serve as the point of attachment of another linear chain, forming a branched structure:

$$\ldots \text{MMMYMMM} \ldots$$
$$\text{M}$$
$$\text{M}$$
$$\text{M}$$
$$\vdots$$

These branches may themselves have branches attached to them, thus forming a bushlike structure. Similarly, two chains may be linked through two such trifunctional groups, with or without an intermediate bridging group, forming a cross-linked polymer:

$$\ldots \text{MMMYMMM} \ldots$$
$$\text{L}$$
$$\ldots \text{MMMYMMM} \ldots$$

This type of link may be *intra*molecular (where the sections linked are part of the same main chain, leading to a cyclic section within the chain) or *inter*molecular. Extensive incorporation of intermolecular cross-links leads to an infinite network comprising the whole sample; the latter then only swells to form a gel in those liquids that are solvents for the purely linear polymer (see Sections 3.3.4 and 3.10.1). By and large, however, for simplicity we shall be concerned almost entirely with the behavior of linear polymers.

An additional important structural feature of polymers of the type

$$-(\text{CH}_2-\text{CRR}')_{\bar{x}}$$

i.e., of vinyl and acrylic polymers (where R' may be a hydrogen atom), is the steric configuration at the substituted carbon atom. Since in general the two sections of polymer chain attached to this atom on either side will be of different length, then there will be two distinct steric configurations at this atom:

$$\begin{array}{ccc}
\text{H} \quad \text{R} & & \text{H} \quad \text{R}' \\
| \quad\ | & & | \quad\ | \\
-\text{C}-\text{C}- & \text{and} & -\text{C}-\text{C}- \\
| \quad\ | & & | \quad\ | \\
\text{H} \quad \text{R}' & & \text{H} \quad \text{R} \\
\text{(I)} & & \text{(II)}
\end{array}$$

If, then, in the sequence of mer units in the chain all these carbon atoms have the same configuration (say, I) then this is termed an *isotactic* chain, while if the configurations are alternately I and II, this is a *syndiotactic* chain; an *atactic* chain results from a random distribution of the two configurations. In general, polymers produced by the simplest methods, such as free-radical addition polymerization at normal temperatures, are atactic; the tactically more ordered configurations have only become important relatively recently with the advent of readily applicable methods (such as the use of coordination catalysts of the Natta or Ziegler types) for the production of these polymers. By and large, whereas atactic polymers are commonly amorphous, the tactic polymers are potentially highly crystalline and hence less soluble than the atactic form. Further, when in solution the relative configurations at adjacent mer units will influence the distance between the consecutive R and R' groups, and will hence influence the interactions between these consecutive groups; this phenomenon needs to be taken into consideration when dealing, for example, with the competitive interactions between polar substituents and the solvent in aqueous solutions. Because of their polarity, hydrophilic polymers are difficult to prepare in such tactically ordered forms, so that these effects have not been studied at all deeply for aqueous systems. However, the development of "clean" methods for the chemical conversion of tactically ordered, less polar polymers to the more hydrophilic ones [such as poly(vinyl acetate) to poly(vinyl alcohol)] has opened up this area for more intensive study.

Further structural possibilities arise for copolymers. Considering a copolymer of two types of mer unit, M and N, for a linear chain, the two commonest arrangements of units are those of the random copolymer

$$\ldots MNMMNNNM\ldots$$

and of the alternating copolymer

$$\ldots MNMNMNMN\ldots$$

On the other hand, by linking together lengths of each of the two homopolymers, one can obtain a block copolymer

$$-(M)_x-(N)_y$$

while the attachment of one homopolymer as a branch to the main chain of the other produces a graft copolymer

$$\ldots MMMYMMM\ldots$$
$$N$$
$$N$$
$$N$$
$$\vdots$$

On this basis, for example, the nucleic acids may be viewed as copolymers of four nucleotides, and the proteins as copolymers of the 20 or so naturally occurring amino acids; however, in both cases the nature and arrangement of the mer units are highly specific to the particular polymer involved. Similarly, the nucleoproteins and glycoproteins may be considered as block or graft copolymers of the two types of polymer chains. In fact, because of the greater complexity of these particular subtances, there is a widespread reluctance to call them polymers, and the generic but vague term "macromolecule" (which should be more strictly "macromolecular substance") has been commonly applied to these and related materials. However, so long as it is realized that the term "polymer" also logically covers "copolymer," then it seems preferable to use this term in place of "macromolecule," particularly since its use emphasises both the chainlike nature of the molecules and the repetition of units of closely (or relatively closely) allied structure.

1.3. Supramolecular Structure in Polymers: Crystalline and Amorphous Components

Even if the molecular structure of the polymer chains is known, this is not sufficient to define the macroscopic properties of the polymer, which also depend to an important extent upon the supramolecular structure of the polymer, i.e., the ordering and arrangement of the polymer chains in the bulk material. Although most synthetic polymers are solids in the common sense of the term, unlike most small-molecule solids, which are usually wholly crystalline or microcrystalline, polymers often have an amorphous component which in many cases is the predominant or even the sole one present. The actual nature of the interrelation between these two components in polymers is a matter of dispute, but one useful model is the "two-phase textural hypothesis"[546] in which the polymer is viewed as composed of closely mixed but still distinct crystalline and amorphous regions. On this picture, such features as mechanical strains and the diffusion of small-molecule components are largely confined to the amorphous regions, while the crystalline regions act as compact reinforcing members which are much less readily deformed and comparatively impermeable to foreign small molecules.

For each of the two component phases there is a characteristic temperature at which its physical properties change fairly abruptly as the temperature is raised. For the amorphous regions there is a *glass transition temperature* T_g characterizing the second-order transition within which it changes from a rigid, brittle solid (glass) to a much more deformable and ductile material (rubber); in molecular terms the transition corresponds to the onset of the rotation of groups or segments on the polymer chains in the amorphous regions. For the crystalline regions there is a *melting point* T_m which corresponds to a first-order transition involving the complete breakdown of the ordering of the chains in these regions. In fact, each of these transitions really takes place over a range of temperatures, and the values quoted in the literature for T_g and T_m are generally the approximate midpoints of these ranges. As might be expected, the value of T_m for a particular polymer having both amorphous and crystalline components is always higher than that of T_g; for many polymers the ratio of the absolute (thermodynamic) values of these temperatures lies in the range 1.7 ± 0.3.

The significance of this duplex structure, which is possessed by many synthetic polymers, is that extent of crystallinity profoundly affects such properties as solubility and chemical and physical reactivity toward small-molecule substances such as water; a polymer at a temperature above T_m will thus have markedly different behavior in these respects from that below this temperature. Similar but less marked differences apply above and below the T_g value, particularly as regards the diffusion of small-molecule substances, although the changes at T_g for the amorphous regions are much less noticeable than they are at T_m for the crystalline regions.

1.4. Division of Polymers According to Origin

A common division applied to polymers, and one reflected in the titles of the chapters in this volume, is according to their origin, i.e., whether natural or synthetic.

Natural polymers comprise many of the materials of common and everyday use, such as wood, cotton, and paper (the first largely and the last two almost entirely cellulose), wool, silk, and leather (all proteins), and *Hevea* (natural) rubber (a polyisoprene); as ultimately used the original natural materials may have been more or less drastically modified (as in the tanning of leather and the vulcanization of rubber). Further, many natural polymeric materials are now treated commercially to yield chemical derivatives quite distinct in behavior from the original material, and much closer in such properties as their solubility in common organic solvents and in their possibilities for fabrication to the purely synthetic polymers. These are the *modified natural polymers*, or *semisynthetic polymers*; a prominent group is that comprising the various derivatives of cellulose, such as cellulose diacetate and triacetate, cellulose nitrate, methylcellulose, and carboxymethylcellulose.

In their native environment many of the natural polymers only have an essentially "inert," structural role (as in the cases of cellulose and keratin), or else they function largely as externally protective agents after secretion from the system (as seems to be the case with rubber, gums, and resins). On the other hand, many of the natural polymers are very closely and intimately involved in the metabolic functioning of the biological system in which they occur; this applies particularly to the nucleic acids, to many of the proteins (such as the enzymes, the serum proteins, and myosin), and also to numerous polysaccharides (such as starch, glycogen, and the mucopolysaccharides) as well as the copolymers of these (nucleoproteins, glycoproteins). This general class of biochemically significant materials has therefore acquired the name of *biopolymers*; since they are structurally quite sensitive materials, to investigate their role in vivo it is necessary to isolate them by as gentle means as possible so that they retain their native characteristics, and then to study them in vitro under conditions as close as possible to those in vivo. The named groups of biopolymers are dealt with specifically in other chapters in this volume; however, it should be emphasized that their behavior as polymeric materials is governed by the same principles as those governing the simpler-structured synthetic polymers dealt with specifically in this chapter.

The class of *synthetic polymers*, the specific subject of this chapter, comprises polymeric materials produced from small-molecule precursors by chemical reactions. There has been a steadily increasing scientific and commercial interest, extending back about 100 years, in the production of these materials; in the early stages of their development they were regarded simply as substitutes for materials of natural origin (as witness terms such as "artificial silk" and "synthetic resins"), but as their development has proceeded they have come to be recognized as new types of materials in their own right. There is now a very large range of synthetic polymers produced commercially; quantitative data on the significant properties of the most important of them are tabulated in the two handbooks[168,1270] already mentioned, while Rodriguez[1268] has listed and dealt in more descriptive detail with a wide range of commercially available synthetic polymer products.

This threefold classification of polymers into natural polymers (including biopolymers), the intermediate group of semisynthetic or modified natural polymers, and the purely synthetic polymers, is a fairly convenient one since it stems from a corresponding division in the aims of many of those working with and studying the particular types. However, it should be emphasized that much is to be gained by linking up these rather distinct divisions into a single unified field of polymer science. In many cases, these divisions are already largely transcended, particularly in the fundamental studies of polymeric materials, where the origin of the polymer is of little concern as long as it is well-characterized and acceptably pure and homogeneous. Similarly, in commercial applications synthetic, semisynthetic, and natural materials compete with each other and serve as a spur for their common development; indeed, in the case of textiles the two types are now commonly used alongside one another, as blends.

For the apparently most contrasting pair, the biopolymers and the synthetic polymers, there are several areas in which the two are interrelated. Synthetic polymers are fairly readily produced with well-characterized structures, and hence serve as the ideal materials for testing theories of polymer behavior to which the more complex biopolymers should also conform; this applies particularly to aqueous systems where the solvent is already so complex as to make it essential to closely define and to simplify the nature of additional components. This is not to say that, for example, particular water-soluble synthetic polymers can always be justifiably proposed as true "model compounds" for the much more complex biopolymers; it should, on the other hand, be possible to obtain information upon the strengths of component interactions (hydrogen bonds, hydrophobic bonds, electrostatic effects, etc.) which take place in aqueous polymeric systems by using simple synthetic polymers of known structures, such quantitative data then being applied to the interpretation of the behavior of related systems containing biopolymers. The main contrast between biopolymers and synthetic polymers is that biopolymers, as a consequence of their more complex molecular structure, have a wider range of possible molecular conformations in solution (as witness the compact shapes of the globular proteins, which may also contain regions of the α-helix, and the double helix of the nucleic acids), whereas, by and large, the molecules of synthetic polymers in solution take up the random coil and related loose, disordered conformations. Nevertheless, this gap is bridged to some extent, for example, by the fact that many proteins when dissolved in reversible denaturing agents such as 6 M guanidinium chloride (plus 2-thioethanol to break the cystine cross-links) take up an essentially random coil conformation;[1465] in the same fashion, poly(vinyl pyridine) quaternized with long-chain alkyl halides (thus forming a "polysoap") can take up a compact conformation in aqueous solution somewhat akin to that of the globular proteins.[1435]

In addition, the experimental techniques for the isolation, purification, characterization and even the in vitro synthesis of biopolymers increasingly involve the use of synthetic polymers. For example, cross-linked polyacrylamide is one of the established materials used for the separatory techniques of gel electrophoresis and gel filtration[31]; similarly, the Merrifield technique for the synthesis of specific polypeptides involves the use of a cross-linked synthetic polymer resin onto which the growing polypeptide chain is attached.[1004,1425] Again, the feasibility of the commercial utilization of enzymes depends very much upon the availability of suitable methods for the attachment of the enzyme molecules onto insoluble supports without at the same time destroying the activity of the enzyme; synthetic polymers are among the most promising of such support materials.[1000]

Finally, materials are now being produced, based upon synthetic polymers, which show marked specific catalytic activity in solution and with the kinetic characteristics (in particular, a dependence of rate upon "substrate" concentration following the form of the Michaelis–Menten equation) closely akin to those of enzymes.[766,827] These "synthetic enzymes" are interesting both in their own right and as possibly useful commercial and laboratory agents, as well as in casting light upon the mode of action of the true enzymes.

Having thus discussed the general features of polymer structure, and compared the different types of polymer classified according to their origin, for the remainder of the chapter the discussion will be largely confined (at least as far as examples are concerned) to the title subject, synthetic polymers, although as already emphasized the principles to be outlined will in most cases apply to polymeric materials in general.

1.5. Classification of Synthetic Polymers

One common and convenient classification of synthetic polymers is according to the type of reaction involved in their formation, i.e., basically a division into addition and condensation polymers.

With addition polymers the chain growth involves the linking of monomer units via the opening of a double bond or a ring, without the elimination of any other molecule; for vinyl and acrylic monomers (of the type $CH_2{=}CRR'$) the growing chain has a reactive end group which may either be free radical, anionic, cationic, or coordinative (to a catalyst surface) in character. In these cases each polymer chain grows very rapidly to its ultimate size, finally "dying" by some process such as, for example, with free-radical polymerization, combination with another such chain. The polymer chains so produced, having a backbone of carbon–carbon bonds, are normally highly stable, particularly to direct hydrolytic attack.

With condensation polymers the polymerization involves the elimination of a small molecule at each successive addition of a monomer unit to the growing chain; a typical example is the formation of a polyester from a diol and a dicarboxylic acid, where the first step of the polymerization would be

$$HO—R—OH + HO_2C—R'—CO_2H$$
$$\rightarrow HO—R—O_2C—R'—CO_2H + H_2O \qquad (1)$$

Unlike addition polymerization, the propagation step in condensation polymerization is often a reversible reaction, and in the case of polyesterification and related processes the production of high-molecular-weight material depends critically upon the removal of the eliminated water from the reaction mixture. A further difference between the two types of process is that with condensation polymerization a polymer molecule still retains its reactive end groups, so that growth can take place by condensation between polymer chains as well as by addition of monomer. Again, in contrast to addition polymers, the chain produced because of the reversibility of its formation reaction is often relatively easily attacked by hydrolytic reagents. Nevertheless, it should be noted that some condensation polymers, such as the nylons and poly(ethylene terephthalate), are in fact hydrolytically quite stable (see Section 4), while hydrolytic instability may be a *desired* property for a particular polymer such as an absorbable surgical suture (see Section 5.6).

This method of classification according to formation reaction is not in fact a completely clear-cut one, since the same polymer may be produced by a variety of routes; thus the chain $-(CH_2CH_2O-)_{\bar{x}}$ may be produced by the addition polymerization of ethylene oxide or by the condensation polymerization of a variety of monomers such as ethylene glycol (with the elimination of water), ethylene chlorohydrin (with the elimination of hydrogen chloride), and so forth. This in turn leads to ambiguities in nomenclature, since polymers are generally named according to the monomer (or monomers) from which they are formed, so that the $-(CH_2CH_2O-)_{\bar{x}}$ chain could be "poly(ethylene oxide)," "poly(ethylene glycol)," "poly(ethylene chlorohydrin)," etc.; in practice, when such ambiguities arise the "addition" name, i.e., poly(ethylene oxide), should be chosen since it leads to the least ambiguity. Even so, the terms "poly(ethylene glycol)" or "polyethylene glycol" are still commonly applied to this polymer, particularly to the shorter-chain polymers and the oligomers which are widely available commercially. An even more systematic but less widely used nomenclature is the direct one applied to the polymer chain, as in "polyoxyethylene" for this specific example.

A further specific point of nomenclature which should be emphasized is that where the name of the monomer involves two (or more) separate words these should be enclosed in parentheses before adding the prefix "poly," i.e., the example cited above is "poly(ethylene oxide)," rather than "polyethylene oxide" which would represent some oxidized form of polyethylene. For single-word monomers such parentheses are not required, hence, "polyethylene." For copolymers it is preferable to use parentheses around each of the monomers involved, even if they are single word, as in "poly(ethylene)co(propylene)."

2. WATER IN RELATION TO THE PREPARATION AND PRODUCTION OF POLYMERS

There are two main areas where the methods of preparation and production of synthetic polymers are significant in the present context.

The first, more general point is that the properties of polymers are much more dependent upon the mode of synthesis, and the extent to which these properties can be altered after synthesis is much more limited, than is the case with small-molecule compounds; this applies particularly to such features of the polymer as the molecular weight distribution, the tacticity (or lack of it), the composition and distribution of different mer units (for copolymers), the presence or absence of branching or cross-links, and the chemical nature of the end groups and of any "minor links" in the polymer chain. The polymer sample as isolated from the synthesis mixture may also contain a variety of small-molecule contaminants from that mixture (monomer, solvent, catalyst or initiator, and so forth).

Purification and fractionation procedures for polymers are available and widely used, e.g., precipitation and exhaustive dialysis for the removal of small-molecule contaminants, and differential solubility for the separation of the polymeric components according to linearity, composition, molecular weight, tacticity, and so forth. However, these techniques (particularly as applied to the separation of the polymeric components) by their very nature seldom give complete separation, while they are also generally very time-consuming and quite troublesome especially where relatively large amounts of the polymer are involved.

For these reasons synthetic polymers should be viewed as almost indelibly stamped by their mode of preparation; a knowledge of this mode (and of any subsequent purification procedures) is therefore necessary in interpreting the behavior of the polymer, including specifically its interactions with water and with aqueous systems. A lack of this knowledge, for example, is a stumbling block in interpreting the isotherms for the uptake of water vapor by hydrophilic polymers (see Section 3.5), where a "toe" observed in an isotherm at low water activities could be definitely ascribed to, say, highly hydrophilic end groups if the nature of these groups and the polymer molecular weight were known.

The second, more specific aspect is that of the roles which water plays in the synthetic procedures. This aspect will also be dealt with in the following sections, where it will be seen that the role can range from the very general one of a fairly inert solvent and diluent for free-radical polymerization, to one involving much more specific chemical effects in ionic polymerization.

2.1. Addition Polymers

In addition polymerization there are four main possible types of reactive intermediates involved in the propagation process: free radical, anionic, cationic, and catalyst-bound (in coordination polymerization); for these four possibilities the respective propagation steps with a vinyl monomer $CH_2{=}CHR$ are

$$P—CH_2—\dot{C}HR + CH_2{=}CHR \rightarrow P—CH_2—CHR—CH_2—\dot{C}HR \quad (2)$$

$$P—CH_2—\overset{\ominus}{C}HR(M^\oplus) + CH_2{=}CHR \rightarrow P—CH_2—CHR—CH_2—\overset{\ominus}{C}HR(M^\oplus) \quad (3)$$

$$P—CH_2—\overset{\oplus}{C}HR(A^\ominus) + CH_2{=}CHR \rightarrow P—CH_2—CHR—CH_2—\overset{\oplus}{C}HR(A^\ominus) \quad (4)$$

$$P—CH_2—CHR—Cat+CH_2{=}CHR \rightarrow P—CH_2—CHR—CH_2—CHR—Cat \quad (5)$$

where P represents the growing chain less the reactive-end mer unit, $M^\oplus$ and $A^\ominus$ are the counterions for the two types of ionic polymerization, and Cat represents the site on the coordination catalyst to which the growing chain is attached.

In each case the overall reaction is essentially the same, being simply (ignoring the end groups) the conversion of x moles of monomer into 1 mol of x-mer:

$$xCH_2{=}CHR \rightleftharpoons {+}(CH_2–CHR{)}_{\overline{x}} \quad (6)$$

This is frequently an irreversible reaction, so that although written for generality as an equilibrium, the equilibrium quotient is so large that the position of the equilibrium (6) will be well to the right; the production of the polymer is then governed largely by kinetic factors such as the rates of the propagation and other reactions. However, from the most general viewpoint this must still be considered as a reversible reaction with the position of the equilibrium governed by the usual thermodynamic principles; in particular, it will have a standard free energy change $\Delta G_p^{\ominus}$ which must be negative for the equilibrium to favor the product, polymer, rather than the reactant, monomer. This free energy change will be related to the corresponding enthalpy and entropy changes by the expression

$$\Delta G_p^{\ominus} = \Delta H_p^{\ominus} - T \Delta S_p^{\ominus} \tag{7}$$

Now since polymerizations of this type are generally both exothermic and exentropic (due to the ordering of monomer units upon polymerization), then there will be a positive absolute temperature T_p, given by

$$\Delta H_p^{\ominus} = T_p \Delta S_p^{\ominus} \tag{8}$$

at which $\Delta G_p^{\ominus}$ changes sign and above which the equilibrium (6) lies to the left and the polymerization is no longer favored on thermodynamic grounds. This value of T_p is known as the "ceiling temperature" for the system, and it thus represents essentially the upper limit at which the particular polymerization reaction can take place.[309] The literature values of T_p and of the other thermodynamic quantities have been tabulated.[716] The general significance of this behavior is that in exploratory studies on novel monomers and/or novel methods for polymerization the failure to produce polymer under "forcing" conditions of relatively high temperatures may arise not from the impossibility of producing polymer by that route but more simply from the fact that the experimental temperature exceeded the value of T_p for the system.

We now discuss the main methods used to carry out addition polymerizations, considering the commonest cases of polymerization in the liquid phase according to the type of diluent or other liquid added (or not added) to the system.

2.1.1. *Bulk (or Block) Polymerization*

In this, the simplest type of polymerization method, a mixture of the pure, undiluted monomer together with a suitably low concentration of an initiator is heated to produce the polymer; alternatively, the monomer alone is irradiated to induce the polymerization. If the polymerization process is carried to completion, the polymer is produced in block form; this method is used widely, for example, for the production of optically clear poly(methyl methacrylate), which is marketed under a variety of trade names (Diakon, Lucite, Perspex, Plexiglas, etc.).

Although simple, the method has a number of disadvantages when structurally well-defined polymers are desired to be produced. In the first place, because of the high exothermicity of vinyl polymerizations,[716] there is always the need for efficient cooling of the polymerizing system if the temperature is to be controlled accurately; this exothermicity is such that for vinyl polymerization carried out adiabatically (i.e., in a fully insulated system) there would be a 2° temperature rise for each 1% conversion. With bulk polymerization, particularly if it is carried to high conversions so that the reaction mixture becomes highly viscous, the removal of this heat is always a difficult practical problem. Second, because of the relatively high polymer concentrations in the reaction mixture (particularly, again, if carried out to high conversions), there is a greater possibility of the formation of branched and cross-linked chains through transfer reactions between "dead" polymer and "living" chain ends. Third, the simplicity of the system means that it is not easy to alter the conditions (e.g., the temperature) so as to obtain products of a variety of average molecular weights while still retaining a convenient rate of polymerization. From these points of view other methods such as solution polymerization would be preferable.

Strictly, so long as the polymerization system only contains pure monomer and initiator, then water cannot be an interactant. However, from the practical viewpoint water may be an accidental component of the system. For example, vinyl monomers generally contain phenolic inhibitors added to stabilize the material during shipping and storage, and these are generally removed before use of the monomer, by washing with aqueous alkali; if this water-saturated monomer is then used directly in the bulk polymerization, the water (being less soluble in the polymer than in the monomer) will separate out in the polymerization and lead to reduced optical clarity and altered dielectric properties of the polymer.

On the other hand, such a polymer, having "pores" or "voids" containing liquid water, would be a material of interest in its own right; in particular, the absorptive properties of the (dehydrated) product toward water vapor would provide useful information on the contentious role of "microvoids" in the corresponding absorptive properties of polymers prepared under (presumably) anhydrous conditions (see Section 3.8).

Similarly, since the "voids" are formed around water molecules, it is possible that the material would show preferential permeation toward water compared with other small-molecule substances; this would be useful in important techniques involving polymer membranes such as reverse osmosis and hemodialysis.

One type of bulk polymerization where water does have an effect is that of radiation-induced polymerization. In the case of stryrene, for example, the rate of the polymerization is increased by rigorous drying of the monomer. Conversely, the addition of water at low concentrations ($1 \text{ m}M$–$1 \text{ }\mu M$) leads to a marked retardation of the polymerization; the fact that other substances of high proton affinity (diethyl ether, ammonia, trimethylamine) have a similar effect indicates that the propagation takes place via free (i.e., not ion-paired) cationic end groups, which has been confirmed by electrical conductivity measurements.[631,1506] The effect of the presence of water on the molecular weight distribution of the product has also been studied.[688]

2.1.2. *Solution Polymerization: Polymer Soluble*

Unless the polymer is specifically required in the block form produced by bulk polymerization, it is more convenient to carry out the polymerization by diluting the monomer with a miscible liquid, which may or may not also be a solvent for the polymer. This has the twofold advantage of both helping to dissipate the heat produced in the polymerization and also of yielding a final mixture which is more easily handled in the subsequent treatment to isolate the polymer in a convenient powder or granular form by, for example, precipitation or freeze-drying. The simplest kinetic behavior results when the diluent used is a solvent for the polymer as well as the monomer; it is useful to distinguish between aqueous and nonaqueous systems.

2.1.2a. *Aqueous Systems.* For free-radical polymerization, water provides an almost ideal solvent medium since, unlike many organic solvents, it is highly resistant to attack by free radicals, so that chain transfer to the solvent is absent and a high-molecular-weight product is more readily obtained than in the generally more reactive organic media; if need be, the molecular weight may be deliberately reduced by the controlled addition of active chain transfer substances to the reaction mixture. This is the type of approach used, for example, by Haas and his colleagues in the synthesis of polyacrylylglycinamide[579] and of polymethacrylylglycinamide,[582] where the product molecular weight was in each case controlled by the addition of isopropanol as the chain transfer agent.

The initiators used for vinyl polymerization in aqueous solution are commonly peroxy compounds such as hydrogen peroxide or peroxydisulfates, which require polymerization temperatures of $50°$ or above to give useful rates of decomposition and hence of polymerization; however for these solutions there is also the possibility of using the mixed "redox" systems, where the free radicals are produced by the reduction–oxidation interaction between the two components, e.g., the simple ferrous ion–hydrogen peroxide system where hydroxyl radicals are produced by

$$Fe^{2\oplus} + H_2O_2 \rightarrow Fe^{3\oplus} + HO^{\ominus} + HO\cdot \tag{9}$$

The advantage of this method is that by adjusting the concentrations of the two reagents it is possible to obtain convenient rates of initiation at or below ambient temperatures; as well as being more convenient experimentally, the use of low temperatures often favors the production of a "cleaner" product in terms of the reduction of side reactions, greater steric definition, and so forth.

In contrast to addition polymerization, water has a marked effect upon most ionic polymerizations, which therefore must be generally carried out in nonaqueous media (see below). An exception are the alkyl α-cyanoacrylates, which appear to polymerize by an anionic mechanism even in aqueous solution (see Section 5.6.3).

2.1.2b. *Nonaqueous Solutions.* As with purely aqueous systems, free-radical polymerizations in nonaqueous solution are not influenced to any great degree by the presence of water, unless the incorporation of water in the medium produces a mixture in which the polymer is insoluble so that it precipitates out during the polymerization (see Section 2.1.3). On the other hand, even traces of water generally have a marked effect upon ionic polymerizations, the sensitivity arising from specific chemical effects. For example, in the low-temperature cationic polymerization of isobutylene catalyzed by boron trifluoride the latter requires a cocatalyst for it to be active; in the case of water as such cocatalyst it is the cation produced by the ionic dissociation of the 1:1 addition compound $BF_3 \cdot H_2O$ which is the actual initiator of the polymerization (Ref. 457, Chapter 5)

$$BF_3 \cdot H_2O \rightleftharpoons HBF_3^{\oplus} + OH^{\ominus} \tag{10}$$

$$HBF_3^{\oplus} + CH_2\!\!=\!\!C(CH_3)_2 \rightarrow BF_3 + CH_3\!\!-\!\!\overset{\oplus}{C}(CH_3)_2, \quad \text{etc.} \tag{11}$$

On the other hand, in the methoxide-catalyzed anionic polymerization of formaldehyde, water acts as a chain transfer agent, in this case serving to reduce the molecular weight of the product (Ref. 1038, p. 14)

$$CH_3O^\ominus + CH_2O \rightarrow CH_3OCH_2O^\ominus, \quad \text{etc.} \tag{12}$$

$$CH_3(OCH_2)_xO^\ominus + H_2O \rightarrow CH_3(OCH_2)_xOH + OH^\ominus \tag{13}$$

$$OH^\ominus + CH_2O \rightarrow HOCH_2O^\ominus, \quad \text{etc.} \tag{14}$$

Finally, in the polymerization of vinyl and diene monomers initiated by lithium alkyls the growth of the chains can be stopped by the addition of a hydroxylic compound such as water, e.g., with styrene and butyl lithium (Ref. 1268, Chapter 4)

$$Bu-(CH_2-CH\phi)_{x-1}-CH_2-\overset{\ominus}{CH}\phi Li^\oplus + H_2O \tag{15}$$

$$\rightarrow Bu-(CH_2-CH\phi)_{\bar{x}}H + LiOH \tag{16}$$

These examples illustrate the diversity of the effects that water can have in ionic polymerizations. An exception to this high sensitivity of ionic polymerization to water is the group of alkyl α-cyanoacrylates, already mentioned above and discussed further in Section 5.6.3 in relation to their use as surgical adhesives.

2.1.3. *Solution Polymerization: Polymer Insoluble*

In general, a polymer tends to be less soluble in a specific liquid than the corresponding monomer, mainly by reason of its higher molecular weight (which gives a lower entropy of solution because of the lower number of molecules contained in a given mass of material), but also because, in the case of vinyl monomers, of the loss of the double bonds which interact favorably with polar solvents. Thus cases often arise in solution polymerization where the polymer separates out of solution as it is formed; in aqueous systems this applies specifically to monomers such as vinyl acetate, methyl methacrylate, and acrylonitrile, which are somewhat water soluble whereas their polymers are completely insoluble.

One feature of this type of system is that the kinetic behavior is abnormal due to the trapping of the growing ends of the polymer chain inside the precipitated particles of polymer; this behavior has been discussed by Bamford and his colleagues.[78]

It is likely that this type of behavior is responsible for the production of almost monodisperse suspensions of polystyrene particles when the monomer is polymerized in suspension with water-soluble initiators and in the absence of any emulsifying agent.[807] Although the mechanism of this polymerization was not discussed by the originators of the method,[807]

the following seems to be a likely explanation. The initiator, peroxydisulfate, produces by its thermal scission sulfate radical ions

$$S_2O_8^{2\ominus} \rightarrow 2SO_4^{\ominus}\cdot \qquad (17)$$

which cannot enter the droplets of styrene monomer and so react instead
with the dissolved styrene present at low concentration in the aqueous
phase; the growing ends of the chains so initiated would then be protected
from termination by burial within the polymeric particle, although still
accessible to the monomer by its diffusion through the particle. On this
basis, each radical ion formed would in turn lead to the production of one
polymer particle. The particles would be prevented from coalescing by their
negatively charged sulfate end groups; the dependence of the stability of
the latices on the presence of these groups was demonstrated by the failure
to produce stable latices when the initiator was changed to hydrogen
peroxide, which would lead only to neutral hydroxyl end groups.[807]

A second interesting feature of this case of solution polymerization
where the polymer is insoluble is that it may be used to detect the presence
of free radicals as intermediates in a reaction system; this applies particularly to aqueous systems because of the wide range of reactivities shown
by the suitable monomers. In this application if a small amount (say, 0.1%
or less) of the monomer is included in the reaction system, so long as the
free-radical intermediates are sufficiently active they will initiate polymerization of the monomer, leading to the production of a readily visible
precipitate of polymer. In fact, by reducing the monomer concentration
to a sufficiently low level only a relatively small proportion of the radicals
will be trapped, so that the main reaction mechanism will not be disturbed,
while at the same time at these low concentration levels the polymer will
remain colloidally suspended in the system to give a turbidity which is
readily detectable by photoelectric means. In addition, the free-radical
intermediates involved should appear as end groups on the precipitated
polymer; thus, the production of thiocyanate radicals (SCN·) during the
photolysis of aqueous solutions of ferric thiocyanate was demonstrated
both by the polymerization of acrylonitrile added to the reaction mixture
and by the detection of thiocyanate end groups on the precipitated polymer.[401]

2.1.4. Suspension Polymerization

In this method the monomer is kept dispersed in a immiscible diluent
during the course of the polymerization. The suspending medium most
commonly used is water, both because it is so widely available and so cheap,

and also because many of the common monomers are essentially insoluble in it; at the same time, because of its high heat capacity, it serves admirably to absorb the heat of the polymerization. In these cases the possible systems are essentially limited to those polymerizing by free-radical mechanisms; for the polymerization to take place within the monomer droplet an oil-soluble initiator (rather than a water-soluble one) must also be used. To keep the polymerizing monomer dispersed, a water-soluble suspending agent such as a polyacrylate is included in the recipe, and the mixture is also stirred. The final product is a suspension or "latex" which may be suitable for direct use (such as a paint or adhesive; see Section 5.4) with only a few formulation additions; alternatively, isolation of the polymer yields a granular or bead form suitable for further modification, such as (with cross-linked polymers) substitution by ionizable groups to produce ion-exchange resins.

It should be borne in mind that, as with bulk polymerization, since the polymer concentration in the suspended particles reaches quite high levels (close to 100% for the common case of complete conversion), the likelihood of branches or cross-links in the molecular structure is increased. This possibility is a factor militating against the use of this method if strictly linear polymers are required.

2.1.5. *Emulsion Polymerization*[160]

In this type of system the reaction mixture comprises water, a water-insoluble monomer, a soap or detergent in an amount well above its critical micelle concentration, and a water-soluble initiator; it is found that this combination can give at the same time both relatively high rates of polymerization and relatively high molecular weights. The key to this behavior lies in the micelles, which dissolve ("solubilize") some of the monomer; free radicals produced in the aqueous phase diffuse into the micelles and there initiate the polymerization, which takes place principally within these micelles. The micelles in fact represent isolated sites for polymerization, since the next radical to diffuse in will immediately terminate the polymer chain, so that at any given time only one-half of the micelles contain growing polymer chains. It is this particular nature of the locus of polymerization which is responsible for the characteristic behavior of the system; this is therefore the most striking example of a physical role played by water in a polymerization system, since it is only in an aqueous system that one would get the three essentially distinct regions: (i) main aqueous phase containing dissolved initiator producing the free radicals, (ii) micelles as the locus of polymerization, and (iii) monomer droplets serving as a reservoir of the material, diffusing into the micelles as it is used up in the propagation reaction.

This type of system also provides examples of the advantages of "redox" initiators over simple peroxy compounds. Thus, this method of polymerization was very early applied to the copolymerization of styrene and butadiene to produce the GR-S or SBR rubbers. Currently this may be carried out either "hot" (at $\sim 50°$) using potassium peroxydisulfate or an organic peroxide as initiator, or "cold" (at $\sim 5°$) with a redox system such as a ferrous salt and an organic hydroperoxide; as compared with the first system, the second gives a product with less chain branching and cross-linking, a greater proportion of the butadiene units trans-1,4 (72%, as compared with 65%), and a lesser proportion cis-1,4 (12%, as compared with 18%; the 16–17% balance in each case are 1,2 units).[1270]

The principal disadvantage of emulsion polymerization, from the viewpoint of preparing pure samples of the polymer, is the likelihood of retention of the emulsifier in the final product, whether simply physically held or else covalently linked by chain transfer reactions.

Emulsion polymerization has been applied recently by several groups of workers[568,731,1127,1562] to prepare monodisperse suspensions of spherical polymer particles for use in fundamental studies on the stability of lyophobic colloids. Although samples of these types of suspensions can be obtained from various commercial sources, many workers prefer to prepare their own. The reason for this is that the particles prepared by such emulsion polymerization have surface charges, which seem to originate either directly from the initiator itself or from the emulsifier via chain transfer reactions, or from both sources; the nature and surface density of these charges therefore depend very much upon the conditions during the preparation, which are seldom precisely known to the purchaser in the case of commercial samples (and often not strictly controlled by the manufacturers themselves). By preparing the suspensions for oneself under known and controlled conditions there can therefore be much closer definition of the nature and the density of these charges, which obviously influence the stability of the colloid to an important extent. Techniques have also been developed to identify the nature of these surface groups; carboxyl groups, for example, can be identified by the dependence of the electrophoretic mobility of the particles upon the pH of the solution, and pH variation can then be used to reduce the surface charge density to a desired level.[568]

The preparation of such suspensions using emulsifier-free systems[807] has already been discussed in Section 2.1.3.

2.2. Condensation Polymers

The commonest examples of condensation polymerization involve reactions where water is the small-molecule component eliminated; a typical simple case is the self-esterification of an ω-hydroxycarboxylic acid

$$x\text{HO—R—COOH} \rightleftharpoons \text{H}(\text{O—R—CO})_x\text{OH} + (x-1)\text{H}_2\text{O} \qquad (18)$$

One limiting factor in the polymerization is that because of its reversible and stepwise nature, the water must be removed efficiently if a reasonably high-molecular-weight polymer is to be produced; thus the reaction must be 99% complete to give a degree of polymerization of 100, and 99.9% complete to give one of 1000. In the case of the commercial production of polyamides (nylons) this is ensured by using temperatures of around 220–280°, the water being "flashed off" at the end of the reaction period.

In certain cases of condensation polymers, water is an actual co-reactant with the monomer; this applies, for example, to the production of silicone polymers through the condensation of disubstituted silyl dichlorides, the water providing the oxygen atoms in the main chain of the polymer

$$x\text{R}_2\text{SiCl}_2 + x\text{H}_2\text{O} \rightarrow (\text{O—SiR}_2)_x + x\text{HCl} \qquad (19)$$

The RTV (room-temperature vulcanizing) silicones similarly depend upon the humidity in the atmosphere for their setting upon exposure, through the hydrolysis of acetate end groups placed upon the ends of the chains followed by the condensation of the silanol (SiOH) groups produced (Ref. 1626, Chapter 11).

An interesting example of condensation polymerization which involves an aqueous solution is that of interfacial polymerization; in this technique a solution of a dichloride in a nonaqueous solvent such as tetrachloroethylene is placed below an aqueous solution of a diamine. A layer of the polyamide is formed at the interface, and can be removed as a continuous "rope" since more polymer is formed as fast as the two layers are brought into contact with one another. The kinetics and mechanism of the process have been the subject of various recent studies.[167,298,957]

2.3. General Considerations on the Production and Preparation of Polymers

The consideration of the actual routes available for the production of a specific polymer has a number of practical aspects in relation to the end use of the polymer, particularly where this involves fundamental investigations of its interactions with water and aqueous systems. If a commercial grade of the polymer is available, then generally this is the one most convenient to use; however, it must be borne in mind that the actual method of production of the polymer is seldom precisely known, and that in commercial practice for economic reasons the polymerization is often carried out up to essentially 100% conversion of the monomer (with consequent extreme variations in monomer and catalyst concentrations), and often with a rising temperature program to ensure such a high conversion. These conditions can only serve to increase the breadth of the molecular weight distribution and to increase the likelihood of chain branching and crosslinking in the final product.

Further, in free-radical addition polymerization, unless oxygen is excluded it acts as an inhibitor, while in smaller amounts it copolymerizes readily with the monomer to produce peroxide groups

$$P—CH_2—\dot{C}HR + O_2 \rightarrow P—CH_2—CHR—O—O \cdot \tag{20}$$

$$P—CH_2—CHR—O—O \cdot + CH_2{=}CHR$$

$$\rightarrow P—CH_2—CHR—O—O—CH_2—\dot{C}HR, \text{ etc.} \tag{21}$$

which represent weak links in the backbone of the polymer. Similarly, a commercial product may contain traces of the other components of the reaction system, either intermixed with it or else covalently linked (e.g., of the emulsifier from emulsion polymerization, by chain transfer processes). Finally the commercial methods for the isolation of polymers often involve rather drastic treatment such as spray- or drum-drying which can lead to further undesirable modification of the polymer.

There is thus much to be said in favor of preparing one's own samples of polymer from the parent monomer under definite, controlled, and reproducible conditions, particularly since this serves to give experience and confidence in the techniques which can then lead to the exploration of other possibilities, such as in the preparation of various types of copolymers. The optimum control is obtained when the polymerization is carried out at a fixed, fairly low temperature on a moderately dilute solution of the monomer in an inert solvent and in the absence of oxygen, and to a relatively small extent of polymerization (say, 10–20%), since this minimizes the width of the molecular weight distribution and reduces the likelihood of side reactions such as cross-linking and chain branching. Pritchard (Ref. 1192, Chapter 1) has emphasized the precautions required both in preparing a reasonably pure sample of poly(vinyl acetate) from vinyl acetate via its free-radical polymerization, and in carrying out the subsequent hydrolysis to produce poly(vinyl alcohol).

From a slightly different viewpoint, a situation may arise where decision has to be made upon what type of polymer is to be produced for a specific purpose; such a purpose could be as for a "model compound" for a particular biopolymer, where it is required to incorporate some specific grouping in a polymer chain of well-defined structure, to investigate the interactions of the group in aqueous polymeric systems. The primary choice here is between polymers of the addition and the condensation types. The advantages of addition polymers are that the polymerization reaction is normally irreversible, with a mechanism of well-defined character; with vinyl and acrylic monomers the carbon–carbon chain of the backbone produces a basically stable molecule of well-defined conformational characteristics, while the "functional" groups may be attached to the monomer (or monomers, for copolymers) in a wide variety of desired fashions. On the other hand, condensation polymerization is often reversible, with frequently a less well-defined mechanism; in the case of polyesters the hydrolytic reactivity would be a disadvantage for studies in aqueous systems, although in fact the polyamides and certain of the polyesters such as poly(ethylene terephthalate) do display high hydrolytic stability.

It is likely that, for example, it was these types of consideration that led Haas and his colleagues to choose polyacrylylglycinamide and polymethacrylylglycinamide for study as potential synthetic polymer substitutes for gelatin[579,582] (see Section 3.10.2).

2.4. Polymer Preparation by Modification of Other Polymers

In certain cases a particular polymer may not be obtainable by direct polymerization, or it may be inconvenient to prepare it directly, so that recourse must be made to preparing it by the chemical modification of some other polymer.

For example, poly(vinyl alcohol) cannot be made by direct polymerization, since vinyl alcohol (CH_2=CHOH) does not exist (attempts to prepare it lead to its isomer, acetaldehyde). This polymer is therefore prepared by the hydrolysis of one of the directly obtainable esters, most commonly poly(vinyl acetate) (Refs. 441; 1192, Chapter 1).

Similarly, in the case of highly polar polymers (particularly polyacids, polybases, and polyelectrolytes) it is difficult to prepare these directly in a specific steric configuration, i.e., as isotactic or syndiotactic forms, so that chemical modification of a sterically ordered precursor polymer must be used. The preparations by this approach of stereoregular poly(styrene sulfonic acid)[69] and of stereoregular poly(methacrylic acid)s[68,888] have been described.

Even where the requirement is simply that of a narrow molecular weight distribution, the technique of chemical modification may be preferred if the other polymer is more conveniently prepared in the monodisperse form. For example, Carroll and Eisenberg[228] have described the preparation of narrow molecular weight distribution poly(styrene sulfonic acid) from an essentially monodisperse polystyrene; here the sulfonation reaction had to be one that produced neither chain degradation nor crosslinking, and yet gave essentially complete para substitution of the benzene rings.

Further possibilities are opened up by the use of copolymers. For example, Nagasawa and Rice[1077] have described the preparation of copolymers of maleic anhydride or diethyl maleate with various vinyl monomers; on hydrolysis these give chains having the adjacent pairs of carboxylic acid groups of maleic acid units, $-CH(CO_2H)-CH(CO_2H)-$ separated by one or more units of the other monomer or its hydrolysis product.

These examples illustrate some of the applications of chemical modification methods in preparing hydrophilic polymers. It should be emphasized that in all these methods the chemical modification reaction used must be as clean and as complete as possible, because of the impracticability of "purifying out" any residual starting material or product of a side reaction.

Finally, the linking of two or more different preformed polymers to give block copolymers also opens up new possibilities for polymer properties and reactivity. Here the preferred method involves parent polymers of defined, narrow molecular weight distribution which are linked via their end groups by way of a clean and irreversible chemical reaction (such as the linking of hydroxyl end groups through urethane formation with a diisocyanate). One example of this is in the production of block copolymers from poly(ethylene oxide) and poly(ethylene terephthalate) by Lyman and his colleagues[934] for use as membranes in hemodialysis.

3. WATER IN RELATION TO THE PHYSICAL PROPERTIES OF POLYMERS

The subject of the physical interactions between polymers and small-molecule components, including water, is a topic of both practical and theoretical interest.

From the fundamental viewpoint, the characterization of a polymer in terms of its molecular weight average or distribution and related properties normally involves studying its behavior (osmotic pressure, viscosity, light scattering, and so forth) in dilute solution,[19,1208] so that the polymer needs to be soluble in some common solvent (i.e., the two must show marked interaction) for it to be so characterizable; once in solution the study of its behavior then gives information upon the interactions between the two components.

More generally, a polymer may not be soluble in a particular small-molecule liquid, but may nevertheless absorb an appreciable amount of it. One practical aspect of this is with water and a fiber-forming synthetic polymer, where for the polymer to be commercially suitable for textiles and fabrics there needs to be a definite degree of interaction between the two—if the polymer is very hydrophobic, there are difficulties both in dyeing the fiber and with the retention of electrostatic charges on its surface, while if it is very hydrophilic, the textiles may be embarrassingly too vulnerable to water and the various forms of aqueous systems that they encounter in use.

In this section we therefore consider the manner in which synthetic polymers interact with small-liquid components in general, and more specifically with water, covering the range from simple absorption by the polymer to cases where the polymer and the liquid mix freely in all proportions. Sections 3.1–3.4 deal with the general features of polymer–liquid systems; the theories and experimental techniques which have been developed to deal with these systems will be briefly outlined. In Section 3.5 we consider more specifically the behavior of the main water-soluble non-electrolyte synthetic polymers in aqueous solution, and the corresponding behavior of synthetic polyelectrolytes is also briefly considered. Multiple-component aqueous systems are then considered, with Section 3.6 concerned with interactions between a synthetic polymer and a small-molecule solute, and Section 3.7 with those between two or more different polymers. In Sections 3.8 and 3.9 the equilibrium (absorption) and kinetic (permeation and diffusion) aspects of the interaction between water-insoluble synthetic polymers and water or aqueous systems are considered. Section 3.10 outlines the characteristics of irreversible and reversible aqueous gels based upon synthetic polymers, and Section 3.11 outlines the behavior of synthetic polymers at aqueous interfaces.

3.1. General Features of Liquid-plus-Polymer Systems

We start by considering the general features of a system composed of a small-molecule liquid L (component 1) and a high polymer P (compo-

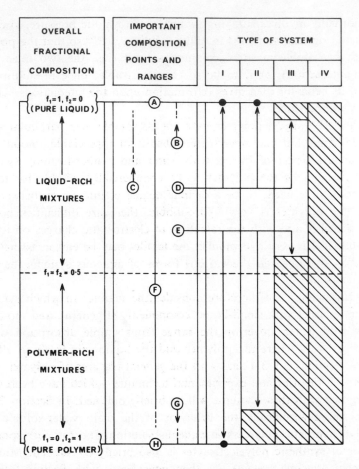

Fig. 1. Important composition points and ranges for mixtures of a small-molecule liquid L (component 1) and a high polymer P (component 2), related to the four main types of system (see text). The "fractional composition" f refers to volume fraction, weight fraction, or the like. The shaded areas represent homogeneous mixtures (solutions).

nent 2), emphasizing particularly the features that distinguish it from a mixture of two small-molecule components; these general features are illustrated diagrammatically in Fig. 1.

It is useful to classify such systems into four main types, numbered in order of increasing degree of interaction between the two components as represented specifically by increasing areas for the composition regions corresponding to *homogeneous* mixtures (solutions) of the two; this is illustrated on the right of Fig. 1. This classification can be specified as follows:

Type I: No (detectable) interaction between the two components; the equilibrium state is pure L in contact with pure P.

Type II: Some interaction between the two; L is absorbed by P to a limited extent, but P is not (detectably) soluble in L.

Type III: Marked degree of interaction; L is strongly absorbed by P, while P is appreciably soluble in L.

Type IV: Strong interactions; L and P are miscible in all proportions.

One special feature to be noted, particularly in the case of types II and III, is the marked asymmetry of the composition regions; this is one feature distinguishing polymeric from nonpolymeric systems.

This fourfold classification is not, of course, a completely clear-cut one. Thus, type I can be linked to the others if we accept that from the thermodynamic viewpoint there must be a definite value for the free energy change of mixing of the two components, however large and positive this value may be; thus the polymer must absorb a finite amount of the liquid, although this may be below the limits of detectability by the techniques currently available. In fact, even such highly hydrophobic polymers as polyethylene, polypropylene, and polytetrafluorethylene absorb small but detectable amounts of water at normal temperatures ($\sim$6 mg/100 g of polymer).[93]*

Similarly, type IV will include cases with a wide range of degrees of interaction, with L ranging from being a very good solvent to being a very poor one.

In one sense, indeed, we could view type III as the most general case; for as the composition points D and E of Fig. 1 are moved (say, by altering the temperature) to the ends of the composition range we obtain first type II and then ultimately type I as special cases, while in the other direction these points will move closer together until they become coincident to produce type IV. However, this simplification is not a useful one, mainly because type III itself is the *least* common one encountered in practice, since it only requires a small alteration in experimental conditions (particularly, temperature) for a type III system to move one way or the other into types II or IV. This is in fact an illustration of the extent to which

---

* However, as will be emphasized later on, the experimental values of water uptake by these relatively hydrophobic polymers are likely to be markedly increased by the presence of even small amounts of hydrophilic contaminants, and the true values for the pure polymers may be much less than the figures quoted.

the solubility of polymers is much more an all-or-none phenomenon than is the solubility of small-molecule substances.

Nevertheless, with these reservations in mind, particularly the need to specify the temperature, we can still consider the four-type classification as a useful one for many practical purposes. In the case of water as the liquid L, we could equate these types with the classification: highly hydrophobic, moderately hydrophobic, moderately hydrophilic, and highly hydrophilic polymer. Both in this specific case and in the more general one it is desirable to put this on a more quantitative basis, using the thermodynamics of polymer solutions.

Following on from this, as also illustrated in the center of Fig. 1, we have a number of points and ranges on the composition scale (labeled by the circled letters A–H) which are important in relation either to the general macroscopic behavior of the systems or to the behavior of the individual components at the molecular level. These can be specified as follows.

(A) $f_1 \to 1$, $f_2 \to 0$: Apparently, this point corresponds simply to the pure liquid; however, it also has significance for the polymer since it is the state of "infinite dilution" or "extreme dilution" of a polymer solution, where we have isolated molecules of P immersed in an infinite medium of L molecules which are unperturbed except for those close to the P molecule or within the volume encompassed by its chain. Experimentally, data for the polymer in this state are typified by the limiting viscosity number (intrinsic viscosity) $[\eta]$ and by related conformational and molecular weight quantities obtained by the extrapolation $f_2 \to 0$.

(B) $0 < f_2 < \sim 0.01$: This is the region of very dilute polymer solutions, where the P molecules interact at most in pairs in the solution; it is thus the region typically covered in viscosity, osmotic pressure, light scattering, and similar experimental techniques. The experimental data for this region, relating to interactions between pairs of polymer molecules, are typified by such quantities as the Huggins viscosity parameter k' and the second virial coefficient A_2 (or B), obtained from the slopes of the lines in the plots for the extrapolation to the extreme dilution state (A).

(C) $f_2 \approx 1/[\eta]$: This is the imprecisely defined region where, as f_2 is increased, there is the onset of overlap and entanglement of the P molecules as they interact with all their nearest neighbors; beyond this region (at higher values of f_2) the solution is essentially uniformly pervaded by solvated and solvent-occluding polymer chains. The location of the region is indicated by such experimentally based methods as the plot of $\log \eta$ (where η is the actual solution viscosity) against f_2,[1190] by the diffusion behavior of small-molecule third components,[1095] and by studies on polymer aerogels produced by solvent sublimation[1532]; less directly, it can be located via calculations for the concentration that would lead to close packing of the polymer molecules, and this is the basis of the value $f_2 = 1/[\eta]$ used as an estimate of its location.

(D) This is the solubility limit for the polymer in the liquid, albeit normally in equilibrium *not* with the pure polymer but with a more or less concentrated solution of the polymer in the liquid at point E.

(E) This is the absorption limit for the liquid in the polymer, i.e., in equilibrium with either the pure liquid or a dilute solution of the polymer at point D.

(F) This is the general region extending from not-too-dilute solutions of the polymer in the liquid, to gels of polymer somewhat swollen by the liquid.

(G) $0 < f_1 < \sim 0.01$: This is the region of the polymer matrix containing molecules of L just sufficiently dense to be interacting with each other either directly or through their effect on the polymer; data for this situation relate experimentally to the first departures from linearity (i.e., from Henry's law) of the absorption isotherm of L vapor into the polymer, and theoretically to clustering functions (see Section 3.8).

(H) $f_1 \to 0, f_2 \to 1$: On the face of it this point simply corresponds to the pure polymer; however, at the same time it corresponds in the case of the liquid to a definite state, that of isolated molecules of L in a polymer matrix which is unperturbed except close to an L molecule; the experimental data for this limiting situation are those from the initial (low pressure) linear section (Henry's law region) of the adsorption isotherm of L vapor into P (see Section 3.8).

More specifically in relation to water as the liquid, we might also expect there to be a point or range where, with increasing water fraction f_1, the solvent shells around individual polymer chains or molecules cease to overlap and regions of "normal" water start to make their appearance. However, for the loose conformations taken up by many linear, flexible-chain synthetic polymers in solution, there is no sharp boundary to the outside of the polymer coil, but rather a roughly Gaussian distribution of chain units about the center of mass; hence, as with region C, for such polymers one cannot expect there to be a sharply defined point at which the regions of "normal" water make their appearance, but instead a range of concentrations over which these regions are gradually formed.

In this introductory discussion, the small-molecule has been taken to be one that exists as a liquid in the pure state, both for simplicity of presentation and also in specific relation to the present context of water and aqueous systems. However, this interacting component may equally well be one that is a gas or vapor in its pure state; indeed, in the case of systems such as type II where a single-point composition results from the equilibration of polymer with the pure liquid, the study of the absorption of the vapor of the liquid over a range of partial pressures up to the saturated vapor pressure gives invaluable information on the mechanism of the interaction between the two components. Gases and vapors as interactants are also important in relation to the rates of their transport across polymer films (see Section 3.9).

The small-molecule interactant may also be a solid in the pure state, as in the case of such solids as camphor and tri-p-cresyl phosphate, which are used as plasticizers, and also in the case of dyes for textile fibers.

3.2. Thermodynamics of Polymer Solutions

In common with all physicochemical studies on interacting systems, in considering the physical interactions between small-molecule liquids and high polymers there are two rather distinct thermodynamic approaches.

First, there is the classical or macroscopic thermodynamic approach, in which data from vapor pressure, heats of mixing, and similar techniques are interpreted on the basis of the free energy, enthalpy, and entropy functions of the system, without reference to its molecular structure.

Second, there is the statistical or molecular thermodynamic approach where these data are interpreted in terms of the molecular structure and the intermolecular interactions in the system.

In practice, these two approaches are necessarily interrelated, if only because the second must depend for its verification upon data provided by the first; further, with polymer systems such techniques as viscosity, osmotic pressure, and light scattering depend very much for their interpretation upon such concepts as the molecular weight and the conformation of the polymer chain. In addition, much of the general interpretation of the behavior of these systems revolves around deciding to what extent it results simply from the chemical structure of the individual repeat units and to what extent it is a more specific result of the concatenation of these units.

3.2.1. Flory–Huggins Theory of Polymer Solutions

The most widely used and firmly based theory of polymer solutions is that initiated independently by Flory[455] and Huggins[693,694] over thirty years ago. Flory in his subsequent monograph[457] has detailed the further developments in the theory, while he[460] has more recently reviewed current work in this field, at the same time modifying certain of the details and also discussing an improved model. In this section the theory is outlined, principally to indicate the assumptions and approximations upon which it is based, so that its applicability to the rather special situation of aqueous polymer solutions can be considered.

The Flory–Huggins theory is centered around the evaluation of the free energy change of mixing ΔG^M of liquid and polymer to form the solution.* We thus start with n_1 moles of a small-molecule liquid (component 1) and n_2 moles of an amorphous high polymer (component 2) which are then mixed to form a homogeneous solution in which the components have volume fractions ϕ_1 and ϕ_2, respectively. (In a recent discussion Flory[460] has indicated that these could perhaps be better replaced by *segment* fractions, but we shall retain volume fractions in this treatment.) The theory is based upon the assumption that at the molecular level the pure liquid, the pure polymer, and the mixture constitute lattices with closely similar and relatively large values for the lattice coordination number z; each molecule of the liquid is taken to occupy one lattice point (although the theory can be readily modified for "longer" molecules of the liquid), while each molecule of the polymer occupies r consecutive lattice points (each containing a so-called segment of the chain), where r is thus the ratio of the molar volume of the polymer to that of the liquid.

The derivation of the expression for ΔG^M then involves the evaluation of two distinct free energy contributions, first the (principally entropic) combinatory contribution, and second the (principally enthalpic) contribution from the interactions between the nearest neighbors (molecules of the liquid and segments of the polymer) in the system.

The first contribution ΔG_{comb} is evaluated assuming random mixing of the molecules of two components, and assuming also that the polymer chain has much the same flexibility in the solution as it has in the pure state; the contribution then takes the form

$$\Delta G_{comb} = RT(n_1 \ln \phi_1 + n_2 \ln \phi_2) \tag{22}$$

For the second contribution ΔG_{inter}, if we represent the three types of nearest-neighbor contacts between the solvent molecules and the polymer segments by [1, 1], [1, 2], and [2, 2], then when the mixing process occurs

---

* For an application of the Flory–Huggins theory to solutions of small molecules see Chapter 1.

a "quasichemical" exchange reaction will take place:

$$\tfrac{1}{2}[1, 1] + \tfrac{1}{2}[2, 2] \rightleftharpoons [1, 2] \qquad (23)$$

One approximation for ΔG_{inter}, applied in the early development of the theory, is to consider only the internal energy change, which for 1 mol of reaction (23) will be

$$\Delta w_{12} = w_{12} - \tfrac{1}{2}(w_{11} + w_{22}) \qquad (24)$$

where the w's represent nearest-neighbor potential energies of interaction between solvent molecules and segments; applying this to the lattice picture with the assumption again of random mixing leads to the internal energy contribution ΔU_{inter} given by

$$\Delta U_{inter} = z n_1 \phi_2 \, \Delta w_{12} \qquad (25)$$

Subsequently it was realized that there is a marked entropic contribution to the value for ΔG_{inter} in the actual systems, which arises partly from the necessarily nonrandom nature of the mixing if $\Delta w_{12} \neq 0$, and also from altered configurational freedom of the individual molecules of solvent and segments with respect to their neighbors; thus for greater generality it is necessary to consider the free energy change Δg_{12} for the exchange reaction (23), and to derive therefrom the contribution to the free energy change of mixing from these interactions:

$$\Delta G_{inter} = z n_1 \phi_2 \, \Delta g_{12} \qquad (26)$$

If we then introduce a dimensionless parameter χ defined by

$$\chi = z \, \Delta g_{12} / RT \qquad (27)$$

eqn. (26) becomes

$$\Delta G_{inter} = RT \chi n_1 \phi_2 \qquad (28)$$

(Since this parameter χ is characteristic of the liquid and the polymer, one should strictly symbolize it by χ_{12}, but the simpler symbolism may be retained except where more than two components are involved.) Flory[460] has proposed the systematic name *residual reduced chemical potential* for χ; generally, however, it is simply referred to as the (Flory–Huggins) *interaction parameter*.* Flory [460] has also defined the division of χ into

---

* In the notation of Huggins[694] this parameter is symbolized μ; however, because of the common use of this in thermodynamics as a symbol for chemical potential, Flory's symbol χ is preferable.

χ_H, the *reduced residual enthalpy of dilution*, or *reduced enthalpy of dilution*, given by

$$\chi_H = -T \partial\chi/\partial T \tag{29}$$

and χ_S, the *reduced residual partial molar entropy*, given by

$$\chi_S = \partial(\chi T)/\partial T \tag{30}$$

such that

$$\chi_H + \chi_S = \chi \tag{31}$$

The quantities χ_H and χ_S correspond respectively to $\varkappa$ and to $\frac{1}{2} - \psi$ of Flory's own previous and still widely used terminology.[457] If we follow the notation of eqn. (27) as applied to free energy, then eqns. (29) and (30) give

$$\chi_H = z\, \Delta h_{12}/RT \tag{32}$$

and

$$\chi_S = -z\, \Delta s_{12}/R \tag{33}$$

It should be noted that the enthalpic contribution (i.e., Δh_{12}) should be considered in terms of $T\chi_H$ rather than of χ_H alone.

Since the complete free energy change of mixing ΔG^M is the sum of ΔG_{comb} and ΔG_{inter}, then addition of eqns. (22) and (28) gives the final expression for ΔG^M

$$\Delta G^M = RT(n_1 \ln \phi_1 + n_2 \ln \phi_2 + \chi n_1 \phi_2) \tag{34}$$

It is instructive to examine the general consequences of eqn. (34) in terms of the effect of χ and of the molecular weight of the polymer upon the thermodynamic feasibility of the mixing process. Since ϕ_1 and ϕ_2 are the volume *fractions*, then the two logarithmic terms on the right-hand side of eqn. (34) are always negative; this allows the value of χ to be positive (up to 0.5, in fact) while still getting complete miscibility (since ΔG^M is then still negative). Considering the effect of molecular weight, for a given *mass* of polymer and of solvent the term $n_2 \ln \phi_2$ will become less negative with increasing molecular weight, finally becoming zero in the limit of infinite molecular weight; thus a higher molecular weight for the polymer makes for a less favorable free energy contribution for the mixing process, leading to the well-recognized inverse relation between the solubility of a polymer and its molecular weight.

In more general terms, from eqn. (28) a positive value of χ corresponds to a positive value of the contribution ΔG_{inter}, which thus disfavors mixing, while the converse is also true. In the great majority of cases the experimental values of χ are positive.[1372] Values of χ up to 0.5 are found for systems showing complete miscibility, while for $\chi > 0.5$ the systems are characterized by only limited miscibility, with higher values of χ corresponding to decreasing extents of interaction of the two components. To the extent, therefore, that χ can be taken to be composition independent, it can be viewed as providing a quantitative basis for the fourfold classification of liquid plus polymer systems outlined in Section 3.1, which gives the following correlation:

Type I: $\chi \gg 0.5$
Type II: $\chi > 0.5$
Type III: $\chi \approx 0.5$
Type IV: $\chi < 0.5$

The quantity χ is clearly a characteristic and important parameter for each liquid plus polymer system at a fixed temperature. In terms of the Flory–Huggins theory, its value ought to be independent both of the molecular weight of the polymer (more specifically, of the molar volume ratio r) and of the composition of the mixture (i.e., of ϕ_1 and ϕ_2). In fact, the latter is most often *not* the case, and the experimentally determined values of χ generally show a marked variation with solution composition. For the most part this may be ascribed to next-nearest-neighbor effects, so that, for example, the contacts [1, 1] and [2, 2] remaining in the solution would not have the same interaction characteristics as in the pure components.

It should be also noted that the Flory–Huggins theory as here outlined starts with the assumption that the polymer is completely amorphous. For polymers that contain crystalline regions the free energy of mixing to form a homogeneous (liquid) solution contains an additional term for the fusion of these regions. In cases where the crystalline regions are unaffected by the solvent, to that only the amorphous regions are involved in absorption of the liquid, the actual effective values for ϕ_1 and ϕ_2 would be those for the amorphous regions; an additional complication is that the polymer should then be viewed as a cross-linked one, the crystallites acting as the cross-links.

One further factor to be considered is that of the lattice coordination number z, which, although it does not specifically appear in the final eqn. (34), nevertheless plays an important role in interpreting experimental data in molecular terms; in particular, if we examine eqns. (27), (32), and (33), which relate χ, χ_H, and χ_S to the molecular interaction contributions, we see that they all specifically contain this coordination number z.

For most small-molecule substances $8 \leq z \leq 12$. On the other hand, for liquid water the effective coordination number is four or somewhat greater (Chapter 8, Volume 1). In the case of carbon-chain and similar polymers, moreover, with each chain group taken to occupy one lattice position, because of the marked preference for one of three fairly distinct orientations (trans, gauche, and antigauche) for the successive links, the coordination number also approximates more closely to a value of four.[459] It therefore appears that such polymers may be able to accommodate themselves better onto the water lattice than onto that of nonaqueous liquids, and that this type of component pair is more appropriately treated by the lattice theory than other liquid plus polymer systems.

On the other hand, in eqn. (34) obtained from the Flory–Huggins treatment, the disappearance of z depends upon the assumption that it is sufficiently large (effectively, infinite). Nevertheless, as Huggins[693] has pointed out, the necessarily finite nature of the value of z is taken into account by making χ an empirical parameter, and this is more specifically indicated by Flory's calling it the *residual* reduced chemical potential [since its function is then to cover any free energy contribution additional to the purely combinatory one given by eqn. (22)].

3.3. Experimental Methods for Studying Liquid/Polymer Interactions

In this section we outline some of the important methods for studying the interactions between a liquid and a polymer. In some cases these explicitly involve the interaction parameter χ, and hence they may be used to evaluate this quantity for a particular system, while in other cases the involvement is implicit and less direct.

3.3.1. *Vapor Pressure of Polymer Solutions*

For a mixture of a reasonably high-molecular-weight polymer and a reasonably low-molecular-weight liquid the vapor pressure above the solution will be solely that of the liquid (component 1). Differentiating eqn. (34) for ΔG^{M} with respect to n_1 (bearing in mind that ϕ_1 and ϕ_2 are also functions of n_1) to obtain the chemical potential of component 1, it then follows that the vapor pressure p_1 is given by

$$\ln(p_1/p_1{}^*) = \ln \phi_1 + (1 - s)\phi_2 + \chi\phi_2{}^2 \qquad (35)$$

where $p_1{}^*$ is the vapor pressure of the pure liquid, and s is the ratio of the molar volume of the liquid to that of the polymer (i.e., the reciprocal

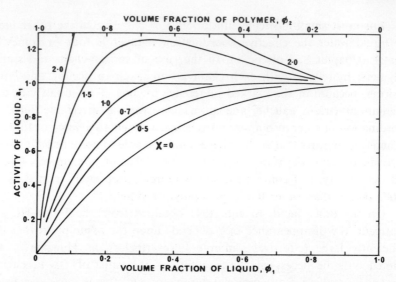

Fig. 2. Plots of the activity a_1 of the liquid (solvent) (component 1) against volume fraction composition ϕ for high polymer (component 2) solutions, as predicted by the Flory–Huggins theory for the indicated values of the interaction parameter χ; the activity values are also equal to the relative partial vapor pressure p_1/p_1^*, where p_1^* is the vapor pressure of the pure liquid.

of r). If pure liquid 1 is taken to be the reference state for this component, then p_1/p_1^* is equal to the activity a_1. Figure 2 shows values of the activity of component 1 as calculated from eqn. (35), plotted against the volume fraction compositions ϕ_1 and ϕ_2 for a series of fairly typical values of χ. In these calculations s has been put equal to zero, equivalent to infinite molecular weight for the polymer; however, calculations for $\chi = 0.5$ using $s = 0.01$ (corresponding to a relatively short-chain polymer) gave a_1 values which were reduced by only between 0.003 and 0.005 for the range $0.1 < \phi_1 < 0.8$. (In fact, it is only for very dilute polymer solutions, where techniques such as osmotic pressure can be applied, that the effect of variations in molecular weight of the polymer on the solvent activity comes into play.) The plots are all tangents to $a_1 = 1.0$ at $\phi_1 = 1.0$ (that for $\chi = 0.5$ is an inflection at this point); for $\chi > 0.5$ the curves cross the $a_1 = 1.0$ line at intersection points representing the composition of the solution (swollen polymer) in equilibrium with the pure liquid. The sections of the curves with $a_1 > 1.0$ have of course no physical significance (except, possibly, for equilibrium with supersaturated vapor).

This correlation between p_1 (or a_1) and ϕ_1 and ϕ_2 can be viewed in its application to actual systems in two ways. First, it can be looked at simply as the manner in which the equilibrium vapor pressure of the liquid over a polymer solution depends upon its composition, i.e., ϕ_1 and ϕ_2 are viewed as the independent variables and p_1 as the dependent variable. Alternatively, it may be viewed in terms of the composition of the solution that is in equilibrium with vapor at a definite value of p_1 or a_1; this then relates to the measurement of the equilibrium uptake (i.e., ϕ_1) of water by an initially solid polymer when placed in an atmosphere of fixed relative partial pressure of the solvent (i.e., in the case of water as the solvent, fixed relative humidity).

3.3.2. *Osmotic Pressure and Light Scattering of Dilute Polymer Solutions**

The osmotic pressure Π of a solution is the excess hydrostatic pressure that must be applied to the solution to raise the activity of the solvent to the same value as the activity of the pure liquid solvent at that temperature. Since the differential of the chemical potential of the solvent with respect to pressure is equal to the partial specific volume of the solvent, which for dilute solutions may be equated to its molar volume V_1, then for a polymer solution showing Flory–Huggins behavior this leads to a relation for Π closely similar to (35):

$$-\Pi V_1/RT = \ln \phi_1 + (1 - s)\phi_2 + \chi\phi_2{}^2 \tag{36}$$

Substituting $1 - \phi_2$ for ϕ_1 and expanding the logarithmic term gives

$$\Pi V_1/RT = s\phi_2 + (\tfrac{1}{2} - \chi)\phi_2{}^2 + \tfrac{1}{3}\phi_2{}^3 + \cdots \tag{37}$$

which upon conversion of ϕ_2 into the polymer concentration c_2 (mass of polymer in unit volume of mixture) gives

$$\Pi/c_2 = (RT/M_2) + RT(\tfrac{1}{2} - \chi)(c_2/V_1\tilde{\varrho}_2{}^2) + O(c_2{}^2) \tag{38}$$

where $\tilde{\varrho}_2$ is the density of the polymer in the mixture. Thus, a plot of Π/c_2 against c_2 should have an initial linear region with an intercept related inversely to the molecular weight (for a polydisperse polymer, the number-average value). The slope of this line, corresponding to the second virial coefficient for the system, will be directly proportional to $\tfrac{1}{2} - \chi$; in the case where $\chi = 0.5$, referred to as the "ideal state" or the "Θ state," the line will be of zero slope.

* See Ref. 1485, Chapters 6 and 10.

Rather similar conclusions, but through less direct application of the theory, apply to the interpretation of light scattering measurements, where the intercept of the corresponding plot is now related directly to the weight-average molecular weight of the polymer, while the slope is related to $\frac{1}{2} - \chi$ as with osmometry. The shape of the scattering envelope, i.e., the angular variation of scattered light intensity, enables the dimensions of the polymer coil to be determined; for systems where $\chi = 0.5$, i.e., the so-called "Θ state," the conformation is that of the unperturbed random flight type (as discussed below in connection with viscosity measurements).

3.3.3. Viscosity of Dilute Polymer Solutions

The measurement of the viscosity of dilute solutions of polymers is a widely used and valuable method for obtaining information on the size, shape, and interactions of the polymer molecules in solution. The theory of the subject has been covered at various levels in the sources cited at the beginning of this chapter.[457,1038,1268,1461] Tompa (Ref. 1485, Chapter 9) has also dealt succinctly with the subject, while Kurata and Stockmayer[835] have reviewed the subject at more length. The practical aspects have been covered by Onyon.[1116] The general basis of the subject will be outlined here as background for the discussion of its applications to synthetic polymers in aqueous solution (Section 3.5).

The specific viscosity η_{sp} of a polymer solution of concentration c_2 is defined by

$$\eta_{\mathrm{sp}} = (\eta - \eta_0)/\eta_0 \tag{39}$$

where η is the viscosity of the solution and η_0 that of the pure solvent. For many dilute polymer solutions the ratio η_{sp}/c_2 is a linearly rising function of c_2, following the equation

$$\eta_{\mathrm{sp}}/c_2 = [\eta] + kc_2 \tag{40}$$

so that a "viscosity plot" of η_{sp}/c_2 against c_2 will be a straight line of ordinate intercept $[\eta]$ and slope k. The quantity $[\eta]$ is termed either the intrinsic viscosity or (in more recently recommended terminology) the limiting viscosity number, which will here be abbreviated l.v.n. (In fact neither of these terms is fully satisfactory since dimensionally $[\eta]$ is neither a viscosity nor a number but has the units of reciprocal concentration.) From eqn. (40) it follows that the l.v.n. relates to the behavior of isolated polymer molecules in solution; from the Einstein relation between η_{sp} and the volume fraction of particles in a dilute suspension it follows that

$$[\eta] = 2.5/\bar{\varrho}_2 \tag{41}$$

where $\bar{\varrho}_2$ is the average density of polymeric material within the volume encompassed by the molecule.

As regards the slope factor k of eqn. (40), this clearly relates to the effect of interactions between pairs of molecules upon the viscosity increment of the solution; generally, for a series of samples of the same polymer type of different molecular weights, the value of k is proportional to $[\eta]^2$, i.e.,

$$k = k'[\eta]^2 \tag{42}$$

where the dimensionless parameter k' is termed the Huggins constant and has a characteristic value (generally ~ 0.5) for that polymer in that solvent.

If values of $[\eta]$ are determined for samples of the same polymer type of differing molecular weight M in the same solvent at the same temperature, then frequently the two are found to be related by the Mark–Houwink–Sakurada equation

$$[\eta] = KM^a \tag{43}$$

where K and a are characteristic parameters for the particular system.

For linear, flexible-chain synthetic polymers in solution the most common type of conformation is a loosely coiled one akin to the true random coil, and for which the experimentally observed values of the exponent a lie within the range 0.5–0.8. The minimum observed value, $a = 0.5$, in fact corresponds to the true or unperturbed random coil conformation, where the deviation from the exact random flight conformation caused by polymer chains not being able to cross their own paths is exactly counterbalanced by the poorness of the solvent favoring segment–segment interactions; thus under these conditions, for the Θ solvent (or Θ temperature for a specific solvent)

$$[\eta]_\Theta = K_\Theta M^{0.5} \tag{44}$$

It should be noted that even in this state there will still be effects from the short-range forces between segments, such as from restricted rotation; the value of K_Θ may thus be used to show the influence of these types of interaction upon the chain conformation.

In general, synthetic polymers in solution do not take up the random coil conformation (unless, that is, they happen to be in a Θ solvent), and they are therefore either in a more or a less expanded conformation. The latter is unusual, since the polymer will then be on the verge of precipitation. In the former, more usual case it is useful to relate the actual conformation to the random coil conformation via a linear expansion coefficient α given by

$$\alpha^3 = [\eta]/[\eta]_\Theta \tag{45}$$

The relation of $[\eta]$, in general, and α to the interactions between segments and molecules of solvents is a complex and disputed topic[835]; however, one useful correlation is that of Flory, which takes the form

$$\alpha^5 - \alpha^3 = 2C_M(\tfrac{1}{2} - \chi)M^{0.5} \tag{46}$$

where C_M is essentially a numerical factor,* and χ is again the Flory–Huggins interaction parameter. On this basis, for $\chi = 0.5$, $\alpha = 1$, and the chain is in the unperturbed random coil conformation; for $\chi < 0.5$, i.e., for better solvents, the value of α from eqn. (46) will be greater than unity. In the limit, for very good solvents (χ large and negative), the term α^3 may be neglected with respect to α^5, so that from eqn. (46) α becomes proportional to $M^{0.1}$, and from eqns. (44) and (45) the exponent a in the Mark–Houwink–Sakurada eqn. (43) becomes 0.8, which is indeed the limiting value generally observed for flexible-chain polymers (see Section 3.5).

Considering the more general types of conformation, i.e., in addition to the loosely coiled forms taken up by synthetic polymers, at one extreme we have the fully compact conformation approached by many proteins in their native states,[1465] as well as by some of the purely synthetic "poly-soaps."[1435] In this limiting case the value of $[\eta]$ will be given by eqn. (41) with $\bar{\varrho}_2$ equal to the density of the solid polymer (i.e., $\sim 1.0 \text{ g cm}^{-3}$), or somewhat less if there is any "solvation" of the polymer molecule. Under these circumstances since $[\eta]$ is independent of M, then the exponent a will be zero, which is thus the absolute minimum value for any specific polymer.

At the other extreme, for a fully extended rigid rod, or for the equivalent shape of a helix, the volume encompassed by the molecule will be the sphere traced out by the molecule as it rotates about its center of mass; from this it follows that $\bar{\varrho}_2$ will be inversely proportional to M^2, so that this gives the value $a = 2$ as the extreme upper limit possible. In fact, the maximal experimental values of a observed for polymers known to conform to this type are generally somewhat less than two, which indicates a certain amount of kinking or bending in the molecule.

3.3.4. Swelling of Cross-Linked Polymers

When a mildly cross-linked polymer is immersed in a liquid the polymer will become swollen to an equilibrium extent depending both upon the density of cross-links in the polymer sample and the value of the inter-

* For the definition of this numerical factor C_M see, for example, Kurata and Stockmayer.[835]

action parameter χ for the pair of components. In practice, the swelling is commonly expressed as the volume swelling ratio or degree of swelling Q_v, which is the ratio of the volumes of the swollen and unswollen systems, i.e.,

$$Q_v = (v_1 + v_2)/v_2 \tag{47}$$

which then becomes in terms of ϕ_2, the volume fraction of polymer in the swollen gel,

$$Q_v = 1/\phi_2 \tag{48}$$

More direct and precise measurements may be made on a weight (mass) basis, giving the weight swelling ratio Q_w, which is the ratio of the weights of swollen and unswollen systems, and which in a similar fashion leads to Q_w being related to the weight fraction of the polymer in the swollen gel by

$$Q_w = 1/w_2 \tag{49}$$

From the thermodynamic viewpoint, at swelling equilibrium the free energy term for mixing of the liquid and the polymer will be balanced by a term for the elastic reaction of the gel, so that application of the polymer elasticity theory of Flory and Rehner[462] gives

$$\ln(1 - \phi_2) + \phi_2 + \chi\phi_2^2 = -N_2 V_1(\phi_2^{1/3} - \tfrac{1}{2}\phi_2) \tag{50}$$

where N_2 is the number of polymer chains per unit volume and V_1 is the molar volume of the liquid. The quantity N_2 may also be written as ϱ_2/M_c, where ϱ_2 is the density of the unswollen polymer and M_c is the molecular weight of the polymer chains between cross-links. Rodriguez[1268] has given a nomograph for solving eqn. (50) in terms of the three variables $N_2 V_1$, ϕ_2, and χ; this nomograph, as well as the equation itself, less directly, shows the manner in which the degree of swelling increases with decrease in the value of χ and with decrease in the extent of cross-linking of the gel (as represented by N_2).

3.4. Correlation of the Thermodynamic Behavior of Polymer Solutions with That of Small-Molecule Systems

One approach which has been widely applied in attempts to relate the behavior of liquid plus polymer systems to the properties of the individual components is to use the concepts of the cohesive energy density and solubility parameter originally developed in the Hildebrand–Scatchard theory of solutions.[662,663] In this section we outline this theory as it applies to wholly small-molecule systems, and then show its application to polymeric systems, in each case examining to what degree it may be extended to highly hydrogen-bonded systems such as those involving water.

3.4.1. Cohesive Energy Density and Solubility Parameter Theory for Small-Molecule Substances

The cohesive energy density of a substance (liquid) is defined as the ratio $\Delta_l^g U/V$, where $\Delta_l^g U$ is the energy of evaporation of a fixed amount of the liquid and V is the volume of this same amount; generally it is a molar quantity that is considered, although since the ratio is an intensive property the actual amount involved is not of consequence. The square root of cohesive energy density is then termed the solubility parameter δ for the liquid, i.e.,

$$\delta = (\Delta_l^g U/V)^{1/2} \tag{51}$$

Table I sets out the values of δ for a number of common liquids; it will be seen that the order is very much that of the "polarity" of the liquids, with the hydroxylic liquids at the upper end of the scale, and with water by far the highest in its value of δ of all the liquids listed. This isolated position for water is a source of one of the main difficulties in applying solubility parameter theory to aqueous systems.

In the application of this quantity to solutions a similar lattice model is used as in the Flory–Huggins theory of polymer solutions. The simplest case to consider is a solution containing two small-molecule components, numbered 1 and 2 (with each molecule occupying one lattice position); on the assumption that the energy of interaction between unlike molecules is the geometric mean of those between like, i.e.,

$$w_{12} = (w_{11}w_{22})^{1/2} \tag{52}$$

then it follows that with random mixing of the molecules the energy change on mixing n_1 moles of component 1 and n_2 moles of component 2 to give volume fractions ϕ_1 and ϕ_2 in the final mixture is given by[663]

$$\Delta U^M = (\delta_1 - \delta_2)^2 V_1 n_1 \phi_2 \tag{53}$$

where V_1 is the molar volume of component 1.

Thus on this basis unless the two liquids have the same cohesive energy density ($\delta_1 = \delta_2$), mixing will always be endothermic, and will be the more so the greater the disparity (in either direction) between δ_1 and δ_2; this

TABLE I. Values of the Solubility Parameter δ for Some Common Liquids at 20-25°C^a

Liquid	δ, cal$^{1/2}$ cm$^{-3/2}$
Perfluorohexane	5.9
Hexane	7.3
Diethyl ether	7.4
Octane	7.6
Dipropyl ether	7.6
Cyclohexane	8.2
Carbon tetrachloride	8.6
Diethylene glycol	9.1
Ethyl acetate	9.1
Tetrahydrofuran	9.1
Benzene	9.2
Chloroform	9.3
Cyclohexanone	9.3
Butanone	9.4
Methylene dichloride	9.8
Acetone	9.8
Dioxan	9.9
Octanol	10.3
Hexanol	10.7
Pentanol	10.9
Butanol	11.1
Propanol	12.0
Dimethylformamide	12.1
Phenol^b	12.5
Ethanol	12.8
Propylene carbonate	13.2
Ethylene glycol	14.4
Methanol	14.6
Ethylene carbonate	14.7
Glycerol	16.5
Water	23.4

^{a} The data quoted are average values from various sources.[17,105,216,662] 1 cal = 4.2 J.
^{b} The value quoted[105] presumably applies to molten phenol, i.e., above 41°.

implies, in turn, that the greater the disparity in δ values, the greater the likelihood of only partial miscibility of the two liquids. In the broadest sense this does indeed apply in practice, for if we examine the order of the liquids in Table I, it is clear that it is the liquids grouped at the top and those grouped at the bottom that most readily exhibit partial miscibility with each other (and these are the pairs for which δ_1 and δ_2 will be most disparate), while the liquids in the middle do tend to be more miscible among themselves and with those at either end of the scale (and for these situations $\delta_1 \approx \delta_2$).

However, the weakness of the theory is that strictly for eqn. (52) to apply, the intermolecular forces involved must be either of the Keesom (dipole–dipole) type or of the London (dispersion) type. Where in particular there are Debye (dipole–induced dipole) type of van der Waals forces, as happens especially with one component having strongly dipolar molecules and the other having highly polarizable molecules, $w_{12} > (w_{11}w_{22})^{1/2}$. Further, for systems where there is the possibility of more specific forces between the molecules of the unlike components, w_{12} will again be high; this would apply, for example, to pairs of liquids such as acetone and chloroform, where the unlike molecules can link via a hydrogen bond but the like molecules can interact only by relatively weaker van der Waals forces.

On the other hand, for more intensively hydrogen-bonded systems the question of the applicability or inapplicability of eqn. (52) is much less clear-cut; it is quite possible, for example, that the strength of a hydrogen bond between unlike molecules may be approximately the geometric mean of that between like molecules, so that as long as the numbers of hydrogen bonds formed remain constant, then eqn. (52) could still apply. Thus, Small[1405] has suggested that for a particular liquid the "proton-donating" and "proton-attracting" (i.e., proton-accepting) powers of the molecules be expressed by two parameters, σ and τ, respectively, with the cohesive energy density then being expressed by the product $\sigma\tau$; for the mixing of two such liquids for which hydrogen bonding is the main cohesive force the energy of mixing will be given by

$$\Delta U^{\mathrm{M}} = (\sigma_1\tau_1 + \sigma_2\tau_2 - \sigma_1\tau_2 - \sigma_2\tau_1)V_1 n_1 \phi_2 \qquad (54)$$

On this basis $\delta_1{}^2 = \sigma_1\tau_1$ and $\delta_2{}^2 = \sigma_2\tau_2$, and thus eqn. (54) reduces to eqn. (53) if $\sigma_1\tau_2 = \sigma_2\tau_1$, i.e., if $\sigma_1/\tau_1 = \sigma_2/\tau_2$; insofar as the hydrogen bond may be viewed as largely electrostatic in nature* and dependent both upon

* As far as water is concerned, this is by now an almost untenable assumption; see Chapters 3, 11, and 14, Volume 1 and Chapter 4, Volume 2. *Editor's note.*

the dipole on the donor X–H bond and upon the fractional charge on the acceptor atom Y, then for a particular substance a high value of σ_1 would indeed be expected to be associated with a high value of τ_1, and similarly for low values, so that the ratio σ/τ could be the same for the liquids. Less generally, a *zero* value for the energy change of mixing ΔU^M would be obtained if either $\sigma_1 = \sigma_2$ or $\tau_1 = \tau_2$ (i.e., for *either* equal "proton-donating" *or* equal "proton-accepting" powers for the two components).

This approach is admittedly speculative, but it suggests that the solubility parameter concept may still be applicable even for highly hydrogen-bonded liquids (especially water and similar substances) at the upper end of the scale of values of the parameter.

3.4.2. *Solubility Parameter Theory for Polymer Solutions*[215]

In the application of the concepts of cohesive energy density and solubility parameter to polymers the first problem that arises is the evaluation of δ for a polymer, since clearly this cannot be obtained from the basic definition of eqn. (51). In practice, δ in these cases is obtained experimentally either by viscosity studies on the polymer in dilute solution or by swelling measurements on a cross-linked sample of the polymer, in each case with a suitable series of liquids; Small[1405] has also outlined the calculation of δ values on the basis of additive group contributions. In either experimental method, since $(\delta_1 - \delta_2)^2$ will be zero for $\delta_1 = \delta_2$ and has positive values (giving endothermic mixing) for $\delta_1 > \delta_2$ and for $\delta_1 < \delta_2$, then both the value of $[\eta]$ and the degree of swelling will be expected to rise to a maximum value (at $\delta_1 = \delta_2$) and then fall off again as the values of δ_1 for the solvents are increased. In the case of the swelling method more precise values of δ_2 are obtained by the use of the Flory–Rehner theory[462]; the application of this latter method has been described by Bristow and Watson[183] and also by Mangaraj and his colleagues,[975,976] while Bristow and Watson[183] have also applied the viscosity method to a series of polymers.

Table II lists δ values that have been obtained for some common synthetic polymers using these types of methods.

Even in these types of measurement the limitations of the basic theory become apparent. Thus, Mangaraj,[975] in swelling measurements on various polymers (natural rubber, polydimethylsiloxane, hydropolybutadiene, and polyurethane), found that aromatic liquids gave anomalous (generally, high) degrees of swelling, so that only the data for the aliphatic liquids were used in the final evaluation of δ for the polymer; a similar restriction was necessary in the subsequent application of the same method by Man-

TABLE II. Values of the Solubility Parameter δ for Some Synthetic Polymers at 20-25°C[a]

Polymer	δ, cal$^{1/2}$ cm$^{-3/2}$
Polytetrafluoroethylene	6.2
Polydimethylsiloxane	7.4 ± 0.2
Polyethylene	8.0 ± 0.2
Polyisobutylene	8.0 ± 0.2
Polyethylene (amorphous)[b]	8.4[17]
Polystyrene	8.9 ± 0.2
Poly(propylene oxide)	9.1 ± 0.4[16]
Poly(methyl methacrylate)	9.2 ± 0.1
Polypropylene	9.3 ± 0.1
Poly(vinyl acetate)	9.4
Poly(ethylene oxide)	10.5 ± 0.5[16]
Poly(ethylene terephthalate)	10.7
Poly(vinyl alcohol)	12.6
Poly(hexamethylene adipamide) (nylon 6.6)	13.6
Polyacrylonitrile	14.0 ± 0.5

[a] Except where otherwise indicated, the values are taken from the "Polymer Handbook"[216] and where no limits are given they represent single estimates. 1 cal = 4.2 J.
[b] Extrapolated from δ values for liquid alkanes.[17]

garaj and his colleagues to polyacrylates and polymethacrylates.[976] This behavior is clearly due to the occurrence of Debye (dipole–induced dipole) forces between the (dipolar) polymer segments and the (polarizable) molecules of the aromatic liquids, giving a higher energy of interaction than would be expected from the individual values of δ_1 and δ_2.

Considering polymer solutions in hydrogen-bonded liquids as solvents, some attempt has been made to fit their behavior into the general pattern by taking into account not only the solubility parameter but also a so-called hydrogen-bonding index (HBI), related to the proton-accepting power of the liquid as given by infrared spectral studies of deuterated methanol (CH_3OD) in the liquid.[105] This two-parameter approach seems to be a useful advance, since it leads, when the two values are plotted as coordinates, to a map with contours representing definite extents of swelling for that polymer.[105]

On the other hand, little attempt seems to have been made, except by Small,[1405] to apply the cohesive energy and solubility parameter concepts to highly hydrogen-bonded systems. Nevertheless, the considerations as outlined at the end of Section 3.4.1 should also apply to polymer solutions. However, this would also require the determination of polymer solubility parameters specifically for these types of systems, a project seemingly still to be attempted; one possible way is outlined in Section 3.7.1.

Finally, even in highly hydrogen-bonded systems the simple solubility parameter concept can still find application in considering the interactions between two or more different solutes. Such a situation arises in the present context with a polymer solute along with a small-molecule solute (see Section 3.6), and also with two different polymers together in solution (see Section 3.7); in the second case, for example, the incompatibility that is frequently observed can be related to the endothermic effect of replacing interactions between like molecules by interactions between unlike ones.

3.5. Synthetic Polymers in Aqueous Solution

In this section we consider the behavior of water-soluble synthetic polymers in their aqueous solutions; we deal first with the nonelectrolyte polymers, and then with polyelectrolytes.

3.5.1. *Nonelectrolyte Polymers*

There are seven common synthetic polymers that are soluble in water at normal temperatures: (i) poly(acrylic acid), (ii) polyacrylamide, (iii) poly(ethylene oxide), (iv) poly(methacrylic acid), (v) polymethacrylamide, (vi) poly(vinyl alcohol), (vii) polyvinylpyrrolidone.

In this section we shall deal first with poly(ethylene oxide), because out of all the seven polymers, this is the one that has been studied most intensively. The remaining six polymers will be considered afterward in order, but with the four related to poly(acrylic acid) [i.e., (i), (ii), (iv), and (v)] discussed together as the "acrylic group."*

---

* It should be emphasized that it is principally in the interests of simplicity and brevity that we have concentrated attention upon the seven named polymers; however, mention has been made of some of the less common polymers of this type [e.g., poly(vinyl methyl ether), under poly(ethylene oxide)] as well as of some of the even more extensive class of copolymers [e.g., poly(vinyl alcohol)co(vinyl acetate), under poly(vinyl alcohol)]. In fact, the information available on these other water-soluble synthetic polymers is at present too fragmentary to make any detailed correlations possible, although more systematic studies should cast much light on this area.

By and large, the discussion will be largely restricted to the behavior of the polymer alone in aqueous solution, with some comparisons with that in nonaqueous solutions; the effects of addition of small-molecule solutes will be dealt with in Section 3.6, and the interactions between two or more polymers in Section 3.7. In each case the behavior of the polymer at finite concentrations is dealt with first, followed by that for extreme dilution.

3.5.1a. *Poly(ethylene oxide)* $+CH_2CH_2O+_{\bar{x}}$. Poly(ethylene oxide) (PEO) is the simplest-structured common synthetic polymer which is water soluble. The very fact of its water solubility is rather remarkable when we consider that for the general class of polyethers $+[CH_2]_m—O+_{\bar{x}}$ neither with $m = 1$ (polyformaldehyde, polyoxymethylene) nor with $m \geq 3$ (poly-trimethylene oxide, etc.) is the polymer at all water soluble. It is also notable that for the methylated derivative of PEO, poly(propylene oxide), $+CH_2CH(CH_3)O+_{\bar{x}}$, only the oligomers are water soluble. On the other and, the isomeric poly(vinyl methyl ether), $+CH_2CH(OCH_3)+$, is water soluble, whereas the corresponding higher alkyl ethers (ethyl, etc.) are not.

It should be borne in mind that, as already indicated, solubility or insolubility is much more of an all-or-none phenomenon with polymers than with small-molecule compounds; in addition, as with small-molecule compunds, solubility involves an equilibrium entailing not only the solute in solution but also the pure solute (or, in the case of polymers the more or less swollen polymer) which is in contact with it. In particular, the forces within the crystalline regions of an expectedly more hydrophilic polymer such as polyoxymethylene will also be an important factor in determining its solubility properties. On the other hand, the "anomaly" of the water solubility of PEO has led to suggestions that it results from the possibility that the polymer may be able to fit better into the water lattice than the other polyethers.[147,1558]

The most extensive thermodynamic study of the water–PEO system is that of Malcolm and Rowlinson,[961] who determined vapor pressure, heat of mixing, phase separation, and other properties for this system at a number of temperatures and over a wide range of compositions. In fact, on the basis of Rowlinson's recent monograph,[1279] this is seemingly the only such study for any synthetic polymer plus water system (the second reference quoted in the monograph[1279] does not in fact deal with this system).

Unfortunately, the data are presented rather unsatisfactorily in some respects; in particular, compositions are given in terms of weight (mass) fractions w, rather than in volume fractions ϕ which as we have already seen are more customary in studies on polymer systems. For this reason, the data have also been interpreted here using w rather than ϕ as the composition parameter; however, since the densities of the two components are not too greatly different, the errors involved are small, and derived data (e.g., for χ) agree with those reported by Malcolm and Rowlinson which were presumably obtained on the more customary volume fraction basis. In addition, the three polymer samples studied, designated PEG 300, PEG 3000, and PEG 5000, where the number indicates the approximate molecular weight of the sample, had respective degrees of polymerization x of 7, 68, and 114; the first is therefore really only an oligomer, while the latter two are around the limit of what would normally be termed a high polymer. Nevertheless, the similarity of, say, the vapor pressure data obtained for the latter two indicates that they can fairly reasonably be considered as high polymers. (It would, of course, be useful to have similar data for polymers of much higher molecular weight, to confirm this point.)

The activity (partial vapor pressure) data for one of the higher molecular weight polymers (PEG 5000) at 65° are plotted in Fig. 3; similar curves were obtained for this and the other polymers at 55 and at 60°. Comparison with Fig. 2 reveals a much greater curvature of the activity plots for this system than expected from the Flory–Huggins theory with a fixed value of χ. This is confirmed in Fig. 4, where the derived values of χ are plotted for the three polymers at several temperatures; in all cases there is a similar marked increase in χ with increase in weight fraction of polymer w_2. [As already stated, the values of χ have here been calculated on the basis of w in place of ϕ as the composition quantity in the Flory–Huggins equation

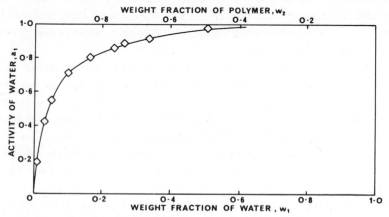

Fig. 3. Solvent activity a_1 versus weight fractions of solvent w_1 and polymer w_2 for the system water + poly(ethylene oxide) PEG 5000 at 65°. Data of Malcolm and Rowlinson.[961]

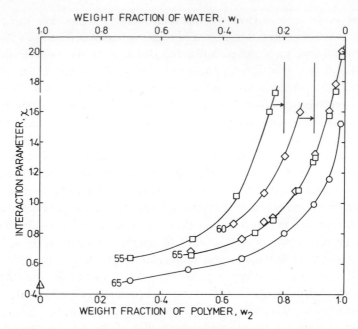

Fig. 4. Values of the interaction parameter χ for poly(ethylene oxide) in water at the indicated temperatures (°C) versus weight fractions of water w_1 and polymer w_2 for PEG 300 (circles), PEG 3000 (squares), and PEG 5000 (diamonds), as calculated from the vapor pressure data of Malcolm and Rowlinson[961]; also shown is the value (0.453) at 27° and $w_2 = 0$ (triangle) from the osmometry data of Schulz.[694] Note that the plots for 60° and 55° have been displaced to the left by 0.1 and 0.2 weight fraction unit, respectively.

(35). However, the derived values agree well with those quoted by Malcolm and Rowlinson; thus, they quote $\chi = 0.75$ at $\phi_2 = 0.66$, and $\chi = 1.24$ at $\phi_2 = 0.9$, compared with the values $\chi = 0.76$ at $w_2 = 0.66$, and $\chi = 1.30$ at $w_2 = 0.90$ obtained here].

Figure 5 shows Malcolm and Rowlinson's data for the heats of mixing ΔH^M (per gram of mixture) measured at 80.3° as a function of the weight fraction composition. The curves are roughly symmetric (but with deviations from exact symmetry which we shall show to be highly significant), with a maximal value for ΔH^M of approximately -8 cal g^{-1} (i.e., exothermic). On the water-rich side of the composition scale the points for the three polymers lie on a common curve, but on the polymer-rich side there is a marked decrease in ΔH^M with increase in molecular weight although the differences between PEG 3000 and PEG 5000 are fairly small.

For many systems, such as those to which the simple lattice theory applies, the heat of mixing is a parabolic function of composition; for

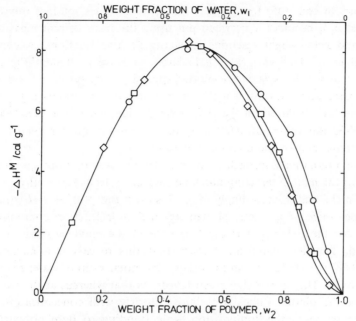

Fig. 5. Heats of mixing ΔH^M for water $+$ poly(ethylene oxide) at 80.3° for PEG 300 (circles), PEG 3000 (squares), and PEG 5000 (diamonds).[961]

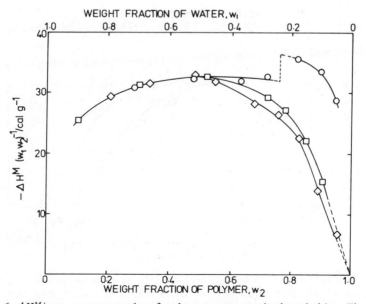

Fig. 6. $\Delta H^M/w_1 w_2$ versus w_1 and w_2 for the same systems (and symbols) as Fig. 5.

example, in eqn. (25) for ΔU_{inter}, for $n_1 + n_2$ moles total of mixture the product $n_1\phi_1$ becomes w_1w_2 when put upon the basis of unit mass of mixture, and with weight fractions replacing volume fractions. Accordingly, the values of $\Delta H^M/w_1w_2$ were calculated, and are plotted in Fig. 6; this shows that for this system the plotted quantity is by no means a constant, there being some variation with composition on the water-rich side, and very marked variation on the polymer-rich side, where for PEG 3000 and PEG 5000 the values of $\Delta H^M/w_1w_2$ are essentially proportional to w_1 at the extreme end of the composition range.

An alternative approach is to consider the system more specifically in terms of the main interacting units, i.e., water molecules and ethylene oxide (monomer) units. Accordingly, Fig. 7 shows the plot of $\Delta H^M/(n_1 + n_2)$ (i.e., per *mole* of mixture) plotted against mole fraction composition y, where for the polymer it is the *base*mole of the ethylene oxide unit that is used. The first significant feature that this reveals is a maximum at $y_1 = 0.68$, $y_2 = 0.32$, i.e., an essentially 2:1 molar ratio of water to ethylene oxide units. The second significant feature is that whereas at the water-rich extreme the plotted values rise almost linearly with concentration, at the other extreme of the composition scale (tending to pure polymer) they seem to approach the concentration axis tangentially, at least for the two "high" polymers PEG 3000 and PEG 5000. Now in interpreting this be-

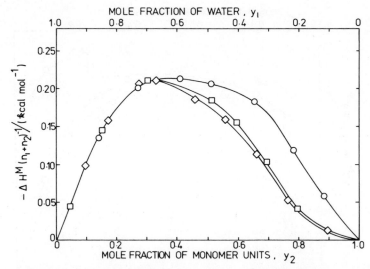

Fig. 7. Heat of mixing per mole of mixture $\Delta H^M/(n_1 + n_2)$ versus mole fractions y_1 and y_2 (where n_2 also refers to the ethylene oxide unit as the molar unit); data source and polymer symbols as Fig. 5.

havior it must be remembered that the value of the heat of mixing is a resultant of bonds broken and interactions overcome between separated molecules of the pure components, compensated for by any new interactions between the unlike components in the mixture. Thus, Fig. 7 demonstrates that at the polymer-rich extreme end, the introduction of the first few (isolated) water molecules into the polymer matrix gives rise to no net heat effect; presumably this is because for each water molecule two hydrogen bonds are broken in removing it from the water lattice and then two of slightly greater strength (compensating for the polymer–polymer interactions overcome) are formed when it is introduced into the polymer, to bridge two separate ethylene oxide units

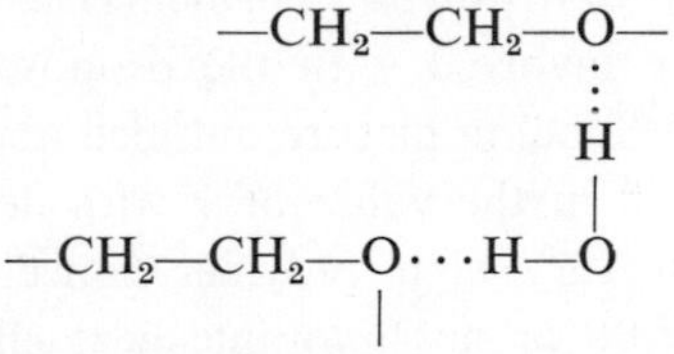

$$-CH_2-CH_2-O-$$
$$\vdots$$
$$H$$
$$|$$
$$-CH_2-CH_2-O \cdots H-O$$
$$|$$

(A similar effect, i.e., zero heat of mixing, is found with water absorption into many polar water-insoluble synthetic polymers; see Section 3.8.) On the other hand, as more water molecules are introduced (i.e., for $y_2 < 1$) there is then a net (exothermic) heat effect, presumably due to cooperative hydrogen bonding of pairs of water molecules onto the same ethylene oxide unit:

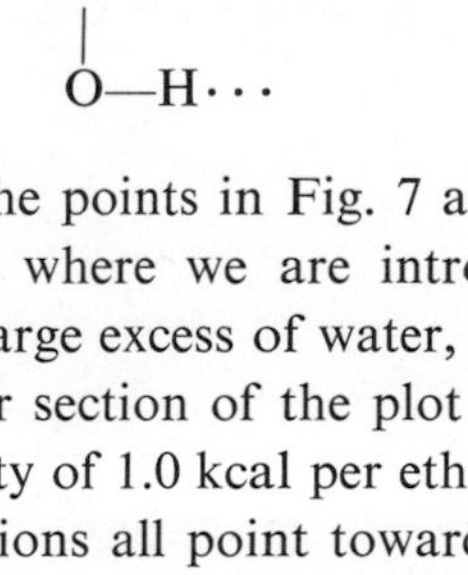

$$O-H \cdots$$
$$|$$
$$H$$
$$\vdots$$
$$-O-CH_2-CH_2-$$
$$\vdots$$
$$H$$
$$|$$
$$O-H \cdots$$

This is also indicated by the points in Fig. 7 at the other (water-rich) end of the composition scale, where we are introducing essentially isolated polymer molecules into a large excess of water, again with a net exothermic heat effect; from the linear section of the plot (Fig. 7) in this region, one can deduce an exothermicity of 1.0 kcal per ethylene oxide unit introduced.

Since these considerations all point toward a stoichiometric 2:1 complex (water:ethylene oxide), we can test this further by plotting

$$\Delta H^M/(n_1 + n_2)y_1^2 y_2$$

as shown in Fig. 8; it will be seen that this quantity is approximately constant for the two "high" polymers, at a value of -1.4 ± 0.2 kcal mol^{-1}. A more precise interpretation would require a knowledge of the equilibrium constant for the complex formation, but nevertheless this is in essential agreement with the value of 1.0 kcal mol^{-1} obtained in the previous paragraph.

Malcolm and Rowlinson[961] correlated their vapor pressure and heat of mixing data to show that the excess entropy of dilution term S_1^E is negative and increases strongly with increasing polymer concentration such that TS_1^E is always greater (i.e., more negative) than the excess enthalpy of dilution term H_1^E; this would be consistent with a degree of ordering of the water molecules (and polymer segments) in the polymer-rich mixtures such as would be involved with the proposed stoichiometric (2:1) complex. The hydrogen bonding picture outlined above would also explain the observed marked fall in the value of χ with decreasing polymer concentration (Fig. 4). At very high polymer concentrations ($w_2 \to 1$) the water molecules give little or no favorable heat effect upon entering the polymer, while their hydrogen-bonded bridging of ethylene oxide units corresponds to an ordering resulting in the high negative value for S_1^E; together, these make the value of χ high. As the polymer concentration is decreased the cooperative bonding of pairs of water molecules onto

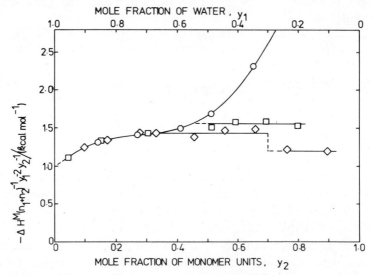

Fig. 8. Plot of $\Delta H^M/(n_1 + n_2)y_1^2y_2$ versus y_1 and y_2 (see Fig. 7), as a test for the formation of a stoichiometric 2:1 (water:ethylene oxide unit) complex; data source and polymer symbols as Fig. 5.

ethylene oxide units gives a favorable heat effect, while the lesser restriction of freedom of the water molecules also makes S_1^E less negative; together, these lead to the observed marked fall in the value of χ (Fig. 4).*

The same pair of workers also studied the phase separation behavior of the mixtures, and showed that PEG 5000 gave a lower critical consolute temperature of 128° (at $w_2 = 0.15$), PEG 3000 gave a closed loop bounded by a lower critical consolute temperature at 170° (at $w_2 = 0.25$) and an upper critical consolute temperature of about 240°, while no phase separation was observed with PEG 300. The occurrence of these *lower* critical consolute temperatures can be related to the predominance of the negative TS_1^E term in the balance between the thermodynamic parameters (entropy and enthalpy) controlling the free energy of mixing, since although at lower temperatures the exothermicity of the mixing process produces a negative value of ΔG^M, at the higher temperatures the increasingly negative TS_1^E contribution ultimately leads to a temperature at which $\Delta G^M = 0$, i.e., at the solubility limit.

The phase separation behavior of poly(ethylene oxide) in water has also been studied by Bailey and Callard,[75] who used higher molecular weight samples and who confirmed the occurrence of lower consolute temperatures. Figure 9 shows the combined lower critical consolute data of Malcolm and Rowlinson[961] and Bailey and Callard[75] presented simply as T_c versus $1/M$; there seems to be no consistency between the trends of the two sets of data. Even when the data are plotted as $(1/x^{1/2}) + (1/2x)$ versus T_c^{-1} (K^{-1}), which on theoretical grounds should give a straight line from which the ideal temperature and the related functions may be obtained (Ref. 457, Chapter 13), no single straight line can be drawn through the scatter of points obtained. It is evident that better experimental data are required on the phase separation phenomena in this system. Further, the fact that Bailey and Callard[75] observed very little difference between the critical temperature for a polymer of molecular weight 2×10^5 (which gave $T_c = 97°$) and one of molecular weight 7×10^6 ($T_c = 96°$) cannot be justifiably interpreted to mean (as they suggest) that this is the Θ point for the system, since the corresponding critical concentrations should then have been very much lower than the values of approximately 1% observed. It should be emphasized that critical solution temperatures, when observed

---

* It should be noted that the heat of mixing measurements were made at a temperature (80.3°) well above the melting point T_m of the polymer (66°). Below T_m the high crystallinity of the pure polymer would greatly modify the heat of mixing behavior, as well as reducing the hydrophilic character at the polymer-rich end of the concentration scale.

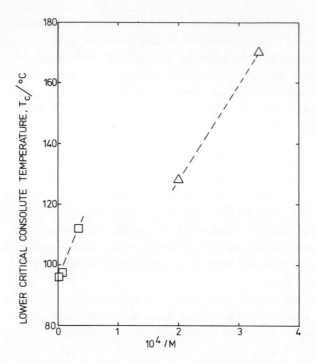

Fig. 9. Dependence of lower critical consolute temperature T_c upon the reciprocal of the molecular weight M for poly(ethylene oxide) in water; triangles, data of Malcolm and Rowlinson;[961] squares, data of Bailey and Callard.[75]

as cloud points, are very sensitive to the breadth of the molecular weight distribution of the material (particularly to the presence of very high-molecular-weight polymer), and should accordingly be carried out using closely fractionated material; the similarity of the values of T_c observed by Bailey and Callard for their two highest molecular weight samples is most probably due to appreciable amounts of much higher molecular weight material in each of these samples.

Thus the most that can be concluded from these two sets of data is that the ideal temperature of the system is approximately 90°.

The thermodynamic behavior observed by Malcolm and Rowlinson[961] for poly(ethylene oxide) in water shows certain similarities to that of other, small-molecule, ether/water mixtures. These same two workers, for example, compared it with dioxan, whose mixtures with water they studied concurrently.[961] More recently Matouš and his colleagues[992,993] have studied the system tetrahydrofuran–water and have shown that like moderately low-molecular-weight poly(ethylene oxide) (such as PEG 300), it has a closed miscibility loop, with a lower critical consolute temperature of 72° and an upper critical consolute temperature of 137°. Rowlinson[1279] has also briefly indicated how the thermodynamic behavior of poly(ethylene oxide) in water can be fitted into the framework of aqueous solutions in general. Further clarification of the thermodynamic behavior of poly-(ethylene oxide)–water systems should come from studies on other, small-molecule ethers, particularly ethylene oxide oligomers; with the latter any hydroxyl end groups should be replaced by alkoxy (preferably methoxy) to make a closer comparison.*

Some further light was cast on the behavior of this polymer in water over the full concentration range by the studies of Liu and Parsons, using the techniques of high-resolution nuclear (proton) magnetic resonance spectroscopy (nmr) and of infrared spectroscopy.[910] The nmr results showed that for PEG 400 in water (D_2O) at 35° the chemical shift δ of the methylene protons (relating to the average segmental environment of the polymer) increased linearly with ϕ_1 from $\delta = 130$ Hz at $\phi_1 = 0$ to $\delta = 138$ Hz at $\phi_1 = 0.5$, but remained essentially constant upon further addition of water; on the other hand, for chloroform ($CDCl_3$) solutions there was a steady linear increase, from $\delta = 130$ Hz at $\phi_1 = 0$ to $\delta = 142$ Hz at $\phi_1 = 1$. These results were interpreted[910] on the basis of *three* molecules of water being required for the hydration of each ethylene oxide unit; however, this interpretation is clearly in error, since for an equivolume mixture ($\phi_1 = \phi_2 = 0.5$) of the two components the molar ratio

$$H_2O:(CH_2CH_2O)$$

---

* It should be noted that a careful choice must be made of the compounds used as small-molecule analogs for the chain units in interpreting the behavior of polymers. Thus in the present case one might be tempted to use ethylene oxide itself as the analog for the poly(ethylene oxide) chain unit. However, as Glew and his colleagues have shown,[532] in the ethylene oxide + water system the ether-molecule oxygen atom is only able to form *one* hydrogen bond with water molecules, whereas in this section it is seen that for poly(ethylene oxide) there are, as expected, two hydrogen bonds formed with each oxygen in the chain. Seemingly, in the case of ethylene oxide monomer, the high strain in the ring affects the hydrogen-bond-accepting properties of the lone pairs on the oxygen atom. The same sort of situation applies to vinyl and acrylic polymers, where the monomer itself (say, $CH_2{=}CHR$) is *not* a suitable small-molecule analog (although surprisingly often used as such), both because the vinyl double bond has different interactions with a polar solvent such as water compared with those of the saturated polymer backbone it produces, and also because this double bond may, through its conjugation with the substituent group R, alter the properties of the latter; a better analog in this case would be the hydrogenated monomer CH_3CH_2R, or better still the homologous isopropyl compound $CH_3{\cdot}CHR{\cdot}CH_3$.

is in fact much closer to *two* (as calculated from molar masses of 18 and 44 g, respectively, and a polymer density of 1.2 g cm^{-3}). This latter result is in agreement with our interpretation of the heat of mixing data. The absence of any anomalies (i.e., such as nonlinearity) at the extreme polymer-rich end of the composition range is probably due to the insensitivity of the method to the exact mode of the bonding of the water molecules, since the nmr technique only looks at the average environment of the methylene protons on the chain. It thus appears from the nmr results that after the "saturation" of the two oxygen lone pairs by hydrogen-bonded attachment of two water molecules there is effectively no further alteration in the chain environment as the solution is diluted further. The fact also that the nmr spectra for the aqueous solutions were not significantly altered over the temperature range 4–80° seems to be a sign of the strength with which the 2:1 "saturation" complex is bonded; because again of the relatively small (albeit measurable) changes in δ that were observed, it is not likely that the data would show up any changes which might be occurring in the average number of water molecules attached to each chain oxygen over this temperature range.

Liu and Parsons[910] also concurrently studied the infrared spectra (2.5–4 and 7–12 μm) of poly(ethylene oxide)s in the crystalline state, and in benzene and aqueous (D$_2$O) solutions at various concentrations (in D$_2$O, 20–80%); on the basis of these spectra and of those of related compounds they suggested that the conformation of the polymer is in a more ordered form in aqueous solution than in benzene solution or in the melt, with a TGT (trans–gauche–trans) preferred conformation for the COCCOC sequence. However, it must be noted that as a consequence of the much lower general infrared transmission of aqueous samples in these regions (even with D$_2$O as solvent) the interpretation of spectra from such samples is much more difficult than it normally is with these spectra, and the conclusions that may be drawn are correspondingly less certain.

Considering the flow properties of water/poly(ethylene oxide) mixtures, Teramoto and his colleagues[1474] have measured the Newtonian viscosities for aqueous solutions of ethylene glycol, tri(ethylene glycol), Carbowax 1000 ($\bar{x} = 28$), and H-20,000 ($\bar{x} = 4 \times 10^2$) over the temperature range 10–90° and over the complete composition range (except for H-20,000, where the maximum polymer concentration studiable was $w_2 = 0.4$); the data were interpreted in terms of free-volume theory. Rodriguez and Goettler[1269] have also studied the viscosity behavior of moderately concentrated aqueous solutions of several poly(ethylene oxide) samples. However, despite the immense practical importance of these properties, the interpretation of the experimental data is still a matter largely of controversy as to the goodness of fit of various empirical mathematical equations,[1268] and does not yet seem to be at a stage where the data can be made to yield information upon what is happening at the molecular level, nor of predicting the macroscopic behavior from known molecular properties.

Turning to the behavior of poly(ethylene oxide) in dilute aqueous solution, as already indicated in Section 3.3.3, viscometry is perhaps the simplest and certainly the most widely used method for studying polymers in dilute solution. In the light, particularly, of the suggestions that the "anomalous" aqueous solubility of poly(ethylene oxide) is due to its close fit into the water lattice,[147] it is therefore timely to compare the rather extensive data available for the Mark–Houwink–Sakurada parameters K and a [compare eqn. (43)] in various solvents. As Elias[394] has shown, for a wide range of polymers plots of K against a for the same polymer in a series of solvents generally show an essentially linear falloff with increase of a in the value of K from its "ideal" value K_Θ at $a = 0.5$ which continues up to $a = 0.7$, the plot then flattening off as the normal limiting value of $a = 0.8$ is approached. Figure 10 shows such an "Elias plot" for poly-

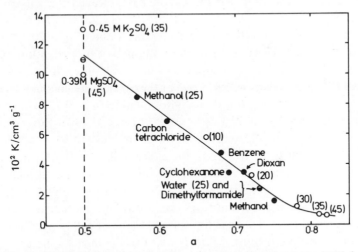

Fig. 10. Mark–Houwink–Sakurada viscosity parameters K and a [eqn. (43)] for poly-(ethylene oxide) in aqueous solution (open circles), in nonaqueous solutions (filled circles), and for the Θ state (barred circle); temperature 20°, except where otherwise indicated in parentheses. Data from Elias[394], Kurata and Stockmayer,[835] Hammes and Schimmel[597] (water at 10°), and Ring et al.[1249] (water at 25°). (We have corrected the accidental transposition of the experimental parameters[1242] for water and methanol at 20° made by Rempp in his original tabulation[1242] and copied by Elias.[394])

(ethylene oxide) in a variety of solvents; also included in the plot is the value of K_Θ quoted by Kurata and Stockmayer,[835] as well as the values for the aqueous Θ solvents 0.45 M potassium sulfate (35°) and 0.39 M magnesium sulfate (45°).[75] (It should be noted that due to the experimental scatter in the double-logarithmic plots used to obtain K and a for these last two systems[75] the ranges of uncertainty for their K_Θ values comfortably cover that of the K_Θ value quoted by Kurata and Stockmayer.[835])*

It will be seen that the points in Fig. 10, both for aqueous and nonaqueous systems, fall about a common curve of the usual form, with the water points perhaps a little higher. Considering for the moment only the nonaqueous solvents: carbon tetrachloride, benzene, cyclohexanone, and dioxan, it is significant that the values of a are in the order of their solubility parameters (compare Table I); in fact the two variables show an almost linear correlation which extrapolates for $a = 0.8$ (the normal limiting value for very good solvents) to $\delta = 11.0\ \text{cal}^{1/2}\ \text{cm}^{-3/2}$, which may be compared with the literature value (Table II) of $10.5 \pm 0.5\ \text{cal}^{1/2}\ \text{cm}^{-3/2}$ for this polymer.

Since for methanol $\delta = 14.6\ \text{cal}^{1/2}\ \text{cm}^{-3/2}$ (Table I), we might expect that since here the δ value for the polymer is exceeded the solvent power and hence the value of a will be less again; this is substantially as observed, except that the marked difference between the value of a at 20° (0.75) and at 25° (0.57) seems more extreme than might be expected (although they are each accompanied by a reasonable value of K), unless it arises from the presence of traces of, say, water in one or the other of the methanol samples. If we include the point for dimethylformamide, $a = 0.73$, which has a solubility parameter δ of 12.1 $\text{cal}^{1/2}\ \text{cm}^{-3/2}$ (Table I), then this again gives a linear plot with the 25° methanol value and $a = 0.80$, $\delta = 11.0\ \text{cal}^{1/2}\ \text{cm}^{-3/2}$. The linearity of the plots on either side of the polymer δ value suggests that the variation of a with the value of δ of the solvent may be a useful method for determining the values of δ for polymers. Unfortunately, the literal extrapolation from the dimethylformamide and methanol (25°) data would indicate that for water ($\delta = 23.4\ \text{cal}^{1/2}\ \text{cm}^{-3/2}$) the value of a should be close to zero, which is that for a compact sphere!

Since the points for water in Fig. 10 lie close to the curve for the nonaqueous systems, this suggests that in this respect water does not pre-

---

* For the rather low molecular weights of the poly(ethylene oxide) samples used in many of the viscosity studies it was necessary to include an additive constant in eqn. (43), which may be neglected for the present purposes.

sent any special features as a solvent for the polymer in these limiting dilutions, compared with nonaqueous liquids. On the other hand, the values of a and K for water do show a marked variation with temperature, although the absence of similar data for the other systems only allows us to suggest that the magnitude of this change is unusual. (Data are, however, available for the precipitation temperatures for some of the systems at finite concentration.[1432]) The steady and almost linear rise shown from 10 to 35° in the value of a demonstrates that water is becoming a markedly better solvent over this temperature range. The relatively much smaller increase in a from 35 to 45° reflects the approach of its limiting value; however, it may also reflect the approach of a maximum in the solvent power of the solvent, for since we have already noted that these systems show a lower critical consolute temperature of $\sim$90°, the solvent power of water must at some point above 45° begin to decrease with increase of temperature. It would be extremely useful to have experimental data on the viscosity of PEO at temperatures above 45°, preferably right up to at least 90°; it is likely that K and a would retrace their path along the Elias plot of Fig. 10, approaching $a = 0.5$ at what would thus definitely be the ideal temperature of the system. Such measurements would be straightforward, given reasonably close fractions of the polymer covering a fairly wide range of molecular weights, since the values of parameters at lower temperatures are so well known that they could safely be applied to the molecular weight calibration of the samples. However, it may be noted from Fig. 10 that the literature values for the parameters at 25° (as determined by Ring and his colleagues[1249]) are somewhat out of line with the others, and this is confirmed by plots (not shown) of the individual parameters against temperature, which give excellent intercorrelations at the other temperatures and which indicate that a more consistent pair of values at 25° would be $a = 0.75$ and $K = 2.2 \times 10^{-2}$ cm³ g⁻¹.

It would also be useful to extend the temperature range downward to 0°, to see whether any anomalous effects occur close to the freezing point of the solvent; the present data show an almost linear correlation between a and temperature, extrapolation of which indicates $a = 0.60$ (with $K = 9 \times 10^{-2}$ cm³ g⁻¹) at 0°, and $a = 0.50$ (with $K = 12 \times 10^{-2}$ cm³ g⁻¹) at $-15°$, the latter thus representing a hitherto unsuspected upper critical temperature (naturally, experimentally unattainable) for the system, but with a value of K_Θ closely similar to that for the lower critical emperature (see Fig. 10).

Some further light is cast on the interactions between water and poly-(ethylene oxide) in dilute aqueous solutions by ultrasonic absorption and

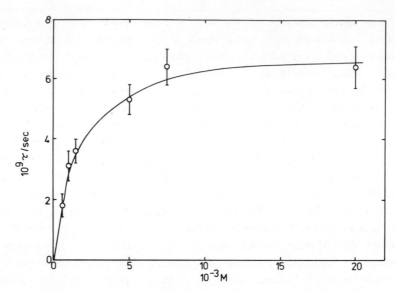

Fig. 11. Dependence of ultrasonic relaxation time τ upon molecular weight M for poly-(ethylene oxide) in water at 2.0 basemolal concentration and 10°; data of Hammes *et al.*[596,597]

velocity measurements carried out by Hammes and his colleagues.[596,597] The measurements were made at frequencies in the range 10–185 MHz on solutions which were mostly at 10° and of 2 basemolal (i.e., molal in ethylene oxide units) (about 9% w/v). The results showed there to be a single ultrasonic relaxation process over the range of conditions used. Figure 11 summarizes the dependence of the relaxation time τ upon the polymer molecular weight M for the above temperature and concentration; the value of τ fell by about 30% when the temperature was raised to 25° and rose by about 10% for $M = 7.5 \times 10^3$ and by about 25% for $M = 2 \times 10^4$ when the concentration was reduced to 1.2 basemolal. The maximum absorption per wavelength was found to be proportional to the polymer concentration.

On the basis of these results Hammes and his colleagues[596,597] suggested that the most likely mechanism for the absorption is a process depending primarily upon interactions between the solvent and the polymer. The ultrasonic sound waves act by perturbing the equilibrium between the water molecules locally interacting with the polymer and those in the bulk of the solvent. The values of the maximum of absorption per wavelength then give information upon this equilibrium which is viewed as a "chemical" process having an associated standard enthalpy change $\Delta H^\ominus$

and standard volume change $\Delta V^{\ominus}$; the data indicate that the quantity $(\Delta V^{\ominus} - \alpha \Delta H^{\ominus})$ (where $\alpha = 2 \times 10^{-4}$ cm^3 cal^{-1}) has a minimum numerical value of ± 0.6 cm^3 basemol^{-1}, so that, for example, if $\Delta V^{\ominus} = 0$, then $\Delta H^{\ominus} = \pm 3$ kcal basemol^{-1} and if $\Delta H^{\ominus} = 0$, then $\Delta V^{\ominus} = \pm 0.6$ cm^3 basemol^{-1}. The $\Delta H^{\ominus}$ value quoted is about the hydrogen bond strength, while the value of $\Delta V^{\ominus}$ is similar in numerical magnitude to that of approximately -1 cm^3 basemol^{-1} associated with the mixing of polymer and solvent.[961] Any attempt at a more precise quantitative interpretation of these results* is clearly vitiated by the ignorance even of the sign of these thermodynamic parameters, which seems to be a basic limitation in the interpretation of experimental data produced by this otherwise promising technique.

Summarizing, water and PEO interact in a manner which can be interpreted reasonably upon the basis of hydrogen bonding between the water molecules and the ether oxygens, up to a limit of a 2:1 molecular ratio. The interaction (starting from the pure liquid components) is exothermic to the extent of between 1 and 2 kcal/mol of ether oxygen, suggesting cooperative bonding of the two water molecules; for polymer-rich systems where only single bonding of water molecules to each ether oxygen is possible the mixing is athermal, which explains the apparent low hydrophilic character of the polymer in this state. The system shows phase separation with lower consolute temperatures, corresponding to a Θ point of approximately 90°. For dilute solutions (extreme dilution) the viscosity parameters show the water to become a progressively better solvent over the temperature range 10–45°, although there must be a reversal in this trend above this temperature at this or a somewhat higher temperature to give the Θ point at approximately 90°. In addition, the reduction of solvent power with fall in temperature from 45 to 10° points to an upper consolute temperature (i.e., Θ point) at about $-15°$.

Considering briefly the other polyethers, for those of the general structure $\left.\left[\left(CH_2\right)_m - O\right]\right._x$ although except for $m = 2$ these are not water soluble, it is pertinent to note that the values of K_{Θ} (which should be independent of solvent) have been determined for the compounds with $m = 1$, 2, 3, 4, 6, 10, and ∞.[418,1599–1601] Some measurements of the thermodynamic properties of water/poly(propylene oxide) systems were carried out by Malcolm and Rowlinson[961] together with their studies on poly(ethylene oxide), although the low water solubility of the polymer meant that only oligomers could be studied, which reduces the utility of the data. Willi-

---

* The interpretation of ultrasound absorption data has been discussed by Blandamer in Chapter 9, Volume 2 of this treatise.

ams and his colleagues[1586] have analyzed the water uptake by polyoxy-methylene ($m = 1$) and obtained a value of ~ 3.9 for the Flory–Huggins interaction parameter χ, and Schneider and his colleagues[1331] have studied the water absorption and transport properties of polyurethanes derived from poly(ethylene oxide), poly(propylene oxide), and poly(tetramethylene oxide).

   3.5.1b. *Poly(acrylic acid), Poly(methacrylic acid), and Their Amides* (*the "Acrylic Group"*). These four related polymers form a very interesting group which in this context it is convenient to term the "acrylic group"; this interest stems from the fact that they are all water soluble (at least at normal temperatures), while as the scheme below shows, their study as a group may serve to demonstrate the effects of chain methylation (PAA → PMAA, PAAm → PMAAm) and of side-group amidation (PAA → PAAm, PMAA → PMAAm) on the chain properties in solution.

$$\left[\begin{array}{c} -CH_2-CH- \\ | \\ CO_2H \end{array} \right]_x \quad \rightarrow \quad \left[\begin{array}{c} CH_3 \\ | \\ -CH_2-C- \\ | \\ CO_2H \end{array} \right]_x$$

Poly(acrylic acid) (PAA)   Poly(methylacrylic acid) (PMAA)

$$\left[\begin{array}{c} -CH_2-CH- \\ | \\ CO \cdot NH_2 \end{array} \right]_x \quad \rightarrow \quad \left[\begin{array}{c} CH_3 \\ | \\ -CH_2-C- \\ | \\ CO \cdot NH_2 \end{array} \right]_x$$

Polyacrylamide (PAAm)   Polymethacrylamide (PMAAm)

In this section we should be concerned with the properties of the two acids in their *undissociated* form, as ensured for dilute solutions by maintaining a low *p*H. The properties of their salts as polyelectrolytes are considered in Section 3.5.2.

   The water uptake and vapor pressure behavior of these polymers have been studied for poly(methacrylic acid) by Katchman and McLaren,[754] for poly(acrylic acid) and partially hydrolyzed poly(methyl acrylate)s, i.e., methyl acrylate/acrylic acid copolymers (as well as their sodium salts) by Hughes and Fordyce,[695] and again for poly(acrylic acid) (and its sodium salt) and for poly(methacrylic acid) by Peterson.[1155] Judging, for example, from the recent (1968) compilation by Barrie,[93] these three papers (which span the years 1951–1958) are the latest work in this area.

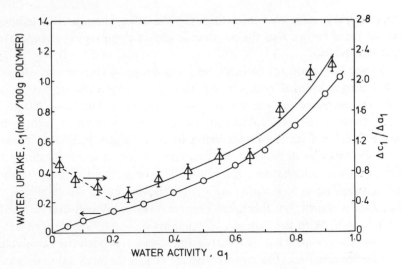

Fig. 12. Equilibrium uptake c_1 (circles) of water by poly(methacrylic acid) at $40°$ versus water activity a_1, together with also the plot of the differential coefficients $\Delta c_1/\Delta a_1$ (triangles); data of Katchman and McLaren.[754]

The most detailed study is the first cited above,[754] where the absorption of water vapor by poly(methacrylic acid) at $40°$ was studied over the wide activity range $0.05 < a_1 < 0.95$; Fig. 12 shows the data obtained for the equilibrium water concentration c_1 as a function of a_1, as well as the differential plot, i.e., $\Delta c_1/\Delta a_1$ versus $\bar{a}_1$. The latter shows up more clearly the "toe", extending up to about $a_1 = 0.2$, in the direct plot; extrapolation indicates that this corresponds essentially to an uptake of 0.05 mol of strongly bound water per 100 g of polymer, i.e., 1 mol of water per 23 or so monomer units. It is likely that this represents strong binding of water by either the end groups in the polymer or some hydrophilic impurity, but the information available is insufficient for a detailed discussion; this serves to emphasize the need for polymers to be fully characterized (even to the nature of their end groups) and purified for studies of this type. For the main part the curve shows an essentially monotonic rise of increasing slope, as would be expected from Flory–Huggins theory (compare with Fig. 2); on the other hand, the curve seems to be tending to a finite intercept of $c_1 = 1.2$ mol water/100 g polymer (which is equivalent, probably coincidentally, to 1 mole per basemole unit), whereas for this water-soluble polymer the curve should continue to rise to approach the $a_1 = 1$ axis asymptotically. However, the uptakes at these higher activities tend to be rather more liable to experimental error, and hence the extrapolation of the curve

in this region is somewhat uncertain. On the other hand, at this temperature (40°) it is known that the polymer is approaching its phase separation
point (see below).

To see how good the fit is to the Flory–Huggins theory, the values of
χ have been calculated both on the basis of the observed uptake c_1 and
upon that corrected for the initial "toe" using $c_1 - 0.05$ (mol/100 g); as
before, these have been calculated using weight fraction w rather than
volume fraction ϕ as the composition quantity, again justifiably because
of the similarity in density of the two components. Figure 13 shows the
results of these calculations; the values of χ calculated directly from c_1
show a steep rise at low values to w_1, corresponding to the "toe" on the
absorption isotherm, but thereafter become remarkably constant at a value
of 1.36 ± 0.01. With the "correct d" quantity $c_1 - 0.05$ the derived values
of χ now fall off somewhat with increasing w_1 but approach the same values
at the higher activities. The curious feature here is the high value of χ even
at the higher water uptakes, which, however, correspond to 85% polymer
and 15% water (by weight); since the polymer is water soluble, the value
of χ must at some point fall off to <0.5 as the solution is diluted further.

The uptake data for poly(acrylic acid)[695,1155] and for the methyl

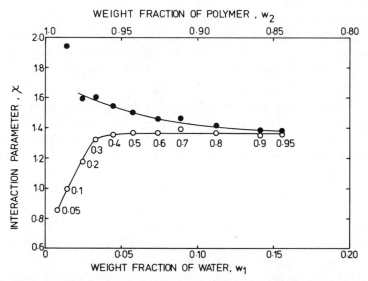

Fig. 13. Flory–Huggins interaction parameter χ for water + poly(methacrylic acid) at
40° versus w_1 and w_2 calculated from water vapor uptake using the measured uptake c_1
(open circles) and the "corrected" uptake $c_1 - 0.05$ (mol/100 g) (filled circles); from the
data of Katchman and McLaren[754] (compare with Fig. 12).

acrylate–acrylic acid copolymers[695] are not presented in a form or sufficient detail to make a similar analysis possible, although in each case they show that for the same activities the uptake is greater than in the case of poly(methacrylic acid). This seems to be an area for fruitful and more detailed study of the water uptake of all the polymers of the acrylic group, covering both the wide range of activities used by Katchman and Mc-Laren[754] as well as a much wider range of temperatures than heretofore. It would also be useful to extend the measurements to much higher water uptakes, i.e., into the water-rich region of compositions, where there may be abrupt changes in χ from the high values in the polymer-rich end to a level below 0.5 in the water-rich end.

Glasel[526] has applied the technique of deuteron magnetic relaxation to the study of the interactions between water (D_2O) and various polymeric solutes, including poly(methacrylic acid), as well as polyvinylpyrrolidone and polyvinylmethyloxazolidone (the results with these last two are discussed below under the first), and also two polypeptides and two polynucleotides. For PMAA the deuteron spin–lattice relaxation time T_1 was determined at $31 \pm 1°$ for a sample with $\bar{M} = 3 \times 10^5$, at pH 1 over the concentration range 0–5%. (Measurements were also made to pH 11, where there will be partial or complete dissociation of the polyacid; these results are therefore discussed in Section 3.5.2.) The presence of the polymer gives rise to a marked fall in the value of T_1 for the solvent, with its reciprocal $1/T_1$ rising linearly with concentration from 1.8 sec^{-1} for pure D_2O to 5 sec^{-1} for 5% polymer solution. These results are interpreted by Glasel in terms of the exchange of the D_2O molecules between the bulk of the liquid and a bound state as "water of hydration," with the observed value of $1/T_1$ being a weighted mean of the values for the two states. The relaxation rate was also measured as a function of frequency for a 1.5% polymer solution, showing that the solvent–polymer interaction was characterized by a single reorientation time of 1.6×10^{-8} sec. Any detailed interpretation of these results is made difficult by the complicating factor of chemical exchange between the hydrogen (deuterium) atoms on the carboxylic groups of the PMAA and in the water; further, as discussed below, we believe that Glasel has incorrectly interpreted the behavior of the systems where little or no change in T_1 is observed (i.e., with PVP, PVMO, and ionized PMAA), so that this must reflect in turn upon any interpretations made where definite effects are observed (as with un-ionized PMAA).

The rheological properties of moderately concentrated aqueous solutions of polyacrylamide have been studied by Rodriguez and Goettler.[1269]

We have already, with poly(ethylene oxide), indicated the limited interpretation that may at present be put upon such data.

Considering the behavior of these polymers in dilute aqueous solution, some particularly interesting light scattering and viscosity measurements on the group have been reported.[1395] Although published more than
fifteen years ago, the measurements do not seem to have subsequently
received the attention they deserve; they are, indeed, particularly valuable
because they represent studies by the same group of workers applying the
same instrumental techniques (light scattering and viscometry) to this
closely related set of polymers, so that the intercomparison of the results
is much less likely to be invalidated by the systematic errors often shown
up when comparing the results from different sets of workers each applying
their individual instrumental techniques.[473] The value of the measurements
is increased by the wide range of temperatures which was covered (maximally
8–80°, somewhat narrower for individual polymers); this enables the thermodynamic effects to be analyzed into the essentially temperature-independent enthalpic and entropic contributions.

The main results are summarized in Table III. The light scattering data
gave in each case essentially temperature-independent values of $\bar{M}_w$, indicating that association (dimerization) is not significant, at least in the limit
of extreme dilution. The second virial coefficient B was found to be linearly
dependent upon $1/T$, hence giving the values for the ideal temperature Θ
and for the enthalpy of dilution parameter $\varkappa$ $(= \chi_H)$ and the entropy
of dilution parameter ψ $(= \frac{1}{2} - \chi_S)$ for each system. It should be noted
that the parameters for PMAA correspond to a *lower* critical consolute
temperature, with $\Theta = 56°$, in agreement with the observation that this
polymer separates out of solution at about 58° or above; for the other
three polymers there is the more usual phenomenon of an *upper* critical
consolute temperature; thus for PAA, $\Theta = 14°$ and the polymer separates
out from solution at about 5° or below. The values of $\varkappa T$ and of ψ are
plotted against one another in Fig. 14, showing a close correlation between
the two quantities, particularly for PAA, PMAA, and PMAAm; the deviation for PAAm may be due at least in part to the uncertainties in the thermodynamic parameters, resulting from the value of Θ $(-38°)$ being well
outside the range of the experimental temperatures. Correlations of this
type between otherwise apparently independent enthalpic and entropic
factors are a fairly common occurrence[871]; in the present case, for example, the positive values of $\varkappa T$ correspond to endothermic dilution, i.e.,
the net effect on dilution is "breaking of bonds," while the positive values
of ψ correspond to endentropic dilution, i.e., the net effect is "greater free-

TABLE III. Light Scattering, Viscosity, and Derived Quantities for Poly(acrylic acid), Poly(methacrylic acid), and Their Amides at Extreme Dilution in Aqueous Solution[a]

Polymer	Molecular weight of sample		Light scattering measurements			Viscosity measurements		
	$10^{-5}\bar{M}_w$	$10^{-5}\bar{M}_v$	Ideal temperature Θ, K	$\varkappa T$, K	$10^2\psi$	$[\eta]_\Theta$, cm³ g⁻¹	$10^{-5}d[\eta]/dT^{-1}$, cm³ deg g⁻¹	$10^2 K_\Theta$,[b] cm³ g⁻¹
PAA[c]	11	4.3	287	18.5	6.47	111	−4.05	14 ± 3
PAAm[d]	3.9	7.1	235	66.0	28.1	45	−2.63	6 ± 1
PMAA[e]	5.9	3.0	329	−6.29	−1.92	17	0.677	2.7 ± 0.4
PMAAm	3.2	2.0	279	1.62	0.58	22	−0.419	4.4 ± 0.5

[a] Maximal temperature range 8–80°. From the data of Silberberg et al.[(1395)]
[b] The limits indicate the differences between values calculated using $\bar{M}_w$ and $\bar{M}_v$ as the molecular weight.
[c] In 0.2 M HCl as solvent.
[d] Due to the Θ temperature being well below the range of the experimental temperatures, the thermodynamic data and values of K_Θ are correspondingly less certain than those for the other polymers.
[e] In 0.02 M HCl as solvent.

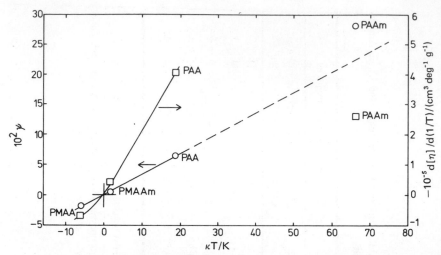

Fig. 14. Entropy of dilution parameter ψ (circles and left-hand scale) and reciprocal absolute temperature coefficient of the limiting viscosity number $d[\eta]/d(1/T)$ (squares and right-hand scale) versus $\varkappa T$ (where $\varkappa$ is the enthalpy of dilution parameter) for the acrylic group of polymers [PAA, poly(acrylic acid); PAAm, polyacrylamide; PMAA, poly(methacrylic acid); PMAAm, polymethacrylamide] at extreme dilution in aqueous solution; maximal temperature range 8–80°. From the data of Silberberg *et al.*[1395]

dom of molecules released." It should be noted that although the correlation line in Fig. 14 for the polymers except PAAm is close to a straight line through the origin, the marked differences between the three corresponding Θ values relate in fact to quite small deviations from exact proportionality. Examining the thermodynamic values of Fig. 14 for correlations with molecular structure, it is evident that, broadly speaking, chain methylation (PAA → PMAA, PAAm → PMAAm) leads to a reduction in the values of both parameters, while amidation (PAA → PAAm, PMAA → PMAAm) leads to somewhat lesser increases. If the values of the parameters are smooth functions of the extents of these two structural modifications, then it should be possible to produce AA–MAA copolymers, and MAA–MAAm copolymers (i.e., partially hydrolyzed PMAAm) which have close to zero values for both ψ and $\varkappa T$; this should produce the intriguing situation where the system is unable to decide whether to have an upper or a lower critical consolute temperature!

There are clearer correlations between Θ and the molecular structure of the polymer, despite the apparent composite basis of Θ (depending as it does both upon ψ and upon $\varkappa$). Thus, with chain methylation, $\Delta\Theta$ is $+42°$ for PAA → PMAA and $+44°$ for PAAm → PMAAm; for amidation, $\varDelta\Theta$ is $-52°$ for PAA $\rightarrow$ PAAm and $-50°$ for PMAA $\rightarrow$ PMAAm. The closeness of the pairs of values obtained for each of the two types of structural change is quite remarkable, the $2°$ differences being well within the experimental limits for the individual Θ values. These correlations, unless they are coincidental, serve both to enhance our confidence in the consistency of the set of experimental data and also act as a pointer toward the prediction of solubility points of polymers from their molecular structure. However, such predictions must naturally also include information on whether it is a lower consolute temperature (as with PMAA) or upper consolute temperature (as with the other three polymers) that is involved.

The data for the limiting viscosity number $[\eta]$ obtained by Silberberg and his colleagues[1395] parallel those of the light scattering studies, especially in regard to the variation of $[\eta]$ with temperature; for PAA, PAAm, and PMAAm the value of $[\eta]$ increases with temperature while with PMAA it decreases. In each case except PAAm the value of $[\eta]$ was found to vary linearly with $1/T$; the slopes of such plots are shown as functions of $\varkappa T$ in Fig. 14 (squares), and again show a close correlation between these two quantities. The correlation is understandable when we recollect that the value of $[\eta]$ represents the hydrodynamic volume of the polymer coil, so that if the dilution is, say, endothermic ($\varkappa T$ positive), then increase of temperature will favor segment–solvent contacts over segment–segment and solvent–solvent contacts, and the polymer coil will thus expand. To this extent, study of the variation of $[\eta]$ with temperature may be viewed as a partial substitute for the technically more difficult light scattering method in thermodynamic studies of this type.

From the variation of $[\eta]$ with temperature and the Θ values obtained from the light scattering studies it is then possible to obtain by extrapolation the values of $[\eta]_\Theta$ (see Table III); as before, the low value of Θ for PAAm as well as the curvature of the plot of $[\eta]$ against $1/T$ makes the derived value of $[\eta]_\Theta$ rather uncertain. Following from this, from eqn. (44) and using the known molecular weights, the K_Θ values can also be calculated. Some more uncertainty is involved here because of the differences between the molecular weights deduced from light scattering ($\bar{M}_w$) and those from viscometry ($\bar{M}_v$); the values of K_Θ in Table III are the means of those obtained by the two methods. The values obtained for PAA, PAAm, and PMAAm are similar to those observed for other carbon chain polymers in solution[835]; it is noticeable that in the case of PAA amidation leads to a halving of K_Θ. Since K_Θ corresponds to the unperturbed conformation of the chain, where the effect of interactions between distantly connected parts of the same chain has been canceled out, any remaining variations must result from interactions between near units on the polymer chain; possibly in the cases of PAAm and of PMAAm, where the values of K_Θ are somewhat low, there is some hydrogen bonding between the amide groups which compacts the conformation. The value of K_Θ for PMAA, however, is very low (being about one-fourth of the expected value), and reflects the compact conformation possessed by this polymer in solution, which also shows itself in the conformational transition that takes place when it is titrated with base (see Section 3.5.2).

The viscosities reported by Silberberg and his colleagues may be compared with those obtained by other groups of workers studying the individual polymers. Thus Scholtan[1339] has studied the behavior of polyacrylamide in aqueous solution at 25° and obtained the Mark–Houwink–Sakurada parameters: $K = 0.63 \times 10^{-2}$ cm^3 g^{-1} and $a = 0.80$; the value of the index a indicates the polymer coil to be highly expanded (at essentially the upper limit for a random coil type conformation) while the low value of K (about one-tenth that of K_Θ—see Table III) is very much that expected at this end of the scale of a. The "goodness" of water as a solvent for polyacrylamide at this temperature which these parameters indicate is reasonable in the light of it being well above the ideal temperature $(-38°)$ for the system (see Table III).

The widely different Mark–Houwink–Sakurada parameters reported subsequently for this same system by Collinson and his colleagues,[270] i.e., $K = 6.8 \times 10^{-2}$ cm^3 g^{-1} and $a = 0.66 \pm 0.05$, do not seem valid by comparison; most probably these values are in error, because they refer to the number-average molecular weights of unfractionated samples of polymer, and also because even these molecular weight values were obtained on the basis of a presumed kinetic mechanism (for the X- or γ-ray-irradiated polymerization in water terminated by acidified ferric perchlorate) rather than being determined directly. It appears that the remarkably high value of 28.0×10^{-2} cm^3 g^{-1} for K_Θ for this polymer quoted in the tabulation of Kurata and Stockmayer[835] is based largely upon this study of Collinson and his colleagues,[270] and hence this value of K_Θ should be disregarded.

For PMAA Katchalsky and Eisenberg[752] have reported viscosity data on fractions at 30° in 0.002 M aqueous hydrochloric acid (their concurrent light scattering measurements showed that the turbidity of the solutions became independent of acid concentration at about this concentration, indicating complete suppression of the polymer acid ionization); the logarithm of the $[\eta]$ values gave the expected linear plot with the logarithm of the degree of polymerization of the polymer samples (Ref. 752, Fig. 7) for which the original authors reported a Mark–Houwink–Sakurada exponent of $a = 0.5$ exactly. Reexamination of this plot suggests that a somewhat higher value of 0.57 for a gives a better fit, and this in turn gives $K = 2.8 \times 10^{-2} \, cm^3 \, g^{-1}$. These are values in reasonable agreement with the data presented in Table III, and they confirm the compactness of the polymer molecules even at this temperature, which is still appreciably below the ideal temperature ($+56°$) for the system.

Summarizing for the acrylic group of polymers, for concentrated aqueous solutions the available water uptake data indicate a high value of χ for PMAA up to the maximum uptake studied, 15% by weight of water. The limited data for this and the other polymers in the group need to be supplemented by more detailed measurements extending well into the water-rich area of the composition range, and also over a wide range of temperatures to enable the enthalpic and entropic contributions to the mixing process to be disentangled. More detailed data are available for extremely dilute solutions, as displayed in Table III. For PAA and PMAAm, water at normal temperatures is a rather poor solvent since the systems have Θ temperatures of $+14$ and $+6°$, respectively. For PMAA the system shows lower critical consolute behavior, with a Θ temperature of $+56°$, while even at normal temperatures ($30°$) water is still a poor solvent for the polymer, whose molecules have a markedly more compact conformation than would be expected at temperatures some distance from the Θ point. For PAAm, water at normal temperatures is a fairly good solvent, since the system has a relatively low Θ value ($-38°$). The Θ values for the four polymers show a close correlation with the structural differences between one polymer and another, chain methylation leading to Θ increasing by $43 \pm 1°$, and amidation of the carboxylic acid group leading to a decrease of $51 \pm 1°$. It is apparent that the study of this group of polymers in aqueous solution should provide a fruitful area for elucidating its interesting but so far enigmatical behavior and relating it to the molecular structures of the polymers.

3.5.1c. *Poly(vinyl alcohol) (PVA) and Related Polymers*

$$\left[CH_2 - CH \underset{\displaystyle OH}{\underset{|}{}} \right]_x \quad (PVA)$$

This polymer is interesting from several aspects. In particular, next to its isomer poly(ethylene oxide), it is the simplest-structured water-soluble synthetic polymer, while the hydroxyl groups confer on it both donating as well as accepting power for hydrogen bonds; at the same time it is of some considerable technological and industrial importance. On the other hand, it is clear from the monograph of Pritchard[1192] that it is rather poorly characterized from many of the more fundamental viewpoints.

Part of this situation seems to stem from the fact that the polymer must necessarily be produced by chemical treatment of another polymer, most commonly by hydrolysis of poly(vinyl acetate); in this specific instance, the hydrolysis is difficult to take to completion, and any residual acetate groups can lead to marked alteration in the polymer's properties.

In addition, such properties of the solid polymer as its water absorption and solubility are complicated by the variable degrees of crystallinity of samples, while the properties in aqueous solution are complicated by gelation and recrystallization phenomena.

The water absorption properties of PVA have been reviewed by Barrie[93] and Pritchard.[1192] Figure 15 shows smoothed values for water uptake at 25° (taken from the compilation of Barrie[93]) measured for a plasticized film of the polymer (of unstated origin, and unstated degree and nature of the plasticization) by Myers and his colleagues.[1075] The isotherm seems to have the form typical of simple solution following the

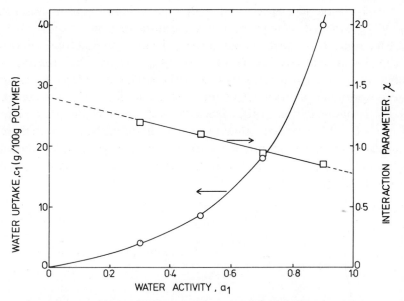

Fig. 15. Equilibrium uptake of water c_1 (circles and left-hand scale) by poly(vinyl alcohol), and derived values of the interaction parameter χ (squares and right-hand scale) versus water activity a_1. From the data of Myers et al.[1075] (see also Barrie[93]).

Flory–Huggins behavior. To test whether this is in fact so, the values of χ have been calculated (as before, on a weight fraction basis) and are also plotted in Fig. 15; this shows that χ is not constant, but falls almost linearly with water activity (the corresponding plot of χ against w_1 is of course much more highly curved at the low uptake end). The extrapolated limiting values for χ in this case are $\chi = 1.4$ for the pure polymer ($a_1 \to 0$) and $\chi = 0.8$ for polymer in contact with water, which corresponds to a finite, albeit large, equilibrium uptake. Considering the temperature variation of absorption, the limited data available[695,1075] would suggest that at 40° the uptake is somewhat less than that for the same water activity at 25°, which means that the solution process (liquid water + polymer $\to$ solution) is somewhat exothermic.

In the above calculations of χ for the polymer it has been taken to be completely amorphous (since the sample was stated to be plasticized) and also the molecular weight has been taken to be effectively infinite. Sakurada and his colleagues[1298] have investigated the water vapor uptake of two PVA samples of known molecular weight, and also of appreciable crystallinity; the values of χ calculated from the data, taking into account the finite values of molecular weight and estimated values of crystallinity, show similar behavior to those in Fig. 15.*

The effects of the nature of the ester group in the parent polymer and the method used for its polymerization on the properties of the derived PVA have been studied by Cooper and his co-workers[280]; the ester groups included acetate, monochloroacetate, dichloroacetate, trichloroacetate, and trifluoroacetate. The parent polymers were prepared by a variety of techniques, mainly either free-radical (benzoyl peroxide) initiated polymerization at 44°, or γ-ray initiated polymerization at 20, -20, or $-80°$. In the case of the parent polymers use of a lower polymerization temperature led to a marked increase in crystallinity; however, for a given ester there were no detectable differences between the rates of hydrolysis for polymers prepared at different temperatures, nor any detectable differences between the degrees of crystallinity of the PVA's produced. On the other hand, the aqueous solubility of the PVA did depend markedly upon both the type of ester grouping in the parent polymer and the temperature of its preparation; the dissolution temperature increased both with increase in the degree of halogenation of the ester methyl and with decrease in the temperature

---

* Valentine[1509] has shown how, by extrapolation from the water absorption behavior of fully amorphous poly(vinyl acetate)s of a range of degrees of hydrolysis, one can estimate the water absorption isotherm for fully amorphous poly(vinyl alcohol).

of polymerization, so that at the extremes, whereas the PVA prepared from "50°" poly(vinyl acetate) has a dissolution temperature of 76°, that from "−20°" poly(vinyl trifluoroacetate) does not dissolve until 126°. These effects for PVA need to be borne in mind not only for this polymer, but also for the other water-soluble polymers such as PAA and PMAA which may be produced by the hydrolysis of various parent ester polymers.

Considering dilute solutions of PVA, Fig. 16 shows the published values of Mark–Houwink–Sakurada viscosity parameters for aqueous solutions, as well as a partially aqueous (water/phenol mixture) and a nonaqueous one (dimethylsulfoxide). The data plotted have been collected in a deliberately rather uncritical fashion from various compilations[834,835, 1192] to show the wide divergences in the published pairs of parameters; a more critical examination suggests that for temperatures in the region 25–30° the "best" pair of values is $K = (6.0 \pm 1.5) \times 10^{-2}$ cm^3 g^{-1} and $a = 0.63 \pm 0.01$; this pair is plotted as the doubled circle in Fig. 16. The values for 80°, since they only represent one study, must be treated with some reserve in the light of the scatter of the literature values for 25 and

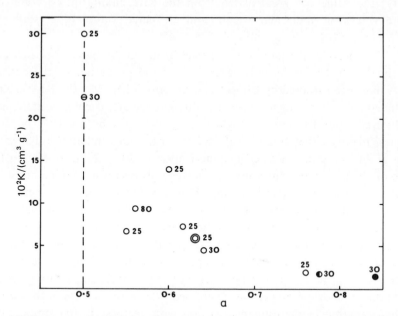

Fig. 16. Literature values of the Mark–Houwink–Sakurada viscosity parameters K and a [eqn. (43)] for poly(vinyl alcohol) in water (open circles), in 15/85 v/v water + phenol (half-filled circle), in dimethylsulfoxide (filled circle), and in the Θ state[835] (barred circle), at the indicated temperatures; the data are taken from various compilations.[834,835,1192] The doubled circle indicates the suggested "best" value for aqueous solutions at 25–30°.

$30°$, while the K_Θ quoted by Kurata and Stockmayer[835] must be similarly viewed since it is apparently based upon the results of the earlier studies. It is noteworthy that in water/phenol and in dimethylsulfoxide the polymer is apparently much more expanded than in aqueous solution; in the latter case the value of a is also markedly higher than the normal limiting value of 0.8. However, it should be clear that it is hazardous to try to draw firm conclusions from these unconfirmed sets of data, particularly in view of the marked divergences between the results obtained for water.

Since PVA is commonly produced by the hydrolysis of poly(vinyl acetate), the product may still contain residual acetate groups; similarly, deliberate partial hydrolysis leads to copolymers of vinyl alcohol and vinyl acetate. The behavior of these copolymers, particularly toward water, is therefore of both practical and theoretical interest.[1270] Considering the variation in water solubility starting with poly(vinyl acetate), up to 50% hydrolysis (i.e., with half of the acetate groups converted into hydroxyls) the material is insoluble in water (cold or hot). At about 70% hydrolysis it is soluble in cold water and in aqueous ethanol and aqueous acetone but insoluble in hot water (i.e., it shows a lower consolute temperature). Between about 75 and 80% hydrolysis the material loses its solubility in the mixed aqueous solvents. Between about 80% and 90% hydrolysis it is soluble both in cold and in hot water. Finally, for 98–100% hydrolysis the product only swells in contact with cold water but dissolves in hot water to form a solution which, if sufficiently concentrated, sets to a gel on cooling.

One of the main methods of chemical modification applied to PVA, particularly to render it water insoluble, is reaction with aldehydes which form acetals with pairs of hydroxyl groups on the polymer. With methoxyacetaldehyde, the polyvinylmethoxyacetals are produced, which have the repeat unit

$$-CH_2-CH-CH_2-CH-$$
$$O \qquad\qquad O$$
$$CH$$
$$CH_2OCH_3$$

and may still be soluble at normal temperatures. Dole and Faller[355] have studied the water absorption of three such polymers, which had 48, 72, and 86% of the hydroxyl groups so replaced; we will call these PVMA 48, PVMA 72, and PVMA 86. The polymers were water soluble at room temperature and down to 0°, but precipitated from solution when heated to 52, 36, and 35°, respectively. Figures 17 and 18 show the values of the interaction parameter χ calculated from the water uptake data, plotted, respectively, against the water activity a_1 and w_1. The first set of plots (Fig. 17) is useful in enabling extrapolations of the values of χ to be made to $a_1 = 1$ (for which $w_1 = 1$, i.e., very dilute polymer solutions); in each case in the method of extrapolation one assumes that there is no marked change in behavior over the extrapolation interval. Certainly, it is curious that for the extrapolation to $a_1 = 1$ it is only for PVMA 72 that the value of χ obtained (0.3) is less than the "critical" complete solubility value of 0.5; even for PVMA 48, which is apparently the most hydrophilic at the lower water activities, the limiting value for χ is 0.65, while that for PVMA 86 is 1.15. Figure 18 then serves to emphasize the relatively low values ($w_1 = 0.35$ and $w_1 = 0.19$) of the maximal uptakes for $a_1 = 0.95$ and the width of the concentration gap spanned in the extrapolation to $a_1 = 1$. The shapes of the plots in Fig. 18, with the value of χ rising relatively sharply to a maximum and then falling off again more gradually, probably represent first the hydration of the residual hydroxyl groups (the weight fractions of water at the maxima, representing complete occupancy of these groups, are in the expected order: PVMA 48 > PVMA 76 > PVMA 86), followed by a similar falloff in χ with water content to that shown

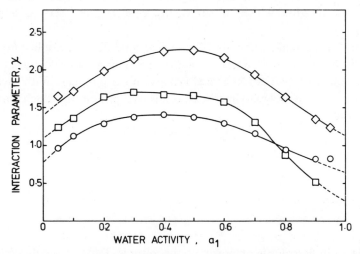

Fig. 17. Interaction parameter χ for water at 25° with the polyvinylmethoxyacetals PVMA 48 (circles), PVMA 72 (squares), and PVMA 86 (diamonds), versus water activity a_1 [the polymer serial numbers indicate the degree of acetylation of the parent poly(vinyl alcohol)—see text]; calculated from the water vapor uptake.[355]

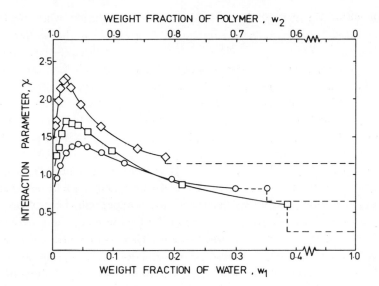

Fig. 18. Interaction parameter values of Fig. 17 plotted against w_1 and w_2; the broken horizontal lines represent the limiting values of χ for $a_1 \rightarrow 1$ (and hence $w_1 \rightarrow 1$) obtained from Fig. 17.

by the other polyether, poly(ethylene oxide) (compare with Fig. 4, where the concentration axes are reversed).

3.5.1d. *Polyvinylpyrrolidone* (PVP)

$$\left[\text{CH}_2\text{—CH—} \right]_x$$

(PVP)

This synthetic polymer is a very significant member of the group of water-soluble polymers considered here, particularly because of its high polarity, arising from the imide group present in the pyrrolidone ring of each repeat unit. The dipole moment of the ring is likely to be much the same as that for N-methylpyrrolidone, which is 4.07 ± 0.04 D.[449,868] Assuming for simplicity that this dipole results from fractional positive and negative charges q centered directly upon the nitrogen and oxygen atoms, then the value of the moment corresponds to 0.4 of an electronic charge on each site:

$$-CH_2-CH-$$

$$N^{q+} \qquad O^{q-}$$

$$CH_2 \qquad C$$

$$CH_2----CH_2$$

The structure of PVP with an amide group (or more strictly, an imide group) in each chain unit together with hydrophobic methylene and methine groups, thus shows certain similarities to the polypeptide chain of proteins, where, however, the amide groups are in the backbone of the polymer and any hydrophobic groups pendant from it. These similarities of structure between the two as well as similarities in behavior such as protective action toward colloids and dispersions (see Section 3.11), and PVP's marked reversible binding power toward small-molecule solutes (see Section 3.6.1), has led to suggestions that it may be viewed as a model compound for the proteins.[729,1030]

One further interesting feature of PVP is the fact that it is soluble not only in water but also in a wide range of organic solvents.[34] Figure 19 illustrates this in the form of a "solubility map," where the solubility parameter δ for the liquid is plotted against the "hydrogen bonding index" (HBI) as used by Beerbower and his colleagues (see Section 3.4). The chain-dotted line in Fig. 19, which delineates the solvent/nonsolvent boundary, shows that solvent power correlates quite well with the two plotted parameters, except that the ethers, esters, and ketones show lower solvent power than might be expected from their HBI values, and the partially chlorinated alkanes show higher solvent power; these anomalies indicate that the solvent power for the polymer is more closely connected with the hydrogen-*donating* ability of the liquid (rather than the hydrogen-*accepting* power with which the value of HBI is seemingly correlated[105]) presumably because of the hydrogen bonding which takes place onto the carbonyl oxygens of the polymer. However, it should be noted that in two-phase systems comprising water and a water-immiscible but hydrogen-bond-donating solvent such as methylene chloride, the polymer is partitioned almost completely into the aqueous layer, demonstrating the high specific solvent power of water.[782]

Further information upon the general interactions between PVP and liquids (including water) was obtained by Breitenbach and Schmidt,[178]

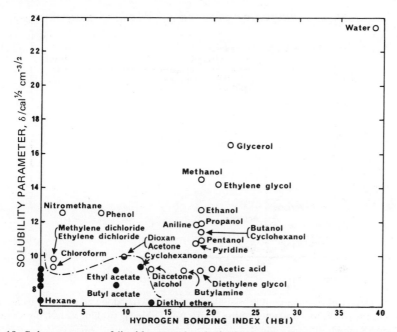

Fig. 19. Solvent powers of liquids toward polyvinylpyrrolidone at normal temperatures (for phenol, at 50°), mapped in terms of δ, the solubility parameter, and HBI, the "hydrogen bonding index."[105] Open circles represent solvents, filled circles nonsolvents, and half-filled circle "borderline" liquids.[34] The four nonsolvent points not labeled at the lower left are for (bottom to top) cyclohexane, toluene, carbon tetrachloride, and benzene.

who studied the swelling of cross-linked samples of the polymer. Figure 20 shows the correlation between the degree of swelling Q (for a polymer of fixed extent of cross-linking) and the solubility parameter of the liquid; the value of Q was taken as the ratio of the mass of swollen polymer plus liquid to that of the unswollen polymer, so that it is then related (see Section 3.3.4) to the weight fraction w_2 of polymer in the swollen gel by

$$Q = 1/w_2 \qquad (55)$$

In place of the normal behavior (see Section 3.4.2), where there is a maximal value of Q at some fairly definite value of δ_1 (which then represents the value of δ for the polymer), Fig. 20 shows that in the present instance there is a threshold value for δ_1 of about 10 cal$^{1/2}$ cm$^{-3/2}$ below which there is little swelling and above which the value of Q rises sharply to reach an essentially constant value of about 18 (equivalent to $w_2 = 0.053$ in the gel) maintained even up to the δ_1 value for water. The only markedly anomalous point on the plot is that for chloroform, where the high value of Q (higher

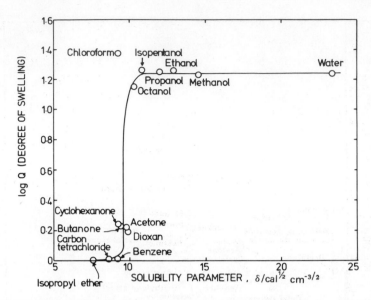

Fig. 20. Degree of swelling Q of cross-linked polyvinylpyrrolidone gels in various liquids, versus δ, the solubility parameter of the liquid (see Table I); swelling data from Breitenbach and Schmidt.[178]

even than that for the alcohols or water but falling into line with them when put on a *volume* rather than *weight* basis) is indicative of its very good solvent power, presumably through hydrogen bonding with the polymer. Judging from the values of Q, therefore, the shorter-chain alcohols (up to pentanol) and also chloroform should have very much the same solvent power, and also value of χ, toward PVP as water has. It should be noted, in the light of the marked changes in χ with composition evident for these types of systems, that the maximal degree of swelling corresponds to $w_2 \approx 0.05$, which represents a fairly dilute polymer solution.

Breitenbach and Schmidt[178] also studied the temperature dependence of the degree of swelling in water and other liquids. For water, Q fell markedly with rise of temperature, from 18.4 ± 0.4 at $20°$ to 14.9 ± 0.3 at $50°$ and to 12.4 ± 0.3 at $70°$, indicating that at these compositions ($w_2 \approx 0.06$) the mixing process is exothermic. A plot of Q against temperature is essentially linear, and the long extrapolation to $Q = 1$ (representing zero swelling) gives a temperature of approximately $170°$, which may be taken as a rough estimate of the corresponding lower critical consolute temperature for an aqueous solution of the polymer.

Considering more specifically the thermodynamics of the interaction of water with the polymer, water absorption isotherms have been determined by several workers[355,727,1457] and the freezing behavior of aqueous solutions of the polymer has also been reported.[726] Of these first three studies, only Dole and Faller[355] have presented data in a tabulated form suitable for further evaluation; Fig. 21 shows their absorption data (for 25°), together with a differential plot which reveals more clearly the initial "toe" extending up to $a_1 = 0.1$. This "toe" represents approximately 0.1 mol of strongly bound water per 100 g of polymer, and most probably arises either from the end groups or other highly hydrophilic impurities in the polymer; in the former case, for example, with two such end groups per polymer chain, it corresponds to a polymer molecular weight of only 10^3, a rather small value but one upon which no further comment can be made because the polymer's molecular weight was not specifically stated.

This anomaly aside, the isotherm of Fig. 21 shows the monotonically rising form of increasing slope which is characteristic of simple solution formation according to the Flory–Huggins theory (compare Fig. 2). To see how well this theory fits, the values of χ have been calculated both from the direct values of the water uptake c_1 and from the values $c_1 - 0.10$ (mol water/100 g polymer) corrected for the initial toe; as Fig. 22 shows, with the direct values there is a steady rise up to a value of 0.58 for χ, while the use of the corrected values has led to some apparent overcom-

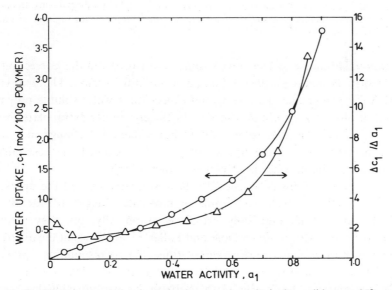

Fig. 21. Equilibrium absorption isotherm for water by polyvinylpyrrolidone at 25°, uptake c_1 (circles and right-hand scale), and differential coefficients $\Delta c_1/\Delta a_1$ (triangles and left-hand scale) versus water activity a_1. Data of Dole and Faller.[355]

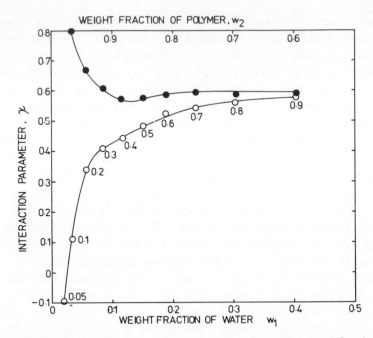

Fig. 22. Interaction parameter χ for water with polyvinylpyrrolidone at 25°, calculated from measured vapor uptake c_1 (open circles) and from "corrected" uptake $c_1 - 0.10$ (mol water/100 g polymer) (filled circles), versus w_1 and w_2; the numbers against the points show the corresponding values of the water activity a_1 (compare with Fig. 21). From the data of Dole and Faller.[355]

pensation, although in fact lower values than 0.10 for the correction do not lead to any more marked flattening in the initial region. The final value of 0.58 for χ is greater than we would expect for a water-soluble polymer; in fact, a plot of χ against a_1 also gives a closely similar extrapolated value for $a_1 = 1$, showing that some marked but unheralded change in χ must occur beyond the limiting uptake of $w_1 = 0.4$ reached in these uptake studies, since the polymer is of course water soluble.

Tadokoro and his colleagues[1457] have similarly studied the uptake of water by the polymer, on a sample of molecular weight 10^6, at 30, 40, and 50°; the results were only presented graphically, but they seem to agree fairly well with those of Dole and Faller[355] at 25° when the difference of temperature is taken into account—thus, the corresponding uptake at $a_1 = 0.9$ was about 80% of that of Dole and Faller.[355] The isotherms also showed a marked "toe," which can probably be attributed in this case to residual traces of the methanol used as solvent in casting the polymer film. The uptake at fixed activity increased with rise in temperature, which in this case indicates *endothermic* mixing (liquid water plus polymer) over the concentration range $w_1 = 0$ to $w_1 = 0.4$; this should be contrasted with the *exothermic* mixing deduced from the swelling data,[178] which correspond to much more dilute systems, $w_2 \approx 0.06$. Within the limitations of conclusions arrived at from these graphically presented data[1457] it seems that over the range $a_1 = 0$ to $a_1 = 0.5$ there is an essentially constant increment of $12(\pm 3)\%$ in the w_1 values (at fixed a_1) for a $10°$ temperature rise; at higher activity values the increment falls off essentially linearly with a_1, extrapolating to a small, possibly negative value for $a_1 = 1$. It is significant that the break in behavior at $a_1 = 0.5$ corresponds to a water uptake of essentially 1 mol/basemol polymer (compare Fig. 21), and that it also corresponds to a sharp change in the trend of χ values with composition (compare Fig. 22). It appears therefore that the transfer of water molecules from the liquid into the solution is endothermic up to the formation of a 1:1 mol (water:vinylpyrrolidone unit) ratio, at which point we would expect the water molecules to be bridging the carbonyl groups of the pyrrolidone rings:

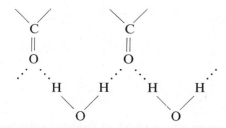

Although one might expect this transfer to be thermally neutral, it seems that the high dipole–dipole forces between the imide groups make it endothermic. Beyond this point further water molecules simply insert themselves into the hydrogen-bonded structure.

Worley and Klotz[1595] included PVP in the series of solutes whose effect on the structure of water they studied by near-infrared spectroscopy. The "water" used was in fact a 6 M solution of HDO in D_2O, for which there were absorption bands in this spectral region at 1.556 μm (λ_1) and 1.416 μm (λ_2); with increase in temperature the absorbance A_1 at λ_1 decreased and that at λ_2, A_2 increased, in such a fashion that the absorbance ratio R ($= A_1/A_2$) was a direct measure of the degree of hydrogen bonding in the liquid. The effect of addition of a solute on the hydrogen bonding may then be characterized in terms of a normalized absorbance ratio $\hat{R}$ given by R_c/R_0, where R_c is the ratio in the presence of the solute at concentration c and R_0 is that for pure solvent. For most inorganic salts

$\hat{R} < 1$, showing that they had a structure-breaking effect; however, for organic salts and particularly those such as sodium propionate and butyrate and tetraethylammonium and tetrabutylammonium salts, $\hat{R} > 1$, showing that they had a net structure-making effect. For PVP, also, $\hat{R} > 1$; thus a 32.8% by weight (4.4 basemolal) solution (for which thus $w_1 = 0.672$) had an $\hat{R}$ value of 1.16 at 25°, comparable to that of the named salts at $2\ m$; this indicates that the polymer has a structure-making effect on the water, which, when compared with the effects of the small-molecule salts, is reasonable in the light of the fact that each monomer unit contains a partial anionic center (the oxygen atom), a substituted partial cationic centre (the nitrogen atom), and also hydrophobic alkyl groups. In terms of the "structural temperature" of the solvent the presence of the polymer at the stated concentration causes a reduction of about 10°.

As mentioned above in the discussion of the "acrylic group" of polymers, Glasel[526] included PVP and also the related polymer polyvinyl-methyloxazolidone (PVMO)

(PVMO)

in the series of polymers whose effect in aqueous solution upon the solvent (D_2O) he studied by deuteron magnetic relaxation measurements. Three different samples of PVP (with $\bar{M} = 1 \times 10^4$, 4×10^4, and 3.6×10^5, respectively) and one of PVMO (with $\bar{M} = 4 \times 10^4$) were used in the studies. At the experimental temperature of $31 \pm 1°$ all four samples gave closely similar results, the reciprocal relaxation time $1/T_1$ showing only a relatively small linear increase with polymer concentration, from 1.9 sec^{-1} for pure D_2O to 2.3 sec^{-1} for 10% polymer. Glasel[526] deduced from this that there is therefore little or no interaction between the polymers and the solvent. This is, of course, a most surprising deduction, particularly in the light of other experimental results presented in this section, and more surprisingly so in the light of both PVP and PVMO being readily water soluble, which would not be so if they did not show marked interactions with the solvent. However, a closer examination of Glasel's paper[526] reveals a serious flaw in his argument; to quote Glasel (Ref. 526, p. 377): "...The extremely low slope of the plot of relaxation rate [i.e., $1/T_1$] vs. concentration indicates vanishingly small interaction of D_2O with these polymers. To be precise, either W [the weight of polymer-associated water per gram of polymer] is near zero, or $(1/T_1)_a$ [the value of $(1/T_1)$ for polymer-associated water] is essentially that of bulk water. Thus, there is no nmr evidence of interaction of water with these polymers. . . ." The crucial point here is the ". . .either. . .or. . ." in the second sentence, where Glasel seemingly misinterprets the second alternative to mean that the water close to the polymer *is* bulk water. On the other hand, in the light of the already highly restricted rotation of water molecules in the bulk, there seems nothing objectionable to the polymer-associated water molecules having much the same degree of rotational restriction; this interpretation, i.e., on Glasel's terminology, $W > 0$ and $(1/T_1)_a \approx (1/T_1)_{\text{free}}$, seems to be a much more reasonable one than Glasel's. Indeed, Glasel states earlier in his paper (Ref. 526, p. 375) that, ". . .the rotational reorientation time of the [polymer-] associated D_2O molecules. . .may be either longer or shorter than for bulk water. . .," so that one could readily arrive at values which are essentially the same by a cancellation or compensation of effects. We have dealt with this point in some detail because it is an important one, since it must reflect upon the interpretation of the results Glasel obtained with the biopolymers that he also studied by this same technique.

Considering now specifically the behavior of PVP in dilute solution, Fig. 23 shows the literature data[168,557] for the Mark–Houwink–Sakurada viscosity parameters K and a. The parameters for water as solvent show a high degree of scatter, even for the same temperature, as witness the differences between the results of the three determinations at 25° and of the two at 30°; even for the three different temperatures (20, 25, and 30°) the parameters should not in fact be greatly different, judging from the effect of temperature upon $[\eta]$ alone; thus Levy and his colleagues[883] found that for a particular sample of the polymer in water the value of $[\eta]$ fell by only 5% when the temperature was raised from 25 to 30°. This general scatter may be attributable in part to the effect of heterogeneity in the molecular weight distribution, but whatever the reason, it shows that viscometry in aqueous solution cannot at present be considered as an accurate means of determining the molecular weights of samples of the polymer. On the other hand, the closeness of the values for chloroform and for methanol is in line with the similarity in the swelling effect of these liquids upon the cross-linked polymer (compare Fig. 20), particularly when this is put on a volume rather than a weight basis; this suggests that one or other of these two would be more suitable than water for molecular weight determinations, although naturally it would be useful to have these

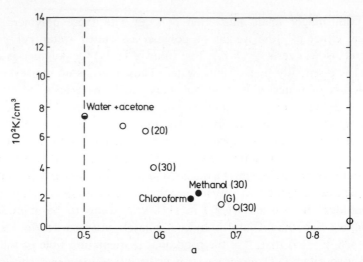

Fig. 23. Literature values of the Mark–Houwink–Sakurada viscosity parameters K and a [eqn. (43)] for polyvinylpyrrolidone in water (open circles), water + acetone mixture (33.2:66.8% v/v) (half-filled circle), and nonaqueous solvents (filled circles); temperature 25°, except where otherwise indicated. The point labeled (G) represents the parameters obtained by Graham:[557] the remaining data are from the compilation in the "Polymer Handbook."[168]

isolated determinations for each solvent medium checked again. The fact, indeed, that water seems, from swelling measurements, also to have much the same solvent power as methanol and chloroform (Fig. 20) suggests that its viscosity parameters should be much the same as those for these two; if a choice has to be made for a pair of parameters to be used, then those of Graham[557] [represented by the point labeled (G) in Fig. 23] would seem to be the best, because of their proximity to those for methanol and chloroform, although even Graham[557] expresses reservations concerning the interpretation of $[\eta]$ in terms of molecular weight for this system.

The osmotic pressure of dilute aqueous solutions of PVP has been studied by Cerny and his co-workers[234]; the results were interpreted on the basis of the theory of Maron, whereby variation of χ with concentration is explicitly taken into account. The results showed that at 30° the infinite-dilution value of χ (symbolized $\mu^0 - \sigma^0$ in the original paper) shows a slight dependence upon the molecular weight of the sample, according to

$$\chi = 0.450 \, M^{0.062} \tag{56}$$

In fact, over the molecular weight range used, $(0.8\text{--}16) \times 10^5$, χ is practically constant at 0.486 ± 0.003 (one s.d.).

Nomura, Miyahara, and their colleagues[989,1100–1102] have measured the partial specific adiabatic compressibilities of a number of polymers (including PVP) in dilute solution by means of ultrasonic velocity measurements; for infinite dilution they obtain the limiting partial specific adiabatic compressibility $\bar{\varkappa}_2^0$, defined as

$$\bar{\varkappa}_2^0 = -(1/\bar{v}_2^0)(\partial v_2/\partial P)^0 \tag{57}$$

where $\bar{v}_2^0$ is the partial specific volume of the solute (polymer), P is the pressure, and the superscript zero refers to infinite dilution. The value of $\bar{\varkappa}_2^0$ for a polymer represents the compressibility contribution made by individual isolated polymer molecules in the solution (including any effect that they may have on the neighboring solvent molecules). For liquids $\varkappa$ has a value of about 100×10^{-12} cm^2 dyn^{-1}. Figure 24 shows the results obtained for the variation of $\bar{\varkappa}_2^0$ with temperature for PVP in water,[1101] for dextran in water,[1102] and also, for comparison, for polystyrene in a number of solvents[1100] [A = dioxan, B = toluene (both good solvents), and C = butanone (a poor solvent)] over the restricted temperature range 30–40°.

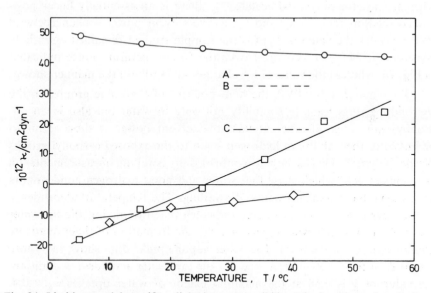

Fig. 24. Limiting partial specific adiabatic compressibility $\bar{\varkappa}_2^0$ of polymers in solution: temperature dependence of $\bar{\varkappa}_2^0$ for polyvinylpyrrolidone in water (squares) and dextran in water (diamonds), compared with the essentially temperature-independent values (horizontal broken lines) for polystyrene in (A) dioxan, (B) toluene, and (C) butanone. The circles represent the temperature dependence of the compressibility $\varkappa_1$ of water. Data of Nomura, Miyahara, and colleagues.[1100–1102]

Unlike the behavior of polystyrene (and indeed, that of most polymers in solution), for both of the aqueous systems the $\bar{\varkappa}_2{}^0$ values are negative throughout the measured temperature range.* Since a value of about 20×10^{-12} cm^2 dyn^{-1} is almost the lower limit for "normal" polymers such as polystyrene and poly(vinyl acetate) in very poor nonaqueous solvents, the numerically lower $\bar{\varkappa}_2{}^0$ values for PVP and dextran in water, and particularly the negative values, indicate the marked "compacting" effect of the polymer, probably through immobilization of the neighboring water molecules. Nomura and Miyahara[1101] have attempted to estimate the amount of water bound by PVP by assuming that the solute is incompressible, and that for the bound water $\varkappa = 18 \times 10^{-2}$ cm^2 dyn^{-1}. Although this method leads to plausible figures at lower temperatures, i.e., 0.458 cm^3 bound water per gram of polymer at 5° and 0.230 cm^3 at 15°, it unfortunately leads to *negative* values at the higher temperatures; for more soundly based calculations it would be necessary to extend the measurements of $\bar{\varkappa}_2{}^0$ to higher temperatures where the value might be expected to level off at that for the "dehydrated" polymer. One further difference between the behavior of these aqueous systems and that of nonaqueous systems is that whereas, say, for poly(vinyl acetate)[989] there is an essentially linear positive correlation between $\bar{\varkappa}_2{}^0$ and $[\eta]$ as the solvent power is changed, with PVP in water the value of $[\eta]$ for the sample studied remained essentially constant (at 24.0 ± 1.5 cm^3 g^{-1}) over the 50° temperature range indicated in Fig. 24, whereas the value of $\bar{\varkappa}_2{}^0$ changes markedly in the manner shown.

To summarize for PVP, the high polarity of the imide groups on the pyrrolidone rings leads to solubility not only in water but also in an unusually wide range of other liquids; the solvent power of these seems to be linked to their ability to hydrogen bond to the exposed carbonyl groups of the polymer. This is largely confirmed by swelling measurements on cross-linked gels, which show that water and good hydrogen-bond donors all give similar, maximal extents of swelling. The temperature dependence of the degree of swelling in water indicates that the mixing (liquid water plus polymer) is somewhat *exothermic* for the formation of these relatively dilute solutions ($w_2 \approx 0.05$). The water vapor uptake data show, like those for other of these polymers, a "toe" at the low water activities; in addition, the value of χ is 0.58 at the maximal extent of water uptake, $w_1 = 0.4$, whereas it has a value of 0.49 for very dilute solutions, indicating that some

---

* However, negative values for $\bar{\varkappa}_2{}^0$ are common for solutes in aqueous solution; see the tabulated values of this quantity for electrolytes given in Chapter 1 of Volume 3 of this treatise.

marked change must occur within the intervening composition range. The temperature dependence of the water vapor uptake indicates that in the corresponding composition region (i.e., $0 < w_1 < 0.4$) the mixing process is *endothermic* to an apparently constant degree up to the formation of a 1:1 molar ratio (water:vinylpyrrolidone unit), subsequently falling with increasing dilution until it becomes either essentially athermal or somewhat exothermic (as indicated by the swelling measurements) for very dilute solutions. The data also indicate, via a rather long extrapolation, that the system should show lower critical consolute behavior at about 170°; this is supported by the data considered below (see Section 3.6.2) on the temperature dependence of the precipitation points for the polymer with glycine and with ammonium sulfate, which indicate values of 130 and 150°, respectively. We may therefore suggest that for this system the ideal temperature is approximately $150 \pm 20°$. Near-infrared spectral studies of the polymer in moderately concentrated aqueous solution show that it has a net structure-making effect upon water, which is reasonable in the light of the nature of the component groups on the polymer and of the known behavior of small-molecule analogs of these groups. Deuteron rotational relaxation measurements on the polymer (and also the related polymer PVMO) in D_2O solution (0–10%) may be most reasonably interpreted in terms of the water molecules bound to the polymer having a rotational reorientation time closely similar to those in the bulk liquid. The dilute solution viscosity data from the literature are too inconsistent to yield any additional information concerning polymer–solvent interactions, and other lines of evidence have to be called in to correlate these viscosity data and to suggest the "best" values for the viscosity parameters for water. Finally, adiabatic compressibility measurements indicate a marked "compacting" effect of the polymer upon water, an effect also shown by the water-soluble natural polymer dextran; further studies are required to clarify these results, which show clear-cut differences between the behavior of polymers in aqueous and in nonaqueous solution.

3.5.1e. *General Comments upon the Experimental Techniques for Studying Synthetic Polymers in Aqueous Solution.* The preceding subsections on the individual polymers have indicated the rather fragmentary state of our knowledge of the properties of their aqueous solutions. Rather than trying to present an overall picture, we will therefore simply emphasize the possibilities for future work, especially in terms of the more general application of techniques so far applied to one or only a few of the polymers.

Considering the basic thermodynamic studies of these systems via vapor pressure measurements and water uptake isotherms, in general the latter conform to the simple BET type III form expected from the Flory–Huggins theory. However, the values of χ calculated on the basis of this theory generally show a marked variation with concentration.

At the low uptake ends of the isotherms the analysis of the data is in many cases complicated by a "toe," evidently due to traces of highly hydrophilic components in the polymer sample, such as the chain end groups, residual solvent, or residual components from the method of preparation. To minimize such effects, so that the true intrinsic absorptive power of the polymer can be studied, the following precautions are recommended:

(a) The end groups should be known to be hydrophobic, or better still the polymer molecular weight should be large enough to make their effect negligible.

(b) The polymer sample should be exhaustively dialyzed to remove residual hydrophilic small-molecule impurities.

(c) The final solvent used, for example, to cast a polymer film should be hydrophobic or highly volatile, or both.

Indeed, the detection of such "toes" with the highly hydrophilic polymers is an indication that similar effects may occur with hydrophobic (water-insoluble) polymers where they would be much more significant because the maximal uptakes are much less; similar precautions should therefore be observed with such polymers in water absorption measurements (see Section 3.8).

Turning to the other end of the water activity scale ($a_1 \rightarrow 1$), one notable general feature is that even for the normal experimental limit of $a_1 = 0.90$ or 0.95 the maximum uptakes are still quite low, seldom corresponding to more than $w_1 = 0.5$ and frequently much less. In fact the region between $w_1 \approx 0.5$ (the upper limit from uptake measurements) and $w_1 \approx 0.99$ (the lower limit for dilute solution measurements) is very much of a terra incognita for these systems. It is evident that this region is troublesome to attempt to study by direct uptake measurements, because of the experimental difficulties involved with such high water activities; it is likely that it would be more useful to apply the isopiestic techniques developed for aqueous solutions of small-molecule solutes. Measurements in this region are all the more important because of the anomalously high values of χ obtained on extrapolating values for the concentration region of the uptake measurements to $w_1 = 1$ or $a_1 = 1$ (which are equivalent for these water-soluble polymers). Since the polymers are water soluble, we expect that $\chi < 0.5$ for this limit of extreme dilution, while the extrapolations generally indicate appreciably higher values; it is likely that there is some relatively abrupt change in χ within the unexplored region of compositions at a point representing either stoichiometry in the polymer hydration or possibly the completion of the buildup of a "disturbed" water structure about the polymer and the initiation of "normal" water structure.*

In addition, these uptake measurements need also to be extended over as wide a range of temperatures as practicable, so that the free energy quantities (such as ΔG^M and χ) can be split up into their enthalpic and entropic contributions. The former may, of course, be more directly and accurately obtained by measurements of heat of mixing via calorimetry, and indeed this alternative approach is strongly recommended wherever practicable; however, in many cases this involves fairly obvious difficulties with these normally highly viscous systems.

For viscosity measurements we have already indicated with poly(ethylene oxide) the limited conclusions which may be drawn from data obtained from concentrated solutions of polymers. For dilute solution measurements, particularly for the limit of infinite dilution, the discussion with poly(ethylene oxide) indicates the power of this relatively simple experimental technique in studying such effects as that of temperature upon solvent power. It is evident, however, that much careful work still needs to be done to better define the Mark–Houwink–Sakurada parameters for these polymers both in aqueous and nonaqueous solution, but that such work would be highly rewarding both in providing better general characterization of this group of polymers and in defining more clearly their interactions with the solvents, especially water.

Ultrasonic velocity and absorption measurements also deserve more widespread application. In the former case the application of the measurement of ultrasonic velocity to the evaluation of the limiting partial specific adiabatic compressibility of PVP (and dextran) in aqueous solution has already been discussed under this polymer. There seem to be distinct differences here between aqueous and nonaqueous systems, particularly in the marked variation of the limiting partial specific compressibility with temperature, and the observation of negative values for this quantity, for polymers in aqueous solution.[†] Extension of the measurements to other polymers may lead to some clarification of these effects. With ultrasonic absorption measurements, we have discussed under poly(ethylene oxide)

---

* See note added in proof (p. 755).
† See, however, the footnote to p. 660.

the limitations in the interpretations of these measurements; once again, studies on other polymers, and particularly the extension over a range of temperatures, may provide more clarification.

Considering the spectroscopic techniques, the most promising for wider application are infrared spectroscopy and nuclear magnetic resonance. In the case of the former the study of polymer spectra (as discussed under PEO) seems to be somewhat unrewarding. However, the study of the near-infrared spectra of the solvent (for H_2O–D_2O mixtures), as discussed under PVP, seems of promise in providing information upon the structure-breaking or -making activity of these types of solutes; of particular interest here is the comparison of the polymer with that of its small molecule analogs. Regarding pmr, as indicated under PEO, this method, using the protons of the polymer chain, can provide valuable information upon the stoichiometry of the chain hydration; it should be emphasized, however, that this was only possible for this particular polymer because of the availability of *liquid* polymer (really oligomer), which meant that the solutions all had low viscosities and hence gave reasonably sharp spectra. Thus extension of the technique to other polymers of the group similarly requires the use of fairly short-chain oligomers, with also a careful control of the end groups (or their elimination by using cyclic oligomers) to ensure they do not give interfering signals.

The use of deuteron magnetic resonance to study the rotational relaxation of the solvent (D_2O) molecules is exemplified by the studies of Glasel applied (as regards synthetic polymers only) to PMAA and to PVP and PVMO. Here the small effects observed with the latter two polymers present some ambiguities in interpretation which have been discussed above.

In conclusion, it is necessary to make a comment upon one specific misuse of the word (and concept of) "hydration" applied to polymers in aqueous solution, which is to equate the degree of hydration of the polymer chain to the number of water molecules within the volume encompassed by the polymer coil. This picture is occasionally used for polymers themselves, but it is often applied to the situation with the micelles of nonionic synthetic detergents where the poly(ethylene oxide) chains form a hydrophilic capsule on the outside of the micelle. Now, with fairly concentrated polymer systems there may, of course, be some justification for considering the water molecules within the encompassed volume to be directly associated with the units of the chain; however, for dilute systems and with a random coil or related loose conformation for the polymer the molecules of solvent within the encompassed volume mostly perform a space-filling role, subject to the limitation that the molecules of solvent adjacent to the chain interact sufficiently strongly with the segments to prevent collapse of the coil through a preference for segment–segment and solvent–solvent contacts. These space-filling water molecules should therefore not be referred to as "hydration," which should be restricted to those molecules adjacent to and interacting with the chain.

3.5.2. *Polyelectrolytes*

When a polymer molecule has charged groups covalently attached to its chain it becomes a polyelectrolyte. With the groups all negatively charged it is a polyanion, with them all positively charged it is a polycation, while with both positive and negative groups present it becomes a polyampholyte. With the exception of polyampholytes containing equal numbers of positive and negative charges there must in all cases be sufficient counterions of the opposite charge present to maintain electroneutrality in the system.

The behavior of polyelectrolytes, particularly toward water and in aqueous solution, is a topic of some considerable scientific importance, not least because many biopolymers (particularly the proteins, nucleic acids, and some polysaccharides) are polyelectrolyte in nature, with their properties strongly dependent upon electrostatic factors. Similarly, certain polyelectrolytes such as poly(acrylic acid) and carboxymethylcellulose are of considerable technological importance. Unfortunately, the study of this important group of substances forms the focal point of three complex and contentious territories of science, i.e., aqueous systems, ionic systems, and polymer systems. It is not possible to do full justice to this triply complex area within the scope of this chapter; we will therefore simply outline some of the main basic quantitative relations governing the properties of polyelectrolytes in solution, while for the fundamental theory the reader should consult more detailed treatments of the subject.[1118,1244]

One characteristic feature of polyelectrolyte salts is their high water solubility when compared with the corresponding equivalent nonionic materials. In broad terms, this higher solubility may be attributed to the more favorable interactions between the charged groups and the water, coupled with the fact that with small counterions there will also be a strongly favorable entropic contribution from these when the complete polysalt is dissociated and dispersed upon solution.

One example of this is the differential solubility of the various types of nylon in hydrochloric acid. Whereas nylon 6,

$$\mathrm{-[NH(CH_2)_5CO]}_{\overline{x}},$$

is soluble in 4 M (and stronger) acid, nylon 6.6,

$$+NH(CH_2)_6NH—CO(CH_2)_4CO+_{\overline{x}},$$

requires 6 M acid for dissolution, and nylon 6.10,

$$+NH(CH_2)_6NH—CO(CH_2)_8CO+_{\overline{x}},$$

is only soluble in acid of this strength upon warming, while nylon 11,

$$+NH(CH_2)_{10}CO+_{\overline{x}},$$

is evidently insoluble even in 11.6 M (concentrated) acid, although in this last case the fibers are converted to a waxlike mass, indicating that the polymer is swollen by the liquid.[1270] This behavior results from the protonation of the polyamide chain (involving its amino end groups and its amide groups), converting it into a polyelectrolyte, acting in opposition to the hydrophobic effect of the methylene groups (for the polymers listed there are respectively five, five, seven, and ten methylene groups per amide group).

Similarly, the "solubilization" of poly(vinyl acetate) by ionic amphiphiles such as sodium dodecyl sulfate and the viscosity behavior of the solutions demonstrate the conversion of the neutral polymer into a polyelectrolyte, presumably by reversible binding of the anions of the amphiphile by the hydrophobic polymer chain.[1305]

However, these examples are rather atypical, and the more usually encountered examples of polyelectrolytes are the simple salts of polyacids or polybases. It is useful to refer to this type as a "simple polyelectrolyte."

The solubility of a simple polyelectrolyte is a function of the nature of the polymer both in the undissolved state and in solution; for this and for more general reasons it is therefore necessary to have information on the conformational state(s) of the polyelectrolyte in aqueous solution, as influenced by the ionic strength of the medium via the interactions between the fixed charges on the polymer chains, the counterions, and any other ions in the system.

If a simple polyelectrolyte is dissolved in water, then so long as the concentration is above the "critical overlap" level the system will be essentially uniformly pervaded by charged polymer chains with the solvent and the counterions moving in the interstices. If the solution is then diluted, the individual chains become progressively separated, and distinct regions containing no polymer chains will appear; the counterions attempt to distribute themselves evenly throughout the system, but are held back preferentially in the polymer regions by the electrostatic attractions of the fixed charges. As the dilution progresses, the loss of the counterions from the polymer coils becomes greater, but can never be complete, except, perhaps, at extreme dilution (where, however, the presence of the hydrogen and hydroxyl ions from the self-dissociation of the water will then become significant).

The effect of this on the conformation is that with increasing dilution the fixed charges are shielded less and less from each other by the counterions, the repulsions between these charges become greater, and the polymer coil accordingly expands. This shows itself most directly in the viscosity of the solutions; whereas with nonelectrolyte polymers the conventional plot of reduced viscosity against polymer concentration c_2 is linear and of small positive slope (compare Section 3.3.3), with simple polyelectrolytes the plot is of negative slope and rises up ever more steeply with decreasing concentration, thus making it impossible to extrapolate directly to zero concentration to obtain the value of $[\eta]$. However, it was shown by Fuoss[504] that viscosity data for polyelectrolytes could be fitted by an empirical equation of the form

$$\eta_{sp}/c_2 = A/(1 + B\sqrt{c_2}) \qquad (58)$$

where A and B are concentration-independent parameters characteristic of the particular system. On the assumption that the equation holds down to $c_2 = 0$ (extreme dilution), then A will correspond to $[\eta]$. One convenient method of applying eqn. (58) is to plot c_2/η_{sp} against $\sqrt{c_2}$, when the plot should be linear with intercept $1/[\eta]$.

If a neutral salt is added to the polyelectrolyte solution, then these anomalous viscosity effects are greatly reduced, and with a sufficiently high salt concentration the viscosity plot becomes of the normal form, i.e., linear with positive slope. This change in behavior results from the screening of the repulsions between the fixed charges on the chains by the counterions from the added salt.

As an alternative to using an excess of neutral salt, the dilution may be carried out at constant ionic strength by using as diluent salt solution of the same ionic concentration (or some fixed fraction of it) as the most concentrated polymer solution; this procedure was originally proposed by Pals and Hermans[1135] and has been termed "isoionic dilution." The value of $[\eta]$ obtained corresponds to the salt concentration of the diluent, since the polyelectrolyte has then been "diluted out." An example of the application of each of these three types of approach to viscosity measurements on polyelectrolytes is the work of Scondac and his colleagues[1352] on partially neutralized copolymers of acrylic acid and vinylpyrrolidone.

For polyelectrolytes in general $[\eta]$ is related to the concentration of added neutral electrolyte above a threshold level of approximately $10^{-4}\ M$ by an equation, derived by Cox,[295] of the form

$$[\eta]/[\eta]_r = (s/s_r)^{-m} \tag{59}$$

where s is the salt concentration and $[\eta]_r$ is the value of $[\eta]$ at some reference salt concentration s_r. For high-molecular-weight polymers $0.55 < m < 0.60$, while for low molecular weights $m \sim 0.4$. At salt concentrations lower than the threshold level the value of $[\eta]$ increases less with decreasing salt concentration and finally attains a limiting value due presumably to maximal extension of the polymer coil.

When the viscosity measurements are extended to samples of different molecular weight then the effect of ionic strength on the more fundamental parameters K and a of the Mark–Houwink–Sakurada relation is revealed. Figure 25 shows such data for sodium polyacrylate in sodium bromide at different concentrations.[834] The plot has a very similar shape to that for

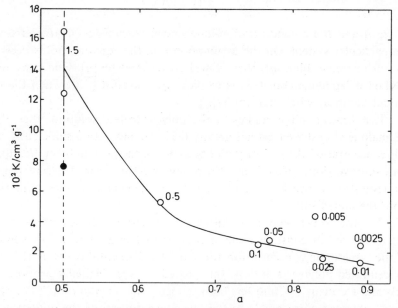

Fig. 25. Mark–Houwink–Sakurada viscosity parameters K and a [eqn. (43)] for sodium polyacrylate at 15° in aqueous sodium bromide at the molar concentrations shown.[834] The filled circle shows K_Θ for un-ionized poly(acrylic acid) in dioxan at 30°.[1038]

nonelectrolyte polymers, except that here the variation in the parameters is caused by variations in the concentration of the added salt. However, the upper limit of a seems to be ~ 0.9, higher than that shown for nonelectrolytes. This occurrence of a higher limiting value of a is also shown more clearly by poly(styrene sulfonic acid), where at 25° in 0.52 M hydrochloric acid or in 0.50 M sodium chloride the exponent a is 1.0 (with K values of 3.4×10^{-4} and 3.1×10^{-4} cm^3 g^{-1}, respectively),[756] and by poly(vinyl sulfuric acid) at 20° in 0.5 M sodium chloride, where the exponent is again 1.0 with $K = 1.03 \times 10^{-3}$ cm^3 g^{-1}.[1141] In each case the values of K are extremely low when compared with those obtained for polymers having smaller values of a (compare with Fig. 25), but nevertheless this follows the trend of the correlations between these two parameters.* The higher values of a are a reflection of the tendency of these chains to acquire a fully extended, rodlike conformation in solution under the influence of the repulsions between the fixed charges, a trend then modified by the effect of the counterions from the added electrolytes.

The association of counterions with the molecule of a polyelectrolyte may be measured directly by electrical transport methods such as conductance and transference[838]; the latter technique, particularly, shows up such association quite unambiguously, e.g., in solutions of sodium polyacrylate an appreciable fraction of the sodium ions are transported with the polyanion toward the anode. In general, with simple monovalent counterions such as the alkali metals for polyanions, and the halide and oxyanions for the polycations, the association may be taken to be of a nonspecific and simple electrostatic character; however, with polycations having quaternary ammonium groups as the cationic centers the solution properties do depend markedly upon the nature of the counterion,[1096] a phenomenon also shown in a similar fashion by micelles of quaternary ammonium amphiphiles[1034] and attributable in each case to specific binding of the counterions. With a polyanion having polyvalent and/or transition metal counterions there may also be specific ion binding, attributable to chelation effects from neighboring anionic centers along the polymer chain.[1039] On the other hand, with organic counterions specific ion binding may arise from interactions between the polymer chain and the organic parts of the ion; for example, aromatic cations show specific and nonstoichiometric association with polyacrylate anions, obeying the form of the Langmuir isotherm and

* However, in each of these cases the polymers were not the pure compounds, since they had been obtained by chemical treatment of other polymers but without complete conversion—the first from polystyrene by sulfonation with only 85% conversion, the second from PVA by sulfation with only about 60% conversion.

with the strength increasing with the size of the aromatic system.[1130,1131] In these cases the behavior tends more to the reversible binding by non-electrolyte polymers discussed in Section 3.6.

Considering more recent work, increasing attention is being paid to the role of the local structure of the polymer chain, and particularly its stereoregularity, in determining the interactions between the polyelectrolyte and its counterions; examples of studies undertaken from this viewpoint include those on the association of the anionic forms of isotactic and syndiotactic PMAA with Cu^{2+} and with Mg^{2+},[1114] of PAA and isotactic and syndiotactic PMAA with bispyridinium cations,[1040] and of isotactic and atactic poly(styrene sulfonic acid) with H^+, Na^+, and cationic dyes.[742]

The general nature of the interactions is also currently being studied by more thorough application of conventional physical techniques, as in the determination of activity coefficients[354] and second virial coefficients[1142] for poly(styrene sulfonate), as well as by the application of less conventional techniques such as dielectric constant and dielectric dispersion.[1219,1285]

One very important aspect of polyelectrolyte chemistry comprises the interconversion of polyacids and their derived polyanions, and polybases and their derived polycations, by protonation/deprotonation reactions. The processes involved are interesting in providing a link between the nonelectrolyte polymers (such as PAA and PMAA, considered in Section 3.5.1) and the polyelectrolytes, while the equilibria involved may be compared with those for the small-molecule equivalents to give information upon polyelectrolyte equilibria in general. These studies are also useful as bases for interpretation of the corresponding titration behavior of such biopolymers as proteins and nucleic acids, which thus casts light upon the nature of the local environment of the groups involved, and upon the conformational flexibility or rigidity of the biological macromolecules.

The titration behavior of polyacids and polybases shows certain marked differences from that of the small-molecule analogs. It we consider, say, the titration of poly(acrylic acid) with a simple strong base such as sodium hydroxide, then the fundamental equilibrium involved is

$$-CH_2-CH- \rightleftharpoons -CH_2-CH- + H^{\oplus} \qquad (60)$$
$$| |$$
$$COOH CO_2^{\ominus}$$

During the titration the values of the degree of neutralization (or more exactly, equivalence) α will be given by the relative amounts of polyacid originally present and of base added. Near the start of the titration, when α is small, the chain is effectively uncharged and hence the pK_a value may be expected to be close to that of a small-molecule analog. However, as the titration proceeds the chain accumulates negative charges, so that it becomes more and more difficult to remove each successive proton and the effective pK_a steadily rises. This means that the measured pH values during the titration no longer conform to the conventional Henderson–Hasselbach equation

$$pH = pK_a + \log[\alpha/(1 - \alpha)] \tag{61}$$

One method of taking this effect into account is to include a term representing the free energy contribution for the removal of the proton from the charged environment of the polymer coil:

$$pH = pK_a{}^0 + \log[\alpha/(1 - \alpha)] + 0.434q\Psi/kT \tag{62}$$

where in the last term, which represents this correction factor, q is the effective charge on the macromolecule, Ψ is the mean electrostatic potential, and k is the Boltzmann constant.

From the more directly useful viewpoint, several equations have been proposed as empirical relations between the apparent dissociation constant pK_{app} [i.e., as calculated by the simple application of eqn. (61)] and pH or α; these include that of Kern[763]

$$pK_{app} = pK_1 + BpH \tag{63}$$

of Kagawa and Tsumura[750]

$$pK_{app} = pK_2 + m \log[\alpha/(1 - \alpha)] \tag{64}$$

and of Katchalsky and Spitnik[753]

$$pH = pK_3 + n \log[\alpha/(1 - \alpha)] \tag{65}$$

As Nagasawa and Rice[1077] point put, these three equations are equivalent, with the constants B, m, and n interrelated by

$$m = B/(1 - B) \tag{66}$$

and

$$n = 1/(1 - B) \tag{67}$$

However, the parameters pK_1, pK_2, and pK_3, although experimentally found to be independant of α or pH for a given titration, are unfortunately not simply or directly related to the true $pK_a{}^0$ of eqn. (62).

The Katchalsky–Spitnik equation (65), in particular, has been useful in showing up a conformational transition between a compact and an expanded form during the base titration of PMAA. By such potentiometric methods, Leyte and Mandel[889] demonstrated that there is an initial section of the titration curve which obeys the Katchalsky–Spitnik equation, a transition region, and then a third and final section where the equation (with different values of pK_3 and n) is obeyed once more; on the basis of this and of previous work they proposed that the first section relates to a compact conformation which is converted over the transition region into the expanded conformation corresponding to the final section of the plot. Figure 26 illustrates how the extent of the transition region (whose limits are fairly sharply defined at either end) depends upon the concentration of added electrolyte (sodium nitrate); it should be noted that the limits did not change appreciably over the polymer concentration range 0.0125–0.073%, showing that the behavior observed is effectively that of isolated independent polymer molecules. From their results Leyte and Mandel[889] deduced that the value of the free energy difference for the transition between the two forms (extrapolated to zero charge density) amounted to 0.12 ± 0.02 kcal basemol^{-1} at 20°; this value of this quantity was essentially independent of ionic strength over the range 0.003–0.1 M. Subsequently Mandel and Stadhouder,[972] from studies of the titration behavior of partially esterified PMAA, showed that the transition depends only on

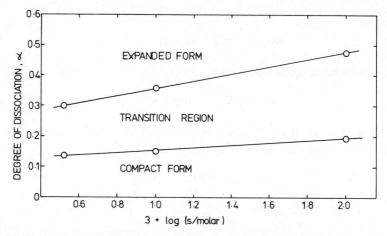

Fig. 26. Conformational transition during the base titration of poly(methacrylic acid) in aqueous solution at 20°: dependence of the extent of the transition region from the compact form to the expanded form upon the concentration s of sodium nitrate. From the data of Leyte and Mandel.[889]

the mean charge density along the cháin, since for different degrees of esterification the same transition curve was obtained as a function of the fraction of ionized groups with respect to the total number of monomer units. Scondac et al.[1272,1353] have similarly investigated copolymers of methacrylic acid with vinylpyrrolidone and with acrylic acid to show the gradual disappearance of these transitions with "dilution" of the methacrylic acid units.

Mandel and his colleagues[971] have further investigated the PMAA transition by potentiometric measurements at 5 and 50° and by spectroscopic measurements in the far ultraviolet (at wavelengths of around 200 nm) and in the infrared (at frequencies around 1700 cm^{-1}). By carrying out these potentiometric studies at these two temperatures they were able to divide the free energy of the transition into its enthalpic and entropic contributions, which, extrapolated to zero degree of dissociation, were equivalent to $\Delta H = 0.30$ kcal basemol^{-1} and $\Delta S = 0.4$ cal deg^{-1} basemol^{-1} (where "mol" refers again to 1 mole of *monomer* units). From the endothermic and endentropic nature of the transition Mandel and his colleagues[971] concluded that the compact form is *not* stabilized primarily by hydrophobic interactions involving the structure of the surrounding solvent, but rather by direct interactions between the neighboring groups on the chain.

Glasel[526] has also followed the conformational transition in PMAA by deuteron magnetic resonance measurements on D_2O solutions at $1 < pH < 11$, using a polymer sample with $\bar{M}_v = 3 \times 10^5$ and at $31 \pm 1°$. At both the extreme pH values quoted the reciprocal relaxation time $1/T_1$ rose linearly with polymer concentration, but while for pH 11 this rise was only from 1.9 sec.$^{-1}$ for pure D_2O to 2.2 sec^{-1} at 10% polymer, at pH 1 the value of $1/T_1$ rose to 5.0 sec^{-1} at 5% polymer, representing a thirtyfold greater effect. Studies at intermediate pH values on a 3% polymer solution showed that the variations in $1/T_1$ with pH mirrored the titration curve of the polymer and the accompanying viscosity changes in the system. Addition of sodium chloride up to 0.5 M caused no change in the $1/T_1$ values of a 3% polymer solution either at pH 1 or at pH 11. We have already criticized in Section 3.5 (under the "acrylic group" and also under polyvinylpyrrolidone) Glasel's interpretation of his deuteron relaxation data. Once again we reject this interpretation upon which, with PMAA at high pH's (i.e., in the anionic form), the relatively small effect observed is stated to be due to the essential absence of any interaction between the polymer and the water; we prefer to regard this as indicating (as with PVP and PVMO) that there is bound water which happens to have much the same rotational correlation time as that in bulk water, although further detailed conclusions cannot be drawn because the effects observed depend both upon the amount bound and its rotational correlation time. It would, indeed, be valuable for this technique to be applied to another polyacid, such as PAA, which is more "normal" in its conformational behavior. In the meantime, however, Glasel's results with various biopolymers that he studied at different pH's, presented in the same paper,[526] need to be reexamined and reinterpreted in the light of the present and the previous discussion.

The pH-dependent transition observed with PMAA is, of course, reversible and is of some interest in relation to reversible transitions of biopolymers between compact and expanded conformations. More recently still, Leyte and his colleagues[888] have demonstrated *irreversible* effects in the potentiometric behavior of isotatctic PMAA.

An interesting type of polyelectrolyte is the form of "polysoap" produced, for example, by quaternizing some of the pyridine groups of poly-4-vinylpyridine with dodecyl bromide and the remainder by ethyl bromide.[1435] Whereas in isopropanol solutions these polymers show normal conformational behavior, in aqueous solution the molecules aggregate (as shown by light scattering measurements). Further, for a polymer having 34% of the quaternizing groups as dodecyl, $[\eta]$ is low and largely independent of concentration of added electrolyte ($[\eta] = 8.9 \pm 0.6 \, \text{cm}^3 \, \text{g}^{-1}$ between 0.05 and 0.20 M potassium bromide); this indicates that the isolated molecule exists essentially as a compact sphere, representing micelle formation within each polymer molecule caused by association of the dodecyl groups.[1435] In a rather similar way Klotz and his colleagues[827] have based their "synthetic enzyme" studies upon polyethyleneimine with 10% of the nitrogen atoms alkylated with dodecyl groups (this provides the compact globular conformation similar to that of native enzymes) and with another 15% of them alkylated with methyleneimidazole (to provide the catalytic sites).

3.6. Interactions between Synthetic Polymers and Small-Molecule Solutes (Cosolutes) in Aqueous Solution

The behavior of a polymer in solution is often influenced by the presence of a small-molecule solute, which it is convenient to term the *cosolute*; this applies particularly to polymers in aqueous solution.

It is useful to distinguish two types of such effects, characterized broadly by the concentration level involved. For low concentrations of the cosolute, say around 100 mM or below, any marked effects normally represent specific interactions of the cosolute with the polymer, i.e., *polymer binding*. On the other hand, at much higher levels of cosolute concentration, say around 1 M and above, the solute becomes comparable in its concentration to that of the solvent, water (which has a concentration of about 55 M for dilute aqueous solutions); effects at these higher levels generally represent not so much specific interactions with the polymer, as more indirect ones through the solvent which we may conveniently refer to as *solvent perturbation* by the cosolute.

It will be noted that we have left a "no-man's land" lying between approximately 100 mM and 1 M which is not specifically allocated to either class. This has been done to emphasize the overlapping nature of the classification; thus bindable cosolutes at higher concentrations may be expected to show solvent perturbation effects, while on the other hand cosolutes showing only the latter effect should begin to act from the lowest concentrations, although the extents of the effects may not be directly detectable by conventional means. The main distinction between the two types of cosolute effect is that binding leads to a lowered thermodynamic activity of the cosolute; it also frequently shows a saturation character, i.e., there is a limiting extent to which the behavior of polymer and cosolute are affected by their interaction, whereas with solvent perturbation there is generally a monotonic increase of the effect with cosolute concentration. Even this again is not a fully clear-cut distinction, since the latter monotonic increase may simply represent the initial part of a binding isotherm extending to much higher cosolute concentrations. Nevertheless, this twofold classification is a useful basis upon which to discuss the subject further, and we shall adopt it in the present discussion.

One additional point is that we shall be mainly considering the effect of cosolutes on polymers which are (by themselves) water soluble. However, the presence of a small-molecule solute can also bring into solution an otherwise water-insoluble polymer; this applies, for example, to poly-(vinyl acetate) in the presence of various synthetic detergents,[1305] which may be attributed to cosolute binding, and to polyacrylonitrile with a variety of salts, particularly divalent-metal iodides and thiocyanates (Ref. 1038, Chapter II).

3.6.1. *Polymer Binding of Cosolutes*

The subject of reversible cosolute binding by synthetic polymers is one with many ramifications; within the scope of this chapter, we can do no more than provide an outline of the subject.

TABLE IV. Reversible Binding of Small-Molecule Cosolutes by Synthetic Polymers in Aqueous Solution

Polymer[a]	Cosolutes	Ref.
PAA	Methylene Blue	1056
PAA	Alkyl and aryl quaternary ammonium compounds	1130
PAA	Arylammonium compounds	1131
PAA	Sodium dodecyl sulfate and benzoic acid	396
PAA	Aliphatic and polycyclic aromatic hydrocarbons	89
PAA	Naphthalene and biphenyl	1109
PAA	Nonionic amphiphiles	1294
PAAm	Sodium dodecyl sulfate and benzoic acid	396
PAAm	Naphthalene and biphenyl	1109
PEI	Naphthalene and biphenyl	1109
PEI	Sodium dodecyl sulfate	1514
PEO	Barbiturates and phenolic compounds	659
PEO	Iodine + iodide system	577
PEO	Phenolic compounds	578
PEO	Pharmaceuticals	840
PEO	Methyl and propyl p-hydroxybenzoates	1027
PEO	Quaternary ammonium compounds	334
PEO	Sorbic acid	148
PEO	Methyl and propyl p-hydroxybenzoates	1143
PEO	Chloroxylenol	180
PEO	Iodine + iodide system	669
PEO	Phenolic compounds	747
PEO	Sodium dodecyl sulfate	739
PEO	Aliphatic anionic and cationic compounds	1288
PEO	Sodium hexadecyl sulfate and hexadecylammonium chloride	1289
PEO	Naphthalene and biphenyl	1109
PEO	Alkylammonium halides and thiocyanates	1296
PESA	Naphthalene and biphenyl	1109
PMAA	Auramine O	144, 1125
PMAA	Sodium dodecyl sulfate and benzoic acid	396
PMAA	Acridine Orange	905
PMAA	Aliphatic and polycyclic aromatic hydrocarbons	89
PMAAm	Sodium dodecyl sulfate and benzoic acid	396
PPA	Naphthalene and biphenyl	1109

TABLE IV. (*Continued*)

Polymer[a]	Cosolutes	Ref.
PPO	Phenolic compounds	578
PPO	Phenolic compounds	747
PPO	Aliphatic anionic and cationic compounds	1288
PPO	Sodium hexadecyl sulfate and hexadecylammonium chloride	1289
PVA/VAc	Sodium dodecyl sulfate	1305
PVA/VAc	Auramine O	1125
PVA/VAc	Aliphatic amphiphiles	1288–1290
PVA/VAc	Sodium dodecyl sulfate	46
PVA/VAc	Chlorazol Sky Blue FF and Orange II	717
PVA/VAc	Naphthalene and biphenyl	1109
PVA/VAc	Alkylammonium halides and thiocyanates	1296
PVA/VAc	Alkylammonium thiocyanates	1291
PVA/VAc	Sodium, potassium, guanidinium, and tetraalkylammonium alkyl sulfates	1295
PVME/MA	Chloroxylenol	180
PVMO	Pharmaceuticals	149
PVP	Azo dyes	179, 1338
PVP	Iodine + iodide system	87
PVP	Pharmaceuticals	658
PVP	Auramine O and Rose Bengal	1125
PVP	Phenolic compounds	578
PVP	Sodium alkyl sulfates (and also other cosolutes)	86
PVP	Orange II and Benzopurpurin 4B	472
PVP	Iodine + iodide system	1394
PVP	Methyl and propyl p-hydroxybenzoates	1027
PVP	Quaternary ammonium compounds	334
PVP	Inorganic and organic compounds	397
PVP	Sodium dodecyl and dodecylpolyoxyethylene sulfates	1287
PVP	Aromatic compounds	1030–1033
PVP	Organic acids and phenolic compounds	1357
PVP	Chlorbutanol, benzyl alcohol, and phenylethanol	73
PVP	Phenolic compounds	180, 747
PVP	Aliphatic anionic and cationic compounds	1288
PVP	Sodium hexadecyl sulfate and hexadecylammonium chloride	1289
PVP	Methyl Orange	792

TABLE IV. (*Continued*)

Polymer[a]	Cosolutes	Ref.
PVP	Sodium fluorescein	1158
PVP	Chlorazol Sky Blue FF and Orange II	717
PVP	Naphthalene and biphenyl	1109
PVP	Hexylresorcinol	1189
PVP	Alkylammonium halides and thiocyanates	1296
PVP	Alkylammonium thiocyanates	1291
PVP	Sodium alkyl sulfates	852
PVP	Mixed aliphatic amphiphiles	1292
PVP	Sodium, potassium, guanidinium, and tetraalkylammonium alkyl sulfates	1295

[a] Abbreviations:

PAA	poly(acrylic acid) and/or its salts.
PAAm	polyacrylamide.
PEI	polyethyleneimine and/or its salts.
PEO	poly(ethylene oxide).
PESA	poly(ethylene sulfonic acid) and/or its salts.
PMAA	poly(methacrylic acid).
PMAAm	polymethacrylamide.
PPA	poly(phosphoric acid) and/or its salts.
PPO	poly(propylene oxide).
PVA/VAc	poly(vinyl alcohol) and/or poly(vinyl acetate) and/or their copolymers.
PVME/MA	poly(vinyl methyl ether)co(maleic anhydride).
PVMO	polyvinylmethyloxazolidone.
PVP	polyvinylpyrrolidone.

Table IV lists some of the studies in this area. The studies are listed alphabetically under the name of the individual polymer concerned, and then in chronological order of the publication for each polymer. The list includes cases where the polymer by itself is water insoluble [e.g., poly-(vinyl acetate)] but becomes soluble in the presence of the cosolute due presumably to the binding of the (hydrophilic) cosolute; also included are cases where binding does *not* take place and where its absence is significant in the light of the structures of the polymer and the cosolute. In addition, examples of binding by polyelectrolytes of organic counterions are also included where it is evident that the nonionic part of the cosolute makes a marked contribution to the binding process.

Broadly speaking, binding shows itself most directly by a lowered thermodynamic activity for the cosolute at that (total) concentration compared with the activity in the absence of the polymer. Thus, when the cosolute is in the pure state, moderately soluble, the binding may be detected by solubility measurements since it leads to increased (total) solubility compared with that in the pure solvent. In the more general case the study of the binding equilibrium involves such techniques as equilibrium dialysis, electrical conductance (ionic cosolutes), potentiometry (if a reversible electrode is available whose potential is sensitive to the cosolute concentration), polarography (for electroreducible or electrooxidizable cosolutes), spectrophotometry (applicable when there are marked spectral charges on binding), and so forth. The aim throughout such studies is to determine separately the concentrations of free and of bound cosolute, over a range of concentrations at a fixed temperature, to define the binding isotherm. In many cases[73,86,472,717,1031,1131,1338] the amount of cosolute bound per unit amount of polymer r is related to the concentration of free cosolute a by the hyperbolic isotherm

$$r = Kan/(1 + Ka) \qquad (68)$$

which is closely analogous in form to the Langmuir isotherm. This type of isotherm indicates that at the molecular level the binding involves the reversible attachment of cosolute molecules onto a fixed number of equally active and independently acting sites, with n corresponding to the site density (in the same units as r) and with K the equilibrium constant for the attachment of a molecule of cosolute onto a site. Extension of the measurements over a suitably wide range of temperatures then enables the standard free energy of binding (calculated from K) to be divided up into the corresponding enthalpic and entropic contributions, thus giving information upon the changes that occur in the energy and the degrees of freedom of the system when binding takes place.[73,87,717,792,1030,1031,1338]

The binding of the cosolute often leads to changes in the conformation of the polymer molecule, which are detectable by such techniques as viscosity and light scattering. Thus, binding of one of the ion types (say the anions) from an ionic cosolute converts the polymer into a polyelectrolyte, and leads to expansion of the polymer coil.[1032,1033] Conversely, cosolutes such as phenols and aromatic acids upon binding cause contraction of the polymer coil, possibly due to hydrogen-bonded cross-links[1032]; at a sufficiently high concentration of the cosolute the polymer precipitates out, which in qualitative terms can be viewed as due to the additional formation of *inter*molecular cross-links.[578,747,1032,1357]

Many of the experimental techniques applied to the study of binding by synthetic polymers are simply more recent applications of those originally developed for use in the even more extensively studied field of protein binding, in particular to such types of cosolutes as drugs,[1008] detergents,[1205] metal ions,[696] and dyes.[1274] Cosolute binding is also shown in vitro by other natural polymers, such as nucleic acids[742] and polysaccharides[227]; indeed, for these two types of polymer, cosolute binding has such directly practical features as the ready staining of nuclei by basic dyes (which led to the coining of the terms "chromatin" and "chromosome"), and the blue complex formed by starch with iodine, which is widely applied in analysis.

Furthermore, for a biopolymer in its native environment, in the great majority of cases a specific ability to reversibly bind particular small-molecule cosolutes forms an essential part of its role in the cell or organism. The systematic study of cosolute binding by synthetic polymers of simple and well-defined structures should do much to cast light on the fundamentals of these types of interactions in aqueous systems; in particular, it should provide basic information upon such component forces between polymers and cosolutes in aqueous systems as ionic interactions, hydrogen bonds, hydrophobic bonds, and so forth, which can then be applied to clarify the generally more complex effects observed with the biopolymers in their native environment.

3.6.2. Solvent Perturbation by the Cosolute

Under this heading we consider effects which take place at relatively high concentrations of the small-molecule substance, i.e., very approximately 1 M and above, where its concentration becomes comparable to that of the solvent itself (for water about 55 M). It is convenient to consider these effects separately for the cosolutes of three types: nonelectrolyte, electrolyte, and the protein denaturing agents (comprising urea, guanidinium salts, and related compounds). We will discuss some of the published work of the effect on water-soluble polymers of high concentrations of cosolutes under this threefold classification; for each type of cosolute the common polymers will be dealt with in the essentially alphabetical order: the "acrylic group" [poly(acrylic acid), poly(methacrylic acid), and their amides]; poly(ethylene oxide); poly(vinyl alcohol) and its copolymers with vinyl acetate; and polyvinylpyrrolidone.

By far the commonest method which has been used for studying the effects of cosolutes at high concentrations is viscometry. If the l.v.n. of the polymer in the absence of cosolute is $[\eta]_0$ and that in the presence of a concentration s molar is $[\eta]_s$, then the viscosity increment $\Delta[\eta]$ is given by

$$\Delta[\eta] = [\eta]_s - [\eta]_0 \qquad (69)$$

Since commonly $\Delta[\eta]$ is proportional to s (particularly at not too high cosolute concentrations), it is convenient to use the molar fractional viscosity increment F_s defined by

$$F_s = \Delta[\eta]/[\eta]_0 s \tag{70}$$

where the "normalization" through the division by $[\eta]_0$ gives a quantity which should be largely independent of the molecular weight of the polymer, and hence applicable in the comparison of the effects of the same cosolute on different types of polymer. The subscript on F_s serves to emphasize that in general its value will depend upon the cosolute concentration; values that are not so-dependent, or which have been extrapolated to zero cosolute concentration, may be denoted by F_0 (limiting molar fractional viscosity increment).

3.6.2a. *Nonelectrolyte Cosolutes.* The most common types of cosolutes in this class which have been used are water-miscible liquids such as the alcohols and acetone; some limited studies have also been carried out with water-soluble solids, such as sucrose. In the first case, that of the liquids, the behavior would in the first instance be expected to be similar to that shown in general for a polymer in a mixed solvent, or solvent plus nonsolvent system.

With this situation in nonaqueous systems the addition of a specific miscible nonsolvent to a solution of the polymer in a particular solvent generally leads to the precipitation of the polymer at a certain critical solvent/nonsolvent ratio, which depends upon the natures of the three components, the molecular weight of the polymer, and the temperature. Studies on the conformation of the polymer coil up to the precipitation point by, say, viscometry, show the polymer coil to be progressively contracted as the concentration of added nonsolvent is increased. Conversely, where the added substance is a better solvent than the initial one, then the viscosity (and in particular limiting viscosity number) may rise. In either case one may, however, get a maximum in the limiting viscosity number due to the polymer having a cohesive energy density value between that of the two liquids.

With aqueous solutions general effects of these types will be occurring alongside those resulting, for example, from any specific structure-making or structure-breaking effects of the cosolute upon the water. Considering *the acrylic group* of polymers, Eliassaf and Silberberg[398] have studied the effects of cosolutes of various types on the behavior of PAA, PAAm, and PMAA in aqueous solution; the studies involved measurements of η_{sp}/c_2

which, since they were made at low polymer concentrations (0.03–0.31%), may be taken (as in this discussion) to be $[\eta]$ values. Eliassaf[395] has supplemented these data with measurements of $[\eta]$ for PMAA in the presence of various cosolutes, studied in each case at only one concentration. The results on the four polymers are summarized in Table V. We may correlate the data obtained with the thermodynamic behavior of the

TABLE V. Limiting Molar Fractional Viscosity Increment F_0 (M^{-1}) for Polymers in Aqueous Solution in the Presence of Cosolutes[a]

Cosolute	Maximum conc. s_{max}, M	Polymer[b]			
		PAA[c]	PAAm	PMAA[d]	PMAAm
Ethanol	6	+0.2	−0.07 ($p. > 5\ M$)	−0.3[e]	$p.$[f]
Acetic acid	3	+0.37	n.s.	−0.24	n.s.
Caprylic acid (octanoic acid)	0.004	0	n.s.	−60	n.s.
Perfluoroctanoic acid	0.014	0	n.s.	0,[g] +120[g]	n.s.
Sodium chloride	1.5(3.3[h])	−0.5 ($p.\ 3.5\ M$)	0	−0.3 ($p.\ 1.5\ M$)	−0.07 ($p. < 5\ M$)
Lithium chloride	1.5(5.0[h])	$-\infty$[i] ($p.\ 2.4\ M$)	0	−0.3 ($p. \sim 3\ M$)	+0.07[j]
Potassium thiocyanate	2.0	n.s.	n.s.	n.s.	+0.4[j]
Urea	3-7 (6.7[h])	+0.04	0	−0.88	+0.2[j]
Guanidinium chloride	0.1	n.s.	n.s.	−25 ($p. < 0.25\ M$)	n.s.

[a] Temperature 28–30°. Abbreviations: n.s. indicates that this polymer + cosolute system had not been studied; $p.$ indicates that the polymer precipitated out at the cosolute concentration quoted. Calculated from the data of Eliassaf and Silberberg[398] and of Eliassaf.[395]

[b] PAA: poly(acrylic acid); PAAm: polyacrylamide; PMMA: poly(methacrylic acid); PMAAm: polymethacrylamide.

[c] In 0.2 M hydrochloric acid as solvent.

[d] In 0.02 M hydrochloric acid as solvent.

[e] The value of $[\eta]$ reaches a minimum at $s = 3\ M$, above which it rises again.

[f] Concentration for precipitation not stated.

[g] $F = 0$ for $0 < s < 8$ mM; $F = +120\ M^{-1}$ for $s > 8$ mM.

[h] For PMAAm.

[i] Unresolved initial "vertical" fall in $[\eta]$ to $\sim 0.15[\eta]_0$ at $s \approx 0.2\ M$, slower fall thereafter.

[j] No precipitation of the polymer at 0° in the presence of the cosolute at its maximum concentration, in contrast with the behavior of this polymer alone in aqueous solution.

four polymers alone in aqueous solution (see Section 3.5.1); all of the polymers except PMAA show upper consolute temperature behavior, with respective Θ temperatures of: PAA, $+14°$; PAAm, $-38°$; PMAAm, $+6°$; while with PMAA the Θ temperature of $56°$ corresponds to a lower consolute temperature. Thus insofar as a cosolute may act by altering the "structural temperature" of the water, then the effect ought to be in the same direction for the group of the first three polymers, but probably less marked for PAAm because its Θ temperature is much lower, while it should be in the opposite direction for PMAA. With ethanol, PAA increases in $[\eta]$, with PMAA there is a decrease followed by a minimum and then an increase, while with PAAm, $[\eta]$ decreases fairly steadily and eventually (at ethanol concentrations greater than $5M$) precipitation occurs; no viscosity data are reported for PMAAm with ethanol, but the fact that precipitation take place at sufficiently high concentrations suggests that reductions in $[\eta]$ take place with this polymer also. These results are consistent with a raising of the structural temperature of the water in the case of the acids, but a lowering in the case of the amides. The use of acetic acid in place of ethanol for the two acids gave similar viscosity results.

Eliassaf and Silberberg[398] also studied the viscosity effects with the cosolutes perfluoroctanoic acid and caprylic (octanoic) acid on PMAA; the interest in the fluorinated compound lies in its activity in preventing denaturation of proteins. With the fluoro acid there was no effect up to a concentration of about 8 mM, above which there was a linear rise in $[\eta]$ with increase in concentration, equivalent to a value of $+120\ M^{-1}$ for F. On the other hand, with caprylic acid itself there was a fall in $[\eta]$ starting from the lowest concentrations, equivalent to $F_0 = -60\ M^{-1}$. The low concentrations involved suggest strongly that it is *polymer binding* rather than *solvent perturbation* that is involved. Since perfluoroctanic acid is a strong acid, it will be present almost entirely as the dissociated form, and the binding of its anions would then give a viscosity increase; the fact that this only takes place above a threshold concentration (8 mM) may result from the compact conformation of this polymer and its resistance to unfolding. With caprylic acid itself, on the other hand, since this is a weak acid, then it will be present mainly as the undissociated form, and binding of this followed by hydrogen-bonded cross-links via the carboxyl groups could give rise to the viscosity reductions observed.

Poly(ethylene oxide) has not been studied very thoroughly for effects of nonelectrolyte cosolutes on its conformation. As regards the alcohols, McGary[949] has shown that the viscosity of a 1% solution of PEO reaches a maximum at about 50% by volume in water/isopropanol mixtures. Blandamer and his colleagues[147] have studied the effect of adding tertiary butanol on the behavior of poly(ethylene oxide) in water; the alcohol was added in concentrations up to 0.1 mole fraction, i.e., 6.2 M. The system showed a very complex series of variations in $[\eta]$ as the alcohol concentration was increased, the value of $[\eta]$ passing successively through a minimum, a maximum, another (poorly defined) minimum, and another (also poorly defined) maximum; the extreme variation in $[\eta]$, however, was less than 10%. The authors[147] interpreted these effects in terms of specific interference by the cosolute with the water structure. However, the results may be due at least in part to binding of the alcohol; it is known, for example, that its trichloroderivative chlorbutanol is bound by another water-soluble polymer, polyvinylpyrrolidone.[73] It is clear that the effect of alcoholic cosolutes on this polymer should provide a fruitful field for further study, particularly in view of the simplicity of the structures of the two components.

With *poly(vinyl alcohol)* Noguchi and Yang[1099] have studied the effect of sucrose on mildly cross-linked films of the polymer (see Table VI). The polymer films were prepared by heat treatment first at 80° and then at 110°, so that they were insoluble in water at normal temperatures, simply swelling on equilibration to increase in length by about 30%; nevertheless,

TABLE VI. Molar Fractional Change in Length $\Delta L/L_0 s$ of Poly(vinyl alcohol) Films Immersed in Aqueous Solutions[a]

Solute	Maximum concentration $s_{\max}$, M	$\Delta L/L_0 s$, M^{-1}
Sucrose	2.5	-0.01[b]
Lithium bromide	3[c]	$+0.011$
Sodium chloride	5	-0.027
Potassium fluoride	10	-0.039[d]
Urea	8	$+0.011$[e]
Guanidinium chloride	3[f]	$+0.023$

[a] Relative to length, L_0, in pure water; from the data of Noguchi and Yang.[1099]
[b] For $S < 1$ M, beyond which the plot of $\Delta L/L_0$ versus s increases in slope and approximates more closely to $\Delta L/L_0 \propto s^2$.
[c] Film dissolved at $s = 4$ M.
[d] For $s < 6$ M, beyond which the plot of $\Delta L/L_0$ versus s decreases in slope.
[e] For $s < 3$ M, beyond which the plot of $\Delta L/L_0$ versus s decreases in slope.
[f] Film dissolved at $s = 5$ M.

the films dissolved fairly rapidly in water at about 65°, indicating that the insolubility is simply due to entanglement of the polymer chains and crystallization during the heat treatment. Over the concentration range up to 2.5 M the presence of sucrose shrank the film by maximally 10% in its linear dimensions; however, the effect is not simply proportional to the sucrose concentration but seems to depend more nearly upon the square of the sucrose concentration. Viscosity studies on the polymer in solution showed that $[\eta]$ remained virtually constant up to 0.5 M sucrose, fell off to 75% of its initial value at 1 M, and then beyond this point began to increase until at concentrations higher than 1.5 M the polymer precipitated out.

For *polyvinylpyrrolidone* Eliassaf and his colleagues[397] have studied, mainly by viscosity measurements, the effect of a range of cosolutes on PVP in aqueous solution. As already indicated, we somewhat arbitrarily take effects shown at the ≤100 mM level to indicate polymer binding (as already included for this polymer in Table IV) while at the 1 M level and above any effects are taken to be due to solvent perturbation by the cosolute. Table VII, where the general data for this latter case are summarized, shows that the nonelectrolytes studied in this way included glycine, leucine (both at pH 6.4, i.e., essentially their isoelectric points), and also acetic acid and propionic acid (both at pH < 3 so that their ionization was supressed). It will be seen that both of the amino acids caused shrinkage of the polymer coil. With glycine a concentration of 2.4 M (at 30°) caused the precipitation of the polymer; additional data presented show that this represents a lower consolute point, in that with 1.9 M glycine the system was clear at 30° but became turbid at 44°, while with 2.7 M the system was turbid at ambient temperatures but cleared on cooling to 10°. Extrapolation of these rather fragmentary results indicates that with the glycine at zero concentration (that is, in pure water) the system should show a lower consolute point at very approximately 130°; this is in general agreement with the deductions made above (see Section 3.5.1) from the effect of temperature on the degree of swelling of cross-linked PVP gels in water. It may be noted that aspartic acid, glutamic acid, and serine all caused marked shrinkage of the polymer coil at the 100 mM level, unlike the other aliphatic amino acids (including glycine and alanine) at these low concentrations; the occurrence of effects at such low concentrations is indicative of *polymer binding* along with cross-linking effects of the bound cosolute.

The marked shrinking effects of glycine and alanine at the molar concentration level, when compared with the relative ineffectiveness of

TABLE VII. Molar Fractional Viscosity Increment F for Polyvinylpyrrolidone at 30° in Aqueous Solution with Cosolutes[a]

Cosolute	s, M	F, M^{-1}
Glycine (pH 6.4)	1.4	−0.14
Glycine (pH 6.4)	2.0[b]	−0.22
Alanine (pH 6.4)	1.8	−0.10
Acetic acid (pH 2.2)	2.4	+0.04
Propionic acid (pH 2.5)	1.3	0
Sodium acetate (pH 6.4–6.6.)	2.0	−0.11
Sodium propionate (pH 6.4–6.6)	1.3	−0.014
Ammonium chloride	1.0	0
Sodium chloride	4.3	−0.03
Sodium perchlorate	0.2	−0.12
Sodium perchlorate	2.0	−0.06
Potassium thiocyanate	2.0	+0.05
Sodium biphospate	0.2[c]	−1.6
Sodium sulfate	0.2[d]	−0.67[e]
Sodium tetraborate	0.1	−0.082

[a] Unless otherwise indicated, the polymer used was of grade K-60, with $\bar{M} = 5.5 \times 10^5$. Data of Eliassaf et al.[397]
[b] Precipitated polymer at $s = 2.4$ M; see text.
[c] Precipitated polymer at $s = 0.3$ M.
[d] Precipitated polymer at $s = 0.7$ M.
[e] With PVP K-90 ($\bar{M} = 9.4 \times 10^5$), $F = -1.1$ M^{-1}.

their deaminated derivatives acetic acid and propionic acid, is evidence of the role of the charged groups on the zwitterionic amino acids in these effects; on the other hand, sodium acetate and sodium propionate have closely similar effects to those of the equivalent amino acids, indicating that the $-NH_3^+$ group on the amino acid is not markedly involved.

With PVP also, Jirgensons[728] showed that in water/propanol mixtures the viscosity (measured as η_{sp}/c_2) shows a maximum (at about $\phi_3 = 0.15$) and a minimum (at about $\phi_3 = 0.60$); the initial addition of propanol thus leads to an increase in η_{sp}/c_2. The same worker[729] studied the acetone precipitation behavior (the liquid is a borderline solvent/nonsolvent for PVP); unlike glycine, the precipitation concentration rises with increase of temperature, although not especially markedly—thus at 5° and for a

0.55% solution of the polymer the volume fraction of acetone at the turbidity point was 0.713, while at 35° it was 0.725.

The rather fragmentary nature of these published results indicates the scope available for further studies of the effect of molar concentrations of nonelectrolyte cosolutes, such as the alcohols, sugars, and the amino acids, on the behavior of synthetic polymers in aqueous solution. Such studies, however, need to be supplemented by other techniques (such as, for example, equilibrium dialysis and cosolute solubility measurements) to ensure that any effects observed are not simply the result of polymer binding.

3.6.2b. *Electrolyte Cosolutes.* Upon addition of an electrolyte (salt) to an aqueous solution of a nonelectrolyte polymer, we may expect in the first instance to get similar types of "salting-in" or "salting-out" effects as observed with small-molecule nonelectrolytes,[915] possibly accompanied by other effects due specifically to the polymeric nature of the solute. One simple theoretical treatment of these former types of effects is that of Debye and McAulay,[330,609] who assumed the ions to be rigid spheres immersed in a continuous medium of bulk dielectric constant ε. The presence of a nonelectrolyte solute at molar concentration c alters the value of ε from that of the pure solvent $\varepsilon°$, the variation generally taking the form*

$$\varepsilon = \varepsilon°(1 - Bc)\qquad(71)$$

where the negative sign indicates that for the majority of solutes (with the notable exeption of such highly dipolar ones as the amino acids[381]) the dielectric constant falls with increasing solute concentration. Since, due to this lowered dielectric constant, the nonelectrolyte present makes the solution a poorer solvent medium for the ions of the salt, then by a reciprocal action the ions present make the solution a poorer solvent medium for the nonelectrolyte, so that its activity coefficient f is increased and it tends to be salted out; this treatment leads then to the Debye–McAulay equation, which when converted to molar terms from the molecular units in which it is more usually quoted, becomes

$$\ln(f/f°) - (BF^2/2RT\varepsilon°N)\sum(c_j z_j^2/r_j)\qquad(72)$$

where F is the Faraday constant, N is the Avogadro constant, c_j is the molar concentration of the jth type of ion, z_j is its charge number, and r_j is its radius. For the simplest case of a uni-univalent electrolyte MA at

---

* The values of B for small-molecule solutes have been tabulated by Hasted in Chapter 7 of Volume 2 of this treatise.

concentration c this becomes

$$\ln(f/f^\circ) = c(BF^2/2RT\varepsilon^\circ N)[(1/r_M) + (1/r_A)] \tag{73}$$

Thus, in general terms, for nonelectrolytes that decrease the dielectric constant of the medium (i.e., B positive), the electrolyte will increase the value of f and hence tend to salt out the nonelectrolyte; the dependence upon z^2 indicates that polyvalent ions should be of much higher efficiency than monovalent ones, while the inverse dependence upon r suggests that the smaller the ion, the greater its salting-out effect. For further discussion of the experimental data and the various theories developed on the effect of electrolytes upon small-molecule solutes the review by Long and McDevit[915] and the monograph by Harned and Owen[609] should be consulted.

We now consider the effect of electrolyte cosolutes upon polymers in aqueous solution. Once again we shall see that the main experimental method used has been that of viscometry, along with the determination of precipitation points.

For *the acrylic group*, viscosity results[398,395] on the effects of lithium chloride and sodium chloride are presented in Table V. The data are difficult to explain fully, but some clarification can be obtained in terms of the Θ temperatures for the four polymers in water (compare Table III), in that a positive value of F may be interpreted as a move of the "structural temperature" away from Θ, and a negative value as indicating that the opposite is taking place. Thus, working upward from the lowest Θ values, for PAAm Θ is -38°; hence at the experimental temperatures the coil, since it is so expanded, would be expected to be insensitive to changes in solvent; this is confirmed by the zero values for F for both NaCl and LiCl. The polymer with the next higher value of Θ is PMAAm, for which Θ is $+6^\circ$; here there is a striking contrast between the effects of the two salts, for LiCl increases both the polymer's coil size and its solubility (so that with $5 M$ salt it is still soluble even at 0°) while NaCl both reduces the coil size and reduces the solubility (the polymer precipitating out at 30° with $5 M$ salt concentration). For PAA, where Θ is $+14^\circ$, there is also a marked contrast between the effects of the two salts, although here only of degree, with even small amounts of LiCl causing an abrupt fall in viscosity while with NaCl the reduction is less sharp. Finally, with PMAA, where Θ is $+56^\circ$ (representing lower consolute behavior), there are closely similar effects, again both reductions in coil size and precipitation at a high enough cosolute concentration, for both salts. Any further interpretation of these effects, in particular on the basis of the Debye–McAulay theory, requires additional data such as, in particular, the dielectric decrements B of these systems.

Considering *poly(ethylene oxide)*, Bailey and Callard[75] have studied a wide variety of inorganic salts, comparing their efficiencies in reducing the precipitation temperature of the polymer and in reducing $[\eta]$ at a fixed temperature. In fact, these two effects run parallel to each other, in that there was a common relation for the salts studied between precipitation temperature and $[\eta]$ at 30°. The reduction in precipitation temperature was also approximately proportional to the concentration of added salt. Table VIII summarizes the reductions (either actual or extrapolated) produced

TABLE VIII. Values of the Molar Reduction ΔT_m (actual or extrapolated) in the Precipitation Temperature (97.5°) of Poly(ethylene oxide)^a on Addition of Salts^b

Salt	Maximum concentration s_{max}, M	ΔT_m, deg
Potassium iodide	1	0.5
Lithium chloride	1	7.5
Potassium bromide	1	11.5
Magnesium chloride	1	14.0
Potassium chlorate	0.5	16
Calcium chloride	1	17.5
Sodium chloride	1	23
Potassium chloride	1	25.5
Sodium acetate	1	30.5
Potassium fluoride	1	46.5
Potassium hydroxide	1	69.5
Lithium sulfate	1	77.5
Zinc sulfate	0.75	84
Magnesium sulfate	0.45	125
Potassium sulfate	0.5	130
Sodium silicate	0.5	132
Potassium carbonate	0.5	138
Sodium carbonate	0.5	148
Trisodium phosphate	0.2	240

^{a} $\bar{M} = 4 \times 10^6$; $c_2 = 0.5\%$.
^{b} From data of Bailey and Callard.[75]

by a 1 M concentration of each particular salt; the values quoted should be taken to represent only a ranking order of the effectiveness of each salt, since for the more effective salts the maximum concentration used was considerably less than 1 M, so that the value quoted in Table VIII was obtained by linear extrapolation from the data at lower concentrations; in addition, as with many of the derived data quoted in this chapter, the values were obtained from results presented only graphically by the original workers. In very broad terms, the results are in accord with the Debye–McAulay theory; thus, as the theory requires, the ΔT_m values for the potassium halides are in order: $I^- < Br^- < Cl^- < F^-$ i.e., inverse order of the ionic radii, while there are also markedly greater values for the divalent anions (sulfate and carbonate), and even more so for the trivalent anion phosphate, than for the salts having monovalent anions. On the other hand, considering the cations, and the alkali metals in particular, the effects seem to be in the reverse order to that expected on the basis of ionic radii, i.e., with the chlorides the order of ΔT_m is $Li^+ < Na^+ < K^+$, and the same order applies to lithium and potassium sulfates; possibly it is the hydrated radii which are the effective quantities, since as is well known from conductance measurements in aqueous solution these follow the order $Li^+ > Na^+ > K^+$. However, it is notable that ΔT_m seems to be much more sensitive to the nature of the ions than would be expected from the Debye–McAulay theory, as witness the remarkably small values of ΔT_m for KI and LiCl.

The effect of the simple salts sodium and ammonium chlorides on PEO in aqueous solution at 10° has been studied by Hammes and Swann,[598] using viscometric and ultrasonic attenuation techniques (the results they obtained with protein denaturing agents are considered below). Both NaCl (at 5.0 m) and NH_4Cl (at 6.0 m) reduced $[\eta]$; the reductions corresponded to F values of 0.08 m^{-1} and 0.05 m^{-1}, respectively. For the ultrasonic measurements it was shown that the single relaxation process observed with poly(ethylene oxide) alone (see Section 3.5.1) is replaced by a combination of longer relaxation times representing new "reactions," which could not be more clearly defined from the data available.

On the other hand, Liu and Anderson[909] found that the addition of $\leq 0.5\ M$ K_2SO_4 had no appreciable effect on the proton magnetic resonance spin–lattice relaxation time T_1 for the methylene groups on PEO in D_2O at 34.5°, although they confirmed that the value of η_{sp}/c_2 (essentially equal to $[\eta]$, since it was measured at $c_2 = 1.0$ g dm^{-3}) was reduced to a very low value in this solution compared with that in pure water. Insofar as nuclear magnetic relaxation is concerned primarily with the relative motions of the neighboring nuclei, then these results show that the local segmental motion of the polymer chain is not affected even by changes which bring the polymer to the point of precipitation.

The effects of salts on PEO in aqueous solution may be viewed alternatively in terms of alterations in the "structural temperature" of the water. Thus, the reductions in precipitation temperature of the polymer caused by the salts imply an increase in the structural temperature; similarly the fact that the polymer shows lower critical consolute behavior, with $\Theta \approx 90°$, implies that the viscosity reductions observed represent a move toward this point, that is, again an increase in the structural temperature. Erlander[415] has discussed these effects on the basis of his "ionic sequence method."

For *poly(vinyl alcohol)* Maeda and his colleagues[958] have studied the effect of the addition of NaOH both on the polymer in aqueous solution (at 30°) and on heat-treated films (at 25 and 50°). In solution, the presence of NaOH caused little change in $[\eta]$ up to about 1.5 M, at which there was a marked reduction by about 30% to another essentially constant level. With the films it was found that the alkali concentration is lower in the gel phase (film) than in the surrounding solution, as would be expected if the phase is of lower dielectric constant; nevertheless, the equilibrium swelling ratio increased with the external NaOH concentration, although showing none of the marked changes observed in $[\eta]$ with a 1.5 M solution. The authors suggest that the observed effects may be explained by breakage of hydrogen bonds involved in junction points between the molecules and in intramolecular links between near-neighbor hydroxyl groups on the polymer chain.

Noguchi and Yang[1099] have similarly studied the effect of LiBr, NaCl, and KF upon PVA in film form (see Table VI). Here, as with most other polymers, the latter two halides shrunk the polymer films; however, LiBr expanded them, while at 4 M and above it dissolved the film. Noguchi and Yang suggest that this difference in the behavior of the alkali halides may be connected with the fact that LiBr is soluble in ethanol, while NaCl is only slightly soluble and KF is essentially insoluble in the alcohol.

The solubility of *polyvinylpyrrolidone* is affected by salts in a similar way to that of PEO, in that is it most readily precipitated out from solution by salts of polyvalent anions such as carbonate, silicate, sulfate, and phosphate.[34] The precipitation point depends upon the molecular weight of the polymer sample, and Na_2SO_4 solutions have been used as precipitants in the turbidimetric titration method for the analysis of molecular weight distribution in PVP.[1340]

Jirgensons[729] has studied the precipitation of PVP from aqueous solution by ammonium sulfate. As with glycine (see above), increase of temperature leads to a decrease in the amount of cosolute required for incipient precipitation; the data indicate an approximately linear relation between the two quantities over the temperature range 4–65°, and a rather long extrapolation suggests that at zero ammonium sulfate concentration the precipitation temperature would be about 150° (for a 1% solution of polymer with $\bar{M} \approx 10^5$), which is similar to the value ($\sim 130°$) obtained for glycine.

More detailed results[397] on the effect of salts on PVP (see Table VII) show that in almost all cases shrinkage of the polymer coil takes place; at sufficiently high concentrations of sodium biphosphate and sodium sulfate the polymer is precipitated out. So effective are these two in particular that the precipitation concentrations are only $0.3\ M$ and $0.7\ M$ respectively, and marked viscosity changes are detectable even at $0.1\ M$. The high activity of salts having polyvalent ions again parallels the effects observed with other polymers such as PEO (compare Table VIII). The exceptions to this general behavior are NH_4Cl, which has no detectable effect even at $1\ M$, and KCNS, which causes a small but measurable expansion; it may be noted that this same cosolute also caused expansion of the coil in the case of polymethacrylamide (see Table V), which may indicate that for each polymer there is some weak binding of this polarizable anion.

3.6.2c. *Denaturing Agents.* We finally consider the effects produced on synthetic polymers by such typical protein-denaturing agents such as urea and other amides, and guanidinium chloride. Because of their specific action on proteins, it is interesting to look for any similarly specific effects when these types of compounds are used as cosolutes with synthetic polymers in aqueous solution, and in particular to see what conformational changes occur and what light any such effects may cast on the denaturation of proteins. Although protein denaturation, in the widest sense of the term, can be induced by a very wide range of chemical reagents as well as by heat, the particular interest in the organic denaturants just named is that their action is often essentially reversible (Ref. 1461, Chapter 9). The most spectacular example of this is guanidinium chloride, which at a sufficiently high concentration ($\sim 6\ M$) and in combination with an agent to cleave any disulfide bonds converts the initially compact conformation of a wide range of proteins into an essentially random coil shape, while upon removal of the guanidinium chloride the native structure of the protein is spontaneously recovered.[1465]

One intriguing point concerning aqueous solutions of urea in particular is that their thermodynamic behavior is approximately ideal, i.e., the vapor pressure of the water is essentially proportional to its mole fraction.[399,1310] The apparent effect of the urea is thus almost to act as an inert diluent which simply reduces the concentration of water molecules in the liquid phase.[485] The effects of urea upon water and its structure have been discussed in Chapters 1 and 5 of Volume 2 of this treatise.

The effects of urea and guanidinium chloride at 28–30° on the *acrylic group of polymers* in aqueous solution are shown in Table V.[395,398] It is notable that in all cases except PAAm the effect of urea is quite clearly to increase the structural temperature of the system. Thus with PAA and PMAAm, which have Θ temperatures of 14 and 6°, respectively (see Table III), the addition of urea increases $[\eta]$, thus moving the polymer further away from its ideal conformation at the stated lower temperatures; with PMAA, which has a Θ temperature of 56°, the addition of urea decreases $[\eta]$, thus moving the polymer toward its ideal conformation at this higher temperature. The absence of any detectable effect on PAAm may be attributable to this system being so far from its Θ temperature ($-38°$) that there are no appreciable residual "long-range" interactions with which the urea might interfere. This may be compared with the similar absence of any effects with either NaCl or LiCl and the relatively small effect with ethanol (Table V). Guanidinium chloride was studied as a cosolute only with PMAA, where it showed approximately 30 times the specific efficiency of urea in shrinking the polymer coil, ultimately precipitating the polymer at $0.25\ M$.

The effect of urea, acetamide, and guanidinium chloride on *poly(ethylene oxide)* in aqueous solution has been studied[597,598] by ultrasonic attenuation and viscosity. Figure 27 shows the results obtained with guanidinium chloride[598]; urea gave closely similar results.[597] The notable feature is that, whereas $[\eta]$ shows a steady but not spectacular rise as the "denaturant" concentration is increased (giving $F = 0.035\ m^{-1}$), the ultrasonic relaxation time τ shows a marked and abrupt fall of about 25% at a cosolute concentration of between 2 and 3 m, the value of τ at the lower and higher cosolute concentrations being essentially independent of concentration. With acetamide, even at 6 m, the value of τ was essentially the same as that in pure water, while the value of $[\eta]$ was only increased somewhat ($F = +0.008\ m^{-1}$).

These results with PEO again demonstrate the differences in the nature of the processes seen by ultrasonic and viscosity measurements, with the former sensitive to the local environment of the chain and its interactions

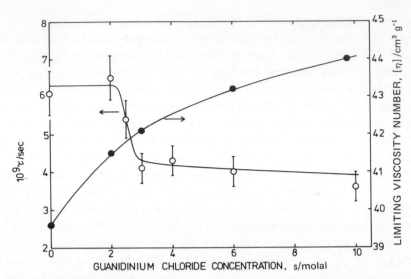

Fig. 27. Effect of guanidinium chloride at concentration s upon the ultrasonic relaxation time τ (open circles and left-hand scale) and upon the limiting viscosity number $[\eta]$ (filled circles and right-hand scale) for poly(ethylene oxide) ($\bar{M} = 2 \times 10^4$) in aqueous solution at 10°; data of Hammes and Swann.[598]

with the solvent, while the latter is more sensitive to the "long-range" interactions, i.e., the contacts between segments only distantly linked as measured along the polymer chain.

In the case of *poly(vinyl alcohol)* Noguchi and Yang[1099] in their studies on heat-treated films of the polymer showed that both urea and guanidinium chloride caused marked expansions of the film (Table VI), the latter being twice as effective as the former at the same concentration and leading indeed to dissolution of the film with concentrations $>5\ M$. These results were paralleled by those from viscometry, where for urea the value of F is approximately $+0.02\ M^{-1}$. On the other hand, Maeda and his colleagues,[958] using polymer samples of fairly closely similar values of $[\eta]$ to those of Noguchi and Yang,[1099] found much higher efficiencies for urea, i.e., $F = +0.10\ M^{-1}$. These differences may possibly be ascribable to differences in the amounts of residual acetate groups in the polymer samples used by the two groups of workers (see the effect of urea on copolymers, discussed below). However, in both cases the effect of the agent was again to raise the structural temperature of the water.

As regards *polyvinylpyrrolidone*, Klotz and Russell[791] have shown that urea had only a minor effect on $[\eta]$ of a sample of the polymer ($\bar{M} = 4 \times 10^4$) in aqueous solution at 25°; in 8 M urea there was a small decrease in $[\eta]$ equivalent to $F = -0.006 \ M^{-1}$. Similarly, Breitenbach and Schmidt[178] found little or no change in the degree of swelling Q of polymer gels in 0.2 or 0.4 M urea compared with that in pure water. Here, if the effect of urea is to raise the structural temperature of the water, then one might expect some reduction in $[\eta]$ or Q in the presence of urea; this is indeed shown in the case of $[\eta]$, while with the swelling measurements the rather low reproducibility in the values of Q and the rather low concentrations of urea used make any final conclusion uncertain. Seemingly, the fact that the polymer is so far away from its ideal state (and we have already deduced, from several lines of evidence, that for this system there is a lower critical consolute temperature at about 150°) means that there are no appreciable "long-range" interactions for a denaturant such as urea to perturb.

On the other hand, Klotz and his colleagues[793,794] have shown that the addition of urea does have an effect on the pK_a of dyes covalently attached to PVP. The primary effect observed in this case is that when such dyes are covalently conjugated either to a protein (such as serum albumin) or to PVP the pK_a of a basic (dimethylanilinium) group on the dye moiety is reduced markedly (from 3.99 to 2.54, in the case of the synthetic polymer); the addition of urea then serves to largely abolish this effect (with 8 M urea the pK_a for the synthetic polymer conjugate is increased to 3.24).[794]

Considering *synthetic copolymers*, some extremely interesting and significant results have been obtained by Saito and Otsuka[1293] on the effects of urea, alkylureas, amides, guanidinium chloride, and tetrabutylammonium bromide on the behavior of a series of such polymers in aqueous solution, in particular on vinylpyrrolidone/vinyl acetate copolymers (PVP/VAc), partially hydrolyzed poly(vinyl acetate)s (PVA/VAc), and also on poly(propylene oxide) (as a low-molecular-weight homopolymer). The two main properties studied were $[\eta]$ and cloud points (i.e., precipitation/dissolution temperatures). The most significant effects were obtained with a PVP/VAc sample (Luviskol VA37, with an approximately 70 mol % acetate content), which, although not soluble in water, could be prepared as an aqueous dispersion. Figure 28 shows the effects of urea and of tetrabutylammonium bromide on these dispersions. With 6 M urea $[\eta]$ was increased up to the value shown by the same polymer in acetone solution (which indeed, since this is a very poor solvent for PVP, would lead to a still rather compact conformation for the copolymer).* The effects observed may be attributed to the break-

---

* I am grateful to Dr. S. Saito for supplying the numerical values of these limiting viscosity numbers from his original data.

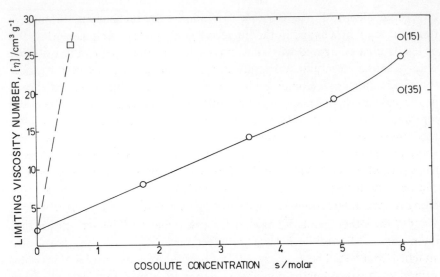

Fig. 28. Effect of urea (circles) and of tetrabutylammonium bromide (square) at concentration s on the limiting viscosity number $[\eta]$ of Luviskol VA37 vinylpyrrolidone–vinyl acetate copolymer in water; at 25° except where indicated. Data of Saito and Otsuka.[1293]

ing of the hydrophobic interactions between the vinyl acetate units of the copolymer chain. It is noticeable that the increase of $[\eta]$ with urea concentration is essentially linear, showing that even in 6 M urea there still remain appreciable hydrophobic interactions (otherwise the plot would at least begin to level off). Further, increase in temperature leads to an appreciable decrease in $[\eta]$ at this urea concentration, and the opposite for decreasing temperature, an effect again symptomatic of hydrophobic interactions. The efficiency of Bu_4NBr is much greater than urea in this respect (Fig. 28), since a concentration of only 0.6 M produces much the same viscosity increase as 6 M urea. Similar effects were noted for these and other polymer–cosolute pairs with the cloud points, which of course are lower consolute solution temperatures; the cloud points of the polymers were generally raised by the addition of the cosolute—thus whereas the cloud point of a 0.23% solution of the same PVP/VAc is about 3°, this is raised to over 60° in the presence of 8 M urea. These and other results are particularly significant because they represent reversible unfolding processes for compact molecules of purely synthetic polymers similar to those observed with the proteins.

Finally, to bring this discussion of cosolute effects full circle and link up with the starting point of polymer binding, we may note that for the three aqueous systems PAA + Methylene Blue,[1056] PMAA + aromatic hydrocarbons,[89] and PVP + Methyl Orange[792] the binding of the co-solute by the polymer is in each case largely abolished upon addition of 6–8 M urea. This common effect suggests that hydrophobic bonds are involved in the binding process in these, and probably also other, cases.

3.7. Interactions between Two or More Polymers in Aqueous Solution

When two different polymers are mixed in a common solvent, unless the polymers are structurally very similar, they generally undergo one or the other of two distinct types of interaction, which in each case leads to phase separation at polymer concentrations above a certain level. In the first and more general case, one of the two phases formed contains almost all of one polymer and very little of the other polymer, with the reverse situation in the second phase; this behavior is generally termed either *incompatibility* or *simple coacervation*. In the second and more special type of case, two phases are again formed but both polymers are concentrated in one of the phases ("precipitate"), while the other phase ("supernatant") may consist of almost pure solvent; this is generally termed *complex co-acervation*. In fact, since the term "coacervation" implies a "heaping to-gether," which is exactly the opposite of what takes place in incompatibil-ity, we shall reserve the unprefixed term *coacervation* for the second type of behavior where such "heaping together" of the polymers does occur. To rationalize these effects, we may extend the previous thermodynamic treatment for the free energy of mixing (Section 3.2) to this three-component case, with the second polymer solute called component 3. On this basis, and now labeling the interaction parameters χ specifically with the pair of components involved, the free energy change of formation of a homo-geneous solution (i.e., of mixing n_1 moles of solvent with n_2 moles of one polymer and n_3 moles of the other) is given by (Ref. 457, Chapter 13)

$$\Delta G^M = RT(n_1 \ln \phi_1 + n_2 \ln \phi_2 + n_3 \ln \phi_3 + \chi_{12}n_1\phi_2 \\ + \chi_{13}n_1\phi_3 + \chi_{23}n_2\phi_3) \tag{74}$$

It should be noted that this introduces not only χ_{13}, the parameter governing interactions of the second polymer with the solvent, but also χ_{23}, which is that governing the interactions between the segments of the two polymers in this particular solvent medium.

3.7.1. Incompatability

Although the phenomenon of incompatability must have been ob-served quite early on in the technical application of polymers particularly in such fields as paints and lacquers, it seems to have been first investigated systematically by Dobry and Boyer-Kawenoki,[353] and later by Kern and Slocombe.[764] These studies confirmed the generality of the phenomenon, the only exceptions (for the nonaqueous systems studied) being polymers of closely related structure such as polystyrene and its ring-methylated derivatives, and the isomeric pair poly(vinyl acetate) and poly(methyl acrylate).

In formal terms, incompatability results from a positive value for χ_{23}, and hence represents unfavorable energetic interactions between the two polymers. At least for moderately nonpolar systems, to which the cohesive energy density theory can be justifiably applied (Section 3.4), this is indeed the behavior expected when we recollect that the interaction energy for this pair depends upon $(\delta_2 - \delta_3)^2$, so that as long as the two polymers are distinctly different then $\delta_2 \neq \delta_3$ and the energy term will be positive whether $\delta_2 > \delta_3$ or $\delta_3 > \delta_2$. When combined with the only small favorable entropy contribution from the polymers in the mixing process (due to the small numbers of molecules for a given mass of the component), this leads to the system's preferring to divide itself into two phases where the polymers of different types are segregated from one another; in fact, for reasonably high polymers this entropy contribution is so small that a correspondingly small, positive value (say, $+0.002$) for χ_{23} would be expected to lead to incipient phase separation (Ref. 1485, Chapter 7).

The phenomenon of incompatability is not restricted to nonaqueous systems but occurs in aqueous systems as well. This area has been investigated in some detail by Albertsson.[10,11] Figure 29 shows one of the phase diagrams (binodials) from his extensive data, in this case for a PEO with a dextran. The thermodynamics of these aqueous systems has been studied by means of their phase equilibria and also via the techniques of sedimentation, osmotic pressure, and viscosity applied to the single-phase systems obtained at concentrations below the phase separation level.[380]

An important characteristic property of a particular system is the critical point, which is the point on the binodial at which the phases would have the same composition, with the tie line then being of zero length; this point may be obtained as the intersection of the locus of the midpoints of the tie lines with the binodial line on the phase diagram, but this latter line is always difficult to delineate close to the critical point, and it would therefore be useful to convert the phase diagram into a linear form by suitable change of coordinates. Toyoshima,[1488] in his discussion on incompatability as it applies to poly(vinyl alcohol) with other polymers in aqueous systems, has suggested (albeit without specific supporting experi-

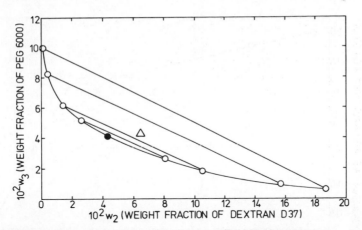

Fig. 29. Incompatability in aqueous two-polymer systems: phase diagram for water (1) + dextran D37 (2) + poly(ethylene oxide) PEG 6000 (3) at 20°, where w represents the weight fraction of the given component. The filled circle indicates the critical point for the system, and the triangle the point of intersection of the two lines in the double logarithmic plot of Fig. 30. Data of Albertsson.[11]

mental evidence) that at the phase separation point for a particular system the volume fractions of the two polymers are related by an equation of the form

$$\phi_2 \phi_3{}^\alpha = \beta \tag{75}$$

For such systems we may again use weight fractions in place of volume fractions, in which case eqn. (75) becomes

$$w_2 w_3{}^\alpha = \beta \tag{76}$$

which we can cast into a linear form:

$$\log w_2 + \alpha \log w_3 = \log \beta \tag{77}$$

Figure 30 shows this double logarithmic plot applied to the phase separation data of Fig. 29; it will be seen that this gives not *one*, but *two* straight lines intersecting at what would be expected to be the critical point for the system. However, the intersection point lies well above the actual critical point for the system, as Figs. 29 and 30 show. Similarly, the direct plots (w_3 versus w_2) for these two lines fall well within the two-phase region. It seems, from Figs. 29 and 30, that the linear plots are only the limiting behavior for systems well away from the critical point. One possible reason for the two-line plot is the heterogeneity of the polymers, whereby the

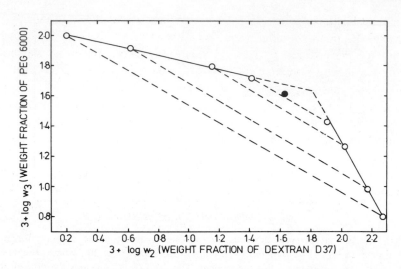

Fig. 30. Double logarithmic plot for data of Fig. 29, according to eqn. (77).

polymer in the phase richer in that particular polymer would be of higher molecular weight than that in the other phase (since the lower molecular weight material tends to distribute itself more evenly than the higher molecular weight material). Certainly, this type of plot (Fig. 30) seems to be a general one among the systems studied by Albertsson,[10,11] although the sharpness of the intersection becomes steadily less marked as the molecular weights of the polymers are reduced. It would be useful to extend these studies to highly fractionated material, both in aqueous and in nonaqueous systems, to see whether the behavior observed is general.

Albertsson has suggested that the study of incompatability in aqueous polymer systems provides a measure of the relative hydrophobic–hydrophilic character of the individual polymers; he has indeed proposed a "hydrophobic ladder" for various water-soluble polymers, where (in order of increasing hydrophobic character) the series is: dextran sulfate; carboxymethyldextran; dextran; hydroxypropyldextran; methylcellulose; poly(vinyl alcohol); poly(ethylene oxide); poly(propylene oxide).[10,11] However, it must be borne in mind that incompatibility is not likely to be governed to an important extent by the solvent–polymer interaction parameters χ_{12} and χ_{13}, which are those that give quantitative expression to the hydrophobic or hydrophilic character, since the polymer is not really altering its degree of interaction with the *solvent* by being in one phase instead of the other; rather, it is the *polymer–polymer* interaction parameter χ_{23} which is the significant quantity here. Indeed, it is likely that Albertsson's ordering should be viewed more as one of the effective solubility parameters for the polymers in the common aqueous environment. There seems to be a potentially fruitful area here for future studies, particularly on the other common synthetic polymers such as those of the acrylic group, as well as PVP; this should enable the effective solubility parameters to be put on a consistent quantitative basis and be related to the molecular structures of the polymers, so that the incompatability behavior of a particular pair of polymers in aqueous solution may be predicted from their molecular structures.

Rather than simply being viewed as an annoying phenomenon, incompatability in aqueous systems has been very usefully applied to give partition systems for the purification of biological cell particles and biopolymers; the technique has been developed (again) principally by Albertsson.[10 11] The most widely used pair of polymers for this purpose is dextran and PEO. The advantage of this type of separation procedure is that the system is largely aqueous and hence mild in its effect upon the often fragile cell particles and biopolymers concerned, while at the same time quite small differences in the nature of the partitioned species can lead to marked changes in the value of its partition coefficient.

3.7.2. Coacervation

With coacervation we are concerned with systems where there is a definite attractive interaction between the two polymeric components. This phenomenon was seemingly first noticed with mixtures of natural polymers such as gum arabic or agar (both polyacids) with gelatin (a polyampholyte) at moderately low pH. The coprecipitation, i.e., "coacervation," which occurs in these mixtures is significant both because it may tend to occur within biological systems themselves, and also because of the possibility that such coacervate droplets may have been the primitive precursors of living cells in the early stages of the evolution of life.[1119,1120] Coacervation of natural polymers arises from electrostatic effects, since the two polymeric components are respectively negatively and positively charged at the pH values used. Similar electrostatic coacervation takes place with synthetic polyelectrolytes in aqueous solution. For example, Michaels and Miekka[1009] have shown that at concentrations <2.0 g dm^{-3} poly(vinylbenzyltrimethylammonium chloride) reacts with sodium poly(styrene sulfonate) to form a water-insoluble but hydrous polysalt; the complex contains equi-equivalent quantities of each polyion irrespective of which polyion is in excess, and it is also essentially free of the small counterions (Na^+ and Cl^-) of the original polyelectrolytes. At much higher initial concentrations

(>6.0 g dm^{-3}) a film of the complex forms at the interface between the two solutions, which then blocks further reaction; these films can be prepared as such, or embedded in filter paper by using this as barrier in a diffusion cell between the two solutions. These polyelectrolyte complexes have been discussed in more general terms by Bixler and Michaels.[145]

With such systems the formal thermodynamic treatment involving interaction parameters χ_{12}, χ_{13}, and χ_{23} can also be applied, although more strictly the polyelectrolyte nature of the two species should be taken into account; phase separation in such polyelectrolyte systems has been discussed by Voorn.[1541] Nevertheless, it is evident that the opposite charges on the two polymer chains must lead to a negative value of χ_{23}, so that the two polymers coaggregate. This effect is increased by the accumulation of interactions between sequences of consecutive units on the two chains; in kinetic terms, once the two chains come together to form one such "salt link" there will be a zippering action as successive units on either side interact with their partners on the other polymer.

Coacervation can also take place between nonionic synthetic polymers in aqueous systems; thus, PEO will form precipitates with PAA (at pH <4), and also with PMAA.[1432] Similarly, PVP will form a precipitate with PAA.[161,1087] The fact that, particularly clearly in the case of the first pair, it is the undissociated form of the acid which is required for precipitation to take place, indicates that hydrogen bonds play an important role in the interaction between the two species; such interchain hydrogen bonding, although it may not be strong between individual units of the chains, particularly in aqueous solution where it is in competition with hydrogen bonds from the water, could nevertheless give a very strong overall interaction. Once again, as with ionic systems, this can be viewed kinetically as the zippering up of the two polymers from either side of the initial junction point.

The coacervation behavior of these nonionic polymers is interesting in providing a synthetic polymer model for the much more specific pairings of nucleotide chains in the double helix of DNA. The role of hydrogen bonding in each case (albeit only presumed with the synthetic polymer) is particularly significant.

Finally, linking up these two distinct types of behavior, incompatability and coacervation, it appears that they each depend upon the polymer–polymer interaction parameter χ_{23} having a suitable value, i.e., $\chi_{23} \gtrsim 0$ for incompatability and $\chi_{23} < 0$ for coacervation. In fact, we must also expect there to be systems where $\chi_{23} \sim 0$, with mildly attractive interactions alongside the small entropic effects so that mixing is possible over the whole range of compositions. It should, indeed, be relatively easy to "design" such a system, i.e., to convert a normally incompatible pair of polymers into such a "neutral" pair by incorporating, by copolymerization, a few anionic groups in one and a few cationic groups in the other; alternatively, hydrogen-bond-donor and -acceptor groups could be used. It is likely that such systems already exist, but have escaped detection because they do not lead to any visible effects. Such completely compatible systems would be of obvious practical interest.

3.8. Equilibrium Aspects of Absorption by Synthetic Polymers

One of the features that distinguishes polymers from other solids is the ease with which they absorb and are penetrated by small-molecule substances. This difference in behavior may be ascribed to the presence of the amorphous (i.e., "liquidlike") component in polymers. In point of fact, there is a close correlation between the amount of amorphous phase in a sample of a polymer and its absorptive capacity; this has been demonstrated clearly in the case of water absorption,[1509] and has been applied as a method to estimate the relative amounts of amorphous and crystalline phases in polymers.

In the broadest sense, the uptake of a small-molecule substance by a polymer can be covered by the general term "sorption," which comprises *ab*sorption (i.e., into the body of the polymer) and *ad*sorption (i.e., onto the interface with its surroundings). The second is of less interest here, since it depends so critically upon the state of subdivision of the material and the presence or absence of adsorbed impurities, and is less readily related to the molecular structure of the polymer, although it is a subject of some practical importance in such applications as adhesion (see Section 5.6). In the present case the discussion will be therefore almost entirely concerned with absorption.*

As is the case with most processes, absorption has both equilibrium and kinetic aspects. Here we shall be concerned with the equilibrium aspect; the kinetics (i.e., permeation and diffusion) will be dealt with in Section 3.9. In practice, these two aspects are so closely interconnected that they are frequently studied in conjunction; for this reason, for example, much information upon equilibrium absorption properties is located in sources ostensibly concerned only with the diffusion or permeation of small-molecule substances in polymers.

---

* The subject of adsorption is covered in Chapters 2 and 5 in Volume 5.

We shall consider first the equilibrium absorption of water by polymers. The general literature up to 1968 has been reviewed by Barrie,[93] while a 1969 issue of the *Journal of Macromolecular Sciences*[1330] contains a number of articles covering more recent work. The more specialized features of the uptake of water by fibers and textiles have been dealt with by Morton and Hearle (Ref. 1046, Part III).

From the experimental viewpoint the general methods of absorption measurement used for polymers with gaseous and fairly volatile liquid absorbates have been outlined by Crank and Park (Ref. 296, Chapter I), while Barrie[93] and Yasuda and Stannett[1609] have considered the special problems that arise with water as absorbate.

We have already discussed some aspects of water uptake by water-soluble polymers in Section 3.6, while uptake by cross-linked polymers is considered briefly in Section 3.10. In the present section attention will therefore be mainly confined to water-insoluble linear polymers.

The equilibrium amount of water absorbed by unit amount of the polymer will normally increase with the partial pressure p_1 of the water vapor; the ratio of these two at a given partial pressure is the solubility coefficient S (for that pressure). The amounts involved are generally put on a volume basis, with the volume of water vapor absorbed being corrected to standard temperature and pressure (stp, $0°$ and 1 atm), so that the common units for S are $cm^3(stp)/cm^3$ cm Hg. For the limiting case of very small values of p_1 the corresponding limiting solubility coefficient is denoted S^0. The partial pressure may alternatively be expressed as its ratio to the saturated vapor pressure, p_1^*. If we ignore relatively minor deviations from perfect gas behavior in the vapor, then $p_1/p_1^* = a_1$; in this case the reference standard state is liquid water (so that, in particular, derived thermodynamic quantities refer to the mixing or solution process in which liquid water is transferred into the polymer). The quantity $100p_1/p_1^*$ is, of course, the relative humidity (RH) on its usual percentage basis. On the other hand, where p_1 alone is considered, then the reference state is the vapor at a standard pressure (and the thermodynamic quantities refer to the sorption process involving the transfer of water vapor molecules into the polymer.)

For many practical purposes it is sufficient to study the uptake at a single value of p_1 or a_1. For fibers and textiles the commonest single-point measurement is the *moisture regain* for 65% RH, which is thus viewed as an average value for the ambient humidity (Ref. 1270, Section 62). For plastics and rubbers it is the *water absorption* corresponding to 100% RH which is important (Ref. 1270, Section 61). However, such single-point values are not sufficient for more fundamental studies, even those limited to the state of the absorbed water at that point. Thus, the single-point value observed could at one extreme simply be one point on a linear (Henry's law) isotherm, with the water thus only relatively weakly bound and hence fairly mobile; on the other hand, at the other extreme it could be the limiting level for a Langmuir-type isotherm corresponding to a "monolayer" of firmly bound and immobile water molecules. A deeper knowledge of the molecular basis of the absorption therefore requires determination of the complete absorption isotherm, from zero to unit activity. Preferably, this should be repeated at a series of temperatures over as wide a range as practicable to enable the enthalpic and entropic contributions to be disentangled.

Figure 31 shows S values taken from the compilation of Barrie[93] for some synthetic polymers, plotted in double logarithmic fashion for display

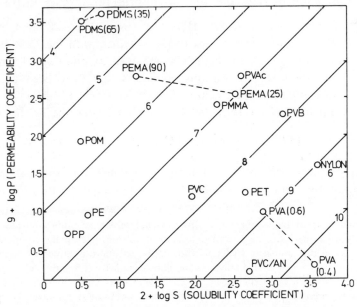

Fig. 31. Solubility coefficient S [in cm³(stp) (cm Hg)$^{-1}$ cm^{-3}] and permeability coefficient P [in cm³(stp) cm (cm Hg)$^{-1}$ cm^{-2} sec^{-1}] for water in synthetic polymers at 25–30° and in the approximate limit of zero water activity, except where otherwise indicated; the parallel lines represent $-\log D$, where D is the corresponding diffusion coefficient (in cm² sec^{-1}). Key to polymers: PDS, polydimethylsiloxane (at 35 and 65°); PE, polyethylene; PEMA, poly(ethyl methacrylate) (at 25 and 90°); PET, poly(ethylene terephthalate); PMMA, poly(methyl methacrylate); POM, polyoxymethylene; PP, polypropylene; PVA, poly(vinyl alcohol) (at $a_1 = 0.4$ and 0.6); PVAc, poly(vinyl acetate) (at 40°); PVB, poly(vinyl butyrate); PVC, poly(vinyl chloride); PVC/AN, poly(vinyl chloride)-co(acrylonitrile). From the compilation of Barrie.[93]

and for comparison against the water permeability coefficient P (see Section 3.9) (except where otherwise indicated the coefficients plotted are essentially their limiting values for zero uptake, S^0 and P^0). The order of solubility coefficients is largely that expected from the structures of the polymer chains, with hydrocarbonlike polymers such as polyethylene, polypropylene, and polydimethylsiloxane lying at the low uptake end, nylon (as well as the water-soluble PVA) lying at the other end, and the ester-containing polymers lying in between. For a more thorough correlation between the absorptive capacity and the molecular structure it is necessary to consider the form of the absorption isotherm in more detail.*

3.8.1. Models for Interpretation of Isotherms for Absorption of Water

We now consider the three main approaches which have been used in the interpretation of isotherms for the absorption of water in polymers.

3.8.1a. *Site-Binding Models.* Here the absorption is treated as if it were *ads*orption onto internal sites in the polymer, with localized binding onto a variety of types of site being accompanied by multilayer formation. Barrie[93] has discussed the various equations that have been proposed based on this model. More recently, D'Arcy and Wall[316] have proposed a general equation of the form (slightly modified from the original)

$$c_1 = [KK'a_1/(1 + Ka_1)] + Ca_1 + [kk'a_1/(1 - ka_1)] \tag{78}$$

where c_1 is the water concentration in the polymer, and the first term in brackets on the right represents a set of strongly binding sites, the second

---

* It should be noted that for these water-insoluble polymers the equilibrium water uptake is likely to be markedly increased by contamination from hydrophilic components or groupings in the sample of material studied. We have already noted in Section 3.5 that with many water-soluble polymers the uptake isotherm shows a "toe" at low water activities which we attributed to the presence of such contaminants; however, the anomalous presence of such a "toe" was only apparent in these cases because the "extra" uptake involved was small compared with the total uptake of the sample. With the present water-insoluble polymers, however, this "extra" uptake could well be comparable with that due to the polymer itself, and hence the experimental data on such samples could give quite erroneous conclusions concerning the relation between uptake and molecular structure of the polymer. For this reason the precautions recommended for sample purification and preparation in Section 3.5 for water-soluble polymers need also to be observed for measurements on water-insoluble ones; in addition, with such highly hydrophobic ones as polyethylene and polypropylene care should be taken that the chains do not contain hydrophilic groups such as $>C=O$ introduced by oxidation reactions during the production and processing of the polymer.

term weakly binding sites, and the last represents multilayer formation on these sites; K, K_1, C, k, and k' will then be concentration-independent parameters characteristic of that polymer at that temperature. The equation was shown to give a very satisfactory fit to the absorption data for 20 natural and synthetic polymers. This type of model is probably most useful with proteins, where, indeed, there are likely to be well-defined sites of different types corresponding to the side groups, the end groups, and the chain amide groups. This is less likely to be the case with synthetic polymers, however, while the inclusion of a simple multilayer adsorption term for processes taking place inside the body of the polymer does not seem physically plausible. In addition, the weakness of any relation like eqn. (78) is that mathematically it is really *too* flexible, since with *five* adjustable parameters almost any experimental isotherm could be fitted satisfactorily. It is therefore essential that any general equation of this type be examined critically, not only from the viewpoint of mathematical fit but also from the viewpoint of its physical and chemical plausibility in terms of the known molecular and supramolecular structure of the polymer.

3.8.1b. *Henry's Law plus Water–Water Interaction Effects.* In this method the initial linear section of the isotherm at $a_1 \to 0$ (Henry's law region) is interpreted in terms of a mechanism for the absorption of the first few water molecules into the polymer; the deviations from Henry's law at higher (i.e., finite) values of a_1 are then interpreted in terms of interactions between absorbed water molecules (in pairs, triplets, and so forth). The method is thus particularly applicable to the common case were the absorption isotherm curves smoothly upward, i.e., type III isotherm on the BET classification.[191]

This approach is well exemplified by recent water absorption studies on poly(alkyl methacrylate)s.[95,96] These studies are particularly interesting, because they not only involved measurements at a wide range of water activities (0.1–0.9) and temperatures (overall, 27–80°), but also were made on four polymers whose ester alkyl groups, i.e., methyl (PMMA), ethyl (PEMA), propyl (PPMA), and butyl (PBMA), form successive members of a homologous series, while less extensive measurements were made on a fifth polymer, poly(propyl acrylate) (PPA), which is an isomer of PEMA; such use of a systematically varied series of polymers greatly assists in relating absorptive behavior to molecular structure. (It may be noted that the sorption and permeation properties of PEMA to a wide variety of gases as well as water vapor had been studied previously by Stannett and Williams.[1416])

The isotherms obtained for the five polymers were in each case of the BET type III; extrapolation of their slopes to $p_1 = 0$ gave the values of S^0. The temperature variation of S^0 was found to follow the usual van't Hoff form

$$S_T^0 = S_\infty^0 \exp(-\Delta H_s^0/RT) \tag{79}$$

where ΔH_s^0 is the limiting heat of sorption, i.e., for the transfer from the vapor phase into the essentially unperturbed polymer matrix.* The derived values of ΔH_s^0 and log S_∞^0, as plotted in Fig. 32, fall on a straight line, with the points in the order of the alkyl chain length in the ester group.[†] Insofar as S_∞^0 may be taken to relate to the entropy change of sorption, then this represents a further example of the commonly encountered correlations between enthalpy and entropy quantities for physicochemical processes.[871]

In the case of PMMA the value of ΔH_s^0 is close to the latent heat of condensation (i.e., 44 kJ mol^{-1}) and the authors therefore suggested that this represents the condensation of water into "microvoids" in the polymer. This is equivalent to an essentially zero value for the limiting heat of mixing ΔH^{M0} (i.e., for the transfer of water from the liquid state into a indefinitely large volume of polymer). The observation, for example, of such a value for ΔH^{M0} in the case of water absorption by a polyurethane

---

* The symbol S_∞^0 is used for the preexponential factor in place of the more common form S_0^0 to emphasize that it represents the extrapolated limiting value of S_T^0 for *infinite*, and not for *zero*, absolute (thermodynamic) temperature. This also applies to to the two other preexponential factors D_∞ and P_∞ introduced in eqns. (89) and (92).

† Frequently, when studies are made on the temperature dependence of an equilibrium property (such as water absorption) or of a kinetic property (such as water permeation or diffusion—see Section 3.9) for a particular system, and this is then interpreted in terms of the van't Hoff or the Arrhenius equation, many workers content themselves with presenting in their published report only the energy quantity (i.e., here ΔH_s) and omit any mention of the preexponential factor (here S_∞). Now, admittedly, the energy quantity is easier to calculate, and also easier to interpret, whereas the preexponential factor requires a further computational step, while its interpretation in the case of water absorption probably involves both the standard entropy change of sorption and the volume density of sites onto which the sorption takes place. Nevertheless, the author would make a strong plea that in all such studies both parameters should be calculated and reported; for, not only do they together represent the complete data in a compact form suitable for interpolation or extrapolation purposes, but also they often show interesting intercorrelations for systems of systematically varied molecular structure [as shown for the poly(alkyl methacrylate)s with their water sorption properties in this section, and for their water permeation properties in Section 3.9].

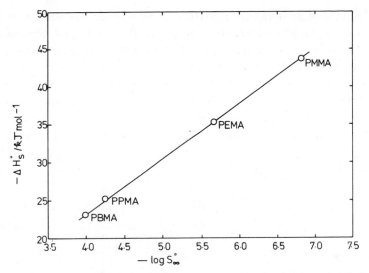

Fig. 32. Van't Hoff parameters [eqn. (79)] for the sorption of water by poly(alkyl meth-acrylate)s in the limit of zero sorption; the units of S_∞^0 are $cm^3(stp)$ $(cm\ Hg)^{-1}\ cm^{-3}$. PMMA, PEMA, PPMA, and PBMA denote the methyl, ethyl, propyl, and butyl esters. From the data of Barrie and Machin.[95]

by Schneider and his colleagues[1332] led them to conclude that "...the interactions of the water in the polymer are dominated by penetrant–penetrant [i.e., water–water] interactions... ." Similar heat of sorption values have been obtained for water for quite a number of polymers, e.g., for nylon interpolymer,[628] poly(vinyl butyral),[628] cellulose acetate,[628] rubber hydrochloride,[628] poly(vinylidene chloride)co(acrylonitrile),[628] poly(vinyl alcohol),[628] polyethylene,[628] (although numerically much lower values of ΔH_s^0 have also been obtained for this polymer[1608]), and nylon 4[684]; correspondingly, the "condensation" interpretation quoted above seems to be that most widely accepted.

It is therefore necessary to state most strongly that this interpretation is quite incorrect. For since the data involved relate to the initial linear section of the absorption isotherm, then in molecular terms they correspond to the absorption of water molecules which become isolated within the polymer matrix (compare point H in Fig. 1). Under these conditions there can be no possibility of the occurrence of, let alone domination by, water–water contacts within the matrix, for otherwise the sorption isotherm would not be linear at the origin; indeed, it seems nonsense to even remotely consider condensation of *liquid* water at these very low vapor pressures.

As a more tenable explanation, it is suggested that insofar as liquid water can be viewed as a four-coordinated, fully hydrogen-bonded structure, then the molar heat of evaporation will represent the breaking of two hydrogen bonds per molecule of water released, so that in a similar fashion an initial heat of sorption value of $\Delta H_s^0 = -44 \text{ kJ mol}^{-1}$ represents the formation of two hydrogen bonds per water molecule entering the polymer matrix. Correspondingly, this explains the essentially zero value for ΔH^{M0}. Evidently, when a water molecule is absorbed it bridges a pair of suitably placed acceptor groups on the chains; for example, in the case of PMMA this would involve one or other of the oxygen atoms in each of a pair of ester groups:

$$
\begin{array}{c}
\text{CH}_3 \\
| \\
-\!\!\!\sim\!\!\sim\!\!-\text{CH}_2-\text{C}-\!\!\sim\!\!\sim\!\!- \\
| \\
\text{C}=\text{O} \\
| \\
\text{H}_3\text{C}-\text{O} \qquad\qquad \text{CH}_3 \\
\vdots \qquad\qquad\qquad | \\
\text{H} \qquad\qquad\qquad \text{O} \\
| \qquad\qquad\qquad | \\
\text{O}-\text{H}\cdots\text{O}=\text{C} \\
\qquad\qquad | \\
-\!\!\!\sim\!\!\sim\!\!-\text{CH}_2-\text{C}-\!\!\sim\!\!\sim\!\!-
\end{array}
$$

Such a model could also explain why the value $\Delta H_s^0 = -44 \text{ kJ mol}^{-1}$ is essentially the upper numerical limit for these and other types of polar polymers.

Considering the results for polymers with larger alkyl groups, it will be seen from Fig. 33 that with increasing alkyl chain length ΔH_s^0 falls off steadily to approach an essentially limiting value of about -22 kJ mol^{-1} for very long alkyl chains. Taking this limiting value to represent the formation of only one hydrogen bond per absorbed molecule, then the variation observed can be interpreted on the basis that with a longer alkyl chain on the ester group the oxygen atoms become more sterically hindered; it then becomes less likely that a water molecule hydrogen-bonded to one ester group would be able to find another group suitably placed nearby for the formation of a second hydrogen bond. The correlated rise in the value of S_∞^0 (Fig. 32) can be interpreted in terms both of the relatively greater freedom of a singly bonded water molecule compared with a doubly bonded water molecule (since this quantity is inversely related to the standard *entropy* change of sorption) and also of the greater numbers

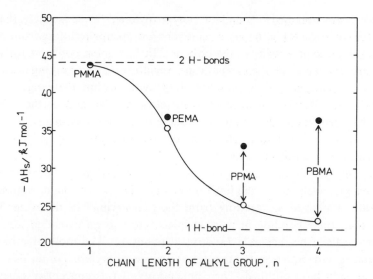

Fig. 33. Enthalpy of water vapor sorption ΔH_s by poly(alkyl methacrylate)s versus the alkyl group chain length n; open circles indicate the limiting values for zero uptake, and filled circles for uptake $c_1 = 4$ cm³(stp) (cm Hg)$^{-1}$ cm^{-3} (the half-filled circle used for $n = 1$ shows the invariance of ΔH_s with respect to c_1 for this polymer). The broken lines labeled "2 H-bonds" and "1 H-bond" represent the levels for the molar enthalpy of condensation of water vapor, and for one-half this value. From the data of Barrie and Machin.[95]

of "lone" ester sites for single hydrogen-bonding compared with the pairs required for bridging.

At finite degrees of absorption ΔH_s for PMMA remains constant at its limiting value ΔH_s^0, while with the longer alkyl chains the numerical value rises with increasing degree of sorption, again apparently toward the same value (i.e., -44 kJ mol^{-1}) (Fig. 33). These variations indicate that with the water molecules already present singly hydrogen-bonded onto the chains, there is then a greater possibility of an entering water molecule being able to form two hydrogen bonds, one with an ester group and another with one of the water molecules already present; the fact that ΔH_s for PMMA does not rise with increasing degrees of coverage confirms that for this polymer ΔH_s^0 already represents the maximal possible value.

Values of ΔH_s^0 of about -22 kJ mol^{-1} are also obtained for a number of other polymers, e.g., polyethylene,[1608] polypropylene,[1608] various silicone rubbers,[94] and poly(vinyl chloride),[1586] indicating that in these cases single hydrogen-bonding is the predominant mechanism. In the cases of polyethylene[1608] (for which in another study a value of about -44

kJ mol^{-1} was obtained[628]) and of polypropylene,[1608] it is likely that the sites for hydrogen bonding are a relatively few groups resulting from oxidation during the processing of the polymer. In fact, most polymers for which data are available for the temperature variation of the limiting water absorption coefficient give values for ΔH_s^0 lying within the range -22 to -44 kJ mol^{-1} (and with most of them close to one end or the other of the range), suggesting that single or double hydrogen bonding, or a mixture of the two, is the main mechanism for the absorption of isolated water molecules into a polymer matrix.

At higher (finite) degrees of absorption, where the amount absorbed rises progressively above the limiting Henry's law value, the deviations are generally explained as resulting from the "clustering" of the water molecules (i.e., their bonding in pairs, triplets, etc.) to an extent greater than that expected from a purely random distribution. The molecular basis of this and its consequences for the poly(alkyl methacrylate)s in terms of ΔH_s have already been outlined. The applications of this concept of clustering have been discussed by Orofino and his colleagues[1124] and by Barrie and Machin.[95] For this situation, however, it seems preferable to take into account the polymeric nature of the absorbent and to apply a solution model for the absorption, as outlined next.

3.8.1c. *Solution Model.* Here the absorption is treated simply as a case of the mixing of the liquid (or its vapor) and the polymer. For amorphous polymers the Flory–Huggins theory may be applied directly, with the water uptake representing the volume fraction of the small-molecule component (liquid) in the solution formed, and the activity of the liquid represented by the relative vapor pressure; this then gives (see Section 3.3.1)

$$\ln a_1 = \ln \phi_1 + (1 - s)\phi_2 + \chi\phi_2^2 \qquad (80)$$

and since for moderately high molecular weights the inclusion of s (the ratio of the molar volume of the liquid to that of the polymer) is found to make little difference, then this simplifies to

$$\ln a_1 = \ln \phi_1 + \phi_2 + \chi\phi_2^2 \qquad (81)$$

This equation has already been plotted out in Fig. 2, for various values of χ; if the axes in Fig. 2 are interchanged, this reveals the expected shape of the isotherm to be that of the BET type III, to which many synthetic polymer systems do indeed conform. In this case, for $\chi > 0.5$ the isotherms cut the $a_1 = 1$ axis at a value of ϕ_1 representing the finite extent of absorption for 100% RH.

The application of this model to absorption of water has also been discussed by Barrie.[93] In the case of partially crystalline polymers the values of ϕ_1 and ϕ_2 which are used must correspond to the amorphous regions of the polymer; this modified model has been applied to nylons,[1417] where excellent fits were obtained to the experimental data for the water uptake at 23° by nylon 6.6 (57% crystalline) with $\chi = 1.46$, and by nylon 6.10 (47% crystalline) with $\chi = 2.18$. Williams and his colleagues[1586] have calculated χ values from absorption measurements for water with poly(vinyl chloride), polyoxymethylene, and poly(ethyl methacrylate) (as well as for some nonsynthetic polymers); in these cases the calculated values of χ were not constant over the isotherm despite the low maximal uptake for these polymers.

The above discussion has been concerned entirely with the thermodynamics of water absorption. It is this actual absorption which is then, of course, responsible for the effects of water (whether liquid or vapor) on the physical properties of polymers. Curiously enough, the effect on the volume of the polymers does not seem to have been studied at all widely, although Starkweather has investigated the anisotropic swelling of oriented nylon 6.6 conditioned to 50 or to 100% RH.[1418] More widely studied are the mechanical properties, such as tensile modulus, and allied dynamical mechanical properties, as measured for nylon 6.12 by Illers,[707] polyurethane by Jacobs and Jenckel,[721] and oriented nylon 6.6 by Starkweather.[1418]

Considering now absorption of a third component in synthetic polymer–water systems, one very important field is the dyeing of textile fibers.[1154,1226]* With the traditional fairly hydrophilic natural polymer fibers, the dyes (or their precursors, in vat dyeing) are applied from aqueous solution. With the relatively hydrophobic synthetic polymers, such as the polyamides, poly(ethylene terephthalate), and polyacrylonitrile, as well as the semisynthetic polymers such as the cellulose acetates, new methods of dyeing had to be developed. One such method is the use of dispersed dyes, which are essentially water insoluble and hence are applied in a finely divided form as their aqueous dispersions; the dyeing equilibrium involves solution of the dye in the fiber, as shown by the fact that with undersaturated systems there is simple partition of the dye between the aqueous and polymer phases. That water does have an effect upon the equilibrium absorption state is shown by the fact that many of the dispersed dyes are sufficiently volatile to be applicable from the dry vapor, when the amounts

---

* See Chapter 3 for a full discussion.

absorbed can be up to three times that obtained with wet dyeing; evidently the small amount of water present in the fiber in the latter case is sufficient to make it markedly less absorptive toward the dye (possibly by filling up sites otherwise used for polymer–dye hydrogen bonds).[1226]

From the more fundamental viewpoint, Zollinger has discussed the contributions made by such interactions as Coulombic forces, van der Waals (London) forces, and hydrophobic bonds in the sorption of anionic dyes from aqueous solution by wool, silk, and polyamides.[1629]

The sorption properties of polyamides toward solutes in aqueous solution are also of medical and pharmaceutical interest; Rodell and his colleagues have studied the sorption (and permeation) of nylon 6 toward benzoic acid, hydroxy-substituted benzoic acids, and the alkyl esters of *p*-hydroxybenzoic acid.[1267]

3.9. Kinetic Aspects of Absorption by Synthetic Polymers: Permeation and Diffusion

The kinetics of the absorption of small-molecule substances is a topic of both practical and of theoretical interest. As far as water and aqueous systems are concerned, it has direct impact in such applications of polymers as packaging, films, containers, fabric coatings, and synthetic leathers, as well as in dyeing, tanning, desalination by reverse osmosis, extracorporeal hemodialysis (artificial kidney machines), and other systems.

Where the kinetic aspects are of particular interest, it is common to term the small-molecule substance concerned the *penetrant*. The standard work on the subject is the multiauthor volume edited by Crank and Park (individual chapters from which have already been cited in the previous section).[296] The more general aspects of transport phenomena in artificial membranes have been reviewed by Lakshminarayanaiah.[846] Certain of the references cited at the beginning of Section 3.8 deal with this subject both in general and more specifically as it applies to water as the penetrant.

There are two distinct but related transport processes to be considered: first, *permeation*, which is the transport of a penetrant from a phase (generally, liquid or gaseous) in contact with one side of a sheet of the polymer, through the sheet and out into the phase on the other side, and second, *diffusion*, which is the process of the transport within the body of the polymer. The interconnection between these two processes, and the absorption equilibrium, is illustrated in Fig. 34, where J is the flux of penetrant through a sheet of thickness t due to a penetrant vapor pressure

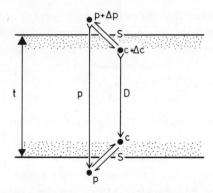

Fig. 34. Interrelations between the processes of sorption (solubility), permeation, and diffusion for the flux of a gaseous or vapor penetrant at pressure differential Δp and concentration differential (in the polymer) Δc across a polymer film of thickness t.

differential Δp.* For this case the permeability coefficient P is given by

$$P = Jt/\Delta p \tag{82}$$

while if the corresponding concentration differential across the bulk of the film is Δc, then the diffusion coefficient D is given by

$$D = Jt/\Delta c \tag{83}$$

Although in the ideal case both P and D would be constants for a given polymer and penetrant at a fixed temperature, in practice P is often pressure dependent and D concentration dependent. Thus for generality one should consider only infinitesimal differentials of p or c across the film, and quote the coefficient for the level of p or c concerned.

For this same case if the absorption isotherm is linear (Henry's law), the solubility coefficient S will be given by both

$$S = c/p \tag{84}$$

and

$$S = (c + \Delta c)/(p + \Delta p) \tag{85}$$

which leads to

$$S = \Delta c/\Delta p \tag{86}$$

so that combining this with eqns. (82) and (83) gives

$$P = DS \tag{87}$$

---

* The flux is the amount of penetrant passing across unit cross-sectional area in unit time.

It should be emphasized that this simple relationship only applies to cases where all three of the parameters are concentration independent; however, it does apply to all systems in the limit of zero concentration, i.e., for *all* systems

$$P^0 = D^0 S^0 \tag{88}$$

The close interrelation between the three coefficients S, P, and D (or at least between their limiting values) indicates that for any particular system only two of them may be regarded as truly independent quantities. With solubility as the firmly based equilibrium property, the choice then lies between diffusion and permeation; in general, diffusion is considered as the more fundamental process, and the directly measured permeation is thus regarded as the resultant of solubility and diffusion effects.

The variation of diffusion coefficient with temperature is normally of the Arrhenius form

$$D_T = D_\infty \exp(-E_D^{\ddagger}/RT) \tag{89}$$

where $E_D^{\ddagger}$ is the activation energy of diffusion and D_∞ is the corresponding preexponential factor; this shows that diffusion has the character of an activated process. The bulk of the current discussion of diffusion and its relation to the molecular structure of polymer and penetrant is based upon the theory of diffusion originally developed for pure liquids and liquid mixtures by Glasstone *et al.*[528] In this treatment an amorphous polymer is considered to have at the molecular level a certain amount of "free volume" or "holes."[825] The critical step in the diffusion process is then taken to be the localized activated jump of a penetrant molecule from one hole to the next. The probability that the molecule will jump is assumed to be controlled in part by the need for cooperative movement of the polymer segments to produce a cylindrical cavity of radius equal to that of the penetrant molecule and of length equal to a characteristic "jump length" λ. On this basis the "Arrhenius parameters" D_∞ and $E_D^{\ddagger}$ are given by

$$D_\infty = (e\lambda^2 kT/h) \exp(\Delta S_D^{\ddagger}/R) \tag{90}$$

$$E_D^{\ddagger} = \Delta H_D^{\ddagger} \tag{91}$$

where $\Delta H_D^{\ddagger}$ and $\Delta S_D^{\ddagger}$ are the standard enthalpy and entropy changes of activation for the "jump", e is the base of natural logarithms, k is the Boltzmann constant, and h is the Planck constant.

As regards the temperature variation of the permeation coefficient, this also normally follows an equation of the Arrhenius form

$$P_T = P_\infty \exp(-E_p{}^\ddagger/RT) \tag{92}$$

From the conventional viewpoint, this may be seen simply as a result of P being the product of S and D [eqns. (87) and (88)], together with their separate dependences [eqns. (79) and (89)].

Considering more specifically the transport of water the published work (up to 1968) has been reviewed by Barrie.[93] Some of these collected data for permeability coefficients of water in synthetic polymers are plotted in a double logarithmic fashion against the corresponding solubility coefficient in Fig. 31; since in most cases the data refer essentially to the limiting values (for $p_1 \to 0$ and $c_1 \to 0$), then from the logarithmic nature of this plot the lines of unit slope represent loci of constant limiting diffusion coefficients D^0, with their negative logarithms as labeled [compare eqn. (88)]. One notable feature shown by this plot is the lack of obvious correlation between any two of the parameters concerned, such as solubility and permeability—a low solubility can equally well go with a low permeability (as in polyethylene and polypropylene), a medium permeability (as in polyoxymethylene), or a high permeability (as in polydimethylsiloxane). This leads us to suspect that solution and permeation or diffusion involve quite distinct and independent sites of action in the polymer matrix.

Now one of the curious features of water transport in fairly polar polymers is that in quite a number of cases, particularly those studied in the most recent and trustworthy work, the diffusion coefficient is found to be markedly concentration dependent (D commonly falling with increase of c_1), whereas the permeability coefficient is found to be essentially pressure independent; this applies, for example, to various silicone rubbers (filled and unfilled)[94] and to the series of alkyl acrylate polymers[95,96] discussed from the equilibrium sorption viewpoint in the previous section. In the formal sense this behavior still fits in with the correlation of these two quantities [eqn. (87)] when the rise also observed in S with increase in pressure (concentration) is taken into account. A considerable amount of mathematical expertise has been devoted to efforts to account for the concentration dependence of the diffusion coefficient, mostly either through the concept of the clustering of the water molecules in the polymer matrix, or alternatively by assuming the existence of two forms of water (mobile and immobile) in the polymer; on the other hand, the constancy of the permeability coefficient has been largely dismissed as coincidental.[93]

There seems, on the other hand, to be a strong case for considering *permeation* to be the more fundamental process, both from the experimental viewpoint of this observed constancy of P with variation of pressure and from the basic thermodynamic viewpoint. The ratio $\Delta p_1/p_1{}^*$ will be the activity difference across the polymer sheet not only in the gas phase on either side but also (in the steady state) within the bulk of the polymer immediately at the surfaces; it is this activity difference which is the true driving force for the transport of the penetrant, so that in the absence of any appreciable "plasticization" by the absorbed penetrant the flux should be proportional to Δp_1. On the other hand, for diffusion, because of the clustering effects which occur particularly with this penetrant (i.e., water), and which lead to curvature in the absorption isotherms, the total concentration of water cannot be taken as proportional to its activity, and since the diffusion coefficient is based upon this total concentration, then its inconstancy is hardly surprising.

The use of this approach, with the activity differential as the driving force for the transport process, may also enable the theory of the process to be put on the secure basis of a true transition state. If we examine the original treatment of rate processes,[528] we find that although in the early part a strictly equilibrium treatment of the activated state was developed for chemical reactions, when viscous flow and diffusion of liquids was considered the authors abandoned this and proceeded on the basis of the molecules "jumping" from one rest position to the next via the activated state. This relapse from the stricter treatment may perhaps have been due to difficulties in applying such a treatment to the multiparticle situation of viscous flow and diffusion in a liquid; it is unfortunate, however, that this same, basically less satisfactory approach has been subsequently applied, in the manner outlined above, to transport of penetrants in polymers. In fact, the strict transition state treatment should be fairly readily applicable to permeation of small-molecule substances through polymers. In such a treatment, there would be taken to be an equilibrium between normal molecules of penetrant A (e.g., those in the gas phase), and activated molecules $A^{\ddagger}$ at critical sites for transport in the polymer:

$$A(g) \rightleftharpoons A^{\ddagger}(p) \tag{93}$$

Where the interior of the polymer is concerned, the concentration (activity) of "normal" A molecules would be given by the number of molecules which would be present, for example, in a small cavity if it were formed at this point; the partial pressure at such a point would presumably be a linear function of the distance across the polymer film. The critical sites would represent "bottlenecks" in the path of the penetrant through the polymer; considering the transport as taking place in the x direction, then for a molecule at such a site the free energy would be a local maximum in the x direction but a local minimum in all directions normal to this. The actual rate of transport would be governed by the rate of ejection of the penetrant molecule from the "bottleneck" into the relatively open area in front of it:

$$A^+(p) \rightarrow A(p) \tag{94}$$

Such a treatment would be expected to lead to relations for P_∞^0 and for E_p^+ similar to eqns. (90) and (91), but with thermodynamic parameters now corresponding to the process (93) for "absorption" by the "bottlenecks," and with λ now relating to the average distance between neighboring "bottlenecks." The prime advantage of the new approach outlined here is that, as already indicated, whereas the property of permeability is an extremely important one from many practical viewpoints, at present it is taken to depend upon two separate factors (i.e., absorption and diffusion) which must be separately evaluated and interpreted in relation to the molecular structures of the penetrant and the polymer; on the other hand, the use of the concept of a true activated state enables permeability to be considered as a process in its own right, with attention then focused upon the nature and properties of the "bottlenecks" within the polymer matrix.

One factor that perhaps deters attempts to make permeation rather than diffusion the more fundamental process is the fact that the experimental values of E_p^+ are frequently small and may even be negative, which seems difficult to square with a true activated process. For example, Fig. 35 shows the Arrhenius parameters [eqn. (92)] for the permeation of water through the poly(alkyl methacrylate)s[95]; here, negative values of E_p^+ are obtained in three cases, although, with the exception of PMMA, the points again show a linear correlation between E_p^+ and $\log P_\infty^0$ (compare with Fig. 32). (It should also be noted that, as indicated by Fig. 35, the temperature dependences of permeation and diffusion frequently show abrupt changes at the glass transition temperature, whereas for absorption there is generally no marked change at this point.)

Such negative values of E_p^+ can be rationalized as arising from actual bonding of the penetrant at the "bottleneck"; the required local minimum in *free* energy is then ensured by a sufficiently unfavorable entropy contribution resulting, say, from a requirement for a specific orientation of the penetrant molecule at this site. This is indeed again demonstrated by Fig. 35, where the more negative the value of E_p^+, the smaller the value of P_∞^0.

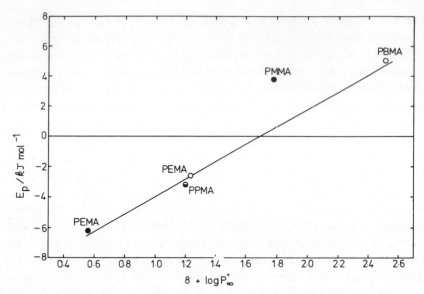

Fig. 35. Arrhenius parameters [eqn. (92)] for the permeation of water through poly(alkyl methacrylate)s, in the limit of zero sorption; the units of $P_\infty{}^0$ are cm³(stp) cm (cm Hg)$^{-1}$ cm^{-2} sec^{-1}. PMMA, PEMA, PPMA, and PBMA denote the methyl, ethyl, propyl, and butyl esters; the open circles represent the parameters for the polymer below the glass transition temperature T_g and the filled circles those above T_g. From the data of Barrie and Machin.[95]

We now briefly consider some examples of the transport of small-molecule third components in synthetic polymer–water systems. The simplest types are the permanent gases; the effect of moisture upon the transport of such gases in various polymer films has been studied by a number of workers.[715,1007,1402]

Dyeing is another industrially important process involving the transport of the third component (dye); the kinetics of these processes have been reviewed by Peters.[1154]

The permeation properties of nylon 6 toward benzoic acid and substituted benzoic acids from aqueous solution have been studied by Rodell and his colleagues,[1267] while a still wider range of penetrants in the same system have been studied by Kostenbauder and his colleagues from the viewpoint of the use of the polymer as an equilibrium dialysis membrane.[804]

In all these three-component systems, as with the simpler two-component systems, it is conventional to regard the directly observed (and practically more important) permeation process as a composite of solution and diffusion factors; however, again it would seem fruitful to consider permeation on the basis of the more fundamental "bottleneck" model already outlined for water itself, as governed by the processes of eqns. (93) and (94).

A somewhat different aspect of permeation in these three-component systems is that the transport of the water is of equal or greater importance to that of the solute, which may in fact be required to be minimized; examples of this are the applications of polymer membranes in reverse osmosis[1412] and also in dialysis (especially hemodialysis). In these cases the requirement of high permeability to water may necessitate using more porous membranes, where the flow may take place less by solution and diffusive flow in a homogeneous medium (as previously assumed) and more by viscous flow through the pores of a sievelike matrix. Thus, in the case of reverse osmosis, the most widely used membrane material is the semi-synthetic polymer cellulose acetate.[1412] Glueckauf and Russell have studied the flow of water through a number of such membranes of differing acetyl content and which had been subjected to various heat treatments; they tentatively interpreted their results in terms of an equivalent pore radius.[537] In the case of dialysis the most widely used polymer is another modified natural polymer, regenerated cellulose, where the assumption of a sponge-like nature may be more tenable than with cellulose acetate. Because of limitations in the selectivity of cellulose, Yasuda and his colleagues have examined the permeation and diffusion behavior of a wide range of hydrophilic polymer membranes toward both water and a variety of such important solutes as sodium chloride, urea, and creatinine; they interpreted their data in terms of a free volume which was taken to be directly proportional to the "hydration" (i.e., the volume fraction of water in the membrane).[1605–1607]

3.10. Gels and Gelation in Water-plus-Synthetic-Polymer Systems

The term "gel" refers to a system formed (in the simplest instance) from one polymer and one liquid, which shows elastic rather than plastic behavior, that is, it is more "solidlike" than "liquidlike." This type of behavior is common among such mixtures that are polymer-rich, but it is useful here to restrict the use of the term to the *liquid-rich* systems which would not normally be expected to behave in this way. The term "gelation" then implies the conversion of an initially viscous or plastic system into a more or less rigid gel; such a conversion usually involves the formation of a cross-linked polymer network. Where the cross-links are covalent bonds an irreversible gel results. On the other hand, in reversible gelation the process takes place typically on cooling a polymer solution, while upon reheating the gel liquefies again, although there are important exceptions to this (see below) where gelation takes place on raising the temperature.

3.10.1. *Irreversible Polymer Gels*

With irreversible gels the cross-links are normally covalent links, either as single bonds or through groupings of atoms between the polymer chains. For the system to be a gel there need to be at least two links between each primary polymer chain and the others in the system, so that the whole mass of the sample becomes essentially one huge single molecule, that is, an infinite network. In practice, the cross-linking may either be induced during the polymerization by the incorporation of a small proportion of a polyfunctional agent along with the main difunctional monomer, or it may be produced subsequently by physical or chemical treatment of the polymer, such as by irradiation or by the vulcanization processes used for rubber. The theory of these gelation processes has been outlined by Flory (Ref. 457, Chapter 9). The first of these two methods is generally preferable, since it enables definite and controllable amounts of cross-links to be introduced; this is important in relation to the properties of the final solvent-swollen gel, for with too few cross-links the gels are so highly swollen as to be fragile and unmanageable, while with too many cross-links the network cannot swell to the desired extent. Physical means for cross-linking, such as irradiation, are less desirable from this viewpoint, since not only is the cross-linking difficult to control and susceptible to interference from small amounts of impurities, but the irradiation can also lead to degradation of the polymer chains.

Polyacrylamide is one water-soluble synthetic polymer for which irreversible cross-linked forms are commercially available; the material Bio-Gel P (Bio-Rad Laboratories, California)[31] is such a product, supplied in bead form for use in such techniques as gel chromatography, in much the same fashion as the cross-linked dextrans which are commercially available (Sephadex, Pharmacia, Sweden[35]). In the case of Bio-Gel P the cross-linking is produced by including the tetrafunctional analog N,N'-methylene-*bis*-acrylamide in the polymerization mixture; the amount of this agent added then controls the equilibrium degree of swelling of the gel by water, which in turn controls the size range of molecules which are partially excluded from the swollen matrix. Thus, at one end of the commercially available range is the most densely cross-linked product Bio-Gel P-2, which has a water regain of 1.5 g water/g dry polymer (corresponding to $w_1 = 0.6$ in the swollen gel); this swollen gel completely excludes peptides with $M > 1.8 \times 10^3$ and freely includes those with $M < 100$, while those of intermediate size are excluded to correspondingly intermediate extents. At the other end of the range is Bio-Gel P-300, with a water regain of 18 g/g (so that $w_1 = 0.95$ in the swollen gel); this gel excludes compact molecules (e.g., globular proteins) with $M > 4 \times 10^5$ and freely includes those with $M < 6 \times 10^4$.[31]

One point of practical importance is that the exclusion and fractionation limits will be seriously altered if the gel swelling is changed such as by the addition of large amounts of solutes (see Section 3.6.2). In fact, polyacrylamide can tolerate relatively large amounts of salts or urea, although it is less tolerant of water-miscible solvents such as ethanol (see Table V).

The cross-linking of a water-soluble polymer also enables its interactions with water and aqueous systems to be studied by novel methods. Thus, with PVP, as already discussed in Section 3.5.1, the swelling properties of its cross-linked gels cast light upon its interactions with various liquids, including water,[178] while use of the same type of material enables the binding properties of the polymer toward cosolutes to be studied more directly than with the linear polymer itself.[179]

Cross-linking of a polyelectrolyte, or substitution of a cross-linked nonionic polymer by ionic groups, gives rise to the commercially available materials widely used as ion-exchange resins.[981,996] By and large, their ion-exchange properties can be explained mainly by fixed-ion/counterion electrostatic effects, as controlled by normal equilibrium behavior. However, the capability of these resins for the separation of the ions of the alkali metal group, and also of the rare earth group, indicates that some specificity of action does exist. Additional specific effects occur with organic ions. For example, Feitelson has studied the interaction of amino acids with a cation-exchange resin (the amino acids were studied around their isoelectric points, so that only the dipolar zwitterionic forms were involved); the aliphatic amino acids were only weakly attracted to the resin, to extents in agreement with the expected dipole–ion interactions, while the aromatic acid phenylalanine was markedly more strongly held, presumably due to dipole–induced dipole interactions involving the benzene ring.[429]

By preparing the original (nonionic) cross-linked polymer in the present of a diluent, the final ionically substituted product has a much higher porosity ("expanded network," "macroporous," or "macroreticular" ion-exchange resins), leading to improved rates of equilibration with the surrounding aqueous medium and also to modified selectivity properties.[826, 1015–1017]

3.10.2. *Reversible Polymer Gels*

Reversible gelation takes place typically with change in temperature. The classic example of this is gelatin itself (see Chapter 5), which is important not only in its familiar culinary applications but even more significantly in the "emulsion" of photographic films. Other natural polymers whose gelling properties in aqueous systems are important include agar and pectin. The work of Johnson[732] on gelatin and agar and of Kohn and Furda on pectin[798] illustrates the contribution that polymer techniques can make to the study of these natural gel systems.

Gelation phenomena in aqueous solution are shown by such polymers as polymethacrylamide, poly(vinyl methyl ether), and PVA. With PMAA [as well as poly(vinyl methyl ether) and some other polymers] the gelation takes place on *increase* in temperature, while with PMAAm (as well as with PVA and most other polymers that gel) it is induced by cooling[1395]; these observations are in line (most evidently for the two "acrylic" polymers) with the changes of temperature for precipitation in *dilute* solutions (see Section 3.5.1). It is apparent, from this correlation, that reversible gelation can be ascribed to the increase in the number of segment–segment interactions, i.e., to the production of labile cross-links, in an exactly analogous fashion to the stable cross-links involved in irreversible gelation.

The commercial importance of gelatin in photographic film, coupled with the uncertainties inherent in its origin and method of manufacture, have led to attempts to develop synthetic polymer substitutes for gelatin.[585] An interesting instance of this is the recent systematic work on the preparation, characterization, and gelation properties of polyacrylylglycinamide (polyacrylamidoacetamide) (PAG) and of polymethacrylylglycinamide (PMG)[579–585]

$$
\left[\begin{array}{c} R \\ | \\ CH_2-C- \\ | \\ CO \\ | \\ NH \\ | \\ CH_2 \\ | \\ CO \\ | \\ NH_2 \end{array} \right]_x
$$

PAG (R = H) and
PMG (R = CH$_3$)

Although not specifically so stated, it is apparent that these polymers had been chosen for investigation because the side chain mimics a di-peptide section of the gelatin chain; this choice was justified by the fact that the polymers do show reversible gelation in aqueous systems.

At concentrations somewhat above 1% (the exact value depending upon the molecular weight of the polymer) PAG forms thermally revers-ible gels in water at around ambient temperatures. From the concentration variation of the gel melting point it was shown that for this polymer the heat of gelation is -8.8 kcal mol^{-1} (per cross-link) compared with a value of about -100 kcal mol^{-1} for gelatin.[584] From equilibrium swelling and elastic modulus measurements it was also shown that for PAG the cross-link involves one group from each of the two chains, and that (unlike the case with gelatin) crystallization plays no important part in the gelation process.[579]

3.11. Synthetic Polymers at Aqueous Interfaces

Adopting the same broad classification as that used elsewhere in this section, where synthetic polymer/water systems are concerned the polymer may be taken either to be fully water soluble (so that in interfacial studies we would be concerned with the interactions at the interface between its aqueous solutions and other phases), or else to be essentially water insoluble (so that the water and the polymer already form such an interface). In this subsection we shall deal mainly with the first case, i.e., that of water-soluble polymers in relatively dilute solution, together with certain examples of water-insoluble polymers spread as films at aqueous interfaces; interfacial interactions between aqueous solutions and solid polymers are significant in a number of applications that cannot be discussed fully here; the second case, however, where the polymer is essentially water insoluble, does have its important aspects, particularly with respect to adhesion in biological and biomedical fields, as discussed by Baier[74] (see also Section 5.6).

Polymers, in common with other solutes, show a variety of types of behavior at the interfaces between their aqueous solutions and other phases. It is most convenient to divide the discussion according to the nature of the second phase, i.e., gas (air), liquid, or solid.

3.11.1. Water–Air Interface

The pioneering work on the study of synthetic polymers at the water–air interface was carried out by Crisp.[300] This work was concerned pri-marily with the properties (especially force–area isotherms) of spread films, mainly of water-insoluble polymers; however, it is interesting to note that stable films could be obtained with such water-soluble polymers as PVA, PAA, and PMAA.[300] Subsequent work in this area has been discussed by Gaines.[505] More recently Zatz and Knowles have studied the force–area isotherms of films of vinylpyrrolidone/vinyl acetate graft copolymers spread on water; the monolayers showed remarkably high compressibility and fluidity right up to the collapse point, probably due to the gradual penetration of the vinylpyrrolidone segments into the aqueous subphase.[1619, 1620]

The equilibrium adsorption of synthetic polymers from aqueous solution onto the air–water interface is a distinct aspect, and one much less studied than that of spread films. Polymers such as PVA show marked adsorption of this type, which eventually leads to a stable surface film even at low polymer concentrations; Pritchard has discussed the surface tension of aqueous PVA solutions, where the interpretation of the data is complicated by the slow diffusion of the polymer to the surface.[1192]

A further interesting aspect of this area is the possible interaction or competition at the interface between a spread film and one adsorbed from the bulk phase below. Zatz and Knowles have demonstrated such effects in the case of poly(vinyl acetate) films spread onto solutions of PVP (at concentrations of about 0.03%); the surface pressure values of the monolayers were increased in the presence of the dissolved polymer, to a greater extent at higher concentrations of the dissolved polymer.[1620] The recorded surface pressure was stable with time, indicating that the system was in a state of equilibrium. With compression of the spread film, the surface pressures approached those obtained in the absence of the dissolved polymer, apparently due to the adsorbed molecules of the latter being squeezed out of the surface. One can view these effects as resulting from a kind of "surface incompatability" between the two polymers.

3.11.2. Water–Liquid Interfaces[2]

The adsorption of solutes at the interface between a pair of immiscible liquids is a subject of general importance because of the part such adsorption can play in the stabilization of emulsions of the two liquids. Where one of the liquids is water, the commercially most important types of synthetic polymer (albeit with oligomeric rather than truly polymeric chains) are the nonionic surfactants which comprise block copolymers of ethylene oxide with an attached polyethylene (alkyl), poly(propylene oxide), or other hydrophobic chain.[1321] With such polymers, where one section of the overall chain is soluble only in one phase and the other section only in the second phase, it is likely that in the adsorbed state the individual chains largely retain their loosely coiled shape within their respective phases (see Chapter 2).

On the other hand, random copolymers of monomers of widely different degrees of water interaction do not seem to have been greatly studied from the viewpoint of their interfacial activity, although this type of copolymer should have some advantages over the block copolymer. A striking example is seen in the behavior of the random copolymers of vinylpyrrolidone and styrene studied by Breitenbach and Edelhauser.[176,177] Figure 36 shows their results for the interfacial tension between water (in which PVP is soluble and polystyrene insoluble) and benzene (for which the converse holds) in the presence of 1 g dm^{-3} of the random copolymer, as a function of molar composition of the copolymer; the minimum in the curve corresponds to a molar ratio of 2:1 (vinylpyrrolidone:styrene), and in fact the polymer with this composition causes spontaneous emulsification of the

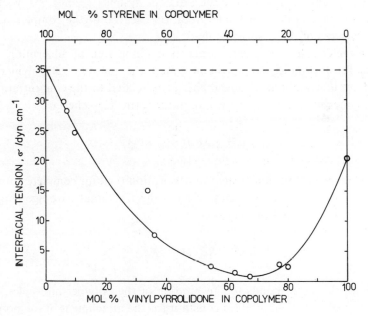

Fig. 36. Variation in the interfacial tension between water and benzene in the presence of 1 g dm^{-3} of random copolymers of vinylpyrrolidone and styrene, as a function of the molar composition of the polymer; the broken line at 35 dyn cm^{-1} indicates the interfacial tension for the pair of pure liquids. Data of Breitenbach and Edelhauser.[176,177]

two liquids, followed by the formation of an interfacial film containing the whole of the polymer. It seems that the high efficiency of this type of polymer results from the randomly placed hydrophilic and hydrophobic units acting to knit the surfaces together along their full length, whereas in the case of block copolymers the sole knitting action occurs at the point of junction of the two chains. This appears to be a promising field for further development, both from the practical and the theoretical viewpoints.

3.11.3. Water–Solid Interfaces

The adsorption of polymers from aqueous solution onto solid surfaces is probably the most important of the three types of interfacial adsorption considered here, principally because of the widespread applications of this phenomenon in such technical processes as the clarification of ground and effluent waters and the dewatering of mineral suspensions and slurries.

The general subject of adsorption of polymers from solution by solids has been briefly discussed by Adamson,[2] while the literature data have been reviewed by Patat et al.[1140] Most of the work has involved non-aqueous systems, and even for the aqueous systems studied many of the results are uncertain in their interpretation because of the complex nature of the surfaces of the adsorbents used.

Nevertheless, in very general terms it is found that at adsorption equilibrium for low-molecular-weight polymers the amount of polymer adsorbed by unit amount of adsorbent, g, is related to the concentration of free polymer in solution c_2 by a Langmuir form of isotherm

$$g = g_s K c_2 / (1 + K c_2) \tag{95}$$

where g_s is the "saturation" extent of absorption (i.e., for complete coverage of the adsorbent surface); denoting by θ the fractional coverage as given by the ratio g/g_s, the isotherm becomes

$$\theta = K c_2 / (1 + K c_2) \tag{96}$$

Insofar as this corresponds to site binding of the polymer molecules, then K will represent the equilibrium constant for the attachment of the polymer molecule onto a site. The isotherm may be rewritten in the form explicit for c_2:

$$c_2 = \theta / \{K(1 - \theta)\} \tag{97}$$

On the other hand, for high-molecular-weight polymers the adsorption data often fit better to the form of the Simha–Frisch–Eirich isotherm:

$$Kc_2 = \{\theta \exp(K_1\theta)/(1 - \theta)\}^\nu \qquad (98)$$

Here the exponent ν represents the number of adhesion points for each polymer molecule on the adsorbent surface, and for high-molecular-weight polymers has an experimental value between two and three, while K_1 characterizes the strength of the interactions between the bound polymer segments. For single-point attachment, $\nu = 1$ and also $K_1 \to 0$, so that the isotherm then reverts to the Langmuir form [compare eqn. (97)]. This behavior, as well as the positive but not marked correlation observed between the amount adsorbed and the molecular weight of the polymer, points to an adsorption mechanism involving a few attached segments separated by loops of the polymer chain. It seems that the polymer coils are in general adsorbed by just a few points of attachment which are most probably on the outer periphery of the coil, while the adsorbed molecule largely retains the shape that it has in solution. The neighboring adsorbed coils, like those free in solution, are unlikely to be able to penetrate one another to any great extent, but would simply become progressively distorted and "flattened" with increasing extent of coverage of the adsorbent as they become more closely packed on its surface.

It should be emphasized that these considerations strictly only apply where the adsorption is freely reversible, and can be considered as an equilibrium process. In cases where the adsorption is irreversible the concept of an adsorption isotherm is no longer applicable since there are no longer definite pairs of values for the concentrations of adsorbed and free polymer.

The fundamental aspects of the use of polymers for effecting either the flocculation or the stabilization of suspended particles in aqueous solution have been treated in a number of papers in a 1966 *Discussion of the Faraday Society*,[1486] and much more recently both theoretical and practical aspects have been considered in a number of papers in a 1972 issue of the *British Polymer Journal*.[269] In the main, the interest has been focused upon inorganic particles in suspension, such as typically arise in the separation and recovery of minerals and in the clarification of water for use; the most widely used polymer for these purposes is PAAm, in the form either of the pure homopolymer or as copolymer with an acidic or basic monomer to give it an anionic or cationic character. The origin of the special activity of PAAm is not fully understood; however, from the observation that while PAAm is both adsorbed by and flocculates aqueous suspensions of freshly prepared thermal silica, but that both these properties are lost as the suspensions "age," Griot and Kitchener have suggested hydrogen bonding onto free surface silanol groups as one possible mode of attachment of the PAAm molecules.[570]

As regards supensions of organic materials, Gregory[568] has studied the flocculation of polystyrene particles with surface carboxyl groups (prepared by emulsion polymerization, as outlined in Section 2.1) with cationic polyelectrolytes of the PAAm type. Here charge–charge interactions seem to play an important although not the sole role in the adsorption.

There are three particularly interesting features to the flocculation of suspensions induced by polymers. First, unlike the case of coagulation induced by simple electrolytes where the sediment finally obtained is compact and hard to filter, with polymers the particles aggregate in loose flocs which do not compact further and which are easy to filter; this is a feature clearly of practical importance in separating the sedimented material from the system. Second, the polymers that are active are often ones, such as PAAm, that are nonionic, unlike the situation with small-molecule coagulants, which are normally electrolytes. Third, there is often an optimum polymer concentration for flocculation, and at higher concentrations the flocculating effect falls off and on the contrary the dispersion may become stabilized ("protected") by the polymer, a feature also of obvious practical importance.

These effects can be attributed to the ability of a polymer molecule to be simultaneously attached to two or more particles, forming a bridge between them; this results from the adsorbed polymer molecule retaining much the same conformation as that free in solution, so that even when adsorbed onto one particle there will be other groups on its periphery available for attachment to a second particle. At the same time the resilience of the polymer molecule prevents the particles from coming into closer contact, and thus leads to the loose flocs produced in the aggregation process. This model also serves to explain the phenomenon of an optimum polymer concentration for flocculation; considering the flocculation process from the kinetic viewpoint, the bridging requires an adsorbed polymer molecule on one particle to come into contact with a vacant site on another particle, so that the joint probability for this will be given approximately by the product $\theta(1 - \theta)$ (where θ is the fractional coverage), which is maximal at $\theta = 0.5$. Thus bridging will take place most readily at the particular polymer concentration that gives essentially half-coverage of the particle surface, and which is then the optimum concentration. At higher polymer concentrations the particle surfaces become progressively covered up by adsorbed polymer, so that the probability for bridging decreases; indeed, because of the presence of the adsorbed layer of polymer the aggregation of the particles is inhibited and the suspension made *more* stable than in the absence of polymer.

This simple model of the action of polymeric flocculants and dispersants is a useful one for many practical purposes. More precise models will require the study of specific systems in greater detail, as has been done by Lyklema and his colleagues[451] in the case of aqueous silver iodide suspensions (whose characteristics are already quite well established) with poly(vinyl alcohol).

Finally, it is interesting to note an application of PVP as a stabilizer of aqueous suspensions, which depends in part upon the special solvent power of the solid polymer. In this application[1456] the material to be dispersed must be one (such as β-carotene, Sudan III, or Chrysarobin) that is soluble in the solid polymer. The technique involves dissolution of the material along with the polymer in chloroform and evaporation of the solvent to form a film in which the material is molecularly dispersed. On dissolving the film in water, a stable dispersion of the material of particle size (with β-carotene) of about 0.1 μm is produced. The method can also be used with other water-soluble polymers such as poly(vinyl methyl ether) or methylcellulose, as well as with the surface-active agent Aerosol OT.

4. WATER IN RELATION TO THE CHEMICAL BEHAVIOR OF SYNTHETIC POLYMERS

The chemical reactivity of polymers is significant in a number of respects in the present specific context; of these the most important instances are the chemical modification of the side groups to convert hydrophobic polymers to hydrophilic ones and vice versa, the hydrolytic stability of the main chain in relation to the serviceability of the polymer, and the general chemical stability of polymers in aqueous solution compared with that in nonaqueous solution or in the solid state. In point of fact, the chemical aspects of synthetic-polymer/water systems are even more diverse than the physical ones, simply because of the greater diversity of possible chemical reactions, and hence within the scope of the present chapter it is only possible to outline a few broad divisions within the field; in particular, we shall discuss some aspects of the above instances in relation to the different modes of chemical attack upon polymer chains, and to the effect of the state of the polymer upon its chemical reactivity.

4.1. Modes of Chemical Attack upon Polymers

Adopting a classification in terms of the extent to which the polymeric nature of the material is modified by the attack, we may distinguish three main modes of chemical attack upon polymers: side-group attack; minor-link main-chain attack; and major-link main-chain attack. Of course, this division is artificial in many respects, since for example any attack upon a side chain must as a result to some extent alter the reactivity of the sequence of atoms forming the main-chain "backbone" of the polymer, while on the other hand much of the so-called "main-chain" attack involves in its initial steps atoms linked to the main chain rather than those in the main chain itself (such as in free-radical attack via hydrogen atom abstraction, and in the aqueous acidic attack on polyesters and the like via hydrogen ion addition to the carboxyl oxygen). Nevertheless, the division is a useful practical one for the present purposes.

4.1.1. Side-Group Attack

The least drastic form of chemical attack upon a polymer is that involving only the side groups. Typical examples of this are the hydrolysis of poly(vinyl acetate) to PVA and that of poly(methyl acrylate) to PAA; these are of interest in the present context because they involve the conversion of an initially relatively hydrophobic polymer into a highly hydrophilic one. The converse then requires, say, the reformation of the ester group in these two cases: alternatively, with PVA, reaction with an aldehyde (RCHO) can give a water-insoluble product through acetal formation with pairs of hydroxyl groups:

$$-CH_2-CH-CH_2-CH- \; + \; R \cdot CHO$$
$$\underset{OH}{|} \qquad\quad \underset{OH}{|}$$

$$\rightarrow -CH_2-CH-CH_2-CH- \; + \; H_2O \qquad\qquad (99)$$
$$\underset{O}{|} \qquad\qquad \underset{O}{|}$$
$$\diagdown \qquad \diagup$$
$$CH_2$$
$$|$$
$$R$$

Many of the aspects of the reactivity of polymers that particularly concern their side groups have been discussed by Morawetz (Ref. 1038, Chapter IX). It should be noted that, as usual, any such chemical reaction will have both equilibrium (thermodynamic) and kinetic aspects; thus, Frank[470] has studied the lactam–amino acid equilibria involved in the ring opening in aqueous solution of PVP and its monomeric analog N-ethylpyrrolidone, while Conix and his colleagues[274,275] have studied the kinetic aspects of this reaction for the amide groups in these and related compounds under highly alkaline conditions where the equilibrium favors the hydrolysis product.

One practical aspect of such side-chain reactions concerns polymer preparation, where it is important that the reaction should proceed essentially to completion, since residual unreacted groups clearly cannot be separated out by any of the conventional techniques which are applied to small-molecule substances (see Section 2.4). Thus in the preparation of PVA from poly(vinyl acetate) the "hydrolysis" is in commercial practice[441] carried out by ester exchange using acidified methanol, with continuous removal of the methyl acetate by distillation:

$$\left[\text{CH}_2\text{—CH} \right] + x\text{CH}_3\text{OH} \rightarrow \left[\text{CH}_2\text{—CH} \right] + x\text{CH}_3\text{CO}_2\text{CH}_3 \quad (100)$$

However, the final product still contains 1 or 2% of acetate groups which must be then removed by aqueous acid hydrolysis at 90–100° (Ref. 1192, Chapter 1).

Even if the side-chain modification is carried out on a strictly linear precursor of known degree of polymerization, then the strictness in linearity and same value of the chain length will also be achieved in the final product only if there has been no main-chain degradation or cross-linking during the reaction. It therefore is advisable to reconvert a portion of the product back to the starting polymer, to check whether its properties are the same as the original; naturally, this reconversion reaction must also be a "clean" and complete one, but if the cycle of reactions leads to no significant alteration, in, say, the $[\eta]$ value of the original polymer then this is a strong indication of the suitability of the pair of reactions involved. This procedure was adopted, for example, by Katchalsky and Eisenberg[752] in their early studies on PAA and PMAA, where the acids were obtained from the polymeric methyl esters by acid hydrolysis, and reconverted using diazomethane.

4.1.2. *Minor-Link Main-Chain Attack*

As a result of the mode of their preparation, many polymers have within their chains minor amounts of "foreign" groups, which may act as weak links by being much more readily broken chemically than the main, normal repeat units of the chain. The presence of small amounts of such groups is much more influential chemically than it would be in the case of a small-molecule compound, since, for example, with a polymer only one such group present in a long polymer chain needs to be broken for the (number-average) molecular weight to be halved. Further, such scission can lead to fragments with active (say, free-radical) end groups from which further degradation of the fragments can propagate.

One example of this is the presence of peroxy groups in vinyl and acrylic polymers produced by free-radical mechanism in the presence of even traces of oxygen [eqns. (20) and (21)]; these groups, because of the ease (particularly by thermal means) with which the O–O bond is broken, act as weak links in the polymer chains. The presence of such peroxy groups may give rise to decreases in $[\eta]$ (and hence, presumably, in the molecular weight) of hydrophilic polymers such as polyacrylylglycinamide, even when in solid form and stored in the dark, and of polyacrylamide when stored in aqueous solution.[581]

Again, the fact that samples of poly(vinyl acetate) frequently show a fall in $[\eta]$ after one cycle of hydrolysis and reacetylation, but that this property thereafter remains stable upon repetitions of the cycle, has been attributed[1001] to intramolecular chain transfer ("backbiting") during the polymerization:

$$\begin{array}{ccccc}
\text{O–CO–CH}_3 & & & \text{O–CO–}\overset{\cdot}{\text{CH}}_2 & \\
| & & & | & \\
\text{–CH}_2\text{–}\overset{\cdot}{\text{CH}}\text{–CH}_2\text{–}\overset{\cdot}{\text{CH}} & \rightarrow & \text{–CH}_2\text{–CH–CH}_2\text{–CH}_2 & & \\
& | & & | & \\
& \text{O} & & \text{O} & \\
& | & & | & \\
& \text{CO–CH}_3 & & \text{CO–CH}_3 &
\end{array} \qquad (101)$$

which leads to hydrolytically labile ester groups within the main chain of the polymer. As already pointed out, the possibility of these or like reactions must be borne in mind when producing a desired type of polymer by chemical modification of another polymer, and checked by "recycling" to ensure that the regenerated polymer has retained its molecular weight and linearity.

The so-called polyurethanes are a class of polymer that contain hydro-lyzable weak links in the form of the urethane groups which are produced in the reaction between the diisocyanate and the preformed hydroxyl-ended polymer (generally, a polyether or a polyester):

HO—P—OH + OCN—R—NCO + HO—P—OH

$\rightarrow$ HO—P—O—CO—NH—R—NH—CO—O—P—OH,   etc.   (102)

Cooper and his colleagues[281] have discussed the significance of the presence of these and other groupings produced in the linking reaction in relation to the hydrolytic stability of polyurethanes.

This example of the production of polyurethanes serves as a reminder that unless a polymer molecule is cyclic, it must contain two end groups chemically distinct (often, widely so) from the main chain unit and hence also viewable as foreign groups in the chain. Their decomposition does not affect the polymer quite so directly as weak links within the main chain, but nevertheless they do form a distinct point of attack on the chain. For example, the formaldehyde polymer polyoxymethylene is unstable due to its tendency to "unzip" from the hydroxyl end groups, and this is therefore prevented in commercial practice by esterifying these groups with acetic acid; however, these acetate groups then provide weak points for the inter-action with water since their hydrolysis may initiate the spontaneous un-zipping reaction. On the other hand, the ease of hydrolysis of such acetate end groups is usefully applied in the room-temperature vulcanizing of special rubbers, which comprise a silicone rubber with end groups of this type; while stable indefinitely out of contact with the atmosphere, when exposed to atmospheric humidity the acetate groups are lost, exposing the silanol (Si–OH) groups at the ends of the chains, which then link up with other such groups, converting the polymer into a solid mass.[1268]

4.1.3. Major-Link Main-Chain Attack

The third and most drastic mode of attack upon a polymer is that involving the main-chain sequence of atoms. Here, for a chain of typical degree of polymerization 10^3, it only requires 0.1% extent of a scission reaction for the breakage of one link (on the average) in each polymer chain, and hence a halving in $\bar{M}_n$.

Considering specifically purely hydrolytic stability, then of all the commercially important synthetic polymers it is the polyamides and the polyesters which are most likely to be susceptible to this form of attack.

Goodman[545] has discussed the molecular structural factors influencing the ease of hydrolysis of synthetic polymers, particularly those used in textile fibers. With the polyamides (i.e., the nylons) the conditions required for hydrolytic cleavage of an amide group are so drastic that no practical problems arise under ordinary conditions of service. On the other hand, with the polyesters there are marked differences in stability which depend on the chemical type of the groupings (R, R') most directly attached to the ester group in the chain: —R—CO—O—R'—; both the completely aliphatic polyesters (with R and R' both alkyl groups) and also the polyesters such as poly(p-phenylene adipate) (where R is alkyl and R' is aryl) are weakened by hydrolysis through continued exposure even to only moist air, whereas poly(ethylene terephthalate) (where R is aryl and R' is alkyl) is unaffected even by lengthy immersion in hot water or contact with aqueous hydrochloric or sulfuric acid. This is, in fact, the same order of reactivity as that shown by small-molecule analogs and it clearly relates to the controlling effect of the group R attached directly to the *carbonyl* group in the ester.[545] Presumably, when R is an aromatic group the conjugation with the carbonyl group stabilizes the ester against attack both by hydrogen ions and by hydroxyl ions. Similar factors explain the parallel hydrolytic behavior of the polyanhydrides.[545] It should be noted, however, that the stability of these polymers, particularly those used for textile fibers, will also depend upon the relative crystalline and amorphous contents of the material (see Section 4.2).

The hydrolytic instability of fully aliphatic polyesters, although it rules them out for use in normal textile fibers, has led to some useful applications, as with poly(glycolic acid), in absorbable surgical sutures[39] (see Section 5.6).

A more general form of chemical attack upon the main chain of synthetic polymers is from atmospheric oxygen via autoxidation, a type of reaction whose rate is also greatly accelerated by light (photoautoxidation). This is a reaction shown by a wide variety of polymers, ranging from the highly hydrophobic ones such as polyethylene and polypropylene across to much more hydrophilic ones such as poly(propylene oxide) and poly-(ethylene oxide). In fact, polymers of the latter type containing oxygen atoms in the main chain are even more susceptible to this autoxidation than those having a backbone of only carbon atoms, which is in line with the greater general susceptibility of ethers to peroxide formation as compared with simple alkanes.[1531] In the case of a polyether such as, for example, poly(ethylene oxide), the autoxidation chain may be taken to be started by traces of free radicals X· already present in the system, the reaction then following the course

$$-CH_2-CH_2-O- + X\cdot \rightarrow -CH_2-\overset{\cdot}{C}H-O- + XH \qquad (103)$$

$$-CH_2-\overset{\cdot}{C}H-O- + O_2 \rightarrow -CH_2-CH-O- \qquad (104)$$
$$\underset{\overset{|}{O_2\cdot}}{}$$

$$-CH_2-CH-O- + -CH_2-CH_2-O-$$
$$\underset{\overset{|}{O_2\cdot}}{}$$

$$\rightarrow -CH_2-CH-O- + -CH_2-\overset{\cdot}{C}H-O- \qquad (105)$$
$$\underset{\overset{|}{O_2H}}{}$$

followed by further cycles of reactions (104) and (105) leading to continued hydroperoxide formation. It the hydroperoxide groups are decomposed (thermally, photolytically, or by the attack of ions of transition metals), then this may lead both to branching of the chain reaction through the generation of fresh radicals (notably, mobile ones such as hydroxyl), and also to the likelihood of scission of the main chain; thus with thermal scission the next two steps could be

$$-CH_2-CH-O-CH_2-CH_2-O-$$
$$\underset{\overset{|}{O_2H}}{}$$

$$\rightarrow -CH_2-CH-O-CH_2-CH_2-O- + HO\cdot \qquad (106)$$
$$\underset{\overset{|}{O\cdot}}{}$$

$$-CH_2-CH-O-CH_2-CH_2-O-$$
$$\underset{\overset{|}{O\cdot}}{}$$

$$\rightarrow -CH_2-CHO + \cdot O-CH_2-CH_2-O- \qquad (107)$$

That this type of degradation mechanism does take place is supported by the marked accelerative effect of light and of transition metal ions upon the degradation of poly(ethylene oxide) in aerated aqueous solution,[1531] and by the marked inhibiting effect of a substance such as isopropanol, which presumably reacts preferentially with the free radicals Y· in the system giving a stable and relatively unreactive radical product[949]

$$CH_3-CH-CH_3 + Y\cdot \rightarrow CH_3-\overset{\cdot}{C}H-CH_3 + YH \qquad (108)$$
$$\underset{\overset{|}{OH}}{} \qquad\qquad\qquad \underset{\overset{|}{OH}}{}$$

This general type of oxidative attack upon polymer chains is clearly important both specifically in regard to the stability of aqueous (and other) solutions of polymers in use, and also more generally in regard to the degradation of most commercially produced synthetic polymers when discarded into the environment (see Section 5.2).

4.2. Effect of Polymer State on Chemical Reactivity

The chemical reactivity of polymers is also markedly influenced by the physical state; further, in the case of solid polymers it will also be influenced by the degree of crystallinity. Considering this latter factor first, the crystalline fraction, because of its regularity and the tightness of packing of the chains, is, as we have already seen (Sections 3.8 and 3.9), essentially impermeable to small-molecule reactants so that external attack is extremely difficult; similarly, any spontaneous decomposition within the phase is also likely to be less easy—thus in the case of a peroxide group held within the phase, even upon its scission the two free radicals would not be able to diffuse very readily away from one another, so that they would be most likely to recombine.

As regards the amorphous phase, although this is far more open to a small-molecule reactant than the crystalline phase, the access is still less easy than for, say, a polymer in dilute solution. In particular, it is much more likely that a reaction in an amorphous phase would be diffusion controlled. This would apply, with respect to water, both directly, to hydrolytic reactions where water is the reactant whose molecules have to permeate into the bulk of the material, and also indirectly, to the effect of water upon diffusion-controlled autooxidation since it is known that the diffusion of simple gases such as oxygen is affected by the ambient humidity (through the presence of the absorbed water in the polymer matrix).[715,1007,1402]

Further, one would expect that an amorphous polymer, or more generally the amorphous component of the polymer, would show a higher reactivity at temperatures above the glass transition temperature T_g (where it is in a "rubbery" state) than below T_g (where it is in a "glassy' state). Water could have a role here because its absorption can lead to the lowering of the glass transition temperature of a polymer.

In the case of polymers in solution the chains will be much more accessible to chemical attack than in the solid state (at least for solvents of low viscosity). However, even in dilute solution polymer molecules possess two features which can cause the reactivity to be markedly different from that of the small-molecule analogs. First, by the very fact of the chainlike nature of the molecule, every chain unit will have others adjacent to it, so that there is a greatly enhanced probability of short-range effects, e.g., neighboring group participation. Second, even for extremely dilute solutions the molecules still form definite regions containing a finite local concentration of the monomer units; this concentration will average very roughly 1%, although the true value will depend very much upon the goodness or poorness of the solvent toward the polymer (and in any case for a random coil type of conformation the distribution of chain units is not uniform but follows an approximately Gaussian distribution about the center of mass). This means that one cannot "dilute out" the longer-range interactions such as electrostatic effects, at least as regards the chain units within the same molecule.

One topic of importance in the chemical reactivity of synthetic polymers in aqueous solution is their radiation stability. This has particular significance both in relation to radiation sterilization of such solutions and in relation to the corresponding sensitivity of biopolymers (particularly the nucleic acids) in their native aqueous environment. In the case of PEO, Borgwardt and his colleagues[156] have studied the kinetics of the pulse radiolysis of the polymer in aqueous solution, where the rate of combination of the macroradicals produced via attack of the HO radicals from the radiolysis of the water could be measured. King and Ward have studied the radiation chemistry of aqueous solutions of PEO from the more general viewpoint; the effects observed are markedly influenced by the presence of trace amounts of oxygen, since the radiation doses required for gelation of apparently well-degassed polymer solutions are greatly reduced when the solutions are *very* efficiently degassed.[768,1556] The radiation stability of polymers in aqueous solution can be greatly increased by the addition of small amounts of specific compounds; thus, Sakurada and Ikada[1297] have shown that thiourea is very effective in inhibiting the gelation of aqueous PVA solutions that is induced by γ rays.

5. WATER IN RELATION TO THE UTILIZATION OF SYNTHETIC POLYMERS

This section is concerned with the roles played by water in connection with the utilization of synthetic polymers, both in general usage (Sections 5.1–5.5) and also in the more specialized medical, pharmaceutical, and allied fields (Section 5.6). Some of these roles have been discussed earlier from their more fundamental aspects, particularly those connected with absorption by polymers (Section 3.8) and adsorption of polymers from solution (Section 3.11). Here we shall be concerned much more with examples where either the factors involved are complex and include a number of interlinked effects, as in the weathering and degradation of plastics, or where there is a complex set of requirements specified for the properties of the polymer (including its interactions with water), as in the medical utilization of synthetic polymers.

5.1. Water–Soluble Polymers

In all applications where the interactions with water and aqueous systems are significant, background knowledge is required on the physical chemistry of such interactions; we have already discussed the behavior of the most important and common of these polymers alone in aqueous solution in Section 3.5, their interactions with small-molecule cosolutes in Section 3.6, and their interactions with other such polymers in Section 3.7, while their gelation behavior has been considered briefly in Section 3.10.2 and their interfacial adsorption behavior considered (particularly with regard to their activity in the flocculation and stabilization of suspensions) in Section 3.11. The applications of these polymers are extremely diverse, and for further details reference should accordingly be made both to the general sources[322,1268] and also to the following more specific sources for the individual polymers indicated: PAA and its homologs,[530] PAAm,[1037] polyethyleneimine,[324] PEO and other polyethers,[507,605,1432] PVA,[48,441,1192] and PVP.[33,34,70]

5.2. Weathering and Degradation of Plastics and Rubbers

Considering water-insoluble polymers, and in particular those used as plastics and rubbers, one important feature determining their suitability for any particular application is their resistance to environmental attack. In service, and particularly in use outdoors, polymers undergo progressive and irreversible physicochemical changes which may be classified as "organic corrosion," during which the polymer is said to "age," or to be undergoing "weathering" or "degradation"; ultimately the material "perishes."[1180,1268] The character of the terms applied to these processes indicates the complex and frequently ill-defined nature of the changes that are taking place. Taken to its extreme, the life of the material could be considered to extend to the time of its complete breakdown; in practice, however, the useful life of the material is generally much shorter than this.

For definiteness, in technical testing and evaluation the effective lifetime is often taken to be the period in which a particular critical physical property (such as impact strength or elongation at break) is reduced to some particular fraction, say, one-half, of its initial value.

5.2.1. Water in Relation to Degradation and Weathering

The three main agents leading to the degradation of plastics and rubbers in service are oxygen, light, and heat. However, water is also a significant agent, since it is known that such materials are frequently more stable in dry atmospheres, even such as that of Arizona, than they are under temperate but more humid conditions.[182] Because of this involvement of water as a degradative agent, in the standard commercially produced equipment for accelerated weathering tests the sample is exposed to intense light both at a specified temperature and at a specified relative humidity, while these may be supplemented by water sprays to simulate the effect of rain.[29,37,38,182,664]

The importance of the three main agents listed above indicates that degradation is usually a photooxidative process, involving peroxide and free-radical intermediates, as discussed in Section 4.1.3. The actual reactions involved will naturally depend upon the specific polymer concerned; most investigations of these processes have been carried out using accelerated methods as indicated above, which thus involve much more extreme conditions than those of the natural environment, and the results so obtained are frequently difficult to extrapolate to give data relating to more normal conditions. Nevertheless, it is possible to see a number of ways in which the presence specifically of water can exert an influence on these processes.

In the first place, to stabilize the polymer against such degradation, it is often formulated with small-molecule additives which act as antioxidants, photoabsorbers, and so forth. In use, contact with liquid water or aqueous solutions (rain, detergent solutions for cleansing, and so forth) can lead to extraction of these agents and hence exposure of the polymer to these same degradation processes.[629] In a similar although less direct fashion, the extraction of other small-molecule additives such as plasticizers and dyes can also lead to lowered usefulness of the polymer formulation.

Second, water adsorbed on the surface of the polymer or absorbed within its bulk is also likely to influence the rates of these photooxidative processes. It is known, for example, that absorbed water affects the rate of diffusion of gases inside polymers;[715,1007,1402] insofar as the photooxidation process is controlled by the diffusion of oxygen from the atmosphere into the bulk of the polymer, then the rate of the degradation process will also be affected by absorbed water. Further, the presence of water is likely to increase the mobility of trace metal ions, and hence to enhance their catalytic effect on the decomposition of hydroperoxides involved as intermediates in the photooxidation.

It is also known that there is a volatile intermediate, possibly H_2O_2, involved in the photooxidation of polymers.[383,384,1347] Sorbed water can play a twofold role here; first, at low concentrations it can enhance the mobility of any H_2O_2 by occupying absorption sites which would otherwise bind the peroxide, while on the other hand at higher concentrations it can form a liquid film which both blocks the voids in the polymer matrix and also absorbs the H_2O_2 itself.

5.2.2. *Water in Relation to the Attack of Ozone on Rubbers*

Polymers are also subject to attack from other and trace components of the atmosphere, such as ozone, oxides of sulfur, and oxides of nitrogen. Ozone, in particular, has a marked degradative effect upon unsaturated polymers such as rubbers. It is known that the rate of uptake of ozone by rubbers is increased by the presence of water vapor; this applies both to natural rubber and also to many synthetic rubbers (with the exception of butyl rubber).[20] Under these conditions the surface degradation is faster than with ozone alone and leads to so-called "frosting" while the degradation products are found to be acetone soluble (unlike those in the absence of moisture). It has been suggested[20] that in the general course of ozonolysis, peroxy zwitterions are formed from the heterolytic decomposition of the ozonide:

$$\begin{array}{c}
\diagdown \quad \diagup \\
C{=}C \\
\diagup \quad \diagdown
\end{array}
+ O_3 \rightarrow
\begin{array}{c}
O_3 \\
\diagdown \quad \diagup \\
C{-\!\!-\!\!-}C \\
\diagup \quad \diagdown
\end{array}
\rightarrow \qquad\qquad (109)$$

$$\begin{array}{c}
O{-}O^{\ominus} \ {}^{\oplus}O \\
\diagdown \mid \quad \mid \diagup \\
C{-\!\!-\!\!-}C \\
\diagup \quad \diagdown
\end{array}
\rightarrow
\begin{array}{c}
\diagdown {}^{\oplus} \\
C{-}O{-}O^{\ominus} \\
\diagup
\end{array}
+ O{=}C \begin{array}{c} \diagup \\ \diagdown \end{array}$$

which in the absence of water combine to form relatively stable polymeric peroxides:

$$x
\begin{array}{c}
\diagdown {}^{\oplus} \quad\quad {}^{\ominus} \\
C{-}O{-}O \\
\diagup
\end{array}
\rightarrow
\begin{array}{c}
\mid \\
{\left[C{-}O{-}O \right]}_{\overline{x}} \\
\mid
\end{array}
\qquad\qquad (110)$$

but that in the presence of water these peroxy zwitterions are converted to hydroxy hydroperoxides:

$$\underset{/}{\overset{\backslash}{C}}\!-\!O\!-\!\overset{\ominus}{O} + H_2O \rightarrow \underset{/}{\overset{\backslash}{C}}\!\overset{\overset{\displaystyle OH}{|}}{}\!-\!O\!-\!OH \qquad (111)$$

which would more readily decompose to form free radicals inducing auto-catalytic oxidation of the rubber.

5.2.3. *Self-Degrading Plastics*

One recent and significant development in relation to the degradation of synthetic polymers is the spreading realization of the disadvantages, in the long-term sense, of commercially producing highly stable synthetic polymer articles; for any such article, unlike one made of a natural organic material, does not readily decay when discarded but tends to survive indefinitely. Indeed, simply because of their durability coupled with their low density (so that they float on water), discarded plastic articles are now among the most unsightly form of litter to be seen polluting the natural environment, while their bulkiness raises problems in their disposal as components of normal household and other waste.[1419] Thus in the present specific context it is evident that serious social and environmental problems arise directly from the interaction (or rather, lack of interaction) between water and these commercial synthetic polymers.

Ironically, therefore, attention is increasingly being directed to the development of plastics that are *self-degrading*, particularly when discarded in an exposed situation. Despite some complacency over the plastics waste problem and despite scepticism as to the feasibility of such developments,[1419] considerable progress has been made in the development of self-degrading plastics, particularly at the University of Aston in Birmingham.[32,1355,1356] Many common synthetic polymers used in plastic articles, particularly polyethylene and polypropylene, are inherently (i.e., thermodynamically) unstable in the atmosphere, and only acquire practicable stability on incorporation of antioxidants (see Section 4.1.3). In the self-degrading formulations an additive is incorporated which absorbs sunlight at wavelengths below about 320 nm and hence photosensitizes the degradation; since window glass absorbs ultraviolet light of these wavelengths, the polymer is stable in an inside environment, but when it is discarded outside the photooxidative degradation sets in rapidly to finally produce a coarsely powdered product which is both greatly more hydrophilic than the original polymer by reason of the carbonyl, hydroxyl, and carboxyl groups produced, and correspondingly more open to further oxidative and biological attack, as well as being in a state where it will be effectively absorbed into the natural environment.

5.3. Textiles

Fibers and the textiles and fabrics which are made from them are a class of polymeric materials with a number of special features placing them somewhat apart from the main run of applied (synthetic) polymers. These special features include a relatively high crystallinity, which results from the deliberately induced orientation of the polymer molecules and which is a requisite for high tensile properties of the fiber, and a high surface area-to-volume ratio, which leads to more direct effects from the atmospheric agents, including moisture; lastly, there is also the fact that the fibers are commonly colored by the incorporation of dyes.

5.3.1. Physical Aspects of Textile–Water Interactions

The general features of water absorption and penetration by polymers, which also apply to textile fibers, have already been outlined in Sections 3.8 and 3.9, along with some related aspects of dyeing of these materials. One particular practical feature of water absorption by textiles is the heat evolved upon absorption, which greatly adds both to the comfort and the utility of the clothing, especially when it is worn in exposed situations.[514] Although the heat evolved per mole of water absorbed is much the same for synthetic and semisynthetic fibers (polyamides, polyesters, cellulose diacetate and triacetate) as it is for natural fibers (wool, cotton, silk), the lower total absorption by the former group of polymers naturally leads to a lower total heat evolution, which is responsible at least in part for the "colder" feel of garments made from these materials. At the same time the lower water uptake also means that, unlike the more hydrophilic natural fibers, even at high humidities the synthetic fiber materials retain their high electrical surface resistivity, so that electrostatic charges acquired by friction take a long time to leak away. This gives rise to various practical difficulties both in such processes as the spinning of the fibers and the fabrication of textiles from them, and also in the comfort for wear of these textiles (Ref. 1046, Part V).

The disadvantages which are thus attendant upon the more hydrophobic character of the synthetic and semisynthetic fibers have led to attempts to develop synthetic polymers for this purpose which are much more hydrophilic; prominent among these is the largely Japanese development of PVA, cross-linked to make it water insoluble by various forms of aldehyde treatment. Nevertheless, as Goodman[546] has commented, "... Truly durable hydrophilic fibers with a high level of mechanical performance still remain to be invented [and] the target of attaining improved hydrophilic properties remains open for future research." In practice, rather than restricting the field to pure homopolymers, the hydrophobic character of the synthetic polymers is mitigated by incorporating small proportions of more hydrophilic monomers into the main chain; further, in actual textiles it is common to use blends of natural and synthetic fibers, which gives an enhanced hydrophilicity while retaining the strength and wear characteristics of the synthetic component.

5.3.2. Chemical Aspects of Textile–Water Interactions

In use, textiles need to be resistant to chemical attack from a variety of aqueous systems; in the case of clothes this involves not only blood, sweat, tears, and other aqueous systems of natural origin, but also the highly hydrolytic and oxidative systems of laundering and bleaching processes. The chemical aspects of the hydrolytic stability of such important fiber-forming synthetic polymers as the polyesters and the polyamides have been outlined in Section 4.1.3; their high crystallinity also serves to enhance the stability of these polymers.

The presence of a dye in a fiber can also play an important part in the stability, and particularly the light stability, of the system. In the first place the dye itself is subject to "photofading," a process which is influenced by the type of polymer matrix involved.[500,906,952] Second, the dye may act as a photosensitizer for the degradation of the fiber, leading to "photo-tendering." Once again these two linked processes are often humidity sensitive, so that, as with plastics, the accelerated test methods involve controlled conditions of light intensity, temperature, and humidity, with water jets where it is required to simulate the effect of rain.[29,37,38,500] Humidity is also a factor in the (dark) fading of dyes on fibers induced by ozone; thus, with a textured nylon fabric dyed with Disperse Blue 3, exposure to dry ozone gave no color change in depth, whereas exposure at higher humidities completely destroyed the dye.[1302]

5.4. Paints and Related Coatings

Polymers used in paints and related coatings such as lacquers and varnishes show two features which distinguish them from polymers in other applications. First, in the original liquid state of the paint the polymer forms only one component of a complex system containing in addition such materials as solvents, thinners, pigments, dispersing and suspending agents, and so forth. Second, in its final state the paint film, simply because it is a film, is much more susceptible to environmental influences than the bulk form of the polymer.

The traditional paints were of course prepared from "drying oils" such as linseed oil, which gradually set and hardened in contact with the air; it is now known that this process is a complex polymerization reaction in which the unsaturated esters of the oil are converted into a cross-linked network via autoxidation reactions with atmospheric oxygen. More recently, these paints have been replaced by alkyd resins, which are products of the transesterification of these same types of oil with polyols, phthalic acid, and so forth. Being rather hydrophobic substances, these oil-based paints can normally only be applied to fairly dry (water-free) surfaces, otherwise the liquid coating will not adhere to the surface; however, paints have been developed which can be applied to steel structures even under water—the basis of one such a paint is a chlorinated rubber in a solvent system which is largely water soluble, together with a long-chain alkyl-amine to give adherence.[163]

Although such common paints do not seem greatly affected in their "drying" by the absence or the presence of water, it is noteworthy that in the case of the traditional Japanese lacquers a high humidity is a prerequisite for giving satisfactory "drying" of the lacquer coat. It seems that this arises from the "drying" being an enzyme-catalyzed process involving an oxidase, *laccase*, which would naturally require a somewhat water-swollen system for proper functioning; heat or the use of metal-based driers does speed up the drying process, but leads to a darker lacquer, while the spontaneous process gives an almost clear finish.[1478] It is interesting to speculate whether, in the light of the current commercial developments in the industrial preparation of enzymes and the current scientific development of "synthetic enzymes," it may not be feasible to produce water-based clear varnishes and lacquers which similarly contain enzymes or synthetic enzymes deliberately added to speed up their "drying."

A relatively recent development in the paint field is that of the so-called "emulsion paints," which are based upon aqueous dispersions of polymers such as poly(vinyl acetate); these have the marked advantage over the oil-based paints that they are readily miscible and thinnable with water, and also that the brushes and rollers used for their application can be washed with water, while it is also not so important for the surfaces to which the paint is being applied to be dry. The formation of the final coherent paint coat takes place by the evaporation of the water and by its penetration into an underlying porous surface, followed by the coalescence of the polymer particles; in most cases a plasticizer ("coalescent") has to be incorporated into the formulation to lower the glass transition temperature of the polymer below the ambient, so that the particles will coalesce readily. The commonest plasticizers used for this purpose are the dialkyl phthalates; unfortunately, these are rather readily leached from the paint film (say, by rainwater, or on washing), so that the film gradually becomes brittle and liable to crack. This effect is minimized by using a mixture of a "fast-acting" plasticizer (e.g., dibutyl phthalate) along with a "slow-acting" one such as tricresyl phosphate, which only gradually diffuses into the polymer particles but is retained much longer because of its much lower volatility.[1561]

The emulsion paint formulation also usually contains polymeric stabilizers to keep the polymer particles in suspension, presumably by adsorption onto the outside of the particles forming a protective layer. Although natural water-soluble polymers such as gum arabic have been used for this purpose, because of both their variability in supply and properties and their susceptibility to biological attack they are being largely supplanted by synthetic materials such as PVA.[1561]

5.5. Polymers in Membranes

Polymer membranes have found increasing applications in the purification and treatment of water and aqueous systems by such processes as electrodialysis and hyperfiltration (reverse osmosis). Glueckauf[536] has outlined the main processes, including the two membrane-based ones named, which are used commercially for seawater desalination, while Sourirajan[1412] has dealt more specifically with reverse osmosis.

In reverse osmosis the aqueous system is placed in contact with a membrane readily permeable to water but not appreciably so to the solutes; if a hydrostatic pressure somewhat greater than the osmotic pressure of the system is applied, then the water will pass through the membrane and the solute will be largely retained by the membrane.

The most widely used membrane material for reverse osmosis is the semisynthetic polymer cellulose acetate; to give a membrane which is both strong and permeable it is generally used in the composite form in which a compact, thin ($\sim$0.25 μm) working layer is backed on one side by a relatively thick ($\sim$100 μm) porous support layer of the material. Cellulose acetate is by no means the ideal membrane material, since it is limited to only mildly acid or mildly alkaline conditions (pH range 3–8, optimum 5.5) due to hydrolysis of the ester groups outside this range, while in addition the thin working layer may be attacked by bacteria which thus destroy the performance of the membrane. This indicates the need for other, more stable polymeric materials for this purpose—while in fact proposals have been made for the use of basically inorganic membranes because of their high stability.[1210] We have already, in Section 2.1.1, pointed out the possible potential for these purposes of materials produced by bulk polymerization of water-insoluble vinyl or acrylic monomers containing deliberately added water.

It is also interesting to note that a process for water purification involving a polymer not in membrane form has been developed by Lazare,[30] whereby the polymer first absorbs the water from the system at a low temperature and then is induced to exude it again by warming it up.

5.6. Medical and Allied Applications of Synthetic Polymers

Synthetic polymers have aroused much interest and optimism as suitable materials for a variety of applications in the medical, pharmaceutical, and allied fields. This has stemmed in part from the seemingly complete inertness of most plastics, and in part from the widespread popular belief that it is possible to synthesize polymers which have any desired property or combination of properties. On the other hand, the requirements that have to be satisfied (in terms of the long-term stability, biosensitivity, and other properties) if a polymer is to be acceptable in the medical field are much more stringent than in most other applications. In the present specific context, interactions between a synthetic polymer and aqueous systems that would be of no great concern for most practical purposes may become of literally vital importance when, say, the synthetic polymer is in the form of a surgical insert and the aqueous system is one of the body fluids (blood, lymph, etc.), or when the polymer is dissolved in an aqueous pharmaceutical preparation which is to be injected into the body.

5.6.1. General Requirements for Suitability

It is in the case of surgical inserts and prostheses implanted inside the body, e.g., as artificial heart valves and artificial joints, that there are the most critical requirements for a synthetic polymer to satisfy if it is to be acceptable. These requirements can be considered under five main categories[907]:

(a) The polymer must be sterilizable in depth.

(b) It must not produce any "foreign-body" reaction in the surrounding tissues, either in the short term or after it has been in the body for some time; even initially smooth polymeric articles can become pitted and crazed after exposure to the body fluids.

(c) If the article is in contact with the blood, e.g., a heart valve, then it must not tend to destroy the blood corpuscles nor must it induce clotting.

(d) The article must not depend for its physical properties or its chemical stability upon small-molecule additives, since these will inevitably be progressively leached out by the body fluids, with deleterious results both to the article and also to the body. This, therefore, rules out polymers, such as poly(vinyl chloride), which rely upon plasticizers to give them the required flexibility for a given purpose; similarly, it rules out a large number of polymers, such as polyethylene and polypropylene, which rely upon small-molecule antioxidants to protect them against the aggressively oxidative and hydrolytic environment of the bloodstream.

(e) The article must be able to withstand applied stresses, both in the short term and after wear in the body.

These stringent requirements stem basically from the need for the article to last a sufficiently long time in the body (preferably, a lifetime), because of the inconvenience of the further surgery which would be required to replace the worn article. It should be realized that, unlike the natural tissues, which are largely self-regenerative, a synthetic polymer can only steadily deteriorate once it has been implanted, and that the most that can be hoped for is that the rate of this deterioration can be kept at an acceptably low level.

One highly hydrophilic synthetic polymer that has been developed specifically for use in implantable protheses is cross-linked poly(hydroxy-

$$
\left[\begin{array}{c} CH_3 \\ | \\ -CH_2-C- \\ | \\ CO \\ | \\ H_2C-O \\ | \\ H_2C-OH \end{array} \right]_x
\qquad \text{(PHEMA)}
$$

ethyl methacrylate) (polyHEMA or PHEMA) where the cross-linking is induced by including small amounts of a tetrafunctional agent such as ethylene glycol dimethacrylate in the polymerization system. This material is, for example, marketed in the U.K. (by Smith and Nephew Ltd.) under the trade name Hydron. The basic properties and clinical applications of the material have been reviewed by Simpson,[1401] while Singh[1400] has described the biological effects of its cardiac implantation. The hydrophilic character of the polymer confers on it the softness in the aqueous environment of the body which makes it suitable for implantation uses in plastic surgery. Synthetic hydrophilic polymers of this type have also been used in contact lenses, where the softness of the water-swollen material gives it an advantage over the more conventional but rather hydrophobic poly-(methyl methacrylate); however, the new types of polymers also have certain disadvantages in this application, particularly a greater difficulty in accurately fabricating the lens to the ophthalmic specification, a greater fragility of the lens in use, and a greater likelihood of bacterial attack than with the more conventional polymer.

5.6.2. Surgical Sutures and Ligatures

The commonest method presently used for closing up wounds and surgical incisions is to stitch the edges together. The traditional surgical suture and ligature material is catgut, which is a modified form of the protein collagen, and which is obtained for this purpose from, for example, the submucosa of the sheep's intestine. Unfortunately, this material has a number of deficiencies related to the fact that it is a complex and non-uniform substance of natural origin, while in addition extreme care has to be taken to ensure that the prepared material is sterile in depth.

In recent times, therefore, synthetic polymers such as the polyamides (particularly nylons 6 and 6.6) and polyesters [particularly poly(ethylene terephthalate)] have been introduced and officially accepted as suture and ligature materials.[737] However, as suture materials these suffer from the disadvantages that, unlike catgut, they are not broken down and absorbed within the wound; in addition, the polyamides rather readily absorb phenols and related substances used widely as germicides. On the other hand, as ligatures (that is, to tie arteries, veins, or the like) these synthetic polymers have definite advantages over catgut, principally because of the greater degree of control over their characteristics and their sterility.

More recently still, synthetic polymer suture materials have been developed and introduced which are broken down readily in the body. For example, poly(glycolic acid) (PGA), with the structure

$$\left[\text{CH}_2 - \underset{\underset{\text{O}}{\|}}{\text{C}} - \text{O} \right]_x$$

has been developed by Cyanamid Ltd., and is marketed for this purpose in the U.K. under the trade name Dexon. This material has the advantages over catgut that it is readily prepared as a sterile material in a range of definite sizes with reproducible properties; at the same time, since it is a simple aliphatic polyester, it is readily hydrolyzed in the body at a convenient rate, with a typical suture lasting for the period of approximately 20 days that is required for it to act while the wound is still healing.[39,941,1003,1022]

5.6.3. Surgical Glues and Adhesives

The suture materials discussed in the previous section also have their own attendant disadvantages, particularly because in the suturing operation the tissues need to be tough enough to hold the stitches. In recent years interest has therefore arisen in the development of polymeric surgical glues and adhesives, the aim thus being to *stick* the edges of the wound together rather than to *stitch* them together. In this case the moist nature of the wound produces problems in making a good adhesive bond between two such areas.[74]

For these purposes an adhesive system must fulfill the following three criteria:

(a) The starting materials, which in practice are generally either a prepolymer plus additives, or else a monomer plus a catalyst, must be noninflammable, nontoxic, and insoluble in the body fluids.

(b) The polymerization must be rapid, even in the presence of the body fluids.

(c) The polymer formed must be tough, flexible, adherent ("tacky"), and immunologically compatible, while although initially it must be resistant to the body fluids, eventually it must be broken down and eliminated.

One approach is to use a preformed polymer which is bonded to the tissue and also self-linked by a suitable small-molecule agent. Examples of this are gelatin linked with resorcinol-formaldehyde, and hydroxyl-ended polyethers or polyesters linked with diisocyanates (to form, in effect, poly-urethanes).[571]

A second, and more fundamental, type of approach, is to use a mono-mer that polymerizes in situ in the wound. One group of such materials used widely for this purpose is that of the alkyl α-cyanoacrylates, which produce polymers of the structure

$$\left[-CH_2-\overset{\overset{\displaystyle CN}{|}}{\underset{\underset{\displaystyle O=C-OR}{|}}{C}}- \right]_x$$

where R is the alkyl group.[134,876,994] The polymerization is notable in ap-parently involving an anionic mechanism even in aqueous solution,[876] i.e., the propagation step is

$$P-CH_2-\overset{\overset{\displaystyle CN}{|}}{\underset{\underset{\displaystyle CO_2R}{|}}{C}}{}^{\ominus} + CH_2=\overset{\overset{\displaystyle CN}{|}}{\underset{\underset{\displaystyle CO_2R}{|}}{C}} \rightarrow P-CH_2-\overset{\overset{\displaystyle CN}{|}}{\underset{\underset{\displaystyle CO_2R}{|}}{C}}-CH_2-\overset{\overset{\displaystyle CN}{|}}{\underset{\underset{\displaystyle CO_2R}{|}}{C}}{}^{\ominus} \qquad (112)$$

It seems that the combined presence of the cyano and the ester groups on the α-carbon atom serves to delocalize the anionic charge from that atom and so inhibit the expected chain-terminating proton transfer reaction with the water:

$$P-CH_2-\overset{\overset{\displaystyle CN}{|}}{\underset{\underset{\displaystyle CO_2R}{|}}{C}}{}^{\ominus} + H_2O \rightarrow P-CH_2-\overset{\overset{\displaystyle CN}{|}}{\underset{\underset{\displaystyle CO_2R}{|}}{C}}H + OH^{\ominus} \qquad (113)$$

while these same two groups on the monomer activate it sufficiently to still give a high rate for the propagation reaction (112). The strong polymer-to-tissue adhesion probably arises from initiation of the polymerization by weakly basic groups on the tissue constituents (proteins, carbohydrates, and so forth), so that the polymer chains become covalently attached to the tissue.*

---

* These cyanoacrylates are marketed also as general adhesives; in the light of their high tissue-bonding power, they should therefore be employed with caution, care being taken particularly to prevent contact with the skin since this can lead to the area contacted becoming bonded to anything else it touches, with obviously painful consequences.

Rather similar problems arise in adhesion, and rather similar polymeric adhesives have been developed, in the fields of dental adhesives and fillings and also of bone adhesives; both the dentine of teeth and bone material have a surprisingly high hydrophilic character which must again be taken into account in the development of bonding agents for such materials.[211,506,869]

5.6.4. Membranes in Hemodialysis

Polymers also find application in the membranes used for the processes of bloodstream purification technically referred to as "extracorporeal hemodialysis," involving equipment popularly referred to as artificial kidney machines. The membrane used for this purpose must be relatively permeable both to water and to body waste-product solutes such as sodium chloride, urea, uric acid, and creatinine, but preferably impermeable to other solutes such as glucose and the amino acids which must be retained in the blood. Regenerated cellulose is the membrane material most commonly used, but it would be desirable to obtain materials with a higher selectivity. Lyman and his colleagues[933–935] have synthesized block copolymers based upon poly(ethylene oxide) and poly(ethylene terephthalate) and investigated their suitability for this purpose, while Yasuda and his colleagues[1605–1607] have studied the permeability of a wide range of hydrophilic polymer membranes toward both water and a variety of relevant solutes in aqueous systems. Once again, the materials proposed in Section 2.1.1, produced by bulk polymerization of water-insoluble vinyl or acrylic monomers containing deliberately added water, may have potential uses in this area.

5.6.5. Polymers As Blood Plasma Substitutes and Related Materials

Because of the natural limitations in the supply and availability of whole blood and of blood plasma, as well as doubts raised about the purity of such materials from the available pooled stocks, a variety of natural, semisynthetic, and synthetic polymers have been proposed from time to time as plasma substitutes.[780] Prominent among the synthetic polymers so proposed is PVP, which was developed in Germany during World War II as a plasma volume expander,[782] and also subsequently used to some extent in the United States for this same purpose. However, this polymer suffers from the disadvantage that much of it is stored indefinitely in the tissues rather than being metabolized or excreted, although no clear harm seems to result from this storage. For this reason, and also

because of the availability as an alternative of the microbiologically produced glucose polymer dextran,[59] which does not seem to suffer to the
same extent from this defect, it is recommended that PVP should only be
used as a plasma substitute if a more suitable alternative is not available.[780]

It may be noted that PVP has also found application as a protective
agent in the freeze-preservation of whole blood; in this connection Ben-
David and Gavendo[109] have studied the protective action of the polymer
toward thermal and mechanical damage of red blood cells.

Within this same general area a material widely clinically used to inhibit the clotting of blood is heparin, which is a mucopolysaccharide bearing sulfate ester groups, prepared from animal tissue. Because of the natural
restrictions in its availability and supply, the likely variability in its properties, and its susceptibility to enzymic breakdown, a purely synthetic
substitute would have several advantages over heparin itself; with this in
mind, Patat and Vogler[1141] have prepared copolymers of vinyl alcohol and
vinyl sulfate by partial sulfation of PVA and have shown them to have
anticoagulant activity.

5.6.6. General Clinical and Pharmaceutical Uses of Synthetic Polymers

Synthetic polymer materials have also found increasingly widespread
application in the general clinical and pharmaceutical fields, particularly
for such items as containers, tubing, and syringes. In all these applications
the possibility should be borne in mind of interactions between the polymer
and various aqueous systems (and also, indeed, nonaqueous systems) with
which they may come into contact; this includes such preparations as
injections, infusions, and pharmaceutical solutions generally. One such
deleterious type of interaction is the leaching out of a small-molecule
component of the polymer (antioxidant, plasticizer, etc.), which both con-
taminates the solution and impairs the properties of the polymer. Conversely, the polymer may absorb and/or be permeable to some crucial
component from the solution; this happens with the conventional rubber
cap-liners, which absorb components such as phenolic preservatives from
pharmaceutical preparations. In this latter instance the problem may be
overcome, for example, by using the commercially available form of cap-
liner which is a composite of a disk of polytetrafluoroethylene (for the side
in contact with the solution) bonded to a disk of a synthetic rubber (to
give the necessary flexibility).[36]

Water-soluble natural polymers have a long and honorable history in
a variety of roles in pharmaceutical preparations, and a number of them
are still in use and officially recognized, e.g., acacia (gum arabic), alginates,

gelatin, and tragacanth. The variability in the supply and properties of these materials, however, is leading gradually to their replacement by semisynthetic materials (such as methylcellulose and carboxymethylcellulose) as well as by the purely synthetic polymers (such as PEO). With all these types of polymers the possibility of interactions between any small-molecule cosolutes and the polymer (see Section 3.6) needs to be considered in devising and evaluating any formulation. Reversible binding of small-molecule drugs or therapeutic substances by polymers, for example, can be put to gainful use in sustained-release and depot therapy, since it can enable more of the active substance to be incorporated in the preparation, or it may allow the same amount to be released at a lower rate. One such example of this is iodine with PVP,[1394] where binding by the polymer greatly reduces the irritant action of the iodine when they are used in combination as a topical germicide. On the other hand, in formulations containing both small-molecule preservative and polymeric components the binding of the preservative by the polymer may reduce the free concentration of the former below the level required for its antimicrobial activity; however, if this level is nevertheless still attained, the binding may serve a useful function in maintaining the free concentration above this level even while the preservative is being lost from the system. In addition, many of these preservatives are phenolic compounds, which can precipitate polymers via the formation of labile cross-links (see Section 3.6.1); again this will be disadvantageous, because it removes the components from solution, and it also prevents a clear homogeneous formulation from being obtained. It should be noted that these phenomena resulting from binding may occur not only in the original formulation of a material, but also more hazardously in clinical practice when two or more preparations are mixed extemporaneously to obtain some desired combination of therapeutic effects.

In a similar fashion the use of two or more different polymers in an aqueous pharmaceutical preparation can lead to phase separation through incompatibility or coacervation (see Section 3.7), with similarly inconvenient, deleterious, or hazardous results.

NOTE ADDED IN PROOF

In section 3.5.1 the values of the Flory–Huggins interaction parameter χ for the water soluble polymers PMAA, PVA, PVMA and PVP were obtained from the literature data for water uptake and solvent activity (relative vapor pressure) using weight fraction, w, as the concentration quantity rather than the more normal volume fraction, ϕ. This method of calculation had to be adopted because the published data are presented in weight terms, and there are no data available for conversion onto a volume basis over the concentration ranges involved. However, it is possible that it was this procedure which was responsible for the observation that in most of these cases the extrapolation $w_1 \to 1$ (i.e., extremely dilute polymer solutions) gave $\chi > 0.5$, whereas one should get $\chi < 0.5$ for this situation with these water soluble polymers.

To examine this possibility, we need only consider the case of these extremely dilute solutions. Using χ_ϕ to represent the value of the parameter calculated on a volume fraction basis, equation (35) (with the molar volume ratio s omitted) reduces for this situation to

$$\ln a_1 = (\chi_\phi - 0.5)\phi_2{}^2 \tag{114}$$

while for concentrations on a weight fraction basis, the corresponding relation is

$$\ln a_1 = (\chi_w - 0.5)w_2{}^2 \tag{115}$$

Equating these two expressions for $\ln a_1$ gives

$$(\chi_\phi - 0.5)/(\chi_w - 0.5) = \bar{v}_1{}^2/\bar{v}_2{}^2 \tag{116}$$

where $\bar{v}$ is the partial specific volume of the component. With $\bar{v}_1 = 1 \text{ cm}^3\text{g}^{-1}$ (water) and $\bar{v}_2 \approx 0.7 \text{ cm}^3\text{g}^{-1}$ (an average value for synthetic polymers in water)

$$(\chi_\phi - 0.5)/(\chi_w - 0.5) \approx 2 \tag{117}$$

Thus if $\chi_w = 0.5$ then also $\chi_\phi = 0.5$, while for values other than 0.5 the deviation of χ_ϕ from this value is essentially twice that of χ_w ($\chi_w = 0.6$, $\chi_\phi = 0.7$; $\chi_w = 0.7$, $\chi_\phi = 0.9$; etc.).

Thus the use of volume fractions in place of weight fractions would not eliminate the anomaly of the extrapolated value of χ for extremely dilute solutions being greater than 0.5, and indeed if anything, it increases the extent of the anomaly. This therefore appears to reinforce the previous conclusion (p. 663) of a relatively abrupt change in χ within the unexplored region between the maximal values of water uptake and the dilute solution region.

It should be noted that for this limiting state of dilute solutions ($\phi_1 \to 1$) it is strictly incorrect to omit the molar volume ratio term s (related inversely to the molecular weight of the polymer) from equation (35);

however, with a high polymer molecular weight and low water uptakes, this would have little effect on the χ values used for the extrapolation. In point of fact it would not have been possible to include this term in the calculations of χ because none of the published data[355,754,1075] included mention of the molecular weight of the polymers studied; this only serves to re-emphasize the need for full characterization of the polymer sample in such studies.

References

1. S. Abrahamsson and I. Pascher, *Acta Cryst.* **21**, 79 (1966).
2. A. W. Adamson, "Physical Chemistry of Surfaces," 2nd ed., Wiley–Interscience, New York (1967).
3. J. E. Adderson and C. G. Butler, *J. Pharm. Pharmacol.* **24**, 130 (1972).
4. J. E. Adderson and H. Taylor, *J. Colloid Sci.* **19**, 495 (1964).
5. J. E. Adderson and H. Taylor, *J. Pharm. Pharmacol.* **22**, 523 (1970).
6. J. E. Adderson and H. Taylor, *J. Pharm. Pharmacol.* **23**, 311 (1971).
7. A. J. Adler and G. D. Fasman, *J. Phys. Chem.* **75**, 1516 (1971).
8. G. Akerlof, *J. Amer. Chem. Soc.* **57**, 1196 (1935).
9. Th. Akermann and H. Rüterjans, *Physik. Chem.* **68**, 850 (1964).
10. P.-Å. Albertsson, *Adv. Protein Chem.* **24**, 309 (1970).
11. P.-Å. Albertsson, "Partition of Cell Particles and Macromolecules," 2nd ed., Wiley, New York, and Almqvist & Wiksell, Stockholm (1971).
12. D. M. Alexander, *J. Chem. Eng. Data* **4**, 252 (1959).
13. D. M. Alexander and D. J. T. Hill, *Austral. J. Chem.* **18**, 605 (1965).
14. D. M. Alexander and D. J. Hill, *Austral. J. Chem.* **22**, 347 (1969).
15. P. Alexander and K. A. Stacey, *Proc. Roy. Soc.* **A212**, 274 (1952).
16. G. Allen, G. Gee, D. Mangaraj, D. Sims, and G. J. Wilson, *Polymer* **1**, 467 (1960).
17. G. Allen, G. Gee, and G. J. Wilson, *Polymer* **1**, 456 (1960).
18. J. Allen, M. C. Phillips, and G. G. Shipley, *Biochem. J.* **125**, 733 (1971).
19. P. W. Allen (ed.), "Techniques of Polymer Characterization," Butterworths, London (1959).
20. J. C. Amberlang, R. H. Kline, O. M. Lorenz, C. R. Parks, C. Wadelin, and J. R. Shelton, *Rubber Chem. Tech.* **36**, 1497 (1963).
21. E. W. Anacker, in "Nonionic Surfactants" (M. J. Schick, ed.), Marcel Dekker, New York (1967), p. 203.
22. E. W. Anacker and R. D. Geer, *J. Colloid Interface Sci.* **35**, 441 (1971).
23. E. W. Anacker, R. D. Geer, and E. H. Eylar, *J. Phys. Chem.* **75**, 369 (1971).
24. E. W. Anacker, R. M. Rush, and J. S. Johnson, *J. Phys. Chem.* **68**, 81 (1964).
25. N. S. Anderson, T. C. S. Dolan, and D. A. Rees, *Nature* **205**, 1060 (1965).
26. R. G. Anderson and M. C. R. Symons, *Trans. Faraday Soc.* **65**, 2550 (1969).
27. S. J. Angyal, *Angew. Chem. Intern. Ed.* **8**, 157 (1969).
28. S. J. Angyal and V. A. Pickles, *Carbohydrate Res.* **4**, 269 (1967).
29. Anonymous, "Atlas Fade-Ometer and Weather-Ometer for Accelerated Light-fastness and Weathering Tests," Bulletin No. 1300, Atlas Electric Devices Co., Chicago (1971).
30. Anonymous, *Chem. Week* **106**, 49 (1970).

31. Anonymous, "Gel Chromatography," Bio-Rad Laboratories, Richmond, California (1971).
32. Anonymous, *Plastics*: *Rubbers*: *Textiles*: *Polymer Age* **3**, 286 (1972).
33. Anonymous, "PVP: An Annotated Bibliography, 1951–1966," Vols. I–III, General Aniline and Film Corp., New York (1967).
34. Anonymous, "PVP: Polyvinylpyrrolidone: Physical, Chemical, Physiological and Functional Properties," Technical Bulletin 7583-033, General Aniline and Film Corp., New York (1964).
35. Anonymous, "Sephadex and other Separation Products," Pharmacia Fine Chemicals, Uppsala, Sweden (1969).
36. Anonymous, "SVL Laboratory Glassware," Sovirel S.A., Levallois-Perret, France (1971).
37. Anonymous, "Xenotest 450 LF Light and Weather Fastness Tester," Leaflet D310 006, Original Hanau Quarzlampen GmbH, Hanau (1970).
38. Anonymous, "Xenotest 150 Light and Weather Fastness Tester," Leaflet D310 013, Original Hanau Quarzlampen GmbH, Hanau (1971).
39. A. R. Anscombe, N. Hira, and B. Hunt, *Brit. J. Surg.* **57**, 917 (1970).
40. G. B. Ansell and J. N. Hawthorne, "Phospholipids," Elsevier Publishing Company, Amsterdam (1964).
41. A. C. Anusiem, J. G. Beetlestone, and D. H. Irvine, *J. Chem. Soc. A* 960 (1968).
42. P. Appel and W. Duane Brown, *Biopolymers* **10**, 2309 (1971).
43. P. Appel and J. T. Yang, *Biochem.* **4**, 1244 (1965).
44. J. Applequist, *J. Chem. Phys.* **38**, 934 (1963).
45. J. Applequist, *J. Chem. Phys.* **50**, 600 (1969).
46. H. Arai and S. Horin, *J. Colloid Interface Sci.* **30**, 372 (1969).
47. R. H. Aranow, L. Witten, and D. H. Andrews, *J. Phys. Chem.* **62**, 812 (1958).
48. C. P. Argaina and E. P. Czerwin, *in* "Water-Soluble Resins," (R. L. Davidson and M. Sittig, eds.), 2nd ed., Reinhold, New York (1968), Chapter 6.
49. D. A. Armitage, M. J. Blandamer, K. W. Morcom, and N. C. Treloar, *Nature* **219**, 718 (1969).
50. E. M. Arnett, W. B. Kover, and J. N. Carter, *J. Amer. Chem. Soc.* **91**, 4028 (1969).
51. E. M. Arnett and D. R. McKelvey, *in* "Solute–Solvent Interactions" (J. F. Coetzee and C. D. Ritchie, eds.), Marcel Dekker, New York, pp. 344–395.
52. A. Arrangton, A. Clouse, D. Doddrell, R. B. Dunlap, and E. H. Cordes, *J. Phys. Chem.* **74**, 665 (1970).
53. M. Arshadi, R. Yamdagni, and P. Kebarle, *J. Phys. Chem.* **74**, 1475 (1970).
54. K. L. Z. Arvan and N. E. Zaitseva, *Opt. Spectr.* **10**, 137 (1961).
55. E. O. Arvidsson, F. A. Green, and S. Laurell, *J. Biol. Chem.* **246**, 5373 (1971).
56. F. Ascoli, C. Botré, and A. M. Liquori, *J. Mol. Biol.* **3**, 202 (1961).
57. R. S. Asquith and A. K. Booth, *J. Soc. Dyers and Col.* **86**, 393 (1970).
58. N. M. Atherton and S. J. Strach, *J. Chem. Soc. Faraday Trans. II*, **68**, 374 (1972).
59. A. Atik, *Anesthesiology* **27**, 425 (1966).
60. D. Atkinson, A. H. Clark, and F. Franks (to be published).
61. D. Atkinson, H. Hauser, G. G. Shipley, and M. Stubbs, *Biochim. Biophys. Acta* **339**, 10 (1974).
62. D. Attwood, *J. Phys. Chem.* **72**, 339 (1968).
63. D. Attwood, P. H. Elworthy, and S. B. Kayne, *J. Phys. Chem.* **74**, 3529 (1970).
64. D. Attwood, P. H. Elworthy, and S. B. Kayne, *J. Phys. Chem.* **75**, 2212 (1971).

65. D. Attwood and L. Saunders, *Biochim. Biophys. Acta* **98**, 344 (1965).
66. R. Auerbach, *Kolloid Z.* **34**, 109 (1924).
67. R. Aveyard and A. S. C. Lawrence, *Trans. Faraday Soc.* **60**, 2265 (1964).
68. N. N. Aylward, *J. Polymer Sci. Part A-1* **8**, 319 (1970).
69. N. N. Aylward, *J. Polymer Sci. Part A-1* **8**, 909 (1970).
70. J. L. Azorlosa and A. J. Martinelli, *in* "Water-Soluble Resins" (R. L. Davidson and M. Sittig, eds.), 2nd ed., Reinhold, New York (1968), Chapter 7.
71. K. Babor and V. Kalac, *Die Stärke* **21**, 202 (1969).
72. H. Baedtker and P. Doty, *J. Phys. Chem.* **58**, 968 (1954).
73. C. K. Bahal and H. B. Kostenbauder, *J. Pharm. Sci.* **53**, 1027 (1964).
74. R. E. Baier, *in* "Adhesion in Biological Systems" (R. S. Manly, ed.), Academic Press, New York and London (1970), Chapter 2.
75. F. E. Bailey and R. W. Callard, *J. Appl. Polymer Sci.* **1**, 56, 373 (1959).
76. R. E. Balanova and M. I. Ushakova, *Chem. Abstr.* **77**, 90396 (1972).
77. H. D. Bale, R. E. Shepler, and D. K. Sorgen, *Phys. Chem. Liquids* **1**, 181 (1968).
78. C. H. Bamford, W. G. Barb, A. D. Jenkins, and P. F. Onyon, "The Kinetics of Vinyl Polymerization by Radical Mechanisms," Butterworths, London (1958), Chapter 4.
79. A. D. Bangham, *Progr. Biophys. Mol. Biol.* **18**, 31 (1968).
80. W. Banks and C. T. Greenwood, *Adv. Carb. Chem.* **18**, 357 (1963).
81. W. Banks and C. T. Greenwood, *European Polymer J.* **5**, 549 (1969).
82. W. Banks and C. T. Greenwood, *Die Stärke* **23**, 452 (1971).
83. W. Banks, C. T. Greenwood, D. J. Hourston, and A. R. Procter, *Polymer* **12**, 452 (1971).
84. I. M. Barclay and J. A. V. Butler, *Trans. Faraday Soc.* **34**, 1445 (1938).
85. J. A. Barker and R. O. Watts, *Chem. Phys. Letters* **3**, 144 (1969).
86. S. M. Barkin, Ph. D. Thesis, Polytechnic Institute of Brooklyn, New York (1957); Publication No. 22,650, Doctoral Dissertation Series, University Microfilms, Ann Arbor, Michigan (1966); *Diss. Abs.* **17**, 1906 (1957).
87. S. Barkin, H. P. Frank, and F. R. Eirich, Proceedings of the International Symposium on Macromolecular Chemistry, IUPAC, Milan/Turin (1954), pub. as suppl. to *Ricerca Sci.* **25**, 844 (1955).
88. G. H. Barlow and H. E. Zaugg, *J. Org. Chem.* **37**, 2246 (1972).
89. G. Barone, V. Crescenzi, A. M. Liquori, and F. Quadrifoglio, *J. Phys. Chem.* **71**, 2341 (1967).
90. G. Barone, E. Rizzo, and V. Vitagliano, *J. Phys. Chem.* **74**, 2230 (1970).
91. M. D. Barratt, F. Franks, and R. Whitehurst (to be published).
92. M. D. Barratt and L. Rayner, *Biochim. Biophys. Acta* **255**, 974 (1972).
93. J. A. Barrie, *in* "Diffusion in Polymers" (J. Crank and G. S. Park, eds.), Academic Press, London and New York (1968), Chapter 8.
94. J. A. Barrie and D. Machin, *J. Macromol. Sci-Phys.* **B3**, 645, 673 (1969).
95. J. A. Barrie and D. Machin, *Trans. Faraday Soc.* **67**, 244 (1971).
96. J. A. Barrie and D. Machin, *Trans. Faraday Soc.* **67**, 2970 (1971).
97. B. W. Barry, J. C. Morrison, and G. F. J. Russell, *J. Colloid Interface Sci.* **33**, 554 (1970).
98. B. W. Barry and G. F. J. Russell, *J. Colloid Interface Sci.* **40**, 174 (1972).
99. C. D. Barry, A. C. T. North, J. A. Glasel, R. J. P. Williams, and A. V. Xavier, *Nature* **232**, 236 (1971).

100. R. Battino and H. L. Clever, *Chem. Rev.* **66**, 395 (1966).

101. R. S. Bear, K. J. Palmer, and F. O. Schmitt, *J. Cell. Comp. Physiol.* **17**, 355 (1941).

102. P. Becher, *in* "Cationic Surfactants" (E. Jungermann, ed.), Marcel Dekker, New York (1970).

103. P. Becher and H. Arai, *J. Colloid Interface Sci.* **27**, 634 (1968).

104. G. Beech, *J. Chem. Soc.* (*A*), 1903 (1969).

105. A. Beerbower, L. A. Kaye, and D. A. Pattison, *Chem. Eng.* **74**, 118 (1967).

106. B. Belleau and J. Lavoie, *Can. J. Biochem.* **46**, 1397 (1968).

107. J. Bello and H. R. Bello, *Nature* **190**, 440 (1961).

108. J. Bello, H. C. A. Riese, and J. R. Vinograd, *J. Phys. Chem.* **60**, 1299 (1956).

109. A. Ben-David and S. Gavendo, *Cryobiology* **9**, 192 (1972).

110. L. Benjamin, *J. Phys. Chem.* **68**, 3575 (1964).

111. L. Benjamin, *J. Colloid Interface Sci.* **22**, 386 (1966).

112. L. Benjamin, *J. Phys. Chem.* **70**, 3790 (1966).

113. A. Ben-Naim, *J. Chem. Phys.* **42**, 1512 (1965).

114. A. Ben-Naim, *J. Phys. Chem.* **69**, 3240 (1965).

115. A. Ben-Naim, *J. Phys. Chem.* **69**, 3245 (1965).

116. A. Ben-Naim, *J. Phys. Chem.* **71**, 4002 (1967).

117. A. Ben-Naim, *J. Chem. Phys.* **54**, 1387 (1971).

118. A. Ben-Naim, *J. Chem. Phys.* **54**, 3696 (1971).

119. A. Ben-Naim and S. Baer, *Trans. Faraday Soc.* **60**, 1736 (1964).

120. A. Ben-Naim and M. Egel-Thal, *J. Phys. Chem.* **69**, 3250 (1965).

121. A. Ben-Naim, J. Wilf, and M. Yaacobi, *J. Phys. Chem.* **77**, 95 (1973).

122. H. P. Bennetto and D. Feakins, *in* "Hydrogen Bonded Solvent Systems" (A. K. Covington and P. Jones, eds.), Taylor and Francis, London (1968).

123. B. C. Bennion and E. M. Eyring, *J. Colloid Interface Sci.* **32**, 286 (1970).

124. B. C. Bennion, L. K. J. Tong, and E. M. Eyring, *J. Phys. Chem.* **73**, 3288 (1969).

125. H. B. Bensusan and S. O. Nielsen, *Biochem.* **3**, 1367 (1964).

126. H. J. C. Berendsen, *J. Chem. Phys.* **36**, 3297 (1962).

127. H. J. C. Berendsen and C. Migchelsen, *Ann. N.Y. Acad. Sci.* **125**, 365 (1965).

128. I. Berezin, N. Kazanskaya, A. Klyosov, and K. Martinek, *Febs Letters* **15**, 121 (1971).

129. R. A. Berg and B. A. Haxby, *Mol. Crystals and Liquid Crystals* **12**, 93 (1970).

130. J. A. Bergeron and M. Singer, *J. Biophys. and Biochem. Cytol.* **4**, 433 (1958).

131. K. Bergman and C. T. O'Konski, *J. Phys. Chem.* **67**, 2169 (1963).

132. T. H. Bevan, D. A. Brown, and T. Malkin, *J. Chem. Soc.* 3495 (1962).

133. T. H. Bevan and T. Malkin, *J. Chem. Soc.* (*London*) 2667 (1951).

134. S. N. Bhaskar, *in* "Adhesion in Biological Systems" (R. S. Manly, ed.), Academic Press, New York and London (1970), Chapter 12.

135. R. L. Biltonen, Ph. D. Dissertation, Univ. of Minnesota (1965).

136. R. L. Biltonen and R. Lumry, *J. Amer. Chem. Soc.* **91**, 4251 (1969).

137. R. L. Biltonen and R. Lumry, *J. Amer. Chem. Soc.* **91**, 4256 (1969).

138. R. L. Biltonen and R. Lumry, *J. Amer. Chem. Soc.* **93**, 224 (1971).

139. R. L. Biltonen, R. Lumry, V. Madison, and H. Parker, *Proc. Nat. Acad. Sci.* **54**, 1018 (1965).

140. R. L. Biltonen, A. T. Schwartz, and I. Wadsö, *Biochem.* **10**, 3417 (1971).

141. B. J. Birch and D. G. Hall, *J. Chem. Soc. Faraday I*, **68**, 2350 (1972).

142. K. S. Birdi and H. Kyllingsbaek, *J. Pharm. Pharmacol.* **23**, 900 (1971).

143. N. J. M. Birdsall, A. G. Lee, Y. K. Levine, and J. C. Metcalfe, *Biochim. Biophys. Acta* **241**, 693 (1971).

144. T. M. Birshtein, E. B. Anufrieva, T. N. Nekrasova, O. B. Ptitsyn, and T. V. Sheveleva, *Vysokomol. Soedin.* **7**, 372 (1965).

145. H. J. Bixler and A. S. Michaels, *in* "Encyclopedia of Polymer Science and Technology" (H. F. Mark, N. Gaylord, and N. M. Bikales, eds.), Wiley–Interscience, New York (1969), Vol. 10, p. 765.

146. J. Blackwell, A. Sarko, and R. H. Marchessault, *J. Mol. Biol.* **42**, 379 (1969).

147. M. J. Blandamer, M. F. Fox, E. Powell, and J. W. Stafford, *Makromol. Chemie* **124**, 222 (1969).

148. S. M. Blaug and S. S. Ahsan, *J. Pharm. Sci.* **50**, 138 (1961).

149. S. M. Blaug and A. G. Rich, *J. Pharm. Sci.* **54**, 30 (1965).

150. D. J. Blears and S. S. Danyluk, *J. Amer. Chem. Soc.* **88**, 1084 (1966).

151. D. J. Blears and S. S. Danyluk, *Biochim. Biophys. Acta* **154**, 17 (1968).

152. H. Block and J. A. Kay, *Biopolymers* **5**, 243 (1967).

153. S. M. Bloom, S. K. Dasgupta, R. P. Patel, and E. R. Blout, *J. Amer. Chem. Soc.* **88**, 2035 (1966).

154. E. R. Blout, J. P. Carver, and J. Gross, *J. Amer. Chem. Soc.* **85**, 644 (1963).

155. J. O'M. Bockris and P. P. S. Saluja, *J. Phys. Chem.* **76**, 2298 (1972).

156. U. Borgwardt, W. Schnabel, and A. Henglein, *Makromol. Chemie* **127**, 176 (1969).

157. C. Botre', V. L. Crescenzi, and A. Mele, *J. Phys. Chem.* **63**, 650 (1969).

158. M. Bourgès, D. M. Small, and D. G. Dervichian, *Biochim. Biophys. Acta* **137**, 157 (1967).

159. F. A. Bovey and F. P. Hood, *J. Amer. Chem. Soc.* **88**, 2326 (1966).

160. F. A. Bovey, I. M. Kolthoff, A. I. Medalia, and E. J. Meehan, "Emulsion Polymerization," Wiley—Interscience, New York (1955).

161. F. Boyer-Kawenoki, *Compt. Rend.* **263**, 203 (1966).

162. J. A. Boyne and A. G. Williamson, *J. Chem. Eng. Data* **12**, 318 (1967).

163. B.P.963906 (1964) to Dofag Co., Vaduz, Liechtenstein.

164. P. Bracha and R. D. O'Brien, *Biochem.* **9**, 741 (1970).

165. E. M. Bradbury and H. W. E. Rattle, *Polymer* **9**, 201 (1968).

166. E. M. Bradbury, C. C. Robinson, H. Goldman, and H. W. E. Rattle, *Nature* **217**, 812 (1968).

167. J. H. Bradbury, P. J. Crawford, and A. N. Hambly, *Trans. Faraday Soc.* **64**, 1337 (1968).

168. J. Brandrup and E. H. Immergut (eds.), "Polymer Handbook," Wiley–Interscience, New York (1966).

169. D. A. Brandt and P. J. Flory, *J. Amer. Chem. Soc.* **87**, 663 (1965).

170. J. F. Brandts, *J. Amer. Chem. Soc.* **86**, 4291 (1964).

171. J. F. Brandts, *J. Amer. Chem. Soc.* **86**, 4302 (1964).

172. J. F. Brandts, *in* "Structure and Stability of Biological Macromolecules," (S. Timasheff and G. Fasman, eds.), Marcel Dekker, New York (1969).

173. J. F. Brandts and L. Hunt, *J. Amer. Chem. Soc.* **89**, 4826 (1967).

174. E. Braswell, *J. Phys. Chem.* **72**, 2477 (1968).

175. G. Brausse, A. Mayer, T. Nedetzka, P. Schlecht, and H. Vogel, *J. Phys. Chem.* **72**, 3098 (1968).

176. J. W. Breitenbach, *Z. Elektrochem.* **59**, 309 (1955).
177. J. W. Breitenbach and H. Edelhauser, Proceedings of the International Symposium on Macromolecular Chemistry, IUPAC, Milan/Turin (1954), publ. as suppl. to *Ricerca Sci.* **25**, 242 (1955).
178. J. W. Breitenbach and A. Schmidt, *Monatsch. Chem.* **85**, 52 (1954).
179. J. W. Breitenbach and E. Wolf, *Makromol. Chemie* **18/19**, 217 (1956).
180. W. B. Breuninger and R. W. Goettsch, *J. Pharm. Sci.* **54**, 1487 (1965).
181. W. S. Brey, T. E. Evans, and L. H. Hitzrot, *J. Colloid Sci.* **26**, 306 (1968).
182. C. A. Brighton, *in* "Weathering and Degradation of Plastics" (S. H. Pinner, ed.), Columbine Press, Manchester and London (1966), Chapter 4.
183. G. M. Bristow and W. F. Watson, *Trans. Faraday Soc.* **54**, 1731, 1742 (1958).
184. A. D. Broom, M. P. Schweizer, and P. O. Ts'o, *J. Amer. Chem. Soc.* **89**, 3612 (1967).
185. C. J. Brown, *J. Chem. Soc. A.* 927 (1966).
186. C. W. Brown, D. Cooper, and J. C. S. Moore, *J. Colloid Interface Sci.* **32**, 584 (1970).
187. F. R. Brown III, Ph. D. Thesis, Harvard University (1970).
188. F. R. Brown III, A. di Corato, G. P. Lorenzi, and E. R. Blout, *J. Mol. Biol.* **63**, 85 (1972).
189. F. R. Brown III, A. J. Hopfinger, and E. R. Blout, *J. Mol. Biol.* **63**, 101 (1972).
190. R. F. Brown, *J. Org. Chem.* **27**, 1202, 3015 (1962).
191. S. Brunauer, "The Adsorption of Gases and Vapors. Volume 1. Physical Adsorption," Oxford University Press, London (1945).
192. W. Bruning and A. Holtzer, *J. Amer. Chem. Soc.* **83**, 4865 (1961).
193. M. Brunori, G. M. Giacometti, E. Antonini, and J. Wyman, *J. Mol. Biol.* **63**, 139 (1972).
194. W. P. Bryan and S. O. Nielsen, *Biochim. Biophys. Acta* **42**, 552 (1962).
195. W. Bubser and H. Eichmanns, *Melliand Textilchem.* No. 1, 23 (1965).
196. W. Bubser and H. Eichmanns, *Palette (Sandoz Limited)* **20**, 20 (1965).
197. C. A. Buch and H. A. Scheraga, *Biopolymers* **7**, 395 (1969).
198. T. J. Buchanan, G. H. Haggis, J. B. Husted, and B. G. Robinson, *Proc. Roy. Soc. (London)* **A213**, 379 (1952).
199. F. Buckely and A. Maryott, "Tables of Dielectric Dispersion Data," Natl. Bur. Std. Circ. 589, Washington, D.C. (1958).
200. H. B. Bull and K. Breese, *Arch. Biochem. Biophys.* **128**, 488 (1968).
201. H. B. Bull and K. Breese, *Arch. Biochem. Biophys.* **128**, 497 (1968).
202. H. B. Bull and K. Breese, *Arch. Biochem. Biophys.* **137**, 299 (1970).
203. H. B. Bull and K. Breese, *Arch. Biochem. Biophys.* **139**, 93 (1970).
204. H. B. Bull and K. Breese, *Arch. Biochem. Biophys.* **149**, 164 (1972).
205. R. R. Bumb and B. Zaslow, *Carbohydrate Res.* **4**, 98 (1967).
206. C. A. Bunton, M. Minch, and L. Sepulvedo, *J. Phys. Chem.* **75**, 2707 (1971).
207. C. A. Bunton and L. Robinson, *J. Phys. Chem.* **73**, 4237 (1969).
208. C. A. Bunton and L. Robinson, *J. Phys. Chem.* **74**, 1062 (1970).
209. L. G. Bunville, E. P. Geiduschek, M. A. Rawitscher, and J. M. Sturtevant, *Biopolymers* **3**, 213 (1965).
210. K. W. Bunzl, *J. Phys. Chem.* **71**, 1358 (1967).
211. M. C. Buonocore, *in* "Adhesion in Biological Systems" (R. S. Manly, ed.), Academic Press, New York and London (1970), Chapter 15.
212. U. Buontempo, G. Careri, and P. Fassella, *Biopolymers* **11**, 519 (1972).
213. R. E. Burge and R. D. Hynes, *J. Mol. Biol.* **1**, 155 (1959).

214. J. Burke, G. Hammes, and T. Lewis, *J. Chem. Phys.* **42**, 3520 (1965).
215. H. Burrell, *Interchem. Rev.* **14**, 3, 31 (1955).
216. H. Burrell and B. Immergut, *in* "Polymer Handbook" (J. Brandrup and E. H. Immergut, eds.), Wiley–Interscience, New York (1966), p. IV-341.
217. C. A. Butler, J. A. Stead, and H. Taylor, *J. Colloid Interface Sci.* **30**, 489 (1969).
218. J. A. V. Butler, *Trans. Faraday Soc.* **33**, 229 (1937).
219. J. A. V. Butler, C. N. Ramchandani, and D. W. Thomson, *J. Chem. Soc.* 280 (1935).
220. S. Cabani, G. Conti, and L. Lepori, *Trans. Faraday Soc.* **67**, 1933 (1971).
221. S. Cabani, G. Conti, and L. Lepori, *Trans. Faraday Soc.* **67**, 1943 (1971).
222. S. Cabani, G. Conti, and L. Lepori, *J. Phys. Chem.* **76**, 1338 (1972).
223. S. Cabani, G. Conti, L. Lepori, and G. Leva, *J. Phys. Chem.* **76**, 1343 (1972).
224. S. Cabani, G. Conti, A. Martinelli, and E. Matteoli, "Proc. 3rd Internat. Conf. Thermodynamics and Thermochemistry, Vienna, 1973."
225. D. S. E. Campbell, D. Cathcart, C. H. Giles, and S. M. K. Rahman, *Trans. Faraday Soc.* **55**, 1631 (1959).
226. F. G. Canepa, P. Pauling, and H. Sörum, *Nature* **210**, 907 (1966).
227. B. Carroll and H. C. Cheung, *J. Phys. Chem.* **66**, 2585 (1962).
228. W. R. Carroll and H. Eisenberg, *J. Polymer Sci., Part A-2* **4**, 599 (1966).
229. E. L. Carstensen, *J. Acoust. Soc. Amer.* **26**, 858 (1952).
230. J. M. Cassel and R. G. Christensen, *Biopolymers* **5**, 431 (1967).
231. B. Casu, M. Reggiani, G. G. Gallo, and A. Vigevani, *Tetrahedron* **22**, 3061 (1966).
232. F. Cennamo and E. Tartaglione, *Nuovo Cimento* **11**, 401 (1959).
233. J. Cerbon, *Biochim. Biophys. Acta* **144**, 1 (1967).
234. L. C. Cerny, T. E. Helminiak, and J. F. Meier, *J. Polymer Sci.* **44**, 539 (1960).
235. S. I. Chan, G. W. Feigenson, and C. H. A. Seiter, *Nature* **231**, 110 (1971).
236. S. I. Chan, C. H. A. Seiter, and G. W. Feigenson, *Biochem. Biophys. Res. Commun.* **46**, 1488 (1972).
237. D. Chapman, *Chem. Rev.* **62**, 433 (1962).
238. D. Chapman, "The Structure of Lipids," Methuen, London (1965).
239. D. Chapman, *in* "Biological Membranes" (D. Chapman and D. F. H. Wallach, eds.), Academic Press, London and New York (1973), Vol. 2, p. 91.
240. D. Chapman, P. Byrne, and G. G. Shipley, *Proc. Roy. Soc. Ser. A* **290**, 115 (1966).
241. D. Chapman, N. F. Owens, M. C. Phillips, and D. A. Walker, *Biochim. Biophys. Acta* **183**, 458 (1969).
242. D. Chapman and N. J. Salsbury, *Trans. Faraday Soc.* **62**, 2607 (1966).
243. D. Chapman and N. J. Salsbury, *Recent Progress Surf. Sci.* **3**, 121 (1970).
244. D. Chapman and D. F. H. Wallach, *in* "Biological Membranes" (D. Chapman. ed.), Academic Press, London and New York (1968), Vol. 1, p. 125.
245. D. Chapman, R. M. Williams, and B. D. Ladbrooke, *Chem. Phys. Lipids* **1**, 445 (1967).
246. J. Charvolin and P. Rigny, *J. de Physique, Colloq. C4* (Suppl. to No. 11–12), **30**, C4, 76 (1969).
247. D. K. Chattoraj and H. B. Bull, *J. Colloid and Interface Sci.* **35**, 220 (1971).
248. M. E. Chevreul, "Recherches chimique sur les corps gras d'origine animale," Levrault, Paris (1823).
249. T. Chiba, *J. Chem. Phys.* **36**, 1122 (1962).
250. T. F. Child, N. G. Pryce, M. J. Tait, and S. Ablett, *Chem. Comm.* 1214 (1970).

251. A. V. Chinnikova, Z. N. Markina, and G. A. Korneeva, *Kolloid. Zh.* **34**, 272 (1972).
252. A. V. Chinnikova-Sineva and Z. N. Markina, *Kolloid Zh.* **34**, 476 (1972).
253. P. Y. Chou and H. A. Scheraga, *Biopolymers* **10**, 657 (1971).
254. N. H. Choulis and L. H. Loh, *Can. J. Pharm. Sci.* **6**, 93 (1971).
255. S. S. C. Chu and G. A. Jeffrey, *Acta Cryst.* **23**, 1038 (1967).
256. S. S. C. Chu and G. A. Jeffrey, *Acta Cryst.* **24B**, 830 (1968).
257. H. S. Chung and I. J. Heilweil, *J. Phys. Chem.* **74**, 488 (1970).
258. A. Ciferri, R. Garmon, and D. Puett, *Biopolymers* **5**, 439 (1967).
259. A. Ciferri and T. A. Orofino, *J. Phys. Chem.* **70**, 3277 (1966).
260. J. Clifford and T. F. Child, *Eur. Biophys. Congr. Proc.* 461 (1971).
261. J. Clifford and B. A. Pethica, *Trans. Faraday Soc.* **61**, 182 (1965).
262. J. S. Clunie, J. F. Goodman, and P. C. Symons, *Trans. Faraday Soc.* **63**, 754 (1967).
263. E. Coates, *J. Soc. Dyers and Col.* **85**, 355 (1969).
264. E. Coates and B. Rigg, *Trans. Faraday Soc.* **57**, 1637 (1961).
265. G. Cohen and H. Eisenberg, *Biopolymers* **6**, 1077 (1968).
266. M. H. Cohen and F. Reif, *Solid State Phys.* **5**, 321 (1957).
267. M. F. Coldman and W. Good, *Biochim. Biophys. Acta* **150**, 194, 206 (1968).
268. H. Coll, *J. Phys. Chem.* **74**, 520 (1970).
269. G. A. Collie (ed.), *British Polymer J.* **4**, No. 3 (1972).
270. E. Collinson, F. S. Dainton, and G. S. McNaughton, *Trans. Faraday Soc.* **53**, 489 (1957).
271. R. Colman, *Biochim. Biophys. Acta* **300**, 1 (1973).
272. G. Conio and E. Patrone, *Biopolymers* **8**, 57 (1969).
273. G. Conio, E. Patrone, and S. Brighetti, *J. Biol. Chem.* **245**, 3335 (1970).
274. A. Conix and G. Smets, *J. Polymer Sci.* **15**, 221 (1955).
275. A. Conix, G. Smets, and J. Moens, Proceedings of the International Symposium on Macromolecular Chemistry, IUPAC, Milan/Turin (1954), publ. as suppl. to *Ricerca Sci.* **25**, 200 (1955).
276. L. Costantino, A. M. Liquori, and V. Vitagliano, *Biopolymers* **2**, 1 (1964).
277. B. E. Conway and R. E. Verrall, *J. Phys. Chem.* **70**, 3952 (1966).
278. B. E. Conway, R. E. Verrall, and J. E. Desnoyers, *Trans. Faraday Soc.* **62**, 2738 (1966).
279. A. Cooper, *J. Mol. Biol.* **55**, 123 (1971).
280. W. Cooper, F. R. Johnston, and G. Vaughan, *J. Polymer Sci. Part A* **1**, 1509 (1963).
281. W. Cooper, R. W. Pearson, and S. Darke, *Industrial Chemist* **36**, 126 (1960).
282. J. M. Corkill and J. F. Goodman, *Adv. Colloid Interface Sci.* **2**, 297 (1969).
283. J. M. Corkill, J. F. Goodman, P. Robson, and J. R. Tate, *Trans. Faraday Soc.* **62**, 987 (1966).
284. J. M. Corkill, J. F. Goodman, and J. R. Tate, *Trans. Faraday Soc.* **60**, 996 (1964).
285. J. M. Corkill, J. F. Goodman, and J. R. Tate, *Trans. Faraday Soc.* **63**, 773 (1967).
286. J. M. Corkill, J. F. Goodman, and J. R. Tate, "Hydrogen-Bonded Solvent System" (A. K. Covington and P. Jones, eds.), Taylor and Francis, London (1968), p. 181.
287. J. M. Corkill, J. F. Goodman, and T. Walker, *Trans. Faraday Soc.* **63**, 768 (1967).
288. J. M. Corkill and T. Walker, *J. Colloid Interface Sci.* **39**, 621 (1972).
289. M. L. Corrin and W. D. Harkins, *J. Amer. Chem. Soc.* **69**, 683 (1947).
290. W. L. Courchene, *J. Phys. Chem.* **68**, 1870 (1964).
291. J. E. Courtois and P. le Dizet, *Carbohydrate Res.* **3**, 141 (1966).

292. J. E. Courtois and P. le Dizet, *Bull. Soc. Chim. Biol.* **52**, 15 (1970).

293. A. K. Covington and P. Jones (eds.), "Hydrogen-Bonded Solvent Systems" Taylor and Francis, London (1968).

294. P. M. Cowan and S. McGavin, *Nature* **176**, 501 (1955).

295. R. A. Cox, *J. Polymer Sci.* **47**, 441 (1960).

296. J. Crank and G. S. Park (eds.), "Diffusion in Polymers," Academic Press, New York and London (1968).

297. B. R. Craven and A. Datyner, *J. Soc. Dyers and Col.* **79**, 515 (1963).

298. P. J. Crawford and J. H. Bradbury, *Trans. Faraday Soc.* **64**, 185 (1968).

299. V. Crescenzi and F. Delben, *Int. J. Prot. Res.* **3**, 57 (1971).

300. D. J. Crisp, *J. Colloid Sci.* **1**, 49, 161 (1946).

301. E. H. Crook, G. F. Trebbi, and D. B. Fordyce, *J. Phys. Chem.* **68**, 3592 (1964).

302. D. M. Crothers and D. I. Ratner, *Biochem.* **7**, 1823 (1968).

303. D. M. Crothers and B. H. Zimm, *J. Mol. Biol.* **9**, 1 (1964).

303a. J. Cullen, M. C. Phillips, and G. G. Shipley, *Biochem. J.* **125**, 733 (1971).

304. C. C. J. Culvenor and N. S. Ham, *Chem. Commun.* No. 15, 537 (1966).

305. B. R. Currell and M. J. Frazer, *RIC Reviews* **2**, 13 (1969).

306. M. Czerniawski, Roczniki Chem. **39**, 1469 (1965); *Chem. Abstr.* **64**, 13430h (1966).

307. S. Dähne, *Naturwiss.* **48**, 715 (1961).

308. S. Dähne, *Z. Chem.* **5**, 441 (1965).

309. F. S. Dainton and K. J. Ivin, *Quart. Rev.* **12**, 61 (1958).

310. R. Damforth, H. Krakaner, and J. M. Sturtevant, *Rev. Sci. Instr.* **38**, 484 (1967).

311. M. D. Danford and H. A. Levy, *J. Amer. Chem. Soc.* **84**, 3965 (1962).

312. J. F. Danielli, *in* "Permeability of Natural Membranes" (H. Davson and J. F. Danielli, eds.), Cambridge Univ. Press, London and New York (1943), p. 64.

313. J. F. Danielli and H. Davson, *J. Cell. Comp. Physiol.* **5**, 495 (1935).

314. I. Danielsson, B. Nielander, S. Sunner, and I. Wadsö, *Acta Chem. Scand.* **18**, 995 (1964).

315. I. Danielsson and P. Stenius, *J. Colloid Interface Sci.* **37**, 264 (1971).

316. R. L. D'Arcy and I. C. Watt, *Trans. Faraday Soc.* **66**, 1236 (1970).

317. A. Darke, E. G. Finer, A. G. Flook, and M. C. Phillips, *Febs Letters* **18**, 326 (1971).

318. A. Darke, E. G. Finer, A. G. Flook, and M. C. Phillips, *J. Mol. Biol.* **63**, 265 (1972).

319. A. J. Darnell and J. Greyson, *J. Phys. Chem.* **72**, 3021 (1968).

320. V. G. Dashevsky and G. N. Sarkisov, *Molec. Phys.* (1973).

321. B. Davidson and G. D. Fasman, *Biochem.* **6**, 1616 (1967).

322. R. L. Davidson and M. Sittig (eds.), "Water-Soluble Resins," 2nd ed., Reinhold, New York (1968).

323. C. W. Davies, "Ion Association," Butterworths, London (1962), p. 128.

324. L. E. Davis, *in* "Water-Soluble Resins" (R. L. Davidson and M. Sittig, eds.), 2nd ed., Reinhold, New York (1968), Chapter 11.

325. R. C. Davis and I. Tinoco, *J. Mol. Biol.* **20**, 29 (1966).

326. R. C. Davis and I. Tinoco, *Biopolymers* **6**, 223 (1968).

327. R. M. C. Dawson, *Biol. Rev. Camb. Phil. Soc.* **32**, 188 (1957).

328. I. C. M. Dea, A. A. McKinnon, and D. A. Rees, *J. Mol. Biol.* **68**, 153 (1972).

329. P. Debye, *J. Phys. Chem.* **53**, 1 (1949).

330. P. Debye and J. McAulay, *Physik. Z.* **26**, 22 (1925).

331. F. Delben, V. Crescenzi, and F. Quadrifoglio, *Int. J. Prot. Res.* **1**, 145 (1969).

332. J. Del Bene and J. A. Pople, *J. Chem. Phys.* **52**, 4858 (1970).
333. C. P. de Loor and F. W. Meijboom, *J. Food Technol.* **1**, 313 (1966).
334. P. D. Deluca and H. B. Kostenbauder, *J. Amer. Pharm. Assoc., Sci. Ed.* **49**, 430 (1960).
335. P. A. Demchenko, *Colloid J. USSR* **22**, 309 (1960).
336. N. C. Deno and C. H. Spink, *J. Phys. Chem.* **67**, 1347 (1963).
337. B. de Nooijer, D. Spencer, S. G. Whittington, and F. Franks, *Trans. Faraday Soc.* **67**, 1315 (1971).
338. A. N. Derbyshire, *Trans. Faraday Soc.* **51**, 909 (1955).
339. A. N. Derbyshire and R. H. Peters, *J. Soc. Dyers and Col.* **72**, 268 (1956).
340. D. G. Dervichian, *in* "Progress in Biophysics and Molecular Biology" (J. A. V. Butler and H. E. Huxley, eds.), Pergamon Press, Oxford (1964), Vol. 14, p. 263.
341. D. G. Dervichian, *Molecular Crystals* **2**, 55 (1966).
342. D. G. Dervichian and M. Joly, *Bull. Soc. Chim. Biol.* **28**, 426 (1946).
343. P. de Santis, E. Giglio, A. M. Ligironi, and A. Ripamonti, *Nature* **206**, 456 (1965).
344. J. E. Desnoyers, G. E. Pelletier, and L. Jolicoeur, *Can. J. Chem.* **43**, 3232 (1965).
345. G. T. de Titta and B. M. Craven, *Nature, New Biology* **233**, 118 (1971).
346. H. J. Deuel, "The Lipids," Interscience, New York (1951), Vol. 1.
347. M. J. Deveney, A. G. Walton, and J. L. Koenig, *Biopolymers* **10**, 615 (1971).
348. C. de Visser and G. Somsen, *J. Chem. Soc. Faraday I* **69**, 1440 (1973).
349. R. M. Diamond, *J. Phys. Chem.* **67**, 2513 (1963).
350. H. O. Dickinson, *Trans. Faraday Soc.* **43**, 486 (1947).
351. A. F. Diorio, E. Lippincott, and L. Mandelkern, *Nature* **195**, 1296 (1963).
352. D. E. Dix and D. B. Straus, *Arch. Biochem. Biophys.* **152**, 299 (1972).
353. A. Dobry and F. Boyer-Kawenoki, *J. Polymer Sci.* **2**, 90 (1947).
354. D. Dolor and H. Leskovšek, *Makromol. Chemie* **118**, 60 (1968).
355. M. Dole and I. L. Faller, *J. Amer. Chem. Soc.* **72**, 414 (1950).
356. F. G. Donnan and A. B. Harris, *J. Chem. Soc.* **99**, 1554 (1911).
357. A. Doren and J. Goldfarb, *J. Colloid Interface Sci.* **32**, 67 (1970).
358. D. E. Dorman and J. D. Roberts, *J. Amer. Chem. Soc.* **93**, 4463 (1971).
359. R. C. Dorrance and T. F. Hunter, *J. Chem. Soc. Faraday Trans. I*, **68**, 1312 (1972).
360. P. Doty, A. Wada, J. T. Yang, and E. R. Blout, *J. Polymer Sci.* **23**, 851 (1957).
361. P. Doty and J. T. Yang, *J. Amer. Chem. Soc.* **78**, 498 (1956).
362. W. F. Dove and N. Davidson, *J. Mol. Biol.* **5**, 467 (1962).
363. M. P. Drake and A. Vies, *Biochem.* **3**, 135 (1964).
364. W. Drost-Hansen, *Ann. N. Y. Acad. Sci.* **125**, 471 (1965).
365. W. Drost-Hansen, *Ind. and Eng. Chem.* **61**, 10 (1969).
366. P. L. Dubin and U. P. Strauss, *J. Phys. Chem.* **74**, 2842 (1970).
367. D. G. Duff and C. H. Giles, *J. Colloid and Interface Sci.* **41**, 407 (1972).
368. D. G. Duff and C. H. Giles, *J. Soc. Dyers and Col.* **88**, 181 (1972).
369. J. Dufourcq and C. Lussan, *Febs Letters* **26**, 35 (1972).
370. R. Durand and Y. Wormser, *Kolloid. Zh.* **33**, 515 (1971).
371. J. Dyckman and J. K. Weltman, *J. Cell. Biol.* **45**, 192 (1970).
372. I. Dzidic and P. Kebarle, *J. Phys. Chem.* **74**, 1466 (1970).
373. D. Eagland and C. Armitage, unpublished data, cited by F. Franks *in* "Physico-Chemical Processes in Mixed Aqueous Solvent Systems" (F. Franks, ed.), Heinemann, London (1967), p. 62.
374. D. Eagland and F. Franks, *Trans. Faraday Soc.* **61**, 2468 (1965).

375. D. Eagland and G. Pilling, *J. Phys. Chem.* **76**, 1902 (1972).

376. D. Eagland and G. Pilling, in preparation.

377. D. Eagland, G. Pilling, R. G. Wheeler, E. G. Finer, F. Franks, M. C. Phillips, and A. Suggett, Faraday Division General Discussion Meeting on Gels and Gelling Processes, 1974.

378. D. Eagland and R. G. Wheeler, in preparation.

379. D. J. Eatough and S. J. Rehfeld, *Thermochim. Acta* **2**, 443 (1971).

380. E. Edmond and A. G. Ogston, *Biochem. J.* **109**, 569 (1968).

381. J. T. Edsall and J. Wyman, "Biophysical Chemistry," Vol. 1, Academic Press, New York (1958), Chapter 6.

382. J. T. Edward, *Can. J. Chem.* **35**, 571 (1957).

383. G. S. Egerton, *J. Soc. Dyers and Col.* **64**, 336 (1948).

384. G. S. Egerton, *J. Textile Inst.* **39**, T305 (1948).

385. G. L. Eichorn, *Advan. Chem.* **62**, 378 (1966).

386. M. Eigen, *in* "The Neurosciences" (G. C. Quarton, ed.), Rockefeller University Press, New York (1967).

387. D. Eisenberg and W. Kauzmann, "The Structure and Properties of Water," Oxford University Press, London (1969).

388. P. Ekwall, *J. Colloid Interface Sci.* **29**, 16 (1969).

389. P. Ekwall and P. Holmberg, *Acta Chem. Scand.* **19**, 455 (1965).

390. P. Ekwall, L. Mandell, and P. Solyom, *J. Colloid Interface Sci.* **35**, 266 (1971).

391. P. Ekwall, L. Mandell, and P. Solyom, *J. Colloid Interface Sci.* **35**, 519 (1971).

392. D. D. Eley, *Trans. Faraday Soc.* **35**, 1242 (1939).

393. D. D. Eley, *Trans. Faraday Soc.* **35**, 1281 (1939).

394. H.-G. Elias, *Kunststoffe: Plastics* **8**, 441 (1961).

395. J. Eliassaf, *J. Appl. Polymer Sci.* **3**, 372 (1960).

396. J. Eliassaf, *J. Polymer Sci. Part B* **3**, 767 (1965).

397. J. Eliassaf, F. Eriksson, and F. R. Eirich, *J. Polymer Sci.* **47**, 193 (1960).

398. J. Eliassaf and A. Silberberg, *J. Polymer Sci.* **41**, 33 (1959).

399. H. D. Ellerton and P. J. Dunlop, *J. Phys. Chem.* **70**, 1831 (1966).

400. B. Ellis, A. S. C. Lawrence, M. P. McDonald, and W. E. Peel, *in* "Liquid Crystals and Ordered Fluids" (J. F. Johnson and R. S. Porter, eds.), Plenum Press, New Yor (1970), p. 277.

401. K. J. Ellis and G. S. Laurence, *Trans. Faraday Soc.* **63**, 91 (1967).

402. O. A. El Seoud, E. J. Fendler, J. E. Fendler, and R. T. Medary, *J. Phys. Chem.* **77**, 1876 (1973).

403. P. H. Elworthy, *J. Chem. Soc.* 5385 (1961).

404. P. H. Elworthy, A. T. Florence, C. B. MacFarlane, "Solubilization by Surface-Active Agents," Chapman and Hall, London (1968)

405. P. H. Elworthy and D. S. McIntosh, *J. Phys. Chem.* **68**, 3448 (1964).

406. E. S. Emerson, M. A. Conlin, A. E. Rosenoff, K. S. Norland, H. Rodriguez, D. Chin, and G. R. Bird, *J. Phys. Chem.* **71**, 2396 (1967).

407. M. F. Emerson and A. Holtzer, *J. Phys. Chem.* **69**, 3718 (1965).

408. M. F. Emerson and A. Holtzer, *J. Phys. Chem.* **71**, 1898 (1967).

409. M. F. Emerson and A. Holtzer, *J. Phys. Chem.* **71**, 3320 (1967).

410. J. Engel, *Arch. Biochem. Biophys.* **97**, 150 (1962).

411. J. Engel, *Z. Physiol. Chem.* **328**, 94 (1962).

412. T. Enns, P. E. Scholander, and E. D. Bradstreet, *J. Phys. Chem.* **69**, 389 (1965).

413. R. F. Epland and H. A. Scheraga, *Biopolymers* **6**, 1551 (1968).

414. J. C. Eriksson and G. Gillberg, *Acta Chem. Scand.* **20**, 2019 (1966).

415. S. R. Erlander, *J. Colloid Interface Sci.* **34**, 53 (1970).

416. L. Espada, M. N. Jones, and G. Pilcher, *J. Chem. Thermodynamics* **2**, 1 (1970).

417. H. C. Evans, *J. Chem. Soc.* 579 (1956).

418. J. M. Evans and M. B. Huglin, *Makromol. Chemie* **127**, 141 (1969).

419. H. Eyring and M. S. Jhon, "Significant Liquid Structures," Wiley, New York (1969).

420. H. Eyring, R. Lumry and J. D. Spikes, *in* "Mechanism of Enzyme Action" (J. McElroy and B. Glass, eds.), John Hopkins Press, Baltimore (1956), p. 123.

421. M. Falk, K. A. Hartman, Jr, and R. C. Lord, *J. Amer. Chem. Soc.* **84**, 3843 (1962).

422. M. Falk, K. A. Hartman, Jr., and R. C. Lord, *J. Amer. Chem. Soc.* **85**, 387 (1963).

423. M. Falk, K. A. Hartman, Jr., and R. C. Lord, *J. Amer. Chem. Soc.* **85**, 391 (1963).

424. M. Falk, A. G. Poole, and C. G. Goymour, *Can. J. Chem.* **48**, 1536 (1970).

425. E. L. Farquhar, M. Downing, and S. J. Gill, *Biochemistry* **7**, 1224 (1968).

426. G. D. Fasman, C. Lindblow, and E. Bodenheimer, *Biochemistry* **3**, 155 (1964).

427. M. E. Feinstein and H. L. Rosano, *J. Colloid Interface Sci.* **24**, 73 (1967).

428. M. E. Feinstein and H. L. Rosano, *J. Phys. Chem.* **73**, 601 (1969).

429. J. Feitelson, *J. Phys. Chem.* **65**, 975 (1961).

430. H. Fellner-Feldegg, *J. Phys. Chem.* **73**, 616 (1969).

431. H. Fellner-Feldegg, *J. Phys. Chem.* **76**, 2116 (1972).

432. G. Felsenfeld and S. L. Huang, *Biochim. Biophys. Acta* **51**, 19 (1961).

433. E. J. Fendler, S. A. Chang, J. H. Fendler, R. T. Medary, O. A. El Seoud, and V. A. Woods, *in* "Reaction Kinetics in Micelles" (Amer. Chem. Soc. Symp.; E. Cordes, ed.), Plenum, New York (1973), p. 127.

434. E. J. Fendler and J. H. Fendler, *Advan. Phys. Org. Chem.* **8**, 271 (1970).

435. J. H. Fendler, E. J. Fendler, R. T. Medary, and O. A. El Seoud, *J. Chem. Soc. Faraday Trans. I* **69**, 280 (1973).

436. E. J. Fendler, J. H. Fendler, R. T. Medary, and O. A. El Seoud, *J. Phys. Chem.* **77**, 1432 (1973).

437. E. J. Fendler, J. H. Fendler, R. T. Medary, and V. A. Woods, *Chem. Commun.* 1497 (1971).

438. J. H. Fendler and L. K. Patterson, *J. Phys. Chem.* **75**, 3907 (1971).

439. M. S. Fernandez and J. Corbon, *Biochim. Biophys. Acta* **298**, 8 (1973).

440. G. E. Ficken, *J. Phot. Sci.* **21**, 11 (1973).

441. C. A. Finch (ed.), "Properties and Applications of Polyvinyl Alcohol," Society of Chemical Industry, Monograph No. 30, London (1968).

442. E. G. Finer, *J. Chem. Soc. Faraday II* **69**, 1590 (1973).

443. E. G. Finer and A. Darke, *Chem. Phys. Lipids* **12**, 1 (1974).

444. E. G. Finer, A. G. Flook, and H. Hauser, *Febs Letters* **18**, 331 (1971).

445. E. G. Finer, A. G. Flook, and H. Hauser, *Biochim. Biophys. Acta* **260**, 49 (1972).

446. E. G. Finer, A. G. Flook, and H. Hauser, *Biochim. Biophys. Acta* **260**, 59 (1972).

447. E. G. Finer, F. Franks, and M. J. Tait, *J. Amer. Chem. Soc.* **94**, 4424 (1972).

448. E. G. Finer and M. C. Phillips, *Chem. Phys. Lipids* **10**, 237 (1973).

449. E. Fischer, *J. Chem. Soc.* 1382 (1955).

450. H. F. Fisher, *Biochim. Biophys. Acta* **109**, 544 (1965).

451. G. J. Fleer, L. K. Koopal, and J. Lyklema, *Kolloid-Z.* **250**, 689 (1972).

452. J. E. Fletcher and J. D. Ashbrook, *Biochemistry* **10**, 3229 (1971).

453. B. D. Flockhart, *J. Colloid Sci.* **16**, 484 (1961).
454. P. J. Flory, *J. Chem. Phys.* **9**, 660 (1941).
455. P. J. Flory, *J. Chem. Phys.* **10**, 51 (1942).
456. P. J. Flory, *J. Chem. Phys.* **13**, 453 (1945).
457. P. J. Flory, "Principles of Polymer Chemistry," Cornell University Press, New York (1953).
458. P. J. Flory, *J. Cell Comp. Physiol.* **49**, 175 (1957).
459. P. J. Flory, "Statistical Mechanics of Chain Molecules," Wiley–Interscience, New York (1969).
460. P. J. Flory, *Disc. Faraday Soc.* **49**, 7 (1970).
461. P. J. Flory and R. R. Garrett, *J. Amer. Chem. Soc.* **80**, 4836 (1958).
462. P. J. Flory and J. Rehner, *J. Chem. Phys.* **11** 512, 521 (1943).
463. P. J. Flory and E. S. Weaver, *J. Amer. Chem. Soc.* **82**, 4518 (1960).
464. J. Formánek and E. Grandmougin, "Untersuchung und Nachweis Organischer Farbstoffe auf Spektroskopischem Wege," 2nd ed., Part 1, Julius Springer, Berlin (1908).
465. T. Förster, *Naturwiss.* **33**, 166 (1946).
466. T. Förster and E. König, *Z. Elektrochem.* **61**, 344 (1956).
467. J. F. Foster, *in* "Starch: Chemistry and Technology" (R. L. Whistler and E. F. Paschall, eds.), Vol. 1, Academic Press, London and New York (1965).
468. K. K. Fox, *Trans. Faraday Soc.* **67**, 2802 (1971).
469. K. K. Fox, I. D. Robb, and R. Smith, *J. Chem. Soc. Faraday Trans. I* **68**, 445 (1972).
470. H. P. Frank. *J. Polymer Sci.* **12**, 565 (1954).
471. H. P. Frank, *J. Coll. Sci.* **12**, 480 (1957).
472. H. P. Frank, S. Barkin, and F. R. Eirich *J. Phys. Chem.* **61**, 1375 (1957).
473. H. P. Frank and H. Mark, *J. Polymer Sci.* **10**, 129 (1953).
474. H. S. Frank, *Z. Physik. Chem. (Leipzig)* **228**, 364 (1965).
475. H. S. Frank, *Fed. Proc.* **24**, S1 (1966).
476. H. S. Frank. *Science* **169**, 635 (1970).
477. H. S. Frank and M. W. Evans, *J. Chem. Phys.* **13**, 507 (1945).
478. H. S. Frank and F. Franks, *J. Chem. Phys.* **48**, 4746 (1968).
479. H. S. Frank and A. S. Quist, *J. Chem. Phys.* **34**, 604 (1961).
480. H. S. Frank and W. Y. Wen, *Disc. Faraday Soc.* **24**, 133 (1957).
481. S. G. Frank and G. Zografi, *J. Colloid Interface Sci.* **29**, 27 (1969).
482. M. D. Frank-Kamenetskii, *Biopolymers* **10**, 2623 (1971).
483. F. Franks, *Ann. N.Y. Acad. Sci.* **125**, 277 (1965).
484. F. Franks, *Nature* **210**, 87 (1966).
485. F. Franks, *in* "Hydrogen Bonded Solvent Systems" (A. K. Covington and P. Jones, eds.), Taylor and Francis, London (1969), p. 31.
486. F. Franks and D. J. G. Ives, *Quart. Rev. Chem. Soc.* **20**, 1 (1966).
487. F. Franks, M. A. J. Quickenden, D. S. Reid, and B. Watson, *Trans. Faraday Soc.* **66**, 582 (1970).
488. F. Franks, J. R. Ravenhill, and D. S. Reid, *J. Solution Chem.* **1**, 3 (1972).
489. F. Franks, D. S. Reid, and A. Suggett, *J. Solution Chem.* **2**, 99 (1973).
490. F. Franks and H. T. Smith, *J. Phys. Chem.* **68**, 3581 (1964).
491. F. Franks and H. T. Smith, *Trans. Faraday Soc.* **63**, 2586 (1967).
492. F. Franks and H. T. Smith, *Trans. Faraday Soc.* **64**, 2962 (1968).

493. F. Franks and B. Watson, *Trans. Faraday Soc.* **63**, 329 (1967).

494. R. D. B. Frazer, B. C. Harrap, T. P. MacRae, F. H. C. Stewart, and F. Suzuki, *Biopolymers* **5**, 251 (1967).

495. A. French and B. Zaslow, *J. Chem. Soc. Chem. Comm.* 41 (1972).

496. D. French, *in* "Chemistry and Industry of Starch" (R. L. Whistler and E. F. Paschall, eds.) Academic Press, New York (1950), Vol. 1, p. 157.

497. D. French, *in* "Symposium on Foods: Carbohydrates and their Roles" (H. W. Schulz, ed.), AVT Publishing Co. (1969).

498. D. French, A. O. Pulley, and W. J. Whelan, *Die Stärke* **15**, 349 (1963).

499. M. E. Friedman and H. A. Scheraga, *J. Phys. Chem.* **69**, 3795 (1965).

500. L. F. C. Friele, *J. Soc. Dyers and Col.* **79**, 623 (1963).

501. A. J. Fryar and S. Kaufman, *J. Colloid Interface Sci.* **29**, 444 (1969).

502. D. Fukushima and J. van Buren, *Cereal Chem.* **47**, 687 (1970).

503. M. E. Fuller and W. S. Brey, *J. Biol. Chem.* **243**, 274 (1968).

504. R. M. Fuoss, *J. Polymer Sci.* **3**, 603 (1948); **4**, 96 (1949).

505. G. L. Gaines, Jr., "Insoluble Monolayers at Liquid–Gas Interfaces," Wiley-Interscience, New York (1966).

506. J. D. Galligan, F. W. Minor, and A. M. Schwartz, *in* "Adhesion in Biological Systems" (R. S. Manley, ed.), Academic Press, New York and London (1970), Chapter 16.

507. N. G. Gaylord, "Polyethers. Part 1. Polyalkylene Oxides and other Polyethers," Wiley–Interscience, New York (1963).

508. K. Gekko and H. Noguchi, *Biopolymers* **10**, 1513 (1971).

509. R. B. Gennis and C. R. Cantor, *J. Mol. Biol.* **65**, 381 (1972).

510. W. L. G. Gent, N. A. Gregson, D. B. Gammack, and J. H. Raper, *Nature* **204**, 553 (1964).

511. S. Y. Gerlsma, *Europ. J. Biochem.* **14**, 150 (1970).

512. R. A. Gibbons, *Nature* **200**, 665 (1963).

513. A. Gierer and K. Wirtz, *Z. Naturforsch.* **A8**, 532 (1953).

514. G. H. Giles, *Chem. and Ind.* 724 (1964).

515. C. H. Giles and A. P. D'Silva, *Trans. Faraday Soc.* **65**, 2516 (1969).

516. C. H. Giles and D. G. Duff, *J. Soc. Dyers and Col.* **86**, 405 (1970).

517. C. H. Giles, I. A. Easton, R. B. McKay, C. C. Patel, N. B. Shah, and D. Smith, *Trans. Faraday Soc.* **62**, 1963 (1966).

518. C. H. Giles and E. L. Neustädter, *J. Chem. Soc.* 918 (1952).

519. C. H. Giles, S. M. K. Rahman, and D. Smith, *Text. Res. J.* **31**, 679 (1961).

520. C. H. Giles and A. H. Soutar, *J. Soc. Dyers and Col.* **87**, 301 (1971).

521. S. J. Gill, M. Downing, and G. F. Sheals, *Biochemistry* **6**, 272 (1967).

522. T. J. Gill, III, C. T. Ladoulis, M. F. King, and H. W. Kunz, *in* "Liquid Crystals and Ordered Fluids" (J. F. Johnson and R. S. Porter, eds.), Plenum Press, New York (1970), p. 131.

523. G. Gillberg, H. Lehtinen, and S. Friberg, *J. Colloid Interface Sci.* **33**, 40 (1970).

524. F. G. R. Gimblett, "Inorganic Polymer Chemistry," Butterworths, London (1963).

525. A. Ginsburg and W. R. Carroll, *Biochemistry* **4**, 2159 (1965).

526. J. A. Glasel, *J. Amer. Chem. Soc.* **92**, 375 (1970).

527. S. Glasstone, *in* "Textbook of Physical Chemistry," MacMillan, London (1962).

528. S. Glasstone, K. J. Laidler, and H. Eyring, "The Theory of Rate Processes," McGraw-Hill, New York and London (1941).

529. D. L. Glaubinger, D. A. Lloyd, and I. Tinoco, *Biopolymers* **6**, 409 (1968).
530. F. J. Glavis, *in* "Water-Soluble Resins" (R. L. Davidson and M. Sittig, eds.), 2nd ed., Reinhold, New York (1968), Chapter 8.
531. D. N. Glew, *J. Phys. Chem.* **66**, 605 (1962).
532. D. N. Glew, H. D. Mak, and N. S. Rath, *Canad. J. Chem.* **45**, 3059 (1967).
533. D. N. Glew, D. H. Mak, and N. S. Rath, *Chem. Comm.* 264 (1968).
534. D. N. Glew, H. D. Mak, and N. S. Rath, *in* "Hydrogen-Bonded Solvent Systems" (A. K. Covington and P. Jones, eds.), Taylor and Francis, London (1968).
535. D. N. Glew and H. Watts, *Canad. J. Chem.* **49**, 1830 (1971).
536. E. Glueckauf, *Nature* **211**, 1227 (1966).
537. E. Glueckauf and P. J. Russell, *Desalination* **8**, 351 (1970).
538. E. D. Goddard and G. C. Benson, *Trans. Faraday Soc.* **52**, 409 (1956).
539. E. v. Goldammer and H. G. Hertz, *J. Phys. Chem.* **74**, 3734 (1970).
540. E. v. Goldammer and M. D. Zeidler, *Ber. Bunsenges. Phys. Chem.* **73**, 4 (1969).
541. J. Goldfarb and L. Sepulveda, *J. Colloid Interface Sci.* **31**, 454 (1969).
542. R. Göller, H. G. Hertz, and R. Tutsch, *Pure and Applied Chem.* **32**, 149 (1972).
543. W. Good, *Nature* **214**, 1250 (1967).
544. F. L. Goodall, *J. Soc. Dyers and Col.* **54**, 45 (1938).
545. I. Goodman, *in* "Fibre Structure" (J. W. S. Hearle and R. H. Peters, eds.), The Textile Institute and Butterworths, Manchester and London (1963), Chapter 4.
546. I. Goodman, "Synthetic Fibre-forming Polymers," Royal Institute of Chemistry, Lecture Series No. 3, London (1968).
547. D. E. Gordon, B. Curnette, Jr., and K. Lark, *J. Mc. Biol.* **13**, 571 (1965).
548. J. A. Gordon and W. P. Jencks, *Biochemistry* **2**, 47 (1963).
549. J. E. Gordon, J. C. Robertson, and R. L. Thorne, *J. Phys. Chem.* **74**, 957 (1970).
550. F. Gormick, L. Mandelkern, A. F. Diorio, and D. E. Roberts, *J. Amer. Chem. Soc.* **86**, 2549 (1964).
551. E. Gorter and F. Grendel, *J. Exp. Med.* **41**, 439 (1925).
552. A. M. Gottlieb, P. T. Inglefield, and Y. Lange, *Biochim. Biophys. Acta* **307**, 444 (1973).
553. E. V. Gouinlock, Jr., P. J. Flory, and H. A. Scheraga, *J. Polymer Sci.* **16**, 383 (1955).
554. E. Graber, J. Lang, and R. Zana, *Kolloid-Z.* **238**, 470 (1970).
555. E. Graber and R. Zana, *Kolloid-Z.* **238**, 479 (1970).
556. D. E. Graham and E. J. A. Lea, *Biochim. Biophys. Acta* **274**, 286 (1972).
557. W. D. Graham, *J. Pharm. Pharmacol.* **9**, 230 (1957).
558. E. H. Grant, *Ann. N.Y. Acad. Sci.* **125**, 418 (1965).
559. E. H. Grant, S. E. Keefe, and S. Takashima, *J. Phys. Chem.* **72**, 4373 (1968).
560. G. T. Grant, E. R. Morris, D. A. Rees, P. J. C. Smith, and D. Thom, *Febs Letters* **32**, 195 (1973).
561. R. A. Grant, R. W. Horne, and R. W. Lox, *Nature* **207**, 822 (1965).
562. W. B. Gratzer and G. H. Beaven, *J. Phys. Chem.* **73**, 2270 (1969).
563. H. B. Gray, Jr., V. A. Bloomfield, and J. E. Hearst, *J. Chem. Phys.* **46**, 1493 (1967).
564. W. E. Gray, W. R. Brewer, and G. R. Bird, *Photogr. Sci. Eng.* **14**, 316 (1970).
565. W. J. Green and H. S. Frank, *J. Solution Chem.* in press.
566. M. Greenhalgh, A. Johnson, and R. H. Peters, *J. Soc. Dyers and Col.* **78**, 315 (1962).
567. J. P. Greenstein and M. Winitz, "Chemistry of the Amino Acids," John Wiley, New York, Vol. 1, p. 528.

568. J. Gregory, *Trans. Faraday Soc.* **65**, 2260 (1969).

569. O. H. Griffith and A. S. Waggoner, *Acc. Chem. Res.* **2**, 17 (1969).

570. O. Griot and J. A. Kitchener, *Trans. Faraday Soc.* **61**, 1026, 1032 (1965).

571. G. A. Grode, R. D. Falb, C. W. Cooper, and J. Lynn, *in* "Adhesion in Biological Systems" (R. S. Manly, ed.), Academic Press, New York and London (1970), Chapter 9.

572. D. W. Gruenwedel, C. H. Hsu, and D. S. Lu, *Biopolymers*, **10**, 47 (1971).

573. E. Grunwald and A. F. Butler, *J. Amer. Chem. Soc.* **82**, 5647 (1960).

574. F. T. Gucker and F. D. Ayres, *J. Amer. Chem. Soc.* **59**, 2152 (1937).

575. E. A. Guggenheim, "Mixtures," Oxford University Press (1952).

576. Yu. V. Gurikov, *J. Struct. Chem.* (*Russian*) **10**, 494 (1969).

577. D. E. Guttman and T. Higuchi, *J. Amer. Pharm. Assoc., Sci. Ed.* **44**, 668 (1955).

578. D. E. Guttman and T. Higuchi, *J. Amer. Pharm. Assoc., Sci. Ed.* **45**, 659 (1956).

579. H. C. Haas, C. K. Chiklis, and R. D. Moreau, *J. Polymer Sci. Part A-1* **8**, 1131 (1970).

580. H. C. Haas, R. L. MacDonald and A. N. Schuler, *J. Polymer Sci. Part A-1* **8**, 1213 (1970).

581. H. C. Haas, R. L. MacDonald, and A. N. Schuler, *J. Polymer Sci. Part A-1* **8**, 3405 (1970).

582. H. C. Haas, R. L. MacDonald, and A. N. Schuler, *J. Polymer Sci. Part A-1* **9**, 959 (1971).

583. H. C. Haas, M. J. Manning, and M. H. Mach, *J. Polymer Sci. Part A-1* **8**, 1725 (1970).

584. H. C. Haas, R. D. Moreau, and N. W. Schuler, *J. Polymer Sci. Part A-2* **5**, 915 (1967).

585. H. C. Haas and N. W. Schuler, *J. Polymer Sci. Part B* **2**, 1095 (1964).

586. E. P. I. Hade and C. Tanford, *J. Amer. Chem. Soc.* **89**, 5034 (1967).

587. A. T. Hagler, H. A. Scheraga, and G. Némethy, *J. Phys. Chem.* **76**, 3229 (1972).

588. D. G. Hall, *Trans. Faraday Soc.* **66**, 1351 (1970).

589. D. G. Hall, *Trans. Faraday Soc.* **66**, 1359 (1970).

590. D. G. Hall, *J. Chem. Soc. Faraday Trans. II* **68**, 668 (1972).

591. D. G. Hall, *J. Chem. Soc. Faraday Trans. II* **68**, 1439 (1972).

592. D. G. Hall and B. A. Pethica, *in* "Nonionic Surfactants" (M. J. Schick, ed.), Marcel Dekker, New York (1967), p. 534.

593. A. Hamabato, S. Chang, and P. H. von Hippel, *Biochemistry* **12**, 1271 (1973).

594. K. Hamaguchi and E. P. Geiderschek, *J. Amer. Chem. Soc.* **84**, 1329 (1962).

595. G. G. Hammes, *Advan. Protein Chem.* **23**, 11 (1968).

596. G. G. Hammes and T. B. Lewis, *J. Phys. Chem.* **70**, 1610 (1966).

597. G. G. Hammes and P. R. Schimmel, *J. Amer. Chem. Soc.* **89**, 442 (1967).

598. G. G. Hammes and J. C. Swann, *Biochemistry* **6**, 1591 (1967).

599. D. J. Hanahan, "Lipide Chemistry," Wiley, New York (1960).

600. D. Hankins, J. W. Moskowitz, and F. H. Stillinger, *J. Chem. Phys.* **53**, 4544 (1970).

601. A. J. Hannaford, *Carbohydrate Res.* **3**, 295 (1967).

602. C. Hansch, *Accounts Chem. Res.* **2**, 232 (1969).

603. C. Hansch, *in* "Drug Design" (E. J. Ariens, ed.), Academic Press, New York (1971), p. 271.

604. C. Hansch and E. Coates, *J. Pharm. Sci.* **69**, 731 (1970).

605. R. H. Harding and J. K. Rose, *in* "Water-Soluble Resins" (R. L. Davidson and M. Sittig, eds.), 2nd ed., Reinhold, New York (1968), Chapter 10.

606. A. C. Hardy and F. M. Young, *J. Opt. Soc. Amer.* **39**, 265 (1949).

607. W. A. Hargraves and G. C. Kresheck, *J. Phys. Chem.* **73**, 3249 (1969).

608. H. S. Harned and B. B. Owen, *in* "The Physical Chemistry of Electrolytic Solutions," 3rd ed., Reinhold, New York (1958).

609. H. S. Harned and B. B. Owen, "The Physical Chemistry of Electrolytic Solutions," 3rd ed., 2nd printing, Reinhold, New York (1963).

610. B. S. Harrap, *Int. J. Protein Research* **1**, 245 (1969).

611. W. F. Harrington, *J. Mol. Biol.* **9**, 613 (1964).

612. W. F. Harrington and G. M. Karr, *Biochemistry* **9**, 3725 (1970).

613. W. F. Harrington and N. V. Rao, *in* "Conformation of Biopolymers" (G. N. Ramachandran, ed.), Academic Press, New York (1967).

614. W. F. Harrington and N. V. Rao, *Biochemistry* **9**, 3714 (1970).

615. W. F. Harrington and M. Sela, *Biochem. Biophys. Acta* **27**, 24 (1958).

616. W. F. Harrington and P. H. von Hippel, *Adv. Protein Chem.* **16**, 1 (1961).

617. G. R. Haugen and E. R. Hardwick. *J. Phys. Chem.* **67**, 725 (1963).

618. H. Hauser, *Biochem. Biophys. Res. Commun.* **45**, 1049 (1971).

619. H. Hauser and M. D. Barratt, *Biochem. Biophys. Res. Commun.* **53**, 399 (1973).

620. H. Hauser, D. Chapman, and R. M. C. Dawson, *Biochim. Biophys. Acta* **183**, 320 (1969).

621. H. Hauser and R. M. C. Dawson, *Biochem. J.* **105**, 401 (1967).

622. H. Hauser and R. M. C. Dawson, *Europ. J. Biochem.* **1**, 61 (1967).

623. H. Hauser and L. Irons, *Hoppe-Seyler's Z. Physiol. Chem.* **353**, 1579 (1972).

624. H. Hauser and M. C. Phillips, *J. Biol. Chem.* **248**, 8585 (1973).

625. H. Hauser and M. C. Phillips, *in* "Berichte vom VI Internationalen Kongress für grenzflächenaktive Stoffe," Carl Hanser Verlag, München (1973), p. 381.

626. H. Hauser, M. C. Phillips, and R. M. Marchbanks, *Biochem. J.* **120**, 329 (1970).

627. M. Hauser and U. Klein, *Z. Phys. Chem.* (*Frankfurt*) **78**, 32 (1972).

628. P. M. Hauser and A. D. McLaren, *Ind. Eng. Chem.* **40**, 112 (1948).

629. W. L. Hawkins, M. A. Worthington, and W. Matreyek, *J. Appl. Polymer Sci.* **3**, 277 (1960).

630. N. Hayama, *Bull. Inst. Chem. Res. Kyoto Univ.* **42**(5), 401 (1964).

631. K. Hayashi, Y. Yamazawa, T. Takagaki, Ff. Williams, K. Hayashi, and S. Okamura, *Trans. Faraday Soc.* **63**, 1489 (1967).

632. C. F. Hazlewood, B. L. Nichols, and N. F. Chamberlain, *Nature* **222**, 747 (1969).

633. J. E. Hearst, *Biopolymers* **3**, 57 (1965).

634. J. E. Hearst and W. Stockmayer, *J. Chem. Phys.* **37**, 1425 (1962).

635. J. E. Hearst and J. Vinograd, *Proc. Nat. Acad. Sci. U.S.* **47**, 1005 (1961).

636. O. Hechter, T. Wittstruck, N. McNiven, and G. Lester, *Proc. Nat. Acad. Sci.* **46**, 783 (1960).

637. D. Hegner, U. Schummer, and G. H. Schnepel, *Biochim. Biophys. Acta* **307**, 452 (1973).

638. K. P. Henrikson, *Biochim. Biophys. Acta* **203**, 228 (1970).

639. R. B. Hermann, *J. Phys. Chem.* **75**, 363 (1971).

640. J. Hermans, Jr., *J. Amer. Chem. Soc.* **88**, 2418 (1966).

641. J. Hermans, Jr., *J. Phys. Chem.* **70**, 510 (1966).

642. J. Hermans, Jr., and G. Acampora, *J. Amer. Chem. Soc.* **89**, 1547 (1967).

643. J. Hermans, Jr., D. Lohr, and D. Ferro, *Nature* **224**, 175 (1969).

644. J. Hermans, Jr., D. Prett, and G. Acampora, *Biochem.* **8**, 22 (1969).

645. K. W. Herrmann and L. Benjamin, *J. Colloid Interface Sci.* **23**, 478 (1967).

646. K. W. Herrmann, J. G. Brushmiller, and W. L. Courchene, *J. Phys. Chem.* **70**, 2909 (1966).

647. T. T. Herskovits, B. Gadegbeku, and J. Jaillet, *J. Biol. Chem.* **245**, 2588 (1970).

648. T. T. Herskovits, H. Jaillet, and B. Gadegbeku, *J. Biol. Chem.* **245**, 2588 (1970).

649. T. T. Herskovits, H. Jaillet, and B. Gadegbeku, *J. Biol. Chem.* **245**, 4544 (1970).

650. H. G. Hertz, *Ber. Bunsenges. Phys. Chem.* **68**, 907 (1964).

651. H. G. Hertz, B. Lindman, and V. Siepe, *Ber. Busenges, Phys. Chem.* **73**, 549 (1969).

652. H. G. Hertz and H. Pfliegel (in press).

653. H. G. Hertz and C. Rädle, *Ber. Bunsenges. Phys. Chem.* **77**, 521 (1973).

654. H. G. Hertz and R. Tutsch, *J. Soln. Chem.* **2**, 000 (1973).

655. R. O. Herzog and A. Polotsky, *Z. Phys. Chem.* **A87**, 449 (1914).

656. M. Hida and T. Sanuki, *Bull. Chem. Soc. Japan* **43**, 2291 (1970).

657. M. Hida, A. Yabe, H. Murayama, and M. Hayashi, *Bull. Chem. Soc. Japan* **41**, 1776 (1968).

658. T. Higuchi and R. Kuramoto, *J. Amer. Pharmaceut. Assoc., Sci. Ed.* **43**, 393, 398 (1954).

659. T. Higuchi and J. L. Lach, *J. Amer. Pharmaceut. Assoc., Sci. Ed.* **43**, 465 (1954).

660. T. Hikota, M. Nakamura, S. Machida, and K. Meguro, *J. Amer. Oil Chem. Soc.* **48**, 784 (1971).

661. J. H. Hildebrand, *J. Phys. Chem.* **72**, 1841 (1968).

662. J. H. Hildebrand, J. M. Prausnitz, and R. L. Scott, "Regular and Related Solutions," Van Nostrand Reinhold Co., New York (1970).

663. J. H. Hildebrand and R. L. Scott, "The Solubility of Nonelectrolytes," 3rd ed. (revised), Dover, New York (1964).

664. D. R. J. Hill, *in* "Weathering and Degradation of Plastics" (S. H. Pinner, ed.), Columbine Press, Manchester and London (1966), Chapter 2.

665. T. L. Hill, "Statistical Mechanics," McGraw-Hill, New York (1956).

666. P. J. Hillson and R. B. McKay, *Trans. Faraday Soc.* **61**, 374 (1965).

667. J. Hindmann, *J. Chem. Phys.* **36**, 1000 (1962).

668. H. Hinz and J. M. Sturtevant, *J. Biol. Chem.* **247**, 6071 (1972).

669. C. F. Hiskey and F. F. Cantwell, *J. Pharm. Sci.* **55**, 166 (1966).

670. M. E. Hobbs, *J. Phys. Chem.* **55**, 675 (1951).

671. A. J. Hodge and F. O. Schmitt, *Proc. Nat. Acad. Sci. U.S.* **46**, 186 (1960).

672. F. Hofmeister, *Arch. Exptl. Pathol. Pharmakol.* **24**, 247 (1888).

673. D. N. Holcomb and K. E. van Holde, *J. Phys. Chem.* **66**, 1999 (1969).

674. F. H. Holmes and H. A. Standing, *Trans. Faraday Soc.* **41**, 568 (1945).

675. W. C. Holmes, *Ind. and Eng. Chem.* **16**, 35 (1924).

676. A. Holtzer and M. F. Emerson, *J. Phys. Chem.* **73**, 26 (1969).

677. T. R. Hopkins and J. D. Spikes, *Biochem. Biophys. Res. Commun.* **28**, 480 (1967).

678. T. R. Hopkins and. J. D. Spikes, *Biochem. Biophys. Res. Commun.* **30**, 540 (1968).

679. W. Hoppe, *Kolloid Z.* **101**, 300 (1942).

680. W. Hoppe, *Kolloid Z.* **109**, 27 (1944).

681. R. A. Horne, *in* "Solute-Solvent Interactions" (J. F. Coetzee and C. D. Ritchie, eds.), Marcel Dekker, New York (1969).

682. R. A. Horne, J. P. Almeida, A. F. Day, and N. T. Yu, *J. Colloid Interface Sci.* **35**, 77 (1971).
683. A. F. Horwitz, W. J. Horsley, and M. P. Klein, *Proc. Nat. Acad. Sci. U.S.* **69**, 590 (1972).
684. P. A. Howell, *J. Phys. Chem.* **73**, 2294 (1969).
685. J. C. Hsia, H. Schneider, and I. C. P. Smith, *Biochim. Biophys. Acta* **202**, 399 (1970).
686. C. Huang, *Biochemistry* **8**, 344 (1969).
687. C. Huang and J. P. Charlton, *J. Biol. Chem.* **246**, 2555 (1971).
688. R. Y. M. Huang and J. F. Westlake, *J. Polymer Sci. Part A*-1, **8**, 49 (1970).
689. W. L. Hubbell and H. M. McConnell, *Proc. Nat. Acad. Sci. U.S.* **61**, 12 (1968).
690. W. L. Hubbell and H. M. McConnell, *Proc. Nat. Acad. Sci. U.S.* **63**, 16 (1969).
691. W. L. Hubbel and H. M. McConnell, *Proc. Nat. Acad. Sci. U.S.* **64**, 20 (1969).
692. W. L. Hubbell and H. M. McConnell, *J. Amer. Chem. Soc.* **93**, 314 (1971).
693. M. L. Huggins, *Ann. N.Y. Acad. Sci.* **43**, 1 (1942).
694. M. L. Huggins, *J. Amer. Chem. Soc.* **64**, 1712 (1942).
695. L. J. T. Hughes and D. B. Fordyce, *J. Polymer Sci.* **22**, 509 (1956).
696. T. R. Hughes and I. M. Klotz, *in* "Methods of Biochemical Analysis" (D. Glick, ed.), Interscience, New York and London (1956), Vol. 3, p. 265.
697. F. Husson and V. Luzzati, *J. Cell Biol.* **12**, 207 (1962).
698. F. Husson, H. Mustacchi, and V. Luzzati, *Acta Crystallogr.* **13**, 668 (1960).
699. J. O. Hutchens, A. G. Cole, and J. W. Stout, *J. Biol. Chem.* **244**, 26 (1969).
700. A. Hvidt and K. Linderstrom-Lang, *Biochim. Biophys. Acta* **14**, 574 (1954).
701. A. Hvidt and S. O. Nielsen, *Adv. Protein Chem.* **21**, 1287 (1966).
702. A. Hybl, R. E. Rundle, and D. E. Williams, *J. Amer. Chem. Soc.* **87**, 2779 (1965).
703. M. Ihnat, A. Szabo, and D. A. I. Goring, *J. Chem. Soc. A* 1500 (1968).
704. T. Iio, *Biopolymers* **10**, 1583 (1971).
705. A. Ikegami, *J. Polymer Sci. A* **2**, 907 (1964).
706. A. Ikegami, *Biopolymers* **6**, 431 (1968).
707. K.-H. Illers, *Makromol. Chemie* **38**, 168 (1960).
708. T. Ingram and M. N. Jones, *Trans. Faraday Soc.* **65**, 297 (1969).
709. I. T. Ingwall, H. A. Scheraga, N. Lotan, A. Berger, and E. Katchalski, *Biopolymers* **6**, 331 (1968).
710. R. B. Inman and R. L. Baldwin, *J. Mol. Biol.* **8**, 452 (1964).
711. R. B. Inman and D. O. Jordan, *Biochem. Biophys. Acta* **42**, 421 (1960).
712. M. N. Inscoe, J. H. Gould, M. E. Corning, and W. R. Brode, *J. Res. Nat. Bur. St.* **60**, 65 (1958).
713. T. Isemura, H. Okabayashi, and S. Sakakibara, *Biopolymers* **6**, 307 (1968).
714. E. Ising, *Z. Physik*, **31**, 253 (1925).
715. H. Ito, *Kobunshi Kagaku* **19**, 158 (1961).
716. K. J. Ivin, *in* "Polymer Handbook" (J. Brandrup and E. H. Immergut, eds.), Wiley–Interscience, New York (1966), p. II–363.
717. S. R. S. Iyer and G. S. Singh, "Proceedings of the Fourth Symposium, Centre for Advanced Study in Applied Chemistry," Dept. of Chemical Technology, University of Bombay (1969), p. 1.
718. R. M. Izat, D. Eatough, R. L. Snow, and J. H. Christensen, *J. Phys. Chem.* **72**, 1208 (1968).
719. C. M. Jackson, *in* "Permeability and Function of Biological Membranes" (L. Bolis *et al.*, eds.), North-Holland, Amsterdam (1970), p. 164.

720. W. M. Jackson and J. F. Brandts, *Biochem.* **9**, 2294 (1970).

721. H. Jacobs and E. Jenckel, *Makromol. Chemie* **47**, 72 (1961).

722. J. J. Jacobs, R. A. Anderson, and T. R. Watson, *J. Pharm. Pharmacol.* **23**, 148 (1971).

723. P. T. Jacobs, R. D. Geer, and E. W. Anacker, *J. Colloid Interface Sci.* **39**, 611 (1972).

724. M. J. Jaycock and R. H. Ottewill, *in* "Chem. Phys. Appl. Surface Active Subst." (Proc. 4th Int. Congr.; J. Th. G. Overbeek, ed.), Gordon and Breach, New York (1967), Vol. 2, p. 545-553; *Chem. Abstr.* **71**, 129040h (1969).

725. E. E. Jelley, *Nature* **138**, 1009 (1936).

726. H. H. G. Jellinek and S. Y. Fok, *Kolloid-Z.* **220**, 122 (1967).

727. H. H. G. Jellinek, M. D. Luh, and V. Nagarajan, *Kolloid-Z.* **232**, 758 (1969).

728. B. Jirgensons, *Makromol. Chemie* **6**, 30 (1951).

729. B. Jirgensons, *J. Polymer Sci.* **8**, 519 (1952).

730. A. Johansson, J. C. Eriksson, and L. O. Andersson, *Chem. Abstr.* **74**, 118097 (1971).

731. G. A. Johnson, S. M. A. Lecchini, E. G. Smith, J. Clifford, and B. A. Pethica, *Disc. Faraday Soc.* **42**, 120 (1966).

732. P. Johnson, *in* "Solution Properties of Natural Polymers" (C. T. Greenwood, ed.), The Chemical Society, Special Publication No. 23, London (1968), p. 243.

733. N. Johnston and S. Krimm, *Biopolymers* **10**, 2597 (1971).

734. C. Jolicoeur and H. L. Friedman, *Ber. Bunsenges. Phys. Chem.* **75**, 248 (1971).

735. C. Jolicoeur and H. L. Friedman, *J. Phys. Chem.* **75**, 165 (1971).

736. C. Jolicoeur and H. L. Friedman, *J. Soln. Chem.* **2**, (1973).

737. S. C. Jolly (ed.), "British Pharmaceutical Codex," The Pharmaceutical Press, London (1973), Part 4.

738. G. Jones and M. Dole, *J. Amer. Chem. Soc.* **51**, 2950 (1929).

739. M. N. Jones, *J. Colloid Interface Sci.* **23**, 36 (1967).

740. M. N. Jones, G. Agg, and G. Pilcher, *J. Chem. Thermodynamics* **3**, 801 (1971).

741. M. N. Jones, G. Pilcher, and L. Espada, *J. Chem. Thermodynamics* **2**, 333 (1970).

742. D. O. Jordan, T. Kurucsev, and M. L. Martin, *Trans. Faraday Soc.* **65**, 598, 612, 616 (1969).

743. J. Josse and W. F. Harrington, *J. Mol. Biol.* **9**, 269 (1964).

744. P. Jost, L. J. Libertini, V. C. Hebert, and O. H. Griffith, *J. Mol. Biol.* **59**, 77 (1971).

745. F. J. Joubert, N. Lotan, and H. A. Scheraga, *Biochem.* **9**, 2197 (1970).

746. E. Jungermann, (ed.), "Cationic Surfactants," Marcel Dekker, New York (1970).

747. B. N. Kabadi and E. R. Hammerlund, *J. Pharm. Sci.* **55**, 1069, 1072 (1966).

748. M. A. Kabayama and D. Patterson, *Canad. J. Chem.* **36**, 563 (1958).

749. M. A. Kabayama, D. Patterson, and L. Piche, *Canad. J. Chem.* **36**, 557 (1958).

750. I. Kagawa and K. Tsumura, *J. Chem. Soc. Japan (Ind. Chem. Sect.)* **47**, 574 (1944); *Chem. Abstr.* **42**, 6205i (1948)

751. L. Kahlenberg, *J. Phys. Chem.* **6**, 1 (1902).

752. A. Katchalsky and H. Eisenberg, *J. Polymer Sci.* **6**, 145 (1951).

753. A. Katchalsky and P. Spitnik, *J. Polymer Sci.* **2**, 432, 487 (1947).

754. B. Katchman and A. D. McLaren, *J. Amer. Chem. Soc.* **73**, 2124 (1951).

755. F. Katheder, *Kolloid Z.* **92**, 299 (1940).

756. M. Kato, T. Nakagawa, and H. Akamatu, *Bull. Chem. Soc. Japan* **33**, 322 (1960).

757. H. Käuffer and G. Scheibe, *Z. Elektrochem.* **59**, 584 (1955).

758. M. V. Kaulgud and K. J. Patil, in press.

759. W. Kauzmann, *Advan. Protein Chem.* **14**, 1 (1959).
760. R. L. Kay and D. F. Evans, *J. Phys. Chem.* **70**, 2325 (1966).
761. P. Kebarle, *J. Chem. Phys.* **53**, 2129 (1970).
762. E. Keh, *Compt. Rend. Ser. C* **272**, 1441 (1971).
763. W. Kern, *Z. Physik. Chem. (Leipzig)* **A181**, 249 (1937/8).
764. R. J. Kern and R. J. Slocombe, *J. Polymer Sci.* **15**, 183 (1955).
765. L. W. Kessler and F. Dunn, *J. Phys. Chem.* **73**, 4256 (1969).
766. H. C. Kiefer, W. I. Congdon, I. S. Scarpa, and I. M. Klotz, *Proc. Nat. Acad. Sci. U.S.* **69**, 2155 (1972).
767. Y. Kim, unpublished observations; see "Structure and Stability of Biological Macromolecules" (S. Timasheff and F. D. Fasman, eds.), Marcel Dekker, New York (1969), p. 82.
768. P. A. King and J. A. Ward, *J. Polymer Sci. Part A-1* **8**, 253 (1970).
769. B. Kingston and M. C. R. Symons, *J. Chem. Soc. Faraday II* **69**, 978 (1973).
770. K. Kinoshita, H. Ishikawa, and K. Shinoda, *Bull. Chem. Soc. Japan* **31**, 1081 (1958).
771. J. G. Kirkwood and F. P. Buff, *J. Chem. Phys.* **19**, 774 (1951).
772. J. G. Kirkwood and F. H. Westheimer, *J. Chem. Phys.* **6**, 506 (1938).
773. E. Kissa, *Text. Res. J.* **39**, 734 (1969).
774. A. Kitahara, *in* "Nonionic Surfactants" (M. J. Schick, ed.), Marcel Dekker, New York (1967), p. 289.
775. A. Kitahara, K. Watanabe, K. Kon-No, and T. Ishikawa, *J. Colloid Interface Sci.* **29**, 48 (1969).
776. A. I. Kitaygorodsky and K. V. Mirskaya, *Tetrahedron* **9**, 183 (1960).
777. D. Kivenson, *J. Chem. Phys.* **33**, 1094 (1960).
778. M. H. Klapper, *Biochim. Biophys. Acta* **229**, 557 (1971).
779. H. B. Klevens, *J. Phys. Chem.* **52**, 130 (1948).
780. A. Kliman, *Anesthesiology* **27**, 417 (1966).
781. N. A. Klimenko, A. Koganovskii, A. E. Nesterov, and V. I. Kofanov, *Chem. Abst.* **76**, 129201 (1971).
782. G. M. Kline, *Modern Plastics* 157 (1945).
783. I. M. Klotz, "Chemical Thermodynamics," Benjamin, New York (1964), p. 243.
784. I. M. Klotz, *Fed. Proc.* **24**, 5 (1965).
785. I. M. Klotz, *Fed. Proc.* **24**, S24 (1966).
786. I. M. Klotz, *J. Amer. Chem. Soc.* **91**, 5885 (1969).
787. I. M. Klotz and S. B. Farnham, *Biochem.* **7**, 3879 (1968).
788. I. M. Klotz and J. S. Franzen, *J. Amer. Chem. Soc.* **84**, 3461 (1962).
789. I. M. Klotz, G. P. Royer, and I. S. Scarpa, *Proc. Nat. Acad. Sci. U.S.* **68**, 263 (1971).
790. I. M. Klotz, G. P. Royer, and A. R. Sloniewsky, *Biochem.* **8**, 4752 (1969).
791. I. M. Klotz and J. W. Russell, *J. Phys. Chem.* **65**, 1274 (1961).
792. I. M. Klotz and K. Shikama, *Arch. Biochem. Biophys.* **123**, 551 (1968).
793. I. M. Klotz, E. C. Stellwagen, and V. H. Stryker, *Biochim. Biophys. Acta* **86**, 122 (1964).
794. I. M. Klotz and V. H. Stryker, *J. Amer. Chem. Soc.* **82**, 5169 (1960).
795. H. Klump and T. Ackermann, *Biopolymers* **10**, 513 (1971).
796. E. Knecht and J. P. Batey, *J. Soc. Dyers and Col.* **26**, 4 (1910).
797. W. S. Knight, Ph. D. Thesis, Princeton University, 1962.

798. R. Kohn and I. Furda, *in* "Solution Properties of Natural Polymers" (C. T. Greenwood, ed.), The Chemical Society, Special Publication No. 23, London (1968), p. 283.

799. J.Konicek and I. Wadsö, *Acta Chem. Scand.* **24**, 1571 (1971).

800. K. Konno and A. Kitahara, *J. Colloid Interface Sci.* **35**, 409 (1971).

801. A. Kornberg, I. R. Lechman, M. J. Bessman, and E. S. Simms, *Biochim. Biophys. Acta* **21**, 197 (1956).

802. G. Körtum, *Z. Physik. Chem.* **B30**, 317 (1935).

803. G. Körtum and M. Seiler, *Angew. Chem.* **52**, 687 (1939).

804. H. B. Kostenbauder, H. G. Boxenbaum, and P. P. Deluca, *J. Pharm. Sci.* **58**, 753 (1969).

805. N. Z. Kostova and Z. N. Markina, *Kolloid. Zh.* **33**, 551 (1971).

806. N. Z. Kostova, Z. N. Markina, P. A. Rebinder, and A. E. Kuzimina, *Kolloid. Zh.* **33**, 79 (1971).

807. A. Kotera, F. Furusawa, and Y. Takeda, *Kolloid-Z.* **239**, 677 (1970).

808. L. Kotin, *J. Mol. Biol.* **7**, 309 (1963).

809. J. J. Kozak, Ph. D. Thesis, Princeton University, 1966.

810. J. J. Kozak, W. S. Knight, and W. Kauzmann, *J. Chem. Phys.* **48**, 675 (1968).

811. H. Krakauer and J. M. Sturtevant, *Biopolymers* **6**, 491 (1968).

812. K. S. Krasnov and G. W. Shilova, *Izv. Vyssh. Ucheb. Zaved. Khim. Khim. Teknol.* **8**, 915 (1965).

813. O. Kratky, I. Pilz, and H. Ledwinka, *Mh. Chem.* **98**, 227 (1967).

814. J. Kraut, *Acta Cryst.* **14**, 1146 (1961).

815. G. C. Kresheck, *J. Phys. Chem.* **73**, 2441 (1969).

816. G. C. Kresheck and L. Benjamin, *J. Phys. Chem.* **68**, 2476 (1964).

817. G. C. Kresheck, E. Hamori, G. Davenport, and H. A. Scheraga, *J. Amer. Chem. Soc.* **88**, 246 (1966).

818. G. C. Kresheck and I. M. Klotz, *Biochem.* **8**, 8 (1969).

819. G. C. Kresheck, H. Schneider, and H. A. Scheraga, *J. Phys. Chem.* **69**, 3132 (1965).

820. C. V. Krishnan and H. L. Friedman, *J. Phys. Chem.* **73**, 1572 (1969).

821. C. V. Krishnan and H. L. Friedman, *J. Soln. Chem.* (in press).

822. T. D. Kroner and W. Tabroff, *J. Amer. Chem. Soc.* **77**, 3356 (1955).

823. K. Kuhn and E. Zimmer, *Z. Naturforsch.* **166**, 648 (1961).

824. T. Kumikatso and K. Ysujii, *J. Colloid Interface Sci.* **41**, 343 (1972).

825. C. A. Kumins and T. K. Kwei, *in* "Diffusion in Polymers" (J. Crank and G. S. Park, eds.), Academic Press, London and New York (1968), Chapter 4.

826. R. Kunin, E. Meitzner, and N. Bortnick, *J. Amer. Chem. Soc.* **84**, 305 (1962).

827. T. Kunitake and S. Shinkai, *Makromol. Chemie* **151**, 127 (1972), and earlier papers in this series.

828. I. D. Kuntz, Jr., *J. Amer. Chem. Soc.* **93**, 514 (1971).

828a. I. D. Kuntz, Jr., *J. Am. Chem. Soc.* **93**, 516 (1971).

829. I. D. Kuntz, Jr. and T. S. Brassfield, *Arch. Biochem. Biophys.* **142**, 660 (1971).

830. I. D. Kuntz, Jr., T. S. Brassfield, G. D. Law, and G. V. Purcell, *Science* **163**, 1329 (1969).

831. I. D. Kuntz, Jr., J. P. White, and C. R. Cantor, *J. Mol. Biol.* **64**, 511 (1972).

832. D. W. Kupke, *Fed. Proc.* **25**, 990 (1966).

833. D. W. Kupke, M. G. Hodgins, and J. W. Beams, *Proc. Nat. Acad. Sci. U.S.* **69**, 2258 (1972).

834. M. Kurata, M. Iwama, and K. Kamada, *in* "Polymer Handbook" (J. Brandrup and E. H. Immergut, eds.), Wiley–Interscience, New York (1966), p. IV-1.
835. M. Kurata and. W. H. Stockmayer, *Fortschr. Hochpolym.-Forsch.* **3**, 196 (1963).
836. K. Kuriyama, *Kolloid-Z.* **181**, 144 (1962).
837. J. Kurtz and W. F. Harrington, *J. Mol. Biol.* **17**, 440 (1966).
838. T. Kurucsev and B. J. Steel, *Rev. Pure Appl. Chem.* **17**, 149 (1967).
839. K. Kusano, J. Suurkunsk, and I. Wadsö (in press).
840. J. L. Lach, K. Ravel, and S. M. Blaug, *J. Amer. Pharmaceut. Assoc., Sci. Ed.* **46**, 615 (1957).
841. B. D. Ladbrooke and D. Chapman, *Chem. Phys. Lipids* **3**, 304 (1969).
842. B. D. Ladbrooke, R. M. Williams, and D. Chapman, *Biochim. Biophys. Acta* **150**, 333 (1968).
843. H. R. Laesser and H. G. Elias, *Kolloid-Z.* **250**, 58 (1972).
844. G. LaForce and B. Sarthz, *J. Colloid Interface Sci.* **37**, 254 (1971).
845. N. Laiken and G. Nemethy, *J. Phys. Chem.* **74**, 3501 (1970).
846. N. Lakshminarayanaiah, *Chem. Rev.* **65**, 491 (1965).
847. M. Lal, *RIC Reviews* **4**, 97 (1971).
848. R. F. Lama and B. C. Y. Lu, *J. Chem. Eng. Data* **10**, 216 (1965).
849. D. Lang, H. Bujard, B. Wolff, and D. Russell, *J. Mol. Biol.* **23**, 163 (1967).
850. J. Lang, J. J. Auborn, and E. M. Eyring, *J. Colloid Interface Sci.* **41**, 484 (1972).
851. J. Lang and E. M. Eyring, *J. Polym. Sci. Part A-2* **10**, 89 (1972).
852. H. Lange, *Kolloid-Z.* **243**, 101 (1971).
853. H. Lange and M. S. Schwuger, *Kolloid-Z.* **223**, 145 (1968).
854. K. Larsson, *Acta Chem. Scand.* **20**, 2255 (1966).
855. K. Larsson, *Zeitschrift Phys. Chem. (Neue Folge)* **56**, 173 (1967).
856. K. Larsson, *Chem. Phys. Lipids* **9**, 181 (1972).
857. K. Larsson and N. Krog, *Chem. Phys. Lipids* **10**, 177 (1973).
858. A. S. C. Lawrence, *Molecular Crystals and Liquid Crystals* **7**, 1 (1969).
859. A. S. C. Lawrence, *J. Soc. Cosmet. Chem.* **22**, 505 (1971).
860. A. S. C. Lawrence and M. P. McDonald, *Molecular Crystals* **1**, 205 (1966).
861. K. D. Lawson and J. T. Flautt, *J. Phys. Chem.* **72**, 2066 (1968).
862. H. W. Leach, *in* "Starch: Chemistry and Technology" (R. L. Whistler and E. F. Paschall, eds.), Vol. 1, Academic Press (1965).
863. J. B. Leathes, *Lancet*, 853, 957, 1019 (1925).
864. H. Lecuyer and D. G. Dervichian, *Kolloid. Z.* **197**, 115 (1964).
865. H. Lecuyer and D. G. Dervichian, *J. Mol. Biol.* **45**, 39 (1969).
866. P. A. Leduc and J. E. Desnoyers, *Can. J. Chem.* **51**, 2993 (1973).
867. A. G. Lee, N. J. M. Birdsall, Y. K. Levine, and J. C. Metcalfe, *Biochim. Biophys. Acta* **255**, 43 (1972).
868. C. M. Lee and W. D. Kumler, *J. Amer. Chem. Soc.* **83**, 4593 (1961).
869. H. L. Lee and M. L. Swartz, *in* "Adhesion in Biological Systems" (R. S. Manly, ed.), Academic Press, New York and London (1970), Chapter 17.
870. J. A. Leermakers, B. H. Carroll, and C. J. Staud, *J. Chem. Phys.* **5**, 878 (1937).
871. J. E. Leffler and E. Grunwald, "Rates and Equilibria of Organic Reactions as Treated by Statistical, Thermodynamic, and Extrathermodynamic Methods," New York and London (1963).
872. R. V. Lemieux and J. D. Stevens, *Can. J. Chem.* **44**, 249 (1966).
873. D. R. Lemin and T. Vickerstaff, *Trans. Faraday Soc.* **43**, 491 (1947).

874. S. Lenher and J. E. Smith, *J. Amer. Chem. Soc.* **57**, 497, 504 (1935).

875. S. Lenher and J. E. Smith, *J. Phys. Chem.* **40**, 1005 (1936).

876. F. Leonard, *in* "Adhesion in Biological Systems" (R. S. Manly, ed.) Academic Press, New York and London (1970), Chapter 11.

877. Y. K. Levine, A. I. Bailey, and M. F. H. Wilkins, *Nature* **220**, 577 (1968).

878. G. S. Levinson, W. T. Simpson, and W. Curtis, *J. Amer. Chem. Soc.* **79**, 4314 (1957).

879. L. V. Levshin and E. G. Baranova, *Opt. Spectros.* (*USSR*) **6**, 31 (1959).

880. L. V. Levshin and V. K. Gorshkov, *Opt. Spectros.* (*USSR*) **10**, 401 (1960).

881. V. L. Levshin and I. S. Lonskaya, *Opt. Spectros.* (*USSR*) **11**, 148 (1961).

882. L. V. Levshin and T. D. Slavnova, *Zh. Prikl. Spektrosk.* **7**, 234 (1967).

883. G. B. Levy, I. Caldas, and D. Fergus, *Anal. Chem.* **24**, 1799 (1952).

884. O. Levy, G. Markovits, and A. S. Kertes, *J. Phys. Chem.* **75**, 542 (1971).

885. S. Lewin, *J. Theoret. Biol.* **17**, 181 (1967).

886. S. Lewin, *Nature New Biology* **231**, 80 (1971).

887. G. N. Lewis, O. Goldschmid, T. T. Magel, and J. Bigeleisen, *J. Amer. Chem. Soc.* **65**, 1150 (1943).

888. J. C. Leyte, H. M. R. Arbouw-van der Veen, and L. H. Zuiderweg, *J. Phys. Chem.* **76**, 2559 (1972).

889. J. C. Leyte and M. Mandel, *J. Polymer Sci. Part A* **2** 1879 (1964).

890. L. J. Libertini, A. S. Waggoner, P. C. Jost, and O. H. Griffith, *Proc. Nat. Acad. Sci. U.S.* **64**, 13 (1969).

891. H. A. Liebhafsky and H. G. Pfeiffer, *J. Chem. Educ.* **30**, 450 (1953).

892. S. Lifson and A. Roig, *J. Chem. Phys.* **34**, 1963 (1961).

893. G. I. Likhtenshtein, *Biofizika* **11**, 23 (1966).

894. P. J. Lillford, to be published.

895. P. J. Lillford and S. Ablett, to be published.

896. G. N. Lin, *in* "Quantum Aspects of Polypeptides and Polynucleotides" (M. Weissbluth, ed.), Interscience, New York (1964), p. 114.

897. I. J. Lin, Is. *J. Technol.* **9**, 621 (1971).

898. I. J. Lin, *Trans. Soc. Mining Eng.*, *AIME* **250**, 225 (1971).

899. I. J. Lin and P. Somasundaran, *J. Colloid Interface Sci.* **37**, 731 (1971).

900. G. Lindblom, B. Lindman, and L. Mandell, *J. Colloid Interface Sci.* **34**, 262 (1970).

901. S. Lindenbaum, *J. Phys. Chem.* **75**, 3733 (1971).

902. S. Lindenbaum and G. E. Boyd, *J. Phys. Chem.* **68**, 911 (1964).

903. B. Lindman and I. Danielsson, *J. Colloid Interface Sci.* **39**, 349 (1972).

904. B. Lindman, S. Forsen, and E. Forslind, *J. Phys. Chem.* **72**, 2805 (1968).

905. A. M. Liquori, G. Barone, V. Crescenzi, F. Quadrifoglio, and V. Vitagliano, *J. Macromol. Chem.* **1**, 291 (1966).

906. A. H. Little and J. W. Clayton, *J. Soc. Dyers & Col.* **79**, 671 (1963).

907. K. Little, *New Scientist* **42**, 118 (1969).

908. R. C. Little and C. R. Singleterry, *J. Phys. Chem.* **68**, 3453 (1964).

909. K.-J. Liu and J. E. Anderson, *Macromolecules* **2**, 235 (1969).

910. K.-J. Liu and J. L. Parsons, *Macromolecules* **2**, 529 (1969).

911. G. Livingstone *et al.* (to be published).

912. G. Loeb and H. A. Scheraga, *J. Amer. Chem. Soc.* **84**, 134 (1962).

913. H. W. Loeb, G. M. Young, P. A. Quickenden, and A. Suggett, *Ber. Bunsenges. Phys. Chem.* **75**, 1115 (1971).

914. F. A. Loewus and D. R. Briggs, *J. Amer. Chem. Soc.* **79**, 1494 (1957).
915. F. A. Long and W. F. McDevit, *Chem. Rev.* **51**, 119 (1952).
916. N. Lotan, A. Yaron, and A. Berger, *Biopolymers* **4**, 365 (1966).
917. R. Lovrien and J. M. Sturtevant, *Biochemistry* **10**, 3811 (1971).
918. M. J. Lowe and J. A. Schellman, *J. Mol. Biol.* **65**, 91 (1972).
919. B. Lubas and T. Wilczok, *Biochem. Biophys. Acta* **120**, 427 (1965).
920. B. Lubas and T. Wilczok, *Biochem. Biophys. Acta* **224**, 1 (1970).
921. B. Lubas and T. Wilczok, *Biopolymers* **10**, 1267 (1971).
922. B. Lubas, T. Wilczok, and O. K. Daskiewicz, *Biopolymers* **5**, 967 (1967).
923. M. Lucas, *Bull. Soc. Chim. France* 2902 (1970).
924. G. Luck, C. Zimmer, G. Snatzke, and G. Sondergrath, *Europ. J. Biochem.* **17**, 514 (1970).
925. R. Lumry, R. L. Biltonen, and J. F. Brandts, *Biopolymers* **4**, 917 (1966).
926. R. Lumry, R. Legare, and W. G. Miller, *Biopolymers* **2**, 489 (1964).
927. R. Lumry and S. Rajender, *Biopolymers* **9**, 1125 (1970).
928. N. Lupu-Lotan, A. Yaron, A. Berger, and M. Sela, *Biopolymers* **3**, 625 (1965).
929. E. S. Lutton, *J. Amer. Oil Chem. Soc.* **27**, 276 (1950).
930. E. S. Lutton, *J. Amer. Oil Chem. Soc.* **42**, 1068 (1965).
931. V. Luzzati, *in* "Biological Membranes" (D. Chapman, ed.), Academic Press, London and New York (1968), Vol. 1, p, 71.
932. V. Luzzati, T. Gulik-Krzywicki, and A. Tardieu, *Nature* **218**, 1031 (1968).
933. D. J. Lyman, *Trans. Amer. Soc. Artif. Int. Organs* **10**, 17 (1964).
934. D. J. Lyman, B. H. Loo, and R. W. Crawford, *Biochem.* **3**, 985 (1964).
935. D. J. Lyman, B. H. Loo, and W. M. Muir, *Trans. Amer. Soc. Artif. Int. Organs* **11**, 91 (1965).
936. J. W. Lyons and L. Kotin, *J. Amer. Chem. Soc.* **86**, 3634 (1964).
937. C. McAuliffe, *Science* **163**, 478 (1969).
938. J. W. McBain and T. H. Liu, *J. Amer. Chem. Soc.* **53**, 59 (1931).
939. W. C. McCabe and H. F. Fisher, *Nature* **207**, 1274 (1965).
940. W. C. McCabe and H. F. Fisher, *J. Phys. Chem.* **74**, 2990 (1970).
941. W. H. McCarthy, *Austral. & New Zeal. J. Surg.* **39**, 422 (1970).
942. P. E. McClain and E. R. Wiley, *J. Biol. Chem.* **247**, 692 (1972).
943. W. F. McDevit and F. A. Long, *J. Amer. Chem. Soc.* **74**, 1773 (1952).
944. C. McDonald, *J. Pharm. Pharmacol.* **22**, 10 (1970).
944a. C. McDonald, *J. Pharm. Pharmacol.* **22**, 774 (1970).
945. C. C. McDonald, W. D. Phillips, and J. D. Glickson, *J. Amer. Chem. Soc.* **93**, 235 (1971).
946. M. P. McDonald and W. E. Peel, *Trans. Faraday Soc.* **67**, 890 (1971).
947. W. McDowall and R. Weingarten, *J. Soc. Dyers & Col.* **85**, 589 (1969).
948. B. G. McFarland and H. M. McConnell, *Proc. Nat. Acad. Sci. U.S.* **68**, 1274 (1971).
949. C. W. McGary, *J. Polymer Sci.* **46**, 51 (1960).
950. R. B. McKay, *Trans. Faraday Soc.* **61**, 1787 (1965).
951. R. B. McKay and P. J. Hillson, *Trans. Faraday Soc.* **61**, 1800 (1965).
952. K. McLaren, *J. Soc. Dyers and Col.* **79**, 618 (1963).
953. W. McMillan and J. Mayer, *J. Chem. Phys.* **13**, 176 (1945).
954. D. W. McMullen, S. R. Jaskunas, and J. R. Tinoco, *Biopolymers* **5**, 589 (1967).
955. E. G. McRae and M. Kasha, *J. Chem. Phys.* **28**, 721 (1958).

956. E. G. McRae and M. Kasha, *in* "Physical Processes in Radiation Biology" (L. Augenstein, R. Mason, and B. Rosenberg, eds.), Academic Press, New York (1964), pp. 23–42.

957. F. MacRitchie, *Trans. Faraday Soc.* **65**, 2503 (1969).

958. H. Maeda, T. Kawai, and S. Seki, *Kobunshi Kagaku* **15**, 719 (1958).

959. B. Magasanik and E. Chargaff, *J. Biol. Chem.* **174** (1948).

960. D. M. Maharajh and J. Walkley, *J. Chem. Soc. Faraday I* **69**, 842 (1973).

961. G. N. Malcolm and J. S. Rowlinson, *Trans. Faraday Soc.* **53**, 921 (1957).

962. W. U. Malik and P. Chand, *J. Electrounal. Chem.* **19**, 431 (1968).

963. W. U. Malik and O. P. Jhamb, *Kolloid-Z.* **242**, 1209 (1970).

964. W. U. Malik and O. P. Jhamb, *J. Amer. Oil Chem. Soc.* **49**, 170 (1972).

965. W. U. Malik and M. Saleen, *J. Oil Col. Chem. Assoc.* **52**, 551 (1969).

966. W. U. Malik, S. K. Srivastava, and D. Gupta. *J. Electro-anal. Chem.* **34**, 540 (1972).

967. W. U. Malik and S. P. Verma, *Kolloid-Z.* **233**, 985 (1969).

968. T. Malkin, *in* "Progress in the Chemistry of Fats and Other Lipids" (R. T. Holman, ed.), Vol. 1, Pergamon Press, Oxford (1952).

969. T. Malkin, *in* "Progress in the Chemistry of Fats and Other Lipids" (R. T. Holman, ed.), Vol. 2, Pergamon Press, Oxford (1964).

970. T. Malkin and M. R. Shurbagy, *J. Chem. Soc.* 1628 (1936).

971. M. Mandel, J. C. Leyte, and M. G. Stadhouder, *J. Phys. Chem.* **71**, 603 (1967).

972 M. Mandel and M. G. Stadhouder, *Makromol. Chemie* **80**, 141 (1964).

973. L. Mandelkern and M. H. Liberman, *J. Phys. Chem.* **71**, 1163 (1967).

974. L. Mandelkern and D. E. Roberts, *J. Amer. Chem. Soc.* **83**, 4292 (1961).

975. D. Mangaraj, *Makromol. Chemie* **65**, 29 (1963).

976. D. A. Mangaraj, S. Patra, and S. Rashid, *Makromol. Chemie* **65**, 39 (1963).

977. C. E. Mangels and C. H. Bailey, *J. Amer. Chem. Soc.* **55**, 1981 (1933).

978. G. Manning, *J. Chem. Phys.* **51**, 924 (1969).

979. G. S. Manning, *Biopolymers* **11**, 937 (1972).

980. P. C. Manor and W. Saenger, *Nature* **237**, 392 (1972).

981. J. A. Marinsky (ed.), "Ion Exchange," Vol. 1, Arnold, London, and Dekker, New York (1966).

982. H. F. Mark, N. G. Gaylord, and N. M. Bikales (eds.), "Encyclopaedia of Polymer Science and Technology," Vols. 1–16, Wiley–Interscience, New York (1964–1972).

983. J. T. Martin and H. A. Standing, *J. Text. Inst.* **40**, 671, 689 (1949).

984. M. D. Maser and R. V. Rice, *Biochim. Biophys. Acta* **74**, 283 (1963).

985. S. F. Mason, *Proc. Chem. Soc.* **119** (1964).

986. S. F. Mason, *J. Soc. Dyers and Col.* **84**, 604 (1968).

987. W. L. Masterton, *J. Chem. Phys.* **22**, 1830 (1954).

988 W. L. Masterton and T. P. Lee, *J. Phys. Chem.* **74**, 1776 (1970).

989. Y. Masuda, T. Hasegawa, and Y. Miyahara, *Nippon Kagaku Zasshi* **82**, 1131 (1961).

990. M. Masuzawa and C. Sterling, *Biopolymers* **6**, 1453 (1968).

991. W. K. Mathews, J. W. Larsen, and M. J. Pikal, *Tetrahedron Letters* **6**, 513 (1972).

992. J. Matouš, J. Hrnčirík, J. P. Novák, and J. Šobr, *Coll. Czech. Chem. Comm.* **35**, 1904 (1970).

993. J. Matouš, J. P. Novák, J. Šobr, and J. Pick, *Coll. Czech. Chem. Comm.* **37**, 2653 (1972).

994. T. Matsumoto, in "Adhesion in Biological Systems" (R. S. Manly, ed.), Academic Press, New York and London (1970), Chapter 13.
995. W. L. Mattice and L. Mandelkern. Biochem. 9, 1049 (1970).
996. P. Meares, in "Diffusion in Polymers" (J. Crank and G. S. Park, eds.), Academic Press, London and New York (1968), Chapter 10.
997. A. B. Meggy, J. Soc. Dyers and Col. 66, 510 (1950).
998. K. N. Mehrotra, V. P. Mehta, and T. N. Nagar, J. Prakt. Chem. 312, 545 (1970).
999. K. N. Mehrotra, V. P. Mehta, and T. N. Nagar, J. Prakt. Chem. 313, 607 (1971).
1000. G. J. H. Melrose, Rev. Pure and Appl. Chem. 21, 83 (1971).
1001. H. W. Melville and P. R. Sewell, Makromol. Chemie 32, 139 (1959).
1002. F. M. Menger and C. E. Portnoy, J. Amer. Chem. Soc. 90, 1875 (1968).
1003. A. T. Mennie (ed.) "Polyglycolic Acid Sutures," Symposium Report, Davis and Geck (Cyanamid of Great Britain Ltd.), London (1972).
1004. R. B. Merrifield, J. Amer. Chem. Soc. 85, 2149 (1963).
1005. J. C. Metcalfe, N. J. M. Birdsall, J. Feeney, A. G. Lee, Y. K. Levine, and P. Partington, Nature 233, 199 (1971).
1006. A. Metzer and I. J. Lin, J, Phys. Chem. 75, 3000 (1971).
1007. J. A. Meyer, C. Rogers, V. Stannett, and M. Szwarc, Tappi 40, 142 (1957).
1008. M. C. Meyer and D. E. Guttman, J. Pharm. Sci. 57, 895 (1968).
1009. A. S. Michaels and R. G. Miekka, J. Phys. Chem. 65, 1765 (1961).
1010. P. F. Mijnlieff, J. Colloid Interface Sci. 33, 255 (1970).
1011. P. F. Mijnlieff and R. Ditmarsch, Nature 208, 889 (1965).
1012. V. A. Mikhailov, J. Struct. Chem. 9, 332 (1968).
1013. V. A. Mikhailov and L. I. Ponomarova, J. Struct. Chem. 9, 8 (1968).
1014. B. Milićević and G. Eigenmann, Helv. Chim. Acta 47, 1039 (1964).
1015. J. R. Millar, D. G. Smith, W. E. Marr, and T. R. E. Kressman, J. Chem. Soc. 218 (1963).
1016. J. R. Millar, D. G. Smith, W. E. Marr, and T. R. E. Kressman, J. Chem. Soc. 2779 (1963).
1017. J. R. Millar, D. G. Smith, W. E. Marr, and T. R. E. Kressman, J. Chem. Soc. 2740 (1964).
1018. J. H. Miller and H. M. Sobell, J. Mol. Biol. 24, 345 (1967).
1019. T. A. Miller and R. N. Adams, J. Amer. Chem. Soc. 88, 5713 (1966).
1020. W. G. Miller and R. E. Nylund, J. Amer. Chem. Soc. 87, 3542 (1965).
1021. F. Millero, Chem. Rev. 71, 147 (1971).
1022. D. C. Miln, J. O'Connor, and R. Dalling, Scot. Med. J. 17, 108 (1972).
1023. J. Mingins and M. M. Standish, Ann. Rep. Prog. Chem. Soc., London 63, 91 (1966).
1024. R. L. Misiorowski and M. A. Wells, Biochemistry 12, 967 (1973).
1025. S. Mitnar, Proc. Nat. Acad. Sci. India 20A, 222 (1951).
1026. M. Miura and M. Kodama, Bull. Chem. Soc. Japan 45, 428 (1972) and references cited therein.
1027. G. M. Miyawaki, N. K. Patel, and H. B. Kostenbauder, J. Amer. Pharmaceut. Assoc. Sci., Ed. 48, 315 (1959).
1028. W. Moffitt and J. T. Yang, Proc. Natl. Acad. Sci. U.S. 42, 596 (1956).
1029. A. Mohammedzadeh-K, R. E. Feeney, R. B. Samuels, and L. M. Smith, Biochim. Biophys. Acta 147, 583 (1967).

1030. P. Molyneux, *in* "The Chemistry and Rheology of Water Soluble Gums and Colloids" (G. Stainsby, ed.), Society of Chemical Industry, Monograph No. 24, London (1966), p. 91.

1031. P. Molyneux and H. P. Frank, *J. Amer. Chem. Soc.* **83**, 3169 (1961).

1032. P. Molyneux and H. P. Frank, *J. Amer. Chem. Soc.* **83**, 3175 (1961).

1033. P. Molyneux and H. P. Frank, *J. Amer. Chem. Soc.* **86**, 4753 (1964).

1034. P. Molyneux, C. T. Rhodes, and J. Swarbrick, *Trans. Faraday Soc.* **61**, 1043 (1965).

1035. A. R. Monahan and D. F. Blossey, *J. Phys. Chem.* **74**, 4014 (1970).

1036. P. Monk and I. Wadsö, *Acta Chem. Scand.* **22**, 1842 (1968).

1037. W. H. Montgomery, *in* "Water-Soluble Resins" (R. L. Davidson and M. Sittig, eds.) 2nd ed., Reinhold, New York (1968), Chapter 9.

1038. H. Morawetz, "Macromolecules in Solution," Wiley–Interscience, New York and London (1965).

1039. H. Morawetz, *Fortschr. Hochpolym.-Forsch.* **1**, 1 (1958).

1040. H. Morawetz and A. Y. Kandanian, *J. Phys. Chem.* **70**, 2995 (1966).

1041. K. W. Morcom and R. W. Smith, *Trans. Faraday Soc.* **66**, 1073 (1970).

1042. J. Morgan and B. E. Warren, *J. Chem. Phys.* **6**, 666 (1938).

1043. E. R. Morris, D. A. Rees, and D. Thom, *J. Chem. Soc. Chem. Comm.* 245 (1973).

1044. T. J. Morrison and F. Billet, *J. Chem. Soc.* (1952), 3819.

1045. T. H. Morton, *Trans. Faraday Soc.* **31**, 262 (1935).

1046. W. E. Morton and J. W. S. Hearle, "Physical Properties of Textile Fibres," The Textile Institute and Butterworths, Manchester and London (1962).

1047. P. Mukerjee, *J. Colloid Sci.* **19**, 722 (1964).

1048. P. Mukerjee, *J. Phys. Chem.* **69**, 4038 (1965).

1049. P. Mukerjee, *Adv. Colloid Interface Sci.* **1**, 241 (1967).

1050. P. Mukerjee, *J. Phys. Chem.* **73**, 2054 (1969).

1051. P. Mukerjee, *Kolloid-Z.* **236**, 76 (1970).

1052. P. Mukerjee, *J. Pharm. Sci.* **60**, 1528 (1971).

1053. P. Mukerjee, *J. Pharm. Sci.* **60**, 1531 (1971).

1054. P. Mukerjee, *J. Phys. Chem.* **76**, 565 (1972).

1055. P. Mukerjee and K. Banerjee, *J. Phys. Chem.* **68**, 3567 (1964).

1056. P. Mukerjee and A. K. Ghosh, *J. Phys. Chem.* **67**, 193 (1963).

1057. P. Mukerjee and A. K. Ghosh, *J. Amer. Chem. Soc.* **92**, 6403 (1970).

1058. P. Mukerjee and A. K. Ghosh, *J. Amer. Chem. Soc.* **92**, 6419 (1970).

1059. P. Mukerjee, P. Kapauan, and H. G. Meyer, *J. Phys. Chem.* **70**, 783 (1966).

1060. P. Mukerjee and K. J. Mysels, "Critical Micelle Concentrations of Aqueous Surfactant Systems NSRDS—NBS 36" Superintendent of Documents, Washington, D.C. (1971).

1061. P. Mukerjee, K. J. Mysels, and C. I. Dulin, *J. Phys. Chem.* **62**, 1390 (1958).

1062. P. Mukerjee, K. J. Mysels, and P. Kapauan, *J. Phys. Chem.* **71**, 4166 (1967).

1063. P. Mukerjee, J. Perrin, and E. Witzke, *J. Pharm. Sci.* **59**, 1513 (1970).

1064. P. Mukerjee and A. Ray, *J. Phys. Chem.* **67**, 190 (1963).

1065. P. Mukerjee and A. Ray, *J. Phys. Chem.* **70**, 2150 (1966).

1066. A. Müller, *Proc. Roy. Soc. A* **114**, 542 (1927).

1067. N. Muller, *in* "Reaction Kinetics in Micelles," *Amer. Chem. Soc. Symp.* (E. Cordes, ed.), Plenum, New York (1953), p. 1.

1068. N. Muller, *J. Phys. Chem.* **76**, 3017 (1972).

1069. N. Muller and R. H. Birkhahn, *J. Phys. Chem.* **71**, 957 (1967).

1070. N. Muller and R. H. Birkhahn, *J. Phys. Chem.* **72**, 583 (1968).

1071. N. Muller and T. W. Johnson, *J. Phys. Chem.* **73**, 2042 (1969).

1072. N. Muller, J. H. Pellerin, and W. W. Chen, *J. Phys. Chem.* **76**, 3012 (1972).

1073. N. Muller and F. E. Platko, *J. Phys. Chem.* **75**, 547 (1971).

1074. N. Muller and N. Simsohn, *J. Phys. Chem.* **75**, 942 (1971).

1075. A. W. Myers, J. A. Meyer, C. E. Rogers, V. Stannett, and M. Szwarc, *Tappi* **44**, 58 (1961).

1076. E. K. Mysels and K. J. Mysels, *J. Colloid Interface Sci.* **38**, 388 (1972).

1077. M. Nagasawa and S. A. Rice, *J. Amer. Chem. Soc.* **82**, 5070 (1960).

1078. T. Nakagawa, *in* "Cationic Surfactants" (E. Jungermann, ed.), Marcel Dekker, New York (1970).

1079. T. Nakagawa and K. Shinoda, *in* "Colloidal Surfactants" (B.-I. Tamamushi and T. Isemura, eds.), Academic Press, New York (1963).

1080. K. Nakanishi, N. Kato, and N. Maruyama, *J. Phys. Chem.* **71**, 814 (1967).

1081. P. K. Nandi and D. R. Robinson, *J. Amer. Chem. Soc.* **94**, 1299 (1972).

1082. P. K. Nandi and D. R. Robinson, *J. Amer. Chem. Soc.* **94**, 1308 (1972).

1083. A. H. Narten, M. D. Danford, and H. A. Levy, *Disc. Faraday Soc.* **43**, 97 (1967).

1084. A. H. Narten and S. Lindenbaum, *J. Chem. Phys.* **51**, 1108 (1969).

1085. J. L. Neal and D. A. I. Goring, *J. Polymer Sci. C* 103, (1969).

1086. J. L. Neal and D. A. I. Goring, *Canadian J. Chem.* **48**, 3745 (1970).

1087. N. Néel and B. Sébille, *Compt. Rend.* **250**, 1052, 1270 (1960).

1088. H. D. Nelson and C. L. de Ligny, *Rec. trav. chim. Pays-Bas* **87**, 623 (1968).

1089. G. Némethy, *Angew. Chem. Internat. Ed.* **6**, 195 (1967).

1090. G. Némethy, *Ann. dell'Istituto Superiore di Sanità* **6**, 492 (1972).

1091. G. Némethy and H. A. Scheraga, *J. Chem. Phys.* **36**, 3382 (1962).

1092. G. Némethy and H. A. Scheraga, *J. Chem. Phys.* **36**, 3401 (1962).

1093. G. Némethy and H. A. Scheraga, *J. Phys. Chem.* **66**, 1773 (1962).

1094. R. A. Newmark and C. R. Cantor, *J. Amer. Chem. Soc.* **90**, 5010 (1968).

1095. Y. Nishijima and G. Oster, *J. Polymer Sci.* **19**, 337 (1956).

1096. I. Noda and I. Kagawa, *Kogyo Kagaku Zasshi* **66**, 1927 (1963).

1097. M. E. Noelken, *Biochem.* **9**, 4117 (1970).

1098. M. E. Noelken, *Biochem.* **9**, 4122 (1970).

1099. H. Noguchi and J. T. Yang, *J. Phys. Chem.* **68**, 1609 (1964).

1100. H. Nomura and Y. Miyahara, *J. Appl. Polymer Sci.* **8**, 1643 (1964).

1101. H. Nomura and Y. Miyahara, *Bull. Chem. Soc. Japan* **39**, 1599 (1966).

1102. H. Nomura, S. Yamaguchi, and Y. Miyahara, *J. Appl. Polymer Sci.* **8**, 2731 (1964).

1103. K. Norland, A. E. Ames, and T. Taylor, *Photogr. Sci. Eng.* **14**, 295 (1970).

1104. Y. Nozaki and C. Tanford, *J. Biol. Chem.* **238**, 4074 (1963).

1105. J. Oakes, *J. Chem. Soc. Faraday Trans. II* **68**, 1464 (1972).

1106. S. Ohki and O. Aono, *J. Colloid Interface Sci.* **32**, 270 (1970).

1107. H. Okabayashi, T. Isemura, and S. Sakakibara, *Biopolymers* **6**, 323 (1968).

1108. M. U. Oko and R. L. Venable, *J. Colloid Interface Sci.* **35**, 53 (1971).

1109. T. Okubo and N. Ise, *J. Phys. Chem.* **73**, 1488 (1969).

1110. T. Okubo and N. Ise, *J. Amer. Chem. Soc.* **95**, 2293 (1973).

1111. E. Oldfield, D. Chapman, and W. Derbyshire, *Febs Letters* **16**, 102 (1971).

1112. E. Oldfield, J. Marsden, and D. Chapman, *Chem. Phys. Lipids* **7**, 1 (1971).

1113. G. S. Omenn and T. J. Gill III, *J. Biol. Chem.* **241**, 4899 (1966).

1114. J. J. O'Neill, E. M. Loebl, A. Y. Kandanian, and H. Morawetz, *J. Polymer Sci. Part A* **3**, 4201 (1965).

1115. L. Onsager, *J. Amer. Chem. Soc.* **58**, 1486 (1936).

1116. P. F. Onyon, *in* "Techniques of Polymer Characterization" (P. W. Allen, ed.), Butterworths, London (1959), Chapter 6.

1117. F. Oosawa, *Biopolymers* **6**, 145 (1968).

1118. F. Oosawa, "Polyelectrolytes," Dekker, New York (1971).

1119. A. I. Oparin, "The Origin of Life on the Earth," 3rd ed., Oliver and Boyd, Edinburgh and London (1957).

1120. A. I. Oparin, K. B. Serebrovskaia, and G. I. Lozovaia, *Doklady Akad. Nauk SSSR* **179**, 1240 (1968).

1121. J. M. O'Reilly and F. E. Karasz, *Biopolymers* **9**, 1429 (1970).

1122. P. J. Oriel and E. R. Blout, *J. Amer. Chem. Soc.* **88**, 2041 (1966).

1123. D. S. Orlander, and A. Holzer, *J. Amer. Chem. Soc.* **90**, 4549 (1968).

1124. T. A. Orofino, H. B. Hopfenberg, and V. Stannett, *J. Macromol. Sci.-Phys.* **B3**, 777 (1969).

1125. G. Oster, *J. Polymer Sci.* **16**, 235 (1955).

1126. S. E. Ostroy, N. Lotan, R. T. Ingwall, and H. A. Scheraga, *Biopolymers* **9**, 749 (1970).

1127. R. H. Ottewill and J. N. Shaw, *Disc. Faraday Soc.* **42**, 154 (1966).

1128. R. J. Owen, L. R. Hill, and S. P. Lapage, *Biopolymers* **7**, 503 (1969).

1129. N. C. Pace and C. Tanford, *Biochem.* **7**, 198 (1968).

1130. A. Packter, *J. Polymer Sci. Part A* **2** 2771 (1964).

1131. A. Packter, *Kolloid-Z.* **202**, 121 (1965).

1132. J. F. Padday, *J. Phys. Chem.* **72**, 1259 (1968).

1133. M. R. Padhye and R. R. Karnik, *in* "Proc. Symp. Dept. Chem. Tech. Univ. of Bombay (1969)," p. 56.

1134. K. J. Palmer and F. O. Schmitt, *J. Cell Comp. Physiol.* **17**, 385 (1941).

1135. D. T. F. Pals and J. J. Hermans, *J. Polymer Sci.* **5**, 733 (1950).

1136. G. D. Parfitt and M. C. Smith, *Trans. Faraday Soc.* **65**, 1138 (1969).

1137. G. D. Parfitt and J. A Wood, *Trans. Faraday Soc.* **64**, 2081 (1968).

1138. C. W. Parker and C. K. Osterland, *Biochem.* **9**, 1074 (1970).

1139. R. C. Parker, L. J. Slutsky, and K. R. Applegate, *J. Phys. Chem.* **72**, 3177 (1968).

1140. F. Patat, E. Killman, and C. Schliebener, *Rubber Chem. Tech.* **39**, 36 (1966).

1141. F. Patat and K. Vogler, *Helv. Chim. Acta* **35**, 128 (1952).

1142. J. R. Patel and R. D. Patel, *Makromol. Chemie* **120**, 103 (1968).

1143. N. K. Patel and N. E. Foss, *J. Pharm. Sci.* **53**, 94 (1964).

1144. D. Pederson, D. Gabriel, and J. Hermans, Jr., *Biopolymers* **10**, 2133 (1971).

1145. E. Peggion, A. Cosani, M. Terbojevich, and G. Borin, *Biopolymers* **11**, 633 (1972).

1146. E. Peggion, A. S. Verdini, A. Cosani, and E. Scaffone, *Macromolecules* **3**, 194 (1970).

1147. L. Pelet-Jolivet and A. Wild, *Kolloid Z.* **3**, 174 (1908).

1148. L. Peller, *J. Phys. Chem.* **63**, 1199 (1959).

1149. J. F. Penin, *J. Phys.* **7**, 1 (1936).

1150. S. A. Penkett, A. G. Flook, and D. Chapman, *Chem. Phys. Lipids* **2**, 273 (1968).

1151. J. B. Peri, *J. Colloid Interface Sci.* **29**, 6 (1969).

1152. L. E. Perlut and K. Amemuja, *Arch. Biochim. Biophys.* **132**, 370 (1969).

1153. J. H. Perrin and L. Saunders, *Biochim. Biophys. Acta* **84**, 216 (1964).

1154. R. H. Peters, *in* "Diffusion in Polymers" (J. Crank and G. S. Park, eds.), Academic Press, London and New York (1968), Chapter 9.

1155. G. H. Peterson, *J. Polymer Sci.* **28**, 458 (1958).

1156. O. G. Peterson, S. A. Tucchio, and B. B. Snavely, *Appl. Phys. Lett.* **17**, 245 (1970).

1157. B. A. Pethica, *in* "Structural and Functional Aspects of Lipoproteins in Living Systems" (E. Tria and A. M. Scanu, eds.), Academic Press, London and New York (1969), p. 37.

1158. R. E. Phares, *J. Pharm. Sci.* **57**, 53 (1968).

1159. P. R. Philip and J. E. Desnoyers, *J. Soln. Chem.* **1**, 353 (1972).

1160. J. N. Philips and K. J. Mysels, *J. Phys. Chem.* **59**, 325 (1955).

1161. M. C. Phillips, *in* "Progress in Surface and Membrane Science," Academic Press, New York and London (1972), Vol. 5, p. 139.

1162. M. C. Phillips and D. Chapman, *Biochim. Biophys. Acta* **163**, 301 (1968).

1163. M. C. Phillips, E. G. Finer, and H. Hauser, *Biochim. Biophys. Acta* **290**, 397 (1972).

1164. M. C. Phillips and H. Hauser, *J. Coll. Interface Sci.* (in press) (1974).

1165. M. C. Phillips, H. Hauser, and F. Paltauf, *Chem. Phys. Lipids* **8**, 127 (1972).

1166. M. C. Phillips, B. D. Ladbrooke, and D. Chapman, *Biochim. Biophys. Acta* **196**, 35 (1970).

1167. M. C. Phillips, R. M. Williams, and D. Chapman, *Chem. Phys. Lipids* **3**, 234 (1969).

1168. V. A. Pickles, M. Sc. Thesis, University of New South Wales (1970).

1169. J. Piercy, M. N. Jones, and G. Ibbotson, *J. Colloid Interface Sci.* **37**, 165 (1971).

1170. R. A. Pierotti, *J. Phys. Chem.* **69**, 281 (1965).

1171. K. A. Piez, *J. Amer. Chem. Soc.* **82**, 247 (1960).

1172. K. A. Piez, *Biochem.* **2**, 58 (1963).

1173. K. A. Piez and A. L. Carillo, *Biochem.* **3**, 908 (1964).

1174. K. A. Piez and J. Gross, *J. Biol. Chem.* **235**, 995 (1960).

1175. K. A. Piez and M. R. Sherman, *Biochem.* **9**, 4129 (1970).

1176. K. A. Piez and M. R. Sherman, *Biochem.* **9**, 4134 (1970).

1177. G. Pilcher, M. N. Jones, L. Espada, and H. A. Skinner, *J. Chem. Thermodynamics* **1**, 381 (1969).

1178. G. C. Pimental and A. L. McClellan, *Ann. Rev. Phys. Chem.* **22**, 347 (1971).

1179. J. M. M Pinkerton, *Nature* **160**, 128 (1947).

1180. S. H. Pinner (ed.), "Weathering and Degradation of Plastics," Columbine Press, Manchester and London (1966).

1181. J. Pitha, R. N. Jones, and P. Pithova, *Can. J. Chem.* **44**, 1045 (1966).

1182. F. Podo, A. Ray, and G. Némethy, *J. Amer. Chem. Soc.* **95**, 6164 (1973).

1183. F. M. Pohl, *Europ. J. Biochem.* **4**, 373 (1968).

1184. F. M. Pohl, *Europ. J. Biochem.* **7**, 146 (1968-69).

1185. F. M. Pohl, *Febs Letters* **3**, 60 (1969).

1186. D. C. Poland and H. A. Scheraga, *Biopolymers* **3**, 283, 305 (1965).

1187. D. C. Poland and H. A. Scheraga, *Biopolymers* **3**, 401 (1965).

1188. D. C. Poland and H. A. Scheraga, *J. Phys. Chem.* **69**, 2431 (1965).

1189. G. P. Polli and B. M. Frost, *J. Pharm. Sci.* **58**, 1543 (1969).

1190. R. S. Porter and J. F. Johnson, *Chem. Rev.* **66**, 1 (1966).

1191. M. J. Povich, J. A. Mann, and A. Kawamoto, *J. Colloid Interface Sci.* **41**, 145 (1972).

1192. J. G. Pritchard, "Poly(vinyl alcohol): Basic Properties and Uses," Macdonald, London (1970).

1193. P. L. Privalov, *Biofizika* **8**, 308 (1963).

1194. P. L. Privalov, *Biofizika* **11**, 23 (1966).

1195. P. L. Privalov, N. N. Khechinashvili, and B. P. Atanasov, *Biopolymers* **10**, 1865 (1971).

1196. P. L. Privalov and D. R. Monaselidze, *Zh. Eksperim. i Teor. Fiz.* **47**, 2073 (1964).

1197. P. L. Privalov, D. R. Monaselidze, G. M. Mrevlishvili, and V. A. Magalddze, *Zh. Eksperim. i Teor. Fiz.* **47**, 2073 (1964).

1198. P. L. Privalov and G. M. Mrevlishvili, *Biofizika* **12**, 19 (1967).

1199. P. L. Privalov and O. B. Ptitsyn, *Biopolymers* **8**, 559 (1969).

1200. P. L. Privalov, I. N. Serbyuk, and E. I. Tiktopulo, *Biopolymers* **10**, 1777 (1971).

1201. P. L. Privalov and E. I. Tiktopulo, *Biopolymers* **9**, 127 (1970).

1202. O. B. Ptitsyn and T. M. Birshtein, *Biopolymers* **7**, 435 (1968).

1203. D. Pugh, C. H. Giles, and D. G. Duff, *Trans. Faraday Soc.* **67**, 563 (1971).

1204. E. Putkey and M. Sundaralingam, *Acta Cryst. B* **26**, 782 (1970).

1205. F. W. Putnam, *Adv. Protein Chem.* **4**, 79 (1948).

1206. O. Quensel, *Trans. Faraday Soc.* **31**, 259 (1935).

1207. E. Rabinowitch and L. F. Epstein, *J. Amer. Chem. Soc.* **63**, 69 (1941).

1208. S. R. Rafikov, S. A. Pavlova, and I. I. Tverdokhlebova, "Determination of Molecular Weights and Polydispersity of High Polymers," Israel Program for Scientific Translations, Jerusalem (1964).

1209. A. Rahman and F. H. Stillinger, *J. Chem. Phys.* **55**, 3336 (1971).

1210. K. S. Rajan, D. B. Boies, A. J. Casolo, and J. I. Bregman, *Desalination* **1**, 231 (1966).

1211. S. Rajender, M. H. Han, and R. Lumry, *J. Amer. Chem. Soc.* **92**, 1378 (1970).

1212. G. N. Ramachandran and R. Chandrasekharan, *Biopolymers* **6**, 1649 (1968).

1213. G. N. Ramachandran and G. Kartha, *Nature* **174**, 269 (1954).

1214. P. S. Ramanathan, C. V. Krishnan, and H. L. Friedman, *J. Soln. Chem.* **1**, 237 (1972).

1215. M. V. Ramiah and D. A. I. Goring, *J. Polymer Sci. C* **27** (1965).

1216. R. P. Rand, D. O. Tinker, and P. G. Fast, *Chem. Phys. Lipids* **6**, 333 (1971).

1217. J.E. B. Randles, *Disc. Faraday Soc.* **24**, 194 (1957).

1218. N. V. Rao and W. F. Harrington, *J. Mol. Biol.* **21**, 577 (1966).

1219. P. S. K. M. Rao and D. Premaswarup, *Trans. Faraday Soc.* **66**, 1974 (1970).

1220. V. S. R. Rao, P. R. Sundararajan, C. Ramakrishnan, and G. N. Ramachandran, *in* "Conformation of Biopolymers" (G. N. Ramachandran, ed.), Academic Press, London (1967), Vol. 2, p. 721.

1221. V. S. R. Rao, K. S. Vijayalakshmi, and P. R. Sundararajan, *Carbohydrate Res.* **17**, 341 (1971).

1222. V. S. R. Rao, N. Yathindra, and P. R. Sundararajan, *Biopolymers* **8**, 325 (1969).

1223. J. Rassing, P. J. Sams, and E. Wyn-Jones, *J. Chem. Soc. Faraday Trans. II* **69**, 180 (1973).

1224. H. Rath, S. Müller, H. Brielmaier, and W. Bubser, *Melliand-Textil-Ber.* **40**, 787 (1959).

1225. H. Rath, J. Rau, and D. Wagner, *Melliand-Textil-Ber.* **43**, 718 (1962).

1226. I. D. Rattee, *Chem. Soc. Rev.* **1**, 145 (1972).

1227. A. Ray, *J. Amer. Chem. Soc.* **91**, 6511 (1969).

1228. A. Ray, *Nature* **231**, 313 (1971).

1229. A. Ray and G. Némethy, *J. Amer. Chem. Soc.* **93**, 6787 (1971).

1230. A. Ray and G. Némethy, *J. Phys. Chem.* **75**, 809 (1971).

1231. D. A. Rees, *Adv. Carb. Chem. Biochem.* **24**, 267 (1969).

1232. D. A. Rees, *J. Chem. Soc. B* **877** (1970).

1233. D. A. Rees, *Biochem. J.* **126**, 257 (1972).

1234. D. A. Rees, *Chem. & Ind.* 630 (1972).

1235. D. A. Rees, *in* "MPT International Review of Science: Organic Chemistry," Series One, Vol. 7, "Carbohydrates" (G. O. Aspinall, ed.), Butterworths (1973).

1236. D. A. Rees, I. W. Steele, and F. B. Williamson, *J. Polymer Sci. C* **28**, 261 (1969).

1237. S. J. Rehfeld, *J. Colloid Interface Sci.* **34**, 518 (1970).

1238. S. J. Rehfeld, *J. Phys. Chem.* **75**, 3905 (1971).

1239. D. S. Reid, M. A. J. Quickenden, and F. Franks, *Nature* **224**, 1293 (1969).

1240. F. Reiss-Husson, *J. Mol. Biol.* **25**, 363 (1967).

1241. F. Reiss-Husson and V. Luzzati, *J. Phys. Chem.* **68**, 3504 (1964).

1242. P. Rempp, *J. Chim. Phys.* **54**, 432 (1957).

1243. G. Rialdi and J. Hermans, *J. Amer. Chem. Soc.* **88**, 5719 (1966).

1244. S. A. Rice and M. Nagasawa, "Polyelectrolyte Solutions: A Theoretical Introduction," Academic Press, London and New York (1961).

1245. A. Rich and F. H. C. Crick, *J. Mol. Biol.* **3**, 483 (1961).

1246. J. Rifkind and J. Applequist, *J. Amer. Chem. Soc.* **86**, 4207 (1964).

1247. J. L. Rigaud, C. M. Gary-Bobo, and Y. Lange, *Biochim. Biophys. Acta* **266**, 72 (1972).

1248. J. L. Rigaud, Y. Lange, C. M. Gary-Bobo, A. Samson, and M. Ptak, *Biochem. Biophys. Res. Commun.* **50**, 59 (1973).

1249. W. Ring, H.-J. Cantow, and W. Holtrup, *European Polymer J.* **2**, 151 (1966).

1250. I. D. Robb, *J. Colloid Interface Sci.* **37**, 521 (1971).

1251. D. C. Robins and I. L. Thomas, *J. Colloid Interface Sci.* **26**, 407 (1968).

1252. D. C. Robins and I. L. Thomas, *J. Colloid Interface Sci.* **26**, 415 (1968).

1253. B. H. Robinson, A. Löffler, and G. Schwarz, *J. Chem. Soc. Faraday Trans. I* **69**, 56 (1973).

1254. C. Robinson, *Proc. Roy. Soc. A* **148**, 681 (1935).

1255. C. Robinson and H. E. Garrett, *Trans. Faraday Soc.* **35**, 771 (1939).

1256. C. Robinson and A. T. Mills, *Proc. Roy. Soc. A* **131**, 576, 596 (1931).

1257. C. Robinson and J. L. Moilliet, *Proc. Roy. Soc. A* **143**, 630 (1934).

1258. C. Robinson and J. W. Selby, *Trans. Faraday Soc.* **35**, 771 (1939).

1259. D. R. Robinson and M. E. Grant, *J. Biol. Chem.* **241**, 4030 (1966).

1260. D. R. Robinson and W. P. Jencks, *J. Amer. Chem. Soc.* **87**, 2462 (1965).

1261. D. R. Robinson and W. P. Jencks, *J. Amer. Chem. Soc.* **87**, 2470 (1965).

1262. R. A. Robinson and R. H. Stokes, *J. Phys. Chem.* **65**, 1954 (1961).

1263. R. A. Robinson and R. H. Stokes, "Electrolyte Solutions," Butterworths, London (1965).

1264. C. H. Rochester and J. R. Symonds, *J. Chem. Soc. Faraday I* **69**, 1267 (1973).

1265. C. H. Rochester and J. R. Symonds, *J. Chem. Soc. Faraday I* **69**, 1274 (1973).

1266. C. H. Rochester and J. R. Symonds, *J. Chem. Soc. Faraday I* **69**, 1577 (1973).

1267. M. B. Rodell, W. L. Guess, and J. Autian, *J. Pharm. Sci.* **55**, 1429 (1966).

1268. F. Rodriguez, "Principles of Polymer Systems," McGraw-Hill, New York (1970).

1269. F. Rodriguez and L. A. Goettler, *Trans. Soc. Rheol.* **8**, 3 (1964).

1270. W. J. Roff and J. R. Scott, "Fibres, Films, Plastics and Rubbers: A Handbook of Common Polymers," Butterworths, London (1971).

1271. K. K. Rohatgi and G. S. Singhal, *J. Phys. Chem.* **70**, 1695 (1966).

1272. A. Roman and I. Scondac, *Makromol. Chemie* **113**, 171 (1968).

1273. H. L. Rosano, A. P. Christodoulou, and M. E. Feinstein, *J. Colloid Interface Sci.* **29**, 335 (1969).

1274. R. M. Rosenberg and I. M. Klotz, *in* "A Laboratory Manual of Analytical Methods of Protein Chemistry (including Polypeptides)," (P. Alexander and R. J. Block, eds.), Pergamon, Oxford (1960), Vol. 2, Chapter 4.

1275. P. D. Ross and R. L. Scruggs, *Biopolymers* **2**, 231 (1964).

1276. J. E. Rothmam, *J. Theor. Biol.* **38**, 1 (1973).

1277. G. Rouser, G. J. Nelson, S. Fleischer, and G. Simon, *in* "Biological Membranes" (D. Chapman, ed.), Academic Press, London and New York (1968), Vol. 1, p. 1.

1278. J. S. Rowlinson, *Trans. Faraday Soc.* **47**, 120 (1951).

1279. J. S. Rowlinson, "Liquids and Liquid Mixtures," Butterworths, London (1969), p. 169.

1280. M. Rudrum and D. F. Shaw, *J. Chem. Soc.* 52 (1965).

1281. R. E. Rundle, *J. Amer. Chem. Soc.* **69**, 1769 (1947).

1282. R. E. Rundle and D. French, *J. Amer. Chem. Soc.* **65**, 1707 (1943).

1283. J. A. Rupley, *J. Phys. Chem.* **68**, 2002 (1964).

1284. A. E. Russell and D. R. Cooper, *Biochem. J.* **113**, 221 (1969).

1285. S. B. Sachs, A. Raziel, H. Eisenberg, and A. Katchalsky, *Trans. Faraday Soc.* **65**, 77 (1969).

1286. H. Saito and K. Shinoda, *J. Colloid Interface Sci.* **35**, 359 (1971).

1287. S. Saito, *J. Colloid Sci.* **15**, 283 (1960).

1288. S. Saito, *J. Colloid Interface Sci.* **24**, 227 (1967).

1289. S. Saito, *Kolloid-Z.* **215**, 16 (1967).

1290. S. Saito, *Kolloid-Z.* **226**, 10 (1968).

1291. S. Saito, *J. Polymer Sci. Part A-1* **8**, 263 (1970).

1292. S. Saito, *Kolloid-Z.* **249**, 1096 (1971).

1293. S. Saito and T. Otsuka, *J. Colloid Interface Sci.* **25**, 531 (1967).

1294. S. Saito and T. Taniguchi, *Kolloid-Z.* **248**, 1039 (1971).

1295. S. Saito, T. Taniguchi, and K. Kitamura, *J. Colloid Interface Sci.* **37**, 154 (1971).

1296. S. Saito and M. Yukawa, *J. Colloid Interface Sci.* **30**, 211 (1969).

1297. I. Sakurada and Y. Ikada, *Bull. Inst. Chem. Res. Kyoto Univ.* **40**, 123 (1963).

1298. I. Sakurada, A. Nakajima, and H. Fujiwara, *J. Polymer Sci.* **35**, 497 (1959).

1299. N. J. Salsbury and D. Chapman, *Biochim. Biophys. Acta* **163**, 314 (1968).

1300. N. J. Salsbury, D. Chapman, and G. Parry-Jones, *Trans. Faraday Soc.* **66**, 1554 (1970).

1301. N. J. Salsbury, A. Darke, and D. Chapman, *Chem. Phys. Lipids* **8**, 142 (1972).

1302. V. S. Salvin, *J. Soc. Dyers and Col.* **79**, 687 (1963).

1303. P. J. Sams, E. Wyn-Jones, and J. Rassing, *Chem. Phys. Lett.* **13**, 233 (1972).

1304. T. S. Sarma and J. C. Ahluwalia, *J. Phys. Chem.* **74**, 3547 (1970).

1305. N. Sata and S. Saito, *Kolloid-Z.* **128**, 154 (1952).

1306. L. Saunders, *Biochim. Biophys. Acta* **125**, 70 (1966).

1307. L. Saunders, J. Perrin, and D. Gammack, *J. Pharm. Pharmacol.* **14**, 567 (1962).

1308. A. B. Savage, *Ind. Eng. Chem.* **49**, 99 (1957).

1309. G. Scatchard and E. S. Black, *J. Phys. Colloid Chem.* **53**, 88 (1949).

1310. G. Scatchard, W. J. Hamer, and S. E. Wood, *J. Amer. Chem. Soc.* **60**, 3061 (1938).

1311. J. Schack and B. S. Bynum, *Nature* **184**, 635 (1959).
1312. I. E. Scheffler and J. M. Sturtevant, *J. Mol. Biol.* **42**, 577 (1969).
1313. G. Scheibe, *Angew. Chem.* **50**, 212 (1937).
1314. G. Scheibe, *Kolloid Z.* **82**, 1 (1938).
1315. G. Scheibe, *Angew. Chem.* **52**, 631 (1939).
1316. G. Scheibe, *Z. Elektrochem.* **52**, 283 (1948).
1317. J. A. Schellman, *Comp. Rend. Lab. Carlsberg, Ser. Chim.* **29**, 230 (1955).
1318. H. A. Scheraga, "The Proteins," Vol. 1 (H. Neurath, ed.) Academic Press, New York (1965).
1319. M. J. Schick, *J. Phys. Chem.* **67**, 1796 (1963).
1320. M. J. Schick, *J. Phys. Chem.* **68**, 3585 (1964).
1321. M. J. Schick (ed.) "Nonionic Surfactants," Marcel Dekker, New York (1967).
1322. M. J. Schick, S. M. Atlas, and F. R. Eirich, *J. Phys. Chem.* **66**, 1326 (1962).
1323. M. J. Schick and A. H. Gilbert, *J. Colloid Sci.* **20**, 464 (1965).
1324. C. Schildkraut and S. Lifson, *Biopolymers* **3**, 195 (1965).
1325. T. Schleich and P. H. von Hippel, *Biopolymers* **7**, 861 (1969).
1326. T. Schleich and P. H. von Hippel, *Biochemistry* **9**, 1059 (1970).
1327. H. Schlenk and D. M. Sand, *J. Amer. Chem. Soc.* **83**, 2312 (1961).
1328. F. O. Schmitt and R. S. Bear, *Biol. Rev. Camb. Phil. Soc.* **14**, 27 (1939).
1329. H. Schneider, G. C. Kresheck, and H. A. Scheraga, *J. Phys. Chem.* **69**, 1310 (1965).
1330. N. S. Schneider (ed.), "Selected Papers on Water Vapor Transport in Polymers" *J. Macromol. Sci.-Phys.* **B3**, No. 4 (1969).
1331. N. S. Schneider, L. V. Dusablon, E. W. Snell, and R. A. Prosser, *J. Macromol. Sci.-Phys.* **B3**, 623 (1969).
1332. N. S. Schneider, L. V. Dusablon, L. A. Spano, H. B. Hopfenberg, and F. Votta, *J. Appl. Polymer Sci.* **12**, 527 (1968).
1333. J. Schnell, *Arch. Biochem. Biophys.* **127**, 496 (1968).
1334. T. J. Schoch, *in* "Symposium on Foods: Carbohydrates and their Roles" (H. W. Schulz, ed.), AVI Publishing Co. (1969), p. 396.
1335. B. P. Schoenborn, *Nature* **208**, 760 (1965).
1336. B. P. Schoenborn, *Nature* **214**, 1120 (1967).
1337. B. P. Schoenborn, R. M. Featherstone, P. O. Vogelhut, and C. Süsskind, *Nature* **202**, 695 (1964).
1338. W. Scholtan, *Makromol. Chemie* **11**, 131 (1954).
1339. W. Scholtan, *Makromol. Chemie* **14**, 169 (1954).
1340. W. Scholtan, *Makromol. Chemie* **24**, 104 (1957).
1341. H. Schonert, *Z. Phys. Chem. (Frankfurt)* **61**, 262 (1968).
1342. H. Schott, *J. Phys. Chem.* **68**, 3612 (1964).
1343. H. Schott, *J. Pharm. Sci.* **60**, 648 (1971).
1344. E. E. Schrier, M. Pottle, and H. A. Scheraga, *J. Amer. Chem. Soc.* **86**, 3444 (1964).
1345. E. E. Schrier and H. A. Scheraga, *Biochim. Biophys. Acta* **64**, 406 (1962).
1346. W. A. Schroeder and L. M. Kay, *J. Amer. Chem. Soc.* **76**, 3556 (1954).
1347. G. Schur, *Kolloid-Z.* **242**, 1212 (1970).
1348. H. P. Schwan, *Advan. Biol. Med. Phys.* **5**, 147 (1957).
1349. H. P. Schwan, *Ann. N.Y. Acad. Sci.* **125**, 344 (1965).
1350. G. Schwartz, *Biopolymers* **6**, 873 (1968).
1351. G. Scibona, P. R. Danesi, A. Conte, and B. Scuppa, *J. Colloid Interface Sci.* **35**, 631 (1971).

1352. I. Scondac, A. Roman, and M. Dima, *Rev. Roumaine Chim.* **11**, 1333 (1966).

1353. I. Scondac, A. Roman, and M. Dima, *Rev. Roumaine Chim.* **12**, 677 (1967).

1354. E. Scott and D. S. Berns, *Biochemistry* **6**, 1327 (1967)

1355. G. Scott, *Plastics*: *Rubbers*: *Textiles* **1**, 361 (1970).

1356. G. Scott, *Waste Disposal* (Official Journal of the National Association of Waste Disposal Contractors, Marylebone Press, Manchester) **5**, 78 (1971).

1357. B. Sébille and J. Néel, *J. Chim. Phys.* **60**, 475 (1963).

1358. J. Seelig, *J. Amer. Chem. Soc.* **92**, 3881 (1970).

1359. D. M. Segal, *J. Mol. Biol.* **43**, 497 (1969).

1360. D. M. Segal and W. Traub, *J. Mol. Biol.* **43**, 487 (1969).

1361. D. M. Segal, W. Traub, and W. Yonath, *J. Mol. Biol.* **43**, 519 (1969).

1362. E. Segerman, *Acta Cryst.* **19**, 789 (1965).

1363. E. Segerman, in "Surface Chemistry" (P. Ekwall, K. Groth, and V. Runnström, eds.) Munksgaard, Copenhagen (1965).

1364. C. H. A. Seiter, G. W. Fergenson, S. I. Chan, and M. C. Hsu, *J. Amer. Chem. Soc.* **94**, 2535 (1972).

1365. J. E. Selwyn and J. I. Steinfeld, *J. Phys. Chem.* **76**, 762 (1972).

1366. L. Sepulveda and F. MacRitchie, *J. Colloid Interface Sci.* **28**, 19 (1968).

1367. J. Setschenow, *Z. Physik. Chem.* (*Leipzig*) **4**, 117 (1889).

1368. L. B. Shaffer, Ph. D. Thesis, University of Wisconsin (1964).

1369. D. O. Shah, *Advances Lipid Res.* **8**, 347 (1970).

1370. J. Shapiro, D. B. Stannard, and G. Felsenfeld, *Biochem.* **8**, 3233 (1969).

1371. L. Shedlovsky, C. W. Jakob, and M. B. Epstein, *J. Phys. Chem.* **67**, 2075 (1963).

1372. C. J. Sheehan and A. L. Bisio, *Rubber Chem. and Tech.* **39**, 149 (1966).

1373. J. C. W. Shepherd and E. H. Grant, *Proc. Roy. Soc.* (*London*) **A307**, 335 (1968).

1374. S. E. Sheppard, *Proc. Roy. Soc.* (*London*) **A82**, 256 (1909).

1375. S. E. Sheppard, *Science* **93**, 42 (1941).

1376. S. E. Sheppard, *Rev. Mod. Phys.* **14**, 303 (1942).

1377. S. E. Sheppard and A. L. Geddes, *J. Amer. Chem. Soc.* **66**, 1995 (1944).

1378. S. E. Sheppard and A. L. Geddes, *J. Amer. Chem. Soc.* **66**, 2003 (1944).

1379. S. E. Sheppard, R. H. Lambert, and R. D. Walker, *J. Chem. Phys.* **7**, 265 (1939).

1380. D. F. Shiao, R. Lumry, and J. Fahey, *J. Amer. Chem. Soc.* **93**, 2024 (1971).

1381. H. Shiio, *J. Amer. Chem. Soc.* **80**, 70 (1958).

1382. H. Shiio, T. Ogawa, and H. Yoshihashi, *J. Amer. Chem. Soc.* **77**, 4980 (1955).

1383. K. Shinoda, *Bull. Chem. Soc. Japan* **26**, 101 (1953).

1384. K. Shinoda, in "Colloidal Surfactants" (B.-I. Tamamushi and T. Isemura, eds.), Academic Press, New York (1963), p. 1.

1385. K. Shinoda, *J. Colloid Interface Sci.* **34**, 278 (1970).

1386. K. Shinoda, M. Hato, and T. Hayashi, *J. Phys. Chem.* **76**, 909 (1972).

1387. K. Shinoda and T. Nakagawa, in "Colloidal Surfactants" (B.-I. Tamamushi and T. Isemura, eds.), Academic Press, New York (1963).

1388. N. Shinozuka, H. Suzuki, and S. Hayano, *Kolloid-Z. Z. Polym.* **248**, 959 (1971).

1389. G. G. Shipley, in "Biological Membranes" (D. Chapman and D. F. H. Wallach, eds.) Academic Press, London and New York (1973), Vol. 2, p. 1.

1390. K. Shirahama, M. Hayashi, and R. Matuura, *Bull. Chem. Soc. Japan* **42**, 1206 (1969).

1391. K. Shirahama, M. Hayashi, and R. Matuura, *Bull. Chem. Soc. Japan* **42**, 2123 (1969).

1392. K. Shirahama and T. Kashiwabara, *J. Colloid Interface Sci.* **36**, 65 (1971).
1393. S. K. Shoor and K. E. Gubbins, *J. Phys. Chem.* **73**, 498 (1969).
1394. S. Siggia, *J. Amer. Pharmaceut. Assoc., Sci. Ed.* **46**, 201 (1957).
1395. A. Silberberg, J. Eliassaf, and A. Katchalsky, *J. Polymer Sci.* **23**, 259 (1957).
1396. J. Simplicio and K. Schwenzer, *Biochem.* **12**, 1923 (1973).
1397. O. Sinanoglu and S. Abdulner, *J. Photochem. Photobiol.* **3**, 333 (1964).
1398. O. Sinanoglu and S. Abdulner, *Fed. Proc. Suppl.* **24**, S. 12 (1965).
1399. S. J. Singer, *in* "Structure and Function of Biological Membranes" (L. Rothfield, ed.), Academic Press (1971).
1400. M. P. Singh, *Bio-Med. Eng.* **4**, 68 (1969).
1401. B. J. Simpson, *Bio-Med. Eng.* **4**, 65 (1969).
1402. V. L. Simril and A. Hershberger, *Modern Plastics* **27**, 95 (1950).
1403. S. R. Sivarajan, *J. Ind. Inst. Sci.* **34**, 75 (1952).
1404. D. M. Small, *J. Lipid Res.* **8**, 551 (1967).
1405. P. A. Small, *J. Appl. Chem.* **3**, 71 (1953).
1406. M. Smith, A. G. Walton, and J. L. Koenig, *Biopolymers* **8**, 173 (1969).
1407. P. J. C. Smith, Ph. D. Thesis, University of Edinburgh (1972).
1408. R. Smith and C. Tanford, *J. Mol. Biol.* **67**, 75 (1972).
1409. R. L. Snipp, W. G. Miller, and R. E. Nylund, *J. Amer. Chem. Soc.* **87**, 3547 (1965).
1410. T. N. Solie and J. A. Schellman, *J. Mol. Biol.* **33**, 61 (1968).
1411. A. T. Sophianopoulos and B. J. Weis, *Biochem.* **3**, 1920 (1964).
1412. S. Sourirajan, "Reverse Osmosis," Logos Press, London (1970).
1413. J. B. Speakman and H. Clegg, *J. Soc. Dyers and Col.* **50**, 348 (1934).
1414. R. D. Spencer and G. Weber, *Ann. N.Y. Acad. Sci.* **158**, 361 (1969).
1415. H. Sprinz, R. Döllstädt, and G. Hubner, *Biopolymers* **7**, 447 (1969).
1416. V. Stannett and J. L. Williams, *J. Polymer Sci. Part C* **3** (Polymer Symposia, No. 10), 45 (1965).
1417. H. W. Starkweather, *J. Appl. Polymer Sci.* **2**, 129 (1959).
1418. H. W. Starkweather, *J. Macromol. Sci.-Phys.* **B3**, 727 (1969).
1419. J. J. P. Staudinger, "Disposal of Plastics Waste and Litter," Society of Chemical Industry, Monograph No. 35, London (1970).
1420. F. Stegner, *Ann. Physik. Chem.* **33**, 577 (1888).
1421. I. Z. Steinberg, W. F. Harrington, A. Bergen, M. Sela, and E. Katchalski, *J. Amer. Chem. Soc.* **82**, 5263 (1960).
1422. P. Stenius and C. H. Zilliacus, *Acta Chem. Scand.* **25**, 2232 (1971).
1423. C. Sterling and M. Masuzawa, *Makromol. Chem.* **116**, 140 (1968).
1424. O. Stern, *Z. Elektrochem.* **30**, 508 (1924).
1425. J. M. Stewart and J. D. Young, "Solid Phase Peptide Synthesis," Freeman, San Francisco (1969).
1426. W. E. Stewart, L. Mandelkern, and R. E. Glick, *Biochem.* **6**, 143 (1967).
1427. F. H. Stillinger and A. Ben-Naim, *J. Phys. Chem.* **73**, 900 (1969).
1428. F. H. Stillinger and A. Rahman, *J. Chem. Phys.* **57**, 1281 (1972).
1429. F. H. Stillinger and A. Rahman, *J. Chem. Phys.* **60**, 1545 (1974).
1430. W. Stoeckenius and D. M. Engelman, *J. Cell Biol.* **42**, 613 (1969).
1431. R. H. Stokes and R. A. Robinson, *J. Phys. Chem.* **70**, 2126 (1966).
1432. F. W. Stone and J. J. Stratta, in "Encyclopedia of Polymer Science and Technology" (H. F. Mark, N. G. Gaylord and N. M. Bikales, eds.), Wiley, New York (1967), Vol. 6, p. 103.

1433. W. H. J. Stork, G. J. M. Lippits, and M. Mandel, *J. Phys. Chem.* **76**, 1772 (1972).

1434. H. Strassmair, J. Engel, and G. Zundel, *Biopolymers* **8**, 237 (1969).

1435. U. P. Strauss and B. L. Williams, *J. Phys. Chem.* **65**, 1390 (1961).

1436. F. C. Strong, *Anal. Chem.* **24**, 338 (1952).

1437. D. S. Studdert, M. Patrone, and R. C. Davis, *Biopolymers* **11**, 761 (1972).

1438. J. M. Sturtevant, S. Rice, and E. P. Denduscheck, *Disc. Faraday Soc.* **25**, 138 (1958).

1439. S. Subramanian and H. F. Fisher, *Biopolymers* **11**, 1305 (1972).

1440. A. Suggett, *in* "Dielectric and Related Molecular Processes," Chemical Society, London (1972), Vol. 1, p. 100.

1441. A. Suggett, *in* "Proceedings of National Physical Laboratory Conference on High Frequency Dielectric Measurements" (J. Chamberlain and G. Chantry, eds.), IPC, London (1973).

1442. A. Suggett, to be published.

1443. A. Suggett, A. H. Clark, and P. A. Quickenden, to be published.

1444. A. Suggett, P. A. Mackness, M. J. Tait, H. W. Loeb, and G. M. Young, *Nature* **228**, 456 (1970).

1445. M. Sundaralingam, *Nature* **217**, 35 (1968).

1446. M. Sundaralingam, *Biopolymers* **7**, 821 (1969).

1447. M. Sundaralingam, *Ann. N.Y. Acad. Sci.* **195**, 324 (1972).

1448. M. Sundaralingam and L. H. Jensen, *Science* **150**, 1035 (1965).

1449. M. Sundaralingam and E. Putkey, *Acta Cryst. B* **26**, 790 (1970).

1450. P. R. Sundararajan and V. S. R. Rao, *Tetrahedron* **24**, 289 (1968).

1451. H. Suzuki and T. Sasaki, *Bull. Chem. Soc. Japan.* **44**, 2630 (1971).

1452. J. Swarbrick and J. Darawala, *J. Phys. Chem.* **73**, 2627 (1969).

1453. C. A. Swenson, *Biopolymers* **10**, 2591 (1971).

1454. C. A. Swenson and R. Formanek, *J. Phys. Chem.* **71**, 4073 (1967).

1455. J. Szeijtli and S. Augustat, *Die Stärke* **18**, 38 (1966).

1456. T. Tachibara and A. Nakamura, *Kolloid-Z.* **203**, 130 (1965).

1457. Y. Tadokoro, S. Okazaki, and H. Yamamura, *J. Sci. Hiroshima Univ., Ser. A-II* **29**, 89 (1965).

1458. M. J. Tait, A. Suggett, F. Franks, S. Ablett, and P. A. Quickenden, *J. Solution Chem.* **1**, 131 (1972).

1459. S. Takashima, *Biopolymers* **1**, 171 (1963).

1460. K. Takeda and T. Yasunaga, *J. Colloid Interface Sci.* **40**, 127 (1972).

1461. C. Tanford, "Physical Chemistry of Macromolecules," Wiley, New York and London (1961).

1462. C. Tanford, *J. Amer. Chem. Soc.* **84**, 4240 (1962).

1463. C. Tanford, *J. Amer. Chem. Soc.* **86**, 2050 (1964).

1464. C. Tanford, *Advan. Protein Chem.* **23**, 121 (1968).

1465. C. Tanford, *in* "Solution Properties of Natural Polymers" (C. T. Greenwood, ed.), The Chemical Society, Special Publication No. 23, London (1968), p. 1.

1466. C. Tanford, *Advan. Protein Chem.* **24**, 1 (1970).

1467. C. Tanford, "Advances in Protein Chemistry" (C. B. Anfinsen, ed.), Academic Press, New York (1970).

1468. C. Tanford, *J. Mol. Biol.* **67**, 59 (1972).

1469. C. Tanford, *J. Phys. Chem.* **76**, 3020 (1972).

1470. C. Tanford, "The Hydrophobic Effect—Formation of Micelles and Biological Membranes," Wiley–Interscience, New York (1973).

1471. C. Tanford and J. G. Burrell, *J. Phys. Chem.* **60**, 1204 (1956).

1472. C. Tanford, R. H. Pain, and N. S. Otchin, *J. Mol. Biol.* **15**, 489 (1966).

1473. J. B. Taylor and J. S. Rowlinson, *Trans. Faraday Soc.* **51**, 1183 (1955).

1474. A. Teramoto, S. Kusamizu, H. Tanaka, Y. Murakami, and H. Fujita, *Makromol. Chemie* **90**, 78 (1966).

1475. D. Thom, Ph. D. Thesis, University of Edinburgh (1973).

1476. E. S. Thomsen and J. C. Gjaldbaek, *Acta Chem. Scand.* **17**, 134 (1963).

1477. R. J. Thorn, *J. Chem. Phys.* **51**, 3582 (1969).

1478. E. Thorpe, "A Dictionary of Applied Chemistry," Longmans Green, London (1928), Vol. 4, p. 3.

1479. J. O. Threlkeld, J. J. Burke, D. Puett, and A. Ciferri, *Biopolymers* **6**, 767 (1968).

1480. A. Tissiers, J. D. Watson, D. Schlessinger, and B. R. Hollingworth. *J. Mol. Biol.* **1**, 221 (1959).

1481. F. Tokiwa, *J. Colloid Interface Sci.* **28**, 145 (1968).

1482. F. Tokiwa and N. Moriyama, *J. Colloid Interface Sci.* **30**, 338 (1969).

1483. F. Tokiwa and K. Ohki, *J. Colloid Interface Sci.* **27**, 247 (1968).

1484. F. Tokiwa and K. Tsujii, *J. Phys. Chem.* **75**, 3560 (1971).

1485. H. Tompa, "Polymer Solutions," Butterworths, London (1956).

1486. F. C. Tompkins (ed.), "Colloid Stability in Aqueous and Non-aqueous Media," The Faraday Society, Discussion No. 42, London (1966).

1487. D. A. Torchia and F. A. Bovey, *Macromolecules* **4**, 246 (1971).

1488. K. Toyoshima, *in* "Properties and Applications of Polyvinyl Alcohol," (C. A. Finch, ed.), Society of Chemical Industry, Monograph No. 30, London (1968), p. 154.

1489. W. Traub, *J. Mol. Biol.* **43**, 479 (1969).

1490. W. Traub and K. A. Piez, *Adv. Protein Chem.* **25**, 243 (1971).

1491. M. Traub and U. Shmueli, *Nature* **198**, 1165 (1963).

1492. W. Traub, A. Yonath, and D. M. Segal, *Nature* **221**, 914 (1969).

1493. H. Träuble and H. Eibl, *Proc. Nat. Acad. Sci. U.S.* **71**, 214 (1974).

1494. H. Träuble and D. H. Haynes, *Chem. Phys. Lipids* **7**, 324 (1971).

1495. P. O. P. Ts'o, G. K. Helmkamp, and C. Sanda, *Proc. Natl. Acad. Sci. U.S.* **48**, 488 (1962).

1496. P. O. P. Ts'o, N. S. Kondo, R. K. Robins, and A. D. Broom, *J. Amer. Chem. Soc.* **91**, 5625 (1969).

1497. T. Y. Tsong, R. L. Baldwin, and E. L. Elson, *Proc. Nat. Acad. Sci.* **68**, 2712 (1971).

1498. T. Y. Tsong, R. L. Baldwin, and P. McPhie, *J. Mol. Biol.* **63**, 453 (1972).

1499. T. Y. Tsong, R. P. Hearn, D. P. Wrathall, and J. M. Sturtevant, *Biochem.* **9**, 2666 (1970).

1500. M. J. B. Tunis and J. Hearst, *Biopolymers* **6**, 1218 (1968).

1501. M. J. B. Tunis and J. E. Hearst, *Biopolymers* **6**, 1325 (1968).

1502. M. J. B. Tunis and J. E. Hearst, *Biopolymers* **6**, 1345 (1968).

1503. M. J. B. Tunis and J. E. Hearst, *Biopolymers* **7**, 1219 (1968).

1504. M. J. B. Tunis-Schneider and M. F. Maestre, *J. Mol. Biol.* **52**, 521 (1970).

1505. H. Uedaira and H. Uedaira, *Bull. Text. Res. Inst. Japan* No. 12, 7 (1964).

1506. K. Ueno, Ff. Williams, K. Hayashi, and S. Okamura, *Trans. Faraday Soc.* **63**, 1478 (1967).

1507. R. J. Urick, *J. Appl. Phys.* **18**, 983 (1947).

1508. P. Urnes and P. Doty, *Advan. Protein Chem.* **16**, 401 (1961).

1509. L. Valentine, *J. Polymer Sci.* **27**, 313 (1958).

1510. E. Valko, *Trans. Faraday Soc.* **31**, 230 (1935).

1511. E. Valko, *in* "Kolloidchemische Grundlagen der Textilveredlung," J. Springer, Berlin (1937).

1512. E. Valko, *J. Soc. Dyers and Col.* **55**, 173 (1939).

1513. E. I. Valko, *Rev. Prog. in Coloration* **3**, 50 (1972).

1514. J. W. A. van den Berg and A. J. Staverman, *Rec. Trav. Chim.* **91**, 1151 (1972).

1515. V. Vand, W. M. Morley, and T. R. Lomer, *Acta Cryst.* **4**, 324 (1951).

1516. L. L. M. van Deenen, *in* "Progress in the Chemistry of Fats and Other Lipids" (R. T. Holman, ed.), Pergamon Press, Oxford (1965), Vol. 8, Part 1.

1517. F. A. Vandenheuvel, *J. Amer. Oil Chem. Soc.* **43**, 258 (1966).

1518. J. H. van der Waals and J. C. Platteeuw, *Advan. Chem. Phys.* **2**, 1 (1959).

1519. K. E. van Holde and G. P. Rosetti, *Biochem.* **6**, 2189 (1967).

1520. L. N. Vauquelin, *Ann. Mus. Hist. Nat.* **18**, 212 (1811).

1521. A. Veis, "The Macromolecular Chemistry of Gelatin," Academic Press, London (1964).

1522. A. Veis, J. Anesey, and J. Cohen, *Arch. Biocyem. Biophys.* **94**, 20 (1961).

1523. Z. Veksli, N. J. Salsbury, and D. Chapman, *Biochim. Biophys. Acta* **183**, 434 (1969).

1524. V. Verezhnikov and L. S. Kotlyar, *Chem. Abstr.* **78**, 86283 (1973).

1525. R. E. Verrall and B. E. Conway, *J. Phys. Chem.* **70**, 3961 (1966).

1526. T. Vickerstaff, *in* "The Physical Chemistry of Dyeing," 2nd ed., Oliver and Boyd, London (1954), Chapter 3.

1527. K. S. Vijayalakshmi ad V. S. R. Rao, *Carbohydrate Res.* **22**, 413 (1972).

1528. B. R. Vijayendran and R. D. Vold, *Biopolymers* **9**, 1391 (1970).

1529. B. R. Vijayendran and R. D. Vold, *Biopolymers* **10**, 991 (1971).

1530. B. Vincent, M. Sc. Thesis, Bristol University (1965).

1531. H. Vink, *Makromol. Chemie* **67**, 105 (1963).

1532. G. V. Vinogradov and L. V. Titkova, *Kolloid-Z.* **239**, 655 (1970).

1533. J. Vinograd, R. Greenwald, and J. E. Hearst, *Biopolymers* **3**, 109 (1965).

1534. W. von Casimii, N. Kaiser, F. Keilmann, A. Mayer, and H. Vogel, *Biopolymers* **6**, 1705 (1968).

1535. P. H. von Hippel, *in* "Treatise on Collagen. I. Chemistry of Collagen" (G. N. Ramachandran, ed.), Academic Press, London (1965).

1536. P. H. von Hippel and W. F. Harrington, *Biochim. Biophys. Acta* **36**, 427 (1959).

1537. P. H. von Hippel and T. Schleich, *Biopolymers* **7**, 861 (1969).

1538. P. H. von Hippel and T. Schleich, *in* "Structure and Stability of Biological Macromolecules" (S. N. Timasheff and G. D. Fasman, eds.), Marcel Dekker, New York (1969), Vol. II.

1539. P. H. von Hippel and K. Y. Wong, *Biochem.* **1**, 664 (1962).

1540. P. H. von Hippel and K. Y. Wong, *J. Biol. Chem.* **240**, 3909 (1965).

1541. M. J. Voorn, *Fortschr. Hochpolym.-Forsch.* **1**, 192 (1959).

1542. M. F. Vuks, L. I. Lisnyanskii, and L. V. Shurupova, *in* "Water in Biological Systems," Consultants Bureau, New York (1971), Vol. 2, p. 79 (transl. from Russian, *Struktura: Rol' Vody v Zhivom Organizme*).

1543. M. F. Vuks and L. V. Shurupova, *Struktura: Rol' Vody v Zhivom Organizme* (1970), Vol. 3, p. 63.

1544. A. Wada, *J. Chem. Phys.* **22**, 198 (1954).

1545. G. Wada and S. Umeda, *Bull. Chem. Soc. Japan* **35**, 646 (1962).

1546. G. Wada and S. Umeda, *Bull. Chem. Soc. Japan* **35**, 1797 (1962).

1547. I. Wadsö, *Acta Chem. Scand.* **22**, 927 (1968).

1548. Ph. Wahl and H. Lami, *Biochim. Biophys. Acta* **133**, 233 (1967).

1549. S. C. Wallace and J. K. Thomas, *Radiat. Res.* **54**, 49 (1973).

1550. G. A. Walrafen, *J. Chem. Phys.* **44**, 3726 (1966).

1551. W. V. Walter and R. G. Hayes, *Biochim. Biophys. Acta* **249**, 528 (1971).

1552. V. K. Walworth. A. E. Rosenoff, and G. R. Bird, *Photogr. Sci. Eng.* **14**, 321 (1970).

1553. R. D. Wanchope and R. Haque, *Can. J. Chem.* **50**, 133 (1972).

1554. J. C. Wang, *J. Mol. Biol.* **43**, 25 (1969).

1555. J. H. Wang, *J. Amer. Soc.* **77**, 258 (1955).

1556. J. A. Ward, *J. Polymer Sci. Part A-1* **9**, 3555 (1971).

1557. D. T. Warner, *Nature* **196**, 1055 (1962).

1558. D. T. Warner, *Ann. N. Y. Acad. Sci.* **125**, 605 (1965).

1559. D. T. Warner, *Ann. Rep. Med. Chem.* 256 (1970).

1560. M. M. Warshaw and I. Tinoco, *J. Mol. Biol.* **20**, 29 (1966).

1561. H. Warson, *Paint, Oil and Colour J.* **142**, 214 (1962).

1562. A. Watillon and A.-M. Joseph-Petit, *Disc. Faraday Soc.* **42**, 143 (1966).

1563. J. G. Watterson and H. G. Elias, *Kolloid-Z.* **249**, 1136 (1971).

1564. J. G. Watterson, H. R. Laesser, and H. G. Elias, *Kolloid-Z.* **250**, 64 (1972).

1565. R. O. Watts, in "Proc. 3rd Internatl. Conf. on Chem. Thermodynamics," Vienna (1973), Vol. 2, p. 88.

1566. S. J. Webb and J. S. Bhorjee, *Can. J. Biochem.* **46**, 691 (1968).

1567. G. Weber, *Advan. Protein Chem.* **8**, 415 (1953).

1568. G. Weber, M. Shinitzky, A. C. Dianoux, and C. Gitler, *Biochem.* **10**, 2106 (1971).

1569. G. Weber and F. W. J. Teale, *Proteins* **3**, 445 (1965).

1571. W. Y. Wen, *J. Sol. Chem.* **2**, 000 (1973).

1572. W. Y. Wen and H. G. Hertz, *J. Sol. Chem.* **1**, 17 (1972).

1572. W. Y. Wen and J. H. Hung, *J. Phys. Chem.* **74**, 170 (1970).

1573. W. Y. Wen, K. Miyajima, and A. Otsuka, *J. Phys. Chem.* **75**, 2148 (1971).

1574. W. Y. Wen and K. Nara, *J. Phys. Chem.* **71**, 3907 (1967).

1575. W. Y. Wen and K. Nara, *J. Phys. Chem.* **72**, 1137 (1968).

1576. M. Wentz, W. H. Smith, and A. R. Martin, *J. Colloid Interface Sci.* **29**, 36 (1969).

1577. W. West and B. H. Carroll, *in* "Theory of the Photographic Process" (C. E. K. Mees and T. H. James, eds.), Macmillan, New York, 1966.

1578. W. West and S. Pearce, *J. Phys. Chem.* **69**, 1894 (1965).

1579. P. G. Westmoreland, R. A. Day, Jr., and A. L. Underwood, *Anal. Chem.* **44**, 737 (1972).

1580. D. B. Wetlaufer, S. K. Malik, L. Stoller, and R. L. Coffin, *J. Amer. Chem. Soc.* **86**, 508 (1964).

1581. P. J. Wheatley, *J. Chem. Soc.* 3245, 4096 (1959).

1582. P. White and G. C. Benson, *Trans. Faraday Soc.* **55**, 1025 (1959).

1583. P. White and G. C. Benson, *J. Phys. Chem.* **64**, 599 (1960).

1584. E. Wilhelm and R. Battino, *J. Chem. Phys.* **56**, 563 (1972).

1585. E. J. Williams and V. F. Smolen, *J. Pharm. Sci.* **61**, 639 (1972).

1586. J. L. Williams, H. B. Hopfenberg, and V. Stannett, *J. Macromol. Sci.-Phys.* **B3**, 711 (1969).

1587. R. M. Williams and D. Chapman, *in* "Progress in the Chemistry of Fats and Other Lipids" (R. T. Holman, ed.), Pergamon Press, Oxford (1970), Vol. 11, p. 1.

1588. A. Wishnia, *J. Phys. Chem.* **67**, 2079 (1963).

1589. A. Wishnia, *Biochem.* **8**, 5070 (1969).

1590. A. Wishnia and T. W. Pinder, *Biochem.* **5**, 1534 (1966).

1591. O. Woerz and G. Scheibe, *Z. Naturforsch. B* **24**, 381 (1969).

1592. D. E. Woessner and B. S. Snowden, *J. Colloid Interface Sci.* **34**, 290 (1970).

1593. D. E. Woessner, B. S. Snowden, and J. C. Chiu, *J. Colloid Interface Sci.* **34**, 283 (1970).

1594. E. Wolfram and E. Borass-Vargha, *Fortschrittsber. Kolloide Polym.* **55**, 143 (1971); *Chem. Abstr.* **77**, 7676 (1972).

1595. J. D. Worley and I. M. Klotz, *J. Chem. Phys.* **45**, 2868 (1966).

1596. H. W. Wyckoff, *J. Biol. Chem.* **242**, 3749 (1967).

1597. J. Wyman, Jr., *J. Amer. Chem. Soc.* **55**, 4116 (1933).

1598. S. H. Yalkowsky and G. Zografi, *J. Colloid Interface Sci.* **34**, 525 (1970).

1599. K. Yamamoto and H. Fujita, *Polymer* (*London*) **7**, 557 (1966).

1600. K. Yamamoto and H. Fujita, *Polymer* (*London*) **8**, 517 (1967).

1601. K. Yamamoto, A. Teramoto, and H. Fujita, *Polymer* (*London*) **7**, 267 (1966).

1602. J. F. Yan, *J. Colloid Interface Sci.* **22**, 303 (1966).

1603. J. F. Yan and M. B. Palmer, *J. Colloid Interface Sci.* **30**, 177 (1969).

1604. N. A. Yaroshenko and P. A. Demchenko, *Ukr. Khim. Zh.* **37**, 723 (1971); *Chem. Abstr.* **76**, 37694 (1972).

1605. H. Yasuda, L. D. Ikenberry, and C. E. Lamaze, *Makromol. Chemie* **125**, 108 (1969).

1606. H. Yasuda, C. E. Lamaze, and L. D. Ikenberry, *Makromol. Chem.* **118**, 19 (1968).

1607. H. Yasuda, A. Peterlin, C. K. Colton, K. A. Smith, and E. W. Merrill, *Makromol. Chemie* **126**, 177 (1969).

1608. H. Yasuda and V. Stannet, *J. Polymer Sci.* **57**, 907 (1962).

1609. H. Yasuda and V. Stannett, *J. Macromol. Sci-Phys.* **B3**, 589 (1969).

1610. T. Yasunaga, S. Fujii, and M. Miura, *J. Colloid Interface Sci.* **30**, 399 (1969).

1611. T. Yasunaga, H. Oguri, and M. Miura, *J. Colloid Interface Sci.* **23**, 352 (1967).

1612. T. Yasunaga, K. Takeda, and S. Harada, *J. Colloid Interface Sci.* **42**, 457 (1973).

1613. D. M. Young and J. T. Potts, Jr., *J. Biol. Chem.* **238**, 1995 (1963).

1614. E. E. Zaev, *Kolloid. Zh.* **34**, 304 (1972).

1615. R. Zana and J. Lang, *Compt. Rend.* **C266**, 893 (1968).

1616. R. Zana and J. Lang, *Compt. Rend.* **C266**, 1347 (1968).

1617. V. Zanker, *Z. Physik. Chem.* **199**, 225 (1952).

1618. V. Zanker, *Z. Physik. Chem.* **200**, 250 (1952).

1619. J. L. Zatz and B. Knowles, *J. Pharm. Sci.* **60**, 1731 (1971).

1620. J. L. Zatz and B. Knowles, *J. Colloid Interface Sci.* **40**, 475 (1972).

1621. B. H. Zimm, *J. Chem. Phys.* **16**, 1093 (1948).

1622. B. H. Zimm, *J. Chem. Phys.* **21**, 934 (1953).

1623. B. H. Zimm and J. K. Bragg, *J. Chem. Phys.* **31**, 526 (1959).

1624. C. Zimmer and H. Venner, *Naturwiss.* **49**, 86 (1962).

1625. H. Zimmermann and G. Scheibe, *Z. Elektrochem.* **60**, 566 (1956).

1626. G. Zografi and S. H. Yalkowsky, *J. Pharm. Sci.* **61**, 651 (1972).

1627. H. Zollinger, *in* "Chemie der Azofarbstoffe," Birkhaüser, Basel (1958).

1628. H. Zollinger, *Chem. and Ind.* (*London*) **21**, 885 (1965).

1629. H. Zollinger, *J. Soc. Dyers and Col.* **81**, 345 (1965).

1630. R. Zsigmondy, *Z. Phys. Chem.* **111**, 211 (1924).

1631. G. Zubay and P. Doty, *Biochem. Biophys. Acta* **29**, 47 (1958).

REFERENCES ADDED IN PROOF

1632. S. Abrahamsson, I. Pascher, K. Larsson, and K.-A. Karlsson, *Chem. Phys. Lipids*
       8, 152 (1972).
1633. D. A. Cadenhead and F. Müller-Landau, in "Protides of the Biological Fluids"
       21st Colloquium (H. Peters, ed.), Pergamon Press, Oxford and New York (1972).
1634. H. L. Friedman and C. V. Krishnan, *J. Soln. Chem.* **2**, 119 (1973).
1635. H. G. Hertz and M. Holz, *J. Phys. Chem.* **78**, 1002 (1974).
1636. O. Kratky and G. Porod, *Rec. Trav. Chim. Pays-Bas* **68**, 1106 (1949).
1637. Y. K. Levine, P. Partington, G. C. K. Roberts, N. J. M. Birdsall, A. G. Lee,
       and J. C. Metcalfe, *Febs Letters* **23**, 203 (1973).
1638. D. L. Melchior and H. J. Morowitz, *Biochemistry* **11**, 4558 (1972).
1639. N. Robinson, Ph.D. Thesis, University of London, p. 41 (1959).
1640. D. M. Small, *Fed. Proc.* **29**, 1320 (1970).

Author Index

803

Subject Index

Absorption
  isotherm, 599, 636, 705, 706, 709, 718
  limit, 599
  of ultrasound, 333, 334, 352, 632
  of water, 584, 585, 647, 704-714
  spectrum of dyes, 185
Activation energy
  of conformational transition, 342, 378
  of diffusion, 24, 716
Activity, 606, 607, 619, 635, 637, 645,
       648, 654, 662, 704, 705, 718
  coefficient, 474, 475, 486, 670
  in terms of virial expansion, 46
Adhesion, 751, 753
Adhesive, 753
Adsorption
  isotherm, 599
  of water vapor by lipids, 255-257
Aerogel, 599
Aggregation
  number of micelle, 142, 143
    effect of nonpolar side chain on, 143
    effect of polar head group on, 143
  of dye, 169, 171
    charge separation effects in, 194, 196
    dispersion force contribution, to, 200,
       201
    maximum overlap of aromatic systems
       in, 194
    nature of the bonding, 198-204
    the role of water in, 204-207
α-Helix
  characteristic vibration frequency in,
       324
  stabilization by hydration interactions,
       355
Alkyl chain fluidity in lipid phases, 245
Amide hydrogen bond, 511
Amphiphilic nature of lipids, 209
Anaesthesia, 83

Angular velocity correlation time
  as determined by e.p.r., 64, 65
  effect of hydrophobic hydration on,
       65, 66
Anionic dyes, 173
  colloidal properties of, 173
  spectral properties of, 181, 182
Anisotropy of molecular motion in lipids,
       235
Annealing of biopolymer, 372
Anomer, 521, 534
Antioxidant, 743, 749
Apparent enthalpy of surfactant in
       micellar solution, 159
Apparent molar volume
  of neutralization, 397, 398
Autoxidation, 736, 738, 746
Axial substitution in sugars, 534

Barclay-Butler rule, 42
Basemole, 571, 622
Base stacking, 354, 355, 361, 362
  in aqueous solution, 356
Beer's law
  deviations by dye solutions, 182-184
BET isotherm, 662, 707, 708
Bilayer, 214
  structure, 238
Binding
  isotherm, 675, 679
  of cosolutes, 675-678
  of hydrocarbons
    thermodynamics of, 517
    to proteins, 516, 517
  of small molecule drugs, 755
  of small molecules to proteins, 483
  site, 88, 89, 706
    and multilayer formation, 707
Binodial, 698
Bridging by adsorbed polymers, 730, 731

Compound Index